ION CHANNEL RECONSTITUTION

ION CHANNEL RECONSTITUTION

Edited by
Christopher Miller

Brandeis University
Waltham, Massachusetts

PLENUM PRESS • NEW YORK AND LONDON

Library of Congress Cataloging in Publication Data

Ion channel reconstitution.

Includes bibliographies and index.
1. Membrane proteins—Purification. 2. Ion channels. I. Miller, Christopher, 1946- . [DNLM: 1. Biochemistry—methods. 2. Ion Channels. QH 601 I6466]
QP552.M44I65 1986 599′.0875 86-4900
ISBN 0-306-42136-4

A Division of Plenum Publishing Corporation
233 Spring Street, New York, N.Y. 10013

Printed in the United States of America

CONTRIBUTORS

Hubert Affolter Department of Pharmacology, University of North Carolina at Chapel Hill, Chapel Hill, North Carolina 27514; present address: Department of Physiology, Yale University School of Medicine, New Haven, Connecticut 06510

W. S. Agnew Department of Physiology, Yale University School of Medicine, New Haven, Connecticut 06510

Osvaldo Alvarez Departamento de Biologìa, Facultad de Ciencias, Universidad de Chile, Casilla 653, Santiago, Chile, and Centro de Estudios Cientificos de Santiago, Casilla 16443, Santiago, Chile

O. S. Andersen Department of Physiology and Biophysics and Department of Anesthesiology, Cornell University Medical College, New York, New York 10021

Robert Anholt Departments of Biology and Physics, University of California San Diego, La Jolla, California 92093; present address: Department of Neuroscience, Johns Hopkins University School of Medicine, Baltimore, Maryland 21205

R. L. Barchi David Mahoney Institute of Neurological Sciences and Department of Neurology, University of Pennsylvania, Philadelphia, Pennsylvania 19104

Joan Bell Department of Physiology and Biophysics, Laboratory of Human Reproduction and Reproductive Biology, Harvard Medical School, Boston, Massachusetts 02115

Roland Benz Department of Biology, University of Konstanz, D-7750 Konstanz, West Germany

Marc B. Blaustein Department of Physiology, University of Maryland School of Medicine, Baltimore, Maryland 21201

Fredic S. Cohen Department of Physiology, Rush Medical College, Chicago, Illinois 60612

Marco Colombini Laboratories of Cell Biology, Department of Zoology, University of Maryland, College Park, Maryland, 20742

Roberto Coronado Department of Pharmacology, University of North Carolina at Chapel Hill, Chapel Hill, North Carolina 27514; present address: Department of Molecular Biophysics, Baylor College of Medicine, Houston, Texas 77030

Tung Ming Fong Department of Biochemistry and Biophysics, University of California, Davis, California 95616

Robert J. French Department of Biophysics, University of Maryland School of Medicine, Baltimore, Maryland 21201

R. E. Furman David Mahoney Institute of Neurological Sciences and Department of Neurology, University of Pennsylvania, Philadelphia, Pennsylvania 19104

Elizabeth D. Getzoff Department of Molecular Biology, Research Institute of Scripps Clinic, La Jolla, California 92037

W. N. Green Department of Physiology and Biophysics and Department of Anesthesiology, Cornell University Medical College, New York, New York 10021

Sherwood Hall Food and Drug Administration, Washington, D.C. 20204

Wolfgang Hanke Lehrstuhl für Zellphysiologie, Ruhr-Universität, D-4630 Bochum, Federal Republic of Germany

Robert Hartshorne Departments of Biology and Physics, University of California San Diego, La Jolla, California 92093; present address: Department of Pharmacology, Oregon Health Sciences University, Portland Oregon 97201

Owen T. Jones Department of Biochemistry and Biophysics, University of California, Davis, California 95616

Peter C. Jordan Department of Chemistry, Brandeis University, Waltham, Massachusetts 02254

Bruce K. Krueger Department of Physiology, University of Maryland School of Medicine, Baltimore, Maryland 21201

Pedro Labarca Departments of Biology and Physics, University of California San Diego, La Jolla, California 92093; present address: Laboratorio de Neurofisiologia, Universidad Catolica de Chile, Casilla 114-D, Santiago, Chile

Ramon Latorre Department of Biology, Faculty of Sciences, University of Chile, and Centro de Estudios Cientificos de Santiago, Casilla 16443, Santiago 9, Chile

Irwin B. Levitan Graduate Department of Biochemistry, Brandeis University, Waltham, Massachusetts 02254

Jon Lindstrom The Salk Institute, San Diego, California 92138

Mark G. McNamee Department of Biochemistry and Biophysics, University of California, Davis, California 95616

Edward Moczydlowski Department of Physiology and Biophysics, University of Cincinnati College of Medicine, Cincinnati, Ohio 45267

Mauricio Montal Departments of Biology and Physics, University of California San Diego, La Jolla, California 92093; present address: Department of Neurosciences, Roche Institute of Molecular Biology, Nutley, New Jersey 07110

Mark T. Nelson Department of Pharmacology, University of Miami School of Medicine, Miami, Florida 33101

William O. Romine, Jr. Department of Physiology, University of Maryland School of Medicine, Baltimore, Maryland 21201

R. L. Rosenberg Department of Physiology, Yale University School of Medicine, New Haven, Connecticut 06510

John A. Tainer Department of Molecular Biology, Research Institute of Scripps Clinic, La Jolla, California 92037

Kenneth K. Tam Department of Biophysics, University of Maryland School of Medicine, Baltimore, Maryland 21201

Michael Tamkun Carnegie Institute of Washington, Baltimore, Maryland 21210

J. C. Tanaka David Mahoney Institute of Neurological Sciences and Department of Neurology, University of Pennsylvania, Philadelphia, Pennsylvania 19104

S. A. Tomiko Department of Physiology, Yale University School of Medicine, New Haven, Connecticut 06510

Akira Uehara Department of Physiology, Fukuoka University School of Medicine, Fukuoka 814–01, Japan

B. W. Urban Department of Physiology and Biophysics and Department of Anesthesiology, Cornell University Medical College, New York, New York 10021

Stephen H. White Department of Physiology and Biophysics, University of California at Irvine, Irvine, California 92717

Jennings F. Worley III Department of Physiology, University of Maryland School of Medicine, Baltimore, Maryland 21201

PREFACE

It is now over 30 years since the idea of ion-conducting pores burst on the electrophysiological scene, 15 years since these were generally realized to be membrane-spanning proteins, and 10 years since the first observations of single ion channels from higher organisms were made. During the past 5 years, several integral membrane channel proteins have been purified in a functionally competent state: the nicotinic acetylcholine receptor, the Na^+ channel, mitochondrial "VDAC," and a variety of porins. The stage is thus set to attack ion channels in the same ways that biochemists have been attacking enzymes for decades: isolation followed by functional analysis in as simple a system as possible, with a view towards understanding the molecular mechanisms of the protein's behavior and how this is related to the underlying molecular structure. This is always a daunting task, all the more so with ion channels because of our still primitive and scanty understanding of channel structures and because of the difficulty in isolating functionally active channel proteins.

In this volume, which can be considered a biochemically slanted companion to Sakmann and Neher's *Single-Channel Recording,* I have tried to present a view of the current landscape of ion-channel reconstitution. These chapters illustrate not only the different approaches and techniques of the major practitioners of ion-channel reconstitution but, as importantly, the varied motivations for doing this kind of work. There are many different reasons for undertaking this radically reductionist approach, but, as the chapters here indicate, questions of molecular mechanism and molecular structure are always present in reconstitution work. Indeed, I can see no justification for ripping a channel protein out of the comfortable home of its native membrane and transferring it to a foreign and "artificial" environment other than to enable the posing of molecular questions about ion-channel behavior, questions that simply cannot be asked in the complicated cellular environment.

Christopher Miller

Waltham, MA

CONTENTS

I. BASICS

Chapter 23

Analysis and Chemical Modification of Bacterial Porins

Roland Benz

Part I

BASICS

Chapter 1

THE PHYSICAL NATURE OF PLANAR BILAYER MEMBRANES

STEPHEN H. WHITE

1. Introduction

A major part of the general strategy for defining the structure–function relationships of ion channels is to reconstitute isolated channel proteins into planar lipid bilayer membranes separating two aqueous compartments. In principle, this part of the strategy allows one to manipulate the lipid and aqueous environments of the protein to elucidate their roles in channel assembly and function *in vivo*. Its success depends on how well one can use the natural physicochemical behavior of the bilayer system to control the reconstitution process and the composition and properties of the lipid bilayer itself. Consequently, the serious student of reconstitution must come to terms with the physical chemistry of these bilayers variously referred to as planar lipid bilayers, black lipid membranes, or thin lipid films.

My aim in writing this chapter is to give the reader a basic understanding of planar bilayers and how to control them. My approach will be largely qualitative because I suspect that most readers are primarily interested in ion channels and only secondarily in the bilayers themselves. Planar bilayers can be incredibly frustrating. All of us who have worked with them have, at one time or another, been reduced to irrational (but satisfying) acts of anger. The frustration usually arises because one is, unknowingly, trying to do something that is thermodynamically unlikely. The number of beakers I sent crashing against the laboratory wall decreased as my understanding of the physical chemistry of the system increased. Here, then, are the essentials of black lipid membranes as I see them.

1.1. Theme of the Chapter

The basic theme of this chapter is that the planar bilayer itself is only one part of a highly heterogeneous physicochemical system. To understand the bilayer, one must understand it in the context of the whole system, which consists

STEPHEN H. WHITE • Department of Physiology and Biophysics, University of California at Irvine, Irvine, California 92717.

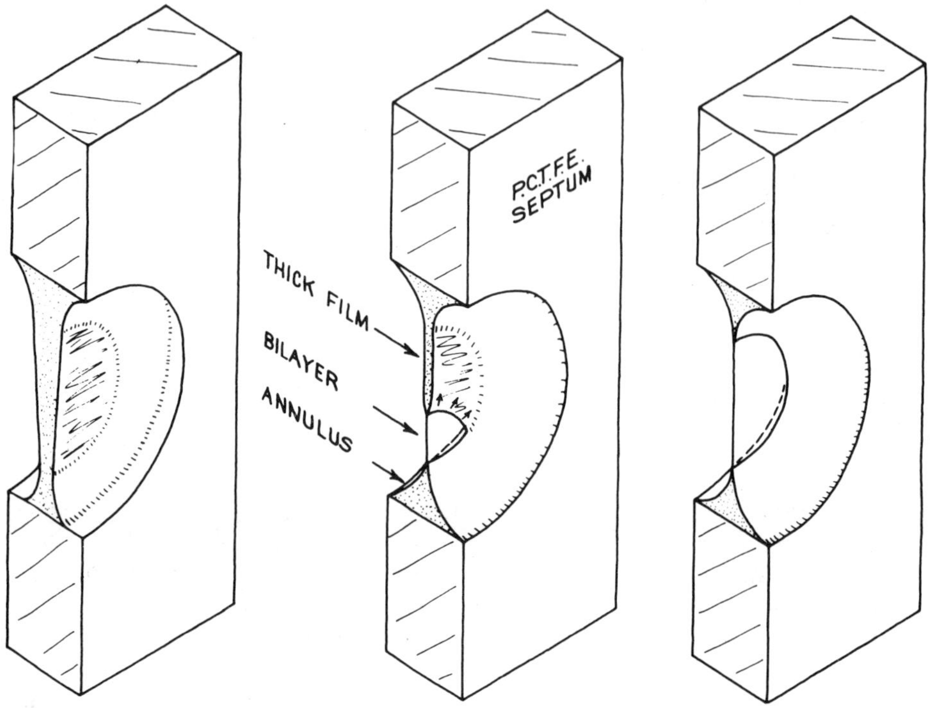

Figure 1. Summary of the process of formation of planar lipid bilayers from lipids dispersed in alkane solvents by the method of Mueller *et al.* (1962). The lipid solution is applied beneath water across an aperture in a nonpolar material such as polychlorotrifluoroethylene (PCTFE). Note that the lipid solution "wets" the septum and forms an annulus (Plateau–Gibbs border) around the bilayer. The initially thick film thins spontaneously because of the Plateau–Gibbs border suction. (From White *et al.*, 1976.)

of several different phases. Consider, for example, the classic black-film system first described by Mueller *et al.* (1962). A bilayer forms spontaneously as in Fig. 1 after a "forming solution" is spread across a small (1–2 mm) aperture in a hydrophobic partition separating two aqueous phases. The forming solution consists of a surface-active lipid "dissolved" in a nonpolar liquid such as *n*-decane. The resulting structure consists of the bilayer film surrounded by an annulus of the bulk forming solution. When viewed under reflected light, the very thin (25–50 Å) bilayer reflects little light relative to its surroundings and thus appears "black"; hence the term black lipid membrane.

The bilayer itself consumes most of our attention, and we thus largely ignore the annular border. Such neglect is not wise if one wishes to control the bilayer. These two parts or phases of the system tend toward chemical equilibrium with one another. Because the annulus has a mass at least a million times that of the bilayer, the chemical potentials of the components of the bilayer are constrained to be the same as those of the annulus. The secret to the control of bilayer behavior and composition is the control of the annulus or any other more massive phases it contacts.

The annulus, then, must receive much of our attention as we seek to use the bilayer as a controlled environment for ion channels. It is not a simple solution of lipid in organic solvent. The lipid is present in the solvent as hydrated aggregates (micelles), which tend toward equilibrium with the bulk aqueous phase. The annulus is a multicomponent system and can exhibit a complex phase behavior. We are most experienced as biologists with the phase behavior of pure lipid dispersed in water. The behavior of hydrated lipid in a nonpolar phase can be quite different.

The planar bilayer system thus consists of several phases that tend to equilibrate with one another. To understand it, one must think about it in terms of the general principles of equilibrium in heterogeneous systems. This can be a confusing business. I have found it to be less so if I organize my thinking around Gibbs's phase rule. This chapter is written, therefore, with the phase rule as its central focus.

1.2. Organization of the Chapter

I proceed as follows. I first discuss the most common strategies for reconstituting channels in planar bilayers to set the context for the material to follow. The discussion is brief because most of this book is devoted to the specific details of reconstitution. It gives me an opportunity, however, to identify topics I believe to be central to the reconstitution problem and to reference key reviews for those topics not reviewed in detail.

I next discuss the general principles of planar bilayer formation and stability. Of particular importance are the differences between those formed by spreading films from bulk solutions by the method of Mueller *et al.* (1962) and those formed by the folding up of monolayers by the method of Montal and Mueller (1972). These principles reveal the complex heterogeneous nature of the planar bilayer. I then discuss the physical chemistry of heterogeneous systems in the context of the bilayer system. This sets the stage for a discussion of the control of bilayer behavior. I close by explaining some of the vagaries of planar bilayers, which have frustrated more than a few biophysicists.

2. Overview of Reconstitution

2.1. Strategies for Reconstitution

Of the many different approaches to reconstitution that have been tried [see reviews by Miller (1983a,b, 1984) and by Montal and his colleagues (1981)], two have emerged as the principal ones (Fig. 2). In the first, planar bilayers are formed either from bulk solutions (Mueller *et al.*, 1962) or from monolayers (Montal and Mueller, 1972). The channels are transferred to the bilayer from vesicles by Ca^{2+}-

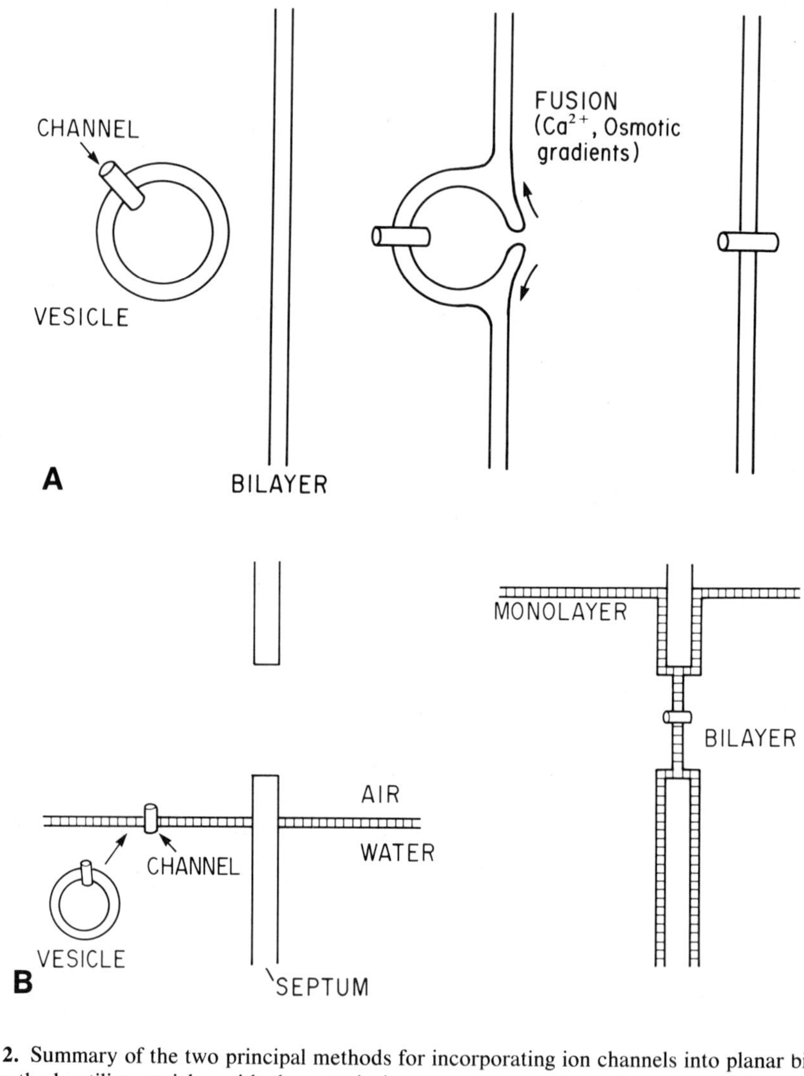

Figure 2. Summary of the two principal methods for incorporating ion channels into planar bilayers. Both methods utilize vesicles with the protein incorporated into their bilayers. (A) The method of Miller and Racker (1976) incorporates the protein by fusing vesicles made with charged lipids to preformed bilayers. The fusion is controlled with Ca^{2+} and the creation of osmotic gradients across the vesicles and planar bilayer. It is assumed that the fusion is complete so that the contents of the vesicles are released *trans* to the side the vesicles are added to. (B) The method of Schindler (1980) involves forming bilayers by folding up monolayers across an aperture in the manner of Montal and Mueller (1972). The monolayers are formed by allowing the protein-containing vesicles to spread spontaneously at the air–water interface. It is assumed that the protein is incorporated into the monolayers.

induced fusion controlled by osmotic gradients across the vesicles and the planar bilayer, which contain negatively charged lipids (Miller and Racker, 1976). The channels are incorporated into the vesicles by dialysis of solutions containing the protein and excess lipid solubilized in detergent. The osmotic stress is viewed as causing the vesicles to swell and rupture in such a way that fusion occurs.

In the second method, lipid monolayers are produced at the air–water interface from the channel-containing vesicles (Schindler, 1980). These monolayers presumably contain the channels, which become incorporated into the bilayer when the monolayers are raised across the aperture by the method of Montal and Mueller (1972).

Both methods can lead to the incorporation of channels into bilayers, but it is not proven that the methods work precisely according to the cartoons shown. Regardless of the accuracy of the cartoons, however, they do illustrate the topics that should attract the interest of the serious student of reconstitution.

2.2. Important Topics for Reconstitution

The two principal methods of forming planar bilayers must be understood thoroughly. These are the subject of the present chapter. In addition, however, the general nature of monolayers at the air–water interface must be appreciated. I highly recommend the reviews of Gershfeld (1974, 1976) and the superb monograph by DeFay and Prigogine (1966). The particularly relevant major issues addressed by Gershfeld concern the phase behavior of monolayers, the equilibrium between bulk lipid phases and adsorbed layers at the air–water interface, and the heterogeneity of surface films. Particular attention should be paid to the question of whether surface films are necessarily monomolecular in thickness and to film penetration by chemical components (e.g., proteins) in the subphase.

Another obvious topic for study is that of the physical chemistry of liposomes and vesicles. Bangham, who is the originator of these systems, has written several reviews that are essential reading (1974, 1980). Israelachvili *et al.* (1980) and Tanford (1980) have summarized the physical principles of liposome and micelle formation. A recent volume edited by Ostro (1983) contains a number of useful papers.

Because the joining together of vesicles with bilayers is central to reconstitution, knowledge of the energetics and mechanics of bilayers in the context of adhesion and deformation is essential. Thorough reviews of these and related topics have been written by Evans and Skalak (1979a,b) and by Parsegian and his colleagues (Parsegian and Rand, 1983; Evans and Parsegian, 1983; Parsegian *et al.*, 1984). The role of osmotic stress in fusion is likely to be more subtle and complicated than simple vesicle swelling. In this regard, the effect of water activity on bilayer mechanics (Parsegian *et al.*, 1984) and the organization of water near the membrane surface (Gruen and Marcelja, 1984) should receive special attention. The molecular packing density of water and the headgroups of phospholipid bilayers are likely to be important also (White and King, 1985).

Finally, the planar bilayer system is best described by the laws of the physical chemistry of surfaces. A familiarity with surface chemistry can be gained from textbooks by DeFay and Prigogine (1966), Adamson (1967), and Aveyard and Haydon (1973).

3. Formation and Stability of Planar Bilayers

Several books and reviews have been devoted to this general subject (Tien and Diana, 1968; Henn and Thompson, 1969; Haydon, 1970; Jain, 1972; Tien, 1974; Finkelstein, 1974; Fettiplace *et al.*, 1975), and I do not intend to be encyclopedic in my approach. Rather, I wish to highlight central ideas that are useful for the later discussion. The most succinct review of the physical chemistry is that of Fettiplace *et al.* (1975).

Planar bilayers formed from either bulk solutions or monolayers are largely indistinguishable from the physicochemical point of view. The major difference is the nature of the bulk phases from which they are formed. Because the first method reveals the general principles better, I begin with it.

3.1. Formation by Spreading from Bulk Solutions

3.1.1. Forming Solution

The first step in the formation of black films by the original method of Mueller *et al.* (1962) is to dissolve—more properly, disperse—a surface-active lipid in a nonpolar solvent. The lipids most commonly used are the monoglycerides (MG) or lipids commonly found in cell membranes such as phosphatidylcholines (PC), phosphatidylethanolamines (PE), or phosphatidylserines (PS), which may be mixed with varying amounts of cholesterol. Phosphatidylserine is particularly useful for conferring fixed charge at the surface, but phosphatidylglycerol (PG) and phosphatidic acid (PA) are also used for this purpose. Films were originally formed from whole lipid extracts of membranes, and these are still used for some purposes. The most common nonpolar solvents these days are alkanes or other long-chain hydrocarbons such as squalene. In the early days, standard lipid extraction solvents (e.g., chloroform:methanol) augmented with alkanes were used, but these are more difficult to use, probably because of time-dependent changes in composition and phase behavior in the presence of excess water.

The lipid in the solvent tends very strongly to form large aggregates (micelles) organized to shield the polar regions of the lipids from the nonpolar solvent (Fig. 3). The simplest micelles are those of the monoglycerides, which tend to form clusters of 25 or so molecules with the acyl chains facing outward (Andrews *et al.*, 1970). Black films cannot be formed unless the aggregates are present; i.e., the concentration of lipid must be above the critical micelle concentration (cmc).

The forming solution remains in contact with the bilayer as the annulus for the bilayer's entire life. In addition, the bilayer and annulus reside in an excess aqueous phase. Because the annulus is so massive compared to the bilayer, the

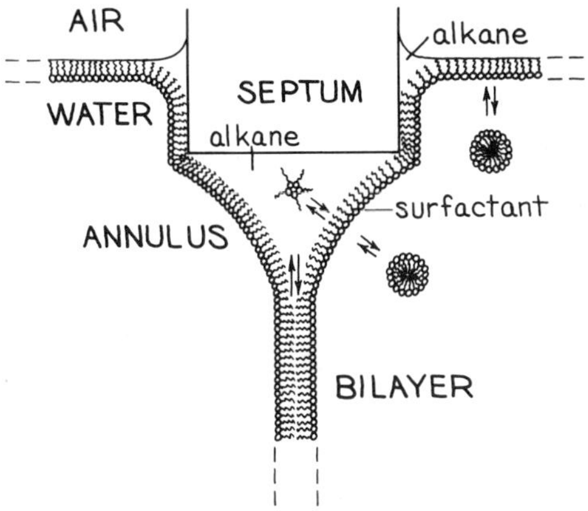

Figure 3. A schematic representation of the black lipid membrane system and the various locations and states of the surface-active lipid. The lipid in the annulus is represented as simple inverted micelles, but other much larger and complex aggregates are possible. The arrows indicate the equilibria between the various phases of the system. The drawing is not to scale. For bilayers formed by spreading from bulk nonpolar solutions, the mass of the annulus is a million times that of the bilayer. Therefore, the chemical potentials of the lipid and solvent (alkane) in the annulus determine the chemical potentials in the bilayer.

chemical potential of the lipid in the aggregates determines the state of the lipid in the bilayer itself. Consequently, the phase behavior of the forming solution in excess water is important to understand. This is not trivial, because one ends up dealing with microemulsions, which are the subject of another whole complex area of physical chemistry (e.g., Ekwall, 1969; Dominquez *et al.*, 1979).

Microemulsions are generally formed from three-component systems consisting of a surfactant, a nonpolar solvent, and a polar solvent. The bulk forming solution has these ingredients when one considers that water, even if not specifically added, must eventually be present in the annulus. I will not not describe the complex behavior of microemulsions; the interested reader can gain entry to the literature from a recent symposium volume edited by Luisi and Straub (1984).

3.1.2. Spreading the Film

A droplet of the bulk forming solution is placed around the rim of an aperture in a Teflon®, Kel-F®, or polyethylene partition submerged in the aqueous phase. Bilayers will not form across the aperture spontaneously; they must be spread with a small brush or, preferably, an air bubble protruding from a fine glass capillary. I make this obvious point because it means that the black film is inherently not an equilibrium system—work must be done to create it. There is always a lower free energy state for the system, which is the bulk solution with no black film; that is, the films tend toward the broken state. The membranes can, however, exist for many hours if other aspects of the complex equilibrium of the system are tended to. It is reasonable to consider them as metastable equilibrium systems (Guggenheim, 1977).

A crucial feature of the black-film system is the interface between the bulk solution of the annulus and the aqueous phase. The lipid adsorbs at the interface from the micelles to a potential energy well resulting from the insolubility of the acyl chains in the aqueous phase and the insolubility of the polar groups in the nonpolar phase. The lipid lowers the interfacial tension γ of the interface (at equilibrium) according to the Gibbs absorption equation

$$d\gamma = -\Gamma \, d\mu \tag{1}$$

where

$$\mu = \mu_0 + RT \ln a \tag{2}$$

In these equations, μ is the chemical potential of the lipid with activity a in the annulus, and Γ is the interfacial concentration (mol/cm^2). μ_0 is the chemical potential in the standard state or a suitable reference state. The activity (a) may be written as ξc_L where c_L is the lipid concentration and ξ the activity coefficient. Equation 1 is the foundation of surface chemistry. It simply states that the decrease in interfacial tension is directly proportional to the change in the chemical potential of the lipid in the annulus. It allows one to determine the interfacial concentration (Γ) of the lipid, which leads in turn to the area/molecule of the lipid in the bilayer (Andrews *et al.*, 1970).

3.1.3. Thinning of the Film

The film formed by spreading the solution across the aperture is initially several micrometers thick but thins spontaneously to the planar bilayer state (Fig. 1). It thins under the influence of several driving forces. The major one in the early stages is the Plateau–Gibbs border suction, which arises from the curvature of the annulus (Plateau–Gibbs border). The curvature is necessary to satisfy simultaneously the contact angles the border must make with the aperture and the film. [These angles are related to the interfacial tensions of the various substances in contact. See Adamson (1967).] Wherever there is a curved interface at mechanical equilibrium, there must be a hydrostatic pressure difference ΔP across it described by the law of LaPlace

$$\Delta P = 2\gamma/R \tag{3}$$

where γ is the interfacial tension and R the radius of curvature. The pressure is always greater on the concave side of the surface so that the pressure in the border must be less than that in the flat region (Fig. 3). Therefore, bulk solution will flow from the flat film into the border, causing the film to thin.

As the film approaches thicknesses of a few hundred Angstroms, a second driving force becomes apparent. It is the London–van der Waals attraction between the aqueous phases separated by the thin film. This force is moderately long ranged and has been described in detail (e.g., Parsegian, 1975; Brooks *et al.*, 1975). Because of it, the free energy change of the film (ΔF) relative to infinite separation is

$$\Delta F = -H/12\pi d^2 \tag{4}$$

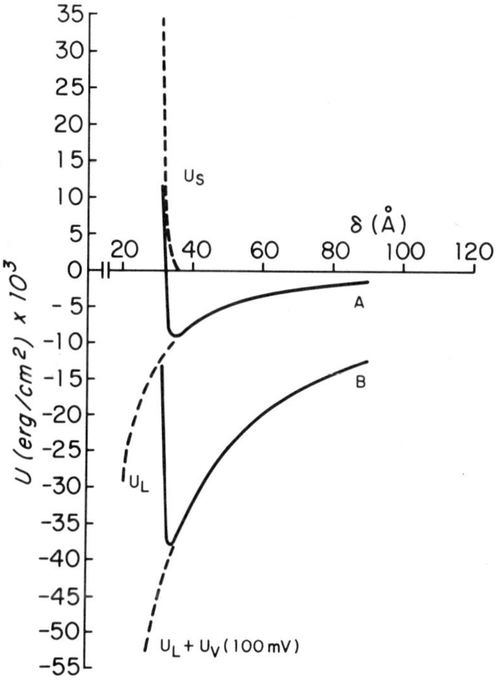

Figure 4. Hypothetical potential energy-versus-thickness curves for planar bilayer membranes. Two opposing forces stabilize the membrane: a London–van der Waals attractive potential (U_L) between the separated aqueous phases (equation 4; $\Delta F = U_L$) and a steric repulsive force (U_S) arising from the interactions of the apposed lipid acyl chains of the bilayer. When an electric field is present, an additional attractive potential (U_V) exists given by equation 6 ($\Delta F_V = U_V$). (From White, 1970.)

where H is the Hamaker coefficient and d the thickness of the film. It is this force that leads to the final thinning and is the major force attempting to thin the membrane at the equilibrium bilayer thickness. The black film (bilayer) usually appears first at the lower margin of the film and propagates rapidly over the surface (Fig. 1). The film is thinner at the lower margin because of the buoyancy of the forming solution.

3.1.4. Stability and Composition of the Film

The bilayer would not be stable unless the van der Waals force were opposed by another set of forces. When the opposing surfaces of the thicker film meet, there is a tendency for the apposed acyl chains on the lipid to interdigitate and to exclude the solvent molecules trapped in the film (Taylor and Haydon, 1966). However, this tendency to "demix" tends to lower the entropy of the bilayer and consequently to raise the free energy of the system, giving rise to an opposing force called the steric repulsive force. The bilayer thus resides in a local free energy well and is stable (Fig. 4).

Another force resisting the thinning of the film, albeit a temporary one, is viscous drag of the solvent flowing out from between the advancing interfaces. When the surfaces are very close, the viscosity of the solvent resists the thinning. As a result, the films are initially somewhat thicker. The excess solvent separates within 15–30 min into microlenses about 1 μm in diameter scattered over the

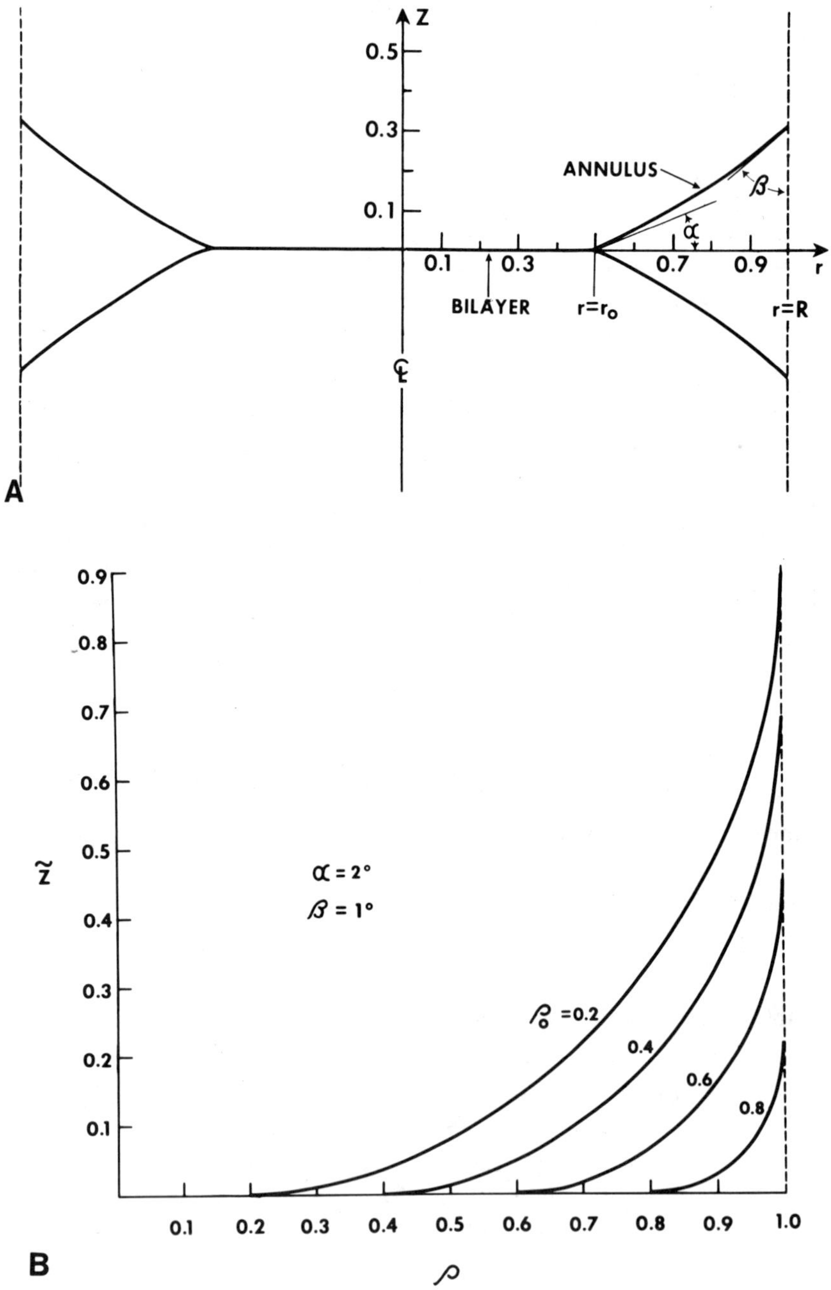

Figure 5. The shape of the annulus surrounding the planar bilayer membrane showing how changes in annulus volume (related to R), the contact angle between the annulus and film (α), and the contact angle between annulus and aperture (β) affect the shape. (A) Overview of system. In B, C, and D, the figures represent the shape in the first quadrant only. Normalized coordinates are used: $\rho = r/R$ and $\tilde{Z} = Z/R$. (B) Variation in shape with changes in ρ. (C) Variation in shape with changes in α. (D) Changes in shape with β. (From White, 1972.)

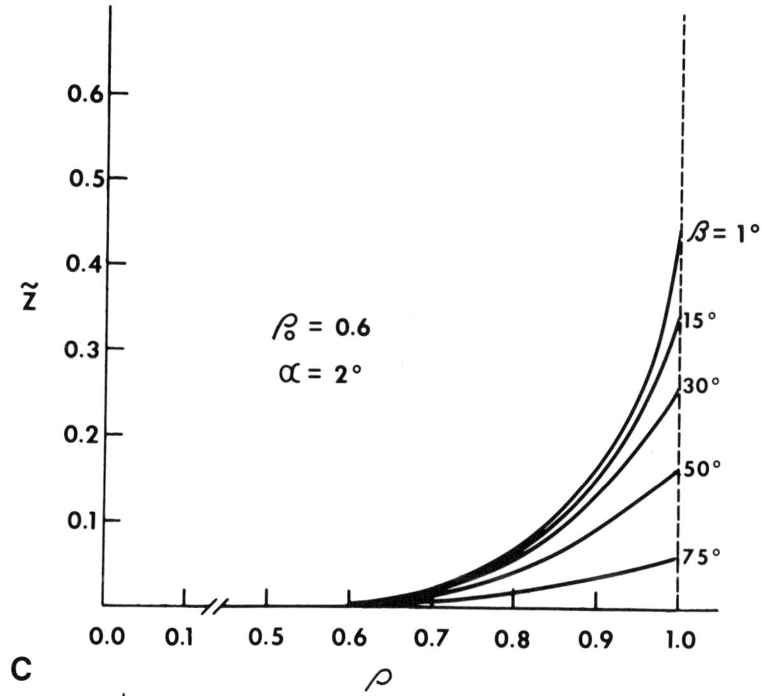

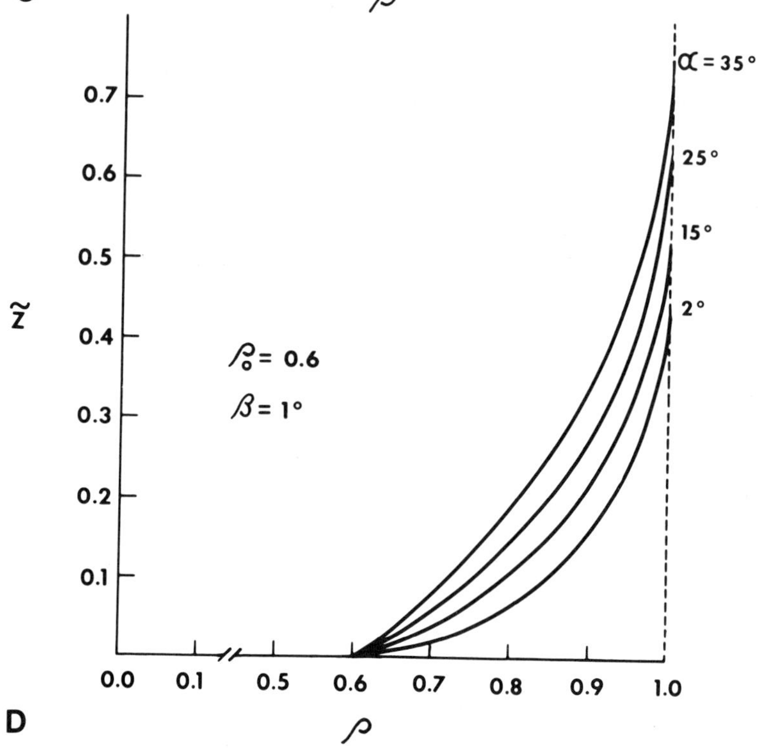

Figure 5. *(continued)*

surface, which are easily seen in reflected light. They do not usually interfere with most electrical measurements and eventually coalesce with the annulus.

The chemical potentials (equation 2) of the solvent and lipid in the bilayer must be equal to those of the annulus when the film reaches its local equilibrium. Because the annulus is much more massive than the bilayer, the solvent in the bilayer tends to come into equilibrium with the annulus. This is equivalent to the situation a single biological cell faces when suspended in a large volume of buffer. We can thus describe an osmotic pressure on the solvent in the annulus trying to drive solvent into the bilayer, which is opposed by the van der Waals force squeezing the bilayer (Andrews, 1970; White, 1980). Most properly then, we must describe the black film's "equilibrium" not as a regular equilibrium (pressure everywhere uniform) but as an osmotic equilibrium (Guggenheim, 1977).

3.1.5. Film Boundary Conditions

An annulus surrounds all black films regardless of method of formation and is essential for the existence of the bilayer (White *et al.*, 1976). It connects the bilayer to the aperture, so its mechanical properties have an effect on one's ability to form and manipulate the bilayer. The influence of the annulus on mechanical stability can be appreciated from a mathematical analysis of the shape of the annulus using the methods of variational calculus (White, 1972). The shape is determined by the contact angle between the forming solution and the aperture (β), the contact angle between the bilayer and the annulus (α), and the volume of solution placed on the aperture (Fig. 5). If one forms the bilayer on an aperture that is the shape of a right circular cylinder, the aspect ratio of the aperture (its length divided by diameter) is likely to be important in achieving mechanical stability because there is a maximum annulus width that can be accommodated by a particular aperture.

One problem with films is that they tend to drift along the length of the cylindrical aperture. Another is that they exchange bulk solution with the surface of the partition. If all surfaces of the forming chamber are thoroughly coated with the bulk forming solution, then bilayer diameter will tend to shrink because solution flows from the septum surface to the annulus as a result of the border suction that led to film thinning in the first place. Both of these problems can be controlled reasonably well by using a countersunk aperture (Fig. 6). The film becomes trapped at the point where the aperture changes diameter. Further, the shelf of the countersink helps to define the volume of the annulus and therefore the bilayer's diameter.

The contact angle (α) between the bilayer and annulus is determined by the difference in free energy (ΔF of equation 4) between a thick film (e.g., the annulus) and a bilayer (Haydon and Taylor, 1968). The relationship between the two is

$$\Delta F = 2\gamma(\cos \alpha - 1) \tag{5}$$

where γ is the annulus–water interfacial tension. Haydon and his colleagues (Haydon and Taylor, 1968; Requena and Haydon, 1975b; Brooks *et al.*, 1975) have

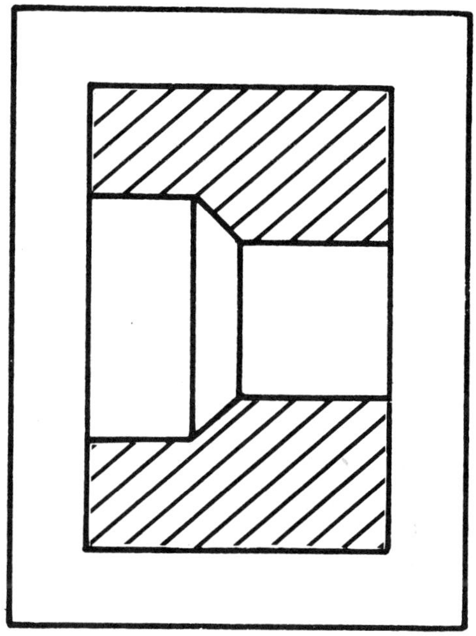

Figure 6. Cross-sectional view of a convenient aperture for forming black lipid films. The countersink keeps the film from drifting along the length of the aperture and helps stabilize the volume of lipid in the annulus to minimize changes in diameter with time.

measured ΔF (the depth of the free energy minimum in Fig. 4) by measuring the contact angle optically.

The application of an electric field to the bilayer adds a compressive force in addition to that of the van der Waals attraction (equation 4) and leads to further thinning of the film (Fig. 4). The change in free energy ΔF_V caused by an electric potential V is

$$\Delta F_V = -\frac{\epsilon_0 \epsilon}{2d} V^2 \tag{6}$$

where ϵ_0 is the permitivity of free space ($8.85 \cdot 10^{-14}$ F/cm^2), ϵ the dielectric coefficient of bilayer, and d the thickness. As implied by equation 5, a change in the contact angle must accompany the voltage-induced change in free energy, as Requena and Haydon (1975a) have verified. This causes the shape of the annulus to change in such a way that the diameter of the bilayer increases (White and Chang, 1981).

3.2. Formation of Black Films from Monolayers

The lipid and some of the solvent from the bulk forming solution will inevitably adsorb at the air–water interface and form either monolayers or possibly

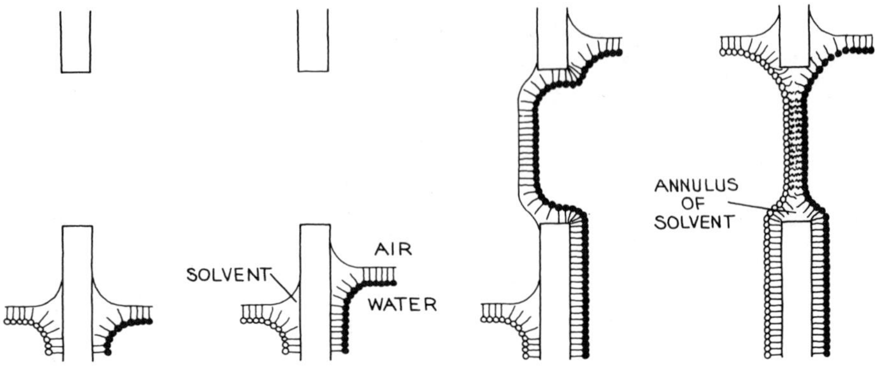

Figure 7. A sequence of sketches showing how a bilayer is produced from monolayers when alkane is present. The alkane causes the acyl chains of the monolayer to have a favorable orientation for bilayer formation. The septum is shown for artistic convenience as being thin; it is orders of magnitude thicker in reality. (From White *et al.*, 1976.)

so-called duplex films (Langmuir, 1933). Another part, then, of the now clearly complex black film sysktem is the lipid and solvent at this interface. Another way of forming black films is to float droplets of the bulk forming solution on the surfaces of the aqueous compartments, which are initially below the level of the aperture. A bilayer can be formed by raising first one and then the other aqueous phase above the aperture, as shown in Fig. 7. There is no fundamental difference between bilayers formed in this way and those formed by painting.

3.2.1. Solvent-Free Bilayers

The basic strategy for forming bilayers by folding up monolayers was introduced by Takagi *et al.* (1965) and perfected by Montal and Mueller (1972). The original rationale was to produce bilayers free of solvent because solvent was nonbiological. Further, because volatile alkanes are anesthetics, the fear arose that solvent-containing bilayers might be anesthetized and thus create a poor environment for ion channels. The main issue is whether or not the bilayer of nerve membranes mediates the anesthetic action. There is strong evidence (Franks and Lieb, 1978, 1981) that it does not, in which case the fear is unfounded.

In any case, Montal and Mueller's goal was to form bilayers with as little solvent as possible. In the original form, lipid was spread from a volatile solvent at the air–water interfaces so that solvent-free monolayers remained after the solvent evaporated. This is the classic method of forming monolayers for study using the Langmuir surface balance (see Gaines, 1966). Prior to the spreading of the monolayers, the initially dry septum was "conditioned" with petroleum jelly applied from a dilute hexane solution. The petroleum jelly provides a small amount of nonpolar lipid for the annulus. Without this material, bilayers cannot be formed as is illustrated in Fig. 8. An annulus of some sort is absolutely required (White *et al.*, 1976). This does not mean that the bilayers necessarily contain very much

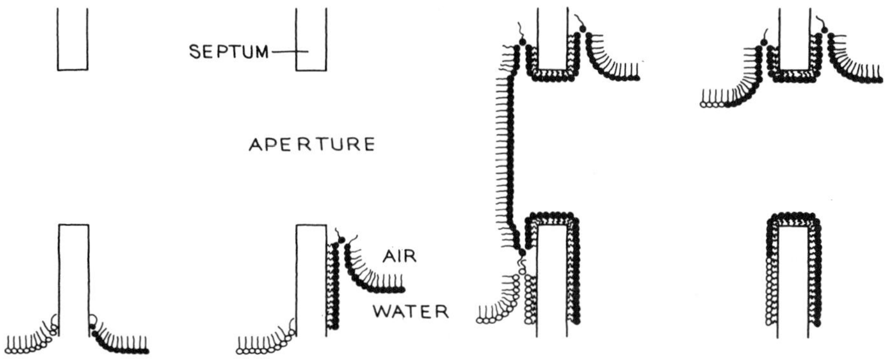

Figure 8. A sequence of sketches showing what happens if one attempts to form bilayers from monolayers in the absence of alkane or other nonpolar solvent. The raising of the left-hand water level "peels off" the monolayer on the right aqueous phase, shown here protruding through the aperture (third and fourth frames). (From White *et al.*, 1976).

solvent. The constituents of the petroleum jelly are probably not very soluble in bilayers.

Solvent-free bilayers can be formed by the painting method simply by using squalene, which does not dissolve in the bilayer but forms a good annulus (White, 1978). Di- or triglycerides (Waldbillig and Szabo, 1979) can be substituted for the squalene. A big advantage of the Montal–Mueller method is that it is possible to form asymmetric bilayers. Another is that one can minimize the total overall effect of the nonpolar solvent so that the properties of the lipid in an aqueous phase rather than a nonpolar phase dominate the system. For example, Pattus *et al.* (1978) and Schindler (1979) showed that one can spread the monolayers from lipsomes rather than from lipid dissolved in solvent. This sort of procedure has actually been used for years by surface chemists studying equilibrium of spreading of monolayers at the air–water interface from bulk lipid (see, e.g., Eriksson, 1971; Phillips and Hauser, 1974; Tajima and Gershfeld, 1978).

3.2.2. The Bulk Phase Distinguishes the Two Methods

The most important difference between the two methods of forming bilayers is the nature of the bulk phase with which the bilayer is attempting to equilibrate. In the first method, one begins with a solution of hydrated lipid in hydrocarbon solvent so that the phase behavior of the lipid–water–hydrocarbon bulk phase determines the state of the bilayer. In the second method, minimum amounts of solvent are present, so that the phase behavior of a predominantly lipid–water bulk phase determines the state of the bilayer.

Figure 9 shows a phase diagram for glycerol monooleate–triolein mixtures in the presence of excess water. The left-hand side represents the situation in which one forms bilayers from bulk nonpolar solutions. At room temperature, the bulk phase is mostly triolein with small amounts of GMO dispersed as micelles.

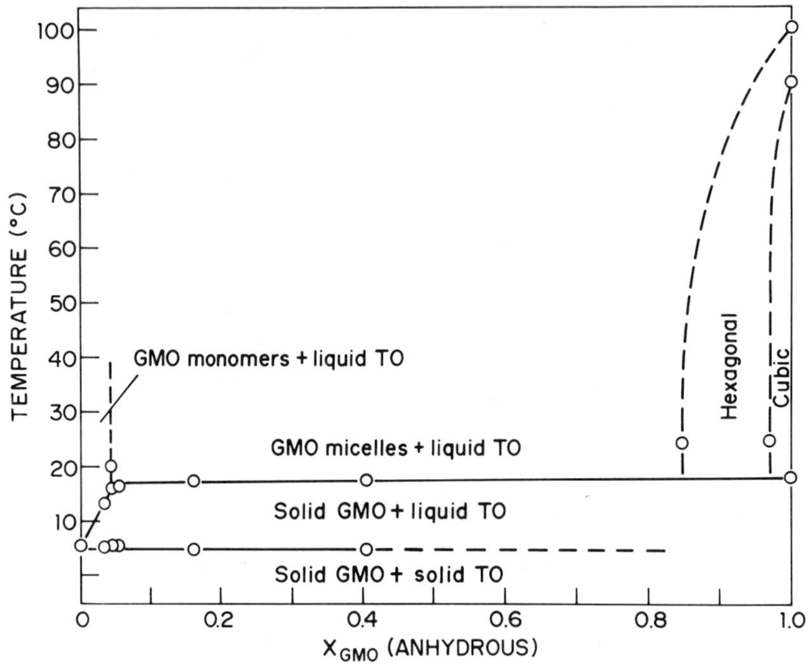

Figure 9. Phase diagram for a mixture of glycerol monooleate (GMO) and triolein (TO) in excess water. This is an example of the type of phase behavior one can expect of the black-film "forming solution" constituting the annulus of the black film. The left-hand side corresponds to the situation for bilayers formed by the spreading method of Mueller *et al.* (1962). The right-hand side corresponds to the situation for bilayers formed by the folding up of monolayers by the method of Montal and Mueller (1972). (Based on unpublished data by S. H. White and T. J. McIntosh, Duke University, Durham, North Carolina.)

On the right is the situation in which one uses minimal amounts of solvent (triolein) as in the formation from monolayers under the conditions of Montal and Mueller. When nearly pure GMO is present, the bulk phase is either a hexagonal or a cubic structure. Thus, if one wishes to understand the differences between bilayers formed by the two methods, one must look to the phase diagram for the system. Because the bulk phases from which the bilayers arise in the two cases are different, the resulting bilayers can be different.

4. Black Films as Hetergeneous Systems

It should be clear by now that the planar bilayer is the smallest part of a complex heterogeneous system. We cannot manipulate it for reconstiution studies unless the combined behavior of the several phases of the whole system is taken into account. To do a proper job, the physical chemistry of heterogeneous systems must be appreciated. Gibbs set down in 1876 the basic rules governing such systems (see Gibbs, 1961). Among the many contributions of his monumental work, his wonderfully elegant phase rule is perhaps the most important.

4.1. Equilibrium in Heterogeneous Systems: The Phase Rule

A heterogeneous system is one comprised of several different phases. If there are no curved interfaces between the various phases and the system is at equilibrium, the phase rule describes the relationship among the number of experimental variables (f, also called degrees of freedom or variance) that can be independently varied, the number of chemical components (c), the number of phases (p), and the number of intensive variables (i) such as temperature, pressure, and electric field. The great utility of the phase rule is that it forces one to think through the physical chemistry of a system in a systematic fashion. The phase rule is

$$f = i + c - p \tag{7}$$

This equation is a consequence of the condition for the equilibrium of a heterogeneous system, which is that the chemical potential (equation 2) of each component be the same in all phases. At equilibrium, in other words, there is no net difference in the partial molar free energy of a chemical species between any of the phases. The major result of this is that the concentrations of the components in the various phases are not independent of one another; the phase rule is the statement of this fact.

The simplest example of an application of the phase rule is the melting of ice at constant pressure. Here, $i = 1$ and $c = 1$, so that if liquid water and solid water coexist, $p = 2$, leading to $f = 0$. This means that the melting of ice is an isothermal event.

4.2. The Applicability of the Phase Rule to Black Films

Because the phase rule is true only at equilibrium, the first question that must be addressed is that of the equilibrium of the system. I noted earlier that the black film system is not inherently an equilibrium system because the free energy of the system is lowered if the membrane breaks. On the other hand, black films can exist for many hours without noticeable changes in properties because they reside in a local potential energy minimum at the higher energy state. We are thus at liberty to treat a black film as a system in a metastable or frozen equilibrium in the sense described by Guggenheim (1977). There are situations, however, in which the system is definitely not at equilibrium. These situations express themselves in two ways. One is that some measured parameter such as specific electrical capacitance may vary strongly with time or be nonreproducible. The other is that films may be difficult to form or may break prematurely. The causes of these frustrating events are discussed in the last section of this chapter.

Another aspect of equilibrium must also be considered because the hydrostatic pressure in the annulus is different from that in the film (section 3.1.3). The phase rule must be modified in such systems to include mechanical equilibrium as described by the Law of LaPlace (equation 3). However, one can show that for black films generated from bulk organic phases using reasonable aperture geometries, the curvature of annulus can be ignored.

Yet another issue concerns the surface phases of the system. They should be included in some way. The phase rule as applied to interfacial systems in general has been discussed elegantly by DeFay and Prigogine (1966). The interfaces of a system appear explicitly in the phase rule, it turns out, only if it is possible to have two or more different states of matter present in a given surface simultaneously. The classic example is the compression of a monolayer from the gaseous state to the liquid expanded state. Over some region of the compression, both phases are present at the air–water interface.

One can show that the complete phase rule for the black film system is

$$f = i + c - p + (s - n) \tag{8}$$

providing the curvature of the annulus can be ignored. In this equation, n is the number of types of surface, and s the total number of surface phases (e.g., liquid expanded, gaseous). The types of surface present in the black film system are the air–water interface, the annulus–water interface, and the bilayer, so that $n = 3$. If each surface type contains only one surface phase, then equation 8 reduces to equation 7. I consider next the application of the phase rules to the films formed by spreading a bulk solution across an aperture.

4.3. Phase Rule Analysis of the Black Lipid Membrane

Specifically, consider a black film formed from lipid dispersed in alkane spread across an aperture immersed in a NaCl solution. Assume that there is only one surface phase per surface so that equation 7 can be used. We may consider first the number of phases present. There are at least a bulk aqueous phase, a bulk air phase (we wish to include the air–water interface), and a bulk organic phase. The bulk organic phase may itself consist of more than one phase depending on how the lipid aggregates in it.

If one uses glycerol monooleate, which forms relatively small aggregates (Andrews *et al.*, 1970), then the activity of the alkane solvent depends rather strongly on the concentration of GMO (Waldbillig and Szabo, 1978) because the aggregates behave as a molecular species. It is generally agreed that under these circumstances, the organic phase should be considered as a single phase (Tolman, 1913). If one uses a phospholipid, on the other hand, very large aggregates result, which have little effect on the activity of the alkane (Fettiplace *et al.*, 1971). In this case, the organic phase should be considered as a two-phase system: bulk alkane plus bulk lipid aggregates. Thus, the bulk forming solution is counted as either one or two phases depending on the behavior of the hydrated lipid in the nonpolar solvent.

There are five components, which include water, NaCl, lipid, alkane, and "air." (Including air is somewhat arbitrary because for most purposes we can ignore it as it appears once in both c and p and thus cancels its presence. I include it for completeness.) The number of intensive variables will be three if one varies temperature, pressure, and the electric field across the system. In general, however, the pressure is constant, and I take $i = 2$. We thus find that either $f = 2 + 5 - 3 = 4$ (GMO) or $2 + 5 - 4 = 3$ (phospholipid). The experimental variables that must be controlled in the case of GMO ($f = 4$) are temperature, electric field, salt concentration, and GMO concentration. If the lipid forms a separate phase, as the phospholipids tend to, then the only variables ($f = 3$) are temperature, electric field, and NaCl concentration. Should the lipid undergo a thermal phase transition at some temperature so that two lipid phases are present (solidus and fluidus), then the number of degrees of freedom would be reduced by one as long as both phases are present. The reduction in f would most likely be revealed by the thermal behavior of the system.

In speaking of the concentrations of salt and lipid as independent variables, it should be kept in mind these statements are equivalent to saying that water and alkane activities are independent variables. Recall that one of the major methods of reconstitution involves osmotic gradients across vesicles and bilayers, suggesting that for some purposes it is more convenient to view the system from the point of view of water activity and its effect on the physical state of the lipid.

This analysis serves to illustrate the major factors affecting the properties of the black-film system. I have not, however, shown explicitly that the degrees of freedom of the system are reflected in the properties and organization of the bilayer. I illustrate this in the next section by considering the control of bilayer properties.

5. Controlling Bilayer Properties

The best method for assessing the state of a bilayer is to measure its specific electrical capacitance (Hanai *et al.*, 1964; White, 1970, 1977; Fettiplace *et al.*, 1971; Requena and Haydon, 1975b), and I thus use this parameter to describe how the bilayer structure changes under various circumstances. The measured specific capacitance of a bilayer, $C_m = C_T/A_m$, is obtained in my laboratory by measuring the total capacitance C_T by standard electrical methods (White and Blessum, 1975) and the membrane area A_m photographically (White, 1970). C_m is determined primarily by the thickness and dielectric coefficient of the hydrocarbon core. I refer to this as the geometric capacitance C_g given by

$$C_g = \epsilon_0 \epsilon_{hc}/d_{hc} \tag{9}$$

where ϵ_{hc} and d_{hc} are the dielectric coefficient and thickness of the hydrocarbon core of the bilayer. The thickness is proportional to the number of lipid molecules n_L per unit area of film and the number of solvent molecules n_S per unit area. Thus,

$$d_{hc} = n_L V_{ac} + n_S V_S \tag{10}$$

where V_{ac} and V_S are the combined molecular volumes of the two acyl chains of the lipid and the molecular volumes of the solvent molecules, respectively (Fettiplace *et al.*, 1971, 1975; White, 1977).

There can also be a contribution C_{dl} to the specific capacitance because of the electric double layers at the bilayer surface (Everitt and Haydon, 1968) and one contributed by the geometric capacitance of the polar headgroups (Coster and Smith, 1974). The latter is not observed, however, without heroic efforts, and I do not discuss this contribution here. The contribution of C_{dl} to C_m can be significant and is accounted for by the equation

$$1/C_m = (1/C_g) + (2/C_{dl}) \tag{11}$$

because the capacitances are in series. C_{dl} has been described in detail by Everitt and Haydon (1968). It is equivalent in form to equation 9 using the dielectric coefficient and Debye length of the double layer. The latter is a strong function of ionic strength and thus electrolyte concentration.

The following sections address in turn how changes in the state of the bulk phases and the intensive variables affect the structure of the bilayer. They focus primarily on black films formed from the monoglycerides dispersed in alkanes because these are the best-studied systems and the ones that are clearly capable of being in a frozen equilibrium.

5.1. The Aqueous Phase

Figure 10 shows the variation in C_m as a function of NaCl concentration for GMO–*n*-decane membranes (White, 1973). The capacitance depends on electrolyte concentration and, except for the 1 M point, shows the expected behavior based on equation 11 (solid curve). The more interesting question, and the one more relevant to the reconstitution issue, is that of the effect of water activity on bilayer structure. The 1 M point in Fig. 10 deviates significantly from the expected curve, and this is probably because of a water activity effect. Andrews *et al.* (1970) made a similar observation and showed the cmc of GMO in alkane solvents to be greatly affected by the salt concentration. This finding makes the important point that aqueous phase hydration affects the phase behavior of the bulk forming solution, which in turn affects the bilayer. It is important to keep in mind that the bilayer will always tend to come into equilibrium with the monolayers adsorbed at the annulus–water interface. If this adsorption changes, the concentration n_L (equation 10) of the lipid in the bilayer changes. Very few measurements of effects of water activity have been reported in the literature. This is unfortunate in face of reconstitution methods that utilize osmotic stress.

The monoglycerides are slightly soluble in the aqueous phase, and it is important that the aqueous phase be in saturation equilibrium with the bulk forming solutions. No rigorous attempt was made to do this (to my embarrassment!) in the data of Fig. 10. As a result, C_g is slightly high (0.41 $\mu F/cm^2$) compared to the

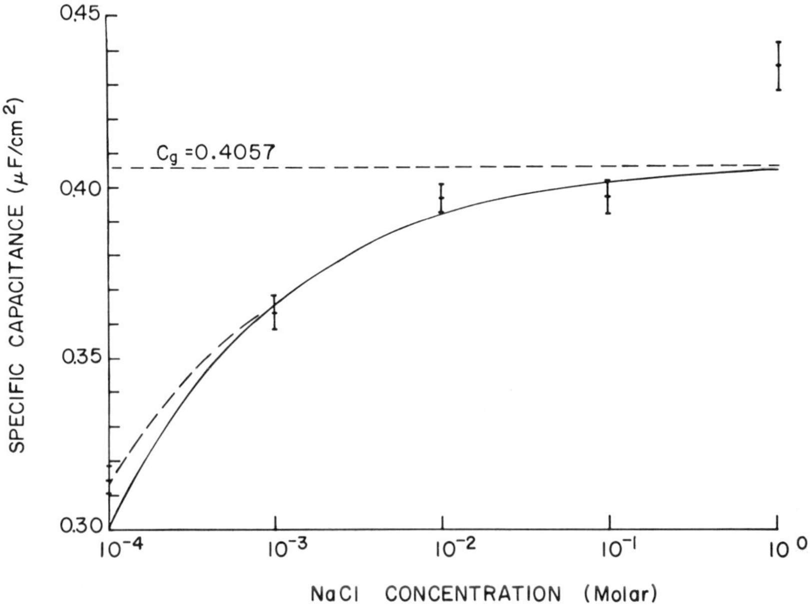

Figure 10. The specific capacitance (C_m) of bilayers formed from glycerol monooleate in NaCl solutions. the solid curve represents the theoretically expected curve (Everitt and Haydon, 1968) caused by the capacitance (C_{dl}) of the double layers at the surface of the membrane assuming the value of geometric capacitance (C_g) shown. The point in 1 M salt is high because of the effect of water activity on the state of the GMO in the annulus (Andrews *et al.*, 1970). (From White, 1973.)

correct value, 0.38 μF/cm^2, obtained when the aqueous phase has been properly equilibrated.

5.2. The Annulus Phase: Alkane Activity

Waldbillig and Szabo (1978) showed that the capacitance of films formed from GMO in *n*-decane depends strongly on GMO concentration, just as expected from the phase rule analysis described in Section 4.3. This observation confirms that the annulus in this case should be considered a single phase.

In addition to its dependence on the activity of the alkane in the annulus, the amount of alkane in the bilayer also depends on the structure of the alkane. Figure 11 shows the solubility of different alkanes in GMO bilayers as reflected by the thickness of GMO membranes (White, 1977). In all cases, the concentration of GMO was 10 mg/ml in the bulk solution so that the differences in solubility of the alkanes reflect primarily differences in molecular interactions with the bilayer itself. The solubility (thickness) depends on the size of the alkane, plotted here as the molecular volume. The figure makes another point that is important: the size parameter that determines solubility is the length of the alkane. This is il-

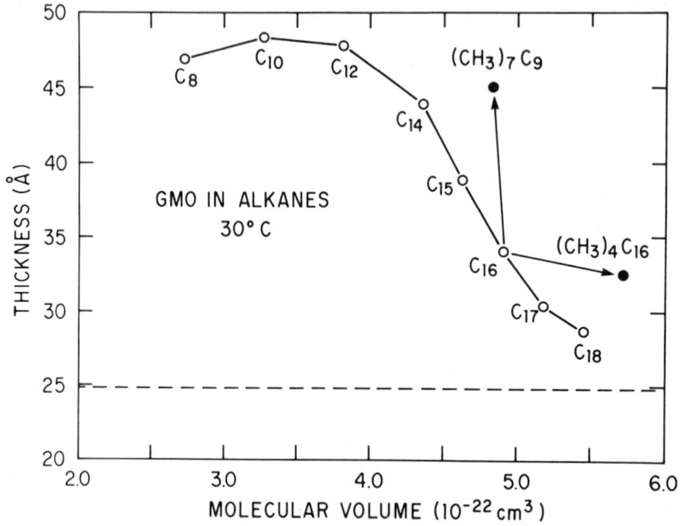

Figure 11. The thickness of planar bilayer membranes formed from glycerol monooleate (10 mg/ml) dispersed in various alkanes at 30°C as a function of alkane molecular volume. The shaded points for heptamethylnonane [labeled $(CH_3)_7C_9$] and tetramethylhexadecane [labeled $(CH_3)_4C_{16}$] demonstrate that the variable that determines alkane solubility in the bilayer is length. The dashed line represents the hypothetical solvent-free bilayer thickness. (From White, 1977.)

lustrated by the hexadecane analogues heptamethylnonane [labeled $(CH_3)_7C_9$] (which has the volume of hexadecane but the length of about decane) and tetramethylhexadecane [labeled $(CH_3)_4C_{16}$] {which has the volume of eicosane [$(CH_3)_2(CH_2)_{18}$] but the length of hexadecane}.

An important implication of Fig. 11 is that with an alkane of sufficient length, one should be able to create a bilayer that contains little solvent by spreading from bulk solution. This possibility was borne out by forming membranes using squalene, which is about 34 carbons long. Solvent-free membranes with a thickness of about 25 Å (Fig. 11) should have a capacitance of about 0.78 $\mu F/cm^2$. Figure 12 shows the capacitance of GMO–squalene membranes as function of temperature (White, 1978). The expected solvent-free capacitance is observed, suggesting that little solvent is present. Waldbillig and Szabo (1979) extended this idea to the use of triglycerides, which also seem to be insoluble in the bilayer.

Also shown in Fig. 12 are measurements reported by Petersen (1983) for GMO–triolein. These membranes are somewhat thinner, suggesting that there may be very small amounts of squalene remaining in the squalene-formed membranes. This point is not entirely clear, however, because the interfacial adsorption from triolein and squalene solutions might be different, leading to different areas/molecule in the bilayer.

The phase rule analysis of Section 4.3 and the findings of Waldbillig and Szabo (1978) noted above show that the activity of the alkane in the annulus determines the concentration in the bilayer. Because Squalene is insoluble in the bilayer but

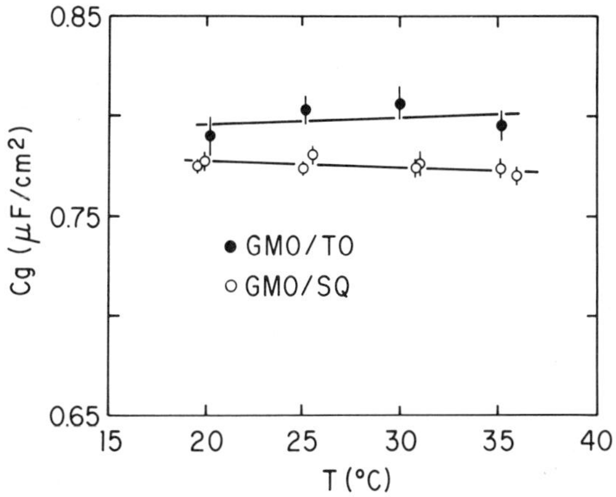

Figure 12. The specific capacitance of "solvent-free" bilayers formed from glycerol monooleate dispersed in squalene (SQ) or triolein (TO). The capacitances correspond closely to those expected of a membrane containing no solvent and having a thickness of 25 Å (Fig. 11). The TO curve is higher than the SQ curve because either some SQ remains in the bilayer or the adsorption of GMO at the SQ–water and TO–water interfaces of the annuli are different. The SQ data are from White (1978), and the TO data from Petersen (1983).

completely miscilbe with alkanes, it is possible to control arbitrarily the amount of alkane in the bilayer by mixing various mole fractions of squalene into the bulk forming solution (White, 1979; Needham and Haydon, 1983). Figure 13 shows the mole fraction of decane in bilayers formed from decane–squalene–GMO solutions with various mole fractions of decane (White, 1979). Here I have plotted mole

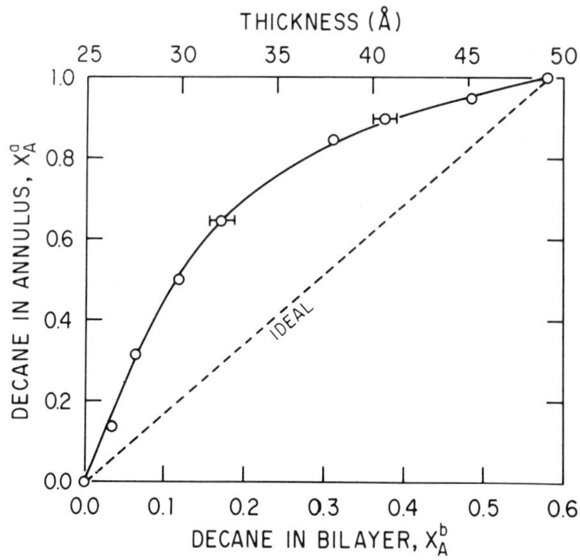

Figure 13. The amount of *n*-decane in GMO bilayers is determined by the activity of *n*-decane in the annulus. $X_A{}^b$ is the mole fraction of *n*-decane in the bilayer, and $X_A{}^a$ is the mole fraction in the annulus resulting from the mixing of squalene with the *n*-decane. The data are plotted in this way because $X_A{}^a$ will be proportional to the vapor pressure of the squalene–*n*-decane mixture; the nonideality of the interaction of the decane with bilayer is thus demonstrated. The deviation of the curve from the ideal case (dashed curve) is positive, indicating a tendency toward phase separation. Data based on those of White (1979).

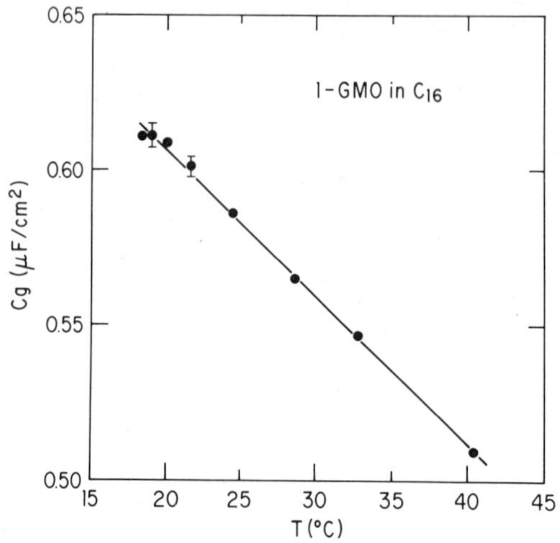

Figure 14. The specific geometric capacitance (C_g) of planar bilayers formed from GMO–*n*-hexadecane as a function of temperature. The solubility of *n*-hexadecane in the bilayer increases with temperature, causing the thickness of the bilayer to increase. Data taken from White (1976).

fraction X_A^a of alkane in annulus (proportional to the vapor pressure of the decane solution) against mole fraction X_A^b of alkane in bilayer to examine directly the ideality of the decane–bilayer solution. If the decane and bilayer mix ideally, then a Raoult's Law straight line relationship would result (dashed curve).

Figure 13 shows two things. First, the amount of decane in the film clearly depends on the activity of the decane in the annulus. Second, it forms a very nonideal solution in the bilayer, showing a large positive deviation from ideality. This nonideal behavior causes the thickness of the bilayer to be strongly dependent on electric field strength (*vide infra*).

The solubility of alkanes in black films can be strongly temperature dependent (White, 1970, 1976). Figure 14 shows the specific capacitance of GMO–hexadecane membranes as a function of temperature under equilibrium conditions. The amount of hexadecane in the bilayer depends very strongly on temperature, with solubility increasing as temperature increases. Because the bilayer is in saturation equilibrium with the annulus, one can determine the enthalpy and entropy of solution of the hexadecane in the bilayer from such measurements (White, 1976). The enthalpy of solution in this case is 3.8 ± 0.1 kcal/mol for the transfer of hexadecane from annulus to bilayer. This means that the binding energy of the hexadecane in the bulk phase of the annulus is considerably larger than in the bilayer. This is strong evidence that the interior of the bilayer is not equivalent to a bulk alkyl liquid. The main point is, however, that temperature can be an important experimental parameter as expected from the phase rule. Only if the enthalpy of solution of the alkane in the bilayer were zero or the alkane were insoluble in the bilayer would the capacitance change with temperature be small, as in Fig. 12.

A surprising result is that the annulus may be frozen without the bilayer breaking (White, 1974, 1975). Figure 15 shows measurements of C_g for GMO–hexadecane films as the temperature is decreased to less than the melting point (T_m = 16°C) of *n*-hexadecane. There is an abrupt increase in C_g at the T_m of

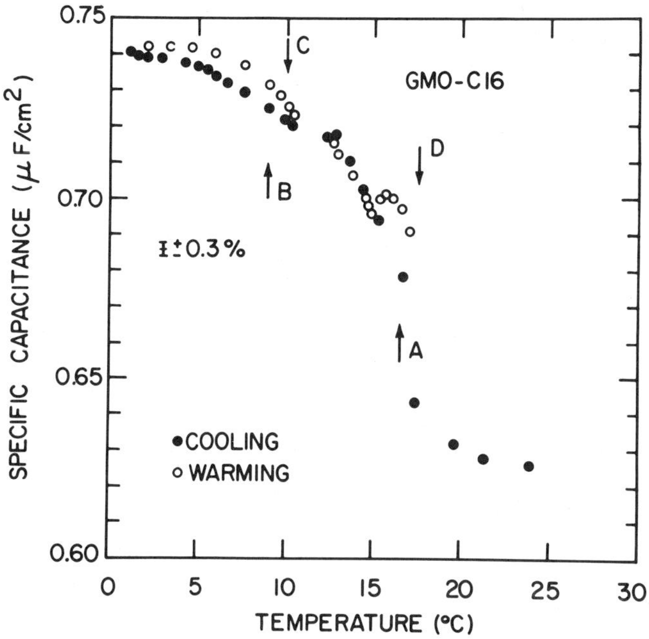

Figure 15. The temperature dependence of the specific geometric capacitance (C_g) of GMO–*n*-hexadecane membranes as the temperature is lowered through the melting point of *n*-hexadecane (16°C). The annulus freezes, but the membrane does not break. There is an abrupt increase in C_g at the freezing point because solid hexadecane is soluble in the bilayer. (From White, 1975.)

hexadecane, apparently because of the freezing out of the hexadecane from the bilayer. This means that hexadecane in the solid state is much less soluble in the bilayer than in the liquid state. It is of interest to apply the phase rule (equation 7) to this system. Assuming no electric field and constant pressure, $i = 1$. There are five components (air, water, NaCl, hexadecane, lipid) and, above T_m, three phases, so that $f = 3$ (water activity, alkane activity, and temperature). Below T_m, the annulus has solidified. If it remains a single phase (a GMO–hexadecane mixture), there are still $f = 3$ degrees of freedom, and the capacitance will depend on the composition of the annulus. If the annulus forms two phases (solid hexadecane and GMO do not mix), then $f = 2$, and C_g will be independent of alkane activity. No experiments have been done to test this issue.

5.3. The Annulus Phase: Lipid Activity

Little work on the effect of lipid-phase behavior on planar bilayer properties has been reported in the literature. The first that I am aware of is that of Krasne *et al.* (1971), but they worked with a complex solution of uncertain composition that was not characterized well physicochemically. Pagano *et al.* (1973) used optical measurements of thickness to study membranes formed from hexadecane solutions containing GMO, glycerol monostearate (GMS), or 1:1 mixtures of GMO–GMS. The GMO and GMS in hexadecane in the presence of excess water have melting isotherms at $T_m = 16°C$ and $T_m = 55°C$, respectively. When $T < T_m$ for GMS-formed membranes, there was a reversible increase in thickness that was consistent with the formation of a gel-state lipid in the bilayer, possibly accompanied by a phase separation of the hexadecane into microlenses. Interestingly, when the GMS–GMO mixture was used and $T < T_m$ for GMS, the bilayer seemed to consist of two phases, judging by the appearance of very bright spots on the surface of the membrane. These were interpreted as GMS crystalline bilayer regions dispersed in liquid–state GMO. The phase rule allows us to decide if this is reasonable.

The phase rule (equation 8) for the situation in which there is more than one surface phase on a given surface can be applied to the experiment. No electric field is present, and pressure is fixed, so $i = 1$. The components are air, water, GMO, GMS, hexadecane, and salt, making $c = 6$. The bulk phases are water, air, and the annulus phases. Above T_m for GMS, the annulus may be counted as a single phase, whereas below T_m it may be counted as two phases because of the precipitated GMS or possibly a GMS–GMO complex. Thus p is taken as 3 for $T > T_m$ and 4 for $T < T_m$. The types of surfaces are the air–water interface, the annulus–water interface, and the bilayer, so that $n = 3$. The major question concerns the number of surface phases. Can the bilayer have liquid GMO and gel GMS on the surface simultaneously?

Consider first the situation above T_m, where I assume $n = s$. There are four degrees of freedom ($f = 1 + 6 - 3 = 4$) which include temperature, NaCl concentration, and the GMS and GMO concentrations. Below T_m, if there are two surface phases on each surface, then $s = 6$ and $f = 1 + 6 - 4 - (6 - 3) = 0$. This could occur only if hexadecane activity ceased depending on GMO concentration in the still-liquid hexadecane and if water activity no longer had an effect on the GMO adsorption. These situations seem unlikely, and I conclude that there are not two surface phases on each surface. If, on the other hand, each surface has only one surface phase, and we attribute the bright spots to GMS crystallized into microlenses, then $s = 3$ and $f = 1 + 6 - 4 - (3 - 3) = 3$ degrees of freedom, which presumably would be temperature, NaCl concentration, and GMO concentration, the GMS having precipitated to form a bulk lipid phase of fixed activity. The point of this exercise has been to indicate that phase rule analysis allows one to think through a complex situation systematically. It further suggests the types of experiments that would have to be done to test various possibilities for the state of the bilayer.

Boheim *et al.* (1980) have done an experiment similar to that of Pagano *et al.* (1973) except they used a mixed-chain PC and formed the bilayers by the method of Montal and Mueller (1972). Hexadecane was used to pretreat the septum. They observed the thickness of the black film to increase 12% above the T_m of the lipid, indicating a structural change in the bilayer accompanying the change in the state of the lipid in the aqueous phase. The phase rule analysis: if T is not equal to T_m, $i = 1$, $c = 5$, $p = 4$ (bulk water, bulk lipid, air, annulus), and $n =$

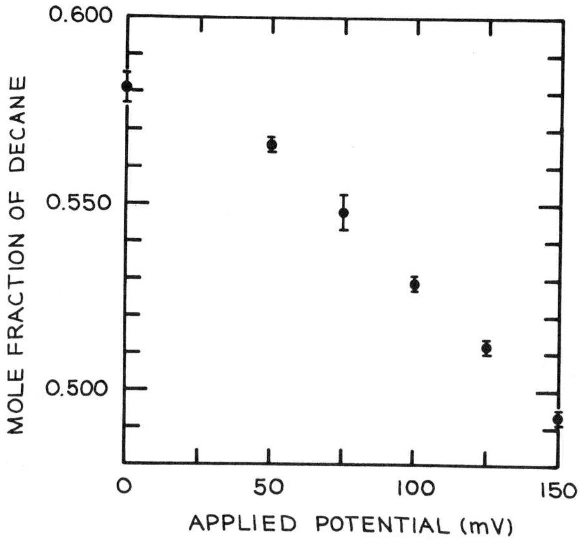

Figure 16. The amount of *n*-decane in a GMO bilayer as a function of the voltage across the membrane. The voltage creates a small pressure (~0.1 atm for 100 mV) in the bilayer, which raises the chemical potential of the decane relative to the annulus. Because the chemical potential of the alkane in the bilayer must be the same as in the annulus, there is a compensatory decrease in amount of decane in the membrane. A pressure of the size induced by the field could not have such a large effect unless the decane formed a very nonideal mixture with the bilayer. Figure 13 shows the nonideality. (From White, 1980.)

s, then $f = 1 + 5 - 4 = 2$, with the variables being water activity and temperature. During the melt there will be two lipid phases present simultaneously so that $p = 5$; the system is isothermal, and $f = 1$ (water activity). It is very interesting that Boheim *et al.* observed significant current fluctuations at the transition temperature as might be expected if two surface phases (gel and liquid crystalline) were present in the bilayer at T_m.

5.4. Electric Field Effects

It was observed very early in their history that electric fields affect C_T of black films (Rosen and Sutton, 1968). White (1970) demonstrated conclusively that this is a result of an increase in C_g resulting from a decrease in thickness. Andrews *et al.* (1970) showed that the extent of the effect depended on the alkane used, decane-formed films being much more sensitive than hexadecane formed films. Figure 16 shows the amount of decane in a GMO bilayer as a function of applied voltage V (White, 1980). This very strong effect is a result of the nonideal behavior of the decane in the bilayer (Fig. 13) (Andrews, 1970; White, 1979, 1980). The electric field causes an increase in pressure in both the annulus and the bilayer through electrostriction (equation 6), which is proportional to V^2. Because the bilayer is very thin compared to the annulus, the pressure in the bilayer is orders of magnitude greater than that in the annulus. As a result, a 100 mV potential will increase the pressure in the bilayer by about 0.1 atm relative to the annulus, which therefore raises the chemical potential relative to the annulus. Because the annulus is so massive compared to the bilayer, the chemical potential of the bilayer must be reduced in some way in the face of the electric field. The reduction comes about by a decrease in the amount of decane in the film, since activity is proportional to concentration. The electric field can have this effect only if the alkane and the bilayer form a highly nonideal solution. If the decane formed an ideal mixture with the GMO of the bilayer, the effect shown in Fig. 16 would be an order of magnitude smaller. Note in Fig. 13 that bilayer thickness is extraordinarily sensitive to the mole fraction (activity) of the decane in the annulus.

The fact that hexadecane-formed bilayers are not very sensitive to voltage can be attributed to two factors. First, there is a much smaller amount of hexadecane in the membrane (Fig. 11). Second, hexadecane forms a more nearly ideal solution with the bilayer, i.e., it does not have the large positive deviation from ideality seen for decane (Fig. 13).

6. The Vagaries of Black Films: Nonequilibrium

The time dependence of the specific capacitance of black films provides a convenient criterion for equilibrium (metastable) because C_g depends strongly on composition. C_g can be measured with a precision of $\pm 0.3\%$ (White and Blessum, 1975); for films formed from monoglycerides and alkanes, C_g does not change within this limit for many hours at a time. Therefore, these films appear to be in equilibrium (metastable). That is not necessarily true for films formed from oxidized cholesterol (White, 1970) or diacylphospholipids (White and Thompson, 1973; Benz and Janko, 1976), whose specific capacitances can change drastically with time and be very nonreproducible.

The changes must occur because the alkane and/or lipid concentrations in the film change. The loss of alkane to the annulus and microlenses is partly responsible for the changes in oxidized cholesterol membranes because the application of a potential for a few minutes leads quickly to membranes with time-independent capacitances (White, 1970). The electric field hastens the movement of the solvent into the annulus and microlenses. Applied voltages have little effect, however, on the time dependence of diacylphospholipid capacitances (White and Thompson, 1973), and for these systems it is probably the adsorption of the lipid at the annulus–water interface and, consequently, the bilayer that is changing.

There are at least four circumstances under which adsorption is likely to change. First, the chemical structure of the lipid might change as a result of oxidation (Huang *et al.*, 1964). Second, the adsorption of the lipid at the interfaces is slow so that a long time is required for equilibrium. Johnson and Saunders (1973) have demonstrated that the interfacial tension at the interface between water and a cyclohexane–lecithin mixture changes from 40 dyne/cm to 2 dyne/cm over a 5-hr period. Third, the aqueous phase is not in equilibrium with the bulk forming solution, so that the lipid desorbs into the aqueous phase, causing the interfacial concentration to be low compared to equilibrium. I indicated earlier that the capacitance of GMO–decane membrane is lowered from about 0.41 to 0.38 $\mu F/cm^2$ when care is taken to equilibrate the two phases. Solubility in the aqueous phase is not likely to be a problem with lecithin. However, one must consider the possibility that lecithin dispersed in excess water has a lower free energy than lecithin dispersed in alkane, leading to a steady loss of lipid to the creation of another phase. Fourth, the aggregated state of the lipid in the annulus may be changing with time because of changes in lipid hydration. Most of the changes observed for the diacylphospholipids (White and Thompson, 1973) are probably caused by this effect. Many of the frustrations of black films can be traced to these nonequilibrium conditions as discussed below.

A common scenario for the formation of black films from lecithins is as follows. A certain amount of lecithin is prepared by lyophilization from a choloform solution. Then alkane is added to make a 1% solution. However, unless the lipid is allowed to absorb some water from the atmosphere or a small amount of methanol is added, it will not disperse and remains as a lump in the bottom of the flask. Immediately after dispersal, the forming solution has the appearance and gross properties of pure alkane. However, if the "solution" is allowed to stand for several hours exposed to moist air (as during summer in Virginia, where I first observed this effect), it assumes the characteristics of a gel. If a small glass rod is dipped into the solution and withdrawn, long strands of it can be pulled from the surface several centimeters. The lecithin apparently continues to take on water and ultimately forms a microemulsion.

If one forms bilayers with the freshly formed solutions, their properties are likely to change drastically with time. Further, one's ability to form the bilayers is time dependent. Often when one attempts to form lecithin films from fresh solutions it is difficult to get them to span the aperture. Even if this step is accomplished, the films are likely to break before reaching the fully black state. If one waits for an hour, the bilayers form without problems, but the specific capacitance is likely to be highly variable. If several additional hours pass, the now well-hydrated forming solution becomes very gellike, and the membranes thin slowly and, once formed, have a stringy appearance with long strands of the annulus stretching into the bilayer region. At the same time the optical reflectance may be highly irregular as though there were patches of multilayers: end of experiment.

7. Epilogue

Finkelstein (1974) has related ". . . some of the folklore surrounding the technique of making lipid membranes, as revealed in both the written literature and the oral tradition." He further comments, ". . . when everything is working properly, bilayers that are stable indefinitely can be made at will. Like any technique, it has its own mysticism." I hope this chapter helps to remove black films from the realms of folklore and and mysticism. There is nothing unusual about them as long as they are accepted as complex heterogeneous systems that operate according to the standard rules of physical chemistry. Biological systems must also obey these rules. In the process of learning how to reconstitute channels into black films whose composition we can control, I suspect we will also learn a good deal about how biological systems use the rules of physical chemistry to accomplish the same goal.

ACKNOWLEDGMENTS. Much of the work described in this chapter was performed in my laboratory over a 10-year span with the help of research grants from the National Science Foundation, the Guggenheim Multiple Sclerosis Foundation, the National Institute of General Medical Sciences, and the National Institute of Neurological Diseases, Stroke, and Communicative Disorders. I was supported by a Research Career Development Award from the last organization from 1976 through 1981. I appreciate the help of Ms. Terri Pouliot and Ms. Jaime Box in preparing this manuscript.

References

Adamson, A. W., 1967, *Physical Chemistry of Surfaces*, ed. 2, Wiley–Interscience, New York.

Andrews, D. M., 1970, Dissertation, Cambridge University, Cambridge.

Andrews, D. M., Manev, E. D., and Haydon, D. A., 1970, Composition and energy relationships for some thin lipid films, and the chain conformation in monolayers at liquid–liquid interfaces, *Special Disc. Faraday Soc.* **1**:46–56.

Aveyard, R., and Haydon, D. A., 1973, *An Introduction to the Principles of Surface Chemistry*, Cambridge University Press, New York.

Bangham, A. D., 1980, Development of the liposome concept, in: *Liposomes in Biological Systems* (G. Gregoriadis and A. C. Allison, eds.), pp. 1–24, John Wiley & Sons, New York.

Bangham, A. D., Hill, M. W., and Miller, N. G. A., 1974, Preparation and use of liposomes as models of biological membranes, in: *Methods in Membrane Biology*, Vol. 1 (E. D. Korn, ed.), pp. 1–68, Plenum Press, New York.

Benz, R., and Janko, K., 1976, Voltage-induced capacitance relaxation of lipid bilayer membranes, effects of membrane composition, *Biochim. Biophys. Acta* **455**:721–738.

Boheim, G., Hanke, W., and Eibl, H., 1980, Lipid phase transition in planar bilayer membrane and its effect on carrier- and pore-mediated ion transport, *Proc. Natl. Acad. Sci. USA* **77**:3403–3407.

Brooks, D. E., Levine, Y. K., Requena, J., and Haydon, D. A., 1975, Van der Waals forces in oil–water systems from the study of thin lipid films, III. Comparison of experimental results with Hamaker coefficients calculated from Lifshitz theory, *Proc. R. Soc. Lond.* **A347**:179–194.

Coster, H. G. L., and Smith, J. R., 1974, The molecular organization of bimolecular lipid membranes, a study of the low frequency Maxwell–Wagner impedance dispersion, *Biochim. Biophys. Acta* **373**:151–164.

DeFay, R., and Progogine, I. 1966, *Surface Tension and Absorption*, Longmans, Green, London.

Dominguez, J. G., Willhite, G. P., and Green, D. W., 1979, Phase behavior of microemulsion systems with emphasis on effects of paraffinic hydrocarbons and alcohols, in: *Solution Chemistry of Surfactants*. Vol. 2 (K. L. Mittal, ed.), pp. 673–697, Plenum Press, New York.

Ekwall, P., 1969, Two types of micelle formation in organic solvents, *J. Colloid Interface Sci.* **20**:16–26.

Eriksson, J. C., 1971, Thermodynamics of surface-phase systems: On the rigorous thermodynamics of insoluble surface films, *J. Colloid Interface Sci.* **37**:659–667.

Evans, E. A., and Parsegian, V. A., 1983, Energetics of membrane deformation and adhesion in cell and vesicle aggregation, *Ann. N.Y. Acad. Sci.* **416**:13–33.

Evans, E. A., and Skalak, R., 1979a, Mechanics and thermodynamics of biomembranes, Part I, *CRC Crit. Rev. Bioeng.* **3**:181–330.

Evans, E. A., and Skalak, R., 1979b, Mechanics and thermodynamics of biomembranes, Part II, *CRC. Crit. Rev. Bioeng.* **3**:331–418.

Everitt, C. T., and Haydon, D. A., 1968, Electrical capacitance of a lipid membrane separating two aqueous phases, *J. Theor. Biol.* **18**:371–379.

Fettiplace, R., Andrews, D. M., and Haydon, D. A., 1971, The thickness, composition and structure of some lipid bilayers and natural membranes, *J. Membr. Biol.* **5**:277–296.

Fettiplace, R., Gordon, L. G. M., Hladky, S. B., Requena, J., Zingsheim, H. P., and Haydon, D. A., 1975, Techniques in the formation and examination of "black" lipid bilayer membranes, in: *Methods of Membrane Biology*, Vol. 4 (E. D. Korn, ed.), pp. 1–75, Plenum Press, New York.

Finkelstein, A., 1974, Bilayers: Formation, measurements, and incorporation of components, in: *Methods of Enzymology*, Volume XXXII, *Biomembranes*, Part B (S. Fleischer and L. Packer, eds.), pp. 489–501, Academic Press, New York.

Franks, N. P., and Lieb, W. R., 1978, Where do general anaesthetics act? *Nature* **274**:339–342.

Franks, N. P., and Lieb, W. R., 1981, Is membrane expansion relevant to anaesthesia?, *Nature* **292**:248–251.

Gaines, G., 1966, *Insoluble Monolayers at Liquid–Gas Interfaces*, Wiley–Interscience, New York.

Gershfeld, N. L., 1974, Thermodynamics and experimental methods for equilibrium studies with lipid monolayers, in: *Methods in Membrane Biology*, Vol. 1 (E. D. Korn, ed.), pp. 69–104, Plenum Press, New York.

Gershfeld, N. L., 1976, Physical chemistry of lipid films at fluid interfaces, *Annu. Rev. Phys. Chem.* **27**:349–368.

Gibbs, J. W., 1961, *The Scientific Papers of J. Willard Gibbs*, Vol. I. *Thermodynamics*, Dover Publications, New York.

Gruen, W. R., and Marcelja, S., 1984, Water structure near the membrane surface, in: *Cell Surface Dynamics: Concepts and Models* (A. S. Perelson, C. DiLisi, and F. W. Wiegel, eds.), pp. 59–91, Marcel Dekker, New York.

Guggenheim, E. A., 1977, *Thermodynamics*, ed. 6, North-Holland, New York.

Hanai, T., Haydon, D. A., and Taylor, J., 1964, An investigation by electrical methods of lecithin-in-hydrocarbon films in aqueous solutions, *Proc. R. Soc. Lond.* **A281**:377–391.

Haydon, D. A., 1970, The organization and permeability of artificial lipid membranes, in: *Membranes and Ion Transport*, Vol. I (E. E. Bittan, ed.), pp. 64–92, Wiley–Interscience, London.

Haydon, D. A., and Taylor, J. L., 1968, Contact angles for thin lipid films and the determination of London–van der Waals forces, *Nature* **217**:739–740.

Henn, F. A., and Thompson, T. E., 1969, Synthetic lipid bilayer membranes, *Annu. Rev. Biochem.* **38**:241–262.

Huang, C., Wheeldon, L., and Thompson, T. E., 1969, The properties of lipid bilayer membranes separating two aqueous phases: Formation of a membrane of simple composition, *J. Mol. Biol.* **8**:148–160.

Israelachvili, J. N., Marcelja, S., and Horn, R. G., 1980, Physical principles of membrane organization, *Q. Rev. Biophys.* **13**:121–200.

Jain, M., 1972, *The Bimolecular Lipid Membrane: A System*, Van Nostrand Reinhold, New York.

Johnson, M. C. R., and Saunders, L., 1973, Time dependent interfacial tensions of a series of phospholipids, *Chem. Phys. Lipids* **10**:318–327.

Krasne, S., Eisenman, G., and Szabo, G., 1971, Freezing and melting of lipid bilayers and the mode of action of nonactin, valinomycin, and gramicidin, *Science* **174**:412–415.

Langmuir, I., 1933, Oil lenses on water and the nature of monomolecular expanded films, *J. Chem. Phys.* **1**:756–776.

Luisi, P. L., and Straub, B. E., 1984, *Reverse Micelles*, Plenum Press, New York.

Miller, C., 1983a, First steps in the reconstruction of ionic channel functions in model membranes, in: *Current Methods in Cellular Neurobiology*, Vol. 3: *Electrophysiological Techniques* (J. L. Barber, ed.), pp. 1–37, John Wiley & Sons, New York.

Miller, C., 1983b, Integral membrane channels: Studies in model membranes, *Physiol. Rev.* **63**:1209–1242.

Miller, C., 1984, Ion channels in liposomes, *Annu. Rev. Physiol.* **46**:549–558.
Miller, C., and Racker, E., 1976, Ca^{++}-induced fusion of fragmented sarcoplasmic reticulum with artificial planar bilayer, *J. Membr. Biol.* **30**:283–300.
Montal, M., and Mueller, P., 1972, Formation of bimolecular membranes from lipid monolayers and a study of their electrical properties, *Proc. Natl. Acad. Sci. U.S.A.* **69**:3561–3566.
Montal, M., Darszon, A., and Schindler, H., 1981, Functional reassembly of membrane proteins in planar lipid bilayers, *Q. Rev. Biophys.* **14**:1–79.
Mueller, P., Rudin, D., Tien, H. T., and Westcott, W. C., 1962, Reconstitution of excitable cell membrane structure *in vitro*, *Circulation* **26**:1167–1171.
Needham, D., and Haydon, D. A., 1983, Tensions and free energies of formation of "solventless" lipid bilayers, *Biophys. J.* **41**:251–257.
Ostro, M. J., 1983, *Liposomes*, Marcel Dekker, New York.
Pagano, R. E., Cherry, R. J., and Chapman, D., 1973, Phase transitions and heterogeneity in lipid bilayers, *Science* **181**:557–559.
Parsegian, V. A., 1975, Long range van der Waals forces, in: *Physical Chemistry: Enriching Topics from Colloid and Surface Science* (H. van Olphen and K. J. Mysels, eds.), pp. 27–72, Theorex, La Jolla.
Parsegian, V. A., and Rand, R. P., 1983, Membrane interaction and deformation, *Ann. N. Y. Acad. Sci.* **146**:1–12.
Parsegian, V. A., Rand, R. P., and Gingell, D., 1984, Lessons for the study of membrane fusion from membrane interactions in phospholipid systems, *Cell Fusion*, pp. 9–27, Pitman, London.
Pattus, F., Desnuelle, P., and Verger, R., 1978, Spreading of liposomes at the air/water interface, *Biochim. Biophys. Acta* **507**:62–70.
Petersen, D. C., 1983, The water permeability of the monoolein/triolein bilayer membrane, *Biochim. Biophys. Acta* **734**:201–209.
Phillips, M. C., and Hauser, H., 1974, Spreading of solid glycerides and phospholipids at the air–water interface, *J. Colloid Interface Sci.* **49**:31–39.
Requena, J., and Haydon, D. A., 1975a, The Lippmann equation and the characterization of black lipid films, *J. Colloid Interface Sci.* **51**:315–327.
Requena, J., and Haydon, D. A., 1975b, Van der Waals forces in oil–water systems from the study of thin lipid films, Part II, *Proc. R. Soc. Lond.* **A347**:161–177.
Rosen, D., and Sutton, A. M. 1968, The effects of a direct current potential bias on the electrical properties of bimolecular lipid membranes, *Biochim. Biophys. Acta* **163**:226–233.
Schindler, H., 1979, Exchange and interactions between lipid layers at the surface of a liposome solution, *Biochim. Biophys. Acta* **555**:316–336.
Schindler, H., 1980, Formation of planar bilayers from artificial or native membrane vesicles, *FEBS Lett.* **122**:77–79.
Tajima, K., and Gershfeld, N. L., 1978, Equilibrium studies of lecithin–cholesterol interactions II. Phase relations in surface films: Analysis of the condensing effect of cholesterol, *Biophys. J.* **22**:489–500.
Takagi, M., Azuma, K., and Kishimoto, U., 1965, A new method for the formation of bilayer membranes in aqueous solution, *Ann. Rep. Biol. Works Fac. Sci.* **13**:107–110.
Tanford, C., 1980, *The Hydrophobic Effect: Formation of Micelles and Biological Membranes*, John Wiley & Sons, New York.
Taylor, J., and Haydon, D. A., 1966, Stabilization of thin films of liquid hydrocarbon by alkyl chain interaction, *Disc. Faraday Soc.* **42**:51–59.
Tien, H. T., 1974, *Bilayer Lipid Membranes (BLM)*, Marcel Dekker, New York.
Tien, H. T., and Diana, A. L., 1968, Bimolecular lipid membranes: A review and summary of some recent studies, *Chem. Phys. Lipids.* **2**:55–101.
Tolman, R. C., 1913, The general principles of equilibria in divided systems, *J. Am. Chem. Soc.* **35**:307–333.
Waldbillig, R. C., and Szabo, G., 1978, Solvent-depleted bilary membranes from concentrated lipid solutions, *Nature* **272**:839–840.
Waldbillig, R. C., and Szabo, G., 1979, Planar bilayer membranes from pure lipids, *Biochim. Biophys. Acta* **557**:295–305.

White, S. H., 1970, A study of lipid bilayer membrane stability using precise measurements of specific capacitance, *Biophys. J.* **10:**1127–1148.
White, S. H., 1972, Analysis of the torus surrounding planar lipid bilayer membranes, *Biophys. J.* **12:**432–445.
White, S. H., 1973, The surface charge and double layers of thin lipid films formed from neutral lipids, *Biochim. Biophys. Acta* **323:**343–350.
White, S. H., 1974, Temperature-dependent structural changes in planar bilayer membranes: Solvent "freeze-out," *Biochim. Biophys. Acta* **356:**8–16.
White, S. H., 1975, Phase transitions in planar bilayer membranes, *Biophys. J.* **15:**95–117.
White, S. H., 1976, The lipid bilayer as a "solvent" for small hydrophobic molecules, *Nature* **262:**421–422.
White, S. H., 1977, Studies of the physical chemistry of planar bilayer membranes using high-precision measurements of specific capacitance, *Ann. N.Y. Acad. Sci.* **303:**243–265.
White, S. H., 1978, Formation of "solvent-free" black lipid bilayer membranes from glyceryl monooleate dispersed in squalene, *Biophys. J.* **23:**337–347.
White, S. H., 1979, Mechanism of the compression of black lipid membranes by an electric field, *Biophys. J.* **25:**9a.
White, S. H., 1980, How electric fields modify alkane solubility in lipid bilayers, *Science* **207:**1075–1077.
White, S. H., and Blessum, D. N., 1975, High precision capacitance bridge for studying lipid bilayer membranes, *Rev. Sci. Instrum.* **46:**1462–1466.
White, S. H., and Chang, W., 1981, Voltage dependence of the capacitance and area of black lipid membranes, *Biophys. J.* **36:**449–453.
White, S. H., and King, G. I., 1985, Molecular packing and the area compressibility of lipid bilayers, *Proc. Natl. Acad. Sci. U.S.A.* **82:**6532–6536.
White, S. H., and Thompson, T. E., 1973, Capacitance, area, and thickness variations in thin lipid films, *Biochim. Biophys. Acta* **323:**7–22.
White, S. H., Petersen, D. C., Simon, S., and Yafuso, M., 1976, Formation of planar bilayer membranes from lipid monolayers, *Biophys. J.* **16:481–489.**

Chapter 2

ION CHANNEL ELECTROSTATICS AND THE SHAPES OF CHANNEL PROTEINS

PETER C. JORDAN

1. Introduction

The accelerating progress that is being made in channel reconstitution will eventually lead to the determination of protein structures, i.e., of channel shapes. These will surely not all be right circular cylinders, and a pore's structure can affect ion passage through it in two ways. Short-range forces govern selectivity and the detailed nature of the potential energy profile for ion permeation. Long-range electrostatic forces control ion access to the channel, influence the relationship of apparent electrical distance to structure, and provide a slowly varying background potential, which modifies channel kinetics.

This chapter treats channel electrostatics and focuses on three major topics. How do various electrical sources, e.g., the permeant ion, an applied voltage, the lipid–water potential difference, affect the potential profile within a pore? How is this profile altered by changes in the shape and charge distribution of the channel protein and in the properties of the bilayer? How can these differences be exploited to provide evidence for specific structural features of the channel proteins?

2. The Electrostatic Model and Its Limitations

Water, pore-forming structures, and phospholipid membranes are complex ensembles containing thousands of interacting functional groups. The various channel systems that have been identified share the same basic purpose—to facilitate and control the transport of a selected solute across the cell membrane. A channel in its permeable state has the general topology illustrated in Fig. 1A. There are four clearly distinguishable dielectric domains, of which two, bulk water and lipid membrane, are relatively easy to characterize; their dielectric constants are ~80 and ~2, respectively. Except for gramicidin (Urry, 1971; Koeppe *et al.*,

PETER C. JORDAN • Department of Chemistry, Brandeis University, Waltham, Massachusetts 02254.

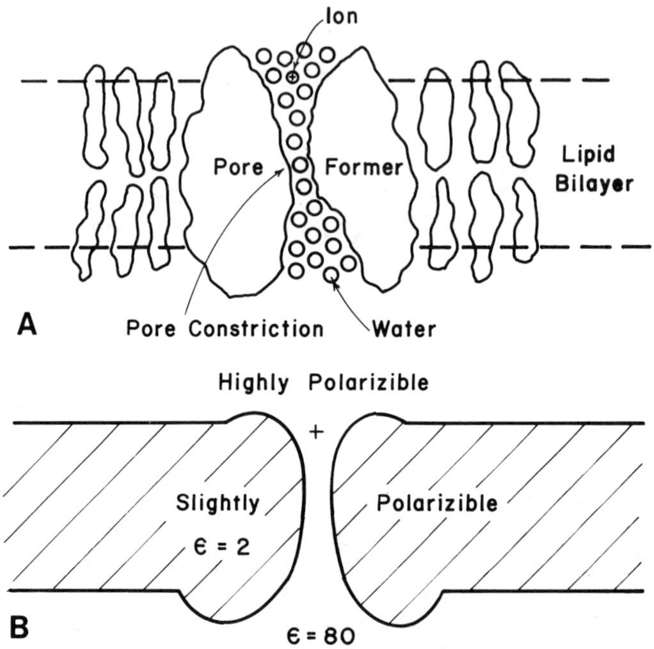

Figure 1. (A) Topology of the ion–water–pore former–lipid ensemble; the discrete nature of the ion–water interaction is emphasized. (B) Simplified two-dielectric, continuum, point-charge description.

1978), for which the conformation of the pore former is known, and the acetylcholine receptor (AChR) channel (Kistler and Stroud, 1981), for which both the pore shape and the amino acid sequence in the five polypeptide subunits are known, detailed structural information about channel formers is absent. All available data are consistent with an intuitively plausible model in which the regions abutting the water are relatively polar and those abutting the lipid are hydrophobic. Consequently, the pore former may be described as mainly lipidlike with thin, highly polarizible surface layers at the water–pore former interfaces (Jordan, 1981).

The dielectric properties of water within the pore are unknown. Since the dielectric constant depends on local solvent structure (Kirkwood, 1939), pore water and bulk water are not equally polarizible. Molecular dynamics simulation of coupled cation–water motion in gramicidinlike pores (MacKay *et al.*, 1984; Lee and Jordan, 1984) show that water in narrow, single-filing pores is structurally very different from bulk water. Consequently, in the constricted region of such highly selective pores as the Na^+ channel (Hille, 1975), the various K^+ channels (Latorre and Miller, 1983), and the Ca^{2+} channel (Hagiwara and Byerly, 1981), uncertainty is introduced by assuming that pore and bulk water are dielectrically equivalent. For relatively open pores such as porins (Benz *et al.*, 1980) and the AChR channel (Maynard *et al.*, 1977; Kistler and Stroud, 1981), the pore water

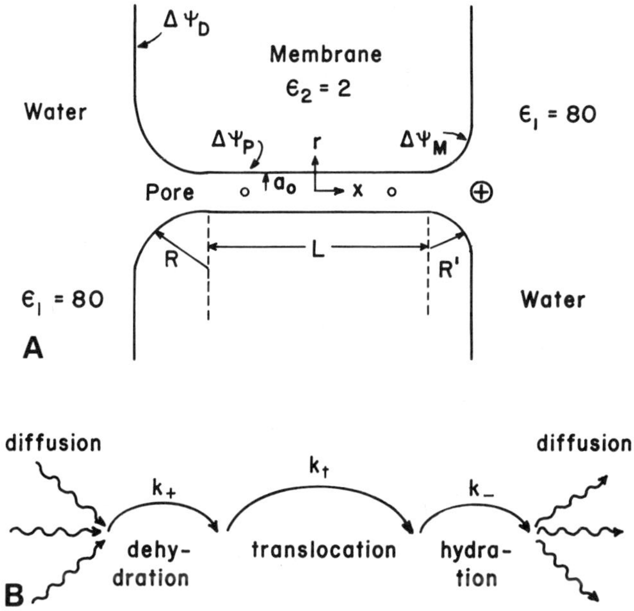

Figure 2. (A) Cross section of a cylindrical pore spanning a membrane of dielectric constant ϵ_2. The bulk water and the pore interior have the same dielectric constant ϵ_1. The constriction length is L; its radius is a_0; and the radii of the mouths are R and R'. Distance, x, is measured from the center of the constriction. The small circles indicate positions of interior binding sites. In addition to the potential generated by the ion at the right-hand pore mouth, there may be potentials caused by the water–lipid potential difference, $\Delta\Psi_D$, by polar groups lining the pore–water interface, $\Delta\Psi_P$, and polar groups at the pore mouth, $\Delta\Psi_M$. (B) Kinetic description of the basic steps in ion permeation. Each of the steps may be composite; if there is only a single binding site in the channel, translocation is absent.

can maintain a structure closer to that of bulk water; the aqueous regions are probably dielectrically quite similar.

A simplified model is illustrated in Fig. 1B. All structural detail has been suppressed. There are only two dielectric continua, one with $\epsilon = 80$ (bulk water, pore water, and the polar surface layer of the pore former) and one with $\epsilon = 2$ (lipid membrane and the hydrophobic part of the pore former). To account for the polarity of the surface layer, the pore's electrical radius is a bit larger than its physical radius (Jordan, 1981). This approximate picture describes an essentially discrete statistical mechanical problem in continuum terms. It cannot account for ion-specific effects. What it does is provide a tool useful for determining the qualitative impact that variation of structural and electrical parameters may have on ion permeation and therefore provide assistance in interpreting experimental data.

The specific model I have analyzed is shown in Fig. 2A; the sizes of the constriction and of the two pore mouths are independently varible. For convenience, the system has cylindrical symmetry. The variable electrical features are the position of an ion within the channel, the applied voltage, the water–lipid potential difference, and dipole potentials caused by the polar nature of the pore former and polar structures lining the channel mouths.

Changing these structural and electrical features significantly influences the basic steps in the ion permeation process illustrated in Fig. 2B. Electrical properties of the pore mouth may control the diffusional process and may have an important effect on the hydration and dehydration steps. The properties of the constriction govern the electrostatic contribution to translocation through the channel.

3. The Image Potential

3.1. Singly Occupied Pores

An ion entering a pore not only interacts with polar moieties at the water–membrane and water–pore former interface but also induces surface charges in those regions where the dielectric properties of the media change rapidly. As the ion is located in the more polarizible region, the charges induced at these electrical phase boundaries have the same polarity as the ion. Thus, the image potential is repulsive. For the model channel geometry of Fig. 2A, it can be represented in the form

$$\Psi = (ze_0/\epsilon_0\epsilon_1 a_0)\,\Phi(\xi,\xi_0;\,\beta,b,b';\,K) \tag{1}$$

where z is the ionic valence, e_0 the electronic charge, ϵ_0 the vacuum permitivity, ϵ_1 the dielectric constant of the bulk water, and a_0 the radius of the pore constriction (Parsegian, 1969; Levitt, 1978; Jordan, 1982). In this form the function Φ is a dimensionless quantity that depends only on scaled properties that describe the *shape* of the pore ensemble, $\beta = L/2a_0$ (the reduced constriction length), $b = R/a_0$, $b' = R'/a_0$ (the reduced mouth radii); the distance the ion is from the center of the constriction, $\xi_0 = x_0/a_0$; the point at which the potential is to be measured, $\xi = x/a_0$; and the dissimilarity of the two dielectrics, $K = \epsilon_1/\epsilon_2$.

The image energy, which represents the interaction of an ion with the induced surface charges, is

$$U = (z^2e_0^2/2\epsilon_0\epsilon_1 a_0)\,\Phi(\xi_0,\,\xi_0;\,\beta,b,b';\,K). \tag{2}$$

Although far less than the energy required to directly solvate an ion in a lipid membrane, which for K^+ in a bilayer 5 nm thick would be ~250 kJ/mol, it is still substantial. For some channels for which structural data are known or may be inferred, peak image energies can be estimated. Results for monovalent cations passing through pore formers incorporated in 5-nm-thick membranes are given in Table I, which also includes the structural parameters used. The channels described span a range of β and b values representative of many actual physical

Table I. Estimated Peak Image Energies of Monovalent Cations in Representative Pore Formers[a]

Channel	Structural parameters[b] (nm)				Scaled parameters			Image energy (kJ/mol)
	a_0	L	R	R'	β	b	b'	
Gramicidin	0.25	2.5	1.25	1.25	5	5	5	~29
Acetylcholine receptor	0.35	4	5.5	1.5	5.7	15.7	4.3	~20–23[c]
Delayed rectifier K^+	0.2	2–3.5	0.8	2.2–0.7	5–8.75	4	11–3.5	~38–44
Large conductance K^+	0.2	1[d]	2	2	2[d]	10	10	~29

[a] In all cases a membrane width of 5 nm is assumed.
[b] Structural parameters are taken from literature values when available (Koeppe *et al.*, 1978; Kistler and Stroud, 1981; Maynard *et al.*, 1977) or are chosen consistent with proposed models (Armstrong, 1975; Miller, 1982). In all cases the electrical radius is slightly larger than the constriction radius to account for polarity (Jordan, 1981).
[c] The range of energies arises from two separate ways of accounting for the fact that the channel former extends beyond the membrane.
[d] Miller's (1982) model presumes a constriction ~1 nm long and ~0.4 nm in radius. It contains a selectivity filter ~0.2 nm in radius. Assuming the filter is minimally 0.3 nm in length yields quoted β value.

possibilities. In all instances, the channel protein's polarity is accommodated by assigning the constriction an electrical radius slightly larger than its physical radius (Jordan, 1981).

It is immediately apparent that the image energy is far more sensitive to the constriction's electrical radius than to any other parameter. The AChR channel, with its long constriction, provides a rather small image barrier to monovalent cation passage because it is really a rather open pore; i.e., a_0 is large. In contrast, the large conductance K^+ channel, which is believed to have a short constriction with an even shorter selectivity filter (Miller, 1982), should have an image barrier comparable to that in gramicidin. These image energies remain a substantial impediment to ion permeation since, under physiological conditions, RT ~2.5 kJ/mol. Without stabilization resulting from interaction with sites in the channel, these barriers would lead to 3000- to 40-million-fold reductions in monovalent cation conductance. The importance of ionic charge is also immediately obvious; for the channels considered, divalent cations would have four times the image energy. Since site binding energy, if ionic in nature, only increases linearly with ionic charge, it is clear that channels adapted for monovalent ion passage discriminate against divalent ion passage. This should be, and is, especially noticeable for the three narrower channels, each of which is blocked by divalent cations at sufficiently high concentrations (Bamberg and Läuger, 1977; Armstrong and Taylor, 1980; Vergara and Latorre, 1983); the AChR channel, in contrast, only exhibits reduced divalent permeability (Lewis and Stevens, 1979).

A more systematic feeling for the effect of structural variation may be obtained from Fig. 3, which illustrates the peak value of the image potential function for a set of symmetrical pores, $\tilde{\Phi}(\beta,b) \equiv \Phi(0,0;\ \beta,b,b;\ 40)$. As constriction length

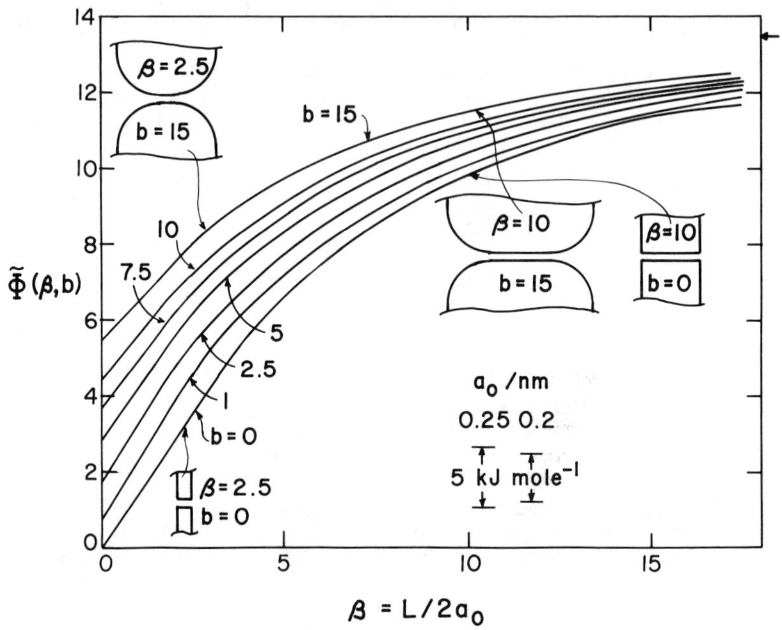

Figure 3. Image potential profile, $\tilde{\Phi}(\beta,b)$ in units of $e/\epsilon_0\epsilon_1 a_0$, for an ion at the center of a cylindrical channel with the geometry illustrated in Fig. 2a. The dielectric ratio, K, is 40. The channel is symmetric ($R = R'$); the parameters are the length-to-diameter ratio, $\beta = L/2a_0$, and the mouth size ratio $b = b' = R/a_0$. Curves for eight different b values are plotted. For the $K = 40$ ratio, the potential in an infinite channel (arrow) is 13.584. To give a sense of scale, an energy of 5 kJ/mol ($2RT$) is indicated for monovalent ions in pores of electrical radius 0.25 nm and 0.2 nm. (Adapted from Jordan, 1984b, with permission.) The cartoons illustrate the shapes of four representative channels.

increases, the peak image potential becomes less dependent on mouth size, finally approaching a common limit. This reflects the observation that the deeper an ion has penetrated a pore, the less it can be influenced by the bulk solvent. The curves for different mouth size parameters have similar shape, which suggests that all the data of Fig. 3 could be simply summarized. The empirical result (Jordan, 1984b) is

$$U^* = 0.87z^2\tilde{\Phi}(\beta^*, 0)/a_0 \tag{3a}$$

$$\beta^* = L/2a_0 + 0.85(R/a_0)^{0.6} \tag{3b}$$

where β^* is an "effective length" greater than the actual length by an amount depending on mouth size; the peak energy, U^*, is given in kilojoules per mole, and distances are in nanometers. For sufficiently short constrictions, $\tilde{\Phi}$ is linear in β, and further simplification is possible,

$$U^* \sim 1.1\, z^2\beta^*/a_0 = 1.1\,(z^2/a_0)[(L/2a_0) + 0.85(R/a_0)^{0.6}],\ \beta^* \leq 6 \tag{4}$$

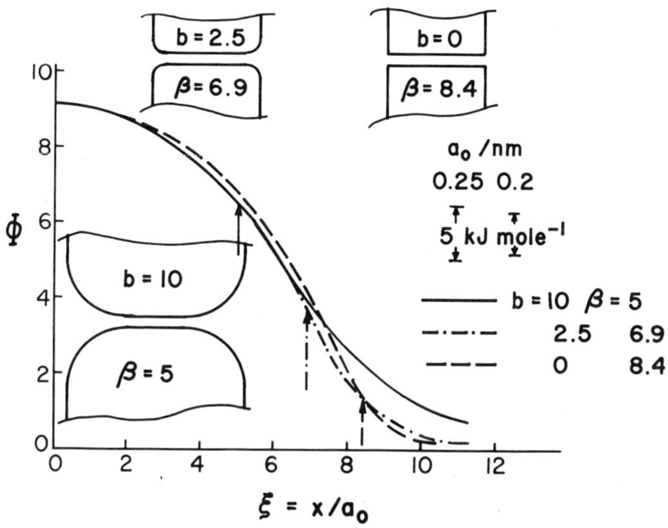

Figure 4. Image potential functions, Φ, for three symmetrical model pores with β* = 8.4. Three mouth sizes are contrasted, b = 10, 2.5, and 0; the cartoons illustrate the shapes of the three channels; the vertical arrows indicate, in each case, the position of the entrance to the constriction. Distance is measured from the constriction center in dimensionless units, $\xi = x/a_0$. To impart a sense of scale, 5 kJ/mol (≡2RT) is indicated for monovalent ions in constrictions with electrical radii of 0.25 nm and 0.2 nm. The dielectric ratio is 40, corresponding to lipid and water.

In addition to being useful, equation 4 emphasizes the varied influences of the different structural features on the maximum image energy; U^* is most sensitive to constriction radius, $\sim a_0^{-2}$, then to constriction length, $\sim L$, and finally to mouth size, $\sim R^{0.6}$.

The similarity between pores with the same value of β* extends to the whole image energy profile as illustrated in Fig. 4. The example chosen (β* = 8.4) contrasts three very different shapes: a short constriction and a large mouth (b = 10), an intermediate constriction and a smaller mouth (b = 2.5), and a longer pore with no mouth (b = 0). Although distinguishable, the profiles differ only trivially within the constricted regions. The result is general, and the image potential profile in the constricted region of any pore may be accurately estimated from the properties of any pore system with the same value of β*.

In the solvent and the channel entrance, the image potential is established by the mouth geometry (see Fig. 6 of Jordan, 1984b). Outside of the constriction, changes of the constriction length have little influence on the image energy. Ions approaching the constriction hardly sense its structure.

The image potential profile in an asymmetric pore can therefore be generated with considerable confidence by combining the three separate profiles: that for each of the two mouth regions and that for an interior region for any pore with the same value of β^* (Jordan, 1984c). For the general pore model of Fig. 2a, the extension of equation 3b is

$$\beta^* = L/2a_0 + 0.425\,[(R/a_0)^{0.6} + (R'/a_0)^{0.6}] \tag{5}$$

It is this procedure that was used to establish the estimates given in Table I.

3.2. Multiply Occupied Pores

An ion entering an already occupied pore must surmount not only the barrier created by its own electrical image but also that caused by the image potential produced by the already resident ion. As the dielectric constant of the lipid is $\sim\frac{1}{40}$ that of water, the interaction between each ion and the surface charges induced by the other ion at the dielectric interface is much stronger than their direct water-mediated coulombic repulsion (Levitt, 1978; Jordan, 1984a,c). It provides an impediment to multiple occupancy but need not be an outright prohibition, even if the pore is narrow (Urry *et al.*, 1980a,b; Hille and Schwartz, 1978) or the permeant ions divalent (Hess and Tsien, 1984; Almers and McCleskey, 1984).

The effect of structural variation on the ionic repulsion energy, ϵ_{12}, can be determined from the image potential, equation 1 (in this application $\xi \neq \xi_0$). At constant binding site separation, the qualitative dependence on a_0, L, and R is similar to that found for singly occupied pores. Energies estimated in this way provide some information from which to draw structural inferences. The ratio of the equilibrium constants for binding the first and second ion in gramicidin A, K_1/K_2, provides the necessary evidence for this assertion. The experimental value is $\sim$100 (Urry *et al.*, 1980a,b); the calculated value is established knowing ϵ_{12}. Since channel length and binding site separation in gramicidin A are known (Koeppe *et al.*, 1978), the model has no adjustable parameters; for a site separation of 2.1 nm, the calculated ratio is $\sim$55 (Jordan, 1984a). The agreement is significant because the calculated K_1/K_2 ratio is very sensitive to this distance. Decreasing it to 1.5 nm would require that $K_1/K_2 \sim 2000$, grossly in error. The model therefore discriminates against an unreasonable assignment of binding site locations.

Other examples are the delayed rectifier K^+ channel and the Ca^{2+} channel from guinea pig heart muscle. The former clearly exhibits multiple occupancy (Hille and Schwartz, 1978); the latter may permit double occupacy (Hess and Tsien, 1984). In order for such behavior to be observed, the free energy for double occupancy, $G_{ij}^{(2)}$ must be comparable to or less than the free energy for occupancy of a single site, g_i. Since

$$G_{ij}^{(2)} = g_i + g_j + \epsilon_{ij} \tag{6}$$

free energy equality implies that $g_i < -\epsilon_{ij}$ (Jordan, 1984c); the pair repulsion energy in a multiply occupied channel is related to the single-site binding energy. Knowing either yields a reasonable estimate of the other. This equivalence provides a basis for informed speculation about channel properties consistent with

Table II. Binding Site Separation, Δ, and Peak Image Energy for Divalent Cation Passage, ϵ^*, Consistent with Three-Barrier, Two-Site Model of Ca^{2+} Channel[a,b] as Functions of Constriction Length, L, and Overall Channel Former Length D

L(nm)	D(nm)	Δ(nm)	ϵ^* (kJ/mol)
1.0	4.0	1.3–1.5	~60
	7.0	1.8	~75
2.0	4.0	1.8	~75
	7.0	2.0	~90
	10.0	2.4	~95
3.0	4.0	2.1	~90
	7.0	2.6	~100
	10.0	3.0	~110

[a] Hess and Tsien, 1984; Almers and McCleskey, 1984.
[b] a_0 is chosen as 0.3 nm, consistent with McCleskey's univalent ion permeation measurements.*

multiple occupancy. An example is the delayed rectifier K^+ channel; I have shown that singly occupied channels cannot contribute significantly to conductance for a large range of constriction geometries and binding site separations (Jordan, 1984c).

Both Hess and Tsien (1984) and Almers and McCleskey (1984) have evidence for double occupancy in different Ca^{2+} channel preparations. The kinetic data of Hess and Tsien (1984) are consistent with a three-barrier, two-site model for which the Ca^{2+} single-site binding energy is ~38 kJ/mol and the barrier to Ca^{2+} translocation is ~22 kJ/mol. McCleskey's studies of monovalent cation permeation on his Ca^{2+} channel preparation suggest that a_0 ~0.3 nm.* If the two channels are structurally similar, the combined observations provide the means to exclude some geometries. Double occupancy means that the site binding energy and the pair repulsion energy, ϵ_{12}, are about equal. The latter is extremely sensitive to the distance between binding sites, Δ. Table II presents estimates of Δ consistent with the Hess and Tsien's (1984) free energy profile for a range of channel structures; L is the constriction length, and D is the overall length of the channel protein. Most significant is the fact that double occupancy precludes very short constrictions. When $L = 1$ nm, no model geometry is compatible with the data since $\Delta > L$, i.e., the binding site would be *outside* the constriction. Except for $L = 2$ and $D = 10$, the other structures are possible. In some cases the binding site may be well within the constriction.

* E. W. McCleskey, private communication. His permeability measurements for a series of univalent cations indicate that in the absence of Ca^{2+}, $(CH_3)_4N^+$ is marginally permeant. From CPK models its ionic radius is ~0.27 nm. The channel's physical radius is then ≥0.27 nm; its electrical radius a_0, would be ~0.3 nm.

To show that none of these structures is excluded on energetic grounds, ϵ^*, the peak of the image energy barrier, is also listed in Table II. It seems large, but this is misleading since the AChR channel is significantly permeable to Ca^{2+} (Lewis and Stevens, 1979). From Table I, AChR has a peak image barrier of ~80–90 kJ/mol for divalent ion passage, comparable to the ϵ^* of Table II. The point of these arguments is simple. Nothing forbids the constriction in this Ca^{2+} channel from being as long as 3 nm; the binding sites may be well inside the constriction. However, the constriction cannot be short; lengths of <1 nm are precluded.

4. Potential Profiles in the Presence of Applied Voltages

Analysis of current–voltage relationships provides information about the fractional voltage drop experienced by an ion at physically interesting sites in the channel systems. These are typically the locations at which the permeant ions bind or at which a blocking ion resides. What is lacking is a way to correlate the measurements of electrical distance with the actual physical position of binding or blocking. The simplest hypothesis is that the total voltage drop takes place over the constricted region of the channel. Although this is a reasonable possibility since the bulk electrolyte is a fairly good conductor, it is not likely to be correct unless the ionic strengths of the bulk aqueous media are very high, >1 M.

The methods used to calculate the image potential can be modified to estimate the profile created by an applied voltage (Levitt, 1978; Jordan, 1982, 1984b,c). Their limitation is the presumption that the bulk water has zero ionic strength, i.e., is a nonconducting medium. However, consideration of the effect of Debye shielding permits estimation of the associated uncertainties. Figure 5 illustrates the zero ionic strength voltage profile for a set of symmetrical pores with different mouth sizes for which β ($= L/2a_0$) is 2.5. The qualitative behavior is typical (see Fig. 8 of Jordan, 1984b). Within the constriction the electric field is constant. As the pore mouth increases in size, more of the voltage drop occurs outside the constriction. If the mouth region is negligibly small ($b = 0$), an ion still senses a significant voltage change before it enters the narrow region. For a longer constriction, a greater fraction of the voltage drop is confined to the immediate vicinity of the pore. However, even for long, narrow constrictions with no exterior mouths, e.g., $\beta = 7.5$, $b = b' = 0$, 10% of the voltage drop takes place outside the constriction. For a short pore with large mouths, e.g., $\beta = 2.5$, $b = b' = 7.5$, only one-third of the voltage drop occurs in the narrow portion of the pore. Interpretation of electrical distance data requires some previous structural knowledge. Without a rough idea of channel shape to limit the possibilities, even sites at which one-third of the voltage drop has taken place may still lie outside the constriction (Jordan, 1984c), at least as long as the ionic strength is low.

As ionic strength increases, the voltage drop tends to be confined to the narrow region of the channel because cations (anions) migrate toward regions of lower (higher) potential. The local charge-density imbalances are compensated

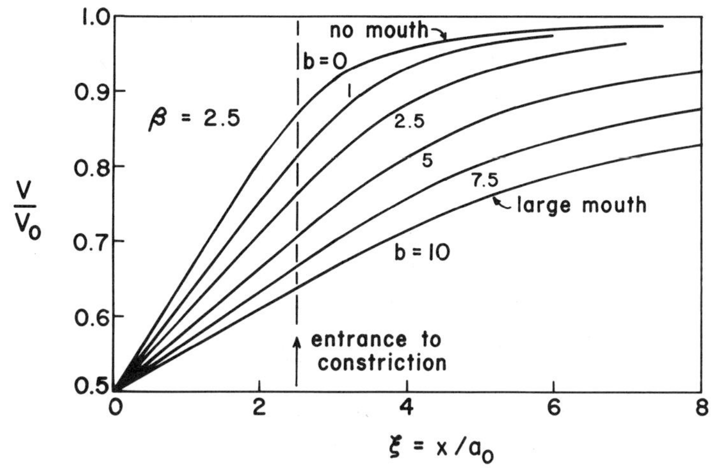

Figure 5. Potential profile on the right-hand cylinder axis of symmetrical channels with the geometry of Fig. 2a. Six mouth sizes, *b*, are contrasted for a pore with a fairly short constriction, $\beta = 2.5$. The vertical line indicates the position of the constriction entrance. The dielectric ratio is 40. (Adapted from Jordan, 1984b, with permission.)

by an increased concentration of counterions. The Debye length determines the range of unscreened potential differences in ionic solutions,

$$\lambda_D = \left\{ \frac{8\pi F^2}{\epsilon_0 \epsilon_1} \frac{I}{RT} \right\}^{1/2} \tag{7}$$

where I is ionic strength, F Faraday's number, R the gas constant, and T the temperature (Wall, 1974). Screening at a distance x into the high-ionic-strength region is proportional to $\exp(-x/\lambda_D)$. It is 65% complete at $x = \lambda_D$ and nearly 90% complete at $x = 2\,\lambda_D$.

Figure 6 illustrates how ionic strength variation alters the voltage profile in gramicidinlike channels incorporated into membranes of variable width. I assume that shielding only takes place in the bulk solution and in the mouth regions. This is plausible since anions cannot enter gramicidin. The profiles are constructed by scaling the zero-ionic-strength voltage profile so that the total potential drop is concentrated within one Debye length of the constriction entrance. This procedure is heuristic; it overestimates screening far from the entrance and underestimates it in the immediate vicinity of the constriction. Nonetheless, it demonstrates the importance of ionic strength. When $I \geqslant 1$ M, $\lambda_D \leqslant 0.3$ nm, and electrical distance is effectively independent of membrane width. At an I of 0.1 M, $\lambda_D = 1$ nm, and the electrical distances at which binding occurs depend on the width of the membrane. The effect is even more pronounced at the lowest ionic strength shown, $I \leqslant 0.01$ M, when $\lambda_D > 3$ nm. Stated differently, *electrical distance measurements may be ionic strength dependent.*

The potential profile of an asymmetric pore can be estimated from those of

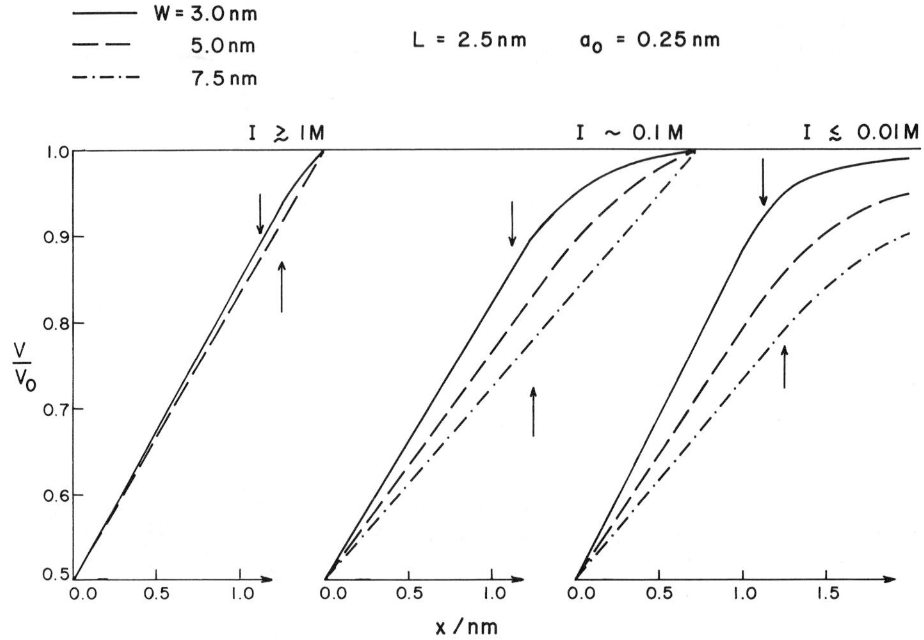

Figure 6. Effect of ionic strength variation on the potential profile in gramicidinlike channels in membranes of variable width at three different ionic strengths. The arrows indicate the positions of the constriction entrance (↑) and of the binding site (↓). Distance is measured from the center of the pore. The method for estimating the profile is described in the text. The dielectric ratio is 40.

symmetrical ones. Just as the image potential in the pore mouth is independent of constriction geometry, the shape of the potential profile created by an applied voltage is completely determined by mouth geometry (Jordan, 1984c). This fact, in conjunction with constancy of the electric field in the constriction, permits construction of voltage profiles for quite general pore geometries.

An example is the delayed rectifier K^+ channel. Inhibition studies with a variety of quaternary ammonium derivatives suggest that the channel has a large interior mouth connected to a narrow, single-file constriction (Armstrong, 1975; Hille and Schwartz, 1978). The block occurs at a fractional electrical distance of $\sim 0.15-0.2$ independent of ionic size for head group diameters between 0.8 nm and 1.2 nm (French and Shoukimas, 1981; Swenson, 1981). Voltage profile calculations (Jordan, 1984c) show that a wide range of reasonable protein structures are totally consistent with these observations. Even were the interior pore mouth just big enough to accommodate the blocker, a significant voltage drop could occur before the ion enters the constriction. Since anions are excluded from the mouth, screening does not significantly affect the interpretation of the data.

The AChR channel, for which structural parameters are given in Table I, is an instance in which ionic strength should influence apparent electrical distance. Because both synaptic and cytoplasmic mouths are large only ~35% of the voltage drop should occur within the constriction when $I = 0$; about half the drop should occur before entering the channel from the synaptic side. The constant-field approximation is totally inadequate. Assuming that anions are excluded from the constriction but can diffuse readily in either mouth, at physiological ionic strengths ($I \sim 0.1$ M) the potential profile is dramatically altered. Only about one-fifth of the voltage drop should take place in the synaptic mouth; about two-thirds should occur in the constriction. Nonetheless, a significant fraction of the voltage drop still occurs outside the narrow part of the channel.

5. Potentials Caused by Surface Dipole Layers

5.1. Lipid–Water Potential Differences

Carrier-facilitated cation conductance is ~100 times greater in membranes formed from glyceryl monooleate (GMO) than in ones formed from phosphatidylcholine (PC) (Hladky and Haydon, 1973). These measurements and studies of lipophilic ion conductance (Pickar and Benz, 1978) indicate that PC is ~120 mV more positive with respect to water than is GMO. Lipid composition has a much smaller effect on conductance in gramicidin B (Bamberg *et al.*, 1976). The conductance roughly doubles in going from PC to GMO. A substantial fraction of the lipid–water potential difference must be shielded in the region of the gramicidin channel that is kinetically rate determining.

The influence of lipid composition can be modeled by assuming that the source of the lipid–water dipole potential difference is a surface layer of point dipoles at the membrane–water interface (Jordan, 1983). The results differ depending on how the pore former is incorporated in the bilayer (P. C. Jordan, unpublished calculations). Two limiting cases contrasted in Fig. 7A, where the pore former projects through the membrane, and Fig. 7B, where it is incorporated into the membrane. In the first case, the source of the membrane potential ends at the pore former–membrane junction. In the second case, increasing pore former thickness requires phospholipid rearrangement in the region of the junction. In both cases, increasing scaled pore former thickness, b, (decreasing β) while keeping scaled membrane width, $\delta = \beta + b$, constant reduces the dipole potential in the pore. The effect is substantially greater when the pore former pierces the membrane.

The profiles illustrated demonstrate clearly the reason that gramicidin B conductance is so slightly affected by lipid variation. The rate-limiting step probably occurs near the constriction entrance (Andersen, 1983). If gramicidin is incorporated into the membrane, the physical situation corresponds to Fig. 7B, and an ion senses less than 25% of the lipid–water potential difference near the channel entrance for $\beta \sim 5$–6, representative of gramicidin. The conductance should less than triple in going from PC to GMO, in reasonable agreement with experiment (Bamberg *et al.*, 1976).

Experiments on gramicidin M^- (Heitz *et al.*, 1982), a chemically altered analogue of gramicidin A for which the central barrier is rate limiting (Heitz *et*

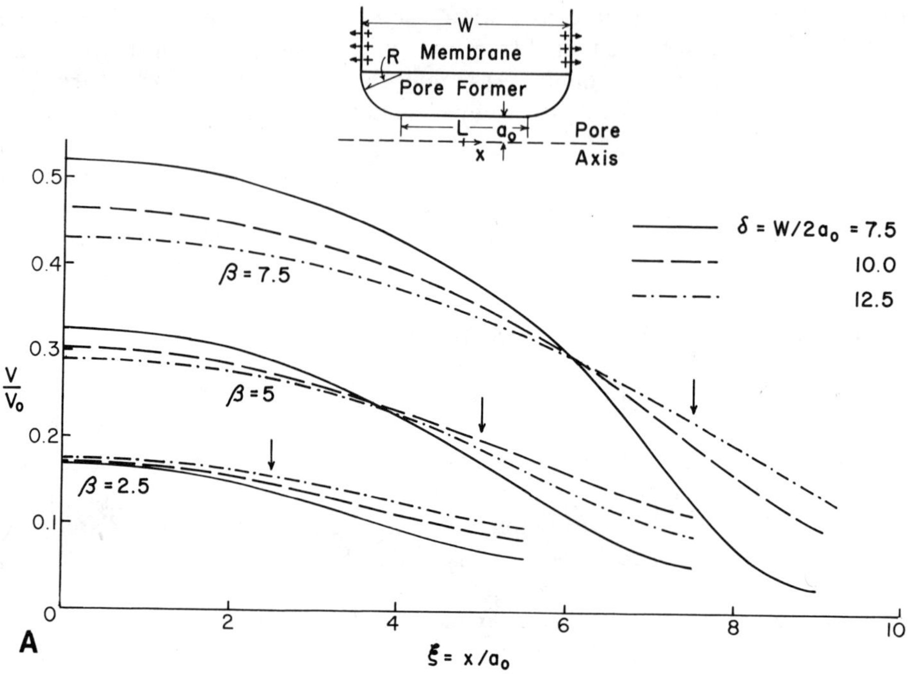
W
Membrane
R
Pore Former
L
a_0
x
Pore Axis
$\delta = W/2a_0 = 7.5$
10.0
12.5
$\beta = 7.5$
$\beta = 5$
$\beta = 2.5$
$\frac{V}{V_0}$
$\xi = x/a_0$
A

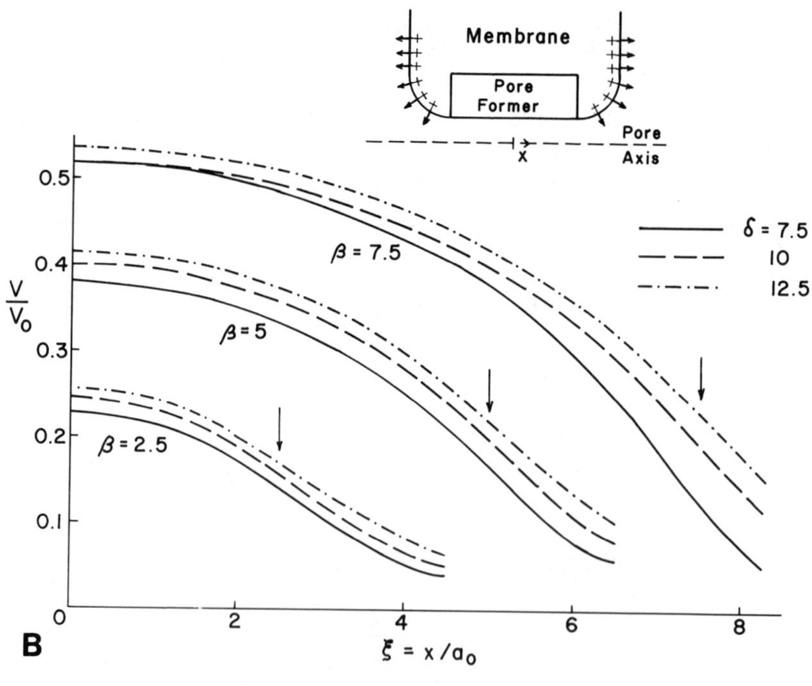
Membrane
Pore Former
x
Pore Axis
$\delta = 7.5$
10
12.5
$\beta = 7.5$
$\beta = 5$
$\beta = 2.5$
$\frac{V}{V_0}$
$\xi = x/a_0$
B

al., 1984), show very different behavior (O.S. Andersen and F. Heitz, personal communication). Here phloretin is used to lower the potential of the lipid phase (Andersen *et al.*, 1976; Andersen, 1978). For a ~100 mV potential change, the conductance of gramicidin M^- increases five- to sixfold. This is completely consistent with the physical model of Fig. 7B, incorporation of the pore former. For membranes between 3 and 5 nm thick, from 38% to 45% of the total lipid–water potential difference is felt at the channel center; the gramicidin M^- conductance is predicted to increase by a factor of 4.4 to 5.8. The model of Fig. 7A, which is clearly inapplicable since the gramicidin dimer is only 2.6 nm long, predicts a smaller effect, in the range of 3.1 to 3.6.

The conclusions are unambiguous. The lipid–water potential difference is greatly shielded within the pore. If the kinetically significant process takes place in or near the channel mouth, lipid variation has little effect on ion permeability. If the interior barrier is rate limiting, a substantial fraction of the potential difference is felt in the kinetically dominant region. The effect is the greatest when a channel protein is incorporated into the membrane.

5.2. Dipole Layers at Pore Mouths

Comparison of Figs. 7A and 7B indicates that there are substantial differences in the shielded dipole potential depending on the details of the charge distribution in the mouth region of a channel protein. Of special interest is the possibility that realistic charge distributions can produce potentials that promote significant increases in cation (or anion) concentration in the mouth of a channel. Such a mechanism could account for the properties of those K^+ channels that exhibit the apparently contradictory properties of high conductance and high selectivity (Coronado *et al.*, 1980; Vergara and Latorre, 1983).

Figure 8 illustrates the axial potential caused by the presence of a dipolar source in the mouth region of a channel protein. As is the case for the image potential (Section 3) and the potential profile arising from an applied voltage (Section 4), the potential in the mouth region depends only on the size of the mouth. It is totally insensitive to the structure of the constriction; the differences are too small to be visible in the scale of the figure for β values between 2.5 and 10. The potential is greatly shielded; even for a system with a large mouth region (b = 10), the potential in the pore mouth is 6% of the potential step, V_0 caused by the dipolar source. However, it remains near its maximum value in a large region.

Figure 7. Axial potential caused by the membrane dipole potential. The source is modeled by point dipoles at the electrical phase boundary with bulk water. The effects of variation of reduced membrane width, $\delta = W/2a_0$, and reduced construction length, $\beta = L/2a_0$, are illustrated. The dielectric ratio is 40. The arrows indicate the position of the entrance to the constriction. (A) The pore former protrudes through the membrane; the source stops at the pore former–membrane junction. (B) The pore former is incorporated into the membrane. The phospholipid molecules reorient themselves to form a mouth region. The dipolar source extends into the mouth.

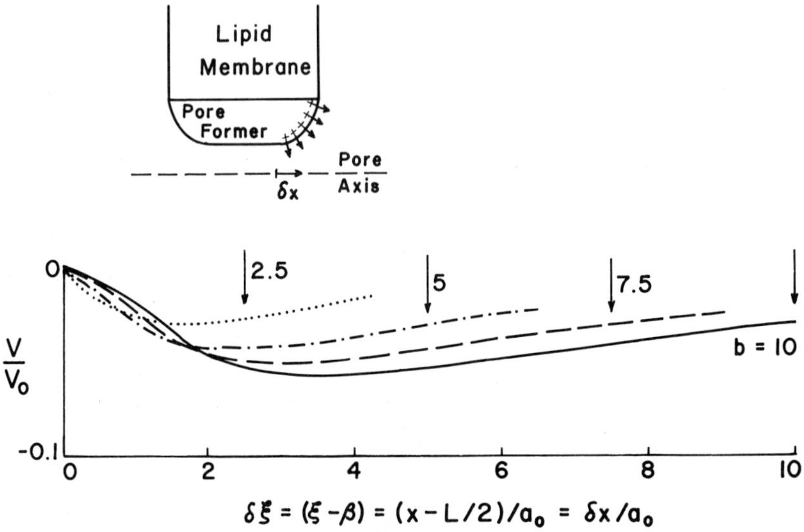

Figure 8. Axial potential in the aqueous region of the pore mouth caused by a dipolar charge distribution in the mouth region of a channel protein. Four mouth sizes are contrasted. The profiles are independent of constriction length. The dielectric ratio is 40. The vertical arrows indicate the limits of the mouth region, $\delta\xi = b$.

When $b = 10$, it is $\geq 0.05\ V_0$ for $\delta\xi$ values between 2 and 6; even at the entrance to the mouth ($\delta\xi = 10$), the shielded potential is half its maximum value. For a channel protein with a large mouth, such as postulated for the K^+ channel from sarcoplasmic reticulum (Miller, 1982) or consistent with the properties of the Ca^{2+}-activated K^+ channel (Latorre and Miller, 1983), charge distributions that produce a 1-V step at the surface of the protein would lead to 30- to 60-mV potentials in the channel mouth. This would increase cation concentration three- to tenfold above that in the bulk of solution and therefore greatly increase channel conductance.

The effect may still be significant even for smaller pore mouths, as is also clear from Fig. 8. Even for the smallest mouth ($b = 2.5$), a 1-V potential step near the protein's surface creates a 30-mV potential in the mouth of the channel. This would be sufficient to enrich the cation (or anion) population threefold above that of the bulk solution.

At low cation concentration, a more quantitative description of conductance in the K^+ channels from sarcoplasmic reticulum (SR) and transverse tubule (TT) is possible. This convergence conductance determines the capture radius of a channel, i.e., the difference between the radius of the entrance to the constriction and the radius of the permeant ion (Läuger, 1976). At low enough concentration, channel conductance is limited not by the interior structure of the constriction but by the ionic concentration in the vicinity of the pore mouth and the rate at which ions enter the mouth region. Table III presents estimates of the capture radius and the constriction radius of the SR and TT K^+ channels assuming various

Table III. Capture and Constriction Radii Determined from Convergance Conductance of K^+ Channels from SR and TT as Functions of Pore Mouth Potential, V_0, and Corresponding Concentration Enrichment Factor

Channel	SR			TT		
Limiting conductivity[a] (nS/M)	3			5		
Pore mouth potential (mV)	0	30	60	0	30	60
Enrichment factor	1	3	10	1	3	10
Capture radius (nm)	0.13	0.04	0.01	0.21	0.07	0.02
Constriction radius (nm)	0.26	0.17	0.14	0.34	0.20	0.15

[a] Coronado *et al.*, 1980; Latorre and Miller, 1983.

charge distributions in the mouth region of the pore; the model geometry parameters are those given in Table I. In the absence of the focusing field created by the dipole layer at the pore mouth, there is no ion enrichment in the pore mouth; the estimated channel radii are 0.26 nm and 0.34 nm for SR and TT, respectively, assuming a K^+ ionic radius of 0.13 nm. These numbers are unreasonably large given the observed selectivities (Coronado *et al.*, 1980; Latorre *et al.*, 1982). A focusing field equivalent to 500 mV increases the ionic concentration in the channel mouth threefold; one of 1 V increases it tenfold. The capture radii are much smaller; the corresponding channel radii are small enough to be consistent with both selectivity and blocking data.

A pore mouth charge distribution equivalent to a voltage step of $\geqslant$500 mV is enough to account for the high permeability of these K^+ channels. Are such structures plausible? Two independent lines of evidence suggest they are. Analysis of the charge distribution at the surface of zinc superoxide dismutase (SOD) has shown that the associated electric field is favorably oriented and is large enough to guide the O_2^- to the catalytic binding site (Getzoff *et al.*, 1983; Getzoff and Tainor, 1985). Is SOD unique? Probably not. If the polar groups lining the channel mouth are only peptide linkages and not polar amino acid residues, the partial charges (Schulz and Schirmer, 1978) indicate that a 1-V potential step may arise if the peptide groups are optimally oriented at a surface density of as little as 1.4 group/nm^2. Assuming gramicidin A to be fairly typical, realistic surface coverages are ~5 group/nm^2; the optimal orientation is not needed. As long as the peptide groups are oriented such that the carbonyl oxygen can face the water, five-sixths of the possible arrangements are consistent with voltage steps of 1 V or more. For a 500-mV step, nine-tenths of the possible arrangements are acceptable. Thus, a wide range of structures is consistent with charge distributions that effectively increase the capture radius without increasing the constriction radius.

ACKNOWLEDGMENT. This work was supported by the National Institutes of Health, Grant GM-28643.

References

Almers, W., and McCleskey, E. W., 1984, Non-selective conductance in calcium channels of frog muscle: Calcium selectivity in a single-file pore, *J. Physiol.* (*Lond.*) **353:**585–608.

Andersen, O. S., 1978, Ion transport across simple membranes, in: *Renal Function* (G. H. Giebiseh and E. F. Purcell, eds.), pp. 71–99, Josiah Macy Foundation, New York.

Andersen, O. S., 1983, Ion movement through gramicidin A channels. Studies on the diffusion controlled association step, *Biophys. J.* **41:**147–165.

Andersen, O. S., Finkelstein, A., Katz, I., and Cass, A., 1976, Effect of phloretin on the permeability of thin lipid membranes, *J. Gen. Physiol.* **67:**749–771.

Armstrong, C. M., 1975, Ionic pores, gates and gating currents, *Q. Rev. Biophys.* **7:**179–210.

Armstrong, C. M., and Taylor, S., 1980, Interaction of barium ions with potassium channels in squid giant axons, *Biophys. J.* **30:**473–488.

Bamberg, E., and Läuger, P., 1977, Blocking of the gramicidin channel by divalent cations, *J. Membr. Biol.* **35:**351–375.

Bamberg, E., Noda, K., Gross, E., and Lauger, P., 1976, Single channel parameters of gramicidin A, B and C, *Biochim. Biophys. Acta* **418:**223–228.

Benz, R., Ishii, J., and Takae, T., 1980, Determination of ion permeability through channels made of porins from the outer membrane of *Salmonella typhimurium* in lipid bilayer membranes, *J. Membr. Biol.* **56:**19–29.

Coronado, R., Rosenberg, R. L., and Miller, C., 1980, Ionic selectivity, saturation and block in a K channel from sarcoplasmic reticulum, *J. Gen. Physiol.* **76:**425–446.

French, R. J., and Shoukimas, J. J., 1981, Blockage of squid axon potassium conductance by internal tetra-N-alkylammonium ions of various sizes, *Biophys. J.* **34:**271–292.

Getzoff, E. D., and Tainer, J. A., 1986, Superoxide dismutase as a model ion channel, in: *Ion Channel Reconstitution* (C. M. Miller, ed.), pp. 57–73, Plenum Press, New York.

Getzoff, E. D., Tainer, J. A., Weiner, P. K., Kollman, P. A. Richardson, J. S., and Richardson, D. C., 1983, Electrostatic recognition between superoxide and copper, zinc superoxide dismutase, *Nature* **306:**287–290.

Hagiwara, S., and Byerly, L. A., 1981, Calcium channel, *Annu. Rev. Neurosci.* **4:**69–125.

Heitz, F., Spach, F., and Trudelle, Y., 1982, Single channels of 9,11,13,15-destryptophyl-phenalanyl gramicidin A, *Biophys. J.* **40:**87–89.

Heitz, F., Spach, F., and Trudelle, Y., 1984, Single channels of various gramicidins. Voltage effects, *Biophys. J.* **45:**97–99.

Hess, P., and Tsien, P. W., 1984, Mechanism of ion permeation through calcium channels, *Nature* **309:**453–456.

Hille, B., 1975, Ionic selectivity of Na and K channels of nerve membranes, in: *Membranes*, Vol. 3 (G. Eisenman, ed.), pp. 255–323, Marcel Dekker, New York.

Hille, B., and Schwartz, E., 1978, Potassium channels as multi-ion single-file pores, *J. Gen. Physiol.* **72:**409–442.

Hladky, S. B., and Haydon, D. A., 1973, Membrane conductance and surface potential, *Biochim. Biophys. Acta* **318:**464–468.

Jordan, P. C., 1981, Energy barriers for the passage of ions through channels. Exact solution of two electrostatic problems, *Biophys. Chem.* **13:**203–212.

Jordan, P. C., 1982, Electrostatic modeling of ion pores. Energy barriers and electric field profiles, *Biophys. J.* **39:**157–164.

Jordan, P. C., 1983, Electrostatic modeling of ion pores. II. Effects attributable to the membrane dipole potential, *Biophys. J.* **41:**189–195.

Jordan, P. C., 1984a, The total electrostatic potential in a gramicidin channel, *J. Membr. Biol.* **78:**91–102.

Jordan, P. C., 1984b, The effect of pore structure on energy barriers and applied voltage profiles. I. Symmetrical channels, *Biophys. J.* **45:**1091–1100.

Jordan, P. C., 1984c, The effect of pore structure on energy barriers and applied voltage profiles. II. Unsymmetrical channels, *Biophys. J.* **45:**1101–1107.

Kirkwood, J. G., 1939, The dielectric polarization of polar liquids, *J. Chem. Phys.* **7**:911–919.

Kistler, J., and Stroud, R. M., 1981, Crystalline arrays of membrane-bound acetylcholine receptor, *Proc. Natl. Acad. Sci. U.S.A.* **78**:3678–3682.

Koeppe, R. C., Hodgson, K. O., and Stryer, L., 1978, Helical channels in crystals of gramicidin A and of a cesium–gramicidin A complex: An X-ray diffraction study, *J. Mol. Biol.* **121**:41–54.

Latorre, R., and Miller, C., 1983, Conduction and selectivity in potassium channels, *J. Membr. Biol.* **71**:11–30.

Latorre, R., Vergara, C., and Hildalgo, C., 1982, Reconstitution in planar lipid bilayers of a Ca^{2+}-dependent K^+ channel from transverse tubule membranes isolated from rabbit skeletal muscle, *Proc. Natl. Acad. Sci. U.S.A.* **79**:805–809.

Läuger, P., 1976, Diffusion-limited ion flow through pores, *Biochim. Biophys. Acta* **455**:493–509.

Lee, W. K., and Jordan, P. C., 1984, Molecular dynamics simulation of cation motion in water-filled, gramicidin-like pores, *Biophys. J.* **46**:805–819.

Levitt, D. G., 1978, Electrostatic calculations for an ion channel. I. Energy and potential profiles and interaction between ions, *Biophys. J.* **22**:209–219.

Lewis, C. A., and Stevens, C. F., 1979, Mechanism of ion permeation through channels in a postsynaptic membrane, in: *Membrane Transport Processes* Vol. 3 (C. F. Stevens and R. W. Tsien, eds.), pp. 133–151, Raven Press, New York.

MacKay, D. H. J., Berens, P., Wilson, K. R., and Hagler, A. T., 1984, Structure and dynamics of ion transport through gramicidin A, *Biophys. J.* **46**:229–248.

Maynard, T., Edwards, C., and Anraku, M., 1977, Permeability of the endplate membrane activated by acetylcholine to some organic cations, *J. Neurobiol.* **8**:173–184.

Miller, C., 1982, Feeling around inside a channel in the dark, in: *Transport in Biological Membranes* (R. Antolini, ed.), pp. 99–108, Raven Press, New York.

Parsigian, V. A., 1969, Energy of an ion crossing a low dielectric membrane: Solution to four relevant electrostatic problems, *Nature* **221**:844–846.

Pickar, A. D., and Benz, R., 1978, Transport of oppositely charged lipophilic ion probes in lipid bilayers having various structures, *J. Membr. Biol.* **44**:353–376.

Schulz, L. E., and Schirmer, R. H., 1978, *Principles of Protein Structure*, p. 30, Springer, New York.

Swenson, R. P., Jr., 1981, Inactivation of potassium current in squid axon by a variety of quaternary ammonium ions, *J. Gen. Physiol.* **77**:255–271.

Urry, D. W., 1971, The gramicidin A transmembrane channel: A proposed $\pi_{L,D}$ helix, *Proc. Natl. Acad. Sci. U.S.A.* **68**:672–676.

Urry, D. W., Venkatachalam, C. M., Spisni, A., Bradley, R. J., Trapani, T. L., and Prasad, K. U., 1980a, The malonyl gramicidin channel: NMR-derived rate constants and comparison of calculated and experimental single-channel currents, *J. Membr. Biol.* **55**:29–51.

Urry, D. W., Venkatachalam, C. M., Spisni, A., Lauger, P., and Khalid, M. A., 1980b, Rate theory calculation of gramicidin single channel currents using NMR-derived rate constants, *Proc. Natl. Acad. Sci. U.S.A.* **77**:2028–2032.

Vergara, C., and Latorre, R., 1983, Kinetics of Ca^{2+}-activated K^+ channels from rabbit muscle incorporated into planar bilayers: Evidence for a Ca^{2+} and Ba^{2+} blockade, *J. Gen. Physiol.* **82**:543–568.

Wall, F. T., 1974, *Chemical Thermodynamics*, ed. 3, W. H. Freeman, San Francisco.

Chapter 3

SUPEROXIDE DISMUTASE AS A MODEL ION CHANNEL

ELIZABETH D. GETZOFF and JOHN A. TAINER

1. Introduction

Detailed information is now available on the biochemistry and physiology of ion channels; however, with certain exceptions such as gramicidin (Wallace, 1984), far less is known about their structures. Moreover, atomic structures of ion channels may continue to be unobtainable for several years. Consequently, it seems worthwhile to consider how existing data on nonmembrane proteins might be used to suggest structural aspects of ion channels and hence aid in the design of relevant biochemical and physiological experiments. Two aspects of known protein structures are particularly pertinent to understanding the structural possibilities for membrane protein ion channels: the existence and nature of deep invaginations in the external molecular surfaces of proteins, and the role of electrostatic forces in the precollision guidance of ions. In this chapter, we use the high-resolution structure of the protein Cu,Zn superoxide dismutase (SOD) as a model to illustrate these features, but the conclusions are representative of other known protein structures. In our analysis we make use of computer graphics techniques, which have been exceptionally successful for the display and evaluation of complex three-dimensional information.

2. Proteins Have a Small Number of Distinct Folds

Before beginning to examine the SOD model for details of protein three-dimensional structure relevant to ion channels, a review of some basic information on protein structure may be helpful. Is it reasonable to use known soluble protein structures as a data base to examine structural implications (topographic and electrostatic) for ion channels? Protein folding is known to result from the linear arrangement of amino acid residues and the nature of the environment; the latter differs considerably between soluble and membrane proteins in terms of hydro-

ELIZABETH D. GETZOFF and JOHN A. TAINER • Department of Molecular Biology, Research Institute of Scripps Clinic, La Jolla, California 92037.

phobicity and hydrogen bonding, at least for the portions of a protein's external surface in contact with the membrane bilayer. Logically, the best answer to this question comes from consideration of known structural information on both soluble and membrane proteins.

Presently about 200 distinct structures of soluble proteins have been solved by X-ray crystallography. Determination of the specific structural arrangement of the amino acid residues in a protein provides substantial insight into understanding their functions at the level of detailed atomic geometry and, in addition, yields information about the important covalent and noncovalent forces that stabilize structural conformation and control activity. Protein structures can be classified according to the internal arrangement of their domains, which are contiguous portions of the polypeptide chain that fold into compact, semiindependent units. A major achievement resulting from the comparative analysis of three-dimensional protein structures has been the discovery that all known structural domains can be grouped into a small number of categories (Rossman and Argos, 1981; Levitt and Chothia, 1976; Richardson, 1977, 1981). Richardson's comprehensive protein structure taxonomy gives four major categories of folding domains: (1) antiparallel α-helix; (2) parallel α-helix/β-sheet; (3) antiparallel β-sheet; and (4) small disulfide- or metal-rich (Richardson, 1981). There is growing evidence that a few principal patterns of interactions characterize each level of structural organization in proteins: side chain to side chain, α-helix to α-helix, β-sheet to α-helices, domain to domain. Examination of this data base of three-dimensional protein structures suggests that these patterns represent structural biases resulting from the stereochemical nature of amino acids, which should, in large part, be equally important for all protein structures.

Thus, the simplest assumption is that membrane protein structures will exhibit general patterns and interactions like those already determined for soluble protein structures as a result of structural biases implicit in their constituent amino acids. Nevertheless, is there any evidence that the types of structures in membrane proteins differ from those found in soluble proteins? Perhaps the best-known membrane protein structure is that of bacteriorhodopsin, which has been solved to 7-Å resolution by electron diffraction and imaging. The fold of bacteriorhodopsin (Henderson and Unwin, 1975) is that of an up-and-down α-helix cluster similar to that of known protein structures such as the hemerythrins (oxygen-binding proteins), the coat protein for tobacco mosaic virus, and the cytochromes c' and b_{562} (Richardson, 1981). The assumption that membrane proteins and soluble proteins share the same structural elements of α-helix and β-sheet is also supported by current data on other membrane proteins such as acetylcholine receptor (Yager *et al.*, 1984; Guy, 1984), cytochrome oxidase (Fuller *et al.*, 1979), gap junction (Makowski *et al.*, 1984), photosynthetic reaction center (Deisenhoffer *et al.*, 1984) and *E. coli* matrix porin (Garavito *et al.*, 1983; Garavito and Rosenbusch, 1980). Based on this hypothesis, it is possible to use the data base of solved protein structures to obtain clues about structural requirements for membrane proteins whose biophysical properties are known. Analysis of these known protein structures may reveal possible features required for the structure and function of ion channels.

3. Superoxide Dismutase Shares Features with Ion Channels

The enzyme Cu,Zn superoxide dismutase (SOD) catalyzes the very rapid two-step dismutation of the toxic superoxide radical ($O_2^- \cdot$) alternately to molecular oxygen and to hydrogen peroxide through the reversible reduction and oxidation of the active-site Cu ion by the following reaction equation (Klug-Roth *et al.*, 1972):

$$\text{SOD-Cu}^{2+} + O_2^- \cdot + H^+ \rightleftharpoons \text{SOD-Cu}^+ + O_2$$

$$\text{SOD-Cu}^+ + O_2^- \cdot + H^+ \rightleftharpoons \text{SOD-Cu}^{2+} + H_2O_2$$

$$2O_2^- \cdot + 2H^+ \rightleftharpoons O_2 + H_2O_2$$

From computer graphics analysis in combination with biochemical information, as specific model was formulated for the enzyme mechanism that requires only minor rearrangements of the ligands during catalysis (Tainer *et al.*, 1983a).

Superoxide dismutase shares constraints imposed on ion-specific channels, specifically, a fast (diffusion-limited) reaction, prohibiting large changes in the protein conformation, and selectivity for ions with respect to both size and charge. Thus, we can examine the structural characteristics of SOD that contribute to features it shares with ion channels.

To evaluate the accuracy of details in the SOD structure, especially those related to solvent-exposed side chains (which often are not well defined in X-ray crystal structures), it is important to know about the level of error. The structure of bovine erythrocyte SOD is now defined to 2 Å resolution with a crystallographic residual error of 19% without waters and 14% with tightly bound waters (Getzoff, 1982). Metal ion positions in the isomorphous derivatives of SOD used for obtaining phase information have been determined as described (Thomas *et al.*, 1974). In addition, complete coordinates for SOD are available from the protein data bank (Tainer *et al.*, 1980, 1982). The crystallographic asymmetric unit contains two enzyme dimers or four identical copies of the SOD subunit structure. These four subunits, fit to their electron density and refined against the observed structure factors independently, yielded a fourfold redundancy of information on the structure of SOD, which makes the accuracy of the molecular surface features of this enzyme reliable at atomic resolution.

4. Protein Surfaces

Protein structure is most commonly represented by skeletal models illustrating the covalent bonding between atoms. However, space-filling models depicting the protein as a solid object enclosed by a surface are more useful for studies of the protein's interaction with other molecules including solvent, substrates, inhibitors, and other macromolecules. The simplest such surface is that constructed using intersecting spheres of van der Waals radii to represent the individual atoms. The smoother molecular surface defined by Richards (1977) results from rolling a probe sphere over the van der Waals surface. Crevices and spaces smaller than the probe sphere are smoothed over and become invisible. The use of a water-sized probe sphere defines those areas of the protein's surface available for interaction with solvent and is particularly helpful for analyzing intermolecular interactions. For computer graphics display, the molecular surface can be depicted as a transparent set of colored points on a vector- or line-drawing display (Connolly, 1981; Connolly and Olson, 1985) or as a thin opaque shell enclosing the protein on a televisionlike raster display (Connolly, 1983a,b, 1985).

Within a protein domain, there are usually subdomains or structural elements that show a high degree of local integrity, such that more stabilizing interactions occur within rather than between subdomains. Different subdomains are characteristic of the individual families of protein folds but in general consist of two or more secondary structural elements (α helices, β strands, or nonrepetitive loops) in preferred packing arrangements. One consequence of this type of integral domain packing is that deep pockets or invaginations in protein surfaces tend to be located between secondary structural elements or subdomains (Connolly, 1981). By analogy, we suggest that ion channels are also more likely to be located between subdomains of a membrane protein.

5. Preexisting Pockets Form Metal Ion Binding Sites

Two fairly simple structural questions can be asked about membrane protein channels. How might an ion channel be designed to allow both efficiency and specificity? Could known structural features of proteins act as gated ion channels, and, if so, how might they work?

To address these questions, we have examined the relationship of six distinct metal ion binding sites to the molecular surface in SOD (Tainer *et al.*, 1983b). In each of these six sites, a pocket or invagination is present in the structure of the native enzyme. These pockets range in depth from less than 2 Å to about 15 Å. The metal ion positions were located during the X-ray crystallographic structure determination of SOD and came from five different derivative preparations selected to be isomorphous with the native protein (Thomas *et al.*, 1974). These binding sites were found inside the protein envelope in the 5.5 Å resolution electron density maps. The metal compounds used (either singly or in pairs) were 3-chloromercuri-2-methoxypropylurea, K_2PtCl_4, K_2IrCl_6, and $HgCl_2$. Some metal binding sites bind more than one type of metal ion and thus represent general positions for localization of metal ions. These pockets occur in the molecular surface of the native enzyme in the absence of any metal ions, yet have an affinity for metal ions. So, these sites might also be expected to bind the metallic compounds used for staining in electron microscopy. Negatively stained periodic specimens may therefore reflect similar irregularities in the molecular surface and

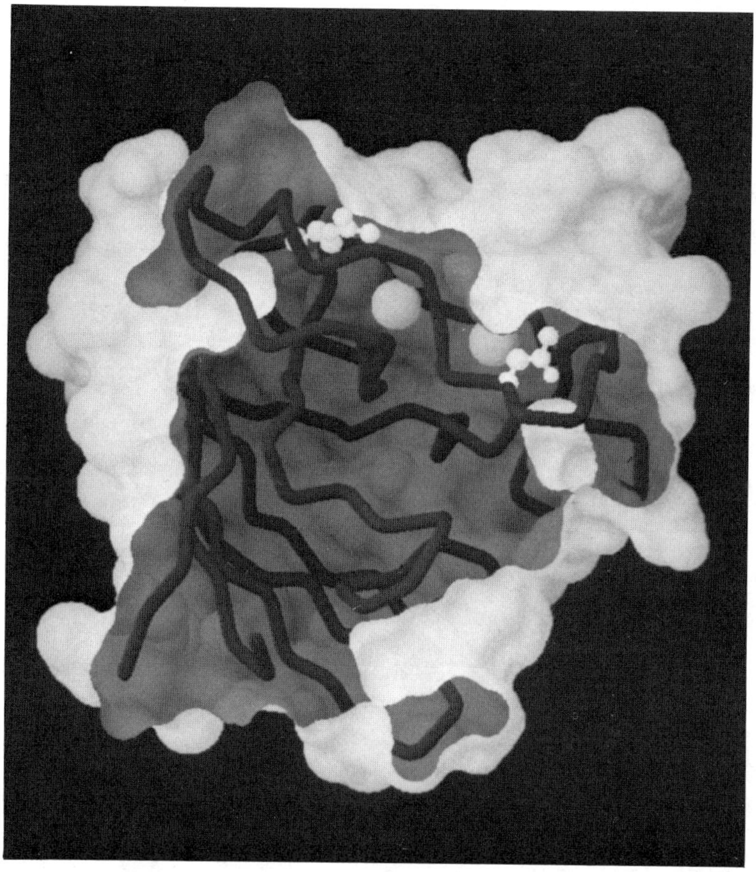

Figure 1. Two deep hydrophilic pockets each closed by a single side chain. The external molecular surface of SOD is shown as a thin, hollow shell shaded white on the outside and gray on the inside. A clipping plane cuts away the front surface of the protein to simultaneously reveal both inside and outside surface topography. The entire α carbon backbone is shown unclipped as a bent wire tube (dark gray). Two light-colored spheres representing the copper (upper center) and zinc (upper right) can be seen inside the external surface underneath the active site. The active-site channel (top center) is eclipsed in this view; the deepest part of the channel is above and touching the copper sphere and can only be seen from inside the protein. Bonds and atoms (white spheres and connections) are shown for the side chains of Arg^{141} (upper left) and Asn^{63} (upper right), which close off the two deep pockets from opening into the active-site channel. A deep invagination of the external surface passes between two segments of the α carbon backbone (upper left) and is closed at the end by the guanidinium group of Arg^{141}. The second pocket similarly passes between two portions of the α carbon backbone (right) and is prevented from forming an open pore only by the side chain of Asn^{63} (upper right). The programs AMS (Connolly, 1983b) and MCS (Connolly, 1985) were used for calculating the surface and the raster image. Adapted from Tainer *et al.*, 1985b.

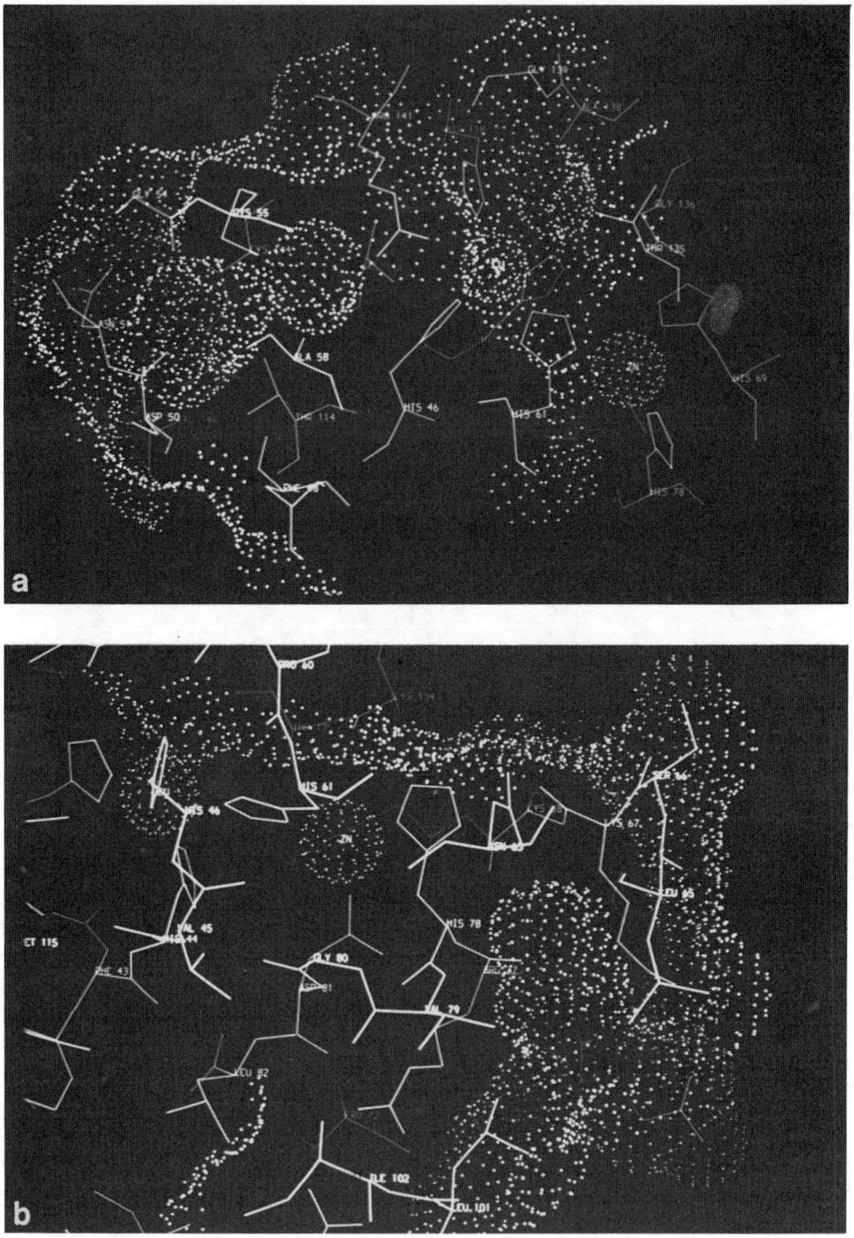

Figure 2. Detailed topography and stereochemistry of two deep pockets in SOD. The external molecular surface is shown as dots, and the atomic model as lines labeled at the α carbon. Deep pockets appear as fingers reaching into the protein from the outside surface. Both sites possess channellike features, and each is prevented from forming an open pore into the active-site channel by a single amino acid side chain. Both the surface and the skeletal model have been sliced by front and back clipping planes to show a thin section. The surface was calculated using the program MS (Connolly, 1983a) and displayed using M. L. Connolly's program HANDLE at the University of California at

provide valuable information about invaginations and channels in macromolecules. All of the sites SOD involve charged or potentially charged amino acid side chains; this finding agrees with the analysis of patterns of negative staining in the electron microscopy of collagen (Piez and Miller, 1974). The charged side chains in the SOD metal binding sites, together with hydrogen-bonding side chains, could provide sufficient ligands for the metal ions without large rearrangements of the polypeptide chain. Surface areas (Connolly, 1981) were calculated for the pits and subdivided into areas contributed by carbon atoms (hydrophobic) and those contributed by noncarbon atoms (hydrophilic). The percentage of hydrophilic surface area ranged from 44% to 80% and tended to be highest for the shallow pockets and lowest for the deep invaginations.

6. Molecular Surface Invaginations as Gated Channel Analogues

Two of the surface invaginations are deep enough to consider in terms of their implications for ion channels in membranes (Fig. 1). Figure 2a illustrates the SOD surface invagination associated with a platinum ion binding site, which is formed by nine residues and closed at the end by the guanidinium group of Arg^{141}. This type of structure has a number of the properties required for an ion channel or gated channel in a membrane protein. It is sufficiently hydrophilic (46% noncarbon-atom surface area) to attract metal ions and has dimensions (about 4 Å diameter) suitable for their access. A single arginine (Arg^{141}) side chain at the bottom of the pit separates it from the active-site channel on the other side. For a gated channel, rotation of single bonds in this side chain into an alternative stable configuration for the Arg guanidinium group would allow the opening of this invagination to form a tunnel through a portion of the protein into the active site. Figure 2b shows the invagination at a mercury ion binding site, which involves five residues and is also fairly hydrophilic (44% noncarbon area). This site possesses the channellike features mentioned above for the platinum binding site; only the side chain of Asn^{63} prevents this pit from forming an open pore through the protein.

Each of the two deep invaginations described above, if opened into a pore, would emerge into the active-site channel of the enzyme (Fig. 1). The addition of this wide mouth to the end of the tunnel increases the analogy to ion channel

---

San Francisco's computer graphics facility. (a) The surface pocket associated with a platinum ion binding site (extended surface, center left) appears as a deep invagination from the external surface that passes between the two sides of the polypeptide chain and underneath the disulfide bond joining Cys^{55} and Cys^{144} (upper left); Arg^{141} (labeled at top) seals off the right-hand end. The Arg^{141} side-chain nitrogen atoms are anchored by hydrogen bonds to the backbone carbonyl oxygen atoms of Cys^{55} and Gly^{59} (Arg^{141} NE–Cys^{55} O; Arg^{141} NH_2–Gly^{59} O). (b) The surface pocket associated with a mercury ion binding site can be seen as a similar indentation passing between two pieces of the polypeptide chain (lower right); Asn^{63} (upper right) blocks the top end of the pocket as shown in this view. The side-chain position of Asn^{63} is stabilized by hydrogen bonds to main chain atoms of Lys^{67} (Asn^{63} OD_1–Lys^{67} N; ASN^{63} ND_2–Lys^{67} O).

structures. Including the active channel cleft, the effective length of the opened pore form of each of these pockets would be about 30 Å. Each gated channel analogue would open near the bottom of the active site cleft, where it would be effectively sequestered from many nonspecific interactions. Thus, the active channel mouth would provide a large access region for ions entering the pore and, as it narrows, a selectivity filter to screen out incorrect ions. Each putative "gate" residue is located near the protein surface, thus allowing sufficient degrees of freedom for the mobility of the "gating" side chain. To complete the ion channel analogy, opening the side-chain gate would require a trigger, such as the binding of an ion, to change the local stereochemical environment. Although the Asn or Arg side chains are inherently flexible, the triggering event would need to break any hydrogen bonds anchoring them to the rest of the protein for gating to occur.

Thus, many related attributes of these two deep invaginations from the SOD structure satisfy potential structural requirements for gated ion channels: ion affinity, hydrophilicity, the presence of charged and hydrogen-bonding side chains to serve as metal ligands, extensive solvent accessibility, sequestering from nonspecific interactions, and the presence of a single flexible but obstructing side-chain gate. Thus, the portions of membrane proteins that function as pores or gated channels in the membrane need not be very different from surface invaginations occurring in the known structures of soluble proteins.

7. Flexibility of Molecular Surfaces

The function and biochemistry of most ion channels suggest that at least some degree of conformational flexibility is required. What information, if any, on local flexibility is available from refined X-ray diffraction structures? Because of the current limitations of theoretical calculations, the empirically determined information about the relative mobilities of different atoms or sites in proteins (available from refined macromolecular structures) is useful for understanding the dynamic aspects of protein structure. This information on mobility is termed the atomic temperature factor and can be given by the equation $B = 8\pi^2\overline{u^2}$, where the mean square displacement $\overline{u^2}$ is averaged over time and the crystal lattice (Debye, 1914). Atomic temperature factors can provide important data about the lower-frequency, concerted motions of groups of atoms, indicating the relative conformational variability of different regions (Swaminathan *et al.*, 1982; Holbrook and Kim, 1984); high temperature factors indicate shallow potential wells (low energy barriers between different conformations) allowing access to multiple conformations at biological temperatures.

We have used temperature factors determined by X-ray structure refinement to examine the mobility of protein surfaces using computer graphics (Tainer *et al.*, 1984; 1985a). Our approach was to map the temperature factors onto the molecular surface (Richards, 1977), which represents the portion of the molecule available for interaction with solvent and other molecules. The resulting pattern of temperature factors was examined together with the underlying stereochemical restraints from hydrogen-bonding and packing interactions. This analysis revealed that protein surfaces characteristically show regions of both high and low mobility; loops and other highly exposed areas are more flexible than well-packed helices and β strands. In simple terms, local molecular topography is highly correlated with mobility such that protrusions are mobile and invaginations are more highly ordered.

In SOD, the enzyme's active channel and the two deep invaginations that form metal ion binding sites all show this pattern of mobility clearly: the inner surface of these indentations is more ordered, whereas the more protruding edges are quite flexible. This same pattern holds for both main-chain and side-chain average temperature factors. Charged side chains, which are often highly exposed around active sites, frequently show multiple conformations in the electron density maps. These residues might thus participate in both the gating and the electrostatic recognition required for membrane ion channels.

Catalysis in SOD requires the enzyme's very rapid interaction with a small substrate. The enzyme's catalytic site, at the bottom of the active site channel, is highly ordered, suggesting that the reaction does not involve significant conformational changes (Tainer *et al.*, 1983a). The observed order can be viewed as a consequence of the number of interactions involving the multiliganded metal ions together with the hydrogen bonds to the ligand side chains. This is consistent with biochemical results suggesting that the active-site structure is more flexible in the absence of the metal ions (Abernethy *et al.*, 1974).

The functional requirements of a gated channel suggest that gating residues must be mobile but that this mobility must be highly specific. Both requirements can be met if the gating residue forms the inner surface of a pocket. A mobile residue on the outer surface would provide a highly flexible gate but little specificity. A residue on the inner surface of a pocket retains some flexibility while gaining a great deal of potential specificity. The number of interlocking interactions involving the gating residue(s) would control the degree of specificity. The influence of topography on surface mobility thus suggests that one should expect movable gates to be associated with surface residues inside a pocket or invagination.

8. Active Channel Topography

The dominant topographical feature of SOD is not the smaller pits, invaginations, and grooves characteristic of all protein surfaces, but the long, deep, active site channel. This Y-shaped channel lies outside of the skeletal core of the protein formed by the repetitive β strand secondary structure. The sides and rim of the channel are constructed of two large, twisted loops joining the β strands. In cross section, the channel funnels downward stepwise from a large shallow depression about 24 Å wide to a deeper well about 10 Å wide and 5 Å deep, then narrows to a <4 Å groove (see Fig. 3). Ligands of the copper and zinc ions form the floor of the narrow part of the channel. The reactive copper ion, which is alternately oxidized and reduced by the superoxide substrate during the enzymatic reaction, is solvent exposed in a small pit at the bottom of the channel. This pit forms a highly complementary position for superoxide anion to bind both the

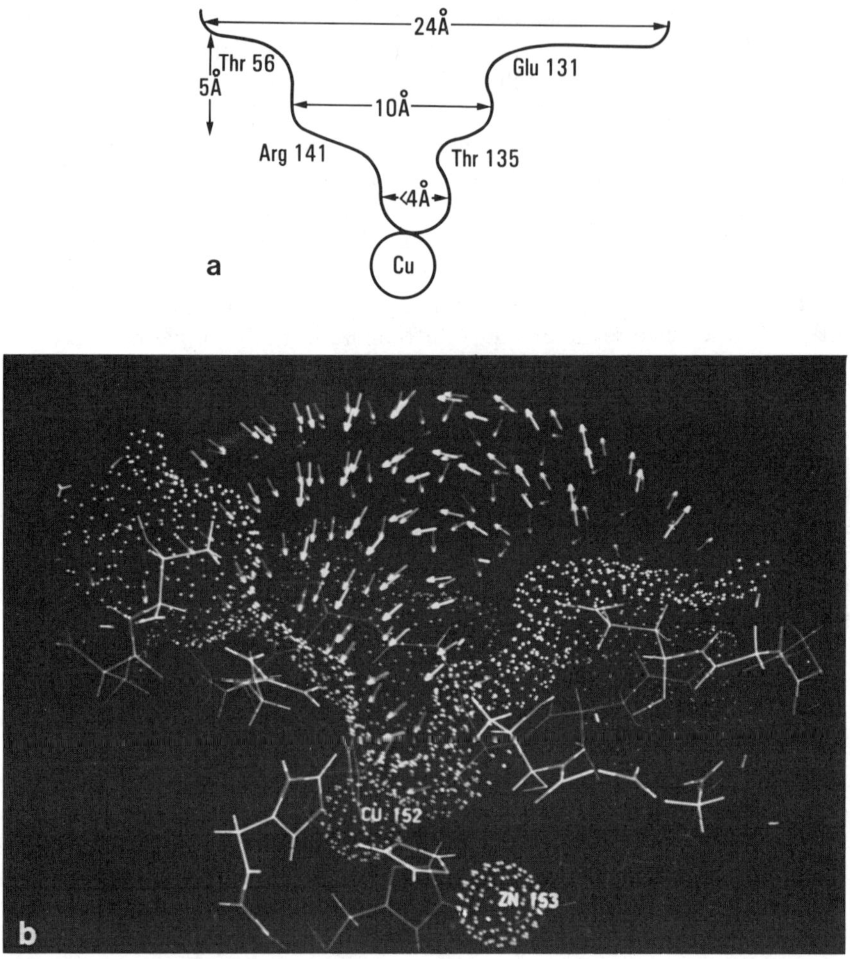

Figure 3. Schematic and computer graphics views of the active site channel. (a) Shape and dimensions of the active site channel in cross-section. The amino acids forming the channel wall are indicated. Lys[134], located forward of the plane of the paper and adjacent to Glu[131], also contributes structurally to channel architecture. (b) A slice through the deep active-site channel of SOD with electrostatic field vectors. The active-site channel is seen in cross section. The vectors were calculated at points outside the molecular boundary on concentric spheres of up to 14 Å radius centered on the axial water ligand of the copper ion. The skeletal model Cu ligands and active-site Arg[141] (left center) can be seen along with the Cu (bottom center) in the most positive part of the active channel. The electrostatic field or electrostatic potential gradient (the magnitude and direction of maximum change in the potential) was calculated by evaluating the partial derivative of the potential with respect to each of the three coordinate axes (Getzoff *et al.*, 1983). By convention, electrostatic field vectors indicate the direction a positive charge would move in the field; we reversed this convention here to show the directional force on the negatively charged superoxide substrate. The surface was calculated using the program MS (Connolly, 1983a) and displayed using the graphics language GRAMPS (O'Donnell and Olson, 1981) and the molecular modeling program GRANNY (Connolly and Olson, 1985). Adapted from Getzoff *et al.*, 1983.

copper ion and activity-important Arg^{141} with the correct geometry. A second small pit forms a water-binding site. The zinc ion is totally inaccessible to our water-sized probe sphere (Tainer *et al.*, 1983a).

The active-site channel structure and topography contribute to SOD's features of ionic selectivity and partial dehydration of bound solvent, which are also necessary to ion channel proteins. The funnellike decrease in size as the wide mouth of the SOD active channel narrows toward the copper ion serves as a selectivity filter to exclude ions larger than a superoxide. The ion selectivity of the enzyme has been demonstrated experimentally by studies of competitive inhibition of catalysis by monovalent anions (Rigo *et al.*, 1977). The results were interpreted as indicating accessibility of the copper ion to Cl^- (3.62 Å diameter) and Br^- (3.90 Å) but not to I^- (4.32 Å). Similar topographic features could be used in ion channel proteins to filter ions on the basis of size. To allow access to and shape complementarity with its binding pocket, the substrate superoxide anion must shed at least some of its hydration shell before binding. The positioning of the side chain of Arg^{141} provides appropriate substitute hydrogen bonding for the ion; while other solvent-accessible hydrogen-bonding atoms in the active site channel interact with solvent water.

In addition to providing size selectivity via surface topography, SOD has evolved to promote substrate specificity through an elaborate system of concerted electrostatic interactions that serve to attract and orient the superoxide substrate into an appropriate docked position for the reaction to take place. The active channel surface topography contributes to these effects by sequestering the reaction from the bulk solvent, thus reducing the effective dielectric constant and increasing electrostatic interactions.

9. Facilitated Diffusion

At physiological pH, SOD has a net negative charge that varies with species: pI values range from 4.6 to 6.8 (Salin and Wilson, 1981). Since the superoxide substrate's net charge is also negative, one would expect an electrostatic barrier to the formation of an enzyme–substrate complex. Instead, our calculation and computer graphics display of the electrostatic potential at the molecular surface of SOD show local electrostatic attraction rather than repulsion of the substrate (Getzoff *et al.*, 1983). The molecular surface of SOD, shaded to indicate electrostatic potential (Fig. 4), reveals no organized pattern except for the striking positive region located over the long, deep channel above the catalytic copper ion. This pattern of positive potential (>7 kcal/mol) extends far beyond the solvent-exposed area of the copper ion (5.2 $Å^2$) to cover most of the active channel, about 9% of the molecular surface (550 $Å^2$) of the protein subunit. Thus, not only is the immediate binding site charge complementary to the negatively charged superoxide anion substrate but additional sequence-conserved amino acids of the active channel (Tainer *et al.*, 1983a) also contribute electrostatically to the attraction of the superoxide anion.

Experimental measurements of the ionic strength dependence of the reaction

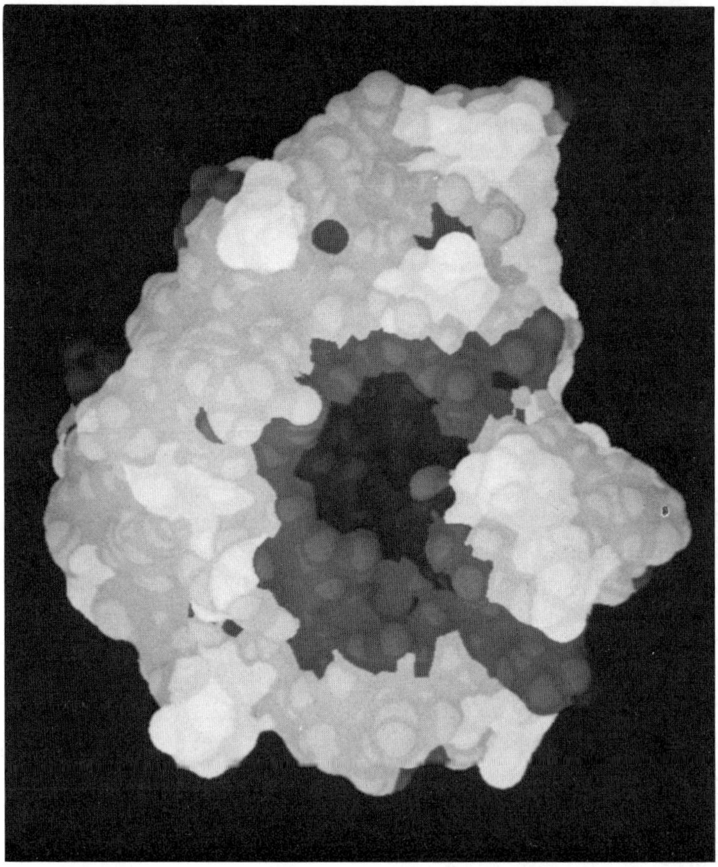

Figure 4. The electrostatically color-coded active-site face of the SOD subunit. The electrostatic potential is divided into five categories coded by a radiating body brightness scale with the most negative potential being white to the most positive potential being black. (white < −21 kcal/mol; light gray −21 to −7 kcal/mol; medium gray −7 to +7 kcal/mol; dark gray +7 to 21 kcal/mol; black > 21 kcal/mol). In SOD, this striking pattern of positive electrostatic potential is unique to the active-site channel. The solid external molecular surface was calculated using M. L. Connolly's programs AMS (Connolly, 1983b) and RAMS (Connolly, 1985) and coded by the electrostatic potentials seen by a water molecule. These electrostatic potentials were calculated from partial charges assigned to the atomic positions of the refined 2 Å crystal structure of the bovine erythrocyte enzyme (Tainer *et al.*, 1983a) using the AMBER data base (Blaney *et al.*, 1982; Weiner and Kollman, 1981). Adapted from Tainer *et al.*, 1985b.

rate confirm that local charges, rather than the net charge, on the protein dominate its electrostatic interactions with the substrate. Raising the ionic strength of the solvent slows catalysis in the native enzyme (Cudd and Fridovich, 1982), implying that the dominant electrostatic forces are attractive (i.e., positive charges on the enzyme interacting with the negative substrate). Reversal of this ionic strength dependence after acylation of seven to eight lysines per subunit (15% residual activity) suggests that local positive charges from one or more lysines are important in providing long-range electrostatic guidance for superoxide anion. Neutralization of active-site Arg^{141} with phenyl glyoxal causes a 90% reduction in activity (Malinowski and Fridovich, 1979) but does not affect the ionic-strength response. These results are compatible with short-range electrostatic stabilization of superoxide anion by Arg^{141} at the binding site, as proposed from computer graphics modeling (Tainer *et al.*, 1983a) and chemical modification studies (Malinowski and Fridovich, 1979).

The SOD dismutation of superoxide anion is extremely rapid (2×10^9 M^{-1} sec^{-1}) (Klug-Roth *et al.*, 1972), about one-fourth the rate of free copper ion in solution (Rabani *et al.*, 1973). Calculations based on collision theory, neglecting steric and orientation factors, suggest that about one in ten random enzyme–substrate collisions are productive (Koppenol, 1982; Getzoff *et al.*, 1983), i.e., the reaction proceeds at about 10% of the maximum calculated diffusion-controlled limit. This enzymatic rate is surprising, since SOD must overcome a large handicap to compete with free Cu; superoxide can react with aqueous Cu ion from any coordinate position but must find the 0.1% of the enzyme's molecular surface that forms a free Cu coordination site. However, this rate could be rationalized by assuming that all random collisions of superoxide anion with the large region of significant positive electrostatic potential covering most of the active channel (about 9% of the molecular surface) are productive. More probably, the attractive electrostatic forces in this region provide nonrandom precollision orientation and/or enhanced diffusion (Getzoff *et al.*, 1983). Dynamics simulations using simple models based on the charge distributions in the crystallographic structure of SOD predict electrostatic enhancement of the diffusion-controlled rate constant (Allison *et al.*, 1985).

10. Electrostatic Orientation

To examine the potential role of electrostatic forces in precollision orientation of the substrate by SOD, we developed a method for the visualization and analysis of an enzyme's three-dimensional electrostatic vector field that enabled the contributions of specific residues to be identified (Getzoff *et al.*, 1983). Long-range guidance effects on approaching superoxide anions were evaluated by the calculation and by the display of electrostatic field vectors within and beyond the SOD active-site channel (Fig. 3b). These vectors indicate that the attraction of superoxide anion is a long-range process occurring even beyond the active-site channel. Judging from the direction of vector flow, a superoxide anion would be repelled from the banks of the channel into the channel itself, where it would be swept along before diving down to dock next to the positive charges of the Cu^{2+} and activity-important Arg^{141} (Getzoff *et al.*, 1983).

Charged residues near the active channel were selectively examined to evaluate their contributions to electrostatic substrate guidance (Getzoff *et al.*, 1983). These residues were individually mathematically neutralized by appropriately reassigning their component partial charges. The field calculated from each modified set of charges was compared with the original field. The largest angular changes in the electrostatic field vector directions identify the amino acid residues which are most important in setting up the field. The importance of an individual charge may also be indicated by examining its sequence variability in the same protein from different species. Amino acid residues that are conserved or conservatively substituted among proteins from widely different species are likely to have important structural or functional roles.

For SOD, charged residues that the electrostatic calculations implicate as important in setting up the enzyme's field are also conserved among the known amino acid sequences. The key charged residues are Arg^{141}, Lys^{134}, and Glu^{131}; Arg^{141} has a local role in binding superoxide anion (Tainer *et al.*, 1983a) and mainly affects the field direction near the Cu ion. The electrostatic potential remains positive farthest from the Cu^{2+} in the direction passing near the positively charged ϵ-amino group of Lys^{134} located 13 Å from the Cu at one end of the active channel. Nonproductive association of superoxide anion with Lys^{134} is reduced by the adjacent negatively charged carboxylate of sequence-conserved Glu^{131}. The strategic placement of this pair of opposite charges helps direct the field vectors into the active channel. The overall electrostatic orientation imposed by these and other residues should increase the number of superoxide anion collisions with SOD that lead to catalysis. Although the calculated magnitude of the electrostatic potential associated with each electrostatic field vector depends on the dielectric model used, potentials along the pathway described above remain greater than the average kinetic energy ($3/2\ kT = 0.9$ kcal/mol) even when damped by the use of the dielectric constant of bulk water ($D = 80$). Thus, the equilibrium concentration of superoxide anion near the active-site channel is significantly greater than that near other regions of the protein, where the potential is more often negative (Getzoff *et al.*, 1983).

11. Implications for Ion Channels

One can envision an ion channel as a deep hydrophilic invagination in the external molecular surface of an integral membrane protein. The diameter of such a channel can change in response to the local environment through simple side-chain rotations and perhaps more extensive structural changes. The channel would certainly contain ordered water molecules, which would allow it to remain open in the absence of ions. Another consequence of ordered water molecules is a decrease in the dielectric constant relative to that for bulk solvent, where the water dipoles can freely align to oppose the electrostatic field. Based on the evidence from structures of enzymes such as SOD, electrostatic forces can be highly significant energetically in guiding the path of ions. Once inside the channel, the ion itself could trigger changes in side-chain conformation, breaking interactions with solvent and between protein side chains. These changes might be expected to affect channel diameter and could also cause the movement of those side chains, forming a gate for ion passage through the channel. For example, binding a positive ion inside an invagination gated at one end by an Arg side chain might trigger the rotation of that positive side chain into an alternative configuration, allowing the relatively negative environment around the original Arg side-chain position to aid ion movement toward the newly opened site. Electrostatic fields around gated channels could improve their efficiency by long-range guidance to the mouth of the channel and by biasing the directional (one-way) movement of ions into the channel. The directional character of an electrostatic field would make ion movement back out of the channel energetically unfavorable.

The hydrophobic core of a membrane bilayer has dimensions of about 30 Å (Cramer *et al.*, 1985), comparable to the thickness of small protein domains. Membrane proteins embedded in the bilayer may have surface topographies that locally shorten this distance significantly. Based on this reasoning, ion channels need not be longer than the two deep metal-binding invaginations in the SOD subunit or than the depth of the Cu ion in the active site (about 15 Å deep), and certainly not longer than the channel formed by opening a deep invagination into the active site channel.

12. Summary

Characteristics of the solvent-exposed molecular surfaces of known protein structures provide clues for understanding ion channels. Examination of the molecular surfaces of highly refined protein structures with computer graphics indicates that protein surfaces characteristically show hydrophilic pockets and invaginations apart from those forming their active sites. For example, SOD has at least six pockets that are known to form metal binding sites. In addition to the deep active-site channel, SOD has two invaginations over 12 Å deep; each is open to the external solvent-exposed surface at one end and separated from the enzyme's active-site channel by a single amino acid side chain at the other. If the blocking side chain were moved, the effective length of these channels through the protein would approach 30 Å including the length of both the metal-binding pocket and the active-site channel. The existence of such deep hydrophilic invaginations, which act as metal binding sites, has implications for the evolution of metal-binding proteins and enzyme active sites and indicates that hydrophilic channels of the type postulated to occur in membrane proteins are compatible with current models of soluble protein structures. The residues lining such channels seem likely to resemble those forming the pockets in SOD. Chemical modification and labeling studies as well as dynamics calculations modeling ion movement through channels might be designed with this information in mind.

Molecular surface topography influences the specificity of the SOD active-site channel for small negative ions. The narrow shape of the active-site channel also plays an important role in reducing the solvent dielectric and thus increasing the effect of electrostatic forces. The electrostatic potential acts to increase specificity by forming a favorable environment for the enzyme's substrate. The electrostatic field (i.e., the derivative of the electrostatic potential) around the protein appears to be crucial to the efficiency and rapid rate of interaction. These same characteristics could be used to provide the efficiency and specificity of ion channels. In SOD, the electrostatic environment appropriate for the enzyme–substrate interaction is localized around the active site; global electrostatic interactions are unfavorable. For ion channels, the local potential of the protein may differ from that of the surrounding membrane bilayer to allow effective electrostatic recognition and guidance to the ion channel.

Perhaps the most important point of this chapter has been to demonstrate that it is possible to use information from existing three-dimensional protein structures to provide insight into ion channel architecture. This mining of structural data is particularly worthwhile if it leads to specific experiments. Furthermore, the expected discoveries ahead in this field will be most useful when placed in the perspective of other structural information.

References

Abernethy, J. L., Steinman, H. M., and Hill, R. L., 1974, Bovine erythrocyte superoxide dismutase: Subunit structure and sequence location of the intrasubunit disulfide bonds *J. Biol. Chem.* **249:**7339–7347.

Allison, S. A., Ganti, G., and McCammon, J. A., 1985, Simulation of the diffusion controlled reaction between superoxide and superoxide dismutase. I. Simple models, *Biopolymers* **24:**1323–1336.

Blaney, J. M., Weiner, P. K., Dearing, A., Kollman, P. A., Jorgensen, E. C., Oatley, S. J., Burridge, J. M., and Blake, C. C. F., 1982, Molecular mechanics simulation of protein–ligand interactions: Binding of thyroid hormone analogues to prealbumin, *J. Am. Chem. Soc.* **104:**6424–6434.

Connolly, M. L., 1981, Protein surfaces and interiors, Ph.D. Dissertation, University of California, Berkeley.

Connolly, M. L., 1983a, Solvent-accessible surfaces of proteins and nucleic acids, *Science* **221:**709–713.

Connolly, M. L., 1983b, Analytical molecular surface calculation, *J. Appl. Crystallogr.* **16:**548–558.

Connolly, M. L., 1985, Depth-buffer algorithms for molecular modeling. *J. Mol. Graphics* **3:**19–24.

Connolly, M. L., and Olson, A. J., 1985, GRANNY, a companion to GRAMPS for the real-time manipulation of macromolecular models, *Computers Chem.* **9:**1–6.

Cramer, W. A., Widger, W. R., Herrmann, R. G., and Trebst, A., 1985, Topography and function of thylakoid membrane proteins, *Trends Biochem. Sci.* **10:** 125–129.

Cudd, A., and Fridovich, I., 1982, Electrostatic interactions in the reaction mechanism of bovine erythrocyte superoxide dismutase, *J. Biol. Chem.* **257:**11443–11447.

Debye, P., 1914, *Ann. Phys.* **43:**4.

Deisenhoffer, J., Epp, O., Miki, K., Huber, R., and Michel, H., 1984, X-ray structure analysis of a membrane protein complex, *J. Mol. Biol.* **180:**385–398.

Fuller, S. D., Capaldi, R. A., and Henderson, R., 1979, Structure of cytochrome *c* oxidase in deoxycholate-derived two-dimension crystals, *J. Mol. Biol.* **134:**305–327.

Garavito, R. M., and Rosenbusch, J. P., 1980, Three-dimensional crystals of an integral membrane protein: An initial X-ray analysis, *J. Cell Biol.* **86:**327–329.

Garavito, R. M., Jenkins, J., Jansonius, J. N., Karlsson, R., and Rosenbusch, J. P., 1983. X-ray diffraction analysis of matrix porin, an integral membrane protein from *Escherichia coli* outer membranes. *J. Mol. Biol.* **164:**313–327.

Getzoff, E. D., 1982, The refined 2 Å structure of copper, zinc superoxide dismutase: Implications for stability and catalysis, Ph.D. Dissertation, Duke University, Durham, North Carolina.

Getzoff, E. D., Tainer, J. A., Weiner, P. K., Kollman, P. A., Richardson, J. S., and Richardson, D. C., 1983, Electrostatic recognition between superoxide and copper, zinc superoxide dismutase, *Nature* **306:**287–290.

Guy, H. R., 1984, A structural model of the acetylcholine receptor channel based on partition energy and helix packing calculations, *Biophys. J.* **45:**249–261.

Henderson, R., and Unwin, P. N. T., 1975, Three-dimension model of purple membrane obtained by electron microscopy, *Nature* **257**:28–32.

Holbrook, S. R., and Kim, S.-H., 1984, Local mobility of nucleic acids as determined from crystallographic data: I. RNA and *B* form DNA, *J. Mol. Biol.* **173**:361–388.

Klug-Roth, D., Fridovich, I., and Rabani, J., 1972, Pulse radiolytic investigations of superoxide catalyzed disproportionation. Mechanism for bovine superoxide dismutase, *J. Am. Chem. Soc.* **95**:2786–2790.

Koppenol, W. H., 1982, On the reactivity of the superoxide anion and the biological function of superoxide dismutase, in *Oxidases and Related Redox Systems* (T. King, M. S. Mason, and M. Morrison, eds.), pp. 127–136, Pergamon Press, Oxford.

Levitt, M., and Chothia, C., 1976, Structural patterns in globular proteins, *Nature* **261**:552–558.

Makowski, L., Casper, D. L. D., Phillips, W. C., Baker, T. S., and Goodenough, D. A., 1984, Gap junction structures, *Biophys. J.* **45**:208–218.

Malinowski, D. P., and Fridovich, I., 1979, Chemical modification of arginine at the active site of the bovine erythrocyte superoxide dismutase, *Biochemistry* **18**:5909–5917.

O'Donnell, T. J., and Olson, A. J., 1981, GRAMPS—A graphics language interpreter for real-time interactive three-dimensional picture editing and animation, *Computer Graphics* **15**:133.

Piez, K. A., and Miller, A., 1974, The structure of collagen fibrils, *J. Supramol. Struct.* **2**:121–137.

Rabani, J., Klug-Roth, D., and Lilie, J., 1973, Pulse radiolytic investigations of the catalyzed disproportionation of peroxy radicals: Aqueous cupric ions, *J. Phys. Chem.* **77**:1169–1175.

Richards, F. M., 1977, Areas, volumes, packing, and protein structure, *Annu. Rev. Biophys. Bioeng.* **6**:151–176.

Richardson, J. S., 1977, β-sheet topology and the relatedness of proteins, *Nature* **268**:495–500.

Richardson, J. S., 1981, The anatomy and taxonomy of protein structure, *Adv. Protein Chem.* **34**:168–339.

Rigo, A., Stevanato, R., Viglino, P., and Rotilio, G., 1977, Competitive inhibition of Cu,Zn superoxide dismutase by monovalent anions. *Biochem. Biophys. Res. Commun.* **79**:776–783.

Rossman, M. G., and Argos, P., 1981, Protein folding, *Annu. Rev. Biochem.* **50**:497–532.

Salin, M. L., and Wilson, W. W., 1981, Porcine superoxide dismutase, *Mol. Cell. Biochem.* **36**:157–161.

Swaminathan, S., Ichiye, T., van Gunsteren, W., and Karplus, M., 1982, Time dependence of atomic fluctuation in proteins: Analysis of local and collective motions in bovine pancreatic trypsin inhibitor, *Biochemistry* **21**:5230–5241.

Tainer, J. A., Getzoff, E. D., Richardson, J. S., and Richardson, D. C., 1980, in, 2SOD:Cu, Zn Superoxide Dismutase Complete Atomic Coordinates (D. C. Richardson and J. S. Richardson, eds.), Brookhaven Protein Structure Data Bank.

Tainer, J. A., Getzoff, E. D., Beem, K. M., Richardson, J. S., and Richardson, D. C., 1982, Determination and analysis of the 2 Å structure of copper, zinc superoxide dismutase, *J. Mol. Biol.* **160**:181–217.

Tainer, J. A., Getzoff, E. D., Richardson, J. S., and Richardson, D. C., 1983a, Structure and mechanism of copper, zinc superoxide dismutase, *Nature* **306**:284–287.

Tainer, J. A., Getzoff, E. D., Connolly, M. L., and Olson, A. J., 1983b, Topography of protein surfaces, *Fed. Proc.* **42**:1998.

Tainer, J. A., Getzoff, E. D., Alexander, H., Houghten, R. A., Olson, A. J., Lerner, R. A., and Hendrickson, W. A., 1984, The reactivity of anti-peptide antibodies is a function of the atomic mobility of sites in a protein. *Nature* **312**:127–133.

Tainer, J. A., Getzoff, E. D., Paterson, Y., Olson, A. J., and Lerner, R. A., 1985a, The atomic mobility a component of protein antigenicity, *Annu. Rev. Immunol.* **3**:501–535.

Tainer, J. A., Getzoff, E. D., Sayre, J., and Olson, A. J., 1985b, Modeling intermolecular interactions: Topography, mobility, and electrostatic recognition, *J. Mol. Graphics* **3**:103–105.

Thomas, K. A., Rubin, B. H., Bier, J. C., Richardson, J. S., and Richardson, D. C., 1974, The crystal structure of bovine $Cu^{2+}Zn^{2+}$ superoxide dismutase at 5.5 Å resolution, *J. Biol. Chem.* **249**:5677–5683.

Wallace, B. A., 1984, Ion-bond forms of the gramicidin A transmembrane channel. *Biophys. J.* **45**:114–116.

Weiner, P. K., and Kollman, P. A. 1981, AMBER: Assisted model building with energy refinement. A general program for modeling molecules and their interactions, *J. Comput. Chem.* **2**:287–303.
Yager, P., Chang, E. L., Williams, R. W., and Dalziel, A. W., 1984, The secondary structure of acetylcholine receptor reconstituted in a single lipid component as determined by Raman spectroscopy, *Biophys. J.* **45**:26–28.

Chapter 4

SINGLE-CHANNEL ENZYMOLOGY

EDWARD MOCZYDLOWSKI

1. An Introduction to Ion Channels as a Class of Enzymes

Until recent years, analysis of ion channel mechanisms belonged to the specialty of electrophysiology. Prerequisites for this work included a background in electronics, microelectrode techniques, and voltage clamping. In this voltage-clamp era, the complete repertoire of biochemical techniques had been difficult to apply to channel proteins because there were no routine assays for purification and functional studies. At present, advances in many fields have salvaged ion channels for biochemistry, as evidenced by the diversity of approaches in this volume. Liposome, planar bilayer, and patch recording studies take their inspiration from test-tube tactics that are familiar to many workers in the biochemical and molecular fields. The classical electrophysiologist now hastens to learn the subtleties of channel cloning, while the molecular biologist pauses to reviews the conclusions of noise analysis.

In such an atmosphere of cross fertilization, the purpose of this chapter is to provide a source of single-channel lore and doctrine that is accessible to students of biochemistry. The premise of the review is that ion channels are a special class of enzymes and that the language necessary for discussing them is a dialect of enzymology. As the author's own training is primarily biochemical, this is a strategy that was adopted in the process of switching from test-tube assays to painting black lipid membranes. Although the material is presented at an elementary level, copious references are given to advanced reviews and primary sources in various areas of channel research. In addition to the present volume and the companion *Single-Channel Recording*, edited by B. Sakmann and E. Neher (Plenum Press, 1983), the channel field now has its first textbook, *Ionic Channels of Excitable Membranes*, by B. Hille (Sinauer Associates, 1984), which provides an extensive coverage of topics introduced here.

To being then, what kind of an enzyme is a channel? Given that enzymes are proteins that enhance reaction rates, the enzymatic function of most channels is to catalyze the transport of inorganic ions across cell membranes. An emphasis on ion transport is a functional emphasis because channels also transport water

EDWARD MOCZYDLOWSKI • Department of Physiology and Biophysics, University of Cincinnati College of Medicine, Cincinnati, Ohio 45267.

and many can transport nonelectrolytes. For example, the flux of urea and formamide mediated by the voltage-dependent Na^+ channel of neuroblastoma cells has been directly measured by radiolabeled tracer flux (Huang *et al.*, 1979). Organic ions are also transported by channels, such as guanidinium ions by the Na^+ channel (Hille, 1971) or molecules as large as ATP by the porin channel of the mitochondrial outer membrane (Colombini, 1980) and the cell–cell gap junction channel (Makowski *et al.*, 1984). However, in regard to electrical excitation, signaling, and secretion phenomena, the function of most ion channels is to promote the selective and rapid flux of any of the inorganic ions, Na^+, K^+, Ca^{2+}, or Cl^-, across a biological membrane. To date, at least 20 different kinds of channels for these various ions have been distinguished by electrical recording and pharmacological methods, and many more are likely to be discovered in the future. Since only two of these ion-selective channels have been purified, as discussed in this volume, ion-channel enzymology is a relatively virgin territory for biochemists.

In order to assess the catalytic ability of channels, three types of solute transport across a membrane may be considered: simple diffusion across a naked lipid bilayer, carrier-mediated transport, and channel-mediated transport. The first of these is commonly expressed as the permeability, P, of a solute in centimeters per second, where the mole flux per unit area, J, is equal to the product of permeability and the solute concentration gradient, ΔC:

$$J = -P\Delta C \tag{1}$$

In contrast to simple diffusion, the kinetic behavior of carrier- and channel-mediated transport is treated quite naturally by standard methods of enzyme kinetics with the observed rate expressed as a saturating function of substrate concentration on both sides of the membrane. Such functions can correspond to a simple Michaelis–Menten hyperbola or exhibit very complex nonlinear behavior. The same is true for the rate laws of enzyme activity versus substrate concentration.

In order to compare rates from a practical viewpoint, the natural measured parameters to compare are the turnover number, k_{cat}, for carriers and the maximal conductance (reciprocal resistance), g_{max}, for single channels. The turnover number of carriers is often measured as V_{max}/E_0, which is the observed maximal transport rate per microgram of protein, V_{max}, divided by the equivalent mass of the functional unit, E_0. The latter parameter is often determined as moles of specific ligand binding sites per microgram of protein. Channel conductance is measured as the slope of single-channel current plotted as a function of applied voltage. This relationship is linear for most channels at low voltages ($< \pm 100$ mV) and under symmetrical salt conditions, indicating ohmic behavior. Since electrical current in amperes can be directly converted to moles of ions per second, conductance values can be used to calculate the number of ions conducted by an open channel per second. Thus, the equivalent expression for the turnover number of a channel is:

$$k_{cat} = g_{max} V N_0 / zF \tag{2}$$

Table I. Comparison of Transport Rates[a]

Transported species	System	P(10^{-12} cm/s) (°C)	Reference[b]
A. Phospholipid bilayer			
Cl^-	PC liposomes	76 (24°)	[1]
Na^+	PC liposomes	0.95 (24°)	[1]
Rb^+	PC liposomes	3.3 (24°)	[1]
Ca^{2+}	PC liposomes	8 (30°)	[2]
D-Glucose	PC liposomes	30 (25°)	[3]
B. Carriers and pumps			
		V_{max}/E_0	
Cl^-	Erythrocyte anion exchanger	5 × 10^4 ions/sec (38°)	[4]
Na^+	Na^+,K^+-ATPase	130–180 ATP/sec (37°)	[5]
K^+	Gastric H^+,K^+-ATPase	20 ATP/sec (37°)	[6]
	Valinomycin	4 × 10^4 ions/sec (23°)	[7]
Ca^{2+}	Sarcoplasmic reticulum Ca^{2+}-ATPase	10 ATP/sec (25°)	[8]
		80–180 ATP/sec (37°)	[9,10]
D-Glucose	Erythrocyte glucose exchanger	1–13 × 10^3 glucose/sec (38°)	[11]
C. Ion-selective channels			
		g (pS)	
Cl^-	*Torpedo* electroplax	10 (22–25°)	[12]
	Glycine receptor	45,31,21 (22–25°)	[13]
	GABA receptor	30,19 (22–25°)	[13]
Na^+	Voltage-activated	12 (9°)	[14]
	Batrachotoxin-activated	20 (25°)	[15]
K^+	Gramicidin	50 (22–25°)	[16]
	Delayed rectifier	17 (6°)	[17]
	Ca^{2+}-activated	240 (18–20°)	[18]
Ca^{2+}	Voltage-activated	30 (20–22°)	[19]

[a] Parameters used to compare the rates of three types of membrane transport processes: P, permeability in cm/s for diffusion through a pure phospholipid bilayer; V_{max}/E_0, turnover number in sec^{-1} for carrier mediated transport; *g*, conductance in pS for channel-mediated ionic current.

[b] References: [1] Mimms *et al.* (1981); [2] Vanderkooi and Martinosi (1971); [3] Brunner *et al.* (1980); [4] Brahm (1977); [5] Jorgensen (1975); [6] Faller *et al.* (1982); [7] Benz and Läuger (1976); [8] Inesi *et al.* (1982); [9] MacLennan *et al.* (1971); [10] Banerjee *et al.* (1979); [11] Brahm (1983); [12] Miller (1982); [13] Hamill *et al.* (1983); [14] Horn *et al.* (1984); [15] Moczydlowski *et al.* (1984a); [16] Eisenman *et al.* (1978); [17] Conti and Neher (1980); [18] Methfessel and Boheim (1982); [19] Brown *et al.* (1984).

where V is the applied voltage, N_0 is Avogadro's number, z is the ionic valence, and F is Faraday's constant = 96,480 coulombs/mol. A voltage of 100 mV is a convenient value to use when comparing channel flux rates, since most cell membranes rarely exceed potentials of this magnitude.

Table I lists the measured permeability of a lecithin bilayer and the turnover numbers of several carriers and channels for the major physiological ions and D-glucose. For an order-of-magnitude comparison of the various types of transport, we may consider the values of ionic flux across an area of bilayer equivalent to that occupied by a large protein molecule (~10^4 Å^2) in the presence of 0.2 M NaCl. This flux would be approximately 10^{-4} ions/sec for Na^+ and 10^{-2} ions/

sec for Cl^- using the data in Table I. Since the Na^+ pump usually exhibits a 3 Na:2K:ATP stoichiometry, one pump occupying this area of lipid would exhibit a flux of 500 Na^+ ions/sec. One of the fastest carriers, the anion exchanger of erythrocytes, transports Cl^- at a rate of 5×10^4 ions/sec. However, the use of equation 2 and Table I shows that many channels operate at rates approaches 10^7 to 10^8 ions/sec at 100 mV of driving force.

The large enhancement of protein-mediated flux over bilayer diffusion does not result from any extraordinary property of membrane proteins. Rather, it is a consequence of the extremely poor partitioning of polar molecules or hydrated ions into the hydrocarbon lipid phase. The task of proteins in evolving transport functions was to provide favorable binding sites for polar molecules during their journey across the membrane. One might then expect that the high flux rates and ohmic behavior of ion channels could be predicted by the physics of simple diffusion through a water-filled pore of atomic dimensions large enough to traverse the bilayer. This expectation has been borne out by modeling studies of cylindrical pores (Hall, 1975; Latorre and Miller, 1983).

Compared to *bona fide* channels, the rates of carrier-mediated transport are at least three orders of magnitude slower. The turnover numbers of carriers fall in the same range as that of typical soluble enzymes. For example, the turnover number of fumarase is 800 sec^{-1} (Tiepel *et al.*, 1968), and that of carbonic anhydrase, one of the fastest enzymes, is 10^6 sec^{-1} (Kernohan, 1965). Carriers and pumps, although matching the efficiency of many enzymes, provide additional impediments to transport with respect to simple pores. Another clue to the difference between carriers and channel is the temperature dependence of their turnover rates (Hille, 1984). The temperature coefficient of conductance for many channels is in the range of $Q_{10} = 1.3–1.6$, close to the value for diffusion in water, $Q_{10} = 1.3$. However, carrier-mediated transport generally exhibits higher values of Q_{10} near 3.0, typical of the higher activation energies of many enzymes.

If channels are indeed now recognized as ion-selective pores, one might argue that the heart of the catalytic mechanism is solved and that they are of no further interest to enzymologists. To the contrary, the problem of ions crossing a pore only begins the description of channel enzymology. A major unanswered question concerns the mechanism of channel gating, which refers to the process of opening or closing the pore. As any new student might guess, a fixed leak or rigid pore would not be of much use to a cell. It could easily be lethal to an unprotected cell, as is known for the case of certain bacterial colicin channels (Schein *et al.*, 1978). Most channels have uniquely specific conditions that favor the opening of their pores. Two common mechanisms are known as voltage-dependent gating and agonist-dependent gating. Voltage-dependent channels activate over a specific range of transmembrane voltage, which can be as narrow as 10–20 mV. The activation of agonist-dependent channels is promoted by the binding of small molecules known as agonists or neurotransmitters. Both types of gated channels can also exhibit inactivation phenomena, which may predominate in a different voltage range or at a higher agonist concentration. Agonist-dependent inactivation is also called desensitization. The voltage-dependent Na^+ channel and the acetylcholine-receptor channel are classic examples of these two types of gating. The time scale of most gating events is in the range of microseconds to seconds. Thus, the duration of a gating event is very similar to the turnover time of a typical soluble enzyme or carrier. Also, the temperature dependence of gating processes is usually higher than that for channel conductance. In these respects, structural rearrangments that underlie the gating of channels appear to share certain features of the conformational transitions of enzymes.

In addition to gating, channels exhibit a variety of inhibition phenomena referred to as channel block or blockade. Block has been found to be induced in various channels by molecules ranging from inorganic ions to anesthetics, toxins, and even agonists. Just as inhibitors are used to deduce enzyme mechanisms, the study of channel block is a useful approach to the structure and mechanism of channels.

Before we leave this introduction, it is instructive to consider the nature of carriers and pumps. Are carriers simply pathetically slow channels, or do they represent an inherently different class of enzymes? Confusion on this point is evident from transport models referring to channels in pumps (Moczydlowski and Fortes, 1981) or references to carrier exchange proteins as channels (Rothstein and Ramjeesingh, 1980). The distinction between carrier and channel is clear in the case of the ionophore models valinomycin and gramicidin. Valinomycin functions as an ion-chelating agent that is soluble within the membrane and transports by diffusion of the complex from surface to surface of the opposed monolayers that form the bilayer (Pressman, 1976). The structure of gramicidin is a head-to-head dimeric helix that spans the membrane. The central cavity of the helix is lined with carbonyl oxygens, which substitute for water in the hydration sphere of cations as they diffuse through the channel (Anderson, 1984). At present, these structural distinctions between carrier and channel cannot be made as easily for integral proteins that span the membrane and certainly do not undergo the kind of mobile transmembrane diffusion shown by valinomyocin. However, carrier and channel mechanisms have been lucidly distinguished by Läuger (1980), who defines channels kinetically as mechanisms with binding sites simultaneously accessible from both sides of the membrane. In contrast, carrier mechanisms involve the alternating exposure of binding sites to either side mediated by a conformational transition that accompanies substrate translocation. However, he has shown that these two cases are really the extreme limits of a common two-state mechanism in which binding sites accessible from both sides undergo reversible transitions of their respective potential energy barriers for ion entry and release. This concept is illustrated by potential energy diagrams in Fig. 1.

The rigid dependence of transport events on conformational coupling causes carriers to display behavior that is foreign to channels. For example, many carriers and pumps exhibit small simple stoichiometries of coupled fluxes, such as 3Na/2K/ATP for Na^+ pump, $1Cl^-/1HCO_3^-$ exchange for the erythrocyte anion transport system, and 1 Na^+/1 substrate for Na^+-coupled glucose or amino acid cotransport systems. The conformational change in carriers can be considered a gating process that actually allows one or a small number of substrate molecules first bound on one side to be released on the other side. In contrast, the confor-

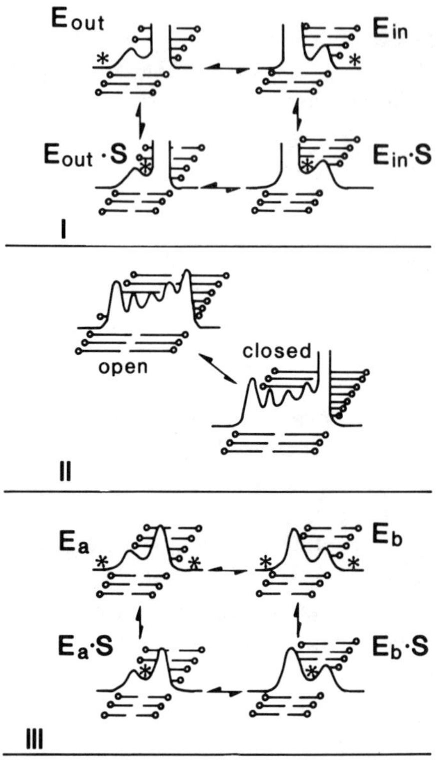

Figure 1. Kinetic distinctions between a carrier and channel based on energy diagrams. Each state is meant to represent the free energy profile of a transported ion or substrate as it crosses the lipid bilayer by a carrier or channel mechanism. Binding sites are represented by energy minima, and transition states correspond to maxima. A broken or open transition state represents an infinitely high barrier. In a pure carrier mechanism (I), an ion can only cross the membrane after a conformational transition of the carrier, as in $E_{in} \rightarrow E_{out}$. In a pure channel mechanism (II), the ion is free to diffuse across the membrane by hopping from site to site in the open channel. In channels, conformational events often involve gating between closed and open states. A unified transport scheme (III) shows how a conformational transition between two states that favor either outward or inward exit can provide a common mechanism for carrier or channel behavior. See Läuger (1980) for a detailed discussion.

mational changes associated with gating in channels serve to open a fixed pore for a time long enough for thousands of ions to diffuse across it. Whatever the mechanism of coupled fluxes in carriers, the degree of coupling can be severe. For example, the cited anion-exchange protein can only pass net Cl^- current at a rate of 5–15 ions/sec (Fröhlich, 1984). This rate is less than 0.01% of the coupled Cl^- exchange reaction and is many orders of magnitude lower than the conductance of even mediocre channels. Ultimately, the relatedness of carriers and channels must be established by their structural homology. Based on sequence data and common structural features, it may one day be possible to classify transport enzymes systematically and trace the path of their molecular evolution (see also Hille, 1984, for a discussion of channel evolution).

2. Mechanisms of Channel Activation

2.1. Analysis of Single-Channel Records

A record of single-channel current is essentially the recorded history of conformational changes of a single macromolecule or functional complex. Although different types of channels exhibit distinct patterns of activity, all such records

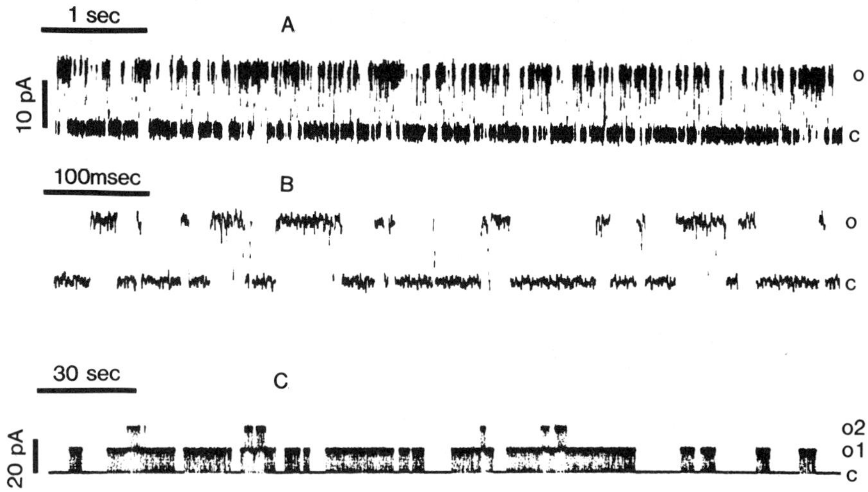

Figure 2. Examples of single-channel behavior. (A) Record of a single Ca^{2+}-activated K^+ channel from rat muscle incorporated in a planar lipid bilayer. Conditions: 1 μM internal Ca^{2+} and 0.2 M symmetrical KCl, +60 mV holding voltage, methods as described in Moczydlowski and Latorre (1983). Closed (c) and open (o) current levels are labeled at the right of each record. (B) A segment of record A is shown at expanded time scale. (C) Record from a bilayer containing two Ca^{2+}-activated K^+ channels in the presence of charybdotoxin, a specific blocker present in *Leiurus quinquestriatus* venom. Conditions: 0.2 M KCl, 100 μM Ca^{2+} internal; 0.01 M KCl, 10 μg/ml *Leiurus* venom external; 0 mV holding voltage. Periods of no channel activity correspond to block of the channel by the toxin (see Miller *et al.*, 1985, for details).

usually exhibit two clearly defined levels of current. The low current level is usually indistinguishable from zero current or the background leak current of the particular recording system. This level is called the closed state. It can also be induced by channel blockers that reside in open channels for a sufficiently long period of time. The highest distinct current level is called the open state and defines the unit conductance of the channel. Closed, open, and blocked states are illustrated by records of a Ca^{2+}-activated K^+ channel shown in Fig. 2. Probably every channel will also be found to dwell in one or more current levels intermediate between open and closed. These intermediate levels are called substates. They may occur seldom or frequently, comprising from less than 1% to a large proportion of all the events. The best-documented examples of this phenomenon are substates of the nicotinic acetylcholine receptor (Hamill and Sakmann, 1981; Auerbach and Sachs, 1983, 1984) and a K^+ channel from frog sarcoplasmic reticulum (Labarca and Miller, 1981). Figure 3 shows examples of substate transitions of the acetylcholine receptor. Other characteristic features of single-channel records are sharp transitions between closed and open states and randomly occurring dwell times in these two levels. Given these features of the data, the objective of single-channel analysis is to interpret such random fluctuations in terms of the structure and conformational transitions of the macromolecule.

To be exact, the single-channel record only reveals the rate-limiting transi-

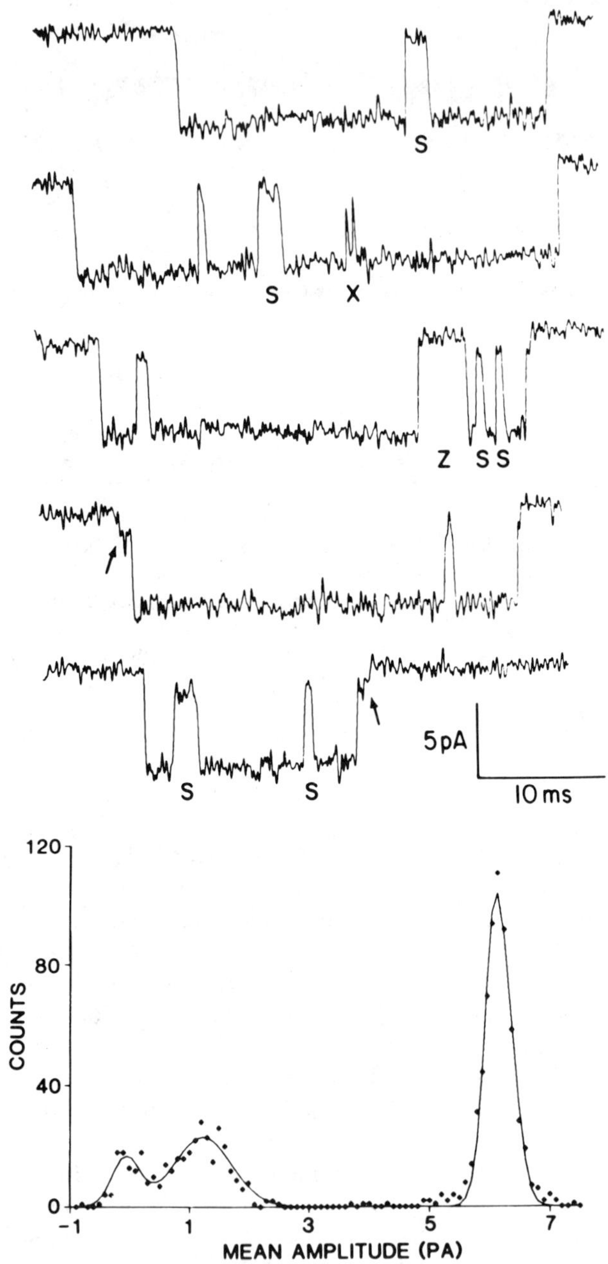

Figure 3. Examples of substate levels of a single channel. Records are opening bursts of acetylcholine receptor channels in the presence of 50 nM acetylcholine, as observed by cell-attached patch-clamp recording of embryonic chick muscle cells. Channel opening is downward. Lower graph is a current amplitude histogram as compiled from single-channel bursts. The 0-pA peak refers to the closed state; the 6.2-pA peak refers to the main open state; the peak at 1.3 pA corresponds to the substate events. (Figures are reproduced from Auerbach and Sachs, 1984, with permission.)

tions between open and closed conformations that occur at the time scale of the recording, usually in the range of 10^{-4}–10^{3} sec. Conformational events not detected are those that are electrically silent (no conductance change) or those occurring outside the limiting time response of the recording. Transitions between open and closed levels appear as square pulses of current because the actual current relaxation between ~0 ions/sec at the closed state to ~10^{6} ions/sec at the open state is too fast to measure. The nearly square time course of the apparent transition step actually corresponds to the time response of the recording or filtering device. The resolved dwell time in one level represents the random waiting time for the slowest steps in a group of states to occur. Thus, single-channel records provide two kinds of molecular information: (1) a direct measurement of the conductance value of the open state, which can be related to the ion permeation mechanism of the channel; and (2) a sampling of resolvable dwell time events in open and closed levels, which can be related to gating or blocking of the channel.

Single-channel analysis can be used to study either steady-state behavior or relaxation behavior following a rapid perturbation (i.e., voltage change). Steady-state recording gives the probability distributions of open and closed events among the various resolvable gating intermediates at equilibrium. Nonstationary analysis gives time-dependent probabilities of channel opening and closing between initial and final equilibrium states. For inactivating channels such as Na^{+} channels, there is essentially no steady-state channel activity at conditions of interest, and, therefore, events must be recorded following a voltage pulse from a resting level at which previously inactivated channels can return to activatible states. For channels for which either method can be applied, kinetic schemes derived from the two approaches must ultimately agree. Relaxation methods can be used to help select among various possible models by testing whether certain pathways are favored by particular initial conditions using prepulse protocols (Aldrich and Yellen, 1984).

An underlying assumption of analysis at steady state is that the probability of the various channel conformations is constant during the length of the record. Some channels do not always exhibit a stable equilibrium. For example, the glutamate-activated channel of locust muscle in cell patches and the Ca^{2+}-activated K^{+} channel in planar bilayers have been shown to exhibit spontaneous variations in opening probability (Patlak *et al.*, 1979; Moczydlowski and Latorre, 1983). This behavior, which is dubbed here the "gearshift" phenomenon, can be readily detected by visual inspection of the records as shown for several examples in Fig. 4. At present, no laboratory has yet attempted a systematic study of the frequency of "gearshift" occurrences, so few conclusions can be drawn. It is not improbable, however, that these shifts may represent very slow conformational changes affecting gating that are superimposed on characteristic rapid processes. Another similar problem is the so called "rundown" phenomenon of Ca^{2+}- or Na^{+}-channel currents. Rundown is a term used to describe slowly disappearing currents or shifting activation parameters after the formation or excision of cell-attached patches (Byerly and Hagiwara, 1982; Fenwich *et al.*, 1982). Such phenomena may not represent an inherent instability of the channels themselves but may reflect cellular or membrane metabolism that is perturbed by removal of the patch from

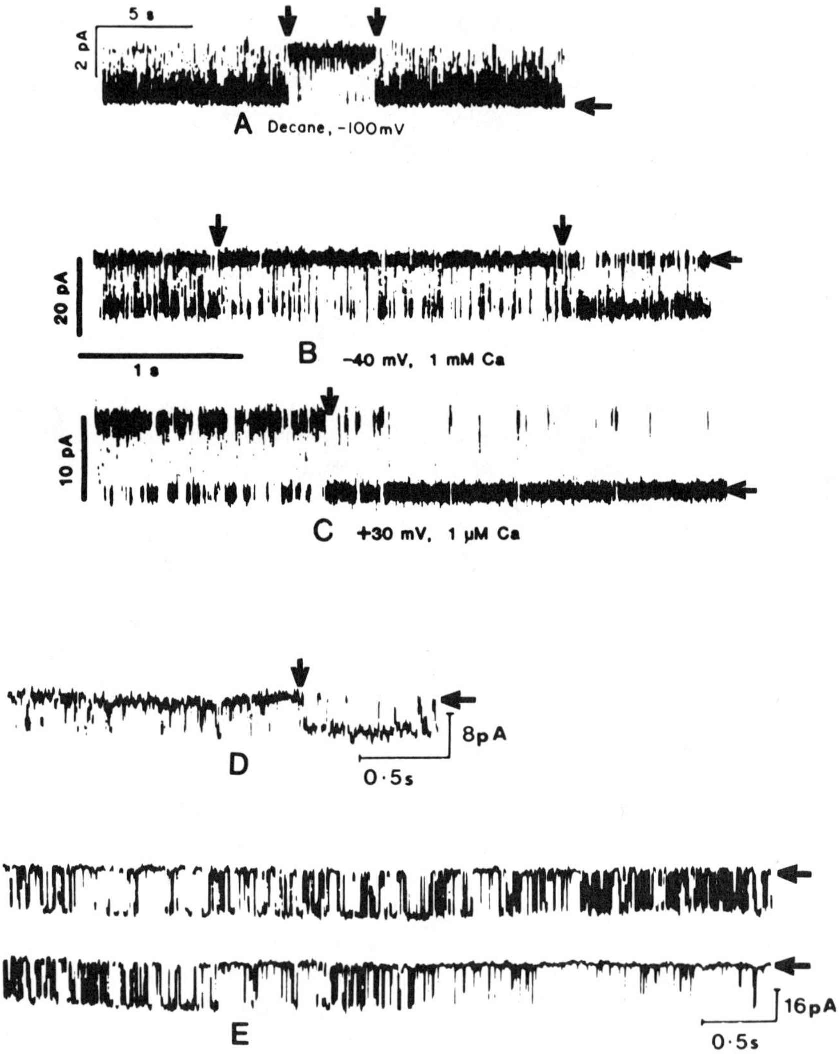

Figure 4. Examples of nonstationary single-channel activity referred to as the "gearshift" phenomenon. (A) Record of a batrachotoxin-activated Na^+ channel in planar bilayers as described by Moczydlowski *et al.* (1984a). (B,C) Records of two different Ca^{2+}-activated K^+ channels in planar bilayers as described by Moczydlowski and Latorre (1983). (D,E) Records of two different glutamate receptor channels from locust muscle in the presence of 0.1 mM L-glutamate as described by Patlak *et al.* (1979). Horizontal arrows at the right of each record mark the zero current level. Vertical arrows mark abrupt changes in opening probability. Figures are reproduced from the cited sources with permission.

the cell. One way to determine if equilibrium or stationary behavior was maintained during the course of an experiment is to check whether the opening probability at some reference voltage and/or agonist concentration is the same at the end of the protocol as when the same channel was first observed.

Since the dwell time in particular closed and open levels is a function of the number of channels present in the patch or bilayer, an exact analysis of dwell times also requires knowledge of the total number of channels contributing to the record. The total number of channels can be determined by measuring the current under conditions in which the opening probability of an individual channel is high, i.e., $P_{open} > 0.7$. Such conditions can often be achieved by using an appropriate voltage or agonist concentration. In cases in which measurements are restricted to low opening probability such as the ACh-receptor channel or the voltage-dependent Na^+ channel, it is sometimes not possible to be certain of the exact number of channels that contributed to the record. In such cases, the maximum number of simultaneous openings only provides a lower limit for the actual number of channels present. Despite this difficulty, it is still possible to rigorously analyze certain types of transitions. An example is the bursting open state of the ACh receptor in many-channel patches, since there is a high probability that such discrete bursts are caused by a single channel in the patch (Colquhoun and Sakmann, 1984). Thus, although *bona fide* singles are the most desirable records to obtain and analyze, a certain amount of useful information can be extracted from multichannel records.

A general-purpose analysis of a typical piece of steady-state single-channel data involves the evaluation of four characteristics of the record: (1) the current amplitude of the open state(s) with respect to the closed state (zero current); (2) the probability of opening averaged over the whole length of the record; (3) the probability distribution of the duration of open-state events; and (4) the probability distribution of the duration of closed-state events. Detailed descriptions and examples of automated analysis of single-channel data have been published (Sachs *et al.*, 1982; Sachs, 1984; Colquhoun and Sigworth, 1984).

2.2. Extracting Rate Constants and Schemes from Single Molecules

Stochastic processes underlie the behavior of populations of molecules. The individual events of a stochastic process are characterized by three features: they have a discrete probability of occurrence, are random, and do not depend on past history, i.e., they have no memory. For example, the simplest possible rate behavior of a population of molecules is to spontaneously transform into a different state, such as the radioactive decay of ^{14}C, the one-step opening of a closed channel, or the dissociation of an inhibitor from an enzyme after washing away the unbound ligand. The analysis of such processes begins with the formulation of a reaction scheme or transition diagram:

$$I \xrightarrow{k_{ij}} J \qquad \text{(S-I)}$$

where k_{ij} is the rate constant in inverse seconds that characterizes the conversion of I to J. The rate law for the unimolecular decay of I is written as a differential equation:

$$d[I]/dt = k_{ij}[I] \tag{3}$$

This equation is directly solved by integration from time 0 to t when $[I] = [I]_0$ at $t = 0$:

$$[I]_t/[I]_0 = \exp(-k_{ij}t) \tag{4}$$

The time course described by equation 4 is the familiar exponential or first-order process characterized by a time constant, $\tau_{ij} = k_{ij}^{-1}$, at which $[I]_0$ decays to e^{-1} of its initial value or by a half-time, $t_{1/2} = (\ln 2)/k$, at which half of the original $[I]_0$ remains. Thus, given a mole of sample I and reliable method of measuring it, a high-school chemistry student would be expected to plot the time course of the sample according to equation 4 and derive a value for k_{ij}.

Suppose, however, in an era of fiscal austerity, that the student in question were only given one molecule of I that could undergo reversible transitions to and from J:

$$I \underset{k_{ji}}{\overset{k_{ij}}{\rightleftharpoons}} J \tag{S-II}$$

How could the student measure the rate constants k_{ij} and k_{ji} by observing the single molecule? The solution to this problem involves a derivation of the exponential probability distribution. This derivation can be found in basic statistics texts and has been previously presented from the viewpoint of pharmacology and channels by Colquhoun (1973) and Colquhoun and Hawkes (1984).

The exponential probability distribution describes the probability $P(t)$ that individual events of residence in state I are greater than a variable of time, t:

$$P(t) = \text{Prob}(\tau_I > t) \tag{5}$$

An expression for this probability can be derived by a straightforward analysis leading to the definition of the derivative of differential calculus (Colquhoun and Hawkes, 1984). The final result is:

$$P(t) = \exp(-k_{ij}t) \tag{6}$$

where $P(0)$ is not defined for $t < 0$.

Equation 6 is one form of the exponential distribution that is commonly encountered in single-channel analysis. It is a cumulative distribution in that it expresses the summed probability of all events that last longer than time t. A second common form of equation 6 is the probability density function, $f(t)$, for which the area under the curve between any two times, t_1 and t_2, is the probability of

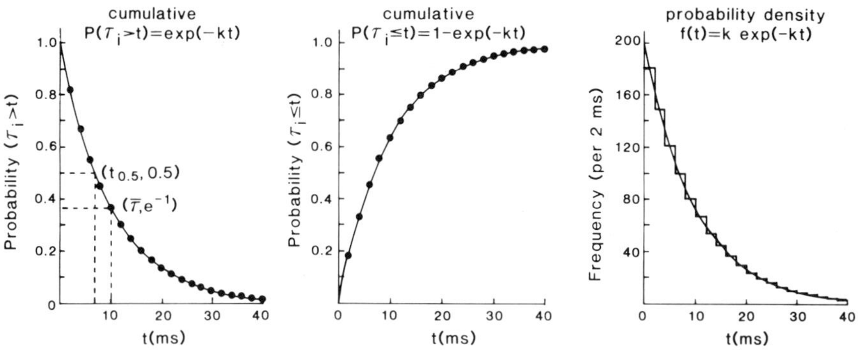

Figure 5. Three representations of dwell-time probability distributions. The cumulative and probability density forms of an exponential distribution are plotted as discussed in the text. The rate constant, k, equals 100 $\sec^{-1}$ for these examples. The theoretical probability density function (p.d.f.) is shown for a bin width of 2 msec and a total of 1000 events. The expected number of events in each bin of the p.d.f. is obtained by multiplying the probability calculated from equation 11 by the total number of events, which in this case equals $100[e^{-kt_1} - e^{-kt_2}]$. The solid curve fitting this p.d.f. is found by multiplying ke^{-kt} by the bin width and the total number of events, which equals $200e^{-kt}$. Note that this latter curve passes through the approximate midpoint of each bin.

events that lasted longer than t_1 but less than t_2. It may be found by taking the derivative of the corresponding cumulative distribution function:

$$P(\tau_1 \leq t) = F(t) = 1 - P(t) = 1 - \exp(-k_{ij}t) \tag{7}$$

$$dF(t)/dt = f(t) = k_{ij} \exp(-k_{ij}t) \tag{8}$$

Statisticians prefer the density function form for fitting points to obtain the rate constant k_{ij} or for testing whether a given process is, in fact, purely exponential. However, for distributions known to be exponential, the cumulative form (equation 6 or 7) is perfectly adequate. The cumulative form also has the advantage that the probability can be read directly from the plot rather than by comparing relative areas, as the density function requires. In any case, the form of the distribution should always be made clear by labeling the vertical axis. "Number of events" is an ambiguous label for the ordinate since it could refer to either type of distribution. Three representations of an exponential distributions are shown in Fig. 5.

Equations 6–8, although given in terms of an exponential distribution, express important relations that can be used to manipulate the various forms of any probability distribution. These relationships for variables defined for positive values, $x \geq 0$, are summarized as follows:

$$f(x) = dF(x)/dx \tag{9}$$

$$\int_0^\infty f(x)\,dx = 1 \tag{10}$$

$$P(t_1 < x < t_2) = \int_{t_1}^{t_2} f(x)\,dx \tag{11}$$

Equations 9 and 11 express the basic derivative–integral relationship between probability density functions, $f(x)$, and cumulative probability functions, $F(x)$. All density functions must be normalized according to equation 10 such that the total area under the function is equal to 100% probability. A probability can be computed by integrating a density function between desired limits (equation 11) or by evaluating the desired cumulative probability, $F(x)$ or $1 - F(x)$, at x.

Two well-known parameters used in describing probability distributions are the population mean, μ, and population variance, σ^2, which are defined as follows (for $x \geq 0$):

$$\mu = \int_0^\infty x f(x)\,\mathrm{d}x \tag{12}$$

$$\sigma^2 = \int_0^\infty (x - \mu)^2 f(x)\mathrm{d}x = \int_0^\infty x^2 f(x)\,\mathrm{d}x - \mu^2 \tag{13}$$

The usual parameter for discussing the spread of values about the mean is the square root of the variance or the standard deviation, denoted as σ. Application of equations 12 and 13 to an exponential probability density function, equation 8, reveals a strikingly simple relationship for the mean and standard deviation of this distribution: $\mu = \sigma = k^{-1}$. Thus, for an exponential distribution, the mean dwell time equals the inverse of the rate constant, which also equals the standard deviation of the population.

In practice, the actual compilation of an exponential distribution often results in an underestimate of the true number of brief events, since a certain proportion of these events are unavoidably missed because of the limited frequency response of the recording system or, in some cases, by the presence of overlapping distributions shorter than the component of interest. In such cases, the observed distributions can be given a consistent time resolution by establishing a certain cutoff limit, t_c, which specifies the minimum duration of measured events. The cutoff limit is set according to the duration of the shortest event that can be accurately measured. Alternatively, it can be set in order to safely exclude an overlapping short component that interfers with the analysis of a long-lived component (e.g., Moczydlowski *et al.*, 1984a). The advantage of using a defined cutoff limit is that this limit can be used to evaluate the error associated with the truncated data. For example, if all openings are well resolved and only brief closings are discarded, the observed closed-time distribution has empty bins from $t = 0$ to $t = t_c$. If the remaining distribution is exponential, its time constant can be obtained by fitting the data at times greater than t_c to an exponential. Extrapolation back to zero time then gives the expected number of missed events. Alternatively, evaluation of the observed arithmetic mean (equation 12) of an exponential distribution truncated by a lower limit t_c, gives the following simple relationship between the observed mean, $\bar{\tau}_c{}^*$, and the true mean, $\bar{\tau}_c$:

$$\bar{\tau}_c{}^* = \bar{\tau}_c + t_c \tag{14}$$

Thus, the observed mean is longer than the true mean by exactly the value of the cutoff limit.

We must also consider the effect of excluding brief closings on the open-time distribution. Whenever a closing is lost because of limited resolution or an imposed cutoff limit, the observed open-time distribution is longer than its true distribution, since many real openings partitioned by short closings are artificially lengthened. If the open-state distribution is exponential, then there also is a simple relationship between the observed mean open time, $\bar{\tau}_o^*$ and the true mean open time, $\bar{\tau}_o$:

$$\bar{\tau}_o^* = \bar{\tau}_o \exp(t_c/\bar{\tau}_c) \tag{15}$$

Thus, the true mean is shorter than the observed mean by the fraction of all closings longer than the closed time cutoff limit, $\exp(-t_c/\bar{\tau}_c)$. The relationships of equations 14 and 15 are simplified cases of more general exact formulas given by Colquhoun and Sigworth (1984) for the relationships between observed, extended mean closed and open times and the true means for exponential distributions when both distributions are modified by short-time cutoff limits. These expressions thus provide a useful method of evaluating errors associated with the unavoidable loss of short events.

Thus far we have considered in detail only a simple unimolecular process, the reversible conversion of state I to state J, scheme S-II. Although such a process is a fundamental building block of reaction mechanisms, the turnover of enzymes often involves many discernible steps. During the course of enzyme catalysis, substrates bind, intermediates form, and end products dissociate while the enzyme undergoes minor and major conformational transitions. Likewise with channels, more complicated kinetic schemes are necessary to account for observed behavior. A full account of every kinetic intermediate is of course an impossible task for a protein. Although some techniques may appear to resolve only a few reaction steps, increased time resolution or a different physical measurement on the same system will continue to yield new intermediates. The primary task of the enzymologist is to characterize enough transient states of a reaction to deduce its underlying principles.

In dealing with reactions mediated by chemical activators or inhibitors, the concentration of such modulators can be varied to identify binding equilibria. For example, the following schemes may be considered:

$$\mathrm{A} + \mathrm{C} \underset{k_{\mathrm{off}}}{\overset{k_{\mathrm{on}}[\mathrm{A}]}{\rightleftharpoons}} \mathrm{A{\cdot}O} \tag{S-III}$$

$$\mathrm{B} + \mathrm{O} \underset{k_{\mathrm{off}}}{\overset{k_{\mathrm{on}}[\mathrm{B};}{\rightleftharpoons}} \mathrm{B{\cdot}O} \tag{S-IV}$$

In S-III an activator or agonist, A, combines with a closed channel, C, to produce an open channel, A·O. In S-IV a blocker, B, combines with an unblocked, open channel, O, to produce a blocked state, B·O. Since these two schemes each contain only one closed or open state, the probability distribution of each state will be described by a simple exponential distribution as discussed previously. However, because the association of the channel with A or B is a bimolecular reaction, each state may be identified by its concentration dependence. The mean open and mean closed times or the observed rate constants are given as follows:

$$\text{Scheme S-III:} \quad \bar{\tau}_{\text{open}} = k_{\text{off}}^{-1};\ \bar{\tau}_{\text{closed}} = (k_{\text{on}}[\text{A}])^{-1} \tag{16}$$

$$\text{Scheme S-IV:} \quad \bar{\tau}_{\text{unblocked}} = (k_{\text{on}}[\text{B}])^{-1};\ \bar{\tau}_{\text{blocked}} = k_{\text{off}}^{-1} \tag{17}$$

Thus, for S-III, the mean open time will be independent of [A], whereas the mean closed time will be inversely proportional to [A]. For S-IV, the opposite behavior is found: blocked or closed events are concentration independent, whereas the open or unblocked events shorten with increasing blocker concentration. For a binding equilibrium, we see that one state corresponds to the dwell time of the ligand on the receptor with a mean of $(k_{\text{off}})^{-1}$, whereas the other state corresponds to the waiting time for an unoccupied receptor to bind ligand, with a mean of $(k_{\text{on}}[\text{L}])^{-1}$. The relationships of equations 16 and 17 are often used to identify binding equilibria in complex reaction schemes.

Instead of the simple one-step schemes of S-III and S-IV, it is often necessary to consider the following two-step processes in the analysis of channel behavior:

$$\text{C} \underset{k_{\text{off}}}{\overset{k_{\text{on}}[\text{A}]}{\rightleftharpoons}} \text{A·C} \underset{\alpha}{\overset{\beta}{\rightleftharpoons}} \text{A·O} \tag{S-V}$$

$$\text{C}_1 \underset{\gamma}{\overset{\delta}{\rightleftharpoons}} \text{C}_2 \underset{\alpha}{\overset{\beta}{\rightleftharpoons}} \text{O} \tag{S-VI}$$

$$\text{C} \underset{\alpha}{\overset{\beta}{\rightleftharpoons}} \text{O} \underset{k_{\text{off}}}{\overset{k_{\text{on}}[\text{B}]}{\rightleftharpoons}} \text{O·B} \tag{S-VII}$$

where C, A·C, and O·B are closed or blocked states, and A·O and O are open states. S-V represents agonist binding to a closed state that can subsequently open. S-VI is a gating mechanism for a channel that is not activated by an agonist but passes through two distinct closed states before opening. S-VII is a two-state closed–open channel in the presence of a blocker that can only bind to the open state and thereby induce a discrete blocked state with conductance indistinguishable from the closed state. In each of these mechanisms there is only one open state. Therefore, the observed open-state distributions will be single exponentials for each scheme. For S-V and S-VI, $\bar{\tau}_{\text{open}} = \alpha^{-1}$, since α is the only pathway by which the open state can close. However, for S-VIII, there are two paths for closing, denoted by α and $k_{\text{on}}[\text{B}]$. Since the probabilities of closing by either path are independent, the mean open time in S-VII is equal to the inverse of the sum, $(\alpha + k_{\text{on}}[\text{B}])^{-1}$. It is a general rule that for any individual state, mean dwell time in that state is the reciprocal of the sum of all the paths for leaving the state.

To deal with the closed-state population of the above schemes, we note that each scheme contains two distinct closed states that cannot be distinguished by conductance value. The solution to the exact form of the probability distribution for two or more indistinguishable states involves systems of differential equations that are best solved by matrix methods. Colquhoun and Hawkes (1977, 1981, 1982,

1984) have described the mathematical methods for handling such problems, including the detailed solutions for the schemes listed above. In general, the unconditional probability distribution for a scheme containing n indistinguishable states is a sum of n exponentials with time constants and amplitudes given by functions of rate constants. For S-V and S-VI, the predicted time constants and amplitudes of the two-exponential distribution of closed events have no simple physical significance and cannot be assigned to a particular step in the mechanism. This is an important rule for cases involving indistinguishable states that are directly connected. However, for mechanism S-VIII, the closed states C and O·B are not in direct communication because they are separated by state O. In such special cases, the observed probability distribution can be interpreted as having two distinct time constants, β^{-1} and $k_{off}{}^{-1}$, with relative amplitudes given by the respective probability of each process, $\alpha/(\alpha + k_{on}[B])$ and $k_{on}[B]/(\alpha + k_{on}[B])$.

In cases in which one reaction is much faster than another, certain simplifications can result. For example, if the binding equilibrium of S-V is much faster than the conformational step, then the two closed states would no longer be kinetically distinguishable. In this case, the closed-state population will approximate a single exponential distribution even though there are two transient closed states.

The unconditional mean closed or mean open time of any multiexponential distribution can always be found by calculating the mean or the expectation integral according to equation 12. Alternatively, Colquhoun and Hawkes (1977) have given a convenient equation for calculating the mean lifetime, $\bar{\tau}$, of any group of n indistinguishable states:

$$\bar{\tau}^{-1} = \sum_{i=1}^{n} p_i k_i; \qquad k_i = \sum_{j=1}^{m} k_j \tag{18}$$

where k_i is the sum of the rate constants of each path, j, by which the ith state can exit the group and p_i is the fraction of the group that is in the ith state at equilibrium. If there are no pathways by which the ith group state can exit the group, then $k_i = 0$. Applying this formula to calculate the unconditional mean open and closed times of S-V to S-VII gives the following results:

Scheme	$\bar{\tau}_{open}$	$\bar{\tau}_{closed}$
S-V	α^{-1}	$\beta^{-1}(1 + K_D/[A])$; $K_D = k_{off}/k_{on}$
S-VI	α^{-1}	$\beta^{-1}(1 + \gamma/\delta[A])$
S-VII	$(\alpha + k_{on}[B])^{-1}$	$(\alpha K_D + \beta[B])/\beta(\alpha K_D + [B]k_{off})$

A consideration of the expected behavior of S-V and S-VII in the limits of zero or infinite [A] or [B] will show that the unconditional mean lifetimes predicted by equation 18 behave as expected in these concentration limits. The extrapolated rate behavior to high or low concentration limits can often be useful in estimating a particular rate constant (e.g., Moczydlowski and Latorre, 1983).

Another important phenomenon of single-channel kinetics is called bursting behavior. Consider what would happen for a mechanism such as S-V, if the time the channel spent in the binding reaction was much longer than the time spent in the opening reaction. This could happen, for example, if the activator [A] concentration was low and if α and β were large compared to k_{off}. Then the channel would dwell in state C a relatively long time before binding agonist. But once A·C was reached, the channel could open and close several times before A dissociates. This would give the appearance of long gaps of no activity interspersed with groups of short openings. Such an obvious clustered group of events is called a burst. Other colorful terminology to describe bursting behavior is often encountered, such as *Nachschlag* (Colquhoun and Sakmann, 1981) or paroxysms (Montal *et al.*, 1984).

To predict the number of openings that should occur in a burst, one begins with the probability that the A·C state will open rather than dissociate, $p = \beta/(\beta + k_{\text{off}})$, and the only other probability of dissociation, $1 - p$. A burst of r open events then consists of r independent events of probability p and one independent event of exit from the burst with probability, $1 - p$. The probability of observing a burst of r openings is then the product of these probabilities:

$$P(r) = (1 - p)p^r \qquad r = 0, 1, 2, \ldots \tag{19}$$

The mean number of openings per burst is given by an infinite series of the product of r and $P(r)$. This series may be recognized as a converging geometric series, and $P(r)$ is a geometric probability distribution. This mean must be corrected by subtracting the probability of a burst with zero openings, since the apparent bursts, those that are actually observed, contain at least one opening. The final result is an apparent mean number of openings per burst of $(1 + \beta/k_{\text{off}})$. The predicted apparent mean burst duration may be found according to a derivation of the probability distribution of the burst length presented by Colquhoun and Hawkes (1981). It is obvious from scheme S-V that this mean length will only be a function of α, β, and k_{off} and will therefore be independent of agonist concentration. By analogy, channels that do not require agonists but rest in at least two closed states before opening (scheme S-VI) can similarly exhibit bursting behavior. Bursting behavior can also be caused by the presence of a blocker (scheme S-VII), as discussed in Section 3.2.

2.3. Voltage-Dependent Processes

A phenomenon unique to integral membrane proteins is the effect of voltage on turnover kinetics. Large transmembrane electric fields of about 100 mV falling over less than 100 Å can provide the energy to move ions and dipoles that may be associated with proteins residing in this environment. In this section, the theory of voltage-dependent rates commonly encountered in the channel literature is reviewed.

The starting point for considering the effect of potential energy on the probability of molecular transitions is the Boltzmann distribution:

$$P_2/P_1 = \exp[-(G_2 - G_1)/kT] \tag{20}$$

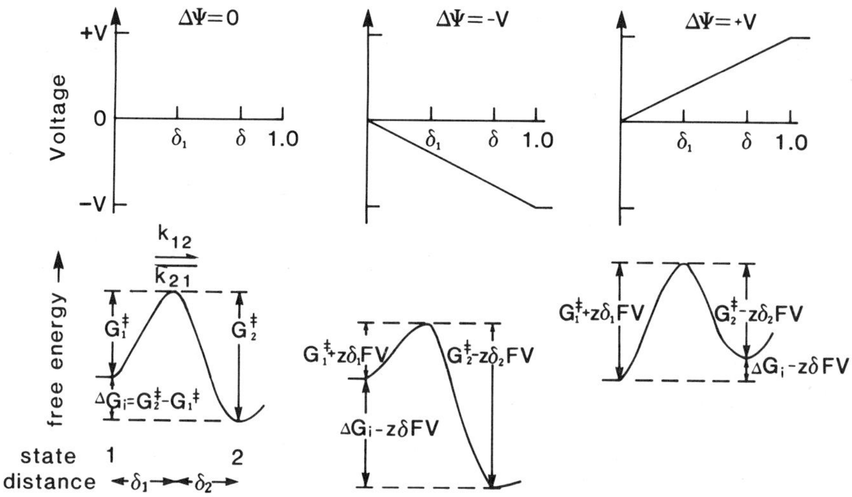

Figure 6. Free energy profile of an ion binding site located in an electric field.

This law describes the ratio of probabilities for a molecule residing in either of two states, P_2/P_1, as an exponential distribution with respect to the energy difference, $G_2 - G_1$, and the absolute temperature, T. The Boltzmann constant, k, is used when the energy per molecule is given, whereas the gas constant, $R = N_0k$, is used when energy per mole is given, N_0 denoting Avogadro's number. For a two-state reversible reaction, the Boltzmann distribution gives the equilibrium constant, $K_{eq} = P_2/P_1$, as a function of the energy difference.

To treat voltage-dependent equilibria in channels, one needs to express the energy differences between various states as functions of voltage. This objective is problematic since the mechanism by which channels transduce voltage is unknown. However, the simplest reasonable notion would be that the electric field does work by moving a charged particle that resides within the field. This particle could either be an ion or a charged amino acid residue of the protein. Both types of "gating particles" have been previously proposed for ion channels. In the absence of detailed information, voltage-dependent equilibria are usually modeled as shown in Fig. 6. Figure 6 shows potential energy diagrams for two positions of a particle within the membrane in the absence of an applied field, $V = 0$, and in the presence of a voltage of either polarity, $-V$ and $+V$. The energy trough of state 1 in this model represents the position of the particle outside the membrane electric field, and state 2 represents a binding site within the membrane field. The peak between troughs 1 and 2 represents the transition state or energy maximum that the ion must transverse between the conformational or binding states 1 and 2. The energy difference between states 1 and 2, ΔG_i, is thus the intrinsic or conformational chemical potential energy between the two states at zero voltage. In Fig. 6 the applied voltage is assumed to fall in a linear fashion with distance across the membrane, resulting in a constant electric field. The potential energy contributed by the voltage is assumed to add to the zero-field profile, resulting in a larger or smaller energy difference between states 1 and 2 depending on the polarity of the voltage.

For a particle bearing one elementary charge, the energy required to move the particle across the whole electric field would be the appropriately signed value of the electron charge, e, times the voltage, V. The energy per mole of charge would be zFV, where z is the valence of the charge and F is Faraday's constant, $F = N_0e$. The energy could, however, be less than zFV if the charged particle was only moved a fraction of the way into the field. In Fig. 6 this fractional distance of the field is represented as δ, the position of site 2 in the field. Thus, if δ varies from 0 to 1, the energy in moving the charge as a function of δ would be $z\delta FV$. The total energy between states 1 and 2, ΔG, as a function of voltage is therefore given by:

$$\Delta G = \Delta G_i + z\delta FV \tag{21}$$

The transition between state 1 and state 2 can be represented as a reversible equilibrium with an equilibrium constant given by the ratio of equilibrium probabilities in the two states or the ratio of rate constants as follows:

$$1 \underset{k_{21}}{\overset{k_{12}}{\rightleftharpoons}} 2 \qquad \text{(S-VIII)}$$

$$K_{eq} = P_2/P_1 = k_{12}/k_{21} \tag{22}$$

The Boltzmann equation can now be used to express equilibrium as a function of voltage:

$$P_2/P_1 = \exp -[(\Delta G_i + z\delta FV)/RT] \tag{23}$$

Voltage-dependent gating is often modeled empirically as a two-state equilibrium according to S-VIII, where state 1 would represent the closed state and state 2 would represent the open state. In this case, a linearized form of equation 23 can be used to fit the voltage-dependent probability of opening, P_o.

$$\ln\left[\frac{P_o}{1 - P_o}\right] = \frac{-z\delta F}{RT}(V - V_0) \tag{24}$$

where $\Delta G_i = -z\delta FV_0$. In this form, V_0 is the applied voltage at which the equilibrium constant equals 1.0 or P_o equals 0.5, the midpoint voltage of the P_o-versus-V relation.

Figure 7 shows theoretical P_o-versus-V relations for the opening probability of several different ion channels in planar bilayers. The steepness or slope of these curves is related to the value of $z\delta$, which is often called the effective valence or the effective gating charge of the reaction. The value of $z\delta$ would thus equal the charge on the hypothetical gating particle if it moved the whole distance of the

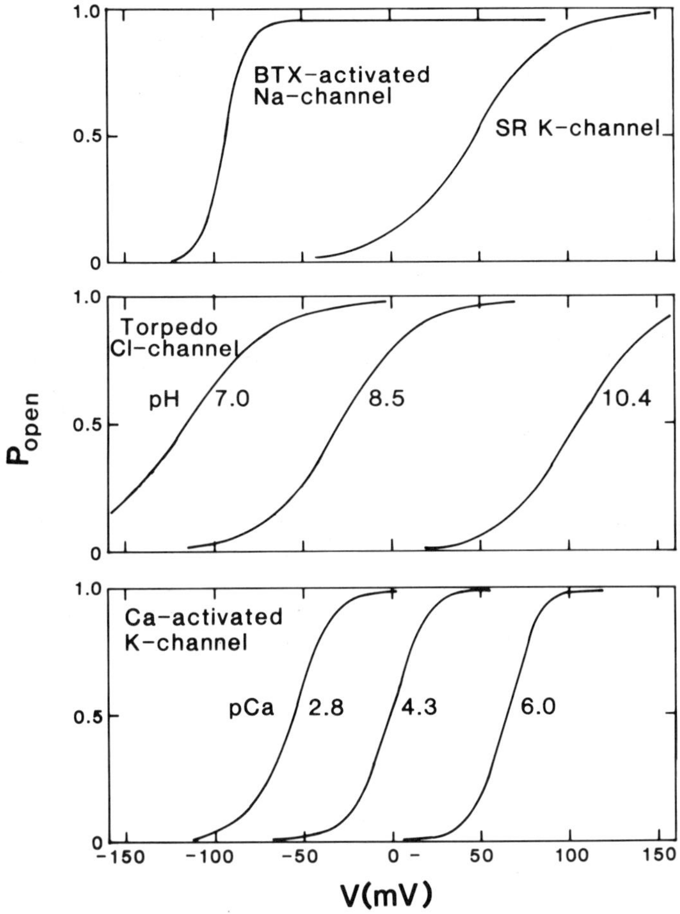

Figure 7. Voltage-dependent opening probability of four different channels observed in planar bilayers. Data for different channels were fitted to equation 24 given in the text. References: BTX-activated Na^+ channel, Moczydlowski *et al.* (1984a). SR K^+ channel Labarca *et al.* (1980); *Torpedo* Cl^- channel, Hanke and Miller (1983); Ca^{2+}-activated K^+ channel, Moczydlowski and Latorre (1983).

field, $\delta = 1$. Typical values of $z\delta$ range from 1 for the sarcoplasmic reticulum K^+ channel to 4.0 for the BTX-modified Na^+ channel. The steep gating region where the probability is most sensitive to voltage occurs at different values of V_0 for different channels, ranging from -95 mV for the Na^+ channel to $+106$ mV for the Torpedo Cl^- channel at pH 10.4. In some channels V_0 is a function of the chemical potential of regulatory ligands such as Ca^{2+} or H^+. For example, the Cl^- channel shifts in V_0 by 60 mV per pH unit (Hanke and Miller, 1983), and the Ca^{2+}-activated K^+ channel shifts V_0 by 35 mV per pCa unit (Moczydlowski *et al.*, 1985).

Although the simple two-state model leading to equation 23 often provides a reasonable fit to many voltage-dependent equilibria, a complete analysis of the kinetics of a process requires that the voltage dependence of individual rate constants be determined. The theory of voltage-dependent rates closely follows the Eyring transition state theory of chemical kinetics. In this theory, statistical mechanics is used to derive the following expression for the rate constant:

$$k_{12} = \frac{\kappa kT}{\mathrm{h}} \exp[-G_1^{\ddagger}/RT] \tag{25}$$

where h is Planck's constant and kT/h has units of $\sec^{-1}$ and represents the frequency of molecular vibrations, sometimes called the universal rate constant. The factor κ is the transmission coefficient, which represents the fraction of molecular attempts that pass over the activation energy barrier to state 2. It cannot be measured independently and is usually disregarded and set to 1.0. In the derivation of equation 25 for any rate constant, the transition from initial to final state is treated as sequential process in which the initial state is in equilibrium with the transition state. Thus, $G^{\ddagger}$ represents the height of the activation energy barrier with respect to the initial state and represents the free energy change of this equilibrium. In the same way as equation 21 was used to give the change in free energy between two states as a function of voltage, equation 26 gives the free energy change between initial states and the transition states for reactions k_{12} and k_{21} as functions of voltage:

$$\begin{aligned} &\Delta G_1^{\dagger}\; G_1^{\dagger}\; z\delta_1 FV \\ &\Delta G_2^{\ddagger}\; G_2^{\ddagger}\; z\delta_2 FV \end{aligned} \tag{26}$$

where $\delta_1 + \delta_2 = \delta$. The corresponding expressions for the voltage-dependent rate constants are now:

$$\begin{aligned} k_{12} &= (\kappa kT/\mathrm{h}) \exp\left[-\frac{G_1^{\ddagger}}{RT}\right] \exp\left[\frac{-z\delta_1 FV}{RT}\right] \\ k_{21} &= (\kappa kT/\mathrm{h}) \exp\left[-\frac{G_2^{\ddagger}}{RT}\right] \exp\left[\frac{z\delta_2 FV}{RT}\right] \end{aligned} \tag{27}$$

Note that k_{12} and k_{21} have opposite voltage-dependent behavior, k_{12} decreasing with voltage and k_{21} increasing with positive voltage. The non-voltage-dependent terms are often combined together as zero-voltage rate constants and an equilibrium constant as follows:

$$\begin{aligned} k_{12}(V) &= k_{12}(0) \exp\left[\frac{-z\delta_1 FV}{RT}\right] \\ k_{21}(V) &= k_{21}(0) \exp\left[\frac{z\delta_2 FV}{RT}\right] \\ K_{\mathrm{eq}}(V) &= \frac{k_{12}(V)}{k_{21}(V)} = K_{\mathrm{eq}}(0) \exp\left[-\frac{z\delta FV}{RT}\right] \end{aligned} \tag{28}$$

Equations 28 thus can be used to describe any rate constant in terms of barrier heights, related to $k(0)$ and a voltage-dependent part related to $z\delta$. Such equations are often used to describe rate constants associated with elementary steps in gating, site-to-site hopping steps associated with ion permeation through the channel, and association or dissociation rate constants for the binding of ligands, such as charged blockers, to a site on the channel protein.

The treatment of a particle moving to a site in a field might be expected to closely approximate the real behavior of a free ion such as a charged blocker or permeant ion moving into the channel from solution, assuming of course that the voltage-driven rates were below the rate of free diffusion up to the entrance to the field. However, in the case of charged amino acid residues on a polypeptide, the electric field cannot be expected to exert such a simple effect since conformational motion of a segment of the chain can be coupled with the rest of the molecule and linked to many other types of chemical interactions that determine the structure of the protein. In the case of channel gating, a somewhat more realistic approach can be taken by treating the effect of voltage as a change in dipole moment of the protein (Stevens, 1978; Neher and Stevens, 1979). In this treatment, one assumes that the gating structure can be modeled as a dipole, or two charges opposite in sign but equivalent in magnitude, q, separated by a distance, d. The dipole moment, μ, is then a vector along an axis joining the two charges with magnitude, qd. The energy of interaction between the dipole and the electric field vector, **E** is then given by the product of **E** and the component of μ parallel to the field.

Thus, in terms of a change in dipole moment instead of a particle moving in a field, the change in free energy now becomes

$$\Delta G = \Delta G_i + \Delta\mu N_0 V/a \tag{29}$$

where the change in molecular dipole moment $\Delta\mu$, given in Debyes (3.3356×10^{-30} coulomb-m), is normalized with respect to the length of the field by the factor a (assumed to equal the membrane thickness, 4 nm). Avogadro's number N_0 is used to convert to a molar free energy. Comparison of equations 21 and 29 shows that the hypothetical $z\delta$ term is equivalent to $\Delta\mu/ae$, where e is the electron charge. Thus, a rather different physical mechanism involving a change in dipole movement can lead to the same form of voltage dependence as that for an ion moving in a field. The slope of a plot of the logarithm of a rate constant or equilibrium constant versus voltage according to equations 28 can thus be interpreted with impartial ignorance as either an effective valence, $z\delta$, or a change in dipole moment $\Delta\mu$. Some of the steepest voltage-dependent processes that have been observed have an e-fold change per 6 mV, which corresponds to a $z\delta$ of 4 and a $\Delta\mu$ of 1000 Debye. For the purpose of comparison, the magnitude of a dipole moment of two elementary charges separated by 1 Å is 4.8 Debye, and reported dipole moments of some ordinary soluble proteins range from 170 to 480 D (VanHolde, 1971).

The use of effective valence, $z\delta$, or effective dipole moment, $\Delta\mu$/a, is convenient because it allows a voltage-dependent interaction to be quantified without detailed knowledge of the actual molecular structures interacting with the field. One simply assumes all the actual charge–field and dipole–field interactions can be summed to a lumped voltage-independent parameter, $z\delta$ or $\Delta\mu$.

3. Mechanisms of Channel Inhibition

3.1. Permeation through a Polypeptide Pore

A wealth of evidence indicates that permeant inorganic ions interact with distinct binding sites during their passage through channels. These ion-binding sites are undoubtedly quite different from the binding sites of organic substrate molecules to enzymes, which permit multiple chemical interactions with various atoms of the ligand. In the case of the pore structure provided by gramicidin (Anderson, 1984), the channel permeation sites could be specified by a few carbonyl oxygens capable of interacting with the ion and providing a local energy minimum in the channel.

The ability of channels to select among inorganic ions arises from the energy of interaction between ions and channel sites. This energy depends on the size and valence of the ion, its hydration chemistry, and electrostatic interactions between the ion and chemical groups at the site. Since the coulomb interaction energy varies inversely with the distance between charges, small ions can exhibit a higher energy of interaction with charged groups and dipoles. For example, small cations such as Li^+ exhibit greater dehydration energies than Cs^+ since Li^+ interacts more strongly with the negative oxygen end of the bond dipole of water. In this view, binding to a site on a protein depends on the competition between ion–water and ion–site interactions. If the ion is able to approach very close to a charged group in a site, then binding to the site will be favored over binding to water, and the site will then select Li^+ over Cs^+. Such a site is termed a "high-field-strength" site. Conversely, a site of low electrostatic interaction energy compared to water is termed a "low-field-strength" site and will favor Cs^+ over Li^+. Such intuitive reasoning formulated in terms of free energy differences leads to the prediction of 11 selectivity sequences of the five alkali cations out of a possible 120 sequences. These predicted sequences are known as the Eisenman selectivity series and are commonly observed for channels and ion activation of enzymes (Eisenman and Horn, 1983).

In addition to ion selectivity, the kinetics of ion permeation through channels also reveals the effects of binding sites. For an inert pore without binding, the current through the pore is predicted to be a linear function of ion activity, and the fluxes in either direction are independent (Läuger, 1980). However, biological ion channels exhibit saturation of channel conductance as the ion activity is increased. Such behavior can be predicted by treating the kinetics of channel transport in the manner of enzyme reactions. This analysis involves drawing a diagram that represents the total number of binding sites and assigning rate constants to the allowed routes of filling empty sites. With the Eyring well-and-barrier approach, an overall rate equation is derived that predicts the current as a function of ion activity and voltage. For the simplest models of a one-site channel or a multisite channel that can only contain a maximum of one ion, channel conductance, g, is predicted to exhibit Michaelis–Menten behavior:

$$g/g_{\max} = a/(a + K_{0.5}) \tag{30}$$

where a is the ion activity and $K_{0.5}$ is the activity at half-maximal conductance (Läuger, 1980). Channels that contain more than one site and can bind two or more ions simultaneously are called multiion channels. The permeation kinetics of such channels can be quite complicated because of multiple states of channel occupancy and possible electrostatic repulsion between neighboring ions. For example, an ion bound at one site could both slow the entry and speed the exit of an adjacent ion, resulting in a lower binding affinity for the second ion. Potassium channels such as the delayed rectifier and inward rectifier are among the best-studied examples of multiion channels (Latorre and Miller, 1983).

Given the nature of the channel as a narrow tube of binding sites, it might be imagined that it would be a simple matter to design an inhibitor for such a mechanism. The requirement for an inhibitor would be to permeate at a much slower rate than the substrate ion or to bind near one end of the pore and occlude the entrance. Such blockers can be found for every type of ion channel. In the case of some K^+ channels, it is difficult to find an organic cation that does not produce some type of blocking. Examples of such blockers are common buffers such as trishydroxymethylaminomethane or quaternary ammonium ions such as choline (Coronado and Miller, 1982). As an alternative to directly occluding the channel, it might also be possible for a molecule binding at site distant from the pore to inhibit binding by an allosteric mechanism. Such a mechanism actually appears to operate in the case of a class of dihydropyridine-based drugs that inhibit Ca^{2+} channels (Hess *et al.*, 1984). However, for many blockers, there is good evidence that the blocker binding site is within the pore. In the case of agonist-activated channels, antagonists can also inhibit by preventing agonist binding as in the case of α-bungarotoxin and *d*-tubocurarine for the acetylcholine receptor of skeletal muscle (Changeux *et al.*, 1984).

By defining the location and properties of blocker binding sites, it is possible to characterize functional sites of channels in much the same way as inhibitors are used to study enzymes. In the following sections, two types of blocking behavior commonly observed at the single-channel level are discussed with emphasis on the evidence for direct occlusion of the pore.

3.2. Kinetics of Slow Block

In the language of enzymology, reversible channel blockers are inhibitory ligands in equilibrium with a binding site. The action of many blockers can be resolved at the single-channel level as discrete blocked periods caused by the residence of an individual molecule on the channel. Such blockers are often called slow blockers, since their dissociation rates are slow enough to be resolved. For the average channel, the resolution limit for short events is about 100 μsec. This limit, which depends on the size of the single-channel current and the recording system, determines the minimum mean dwell time necessary for slow block to be observed. As discussed in Section 2.2, scheme S-VII forms the initial basis for analysis of blocker kinetics:

$$C \underset{\alpha}{\overset{\beta}{\rightleftharpoons}} O \underset{k_{off}}{\overset{k_{on}[B]}{\rightleftharpoons}} O{\cdot}B \qquad \text{(S-VII)}$$

This scheme assumes that the blocker, B, is in binding equilibrium with the open state of the channel, O, where the equilibrium dissociation constant of the blocker is equal to $K_D = k_{off}/k_{on}$. Colquhoun and Hawkes (1984) have presented derivations of the predicted behavior of the above scheme. Here this theory is summarized and compared with actual results for various channels.

The single open state of scheme S-VII gives rise to the following simple expressions for the cumulative probability distribution and mean length of openings or unblocked events:

$$P(\tau_o > t) = \exp[-(\alpha + k_{on}[B])t] \qquad (31)$$

$$\bar{\tau}_o = (\alpha + k_{on}[B])^{-1} \qquad (32)$$

Since the closed and blocked states cannot be distinguished by conductance, the distribution of all shut periods follows a sum of two exponentials:

$$P(\tau_s > t) = \left(\frac{\alpha}{\alpha + k_{on}[B]}\right)\exp(-\beta t) + \left(\frac{k_{on}[B]}{\alpha + k_{on}[B]}\right)\exp(-k_{off}t) \qquad (33)$$

$$\bar{\tau}_s = (\alpha K_D + \beta[B])/\beta(\alpha K_D + k_{off}[B]) \qquad (34)$$

The particular form of the unconditional shut-state distribution of scheme S-VII has an especially simple interpretation because the two shut states are not in direct communication. It may be noted that the time constant of each exponential in equation 33 corresponds to either of two rate constants, β or k_{off}. When the blocker concentration is set to zero, equation 33 reduces to the behavior of the simple closed–open gating reaction. Therefore, in this case one can uniquely identify the two components of the closed-state distribution with either β of the gating reaction or k_{off} of the blocker. The analysis of slow block is straightforward in either of two extremes, $k_{off} \gg \beta$ or $k_{off} \ll \beta$.

The case in which $k_{off} \gg \beta$ has been encountered in QX-222 block of the acetylcholine receptor (Neher and Steinbach, 1978) or decamethonium block (see Fig. 10) of the rabbit SR K^+ channel (Coronado and Milller, 1980; Miller, 1982). In these examples, the individual blocked states are shorter than the normal closed events of the channel in the absence of blocker. At blocker concentrations at which many blocking events can occur per channel opening, the channel records present an appearance of bursts of openings in which the gaps within the bursts have a mean of $1/k_{off}$ and the gaps between bursts have a mean of 1/β. The mean burst length can be derived simply by grouping the open and blocked states and applying equation 20 to obtain:

$$\bar{\tau}_{\text{burst}} = \alpha^{-1}(1 + [\text{B}]/K_{\text{D}}) \tag{35}$$

Colquhoun and Hawkes (1984) have shown that $\bar{\tau}_{\text{burst}}$ is actually comprised of the sum of the mean total open time per burst, $1/\alpha$, and the mean total blocked time per burst $[\text{B}]/\alpha K_{\text{D}}$. Thus, k_{off} can be obtained from the mean of gaps within bursts, k_{on} from the unconditional mean open time, and K_D independently from the burst length.

The case in which $k_{\text{off}} \ll \beta$ has been encountered for Ba^{2+} and charybdotoxin (Fig. 2) block of the Ca^{2+}-activated K^+ channel (Vergara and Latorre, 1983; Miller *et al.,* 1985) and also for saxitoxin and tetrodotoxin block of the batrachotoxin-modified Na^+ channel (French *et al.,* 1984; Moczydlowski *et al.,* 1984a). In these examples, the blocked states correspond to long-lived shut states that punctuate periods of normal gating activity. Thus, the gating activity occurs as a burst, but the blocking events comprise intraburst gaps with mean k_{off}^{-1}. In this case, the mean burst length is given as:

$$\bar{\tau}_{\text{burst}} = (1 + \alpha/\beta)(k_{\text{on}}[\text{B}])^{-1} \tag{36}$$

Equation 36 predicts that the burst length is inversely proportional to blocker concentration and is also a function of the gating equilibrium α/β. This latter effect provides a sensitive test of whether the blocker in fact binds only to the open state. If the gating equilibrium can be shifted toward the closed state, for example, by changing the voltage or reducing agonist concentration, the mean burst length should increase. Vergara *et al.,* Latorre (1983) actually found a sixfold increase in the mean burst length of the Ca^{2+}-activated K^+ channel in the presence of Ba^{2+} as the opening probability was lowered to $\alpha/\beta = 5.7$, which closely follows the predictions of scheme S-VII.

3.3. In the Limit of Fast Block

If the dissociation rate of the blocker is fast, $k_{\text{off}} > 10^5\ \text{sec}^{-1}$, its residence time will not be distinctly resolved by single-channel recording. This range of dissociation rates is typical of many low-affinity ligands of proteins. Since the short-lived blocked states cannot be resolved, the very fast "noise" caused by blocker binding and unbinding will be filtered by the amplifier and result in a lower apparent single-channel conductance. Such an effect can be demonstrated simply by heavilty filtering (1–10 Hz) a single-channel record taken in the presence of a blocker whose binding rates are fast but can be resolved at high frequency response (1–10 kHz, e.g., see Fig. 7 of Yellen, 1984a). This effect of filtering can be used to derive a theory for measuring rate constants by fitting the amplitude distribution of records taken in the presence of noisy or "flickery" blockers whose off rates cannot be cleanly resolved by dwell-time analysis (Fitzhugh, 1983; Yellen, 1984a). However, for very fast blockers that cause a lower apparent single-channel current, the equilibrium dissociation constant of the blocker can still be obtained by analyzing the blocker titration curve.

The basis for this analysis is that the ratio of single-channel current in the presence of blocker to that in its absence, i/i_0 equals the fraction of unblocked channels or the fraction of time that a single channel is not occupied by blocker. For the following equilibrium involving the binding of n blocker molecules to an open channel, the following equations can readily be derived:

$$n\text{B} + \text{O} \overset{K_\text{B}}{\longleftrightarrow} n\text{B}\cdot\text{O} \tag{S-IX}$$

$$i/i_0 = (1 + [\text{B}]^n/K_\text{B})^{-1} \tag{37}$$

$$\log(i_0/i - 1) = n \log[\text{B}] - \log K_\text{B} \tag{38}$$

Equation 37 describes the expected behavior of the blocker titration curve; its reciprocal is a linear transformation that gives the K_B as the abscissa intercept. Equation 38 is a Hill plot that can be used to evaluate the Hill coefficient, n. If $n \neq 1$, this is an indication that the mechanism may involve more than one blocker binding site.

The kinetics of fast blockade of a single open state can be explained in terms of the previous discussion of slow block in Section 3.2 for the case, $k_{\text{off}} \gg \beta$. The difference is that there are no gaps within bursts caused by discrete blocking events. Rather, the mean open time is increased since the length of the apparent open state really consists of the normal open time of the unblocked channel and the total time that the open channel is occupied by blocker. Thus, the apparent mean open time in the presence of a fast blocker is equal to the burst length given by equation 35, which increases linearly with blocker concentration. Figure 8 shows an example of actual behavior following equation 35 as reported by Coronado and Miller (1982) for fast block of the SR K^+ channel by bis[tris(hydroxymethyl)methylamino]propane (BTP). Since both the reciprocal of apparent channel conductance (equation 37) and the apparent mean burst length (equation 35) are linear functions of the term $(1 + [\text{B}]/K_D)$, Coronado and Miller (1982) noted that the product of these observed parameters is predicted to be constant at all blocker concentrations according to:

$$(i/i_0)\bar{\tau}_{\text{burst}} = \alpha^{-1} \tag{39}$$

Behavior following the relationship of equation 39 was also demonstrated in the BTP blocking study. Given the observed agreement with the predictions of scheme S-VII, the cited examples of block of the acetylcholine receptor by QX-222, of the SR K^+ channel by BTP, and of the Ca^{2+}-activated K^+ channel by Ba^{2+} provide good evidence from single-channel data that some blockers bind preferentially to the open state.

The mechanistic basis for preferred binding to the open state has not been fully explained. One possibility is that blocker binding affinity is a function of the

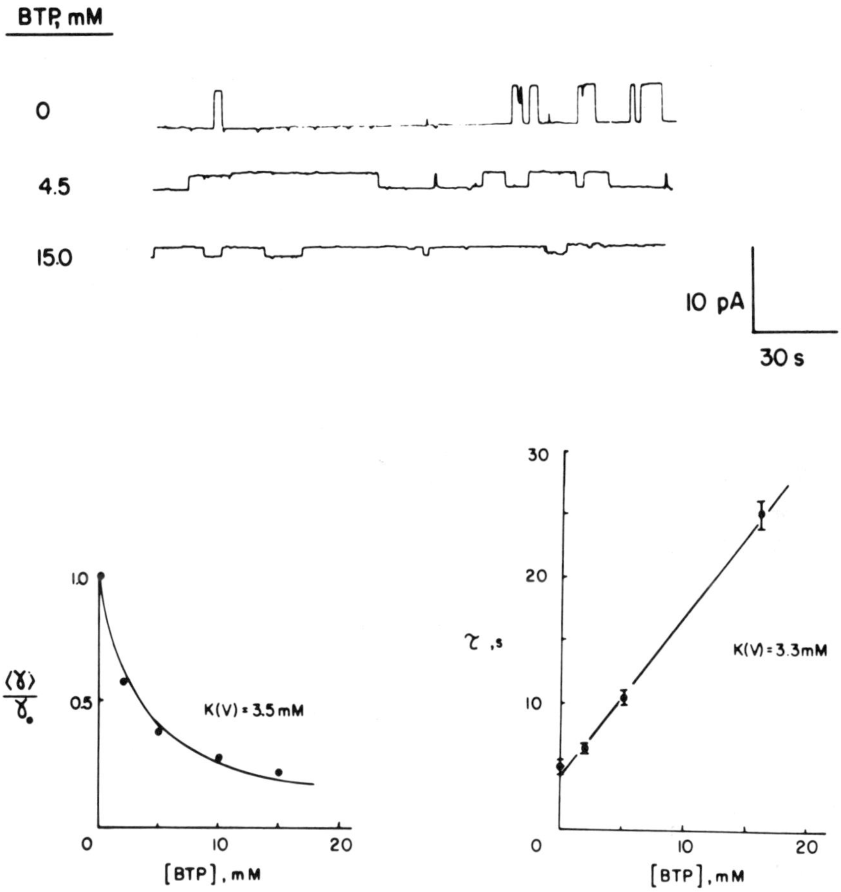

Figure 8. Example of an analysis of a fast blocker. Upper current records show the behavior of an SR K^+ channel in the absence and in the presence of two different concentrations of BTP at a voltage of −30 mV. The lower graphs plot the relative single-channel conductance (left) and the apparent mean open time (right) versus blocker concentration according to equations 37, and 35 given in the text. Figures are taken from Coronado and Miller (1982) with permission.

conformational state of the channel, with the closed state exhibiting a very low binding affinity. Alternatively, if the blocking site is actually in the permeation path, both ion permeation and blocker binding would be prevented by channel closing. In contrast to the preferred binding of a blocker to the open state, suppose instead that the closed state of the channel binds a blocker and that this closed–blocked state either cannot open or opens slowly. Then, the mean closed time would be predicted to lengthen with blocker concentration. Such behavior has recently been observed for an effect of the local anesthetic procaine on the batrachotoxin-activated Na^+ channel from rat muscle (see Moczydlowski *et al.*, Chapter 16, this volume).

3.4. Relief of Block by Permeant Ions

The effects of a blocker can often be overcome by raising the permeant ion concentration. The following scheme is commonly used to model such behavior:

$$\begin{array}{ccc} \mathrm{B} + \mathrm{O} & \underset{k_{\mathrm{off}}}{\overset{k_{\mathrm{on}}[\mathrm{B}]}{\rightleftharpoons}} & \mathrm{O{\cdot}B} \\ + & & \\ \mathrm{X} & & \\ \updownarrow K_x & & \\ \mathrm{O{\cdot}X} & & \end{array} \tag{S-X}$$

In this scheme, O and O·X are open, conducting states of a channel that has one site that can either be occupied by the permeant ion X or the blocker B.

For the case of slow block, where the blocker rate constants can be measured, scheme S-X makes the simple prediction that only the apparent association rate constant, k_{app}, will depend on the concentration of the competitor, X, as follows:

$$k_{\mathrm{app}}^{-1} = k_{\mathrm{on}}(0)^{-1}(1 + [\mathrm{X}]/K_{\mathrm{x}}) \tag{40}$$

where $k_1(0)$ is the value of the rate constant extrapolated to zero competitor concentration. Equation 40 is given in the form of a convenient linear transformation that can be used to obtain $k_1(0)$ from the ordinate intercept and K_{x} from the abscissa intercept. Several examples of slow blocking reactions that have been shown to exhibit this behavior are block of the SR K^+ channel by decamethonium (Miller, 1982), block of the Ca^{2+}-activated K^+ channel by Ba^{2+} (Vergara and Latorre, 1983), and block of the Na^+ channel by tetrodotoxin (Moczydlowski *et al.*, 1984a). For the case of fast block, scheme S-X yields the following modification of equation 37:

$$i_0/i = 1 + \frac{[\mathrm{B}]}{K_{\mathrm{B}}(1 + [\mathrm{X}]/K_{\mathrm{x}})} \tag{41}$$

The abscissa intercept of a plot of i_0/i versus [B] gives an apparent dissociation constant for blocker:

$$K_{\mathrm{app}} = K_{\mathrm{B}}\,(1 + [\mathrm{X}]/K_{\mathrm{x}}) \tag{42}$$

The values of K_{app}' measured at several concentrations of X can be replotted according to equation 42 to determine the equilibrium dissociation constants for competitor, K_{x}, and that for the blocker at zero X concentration, K_{B}. Such an analysis has been carried out for the case of fast block of the SR K^+ channel by BTP (Coronado and Miller, 1982).

The observation of simple competition in the case of the cited channel-blocking reactions is evidence that the binding of blocker and of the permeant ion are mutually exclusive. Competition alone does not reveal whether the blocker binding site occupies part of the permeation path, since the permeant cation could also bind to other sites on the protein that do not form part of the pore. However, the topology of the channel can sometimes be used to identify the blocking sites as part of the pore. In some cases, block can be relieved by raising the permeant ion concentration on the side of the channel opposite (*trans*) to that from which the blocker acts. This effect could be explained by the accessibility of the permeant ion to the blocking site through the pore itself. For example, relief of block by raising the *trans* (external) K^+ ion concentration has been observed in the case of internal block of the delayed rectifier by C_9 (Armstrong, 1975) and internal block of the Ca^{2+}-activated K^+ channel by Ba^{2+} (Vergara and Latorre, 1983) and Na^+ (Yellen, 1984b).

Such *trans* effects on blocker binding constants can occur by either a decrease in the association rate constant or an increase in the dissociation rate constant of the blocker. For the first case, a one-site occupancy competition would predict this effect; for the second case, this two-site noncompetitive effect could occur by ion repulsion. A multisite channel with multiple occupancy could permit a permeant ion to occupy a site in the pore adjacent to the blocker ion. If there is repulsion, the exit rate of the blocker could be enhanced. This latter effect is sometimes termed a "knock-out" or "knock-off" mechanism and has been observed in example of C_9 block of delayed rectifier and Na^+ block of the Ca^{2+}-activated K^+ channel, as cited above.

In contrast, a purely competitive occupancy effect by *trans* K^+ has been reported for Ba^{2+} block of the Ca^{2+}-activated K^+ channel (Vergara and Latorre, 1983). A knock-off effect on the dissociation rate constant of a blocker is considered to be good evidence for a multiion channel mechanism; however, a purely competitive effect on the association rate constant does not necessarily rule out a multiion channel.

3.5. Summary of the Characteristics of Block by Occlusion

Most examples of slow and fast block studied at the single-channel level exhibit voltage-dependent binding constants. The usual analysis of such voltage dependence follows the Eyring treatment of a binding site in an electric field described in Section 2.3. This mechanism is often referred to as the "Woodhull" mechanism after Woodhull's (1972) paper on proton blockade of the Na^+ channel. In this treatment, the apparent rate and equilibrium binding constants of the blocker are fit to exponential functions of voltage. For example, Fig. 9 shows actual examples of voltage-dependent block of the SR K^+ channel by both fast and slow blockers as analyzed by equations 37 and 28 (where $n = 1$). Also, Fig. 10 shows the measured voltage dependence of the individual binding rate constants of decamethonium to the SR K^+ channel, plotted according to equation 28.

A primary requirement of a mechanism involving work performed by moving a charged particle in an electric field is that the utilized energy be proportional to the charge valence or dipole moment of the particle. Table II contains several

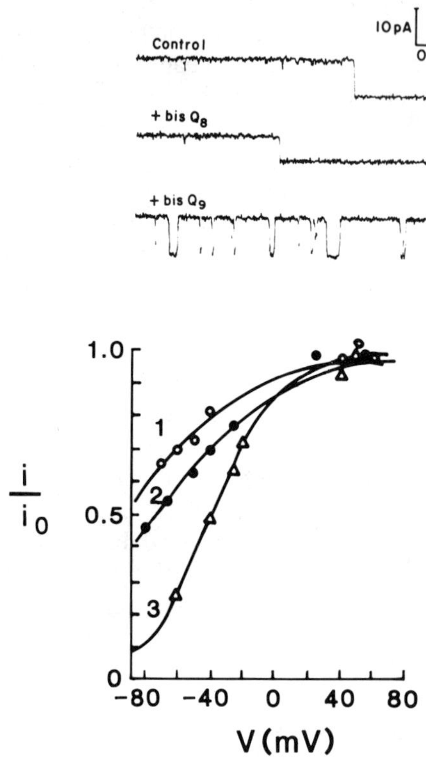

Figure 9. Blocking effects of trimethylammonium derivatives on the SR K^+ channel. Blockers used in Figs. 9 and 10 belong to the "Q_n" and "bisQ_n" series, where Q_n refers to an *n*-carbon linear alkane with a trimethylammonium group on one end and "bisQ_n" has a trimethylammonium group on each end. Top record, closing event in the absence of blocker; middle record, closing event in the presence of 0.7 mM bisQ_8 shows a 40% reduction in the apparent conductance; bottom record, discrete blocking events induced by 0.1 mM bisQ_9. For this channel, fast block is observed for blockers of the monovalent Q_n series and the divalent bisQ_n series up to bisQ_8. For the divalent blockers from bisQ_9 to bisQ_{12}, slow block is observed. The lower graph is a plot of the probability of being unblocked versus voltage (according to equations 37 and 28 in the text) in the presence of (1) 4 mM Q_6, (2) 4.5 mM bisQ_6, and (3) 0.16 mM bisQ_{11}. Note that the steepness of the curves corresponds to $z\delta = 0.66$ for Q_6 and bisQ_6; $z\delta = 1.26$ for bisQ_{11}. (From Miller, 1982, with permission.)

examples where such a charge dependence is apparent and one case where it is not. For the Ca^{2+}-activated K^+ channel, block by Ca^{2+} and Ba^{2+} has a higher effective valence than block by Na^+ or TEA^+ (Vergara and Latorre, 1983; Vergara *et al.*, 1984; Yellen, 1984a). For the SR K^+ channel, the effective valence for BTP is higher at pH 5.9 than at pH 8.5, correlating with the protonation of a second group on this diamino compound (Coronado and Miller, 1982). Also, for this latter channel, certain divalent bisquarternary compounds exhibit about twice the effective valence as do monovalent blockers. For the acetylcholine receptor channel, the observed voltage dependence of block also correlates with charge: natural benzocaine exhibits no voltage dependence (Ogden *et al.*, 1981), whereas the voltage dependence of gallamine^{3+} block is greater than that of QX-222$^+$ (Neher and Steinbach, 1978; Colquhoun and Sheridan, 1981). Such examples of correlation with net charge support the view that the voltage dependence of block in these channels arises from direct entry of the blocker into the electric field that spans some region of the pore. The interpretation that the δ term of $z\delta$ gives a measure of the relative depth of field penetration may well be justified in these cases, since the charge dependence satisfies the basic requirement of the Woodhull mechanism.

In some cases there is dramatic evidence that the blocker actually enters the pore in the observation of relief of block at high voltage associated with actual

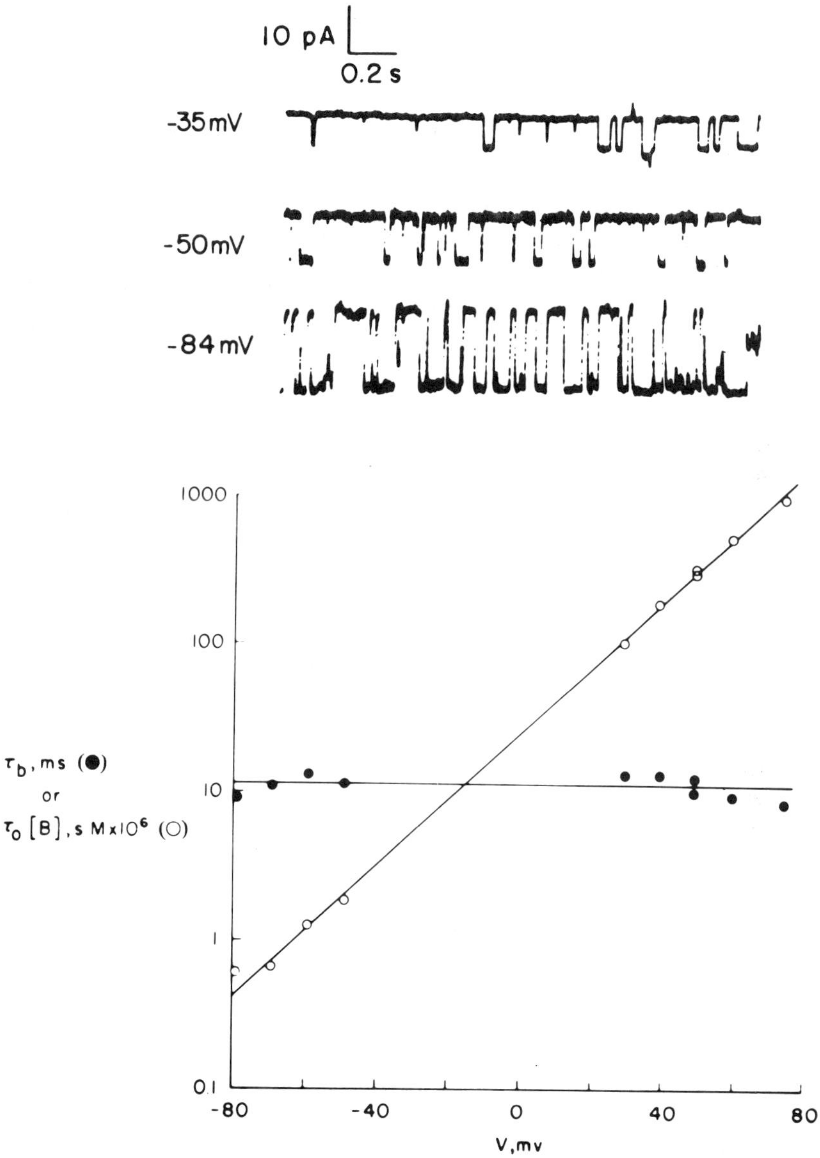

Figure 10. An example of the voltage dependence of a slow blocker. Current records show discrete blocking events (downward) induced in a single SR K^+ channel by 2 μM bisQ_{12} at three different voltages. The lower graph plots the reciprocal of the dissociation rate constant (●) and association rate constant (○) of bisQ_{10} (decamethonium) on a logarithmic scale versus voltage. Note that for this example, the dissociation rate is independent of voltage. (From Miller, 1982, with permission.)

Table II. Voltage Dependence of Block Observed for Single Channels[a]

Reference[a]	Side[b], blocker[c]	$k_{off}(0)$ (sec^{-1})	$z\delta_1$	$k_{on}(0)$ (sec^{-1} M^{-1})	$z\delta_2$	$K_D(0)$ (M)	$z\delta$
A. Ca^{2+}-activated K^+ channel							
[1]	i,Ba^{2+}	3×10^{-1}	0.33	8.3×10^{3}	1.33	3.6×10^{-5}	1.66
[1]	i,Ca^{2+}	2.4×10^{-1}	0.31	9×10^{-1}	1.02	3×10^{-1}	1.33
[2]	i,Na^+	1×10^{5}	~0	1.3×10^{6}	0.86	7.7×10^{-2}	0.86
[3]	i,TEA^+	—	—	—	—	3.5×10^{-2}	0.34
[4]	i,C_9^+	5×10^{3}	~0	2.6×10^{8}	0.1	1.9×10^{-5}	0.1
[1]	o,Ba^{2+}	3.5×10^{-1}	~0	1.9×10^{2}	0.71	1.8×10^{-3}	0.71
[3]	o,TEA^+	—	—	—	—	2.9×10^{-4}	~0
B. SR K^+ channel							
[5]	TMA^+	—	—	—	—	5.0×10^{-1}	0.68
[5]	TEA^+	—	—	—	—	8.5×10^{-2}	0.62
[5]	$choline^+$	—	—	—	—	2.3×10^{-1}	0.61
[5]	BTP^+, pH 8.1	—	—	—	—	1.4×10^{-2}	0.57
[5]	BTP^{2+}, pH 5.9	—	—	—	—	2.0×10^{-2}	1.1
[6]	$bisQ_2$	—	—	—	—	3.0×10^{-1}	1.0
[6]	$bisQ_{5,6,7,8}$	—	—	—	—	—	0.65
[6]	$bisQ_9$	1.7×10^{2}	~0	2.4×10^{4}	1.06	7.1×10^{-3}	1.06
[6]	$bisQ_{10}$	8.4×10^{1}	~0	3.4×10^{4}	1.25	2.5×10^{-3}	1.25
[6]	$bisQ_{11}$	7.3×10^{1}	~0	5.6×10^{4}	1.23	1.3×10^{-3}	1.23
[6]	$bisQ_{12}$	3.0×10^{1}	~0	2.0×10^{5}	1.22	1.5×10^{-4}	1.22
C. Nicotinic acetylcholine receptor							
[7]	o,$Benzocaine^0$	3×10^{2}	~0	1.8×10^{6}	~0	1.1×10^{-4}	~0
[8]	o,$QX\text{-}222^+$	4.5×10^{3}	0.41	3.6×10^{6}	0.34	1.2×10^{-3}	0.75
[9]	o,$Acetylcholine^+$	—	—	—	—	5×10^{-2}	0.78
[9]	o,$Suberyldicholine^{2+}$	1×10^{4}	0.42	2.2×10^{6}	0.37	5.2×10^{-3}	0.79
[10]	o,$Gallamine^{3+}$	1.1×10^{4}	0.69	1.2×10^{7}	0.27	7.8×10^{-4}	0.96
D. Tetrodotoxin-sensitive Na^+ channel							
[11]	i,9-$Aminoacridine^+$	2.1×10^{2}	0.63	1.0×10^{7}	0.05	2.1×10^{-5}	0.68
[12]	o,STX^{2+}	6.1×10^{-2}	0.38	1.4×10^{7}	0.24	4.3×10^{-9}	0.62
[12]	o,TTX^+	1.2×10^{-1}	0.33	4.5×10^{6}	0.26	2.7×10^{-8}	0.59
[12]	o,Cl^0	1.1	0.41	4.4×10^{5}	0.25	2.5×10^{-6}	0.67

[a] The table summarizes selected examples of voltage-dependent blocking equilibria that were fit to equations 28 as discussed in the text. References: [1] Vergara and Latorre (1983); [2] Yellen (1984a); [3] Vergara *et al.* (1984); [4] Yellen (1984b); [5] Coronado and Miller (1982); [6] Miller (1982); [7] Ogden *et al.* (1981); [8] Neher and Steinbach (1978); [9] Sine and Steinbach (1984); [10] Colquhoun and Sheridan (1981); [11] Yamamoto and Yeh (1984); [12] Moczydlowski *et al.* (1984b).

[b] The side of the blocker's action is denoted as i, intracellular face, or o, extracellular face. For the SR K^+ channel, listed blockers act on the *trans* side, bilayer convention (see references 5, 6).

[c] Abbreviations: TEA, tetraethylammonium; C_9, nonyltriethylammonium; TMA, trimethylammonium; BTP, 1,3-bis[tris(hydroxymethyl) methylamino] propane; "$bisQ_n$" refers to an *n*-carbon linear alkane with a trimethylammonium group on each end; QX-222 is a quaternary ammonium derivative of lidocaine; STX, saxitoxin; TTX, tetrodotoxin; C1 is a doubly sulfated derivative of saxitoxin.

permeation by the blocker. This effect has been observed in the case of Cs^+ block of the delayed rectifier (French and Shoukimas, 1985) and for block of the ACh receptor by acetylcholine (Sine and Steinbach, 1984).

In the absence of detailed information, voltage-dependent reactions are often modeled using a simple potential energy diagram with a symmetrical activation energy barrier located half-way between two wells representing the initial and final states as depicted in Fig. 6. Such a scheme implies equal voltage dependence for binding association and dissociation reactions if the electric field is constant. In contrast to this picture, the results of Table II show that either the association or dissociation step can express most of the voltage dependence of the equilibrium for various blocking reactions. For example, as shown in Fig. 10, for the case of slow block of the SR K^+ channel by bisQ_{10}, all of the voltage dependence resides in the association step. In contrast, for block of the Na^+ channel by 9-aminoacridine (Yamamoto and Yeh, 1984), most of the voltage dependence resides in the dissociation step. According to the site-in-a-field paradigm, such asymmetry can have two possible meanings: (1) the physical location of the energy barrier can vary from the field entrance to the site; (2) there can be a charge difference in the species participating in the rate-limiting step for the forward or reverse reactions; e.g., a tightly complexed ion pair formed between a cationic blocker and a carboxylate group in the site might give rise to a voltage-independent dissociation rate.

Several anomalies in observed values of $z\delta$ deserve particular comment. For block of the delayed rectifier by external Cs^+, $z\delta$ varies with Cs^+ concentration and may approach values as high as 1.3 (Adelman and French, 1978). The classical interpretation of this result is that the delayed rectifier is a multiion channel and the anomalously high $z\delta$ values represent the entry of two or more Cs^+ ions into the field. This interpretation is reinforced by the observation that the titration curve of Cs^+ block exhibits a Hill coefficient greater than 1. In the case of blockade by long-linear-chain compounds with charges at either end, a lack of an effect on $z\ \delta$ by the second charge has been observed for blockers in the series bisQ_4 to bisQ_7 for the SR K^+ channel (Miller, 1982) and in the case of suberyldicholine versus acetylcholine for the ACh receptor (Sine and Steinbach, 1984). To explain such results, Miller (1982) has presented the reasonable interpretation that only one charge enters the field in such molecules, with the second charge residing outside the field. This effect has been proposed as a novel method for actually measuring the length of the field, which was determined as 13 Å for the SR K^+ channel. In this same blocker series, a sudden increase in $z\delta$ was observed for bisQ_9 through bisQ_{12}. This result was explained as the binding of both charges in a particular "horseshoe" conformation of the blocker that is unfavored for the lower chain members of the bisQ_n series. A final anomaly is the case of Na^+-channel block by a series of saxitoxin derivatives with net charges of 0, 1^+, and 2^+, which show no apparent dependence on charge (Moczydlowski *et al.*, 1984b). This result is taken to mean that for this case the toxin binding site does not lie within the transmembrane electric field and the voltage dependence arises by indirect allosteric coupling with a distant dipole.

In summary, particular characteristics of certain blockers confirm that some blocker binding sites form part of the channel passageway for ions. Most of these characteristics have been cited by French and Shoukimas (1985) as evidence for a deep channel location of certain blocker sites of the squid axon delayed rectifier. These characteristics include preferential binding to the open state, voltage dependence of binding, relief of block at high voltage as a result of blocker permeation, correlation of voltage dependence with net charge of the blocker, apparent stoichiometries greater than 1 for multiion channels, and relief of block by *trans* permeant ions through competition or knockoff mechanisms. In contrast, for some blockers, there is little evidence that argues in favor of a site location within the pore. Examples are voltage-independent block of the Ca^{2+}-activated K^+ channel by external TEA (Vergara *et al.*, 1984), weakly voltage-dependent block of the delayed rectifier by large organic cations (French and Shoukimas, 1985), and charge independence of voltage-dependent of block the Na^+ channel by guanidinium toxins (Moczydlowski *et al.*, 1984b). In such cases, block may nevertheless occur by an occlusion mechanism at the outer mouth of the channel. In contrast to "deep" block, this type of block has been termed "shallow" block by French and Shoukimas (1985).

ACKNOWLEDGMENTS. The author was supported by the Elma Margaret Lapp Fund of the University of Cincinnati College of Medicine and an Established Investigatorship from the American Heart Association. Anita Tolle provided invaluable secretarial assistance in preparing the manuscript.

References

Adelman, W. J., and French, R. J., 1978, Blocking of squid axon potassium channel by external caesium ions, *J. Physiol. (Lond.)* **276:**13–25.

Aldrich, R. W., and Yellen, G., 1984, Analysis of nonstationary channel kinetics, in: *Single-Channel Recording* (B. Sakmann and E. Neher, eds.), pp. 287–299, Plenum Press, New York.

Anderson, O. S., 1984, Gramicidin channels, *Annu. Rev. Physiol.* **46:**531–548.

Armstrong, C. M., 1975, Potassium pores of nerve and muscle membranes, in: *Membranes: A Series of Advances*, Vol. 3 (G. Eisenman, ed.), pp. 325–358 Marcel Dekker, New York.

Auerbach, A., and Sachs, F., 1983, Flickering of a nicotinic ion channel to a subconductance state, *Biophys. J.* **42:**1–10.

Auerbach, A., and Sachs, F., 1984, Single-channel currents from acetylcholine receptors in embryonic chick muscle: Kinetic and conductance properties of gaps within bursts, *Biophys. J.* **45:**187–198.

Banerjee, R., Epstein, M., Kandrach, M., Zimniak, P., and Racker, E., 1979, A new method of preparing Ca^{2+}-ATPase from sarcoplasmic reticulum: Extraction with octylglucoside, *Memb. Biochem.* **2:**283–297.

Benz, R., and Läuger, P., 1976, Kinetic analysis of carrier-mediated ion transport by the charge-pulse technique, *J. Membr. Biol.* **27:**171–191.

Brahm, J., 1977, Temperature-dependent changes of chloride transport kinetics in human red cells, *J. Gen. Physiol.* **70:**283–306.

Brahm, J., 1983, Kinetics of glucose transport in human erythrocytes, *J. Physiol. (Lond.)* **339:**339–354.

Brown, A. M., Kunze, D. L., and Yatani, A., 1984, The agonist effect of dihydropyridines on Ca channels, *Nature* **311:**570–572.

Brunner, J., Graham, D. E., Hauser, H., and Semenza, G., 1980, Ion and sugar permeabilities of lecithin bilayers: Comparison of curved and planar bilayers, *J. Membr. Biol.* **57:**133–141.

Byerly, L., and Hagiwara, S., 1982, Calcium currents in internally perfused nerve cell bodies of *Limnea stagnalis, J. Physiol. (Lond.)* **322:**503–528.

Changeux, J. P., Devillers-Thiery, A., and Chemoulli, P., 1984, Acetylcholine receptor: An allosteric protein, *Science* **225:**1335–1345.

Colombini, M., 1980, Pore size and properties of channels from mitochondria isolated from *Neurospora crassa, J. Membr. Biol.* **53:**79–84.

Colquhoun, D., 1973, *Lectures on Biostatistics*, Clarendon Press, Oxford.
Colquhoun, D., and Hawkes, A. G., 1977, Relaxation and fluctuations of membrane currents that flow through drug-operated channels, *Proc. R. Soc.* (*Lond.*) [*Biol.*] **199:**231–262.
Colquhoun, D., and Hawkes, A. G., 1981, On the stochastic properties of single ion channels, *Proc. R. Soc.* (*Lond.*) [*Biol.*] **211:**205–235.
Colquhoun, D., and Hawkes, A. G., 1982, On the stochastic properties of bursts of single ion channel openings and of clusters of bursts, *Phil. Trans. R. Soc. Lond.* [*Biol.*] **300:**1–59.
Colquhoun, D., and Hawkes, A. G., 1984, The principles of stochastic interpretation of ion-channel mechanisms, in: *Single-Channel Recording* (B. Sakmann and E. Neher, eds.), pp. 135–175, Plenum Press, New York.
Colquhoun, D., and Sakmann, B., 1981, Fluctuations in the microsecond time range of the current through single acetylcholine receptor ion channels, *Nature* **294:**464–466.
Colquhoun, D., and Sakmann, B., 1984, Bursts of openings in transmitter-activated ion channels, in: *Single-Channel Recording* (B. Sakmann and E. Neher, eds.), pp. 345–364, Plenum Press, New York.
Colquhoun, D., and Sheridan, R. E., 1981, The modes of action of gallamine, *Proc. R. Soc.* (*Lond.*) [*Biol.*] **211:**181–203.
Colquhoun, D., and Sigworth, F., 1984, Fitting and statistical analysis of single channel records, in: *Single-Channel Recording* (B. Sakmann and E. Neher, eds.), pp. 191–263, Plenum Press, New York.
Conti, F., and Neher, E., 1980, Single channel recordings of K^+ currents in squid axons, *Nature* **285:**140–143.
Coronado, R., and Miller, C., 1980, Decamethonium and hexamethonium block K^+ channels of sarcoplasmic reticulum, *Nature* **288:**495–497.
Coronado, R., and Miller, C., 1982, Conduction and block by organic cations in a K^+-selective channel from sarcoplasmic reticulum incorporated into planar phospholipid bilayers, *J. Gen. Physiol.* **79:**529–547.
Eisenman, G., and Horn, R., 1983, Ion selectivity revisited: The role of kinetic and equilibrium processes in ion permeation through channels, *J. Membr. Biol.* **76:**197–225.
Eisenman, G., Sandblom, J., and Neher, E., 1978, Interactions in cation permeation through the gramicidin channel: Cs, Rb, K, Na, Li, Tl, H and effects of anion binding. *Biophys. J.* **22:**307–340.
Faller, L., Jackson, R., Malinowska, D., Mukidjam, E., Rabon, E., Saccomani, G., Sachs, G., and Smolka, A., 1982, Mechanistic aspects of gastric (H^+–K^+)-ATPase, *Ann. N.Y. Acad. Sci.* **402:**146–163.
Fenwick, E. M., Marty, A., and Neher, E., 1982, Sodium and calcium channels in bovine chromaffin cells, *J. Physiol.* (*Lond.*), **331:**559–635.
FitzHugh, R., 1983, Statistical properties of the asymmetric random telegraph signal, with applications to single-channel analysis, *Mat. Biosci.* **64:**75–89.
French, R. J., and Shoukimas, J. J., 1985, An ion's view of the potassium channel: The structure of the permeation pathway as sensed by a variety of blocking ions, *J. Gen. Physiol.* **85:**669–698.
French, R. J., Worley, J. F., and Krueger, B. K., 1984, Voltage-dependent block by saxitoxin of sodium channels incorporated into planar lipid bilayers, *Biophys. J.* **45:**301–310.
Fröhlich, O., 1984, How channel-like is a biological carrier? Studies with the erythrocyte anion transporter, *Biophys. J.* **45:**93–94.
Hall, J. E., 1975, Access resistance of a small circular pore, *J. Gen. Physiol.* **66:**531–532.
Hamill, O. P., and Sakmann, B., 1981, Multiple conductance state of single acetylcholine receptor channels in embryonic muscle cells, *Nature* **294:**462–464.
Hamill, O. P., Bormann, J., and Sakmann, B., 1983, Activation of multiple-conductance state chloride channels in spinal neurones by glycine and GABA, *Nature* **305:**805–808.
Hanke, W., and Miller, C., 1983, Single chloride channels from *Torpedo* electroplax. Activation by protons, *J. Gen. Physiol.* **82:**25–45.
Hess, P., Lansman, J. B., and Tsien, R. W., 1984, Different modes of Ca channel gating behavior favored by dihydropyridine Ca agonists and antagonists, *Nature* **311:**538–544.

Hille, B., 1971, The permeability of the sodium channel to organic cations in myelinated nerve, *J. Gen. Physiol.* **59**:637–658.

Hille, B., 1984, *Ionic Channels of Excitable Membranes*, Sinauer Associates, Sunderland, Massachusetts.

Horn, R., Vandenberg, C.A., and Lange, K., 1984, Statistical analysis of single sodium channels. Effects of N-bromoacetamide, *Biophys. J.* **45**:323–335.

Huang, L. M., Catterall, W. A., and Ehrenstein, G., 1979, Comparison of ionic selectivity of batrachotoxin-activated channels with different tetrodotoxin dissociation constants, *J. Gen. Physiol.* **73**:839–854.

Inesi, G., Watanabe, T., Coan, C., and Murphy, A., 1982, The mechanism of sarcoplasmic reticulum ATPase, *Ann. N.Y. Acad. Sci.* **402**:515–534.

Jorgensen, P. L., 1975, Isolation and characterization of the components of the sodium pump, *Q. Rev. Biophys.* **7**:239–274.

Kernohan, J. C., 1965, The pH–activity curve of bovine carbonic anhydrase and its relationship to the inhibition of the enzyme by anions, *Biochim. Biophys. Acta* **96**:304–317.

Labarca, P. P., and Miller, C., 1981, A K^+-selective, three-state channel from fragmented sarcoplasmic reticulum of frog leg muscle, *J. Membr. Biol.* **61**:31–38.

Labarca, P., Coronado, R., and Miller, C., 1980, Thermodynamic and kinetic studies of the gating behavior of a K^+-selective channel from the sarcoplasmic reticulum membrane, *J. Gen. Physiol.* **76**:397–424.

Latorre, R., and Miller, C., 1983, Conduction and selectivity in potassium channels, *J. Membr. Biol.* **71**:11–30.

Läuger, P., 1980, Kinetic properties of ion carrier and channels, *J. Membr. Biol.* **57**:163–178.

MacLennan, D. H., Seeman, P., Iles, G. H., and Yip, C. C., 1971, Membrane formation by the adenosine triphosphatase of sarcoplasmic reticulum, *J. Biol. Chem.* **246**:2702–2710.

Makowski, L., Caspar, D., Phillips, W., Baker, T., and Goodenough, D., 1984, Gap junction structures: Variation and conservation in connexon conformation and packing, *Biophys. J.* **45**:208–218.

Methfessel, C., and Boheim, G., 1982, The gating of single calcium-dependent potassium channels is described by an activation–blockage mechanism, *Biophys, Struct. Mech.* **9**:35–60.

Miller, C., 1982, Bis-quaternary ammonium blockers as structural probes of the sarcoplasmic reticulum K^+ channel, *J. Gen. Physiol.* **79**:869–891.

Miller, C., Moczydlowski, E., Latorre, R., and Phillips, M., 1985, Charybdotoxin, a high-affinity inhibitor of single Ca^{2+}-activated K^+ channels from mammalian skeletal muscle, *Nature* **313**:316–318.

Mimms, L. T., Zampighi, G., Nozaki, Y., Tanford, C., and Reynolds, J. A., 1981, Phospholipid vesicle formation and transmembrane protein incorporation using octylglucoside, *Biochemistry* **181**:833–840.

Moczydlowski, E. G., and Fortes, P. A. G., 1981, Inhibition of sodium and potassium adenosine triphosphatase by 2′,3′-O-(2,4,6-trintro-cyclohexadienylidine) adenine nucleotides, *J. Biol. Chem.* **256**:2357–2366.

Moczydlowski, E., and Latorre, R., 1983, Gating kinetics of Ca-activated K-channels from rat muscle incorporated into planar lipid bilayers: Evidence for two voltage-dependent Ca^{2+} binding reactions. *J. Gen. Physiol.* **82**:511–542.

Moczydlowski, E., Garber, S. S., and Miller, C., 1984a, Batrachotoxin-activated Na^+ channels in planar lipid bilayers: Competition of tetrodotoxin block by Na^+, *J. Gen. Physiol.* **84**:665–686.

Moczydlowski, E., Hall, S. Garber, S. S., Strichartz, G. R., and Miller, C., 1984b, Voltage-dependent blockade of muscle Na^+ channels by guanidinium toxins: Effect of toxin charge, *J. Gen. Physiol.* **84**:687–704.

Moczydlowski, E., Alvarez, O., Vergara, C., and Latorre, R., 1985, Effect of phospholipid surface charge on the conductance and gating of a Ca^{2+}-activated K^+-channel in planar lipid bilayers, *J. Membr. Biol.* **83**:273–282.

Montal, M., Labarca, P., Fredkin, D. R., and Suarez-Isla, B. A., 1984, Channel properties of the purified acetylcholine receptor from *Torpedo californica* reconstituted in planar lipid bilayer membranes, *Biophys. J.* **45**:165–174.

Neher, E., and Steinbach, J. H., 1978, Local anaesthetics transiently block currents through single acetylcholine receptor channels, *J. Physiol.* (*Lond.*) **277:**153–176.

Neher, E., and Stevens, C. F., 1979, Voltage-driven conformational changes in intrinsic membrane proteins, in: *The Neurosciences Fourth Study Program* (F. O. Schmitt and F. G. Worden, eds.), pp. 623–629, MIT Press, Cambridge.

Ogden, D. C., Siegelbaum, S. A., and Colquhoun, D., 1981, Block of acetylcholine-activated ion channels by an uncharged local anesthetic, *Nature* **289:**596–598.

Patlak, J. B., Gration, K. A. F., and Usherwood, P. N. R., 1979, Single glutamate-activated channels in locust muscle, *Nature* **278:**643–645.

Pressman, B. C., 1976, Biological applications of ionophores, *Annu. Rev. Biochem.* **45:**501–530.

Rothstein, A., and Ramjeesing, M., 1980, The functional arrangement of the anion channel of red blood cells, *Ann. N. Y. Acad. Sci.* **358:**1–12.

Sachs, F., 1984, Automated analysis of single-channel records, in: *Single-Channel Recording* (B. Sakmann and E. Neher, eds.), pp. 265–285, Plenum Press, New York.

Sachs, F., Neil, J., and Barkakati, N., 1982, The automated analysis of data from single ionic channels, *Pflügers Arch.* **395:**331–340.

Sakmann, B., and Neher, E. (eds.), 1983, *Single-Channel Recording*, Plenum Press, New York.

Schein, S. J., Kagan, B. L., and Finkelstein, A., 1978, Colicin K acts by forming voltage-dependent pores in phospholipid bilayer membranes, *Nature* **276:**159–163.

Sine, S. M., and Steinbach, J. H., 1984, Agonists block currents through acetylcholine receptor channels, Biophys. J. **46:**277–284.

Stevens, C. F., 1978, Interactions between intrinsic membrane protein and electric field. An approach to studying nerve excitability, *Biophys. J.* **22:**295–306.

Tiepel, J. W., Hass, G. M., and Hill, R. L., 1968, The substrate specificity of fumarase, *J. Biol. Chem.* **243:**5684–5694.

Vanderkooi, J. M., and Martonosi, A., 1971, Sarcoplasmic reticulum. XVI. The permeability of phosphatidylcholine vesicles for calcium, *Arch. Biochem. Biophys.* **147:**632–646.

VanHolde, K. E., 1971, *Physical Biochemistry*, Prentice Hall, Englewood Cliffs, New Jersey.

Vergara, C., and Latorre, R., 1983, Kinetics of Ca-activated K-channels from rabbit muscle incorporated into planar bilayers: Evidence for a Ca^{2+} and Ba^{2+} blockage, *J. Gen. Physiol.* **82:**543–568.

Vergara, C., Moczydlowski, E., and Latorre, R., 1984, Conduction, blockage and gating in a Ca^{2+}-activated K^{+} channel incorporated into planar lipid bilayers, *Biophys. J.* **45:**73–76.

Woodhull, A., 1972, Ionic blockade of sodium channels in nerve, *J. Gen. Physiol.* **61:**687–708.

Yamamoto, D., and Yeh, J. Z., 1984, Kinetics of 9-aminoacridine block of single Na channels, *J. Gen. Physiol.* **84:**361–377.

Yellen, G., 1984a, Ionic permeation and blockade in Ca-activated K-channels of bovine chromaffin cells, *J. Gen. Physiol.* **84:**157–186.

Yellen, G., 1984b, Relief of Na^{+} block of Ca^{2+}-activated K^{+} channels by external cations, *J. Gen. Physiol.* **84:**187–199.

Chapter 5

HOW TO SET UP A BILAYER SYSTEM

OSVALDO ALVAREZ

1. Introduction

In this chapter I explain how to set up a planar bilayer system. The aim is to guide investigators taking their initial steps in the bilayer field in making their first setup. The intention is to give practical guidance rather than a thorough theoretical treatment of the different subjects. I assume that the reader has minimal knowledge of electronics and is planning to build the equipment.

The exposition is divided into several sections, each describing a small part of the whole system. The sections are: (1) mechanical hardware, describing chambers, electrodes, and interference shielding; (2) electrical hardware, covering pulse generators, current-to-voltage converter, and signal conditioning; and (3) procedures for membrane formation.

2. Mechanical Hardware

2.1. Membrane Chambers

Planar bilayers are formed using two main techniques, painted membranes or folded membranes.

Painted membranes are formed by placing a small amount of a lipid solution, usually disolved in decane, onto an aperture in a wall separating two aqueous solutions. As the lipid solution drains to the border of the aperture, a film is formed in the central part of the aperture. This film gets thinner and finally becames a bilayer. A membrane chamber for painted bilayers is shown in Fig. 1. It consists of two parts. One is a plastic cup with a sector of the side wall thinned and perforated. The other is a Teflon® block with two intersecting holes. The cup is fitted into one of these holes. Draining of the solvent depends on the material the cup made of. Teflon® cups are adequate for large membranes (1 mm^2); for smaller membranes (down to 0.1 mm^2), membranes do not thin to bilayers in Teflon®

OSVALDO ALVAREZ • Departamento de Biologìa, Facultad de Ciencias, Universidad de Chile, Casilla 653, Santiago, Chile, and Centro de Estudios Cientificos de Santiago, Casilla 16443, Santiago, Chile.

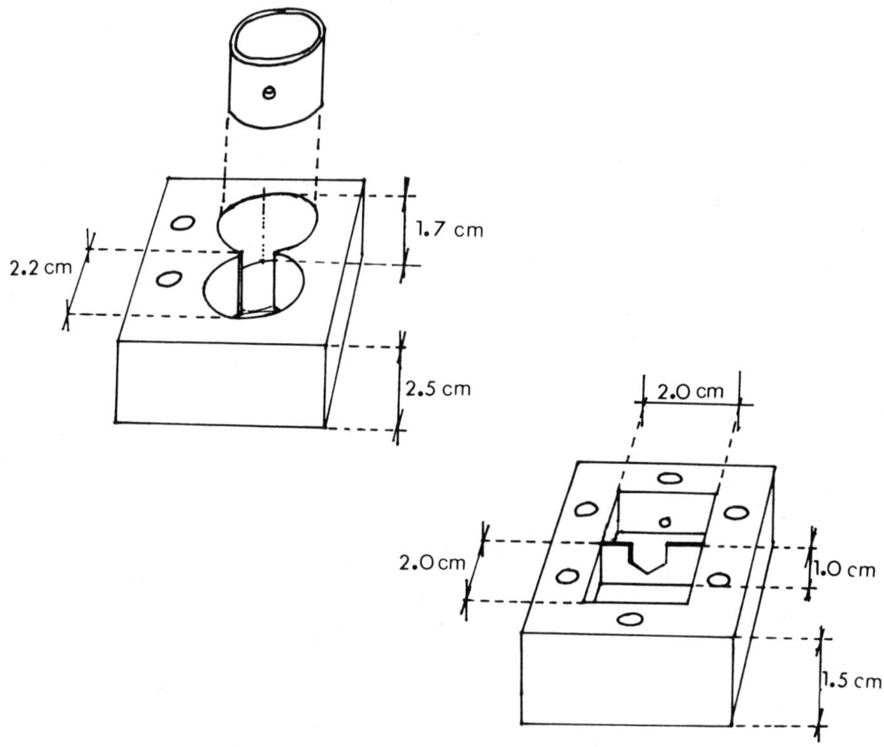

Figure 1. Membrane chambers.

cups, and polycarbonate should be used. Lipids can be applied to the hole in the cup using either a small camel hair brush, a small Teflon® spatula, or a fire-polished glass rod.

Folded membranes are formed by apposition of two monolayers. A chamber for this type of bilayer can be constructed from a Teflon® block, as shown in Fig. 1. Two rectangular cavities are machined, leaving a thin separating wall. A V-shaped piece of the wall is cut to allow the compartments to communicate. A thin (60-μm) Teflon® film is affixed to this wall using high-vacuum silicone grease for electric insulation. A hole has been cut from the film using a hypodermic needle that has been sharpened as a cork borer. To form the bilayer, a small amount of electrolyte solution is placed into each pool and covered with a monolayer of lipids. The liquid levels are then raised on both sides simultaneously by adding solution. The monolayers cover the Teflon® film on both sides with the hydrophobic tails facing the partition (water, lipids, teflon, lipids, water). The bilayer is formed by apposition of the monolayers at the hole in the Teflon® film (water, lipids, lipids, water). Two 2-ml disposable syringes fitted to two holes in the chamber are used for filling.

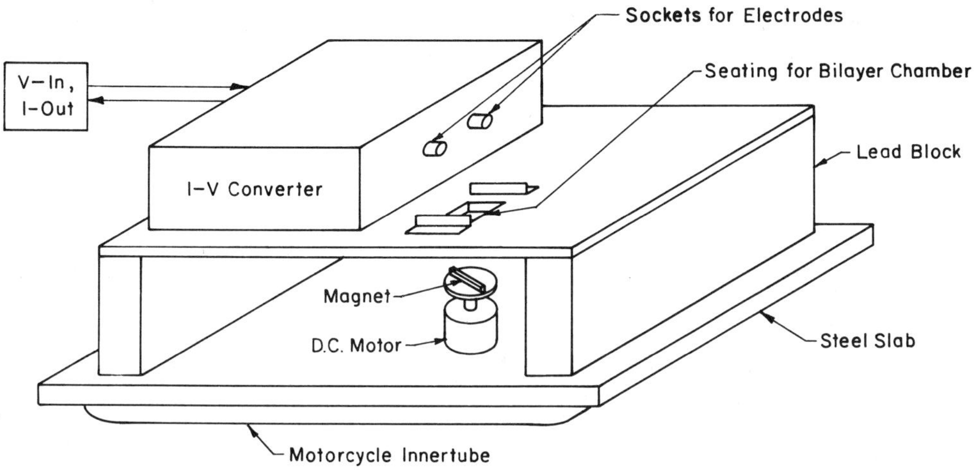

Figure 2. Mounting of the membrane chamber and current monitor on a vibration-proof surface.

2.2. Electrodes

Electrical connection of the membrane chamber to the measurement equipment is done by reversible electrodes. Silver–silver chloride electrodes are generally used because they are stable and easy to prepare. Silver chloride can be deposited by placing the electrode under preparation and another silver wire into 0.1 M HCl and passing a current of 10 mA for 2 min through the silver wires. Silver chloride is deposited on the anode. Electrodes can also be covered with silver chloride by immersion in Clorox® bleach.

For many experiments, just placing the silver electrodes into the membrane solutions is adequate. For experiments in which solutions do not contain Cl^- or when solutions at both sides of the membrane have different Cl^- concentrations, salt bridges are needed.

2.3. Interference Problems

Alternating current pickup from power lines is the most serious source of interference. To avoid this problem, the membrane chamber and the current-measuring circuit must be enclosed in some sort of Faraday cage.

Vibrations transmitted to the membrane chamber can cause serious interference when measuring low-level currents and must be avoided. Floor vibrations can be attenuated by placing the chamber on a heavy platform seated on an inflated rubber innertube of a pushcart wheel (see Fig. 2). Vibrations coming through the air such as air conditioning noise or conversations can also disturb and are more difficult to foil. I have used the upper compartment of a discarded two-door refrigerator as a sound-proof chamber, substantially improving the signal-to-noise ratio of the low-level current recordings.

3. Electrical Equipment

The electrical equipment consists of a waveform generator, a current-to-voltage converter, a voltage amplifier, a low-pass filter, and a step response correction. In Fig. 3, a block diagram of the system is presented. The bilayer is connected to a function generator and a current-to-voltage converter. The electrical potential at one side of the membrane is the output of the function generator, and the other side is actively held at ground by the current-to-voltage converter. Current passing through the membrane is transduced into an analogue voltage signal by the current-to-voltage converter. The sensitivity of this converter is usually 1–10 mV/pA, and its output must be amplified by the signal-conditioning circuit before being displayed on an oscilloscope or recorded on tape. High-frequency components of the current are attenuated by the current-to-voltage converter and are restored in the signal-conditioning unit using a high-frequency booster. Very-high-frequency components are removed in this unit to improve signal-to-noise ratio. When membrane potential is changed as a step function, a large current spike arises from the charging of the membrane capacitance. This current can be subtracted by the inverting amplifier and the capacitor as shown in the figure.

3.1. Power Supply

Batteries must be used to avoid alternating current interference in the most critical circuits, which are the waveform generator and the current-to-voltage converter. Rechargeable 7-V Ni–Cd ("transistor radio") batteries are small and easy to use, and a power supply for an operational amplifier takes only two of these batteries. For other circuits, regular power supplies can be used.

3.2. Pulse Generator

For single-channel recording in a planar bilayer, a waveform generator capable of delivering DC, pulses, or triangle functions is needed. Many commercial generators are available, but, in my experience, they are not suitable for this type of work. The problems I have found are mainly excessive noise and AC interference. Low-noise pulse generators can easily be built at a very low cost.

The source for DC voltages can be a 1.5-V battery connected to a voltage divided to produce a total voltage span of −200 mV to +200 mV controlled by a 10-turn potentiometer with a calibrated counting dial. The DC voltage is turned on and off and polarity selected by a three-position switch.

The bilayer system must include a rapid method of estimating membrane capacitance. Capacitance measurements using triangle waveforms are convenient since the amplitude of the current is proportional to the membrane capacitance and can be measured directly using an AC voltmeter. A triangle generator can

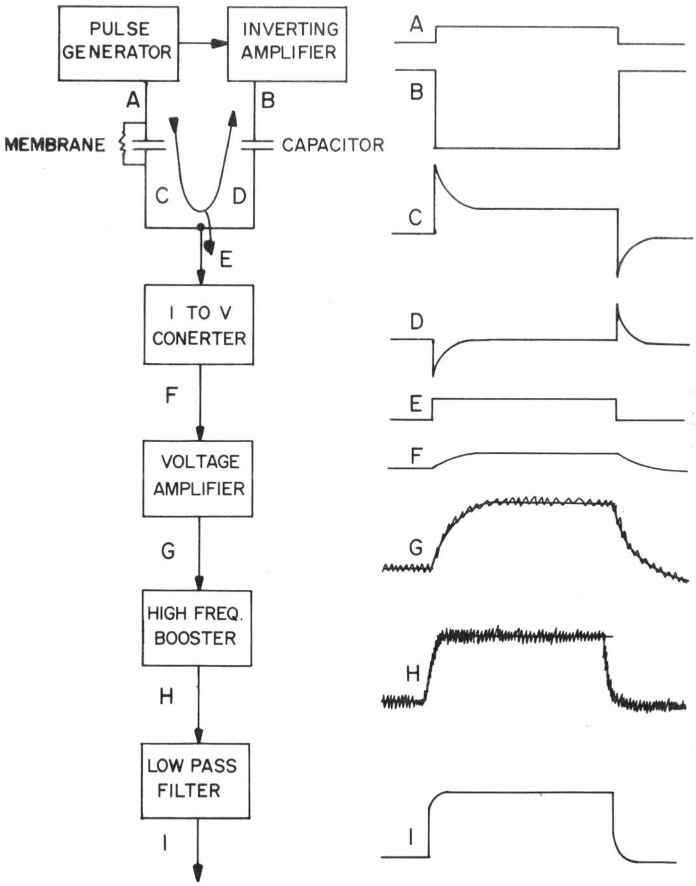

Figure 3. Block diagram of electrical circuits. Voltage from the pulse generator, A, is fed to the membrane; an inverted and amplified version, B, is fed to the compensating capacitor. If this capacitor has a capacitance one-tenth the membrane capacitance, then the inverted pulse should be ten times as large as the original pulse. Waveform C is the current passing through the membrane, which is the sum of the current through membrane resistance and capacitance. Waveform D is the current passing through the capacitor. Since the capacitive currents are equal and of opposite sign, they circulate from the membrane capacitor to the compensating capacitor, and only the membrane resistor current enters the current-to-voltage converter, E. The output of the current-to-voltage converter is waveform F. The amplitude of this signal is brought from the millivolt range to the volt range by the voltage amplifier, G. The rounded edges are squared by the high-frequency booster, waveform H. High-frequency noise is removed by the low-pass filter, and the final processed signal, I, is ready for display. Waveforms are idealized drawings.

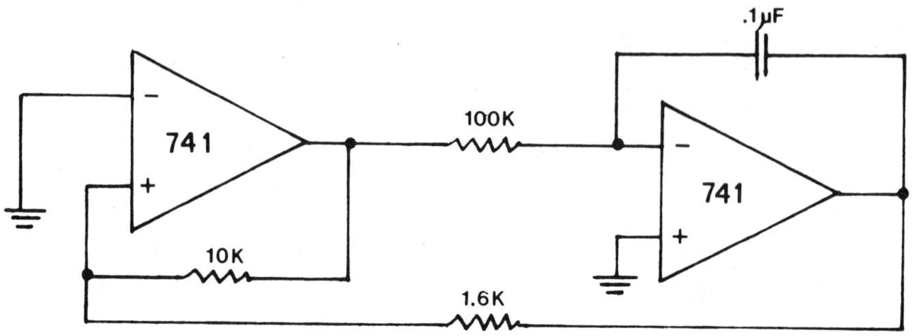

Figure 4. Triangle- and square-wave generator. This simple two-operational-amplifier circuit can be used to generate triangular waves, which are useful to measure membrane capacitance and to calibrate high-frequency boosters. It consists of an integrator and a voltage comparator. The integrator produces a ramp of positive slope for a constant negative voltage at the input and one of negative slope for positive input. The voltage comparator looks at the integrator output, and its output is in turn the input of the integrator. When the integrator voltage reaches a certain value, the output of the comparator changes sign, and the integrator ramp changes slope. This is reversed at the other end of the voltage span. The amplitude of the triangle is determined by the hysteresis of the voltage comparator, which depends on the input and feedback resistors of the comparator. The triangle slope is determined by the input resistor and the feedback capacitor of the integrator.

be constructed using two 741 operational amplifiers, one connected as an integrator and the other as a voltage comparator, as shown in Fig. 4.

3.3. Current-to-voltage Converter

3.3.1. Theory

The basic current-to-voltage converter is an operational amplifier connected as illustrated in Fig. 5A. The input capacitor, C_i, represents the membrane capacitance, and C_f is the stray capacitance associated with the feedback resistor. Assuming an ideal amplifier of infinite gain and that no current goes into the input, all the incoming current, I, flows through the feedback resistor, R_f, into the amplifier's output, and the voltage at the input is zero. With these assumptions, the voltage at the output, V_o, is given by equation 1.

$$I = -V_o/R_f \tag{1}$$

To measure low-level currents, of the order of picoamperes, large feedback resistors are used, of the order $10^9\ \Omega$. Capacitance of these high-megohm resistors becomes significant and must be considered. If capacitance is included in the circuit analysis, then the incoming current circulates through the capacitance and resistance of the feedback element. V_o is now described by equation 2.

$$I = -V_o/R_f - C_f(dV_o/dt) \tag{2}$$

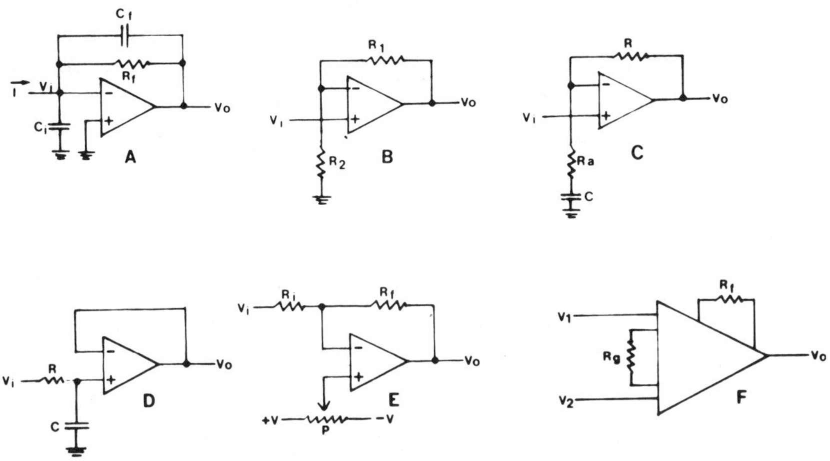

Figure 5. Operational amplifier circuits. These are several connections for amplifiers that are used as building blocks for the bilayer system. (A) Current-to-voltage converter. (B) Voltage follower with gain. (C) High-frequency booster. (D) Low-pass filter with buffer. (E) Inverting amplifier with offset. (F) Differential amplifer. Circuits A to E use operational amplifiers; circuit F uses an instrumentation amplifier, e.g., Analog Devices AD 521.

When this is solved for V_o (see equation 3), it is apparent that for a step of current of amplitude I_o, V_o rises to the steady value $-I_oR_f$, following an exponential curve with a time constant, T_f, equal to R_f times C_f.

$$\begin{aligned} V_o &= -R_fI[1 - \exp(-t/T_f)] \\ T_f &= C_fR_f \end{aligned} \tag{3}$$

For R_f of $10^9\ \Omega$ C_f of 10^{-12} F, the time constant of the circuit is 10^{-3} sec. Decreasing C_f should speed up the system according to equation 3, but this is not always true because this equation was derived assuming that the operational amplifier keeps the voltage at the input, V_i, at zero all the time. This is equivalent to assuming that the amplifier is infinitely fast and instantaneously changes V_o to satisfy equation 2. Real amplifiers change V_o at a finite rate that is proportional to V_i and proportional to the bandwidth of the amplifier, as in equation 4.

$$\begin{aligned} dV_o/dt &= -V_i/T_a \\ T_a &= 1/6.28 \text{ gain–bandwidth product} \end{aligned} \tag{4}$$

The amplifier time constant, T_a, is the reciprocal of the amplifier's bandwidth in radians per second. The factor in its definition in equation 4, 2π, permits the use of the bandwidth expressed in cycles per second as it is in the amplifier specifications.

Since V_o does not change instantly, V_i is not always zero, i.e., V_i jumps and then is returned to zero by the amplifier when a step of current is applied to the circuit. Input capacitance, C_i, is charged when V_i changes, and part of the incoming current goes through C_i to ground. The complete expression for the current, considering that V_i changes and that current goes through three routes, is given in equation 5.

$$I = [(V_i - V_o)/R_f] + C_f(dV_i/dt - dV_o/dt) + C_i(dV_i/dt) \tag{5}$$

Solving for V_o is more complex now. Equation 4 allows V_i to be eliminated from the expression, but the second derivative of the output voltage appears (see equation 6).

$$I = -V_o/R_f - (C_f + T_a/R_f)(dV_o/dt) - (C_f + C_i)T_a(d^2V_o/dt^2) \tag{6}$$

Solving equation 6 is beyond the scope of this book and can be found in Sigworth (1983) and Alvarez *et al.* (1985). For the case of a bilayer system, some simplifications can be made. C_i is the membrane capacitance, *ca.* 500 10^{-12} F; C_f, the capacitance of the feedback resistor, is 10^{-12} F or less, so C_f can be eliminated from the third term of equation 6. If T_a is 10^{-6} sec (i.e., the amplifier has a gain–bandwith product of *ca.* 6 MHz) and R_f is 10^9 Ω, then T_a/R_f can be eliminated from the second term. With these simplifications, equation 6 is reduced to:

$$I = -V_o/R_f - C_f(dV_o/dt) - C_iT_a(d^2V_o/dt^2) \tag{7}$$

Now, if C_i and T_a are small values, and C_f is large enough, then the term with the second derivative can be ignored, equation 6 reduces to equation 2, and the step response is just an exponential function. If the term with the second derivative is significant, then the step response shows oscillations, since the first and second derivatives take values of opposite sign (V_o increases but at a decreasing rate), so there can be several values of V_o as a function of time that satisfy equation 6. This situation should be avoided by using a fast amplifier and small membranes and a reasonable value for C_f. The complete mathematical treatment to calculate the values of the different design parameters will not be given here, but in Section 3.3.7 a conservative current-to-voltage converter design that will work is described.

3.3.2. Feedback Resistor

Most of the problems in making a fast and stable current-to-voltage converter arise from the use of large feedback resistors. If R_f has a small value, only the first term in equation 6 would be significant, and the current-to-voltage converter would behave ideally. At first glance it would appear convenient, then, to use a smaller feedback resistor and to compensate for the decrease in sensitivity by using voltage gain stages after the current-to-voltage converter. The reason to use large feedback resistors is that the system has less noise. This is because even though the voltage noise (thermal noise) from the resistor increases with the square root of R, the current-to-voltage converter sensitivity increases as a direct function of R. For a bandwith of 1 kHz, the contribution of the noise from the feedback resistor is 13 pA for a 10^5-Ω resistor, 1.3 pA for a 10^7-Ω resistor, or 0.13 pA for a 10^9-Ω resistor. This tells us that for low noise, a large feedback resistor must be used; a 10^9-Ω resistor is a good choice.

3.3.3. Amplifier Gain–Bandwidth Product

To have a current-to-voltage converter with a fast step response that is free of oscillations, a fast operational amplifier has to be used. To calculate how fast the amplifier must be, we must solve equation 7. This is beyond the scope of this book, but for practical purposes, a lower limit of 1 MHz can be set.

3.3.4. Amplifier Noise

Noise is another important amplifier characteristic to consider. The relevant parameter to look at on the amplifier's data sheet is the input voltage noise. Any amplifier with less than 30 nV/Hz^{-1} can be used. See the *Burr-Brown Product Data Book* under OPA 101 application information notes for a good discussion of this subject.

3.3.5. Feedback Capacitance

To have a reasonably fast current-to-voltage converter, feedback capacitance must be kept low, of the order of tenths of a picofarad. Feedback capacitance is the sum of the capacitances of the feedback resistor and the connecting wires and the internal capacitance of the operational amplifier. Stray capacitances can be significant if special care in the layout of the circuit is not taken. Connecting wires carrying the output of the amplifier must be routed away from the input cables to minimize C_f. Keeping the input wires as short as possible also helps in reducing capacitative coupling to the output wires.

3.3.6. Input Capacitance

The presence of oscillations in the step response of the current-to-voltage converter depends on the input capacitance. The smaller the input capacitance (membrane area), the less likely it will be to get oscillations. Input capacitance also affects the noise of the system. Random noise increases as C_i increases. Voltage at the input of the amplifier moves around the average value, and this causes a current through the input capacitance. This is as if a current were injected into the converter and appears as a voltage at the amplifier's output. The magnitude of this current depends on C_i. This means that the membrane capacitance must be kept low to minimize system noise.

To keep C_i low, it is advisable to place the current-to-voltage converter close to the membrane chamber and to connect the current electrode directly to the box containing the amplifier, as shown in Fig. 2. If shielded cables are used to connect the electrode to the amplifier, cable capacitance can contribute significantly to C_i, and this should be avoided.

3.3.7. A Practical Design

A current-to-voltage converter that is almost guaranteed to work well as a first approach can be constructed using an OPA 111 (Burr-Brown, Tucson, AZ) operational amplifier and a 10^9-Ω resistor (Eltec Instruments Inc., Daytona Beach, FL, Model 104). This current-to-voltage converter will give no overshoot or oscillations with membranes up to 500 pF. The sensitivity is 1 mV/pA. The step response is nearly exponential and has a response time of 300 μsec and an average noise of less than 1 pA at a 1-kHz bandwith. These data are for the abovementioned feedback resistor; using other models or makes can result in a slower system if the resistor capacitance is larger or oscillations if smaller.

3.3.8. Capacitance Compensation

Because planar bilayers have large capacitances (~200 pF), a change of applied voltage (such as a step) will give very large current transients, and the head-stage amplifier will saturate. To avoid this, an easy solution is to subtract an identical current through a "dummy" membrane before the current reaches the head-stage amplifier. This is done as shown in the block diagram (Fig. 3). In principle, the voltage command applied to the bilayer is simultaneously inverted and applied to a capacitor that has exactly the same value as the membrane capacitance. In this way, the "positive" capacitive current of the planar bilayer is exactly cancelled by the "negative" current from the dummy capacitor. In practice, however, it is not always possible to use a dummy capacitor exactly the same size as the bilayer capacitance. Therefore, the voltage command is inverted with a variable gain (about two- to tenfold), and the dummy capacitor is chosen to be ~40 pF, i.e., approximately fivefold lower than the bilayer value. The gain of the voltage inverter can then be adjusted until the bulk of the capacitive spike is nulled out.

3.4. Signal Conditioning

The current passing through the membrane is converted into a voltage by the current-to-voltage converter. The output of that stage can be of the order of millivolts or fractions of a millivolt. In most cases, this low-level voltage must be further amplified before being displayed, recorded, or processed. Furthermore, we may need to add an offset to the voltage in order to make better use of the dynamic range of a tape recorder, reduce the bandwidth before signal digitization, or correct the step response to improve time resolution. All of these operations are called signal conditioning.

3.4.1. Gain and Offset

Adding gain and offset can readily be done using an instrumentation amplifier such as the AD 521 (Analog Devices, Norwood, MA), as shown in Fig. 5E. Gain can be controlled by changing the gain resistor of the amplifier and should be 1, 10, or 100. Offset can be added directly at the output stage of the amplifier, so it will stay constant when the gain is changed. It is advisable to mount the voltage-gain amplifier in the same box where the current-to-voltage converter is mounted. Great care must be exercised in insulating the output wires of this amplifier from the input of the current-to-voltage converter. If they are close, negative-capacitative feedback will significantly slow the system response, especially at high gain. A metallic partition inside the box separating the two amplifiers will help. A battery power supply should be used, and power supply lines decoupled using capacitors near the amplifiers. The signal at this point will have an amplitude of 1 volt per 10 pA, and it can now be passed to other circuits located elsewhere without danger of noise pickup.

3.4.2. Low-Pass Filter

The signal coming from the voltage amplifier may include extraneous noise. Since noise extends over the entire frequency spectrum, it can be reduced by narrowing the system bandwidth. This can be done using a low-pass filter. A minimal low-pass filter can be constructed using a resistor–capacitor combination and a voltage follower to keep output impedance low.

3.4.3. Step Response Correction

When a step of current is applied to a current-to-voltage converter, the output voltage is not a step function but rather an exponential function with a time constant equal to the *RC* combination on the feedback loop. This time constant can limit the overall time resolution of a planar bilayer system, and it can be improved by using a circuit such as the one shown in Fig. 5C, sometimes called a high-frequency booster. This circuit is similar to the voltage follower with gain (Fig. 5B). In this connection, the amplifier places a voltage at the output, V_o, such that the voltage at the inverting input, V_a, is equal to the voltage at the noninverting input, V_i, which is the input to the circuit. Disregarding the small resistor from the capacitor to ground, the voltage V_a will rise exponentially to a final value equal to V_o when V_o jumps from zero to some constant value. This can easily be seen, since V_a is just the charge of a capacitor. If the voltage at the point V_i is an exponential function, since it is the output of the current-to-voltage converter, then the voltage at the output of the amplifier must be a step function, provided that the *RC* product in this circuit is adjusted to be the same as that of the current-to-voltage converter. The small resistor between the capacitor and ground is used to limit the high-frequency gain and to stabilize the circuit.

The use of this step-response correction deserves some comments. This step-response correction works only if the step response of the current-to-voltage converter is a perfect exponential function. Minimal deviations from this ideal step response can be obtained by using a very fast amplifier with small input capacitance and selecting feedback resistors to get the best response. Kobra resistors (K & M Electronics, W. Springfield, MA) are excellent for this purpose.

The high-frequency booster has a gain that increases with frequency. This means that the high-frequency noise content of the signal will be amplified, and the signal-to-noise ratio will deteriorate. To keep noise low, it is convenient to filter out the high-frequency noise before the booster. However, care must be taken not to distort the exponential shape of the step response before it is fed to the high-frequency booster.

3.5. A Minimal System to Get Started

For a minimal bilayer system, not all the circuits described are necessary. As a pulse generator, the proposed triangle oscillator can be used to monitor membrane capacitance during membrane formation, and DC voltages can be obtained from a battery and a voltage divider. A current-to-voltage converter without capacitance compensation can be constructed using a Burr-Brown OPA 111 amplifier and a 1-GΩ resistor. Minimal signal conditioning may be a simple RC low-pass filter and an amplifier to increase the gain to 100 pA/V. Current can be displayed using an oscilloscope, and higher-level signals can be picked up for a tape or paper recorder from the vertical amplifier of the oscilloscope, which is generally available.

For this system the main components are:

1. 741 operational amplifiers: two for triangle generator. These are inexpensive and are available in any electronics store.
2. OPA 111 operational amplifier for current-to-voltage converter (Burr-Brown, P.O. Box 11400, Tucson, AZ). A special mating Teflon® socket is needed for low leakage.
3. AD 521 instrumentation amplifier for added sensitivity (Analog Devices, P.O. Box 280, Norwood, MA).
4. 1-GΩ resistor (Model 104, Eltec Instruments, Inc., Daytona Beach, FL).
5. Precision 1% resistors: 1 kΩ, 10 kΩ, and two 100 kΩ for the AD521 amplifier. One 100-kΩ resistor is used in the feedback loop, and the other resistors are for selecting the gain: 1×, 10×, and 100×.
6. Precision 10-turn potentiometer and counting dial for DC voltage generator.
7. Miscellaneous: two aluminum boxes, switches, rechargeable 7-V batteries and matching clips, BNC connectors, BNC patch cables, PC boards, etc. . . .

The DC and triangle generators are mounted in one box, and the current-to-voltage converter and gain amplifier in another box. The latter should be placed in the immediate vicinity of the membrane chamber and enclosed in a Faraday cage and on a vibration-proof platform. Connection from the pulse generator to the current-to-voltage box is made using shielded coaxial cable with BNC connectors. The current-to-voltage box has four BNC connectors. Two of these connectors must have Teflon® insulation and are used to accept the membrane electrodes. One is connected to the voltage input, a third BNC connector, and the other to the input of the current-to-voltage converter. The fourth BNC connector is the output of the gain amplifier and goes to the oscilloscope. To prevent ground loops, do not connect the pulse generator box to ground but through the coaxial cable going to the current detector box. If voltage is monitored on the oscilloscope, use a shielded cable but cut the ground cable close to the oscilloscope.

4. Procedures for Membrane Formation

4.1. Cleaning the Chambers

To clean Teflon® chambers, first wash them with running water to remove all salts. The vacuum grease is removed using a paper tissue. The chamber is then soaked overnight in Na dichromate–sulfuric acid cleaning solution. The cleaning solution is discarded, and the chamber is soaked for 4 hr in running tap water. The chambers are then washed several times with double-distilled water, washed with ethanol and then chloroform, and dried in a stream of nitrogen. Chambers are stored in a dust-proof container. Should the chamber get stained, it must be treated with a boiling solution of NaOH in alcohol in a reflux apparatus. Detergents should not be used.

4.2. Water Purification

Purity of the water used in bilayer experiments is important to obtain stable membranes. I use double-distilled water with the second distillation done in an all-glass apparatus, and water is distilled from alkaline potassium permanganate. Teflon® boiling stones should be used to prevent bumping.

4.3. Lipid Solutions

Lipids are to be stored in a deep freezer to prevent decomposition. Lipids are usually purchased in chloroform. Hexane solutions are used for folded bilayers, and decane solutions for painted bilayers. These solutions must be freshly prepared at a concentration of 10 mg/ml each day. The appropriate amount of chloroform solution is measured into a vial, and the solvent evaporated under a stream of nitrogen. The final solvent is then added.

For painted bilayers, the hole must be preconditioned with a little lipid to achieve easy membrane formation. A drop of decane–lipid solution is smeared around the hole, and the solvent is allowed to evaporate under nitrogen. The cup

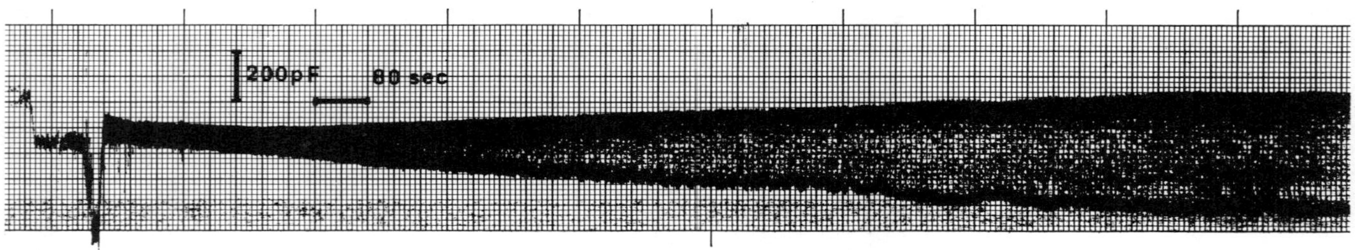

Figure 6. Time-dependent increase of membrane capacitance. This is an actual record taken from the thinning process of a bilayer made from a decane solution of phosphatidylethanolamine, showing the time evolution of membrane capacitance as measured using a triangular voltage wave.

is then immediately immersed under the aqueous solutions. The hole need not be recoated more than once or twice during a day's work.

4.4. Membrane Formation

To monitor membrane formation a triangular wave of 20 mV peak to peak and 100 Hz frequency is used. The gain is set to 1000 pA/V, and current is measured on the oscilloscope at 1 V/cm. A capacitor of 100 pF is connected in place of the bilayer to calibrate the amplitude of the square wave of current in picofarads. It should be 400 pA for 100 pF. For painted bilayers, the compartments of the chamber are filled with the appropriate aqueous solutions. Current is very large at this moment, and the oscilloscope beam disappears. Lipids are applied to the hole, and the oscilloscope beam returns. This indicates that the hole is plugged with lipids. A small square wave of current should be seen on the oscilloscope, which corresponds to the capacitance of the cup wall. As time passes (see Fig. 6), the current amplitude increases, indicating the thinning of the bilayer. The final capacitance should be *ca.* 0.4 $\mu F/cm^2$ of bilayer plus the initial capacitance. The time it takes for the final value to be reached depends on the hole material and the lipid concentration. It is faster with dilute solutions.

For folded bilayers, the procedure varies somewhat. Here we start with the monolayer-coated solutions below the hole level. At this time, a very small capacitance is measured. As the levels are raised, the capacitance increases as monolayers are apposed. The completion of membrane formation is indicated by a sudden decrease in capacitance. Now the levels can further be raised to cover the bilayer completely. Final capacitance is *ca.* 0.7 $\mu F/cm^2$.

4.5. Checking for Leakage

Once the membrane is formed, the triangle generator should be turned off and DC voltage applied. To measure conductance, the gain is increased to 1 pA/V. A good membrane should pass less than 1 pA at 100 mV. Channels can now be added.

ACKNOWLEDGMENTS I acknowledge the able assistance of Mr. Juan Espinoza in the preparation of the illustrations of this chapter and the construction of the systems described. This work was supported by Universidad de Chile Grant B-1985-8413 from Departamento de Desarrollo de la Investigación and Grant 1299 from Fondo Nacional de Investigación Científica (Chile).

References

Alvarez, O., Benos, D., and Latorre, R., 1985, The study of ionic channels in planar lipid bilayer membranes, *J. Electrophysiol. Tech.* **12**:159–178.

Sigworth, F., 1983, Electronic design of the patch clamp, in: *Single-Channel Recording* (B. Sakmann and E. Neher, eds.), pp. 3–35, Plenum Press, New York.

Handbooks

Burr-Brown Product Data Book, 1984, Burr-Brown, Tucson, Arizona.
Data Acquisition components and Subsystems, 1980, Analog Devices, Norwood, Massachusetts.
Signetics Analog Manual, 1976, Signetics Corporation, Sunnyvale, California.

Chapter 6

FUSION OF LIPOSOMES TO PLANAR BILAYERS

FREDRIC S. COHEN

1. Introduction

During the late 1970s and early 1980s, methods to reconstitute integral membrane channel-forming proteins into planar membranes were developed. There are now two basic strategies. The first is to fuse vesicles that contain channels to preformed phospholipid planar membranes (Miller and Racker, 1976; Cohen *et al.*, 1980; Akabas *et al.*, 1984). The vesicles can be those obtained by the usual methods of cell fractionation—homogenization and centrifugation— and are often referred to as "native" vesicles, or they can be phospholipid vesicles that have had a purified channel incorporated into the vesicular membrane, known as "artificial" vesicles. The second strategy is to incorporate membrane channels into monolayers derived from vesicles (Verger and Pattus, 1976; Pattus *et al.*, 1981; Schindler, 1979) and to form bilayers with functional channels from these monolayers (Schindler and Rosenbusch, 1978; Schindler and Quast, 1980; Nelson *et al.*, 1980; Vodyanoy and Murphy, 1982; Coronado and Latorre, 1983). Either native or artificial vesicles can be used to form the monolayers.

As native vesicles are easier to obtain than artificial vesicles, the former have usually been used for reconstitution via fusion methods. Channels reconstituted by this method are numerous and include a K^+-selective channel from muscle sacroplasmic reticulum (Miller and Racker, 1976; Labarca *et al.*, 1980), a Cl^--selective channel from *Torpedo* electroplax (White and Miller, 1979; Hanke and Miller, 1983), Ca^{2+}-activated K^+ channels from muscle (Latorre *et al.*, 1982; Moczydlowski and Latorre, 1983), K^+ channels from lobster axolemma (Coronado *et al.*, 1981, 1984), TTX-sensitive voltage-dependent Na^+ channels from brain (Krueger *et al.*, 1983; French *et al.*, 1984; Moczydlowski *et al.*, 1984), an amiloride-sensitive sodium channel from A6 cells (Sariban-Sohraby *et al.*, 1984), voltage-dependent Ca^{2+} channels from brain (Nelson *et al.*, 1984), K^+ and Cl^- channels from cardiac sarcolemma (Coronado and Latorre, 1982), and cation-selective channels from rod outer segment (Hanke and Kaupp, 1984). Fusion of artificial vesicles to planar membranes for the purpose of reconstitution has not been nearly as common because purified channel-containing vesicles were not routinely avail-

FREDRIC S. COHEN • Department of Physiology, Rush Medical College, Chicago, Illinois 60612.

able. Such preparations are, however, being developed and the method is now being used for artificial vesicles, as illustrated by the reconstitution of purified voltage-dependent Na^+ channels into planar membranes (Hartshorne *et al.*, 1985).

Attention in this chapter is directed toward fusing artificial vesicles to planar membranes. Minor modifications used with native vesicles are pointed out.

2. Mechanism of Fusion

When vesicles are added to one of the aqueous compartments bathing a planar membrane (*cis* side), proteins can, in principle, insert into the planar membrane by either of two processes: (1) the proteins can transfer from the vesicular membrane to the planar membrane; or (2) the vesicle itself can fuse with the planar membrane, and all the proteins in the vesicular membrane are thereby inserted into the planar membrane.

Clearly, if many channels incorporate simultaneously into the planar membrane, this is an indication that protein insertion is occurring via fusion of the entire vesicular membrane into the planar membrane rather than by transfer of individual channel-forming proteins from one membrane to the other. Simultaneous incorporation has thus been used as the criterion to determine whether fusion has in fact occurred (Miller and Racker, 1976; White and Miller, 1979; Cohen *et al.*, 1980, 1984). In one study, release of vesicular contents across the planar membrane (i.e., into the *trans* compartment) was shown to take place under the same conditions and with the same time course as did group insertion of vesicular channels into the planar membrane (Zimmerberg *et al.*, 1980a). Therefore, it was demonstrated that these channels are reliable indicators of fusion. The use of such markers has allowed the conditions necessary for fusion to be elaborated (Cohen *et al.*, 1980, 1984; Akabas *et al.*, 1984).

Fusion of artificial vesicles has been shown to consist of two experimentally distinguishable steps. In the first, a very tight, close, and irreversible association between the vesicles and planar membrane is formed, a configuration called the "prefusion" state. If the vesicular and/or planar membrane contains negatively charged lipids, addition of divalent cations to the *cis* side promotes the formation of this state. This association between vesicle and planar membrane is not a point contact but rather an adhesion of a substantial fraction of the vesicle with the bilayer (Akabas *et al.*, 1984). In the second step, the actual fusion of vesicular and planar membranes takes place. For this second fusion step to occur, it is necessary that an intravesicular hydrostatic pressure develop within those vesicles that are in the prefusion state. This pressure is routinely induced by osmotically swelling the vesicles. When lysis occurs in the region of contact between the vesicle and planar membrane, fusion occurs. The picture we envision is that of a vesicle tightly abutted to the planar membrane. When the vesicle swells, the vesicular and planar membranes stretch, exposing hydrophobic regions of the acyl chains, which leads to phospholipid rearrangements and to the melding of the two membranes (Akabas *et al.*, 1984). Although certainly not proven, this picture of the process is a useful shorthand that aids one in correctly manipulating conditions to promote fusion and hence reconstitution.

3. Method of Fusion

3.1. Binding

To fuse vesicles and planar membrane, it is necessary to bring about an intimate contact between them. Vesicles are added to the *cis* side of a planar membrane. When negative lipids are present in either the planar or vesicular membrane, addition of millimolar levels of divalent cations to the *cis* side is a simple way to accomplish this intimate contact: Ca^{2+}, Mg^{2+}, Co^{2+}, Ba^{2+}, and Mn^{2+} all promote the prefusion state, although not with quite the same efficacy. Calcium, for instance, works a bit better than Mg^{2+}; La^{3+} is also effective in promoting the prefusion state and does so at lower concentrations than do the divalents. The exact concentrations needed vary with the lipids used. Proteins such as fibronectin (Young and Young, 1984) can also induce adhesion. For native vesicles, divalents have been used exclusively. With the solvent-containing black lipid membranes (BLMs) and artificial phospholipid vesicles, 10 mM Ca^{2+} is invariably sufficient. Solvent-free membranes (SFMs) require more Ca^{2+} for binding, with 15 mM Ca^{2+} usually adequate. These are useful concentrations to start with, but lower concentrations can often be used. After fusion, the *cis* side of the chamber can be perfused to establish any ionic conditions desired. When only neutral lipids are present in both the planar and vesicular membranes, the prefusion state can form spontaneously depending on the lipids and type of planar membrane used. Divalent cations do not promote the prefusion state for an all-neutral lipid system.

In choosing divalent cation concentrations, adsorption of vesicles to the planar membrane must be weighed against vesicle–vesicle aggregation. Divalent cations induce vesicle aggregation as well as planar membrane adhesion, and aggregated vesicles fuse poorly, if at all. Therefore, although increasing divalent (or trivalent) cation concentrations better promotes the desired adhesion, it might also better promote vesicle aggregation and thereby decrease the number of vesicles adsorbed. If these simple considerations are kept in mind, adjustments at the setup will establish convenient conditions.

With native vesicles, the same considerations apply. In practice, lower divalent concentrations are usually sufficient. Exact conditions will depend on the source of the vesicles.

3.2. Swelling

The addition of divalent cations alone is ineffective in promoting fusion of artificial vesicles to planar membranes. As stated, vesicles in the prefusion state need to swell for fusion to occur. Two methods have been used to achieve this swelling.

In the first method, after vesicles are in the prefusion state, a solute on the *cis* side that is impermeant through the vesicle is replaced with a permeant solute. The solute enters the vesicle with water following, thereby maintaining osmolarity, and the vesicle swells and fuses (Cohen *et al.*, 1982; Akabas *et al*,. 1984). Although conceptually straightforward, this method is not technically convenient and therefore not really practical for reconstitution.

The other method is to establish an osmotic gradient across the planar membrane, *cis* side hyperosmotic with respect to the *trans* side. Water then flows *trans* to *cis*, and a fraction of it enters vesicles in the prefusion state. Mechanisms for this phenomenon have been analyzed (Zimmerberg *et al.*, 1980b; Akabas *et al.*, 1984). As a result of this water flow into adsorbed vesicles these vesicles swell, and in the steady state a pressure develops in them that causes some to burst and fuse. This method has been successfully used for fusion of artificial vesicles (Cohen *et al.*, 1980; Akabas *et al.*, 1984; Hartshorne *et al.*, 1985) and a variety of native vesicle (Coronado and Latorre, 1982; Latorre *et al.*, 1982; Moczydlowski and Latorre, 1983; Sariban-Sohraby *et al.*, 1984; Bell and Miller, 1984; Coronado *et al.*, 1984).

Osmotic-gradient-induced fusion has the great advantage of being easy to achieve experimentally, and it can be implemented in a variety of ways. An osmoticant (an invented but useful word to indicate a solute that, when added to a solution, raises the osmolarity of that solution) can be added to the *cis* side after vesicles are added, or vesicles can be added to a preformed gradient. The gradient can also be established before or after addition of divalents. The order of adding vesicles, divalent, and osmoticant is basically unimportant and can be chosen to suit experimental convenience.

In selecting an osmoticant, it is important to keep in mind that its ultimate role is to promote swelling of the vesicles. Therefore, it is necessary that the vesicle membrane be permeable to that osmoticant. Addition of a vesicle-impermeant osmoticant will cause net shrinkage of the vesicle and will not promote fusion. But the osmoticant must not be too permeant through the planar membrane. If the osmoticant is so permeant that the gradient is dissipated across the unstirred layers adjacent to the membrane, there will not be adequate net movement of molecules into the vesicle to cause swelling.

Urea and glycerol are, for example, good osmoticants. They are sufficiently permeant through vesicle membranes to enter rapidly and thereby prevent vesicle shrinkage as a result of solute and water movement between the vesicles and bathing *cis* compartment. But their permeabilities through lipid bilayers are low enough to promote water movement into adsorbed vesicles. Ethylene glycol and formamide, which have larger permeabilities than urea and glycerol, are not, in practice, effective osmoticants.

If the vesicular membrane has channels that allow salts to enter the vesicle interior, then permeant salts can also be used as osmoticants. For artificial vesicles in contact with BLMs, a 250 mOsm gradient across the planar membrane will induce fusion. When SFMs are used, a 450 mOsm gradient should be adequate.

Usually, however, fusion can be obtained with smaller gradients. Larger gradients lead to more fusion events, as expected.

4. Solvent-Free versus Solvent-Containing Planar Membranes

Conceptually, the method for fusing vesicles to planar membranes is the same for solvent-free (SFM) and solvent-containing (BLM) membranes. But there are some phenomenological differences to be aware of.

The most dramatic—and at this time still mysterious—difference is that stirring of the *cis* side is required for fusion of vesicles to BLMs but totally irrelevant when fusing them to SFMs. When vesicles are fusing to BLMs, halting the stirring results in a substantial reduction of the rate of fusion and often in complete abolition of fusion, usually within 30 sec. Resumption of stirring leads to the previous high rate of fusion. Although the mechanism for this stirring effect is unknown, we do know that stirring promotes the adsorption of vesicles to BLMs (Akabas *et al.*, 1984).

As already mentioned, in general, higher concentrations of cationic divalents and larger transplanar membrane gradients are required with SFMs than with BLMs for successful fusion.

With SFMs, phosphatidylethanolamine (PE) is a better phospholipid than phosphatidylcholine (PC) for promoting fusion. For instance, artificial vesicles with a PC:PE ratio of 1:1 fuse better to SFMs composed of PE than to SFMs of PC. With BLMs, however, there are no apparent differences between using PE and PC. PC vesicles fuse very well to PC BLMs but not at all to PC SFMs. These differences in the abilities of PE and PC to promote fusion are related to differences in their ability to promote the formation of the prefusion state. With SFMs, PE is better in promoting the prefusion state, whereas with BLMs either phospholipid works equally well.

Generally speaking, it is easier to fuse vesicles to BLMs than to SFMs. The conditions are less stringent for the former, and higher rates of fusion are obtained. If the effects of the solvent on the protein can be tolerated, it will probably prove easier for the practitioner who is using the fusion method to start reconstitution with BLMs.

5. Additional Considerations

In fusing vesicles to planar membranes some simple adjustments of conditions other than those already mentioned can improve the chances for success.

5.1. Hydrostatic Pressure Considerations

Vesicle interiors do not have hydrostatic pressures until their membranes are taut. A shrunken vesicle will swell appreciably before building up an intravesicular hydrostatic pressure. It is therefore important that the vesicles free in solution be as fully swollen as is possible without bursting. In practice, water flow resulting from the osmotic gradient across the planar membrane will probably not be sufficient to fully swell a shrunken vesicle, and fusion will not occur. The simplest way to swell the vesicles is to add them to a *cis* side that is hypoosmotic with respect to the vesicular contents. With natural vesicles that are loaded with sucrose as a result of being centrifuged in a sucrose gradient, this is often accomplished by adding them to a sucrose-free (hypoosmotic) *cis* side of the planar membrane. To determine the maximum transvesicular gradient the vesicles can withstand without rupturing, any standard vesicle lysis assay can be used. A particularly convenient one is to use a macroscopic ion-selective electrode that is sensitive to an ion in high concentration in the vesicle. In adding vesicles to media free of that ion, varying the osmotic strength of the media determines the amount of released ion at any given osmotic strength. Total quantity of vesicle-trapped ion can be determined by adding sufficient detergent to vesicles to release all ions.

5.2. Establishing the Osmotic Gradient

It sometimes proves convenient to add vesicles to a chamber that already has an osmotic gradient across the planar membrane, *cis* side hyperosmotic. When the vesicular interior is hyperosmotic with respect to the *cis* chamber, then vesicles become fully swollen. Addition of divalents causes the vesicles to come into intimate contact with the planar membrane. Some of the water flowing from the *trans* into the *cis* compartment will flow into the vesicles, causing further swelling and fusion. (As mentioned, if an osmoticant is added after the vesicles, the osmoticant should be permeant through the vesicle membrane.)

5.3. Handling Vesicles

Channels can be incorporated into a number of types of phospholipid vesicles, e.g., small sonicated vesicles, freeze–thaw vesicles, detergent-dialyzed, and multilamellar vesicles. Independent of the exact nature of the vesicles and the method used to incorporate channels into them, the same considerations for fusing them to planar membranes apply for all.

Unless there is a sufficient concentration of vesicles in the chamber, there will not be enough collisions between vesicles and planar membrane to form the prefusion state. How many vesicles are sufficient depends not only on the volume of the *cis* compartment but also on the surface-to-volume ratio of the chamber. Notably, with SFMs, which have supporting monolayers, shallow chambers with much surface area tend not to work as well as deeper chambers with the same volume but less surface area. Presumably, vesicles are adsorbing to the monolayers in sufficient quantity to effectively reduce the concentration of those free in solution. Typically, 10^{11} vesicles/ml is a good starting concentration.

With time, vesicles in the stock solution sometimes aggregate in the holding tube. A very light sonication of the vesicles (a few seconds in a bath sonicator) just before adding them to the chamber helps dissociate them, and greater fusion is obtained.

Finally, it is easy to stop fusion. Just abolish the osmotic gradient by adding osmoticant to the *trans* compartment. With BLMs, stopping the stirrers also halts further fusion.

6. Other Methods

There are other methods to reconstitute proteins by fusionlike mechanisms.

By making solvent-free planar membranes from 1-stearoyl-3-myristoyl-glycero-2-phosphocholine, which is a mixed-chain lecithin, fusion of vesicles can be induced by simply adding calcium and holding the membrane at or below the phase-transition temperature (Hanke *et al.*, 1981). But this method is restricted; it requires the use of certain special phospholipids. They must have experimentally convenient phase-transition temperatures, and the planar membranes made from them must not break or become leaky when frozen. Membranes made from mixtures of phospholipids do not have a well-defined transition temperature and thus are not suitable for reconstitution by this method.

It should not be inferred that, in general, phospholipid bilayers held below or at the phase-transition temperature will fuse better than fluid bilayers or that frozen membranes require only divalent cations for fusion to proceed. If the planar membrane remains fluid, and artificial vesicles with well-defined phase-transition temperatures (T) are used, then the rate of fusion is small below T, high above T, and increases in a steplike manner at T. Fusion is not greater at or near the transition temperature than at temperatures above T (Cohen and Akabas, 1982). In all cases, an osmotic gradient across the planar membrane is necessary to promote fusion.

With native vesicles, divalent cations alone can induce channel incorporation into the planar membrane, especially into BLMs. For some preparations, incorporation takes place via a fusion mechanism as evidenced by simultaneous insertion of several channels (Miller and Racker, 1976; White and Miller, 1979). In these cases, the aqueous solutions are such that the vesicles are fully swollen; presumably, some of the vesicles are maintaining a pressure that is sufficient to cause them to fuse when brought into contact with the planar membrane. In such instances, an osmotic gradient across the planar membrane causes the rate of fusion to increase by more than an order of magnitude.

For some native vesicle preparations, however, protein incorporation after addition of divalent cations may not be occurring via fusion. Insertion of the Ca^{2+} channel from cilia of paramecium is an example in which incorporation does not depend on vesicle swelling, and channels insert one at a time rather than several simultaneously (Ehrlich *et al.*, 1984).

In conclusion, the fusion process can be conceptually divided into two distinct phases: formation of the prefusion state and subsequent osmotic swelling of the vesicles. This situation is useful and allows well-considered variation in experimental conditions. Furthermore, the transbilayer osmotic gradient method works well for both crude native vesicles and artificial vesicles with incorporated purified channels.

ACKNOWLEDGMENTS. I wish to thank Drs. Robert Eisenberg and Walter Niles for critical readings.

References

Akabas, M. H., Cohen, F. S., and Finkelstein, A., 1984, Separation of the osmotically driven fusion event from vesicle–planar membrane attachment in a model system for exocytosis, *J. Cell. Biol.* **98:**1063–1071.

Bell, J., and Miller, C., 1984, Effects of phospholipid surface charge on ion conduction in the K^+ channel of sarcoplasmic reticulum, *Biophys. J.* **45:**279–287.

Cohen, F. S,. and Akabas, M. H., 1982, The nature of the region of contact between fusing membranes, *Biophys. J.* **37:**26a.

Cohen, F. S., Zimmerberg, J., and Finkelstein, A., 1980, Fusion of phospholipid vesicles with planar phospholipid bilayer membranes. II. Incorporation of a vesicular membrane marker into the planar membrane, *J. Gen. Physiol.* **75:**251–270.

Cohen, F. S., Akabas, M. H., and Finkelstein, A., 1982, Osmotic swelling of phospholipid vesicles causes them to fuse with a planar phospholipid bilayer membrane, *Science* **217:**458–460.

Cohen, F. S., Akabas, M. H., Zimmerberg, J., and Finkelstein, A., 1984, Parameters affecting the fusion of unilamellar phospholipid vesicles with planar bilayer membranes, *J. Cell Biol.* **98:**1054–1062.

Coronado, R., and Latorre, R,. 1982, Detection of K^+ and Cl^- channels from calf cardiac sarcolemma in planar lipid bilayer membranes, *Nature* **298:**849–852.

Coronado, R., and Latorre, R., 1983, Phospholipid bilayers made from monolayers on patch-clamp pipettes, *Biophys. J.* **43:**231–236.

Coronado, R., Huganir, R., and Mautner, G., 1981, A K^+-selective conductance sensitive to cholinergic antagonists obtained by fusion of axonal membrane vesicles to planar bilayers, *FEBS Lett.* **131:**355–358.

Coronado, R., Latorre, R., and Mautner, H. G., 1984, Single potassium channels with delayed rectifier behavior from lobster axon membranes, *Biophys. J.* **45:**289–299.

Ehrlich, B. E., Finkelstein, A., Forte, M., and Kung, C., 1984, Voltage-dependent calcium channels from *Paramecium* cilia incorporated into planar lipid bilayers, *Science* **225:**427–428.

French, R. J., Worley, J. F. III, and Krueger, B. K., 1984, Voltage-dependent block by saxitoxin of sodium channels incorporated into planar lipid bilayers, *Biophys. J.* **45:**301–310.

Hanke, W., and Kaupp, U. B., 1984, Incorporation of ion channels from bovine rod outer segments into planar lipid bilayers, *Biophys. J.* **46:**587–595.

Hanke, W., and Miller, C., 1983, Single chloride channels from *Torpedo* electroplax: Activation by protons, *J. Gen. Physiol.* **82:**25–45.

Hanke, W., Eibl, H., and Boheim, G., 1981, A new method for membrane reconstitution: Fusion of protein-containing vesicles with planar bilayer membranes below lipid phase transition temperature, *Biophys. Struct. Mech.* **7:**131–137.

Hartshorne, R. P., Keller, B. U., Talvenheimo, J. A., Catterall, W. A., and Montal, M., 1985, Functional reconstitution of the purified brain sodium channel in planar lipid bilayers, *Proc. Natl. Acad. Sci. U.S.A.* **82:**240–244.

Krueger, B. K., Worley, J. F. III, and French, R. J., 1983, Single sodium channels from rat brain incorporated into planar lipid bilayer membranes, *Nature* **303:**172–175.

Labarca, P., Coronado, R., and Miller, C., 1980, Thermodynamic and kinetic studies of the gating behavior of a K^+-selective channel from the sarcoplasmic reticulum membrane, *J. Gen. Physiol.* **76**:397–424.

Latorre, R., Vergara, C., and Hidalgo, C., 1982, Reconstitution in planar lipid bilayers of a Ca^{2+}-dependent K^+ channel from transverse tubule membranes isolated from rabbit skeletal muscle, *Proc. Natl. Acad. Sci. U.S.A.* **79**:805–809.

Miller, C., and Racker, E., 1976, Ca^{2+}-induced fusion of fragmented sarcoplasmic reticulum with artificial bilayers, *J. Membr. Biol.* **30**:283–300.

Moczydlowski, E., and Latorre, R., 1983, Gating kinetics of Ca^{2+}-activated K^+ channels from rat muscle incorporated into planar lipid bilayers. Evidence for two voltage-dependent Ca^{2+} binding reactions, *J. Gen. Physiol.* **82**:511–542.

Moczydlowski, E., Garber, S. S., and Miller, C., 1984, Batrachotoxin-activated Na^+ channels in planar lipid bilayers. Competition of tetrodotoxin block by Na^+, *J. Gen. Physiol.* **84**:665–686.

Nelson, N., Anholt, R., Lindstrom, J., and Montal, M., 1980, Reconstitution of purified acetylcholine receptors with functional ion channels in planar lipid bilayers, *Proc. Natl. Acad. Sci. U.S.A.* **77**:3057–3061.

Nelson, M. T., French, R. J., and Krueger, B. K., 1984, Voltage-dependent calcium channels from brain incorporated into planar lipid bilayers, *Nature* **308**:77–80.

Pattus, F., Rothen, C., Streit, M., and Zahler, P., 1981, Further studies on the spreading of biomembranes at the air/water interface. Structure, composition, enzymatic activities of human erythrocyte and sarcoplasmic reticulum membrane films, *Biochim. Biophys. Acta* **647**:29–39.

Sariban-Sohraby, S., Latorre, R., Burg, M., Olans, L., and Benos, D., 1984, Amiloride-sensitive epithelial Na^+ channels reconstituted into planar lipid bilayer membranes, *Nature* **308**:80–82.

Schindler, H., 1979, Exchange and interactions between lipid layers at the surface of a liposome solution, *Biochim. Biophys. Acta* **555**:316–336.

Schindler, H., and Quast, U., 1980, Functional acetylcholine receptor from *Torpedo marmorata* in planar membranes, *Proc. Natl. Acad. Sci. U.S.A.* **77**:3052–3056.

Schindler, H., and Rosenbusch, J., 1978, Matrix protein from *Escherichia coli* outer membranes forms voltage-controlled channels in lipid bilayers, *Proc. Natl. Acad. Sci. U.S.A.* **75**:3751–3755.

Verger, R., and Pattus, R., 1976, Spreading of membranes at the air–water interface, *Chem. Phys. Lipids* **16**:285–291.

Vodyanoy, V., and Murphy, R. B., 1982, Solvent-free lipid biomolecular membranes of large surface area, *Biochim. Biophys. Acta* **687**:189–194.

White, M. M., and Miller, C., 1979, A voltage-gated anion channel from the electric organ of *Torpedo californica*, *J. Biol. Chem.* **254**:10161–10166.

Young, T. M., and Young, J. D., 1984, Protein-mediated intermembrane contact facilitates fusion of lipid vesicles with planar bilayers, *Biochim. Biophys. Acta* **775**:441–445.

Zimmerberg, J., Cohen, F. S., and Finkelstein, A., 1980a, Fusion of phospholipid vesicles with planar phospholipid bilayer membranes. I. Discharge of vesicular contents across the planar membrane, *J. Gen. Physiol.* **75**:241–250.

Zimmerberg, J., Cohen, F. S., and Finkelstein, A., 1980b, Micromolar Ca^{2+} stimulates fusion of lipid vesicles with planar bilayers containing a calcium-binding protein, *Science* **210**:906–908.

Chapter 7

INCORPORATION OF ION CHANNELS BY FUSION

WOLFGANG HANKE

1. Introduction

From the time when Mueller and colleagues (1962) first developed the planar lipid bilayer, a major goal has been to incorporate biological membrane transport proteins into these electrically accessible model membranes. One important group of transporters is that of the ion-channel proteins. Many different ion channels have been identified in biological membranes, and it is important to investigate them in single-channel experiments. Just as patch recording (Hamill *et al.*, 1981) is the only method for observing single channels in their native membranes, the planar bilayer has thus far proven to be the system of choice for studying single channels in model membranes of defined composition. The insertion of channel proteins into planar bilayers is thus the major problem that must be solved before such reconstitution experiments can be undertaken.

Several methods have been developed to incorporate proteins into planar lipid bilayers. Proteins that are soluble both in water and in lipid can be directly added to the aqueous solution around the bilayer and will diffuse into it. Ion channels from the membranes of higher organisms, however, are not water soluble. Biochemical isolation procedures usually leave them incorporated into the membranes of lipid vesicles. From these membranes they must be transferred into the planar lipid bilayer. Two ways are presently available to do this. Schindler and Rosenbusch (1978) developed a technique to incorporate proteins into planar lipid bilayers via monolayers at the air–water interface in equilibrium with vesicle suspensions. Two such monolayers having the proteins already in them are folded over a hole in a septum to form a bilayer, similar to the method of Montal and Mueller (1972). I discuss here another technique of protein incorporation into planar lipid bilayers: the fusion of channel-containing vesicles with preformed planar lipid bilayers (Miller and Racker, 1976).

WOLFGANG HANKE • Lehrstuhl für Zellphysiologie, Ruhr Universität, D-4630 Bochum, Federal Republic of Germany.

2. Methods of Bilayer Formation

Before going detailing the fusion of vesicles into bilayers, I comment briefly on the different methods of bilayer formation available now. The first method was developed by Mueller and co-workers (Mueller *et al.*, 1962), who "painted" a membrane from a solution of lipids in *n*-decane onto a small hole in a partition between two chambers of aqueous solution. These painted bilayers contain large amounts of decane.

Another method was developed by Montal and Mueller (1972) using monolayers on top of aqueous solutions. These monolayers were spread from lipids disolved in a highly volatile organic solvent such as pentane. Two such monolayers are combined into a planar lipid bilayer containing only small amounts of organic solvent. As most of the solvent is allowed to evaporate before bilayer formation, the bilayer is left virtually solvent-free. (A small amount of nonvolatile solvent such as squalene, which does not partition into the bilayer, must be applied to the septum before the bilayer is formed.) Alternatively, Schindler and Rosenbusch (1978) formed planar bilayers from monolayers formed not from a pentane solution of lipid but from a suspension of lipid vesicles. The bilayers made according to this method contain only very small amounts of organic solvent, originating again from the preconditioning of the septum.

3. Methods of Fusion

Bilayers made according to any of the above methods can be the starting point of a fusion experiment. All fusion experiments, however, have some technical details in common. The vesicles containing the ion-channel-forming proteins in their membranes are added to the aqueous solution on one side of the planar lipid bilayer (*cis* side). This side usually is vigorously stirred. The *cis* side should be hyperosmotic to the *trans* side, or the vesicles should be hyperosmotic to the compartment solutions used.

To know whether vesicles have fused with the planar bilayer one has to monitor the occurrence of fusion events. This is most easily done by applying a constant potential across the planar lipid bilayer and measuring the current flowing. When a vesicle carrying channels fuses with the bilayer under such conditions, the current increases in discrete jumps. Figure 1A shows a series of such fusion events under different conditions, as indicated in the figure legend. Figure 1B shows the "fine structure" of the fusion events of Fig. 1A. Each fusion event inserts some potassium-selective ion channels from the sarcoplasmic reticulum into the planar bilayer, thus giving an increased fluctuating current across the bilayer at constant potential. The fusion depends on the amount of hyperosmolarity, the area of the planar lipid bilayer, and the number of vesicles in the aqueous solution.

After some vesicles (or only one) have fused with the bilayer, the *cis* side is usually perfused with the solution of the *trans* side to reestablish symmetrical conditions and to suppress further fusion events. Then the incorporated ion channels can be investigated in detail.

Besides the conditions described above, additional technical effort is necessary to get reproducible fusion rates. The following section describes prerequisites that are useful for controlling the fusion process. Together with each method, commentary follows on the type of bilayer used and on present examples of ion-channel fluctuation traces obtained with it. The methods are distinguished by the maneuvers used to control the fusion rate.

3.1. Fusion of Vesicles with Planar Lipid Bilayers Containing Negatively Charged Lipids in the Presence of Calcium in the Aqueous Solution

One of the first methods used to fuse vesicles with planar lipid bilayers used bilayers containing negatively charged lipids [phosphatidylserine (PS), phosphatidic acid (PA), phosphatidylglycerol (PG)]. In addition to this and to the general prerequisites of fusion experiments mentioned above, millimolar amounts of calcium have to be present in the aqueous solution of the *cis* side (Miller and Racker, 1976; Miller, 1978; Labarca *et al.*, 1980). The fusion rate in such experiments can be controlled by the amount of negatively charged lipids in the bilayer, by the calcium concentration, and by the parameters mentioned above. Figure 1 shows the fusion of vesicles from the sarcoplasmic reticulum containing a potassium-selective ion channel with a planar bilayer under such conditions. In part B of this figure, the fluctuation of the incorporated ion channels is shown in some detail. It is demonstrated furthermore that the amount of solvent present in the planar bilayer does not necessarily affect the ion channel in a dramatic way.

Besides the example given, many other ion channels have been reconstituted by this method, mainly using decane-containing painted bilayers. In some experiments "solvent-free" folded bilayers have been used, too. An example is shown in Fig. 1B, trace B, for the potassium-selective channel from the SR. As pointed out, the channel itself was not significantly affected by the type of bilayer used, but the kinetics of the channel gating were slightly faster in the virtually solvent-free case (Labarca *et al.*, 1980). The fusion rates, however, were found to be significantly smaller when folded bilayers were used. A detailed analysis of the effects of organic solvent on the channel proteins and on the fusion process itself is still needed (Cohen *et al.*, 1984), as is discussed in Chapter 6.

It is difficult to develop mechanistic models of the fusion process itself. It is known that calcium binds to phosphatidylserine bilayers and leads to a phase separation (Papahadjopoulos *et al.*, 1977). The phase boundaries thus formed may be significant to create fusogenic spots when the water between bilayer and vesicle membrane has been shifted away. Alternatively, positively charged polypeptides have been used instead of Ca^{2+} to promote fusion in such experiments with good results. It has been shown too that micromolar calcium stimulates fusion of vesicles with a planar bilayer containing a high-affinity calcium-binding protein (Zimmerberg *et al.*, 1980a).

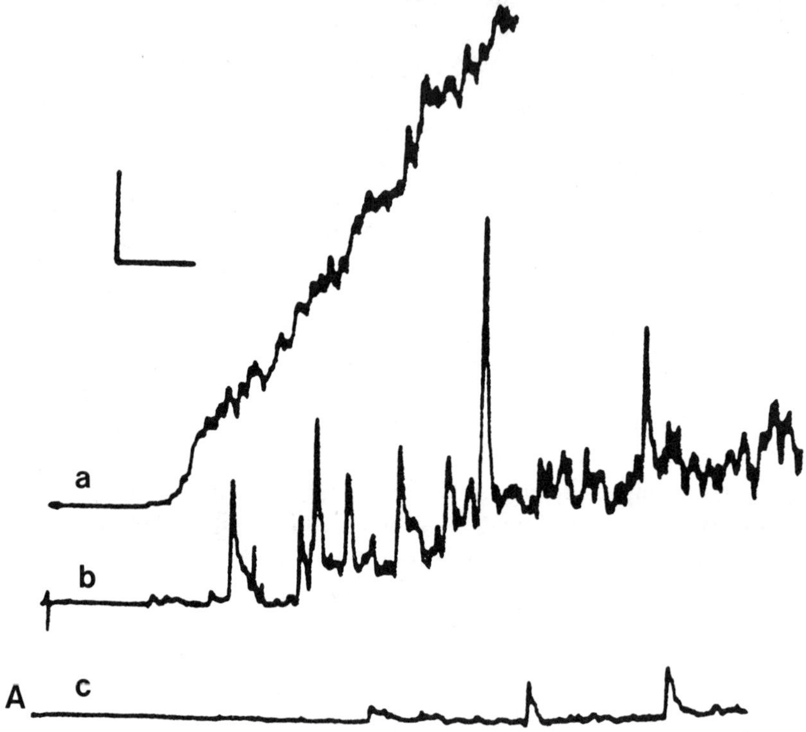

Figure 1. (A) The fusion of many SR membrane fragment vesicles with a planar bilayer. The vesicles contain a potassium-selective channel in their membranes, increasing the membrane permeability to potassium after each fusion event in an aqueous solution containing 0.1 M potassium. Each fusion event is accompanied by a current transient, which is to be seen best in trace (c) at low fusion rates. In all cases the bilayer was made of 70% washed asolectin and 30% DPG. The membrane current was monitored at a potential of −50 mV. Vesicles were added at the beginning of each trace, and the fusion is shown at different conditions: (a), highly favorable fusion conditions with 1 mM calcium, 200 mM and glycerol, and SR (50 μg/ml) added to the *cis* side (scale bars are 10 $\mu S/cm^2$ and 20 sec); (b), normal fusion conditions with 0.5 mM calcium and SR (10 μl/ml) added to the *cis* side (scale bars are 0.5 $\mu S/cm^2$ and 20 sec); (c), reduced fusion conditions with only SR (1 μg/ml) added to the *cis* side, otherwise as in (b). The dependence of the fusion rates on calcium and vesicle concentration can be seen clearly from these traces.

3.2. Fusion of Vesicles with Planar Lipid Bilayers Containing Large Amounts of Phosphatidylethanolamine Lipids

It was found that vesicles fuse with planar bilayers only if these contain large amounts of phosphatidylethanolamine (PE) lipids (Miller and Racker, 1976; Latorre *et al.*, 1982; Hanke, 1984). The fusion rates here depend on the amount of PE lipids in the bilayer in addition to the general factors mentioned above. To enhance the stability of the bilayer, a small amount, 10–20%, of phosphatidylcholine (PC) is included in the membrane-forming solution. Experiments with such bilayers have been carried out with both painted and folded planar bilayers

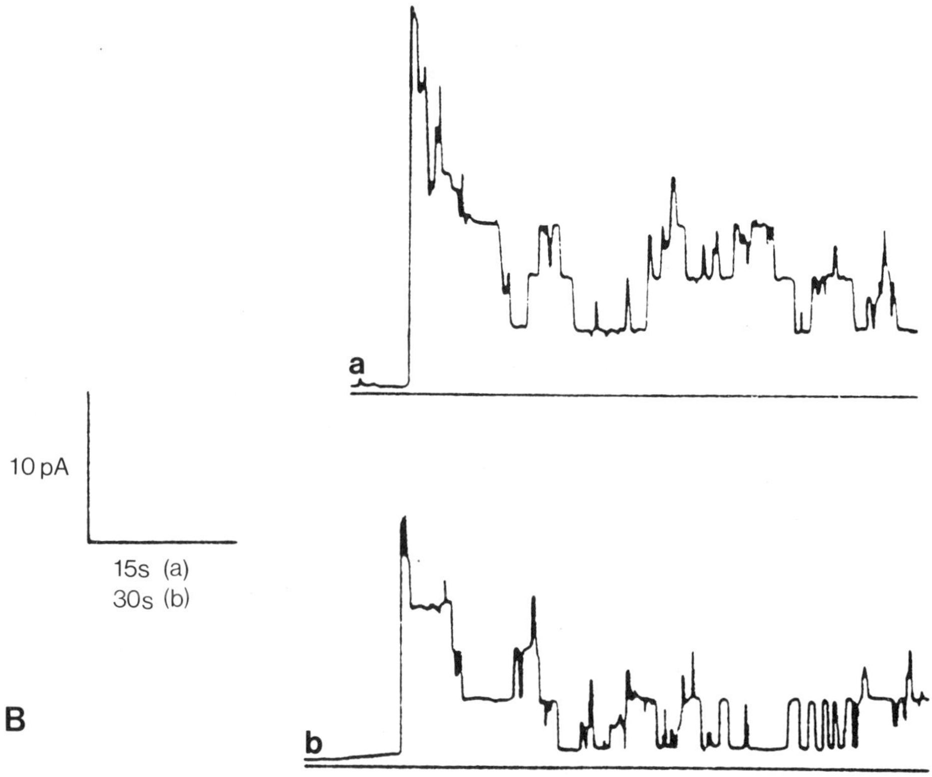

Figure 1. (B) The fine structure of fusion spikes is shown in more detail in folded (a) and painted (b) bilayers. The fluctuation of single channels after the fusion transient is clearly resolved. Both membranes were made from 70% washed asolectin and 30% PA. Fusion was monitored at +20 mV under conditions otherwise as given in part A. The channel fluctuations are nearly independent of the type of bilayer used, but the kinetics are slightly faster in folded bilayers. (Both parts reproduced from Labarca *et al.*, 1980.)

(Hanke, 1984; Moczydlowski *et al.*, 1984). Very often, the reconstituted ion channels were not significantly affected by the type of bilayer; this does not hold for all ion-channel-forming proteins known, however. The fusion rates were much smaller when folded bilayers were used in any case (Cohen *et al.*, 1984).

Figure 2 shows traces from the same type of ion channel reconstituted into both painted and folded bilayers. It is the sodium channel protein purified from rat brain (Hanke *et al.*, 1984a). During the preparation, the channel has apparently lost its inactivation process, but the other properties of this channel as known from electrophysiology have been conserved, giving us a convenient way to study it. In both types of traces shown in Fig. 2, membranes made from 75% egg PE, 20% 1-palmitoyl-2-oleoyl-3-glycero-PC (POPC), and 5% cholesterol were used together with an aqueous solution containing 150 mM sodium plus other ions. In both types of planar bilayers, the channel exhibits about the same conductance and kinetic properties. Again, the ion-channel-forming protein was found not to

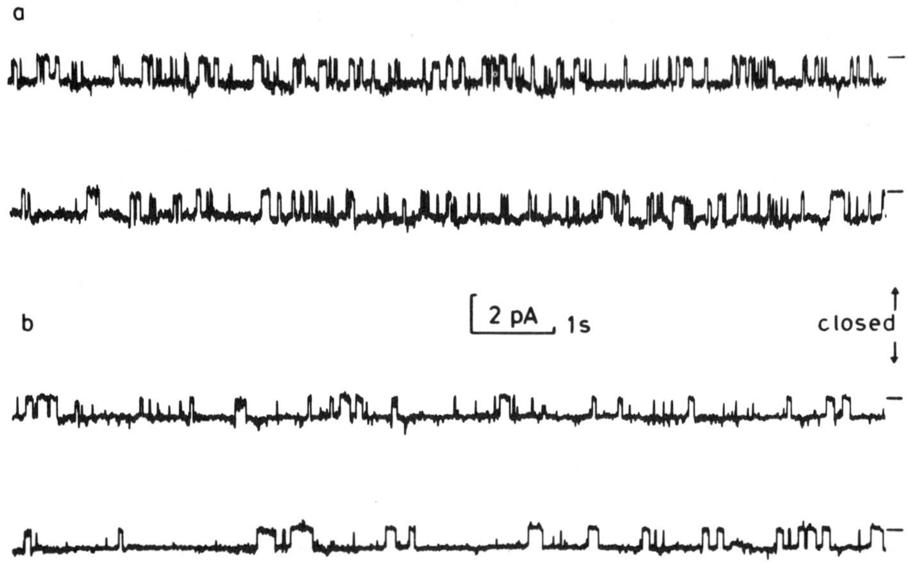

Figure 2. (a) The fluctuation of a single sodium channel purified from rat brain is shown after incorporation into a folded bilayer made from 75% egg PE, 20% POPC, and 5% cholesterol. The aqueous solution contained 150 mM sodium plus other ions. The vesicles were hyperosmotic to the aqueous solution with an additional 200 mM sucrose. A potential of −35 mV was applied to the membrane, and the channel type shown here has a conductance of 28 pS. (b) Under conditions identical to those given in a, fluctuations of the same channel type are shown in a painted bilayer. The applied potential was −30 mV. No significant changes in the channel properties were found compared to a except that the channel kinetics were slightly faster in b. The closed state is indicated for each trace. For details of this channel see Hanke *et al.* (1984a).

be significantly affected by the amount of organic solvent present in the planar lipid bilayer, but the fusion rates, including calcium-induced fusion, were much higher in the painted bilayers.

Not very much is known about the mechanism by which PE lipids induce fusion. From P-NMR, calorimetric, and structural studies, it is known that PE lipids may undergo a phase transition from a lateral phase to a hexagonal, nonbilayer structure. If this happens in planar bilayers, areas of nonbilayer structure may create "fusogenic" instabilities and phase boundaries. When the bilayer is stabilized by smaller amounts of lipid in the lateral structure, both lateral and hexagonal phases may coexist. The hexagonal structures or so-called lipidic particles created by PE lipids in planar bilayers thus may be the reason for the increase of fusion rates found with these types of bilayers. Evidence for this has been found in pure vesicle fusion experiments (Verkleij *et al.*, 1980; Verkleij, 1984).

3.3. Fusion of Vesicles with Bilayers in the Frozen State

In vesicle–vesicle fusion experiments, it has been shown that fusion of membranes is highly favored when the membranes of these vesicles are made from

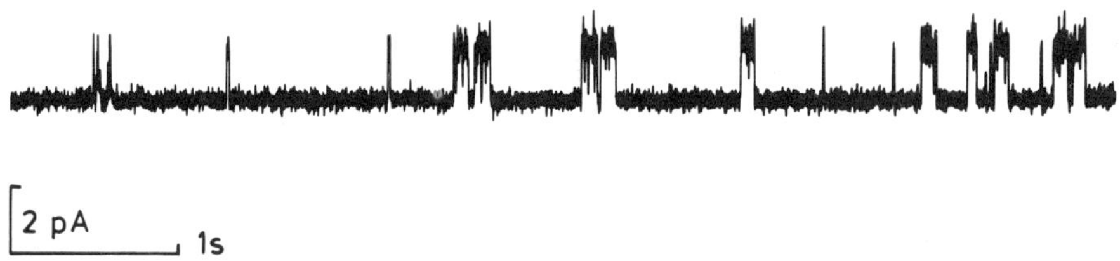

Figure 3. Fluctuations of a potassium-permeable channel from *Paramecium* cilia membranes are shown in a folded bilayer made from 1,3-SMPC (T_0 = 30°C). The temperature of the experiment was 22°C, well below T_0, and the applied potential was +100 mV. The aqueous solution contained 100 mM potassium plus other ions, and the vesicles were hyperosmotic with an additional 500 mM sucrose. A conductance of about 15 pS was found for the channel. For details of this channel see Boheim *et al.* (1982).

pure lipids in the range of the phase-transition temperature (Schullery *et al.*, 1980). For a long time it was technically impossible to form planar bilayers in the range of the phase-transition temperature of the lipid used. However, Boheim and co-workers (Boheim *et al.*, 1980) succeeded in forming stable planar bilayers in the frozen state by employing synthetic mixed-chain (saturated) PC lipids to form folded planar lipid bilayers below the phase transition temperature. These bilayers were stable in a wide temperature range above, at, and below the phase transition. Only under these specific conditions (i.e., folded bilayers and saturated mixed-chain PC lipids) have such experiments been successfully carried out.

It was shown later that it is possible to fuse protein-containing vesicles with bilayers in the frozen state (Hanke *et al.*, 1981; Boheim *et al.*, 1982), thus leading to the incorporation of proteins into frozen planar bilayers. After fusion, the physical state of the planar bilayer is, of course, no longer a well-defined single-component system, since the entire content of the vesicle membranes is incorporated into it.

In Fig. 3 traces of a cationic channel from paramecium ciliary membranes (Hanke *et al.*, 1981; Boheim *et al.*, 1982) are shown after reconstitution of the ion-channel-forming protein into a planar bilayer made from 1-stearoyl-3-myristoyl-2-glycero-PC (1,3-SMPC) at 22°C. The phase transition of this lipid is about 30°C. It is surprising that these ion channels exhibit nearly unchanged gating in a frozen lipid matrix. This can be explained best by the assumption of a still-fluid boundary lipid around the proteins or by effects of the incorporated protein on the physical state of the boundary lipids. On the other hand, it is possible that the gating of these channels does not involve a gross interaction with membrane lipids, but merely a small movement of groups within the channel protein.

The fusion process itself may be explained by the coexistence of fluid and frozen phases in the planar bilayer in the range of the phase-transition temperature (T_c) and somewhat below it. Again, phase boundaries are postulated to create instabilities and fusiogenic spots, which are a prerequisite for fusion. This idea is supported by the finding that the fusion rates in all experiments are high at T_c, become smaller with decreasing temperature, and are negligible above T_c. Iden-

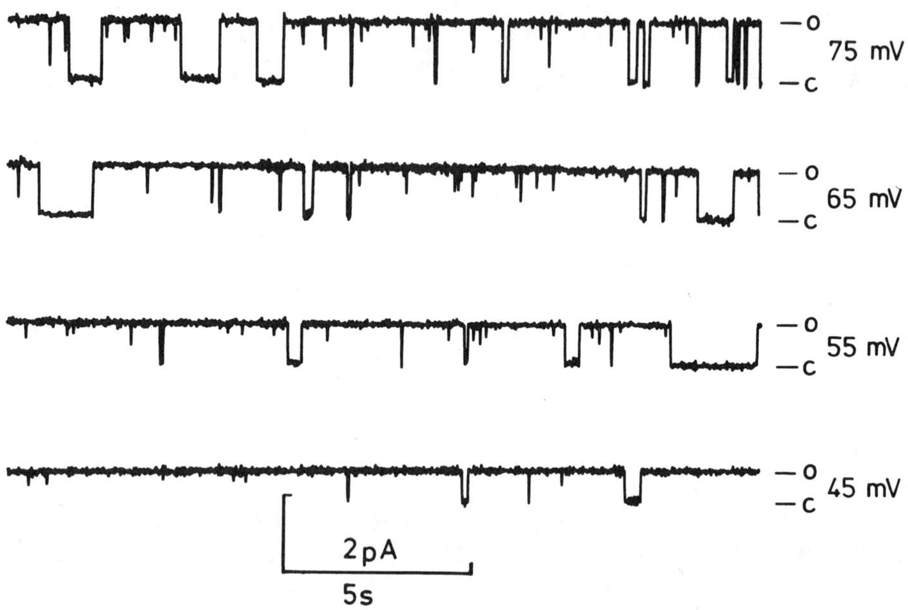

Figure 4. Traces of a sodium-permeable channel from bovine rod outer segment membranes are given after fusion of membrane fragment vesicles with a planar bilayer made from 50% 1,3-SMPC and 50% egg PE at 22°C; 150 mM sodium was present in the aqueous solution, and the vesicles were hyperosmotic with an additional 600 mM sucrose. For details see Hanke and Kaupp (1984).

tical results were found in vesicle–vesicle fusion experiments (see, for example, Schullery *et al.*, 1980).

3.4. Mixed Fusion Methods

Of course it is possible not only to use the purer methods described above but to combine some of them to make use of the advantages of more than one. This is done when special requirements of the lipid composition apply or when the fusion rates must be increased by any possible trick. Very often in a series of experiments the researcher has found a special set of parameters that work best with his preparation simply to be a mixture of different fusion methods. This way to solve problems results from a lack of sufficient understanding about the fusion process and about the lipid dependence of channel function, as is pointed out later.

Presented here is only one example of such a finding. Recently, Hanke and Kaupp (1984) reconstituted channels from bovine rod outer segment membranes into planar bilayers. The method they used was that of fusing membrane fragment vesicles with preformed folded bilayers. The experiments were found to work best with bilayers made from 50% egg PE and 50% 1,3-SMPC (to about 30°C), so they used a mixture of the methods of Sections 3.2 and 3.3. Of course, nothing detailed can be said about the physical state of the bilayers they used because they were made as mixtures of fluid and frozen lipids and additional protein was incorporated as a consequence of the fusion of vesicles with the bilayer. Indeed, frozen, fluid, and hexagonal phases may all coexist in one planar bilayer.

Fig. 4 presents some traces at different potentials of sodium-permeable channels from bovine rod outer segment membranes reconstituted into such planar bilayers (Hanke and Kaupp, 1984). In some of their parameters, these channels resemble those shown in Fig. 2 (conductance, selectivity), whereas in others, they are totally different from those (kinetics, pharmacology).

4. General Remarks on the Incorporation of Ion Channels by Fusion

4.1. Comparison of the Different Fusion Methods

As pointed out above, such fusion method has its own advantages and disadvantages. In any case, the question has to be addressed whether one should use painted or folded bilayers. Although decane increases the fusion rates significantly, organic solvents are not part of biological membranes, and the study of their effects of ion-channel function is usually not the goal of an experiment. So the following conclusion can be made: whenever possible, the amount of organic solvents in the planar bilayer should be kept small; otherwise it has to be demonstrated in each case whether the organic solvent affects the protein or not.

4.2. How to Control Fusion Rates

In all ion-channel reconstitution experiments using the fusion method, the most critical point is to control the fusion process itself. We know that fusion rates depend on some general experimental parameters, as listed:

1. Vesicle concentration in the aqueous solution
2. Area of the planar bilayer
3. Amount of hyperosmolarity of the *cis* side or the vesicles
4. Amount of organic solvent in the planar bilayer.

If any of these parameters is increased, the fusion rates are increased (Zimmerberg *et al.*, 1980b; Cohen *et al.*, 1980, 1984). Additional parameters to control and increase the fusion rates are:

5. The amount of negatively charged lipid in the planar bilayer together with the concentration of calcium in the aqueous solution
6. The amount of PE-lipids in the planar bilayer
7. The temperature when a planar bilayer in the frozen state is used.

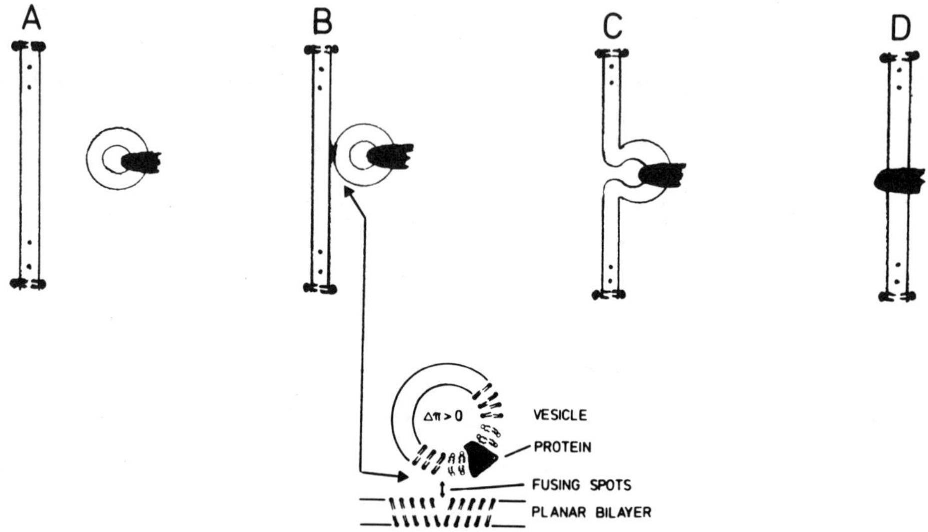

Figure 5. A cartoon of the fusion of a vesicle with a planar bilayer. First, vesicle and planar bilayer coexist well separated. (A); they come into close contact and remain associated for a period of time (B). Then they fuse (B → C), and the vesicle membrane becomes an omega-shaped invagination of the planar bilayer (C). The bilayer becomes a flat membrane again (D). At this time, the entire contents of the vesicle membrane have become part of the planar bilayer, i.e., all proteins and lipids of the vesicle membrane are incorporated into the planar bilayer.

All these control mechanisms, however, only give a qualitative control of fusion rates, and it remains an unhappy circumstance that quantitative manipulation of vesicle fusion into planar bilayers is still an elusive goal (Zimmerberg *et al.*, 1980b; Cohen *et al.*, 1980, 1984).

Another important problem that only rarely has been addressed is the question of to what extent the vesicle preparation itself affects the fusion rates. Many researchers have reported that under otherwise identical conditions one preparation fuses quite well but another does not. This finding is not surprising, as we have shown that the composition of the planar bilayer affects the fusion rates significantly, and in principle the problem is symmetric with respect to the vesicle membrane composition.

4.3. Models of the Fusion Process Itself

It is one thing to know how to use a technique but something totally different to understand the basic mechanisms involved in it. This is the situation for the incorporation of ion-channel proteins into planar bilayers by fusion. The technique is known and has been used with success. The fusion process itself, however, is not understood.

In Fig. 5 I present a cartoon of our view of the fusion of a vesicle with a planar bilayer. After addition of the vesicles to the aqueous solution, bilayer and vesicle coexist well separated (A). They aggregate, and for a certain time the two bilayers coexist in close contact (B). By a process we do not understand, the two bilayers fuse, and the membrane of the vesicle becomes an invagination of the planar bilayer (C). The bilayer then flattens out to minimize surface energy. All of these events are easy to understand except the moment when the two bilayers fuse. Available data suggest that instabilities in the planar bilayer are required to start the fusion process (phase boundaries, hexagonal structures, clustered negatively charged lipids, or others) (Hoeckstra, 1982). Additionally, it seems to be necessary to exclude the water from between the two associated membranes before fusion can occur (Markin and Kozlow, 1983; Kozlow and Markin, 1983). Recently, a new technique has been published (Markin and Kozlow, 1983; Kozlow and Markin, 1983) to study directly the fusion of two planar bilayers. This may help to get some deeper understanding of the principles involved.

5. Conclusion and Discussion

The above short outline has addressed some of the problems and possibilities of ion-channel reconstitution by fusion. Since most ion-channel protein preparations are isolated as vesicles with the proteins incorporated in their membranes, these proteins have to be transferred into a planar lipid bilayer somehow. The fusion method to incorporate proteins into planar bilayers starts directly with the vesicles in which the proteins normally reside. Since fusion itself probably does not affect the proteins, this method seems to be the gentlest way to bring about protein insertion. One has to live, however, with some disadvantages of the method. The choice of the lipid composition of the planar lipid bilayer is not free. The fusion rates can only be controlled qualitatively, and the underlying process is not understood. If one is only interested in exploiting this method for studying ion channels in planar bilayers, one can live with these problems. Fusion of membranes itself, however, is a much more general process in biology. The study of the fusion process, therefore, should be one of the major goals of future work, as it would help to better control the technique described, and it would bring some deeper understanding of numerous biological problems.

References

Boheim, G., Hanke, W., and Eibl, H., 1980, Lipid phase transition in planar lipid bilayer membrane and its effects on carrier and pore mediated transport, *Proc. Natl. Acad. Sci. U.S.A.* **77**:3403–3407.

Boheim, G., Hanke, W., Methfessel, C., Eibl, H., Kaupp, U. B., Maelicke, A., and Schultz, J. E., 1982, Membrane reconstitution below lipid phase transition temperature, in: *Transport in Biomembranes* (P. Spach ed.), pp. 87–97, Raven Press, New York.

Cohen, F. S., Finkelstein, A., and Zimmerberg, J., 1980, Fusion of phospholipid vesicles with planar phospholipid bilayer membranes. II. Incorporation of a vesicular membrane marker into the planar membrane, *J. Gen. Physiol.* **75**:251–270.

Cohen, F. S., Akabes, M. H., Zimmerberg, J., and Finkelstein, A., 1984, Parameters effecting the fusion of unilamellar phospholipid vesicles with planar lipid bilayers, *J. Cell Biol.* **98:**1054–1062.

Coronado, R., and Latorre, R., 1983, Formation of phospholipid bilayers on patch-clamp pipettes, *Biophys. J.* **43:**231–236.

Duezguenz, N., and Ohki, S., 1980, Fusion of small unilamellar liposomes with phospholipid planar bilayer membranes and large single bilayer vesicles, *Biochim. Biophys. Acta.* **640:**734–747.

Hamill, O. P., Marty, A., Neher, E., Sakmann, B., and Sigworth, F. J., 1981, Improved patch-clamp technique for high resolution current recording from cells and cell-free patches, *Pfluegers Arch.* **391:**85–100.

Hanke, W., 1984, Reconstitution of channel proteins from *Torpedo* electroplax into virtually solvent free planar lipid bilayers, *Biochem. Bioeng. J.* **12:**341–151.

Hanke, W., and Kaupp, U. B., 1984, Incorporation of ion channels from bovine rod outer segments into planar lipid bilayers, *Biophys. J.* **46:**587–595.

Hanke, W., Eibl, H., and Boheim, G., 1981, A new method for membrane reconstitution: Fusion of protein containing vesicles with planar lipid bilayers below lipid phase transition temperature, *Biophys. Struct. Mech.* **7:**131–137.

Hanke, W., Boheim, G,. Barhanin, J., Pauron, D., and Lazdunski, M., 1984a, Reconstitution of highly purified saxitoxin sensitive sodium channels into planar lipid bilayers, *EMBO J.* **3:**509–515.

Hanke, W., Methfessel, C., Wilmsen, H. U., and Boheim, G., 1984b, Ion channel reconstitution into planar lipid bilayers on glass pipettes, *Biochem. Bioeng. J.* **12:**329–339.

Hartshorne, R. P., Keller, B. U., Talvenheimo, J. A., Catterall, W. A., and Montal, M., 1984, Functional reconstitution of the purified sodium channel into planar lipid bilayers (in press).

Hoeckstra, D., 1982, Role of lipid phase transition and membrane hydration in phospholipid vesicle fusion, *Biochemistry* **21:**2833–2840.

Kazlow, M. M., and Markin, V. S., 1983, Cell biophysics, possible mechanism of membrane fusion, *Biophysics* **2:**255–261.

Labarca, P., Coronado, R., and Miller, C., 1980, Thermodynamic and kinetic studies of the gating of a potassium selective channel from sarcoplasmic reticulum, *J. Gen. Physiol.* **76:**397–424.

Latorre, R., Vergara, C., and Hidalgo, C., 1982, Reconstitution in planar lipid bilayers of a calcium dependent potassium channel from tranverse tubule membranes isolated from rabbit skeletal muscle, *Proc. Natl. Acad. Sci. U.S.A.* **79:**805–809.

Markin, W. S., and Kozlow, M. M., 1983, Primary act in the process of membrane fusion, *Biophysics* **28:**73–78.

Miller, C., 1978, Voltage gated cation conductance channel from fragmented sarcoplasmic reticulum: Steady state electrical properties, *J. Membr. Biol.* **40:**1–23.

Miller, C., and Racker, R., 1976, Calcium induced fusion of fragmented sarcoplasmic reticulum with artificial planar bilayers, *J. Membr. Biol.* **30:**283–300.

Miller, C., and Rosenberg, R. C., 1979, Modification of a voltage gated potassium channel from sarcoplasmic reticulum by a pronase derived specific endopeptidase, *J. Gen. Physiol.* **74:**457–478.

Moczydlowski, E., Garber, S. G., and Miller, C., 1984, Batrachotoxin activated sodium channels in planar lipid bilayers. Competition of tetrodotoxin by sodium, *J. Gen. Physiol.* **84:**665–686.

Montal, M., and Mueller, P., 1972, Formation of bimolecular membranes from lipid monolayers and a study of their electrical properties, *Proc. Natl. Acad. Sci. U.S.A.* **69:**3561–3566.

Mueller, P., Rudin, D. O., Ti Tien, H., and Wescott, W. C., 1962, Reconstitution of cell membrane structure *in vitro* and its transformation into an excitable system, *Nature* **194:**979–980.

Papahadjopoulos, D., Vail, W. J., Newton, G., Nir, S. M., Jacobson, K., Poste, G., and Lazo, R., 1977, Studies on membrane fusion. III. The role of calcium induced phase changes, *Biochim. Biophys. Acta* **465:**579–598.

Schindler, H., and Rosenbusch, J. P., 1978, Matrix protein from *Escherichia coli* outer membrane forms voltage-controlled channels in lipid bilayers, *Proc. Natl. Acad. Sci. U.S.A.* **75:**3751–3755.

Schullery, S. E., Schmidt, C. F., Felgner, P., Tillack, T. W., and Thompson, T. E., 1980, Fusion of dipalmitoylphosphatidylcholine vesicles, *Biochemistry* **19:**3919–3923.

Tank, D., Miller, C., and Webb, W. W., 1982, Isolated patch recording from liposomes containing

functionally reconstituted chloride channels from electroplax, *Proc. Natl. Acad. Sci. U.S.A.* **79**:7749–7753.
Verkleij, A. J., 1984, Lipidic intramembraneous particles, *Biochim. Biophys. Acta* **779**:43–63.
Verkleij, A. J., Van Echtfeld, C. J. A., Gerritsen, W. J., Cullis, P. R., and DeKruiff, B., 1980, The lipidic particle as an intermediate structure in membrane fusion process and bilayer to hexagonaL HII transitions, *Biochim. Biophys. Acta* **600**:620–624.
Wilmsen, H. U., Methfessel, C. Hanke, W., and Boheim, G., 1983, Channel current fluctuation studies with solvent free planar lipid bilayers using Neher–Sakmann pipettes, in: *Physical Chemistry of Transmembrane Ion Motion* (G. Spach, ed.), pp. 479–485, Elsevier, Amsterdam.
Zimmerberg, J., Finkelstein, A., and Cohen, F. S., 1980a, Micromolar calcium stimulates fusion of lipid vesicles with planar bilayers containing a calcium-binding protein, *Science* **210**:906–908.
Zimmerberg, J., Finkelstein, A., and Cohen, F. S., 1980b, Fusion of phospholipid vesicles with planar phospholipid bilayer membranes. I. Discharge of vesicular contents across the planar membrane, *J. Gen. Physiol.* **75**:241–250.

Part II

NICOTINIC ACETYLCHOLINE RECEPTOR

Chapter 8

THE RECONSTITUTED ACETYLCHOLINE RECEPTOR

MAURICIO MONTAL, ROBERT ANHOLT, and PEDRO LABARCA

1. Introduction

The nicotinic acetylcholine receptor (AChR) is the postsynaptic membrane protein that transduces the binding of acetylcholine (ACh) into the transient opening of a cation-selective channel. This signal-transduction event is the fundamental process in the transfer of information across synaptic junctions in excitable tissues (reviewed by Changeux, 1981; Changeux *et al.*, 1984; Anhold *et al.*, 1984; Karlin, 1980; Conti-Tronconi and Raftery, 1982).

Classical protein biochemistry together with the application of monoclonal antibody technology and recombinant DNA technology have contributed greatly to our present understanding of the molecular structure of the AChR. The salient structural features of the AChR are summarized schematically in Fig. 1 and considered in some detail by our collaborator Jon Lindstrom in Chapter 9. Briefly, the AChR from *Torpedo* is a ~250,000-dalton pentamer composed of four homologous glycopeptides present as an $\alpha_2\beta\gamma\delta$ complex (Reynolds and Karlin, 1978; Lindstrom *et al.*, 1979; Raftery *et al.*, 1980, 1983; Noda *et al.*, 1983b). Each AChR monomer contains two nonequivalent ACh binding sites distinguishable by their affinity for antagonists and their reactivity with affinity labels such as the agonist analogue bromoacetylcholine (Damle *et al.*, 1978) or the antagonist 4-(N-maleimido)benzyltrimethyl ammonium iodide (MBTA) (Weill *et al.*, 1974, Karlin *et al.*, 1975), which attach to the α subunits (Karlin, 1980; Karlin *et al.*, 1983; Changeux *et al.*, 1984; Kao *et al.*, 1984). In the native membrane of *Torpedo*, the AChR occurs predominantly as a dimer formed by a disulfide bridge between the δ subunits of adjacent monomers (Chang and Bock, 1977; Hamilton *et al.*, 1979).

All four AChR subunits are transmembrane polypeptides (Tarrab-Hazdai *et*

MAURICIO MONTAL, ROBERT ANHOLT, and PEDRO LABARCA • Departments of Biology and Physics, University of California San Diego, La Jolla, California 92093. Present addresses: M. M., Department of Neurosciences, Roche Institute of Molecular Biology, Nutley, New Jersey 07110; R. A., Department of Neuroscience, Johns Hopkins University School of Medicine, Baltimore, Maryland 21205; P. L., Laboratorio de Neurofisiologia, Universidad Catolica de Chile, Casilla 114-D, Santiago, Chile.

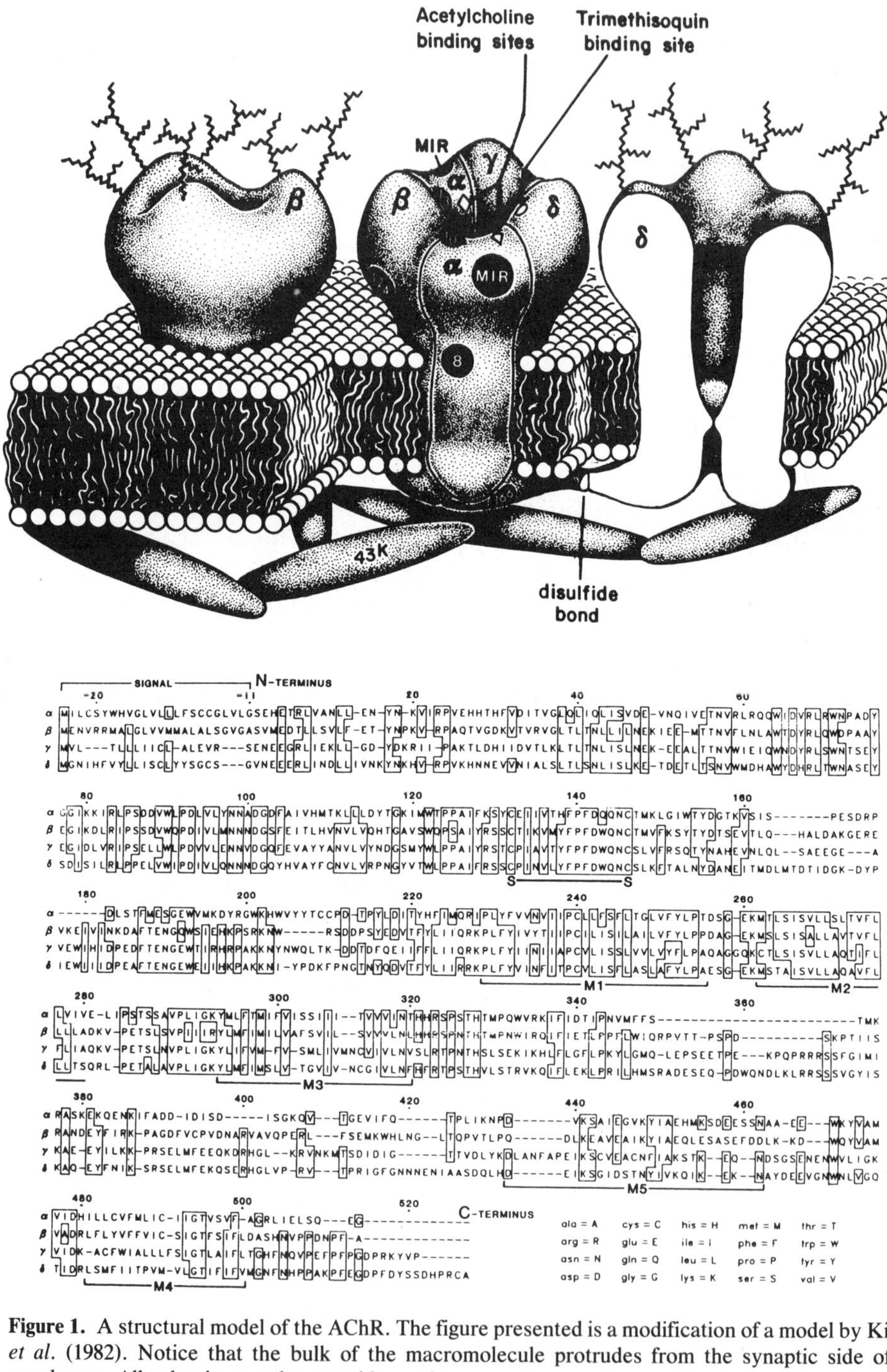

Figure 1. A structural model of the AChR. The figure presented is a modification of a model by Kistler *et al.* (1982). Notice that the bulk of the macromolecule protrudes from the synaptic side of the membrane. All subunits are glycopeptides and traverse the membrane at least once. Receptor monomers are linked as dimeric units via a disulfide bond between the δ subunits. The α subunits are shown to contain the ACh binding sites (Weill *et al.*, 1974), main immunogenic region (MIR), and

al., 1978; Strader *et al.*, 1979; Wennogle and Changeux, 1980; Strader and Raftery, 1980; Froehner, 1981), and the complete amino acid sequences of all four subunits have been deduced from the nucleotide sequences of cDNA clones (Fig. 1) (Noda *et al.*, 1982, 1983a,b; Claudio *et al.*, 1983; Devilliers-Thiery *et al.*, 1983). Analysis of hydrophobic sequences has led to the identification of several putative transmembrane segments for each of the subunits (Fig. 1, Noda *et al.*, 1983b; Claudio *et al.*, 1983; Devilliers-Thiery *et al.*, 1983; Guy, 1984; Fairclough *et al.*, 1983; Finer-Moore and Stroud, 1984; Young *et al.*, 1985; Criado *et al.*, 1985). Since the carboxyl and amino termini of the AChR subunits appear to be exposed on opposite sides of the membrane, each subunit must traverse the membrane an odd number of times (Lindstrom *et al.*, 1984; Finer-Moore and Stroud, 1984; Guy, 1984; Young *et al.*, 1985). The exact transmembrane quaternary arrangement of the five subunits as well as the precise conformation of the cation channel and its mechanism of agonist-induced gating remain unclear (Lindstrom, Chapter 9).

Functional reconstitution of the nicotinic AChR in model membranes enables investigation of its function *in vitro* under conditions in which both the protein and lipid composition of the reconstituted membrane can be varied in a controlled manner along with the concentrations of agonist and ions in the medium. Furthermore, the isolated AChR can be manipulated and modified biochemically prior to or after reassembly, and the effects of such treatments on AChR function can then be assessed in the reconstituted membrane. Thus, a reconstituted system allows the investigation of structure–function relationships within the receptor molecule and may ultimately contribute to a complete understanding of the molecular events that underlie ACh-regulated opening of the postsynaptic channel.

---

binding sites for monoclonal antibodies 10 and 13 (Tzartos and Lindstrom, 1980; see also Chapter 9 by Lindstrom). The position of monoclonal antibody 124 at a sterically masked location is tentative. The binding site of monoclonal antibody 8 near the membrane interphase is based on Gullick *et al.* (1981), and the intracellular position of the antigenic determinant to which monoclonal antibody 166 binds is according to Anderson *et al.* (1983) and Gullick and Lindstrom (1983). A local anesthetic binding site has been indicated on the extracellular surface of the δ subunit according to Wennogle *et al.* (1981). Subunit topography is indicated as proposed by Kistler *et al.* (1982). Relative dimensions are based on Ross *et al.* (1977) and Wise *et al.* (1979). It should be noted that the shapes of the subunits are merely schematic and hypothetical. The positions of antigenic regions and ligand binding sites are, similarly, only schematic illustrations and are not intended to indicate precise topographical locations. The amino acid sequences of the four subunits deduced by Noda *et al.* (1983b) from the corresponding nucleotide sequences of cloned cDNAs are presented in the lower portion of the figure. The alignment for maximum homology is according to Noda *et al.* (1983b). The boxes indicate the occurrence of identical amino acids at homologous positions. Notice that each subunit possesses a signal peptide. In this alignment, the proposed active-site disulfide would be formed between cysteines 132 and 146 of the α subunit, and the proposed δ–δ disulfide might be formed via the cysteine at position 526. Four transmembranous hydrophobic α-helical domains proposed by Noda *et al.* (1983b), Claudio *et al.* (1983), and Devilliers-Thiery *et al.* (1983) are indicated M1, M2, M3, and M4. In addition, Finer-Moore and Stroud (1984) and Guy (1984) have proposed a fifth transmembranous domain in each subunit, indicated in the figure as M5. This domain consists of an amphipatic α helix that contains charged amino acids on one side and apolar amino acids on the other side, such that alignment of these corresponding amphipatic domains of the five subunits would form a transmembrane channel with a hydrophilic core (Finer-Moore and Stroud, 1984; Guy, 1984; see also Mishina *et al.*, 1985). In this scheme, the N termini would be located on the synaptic side of the AChR, whereas the C termini would be intracellular. (From Anholt *et al.*, 1984.)

The aim here is to review the techniques and concepts that have evolved during the study of the reconstituted AChR.

2. Reconstitution of Acetylcholine Receptors in Lipid Vesicles

2.1. Preservation of the Functional Integrity of the Agonist-Regulated Cation Channel of the AChR during Solubilization of the Electric Organ Membrane

Although once a challenging problem, solubilization of the AChR and its reincorporation into lipid vesicles with full retention of its functional properties is now a routine procedure (for reviews, see Anholt, 1981; McNamee and Ochoa, 1982; Anholt *et al.,* 1983).

The main principle for the successful reconstitution of functional AChR is the notion that the integrity of its agonist-regulated ion channel is critically dependent on the continuous presence of stabilizing lipid (Epstein and Racker, 1978; Huganir *et al.,* 1979; Huganir and Racker, 1982; Changeux *et al.,* 1979; Lindstrom *et al.,* 1980; Anholt *et al.,* 1981; Popot *et al.,* 1981; Wu and Raftery, 1979). This lipid dependence is illustrated in Fig. 2. Only about 10% of the initial cation channel activity survives when extracts of native membranes (1.67 mg/ml protein) are incubated in 2% cholate. This residual activity results from protection by endogenous lipids in the extract. However, the presence of relatively low amounts of supplementary soybean lipids preserves channel function. Similarly, Heidmann *et al.* (1980) showed that the characteristic allosteric properties of the membrane-bound AChR are preserved by the presence of protective lipids in the micellar solution, whereas the capacity of the AChRs to bind $[^{125}I]\alpha$-bungarotoxin (α-BGT) is not affected.

As shown in Fig. 3A, extraction of native membranes is maximal at 2% cholate and amounts to a recovery of about 68% of the initial AChRs in the extract at cholate/lipid ratios of 10:1 (w/w). A class of M_r 43,000 proteins anchors AChRs to the cytoskeleton (Sobel *et al.,* 1977; Cartaud *et al.,* 1981; Sealock, 1982; Froehner *et al.,* 1981; Gysin *et al.,* 1981; St. John *et al.,* 1982). Alkaline extraction methods, which remove these proteins (Neubig *et al.,* 1979; Moore *et al.,* 1979; Barrantes *et al.,* 1980) do not increase the extraction yield.

When detergent concentrations above 3% cholate are used, a strong irreversible inactivating effect of the detergent on the channel becomes evident, even though the same 10:1 (w/w) cholate/lipid ratio is maintained (Fig. 3B). This channel inactivation observed at elevated detergent concentrations can be accounted for by the high concentration of lipid-free cholate micelles, which displace the lipid protecting the AChR channel. In view of the 20-fold molar excess of cholate over lipid in the micellar solution under optimal conditions (2% cholate, 2 mg/ml of soybean lipid), it is evident that the affinity of the lipids for the AChR

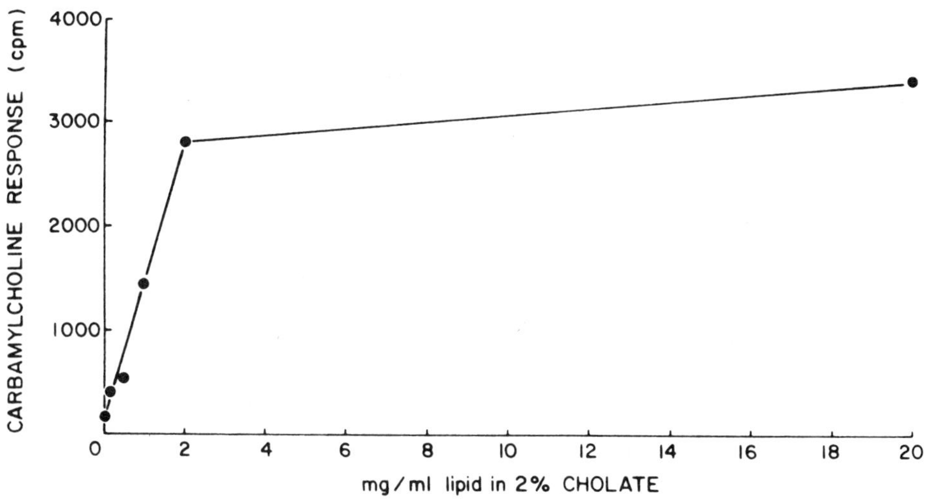

Figure 2. Protection of ion channel function during solubilization of AChRs in the presence of supplementary soybean lipids. Pelleted AChR-rich electric organ membranes (1.67 mg of protein) were solubilized in 0.5 ml of 2% cholate, 100 mM NaCl, 10 mM Na-phosphate buffer, pH 7.5, in the presence of the indicated amounts of soybean lipid added as a sonicated dispersion in distilled water. After gentle agitation for 16 hr at 4°C, the lipid in all samples was adjusted to 25 mg in a final volume of 1 ml. The samples were agitated for 10 min and centrifuged for 30 min at 100,000 × *g*. Reconstituted vesicles were formed via removal of the cholate by dialysis (Brunner *et al.*, 1976). The cumulative $^{22}Na^+$ uptake in response to 10^{-4} M CCh was assayed. (From Lindstrom *et al.*, 1980.)

must be significantly higher than the affinity of cholate for the AChR. This suggests that the AChR channel must be shielded from the detergent molecules and the aqueous medium by a protective annulus of lipids (as illustrated in Fig. 4). Cholate forms expandable micelles in which the planar cyclopentenophenanthrene rings are apposed and the hydroxyl groups attached to the rings interact with the aqueous environment. This peculiar micellar structure of cholate may enable it to carry AChRs as lipoprotein complexes in a way in which the ion channel is protected by a lipid annulus (Fig. 4) (Anholt *et al.,* 1981).

Although the use of other detergents, such as the nonionic dialyzable detergent octylglucoside, has also been reported (Gonzalez-Ros *et al.,* 1980), cholate dialysis is the most widely used technique for the incorporation of functionally intact AChRs in lipid vesicles, since a general consensus exists that AChR function is recovered after exposure to mixed micelles of cholate and lipid.

2.2. Formation of Reconstituted Vesicles

Although low concentrations of lipids are sufficient to protect the cation channel during solubilization in cholate, approximately tenfold higher lipid

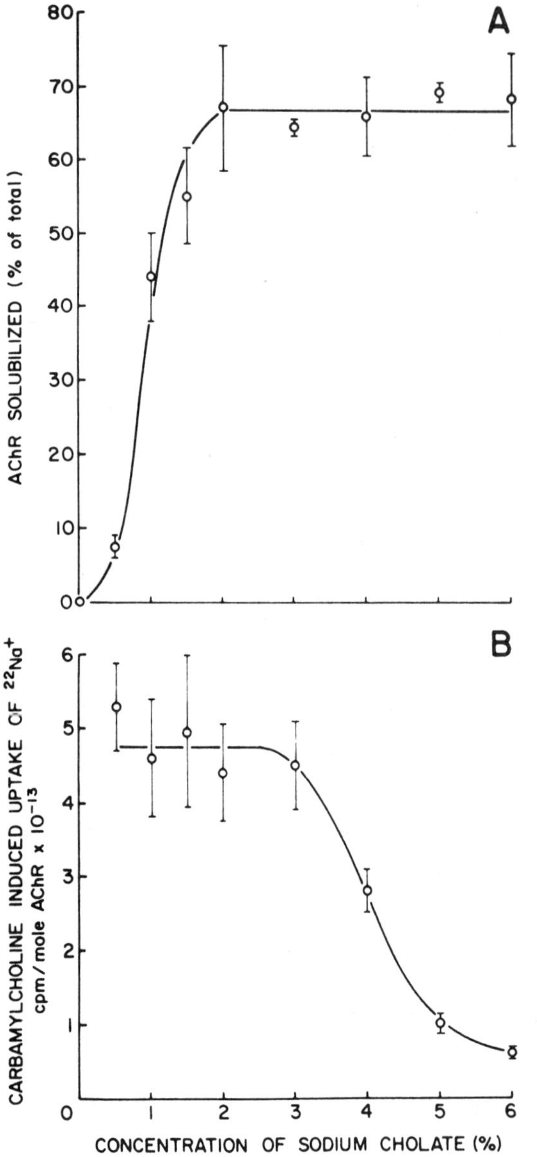

Figure 3. Solubilization of AChRs from AChR-rich electric organ membranes (A) and preservation of AChR channel function in mixed micelles of sodium cholate and soybean lipids (B). Aliquots of AChR-enriched membranes were centrifuged for 15 min in an Eppendorf microfuge. The pellets were suspended at a concentration of 4 mg/ml of protein in 100 mM NaCl, 10 mM Na-phosphate buffer, pH 7.5, supplemented with different concentrations of sodium cholate and soybean lipids at a constant weight ratio of 10:1. After 18–24 hr of shaking, the cholate and lipid concentrations in all samples were adjusted to 2% and 25 mg/ml, respectively. After an additional incubation period of 30 min of slow shaking, the samples were centrifuged in an Eppendorf microfuge as before; the supernatants were collected, assayed for $[^{125}I]\alpha$-BGT binding activity, and dialyzed in order to form reconstituted vesicles. The resulting vesicles were then assayed for CCh-dependent uptake of $^{22}Na^+$ at 10^{-4} M CCh. Data were obtained from three sets of triplicate measurements. (From Anholt *et al.*, 1981.)

concentrations are required to preserve the integrity of the channel during the reassembly of lipid and AChR into vesicles. When AChRs are reconstituted under optimal conditions (lipid/protein $\geq$ 16:1), they reassemble into lipid vesicles at a constant protein/lipid ratio such that the AChR packing density is fivefold lower than that in the native electric organ membrane (Fig. 4). Reconstitution under suboptimal lipid/protein ratios leads to a denser packing of AChRs in the reconstituted membranes and a corresponding irreversible inactivation of a fraction of

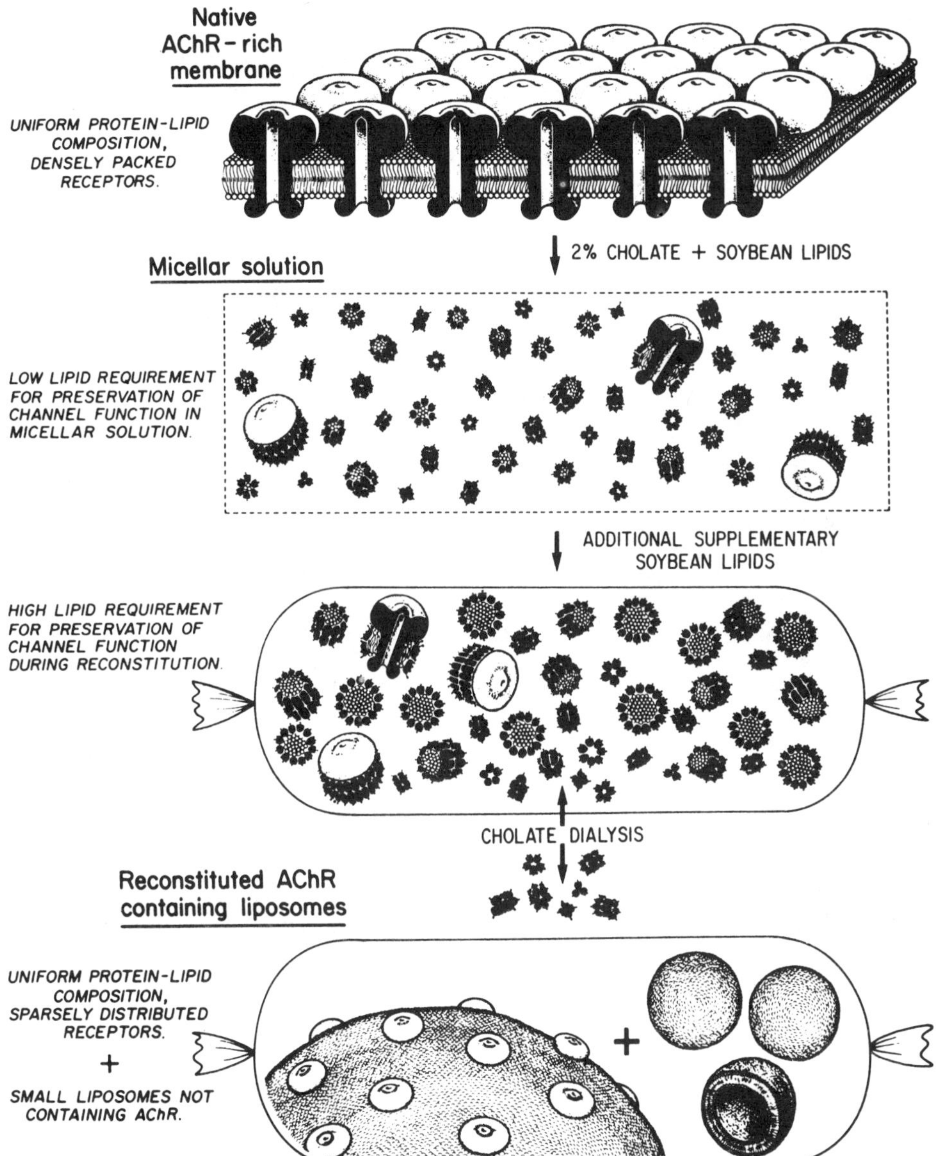

Figure 4. Schematic representation of AChR–lipid interactions during the solubilization and reconstitution processes. (From Anholt *et al.*, 1981.)

the AChR channels (Anholt *et al.*, 1981). Freeze–fracture replicas of reconstituted preparations reveal the presence of two populations of vesicles: large vesicles containing intramembranous particles that correspond in size to AChRs and smaller liposomes free of particles (Fig. 5). About 70% of the AChRs are oriented with their agonist-binding sites on the external surface of the vesicles. Immune precipitation of reconstituted vesicles with monoclonal antibodies provided evi-

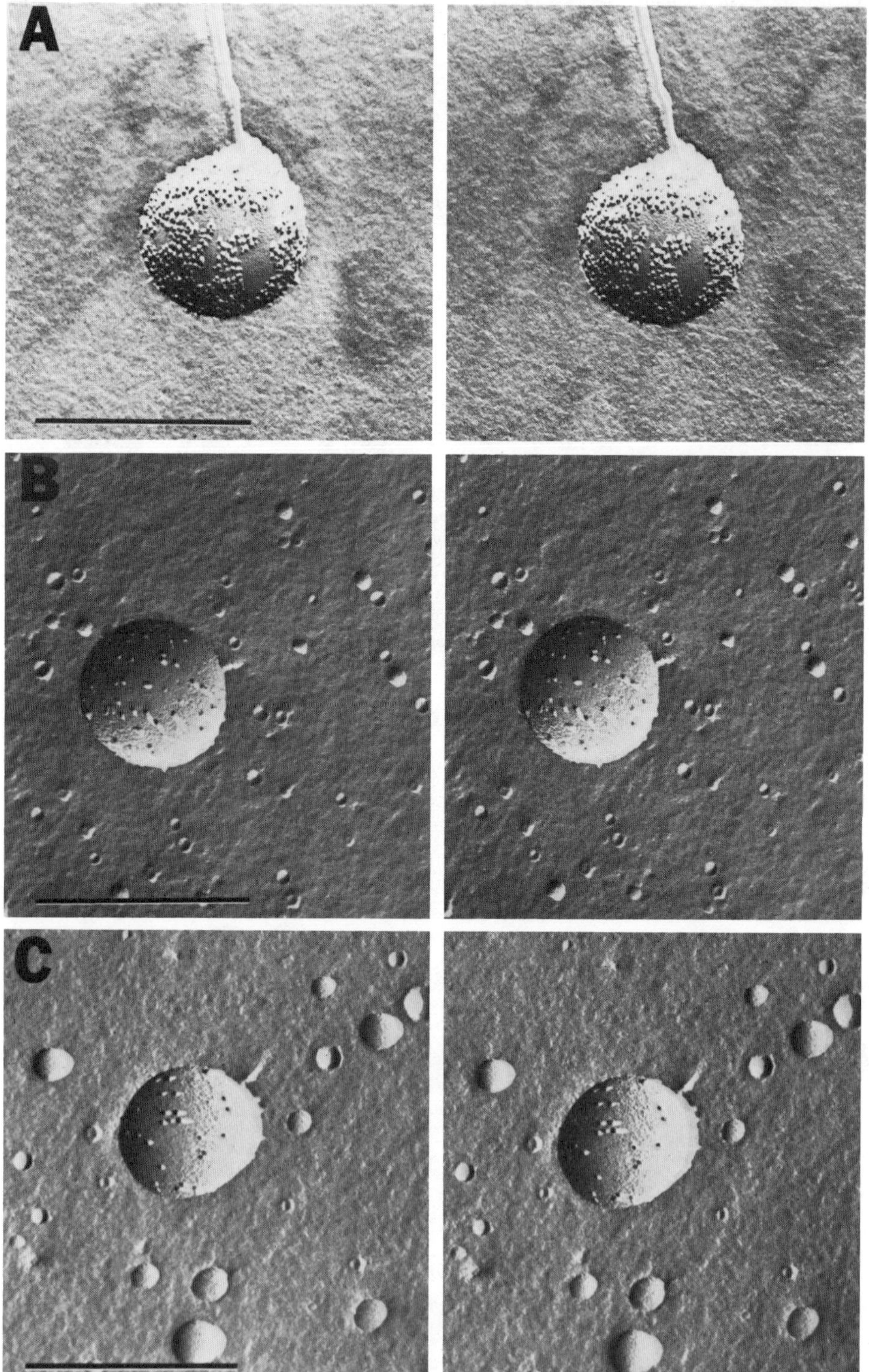
A
B
C

dence that receptors in a single vesicle are oriented either all right-side out or all inside out (Anholt *et al.*, 1981).

Since AChRs make up about 7% by weight of the reconstituted membranes (Anholt *et al.*, 1981), from knowledge of the partial specific volume of the AChR (Reynolds and Karlin, 1978), the average area occupied by a phospholipid molecule (Shah and Schulman, 1967), and the isopycnic density of liposomes without AChRs (Anholt *et al.*, 1981), one can estimate that a 2000-Å vesicle contains approximately 100 AChR monomers (Anholt *et al.*, 1982).

In order to reduce limitations of the intravesicular volume on the agonist-activated radioisotope uptake measured at equilibrium (discussed in detail below), attempts have been made to prepare vesicles with a large internal volume per AChR. Popot *et al.*, (1981) reported the preparation of large vesicles (mean diameter 950 ± 550 Å by partial cholate dialysis and subsequent gel filtration. We reported that subjecting reconstituted vesicles to a freeze–thaw cycle enhances the sealing of the vesicles and causes vesicle fusion. Inclusion of cholesterol (20% w/w) during the reconstitution procedure greatly enhances vesicle fusion during a subsequent freeze–thaw cycle. This, however, results in a more random orientation of AChRs in the reconstituted vesicles (Anholt *et al.*, 1982).

Size analysis of the reconstituted vesicles, performed by electron microscopy after negative staining, established three general patterns of vesicle organization, as summarized in Fig. 6. (1) Liposomes formed in the presence of AChR have larger mean diameters than the corresponding preparations formed in the absence of AChR. (2) Freezing and thawing of the preparations shifts the mean value of the diameter to larger values and causes the appearance of a large number of vesicles greater than 1000 Å in diameter. (3) Cholesterol-supplemented liposomes containing AChRs undergo extensive fusion during freezing and thawing, and the resulting vesicles have a mean diameter of 760 Å; 25% of the vesicles are now larger than 1000 Å in diameter, and 50% of the total included volume is contained in the 3.5% of the vesicles with diameters larger than 1400 Å (Anholt *et al.*, 1982). Fusion with liposomes can be expected to lower the packing density of AChRs in the reconstituted membranes, leading to a further increase in the available

Figure 5. Stereoscopic electron micrographs of freeze-fracture replicas of native electric organ membranes and reconstituted vesicles. (A) Native electric organ membrane; (B) regular reconstituted vesicles after a freeze–thaw cycle; (C) cholesterol (20% w/w)-supplemented reconstituted vesicles after a freeze–thaw cycle. In panel A, a native membrane vesicle containing densely packed AChRs is presented. The AChRs appear as 90-Å particles that protrude at least 45 Å from the plane of the membrane, as measured from the projected shadow of particles at the top of the vesicle. In panel B, a reconstituted vesicle containing sparsely distributed 90-Å particles can be seen surrounded by a homogeneous population of smaller vesicles devoid of particles. In panel C, a similar-size vesicle containing particles is surrounded by particle-free liposomes, which appear larger than their counterparts in panel B. The large vesicles in panels B and C contain about 40 and 26 particles, respectively, on their visible area (about half of the vesicle surface). The stereo pairs can be viewed with a conventional three-dimensional viewer (Ted Pella, Inc.). The bar indicates 500 nm. (From Anholt *et al.*, 1982.)

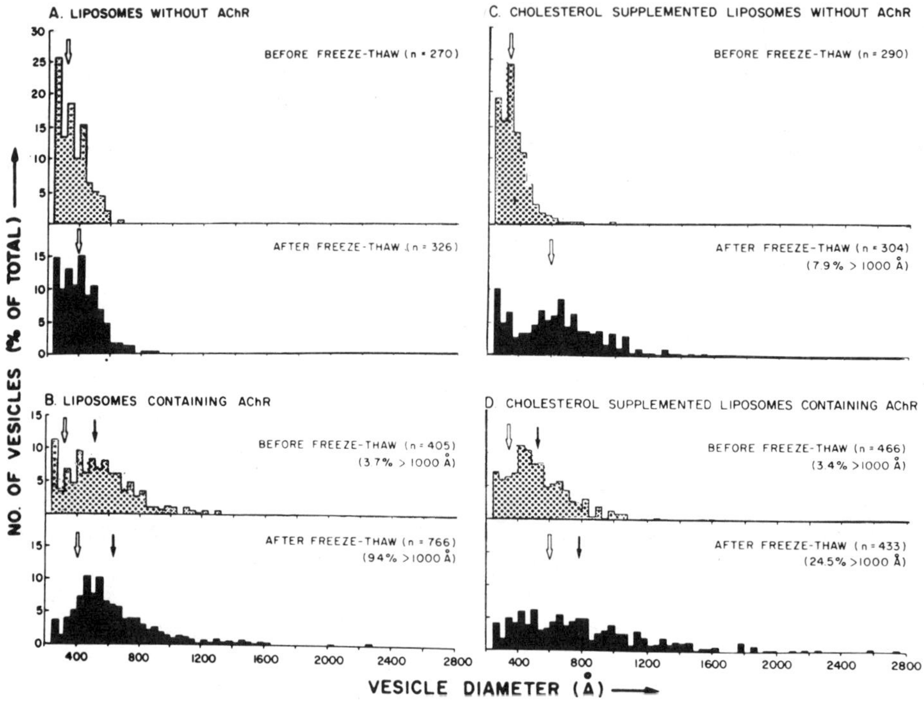

Figure 6. Size distribution of reconstituted vesicles. Electron micrographs of negatively stained vesicles were prepared. Vesicles were measured according to their longest diameter to the nearest 40 Å. Occasional structures appearing smaller than 240 Å (the approximate limiting diameter for a minimal-size liposome according to Brunner *et al.*, 1976) were discarded from the analysis. Sampling was continued until the histogram appeared to stably build up without significantly changing in terms of distribution. All liposomes in a given field were measured. Open arrows indicate the mean diameter of liposomes formed in the absence of AChR. Closed arrows indicate the mean diameter of liposomes formed in the presence of AChR. The number of vesicles with diameter larger than 1000 Å is given as an arbitrary indicator of shifts in size distribution as a result of vesicle fusion. (From Anholt *et al.*, 1982.)

internal volume per AChR for cation accumulation during the agonist-induced permeability response (discussed in more detail in Sections 2.3.2 and 2.5).

It is evident from the above discussion that the number of AChR-containing vesicles, the density of AChRs per vesicle, the internal volume per AChR in the reconstituted vesicles, the orientation of the reconstituted receptors, and the dynamics of vesicle fusion during the freeze–thaw cycle are all variable parameters that can, among other factors, be influenced by the lipid composition of the membranes. It should therefore be emphasized that experiments in this model system have to be interpreted with caution, since alterations in radioisotope flux measurements may result from alterations in the structure of the reconstituted membranes instead of reflecting a change in AChR function.

2.3. Assessment of the Pharmacological Integrity of the Reconstituted AChR

2.3.1. State Transitions

It is essential with any reconstituted system to verify the pharmacological integrity of the component of interest in order to justify the use of the system as a model for the function of its elements *in vivo*. Reconstituted vesicles containing pharmacologically intact AChRs should exhibit an agonist-activated, dose-dependent translocation of cations that can be blocked by known inhibitors such as snake neurotoxins, *d*-tubocurarine, and local anesthetics. Furthermore, preincubation with neurotransmitter should prevent channel opening, indicating the occurrence of desensitization.

There is widespread agreement that AChRs can exist in at least three different conformational states: a "resting state" in which the cation channel is closed, an activated state in which the cation channel is open, and a "desensitized state" in which the cation channel remains closed in the presence of agonist (Changeux, 1981; Changeux *et al.,* 1984; Conti-Tronconi and Raftery, 1982; Anholt *et al.,* 1984). These state transitions are illustrated in Fig. 7. Ligand binding to the AChR in its resting state results in opening of the channel, which remains open for several milliseconds, allowing the passage of cations. Prolonged exposure to the agonist will stabilize the AChR in its desensitized state, which is characterized by a higher affinity (by two orders of magnitude) for the ligand than the resting state (Weber *et al.,* 1975; Lee *et al.,* 1977; Weiland and Taylor, 1979; Sine and Taylor, 1979; Boyd and Cohen, 1980a,b; Cash *et al.,* 1981; Moore and Raftery, 1980). Evidence suggests that the desensitization process occurs in two stages, a fast step ($t_{1/2} \cong$ 300 msec) that initiates the desensitization process followed by a slow step ($t_{1/2} \cong$ 6–7 sec) that stabilizes the desensitized conformation (Sakmann *et al.,* 1980; Neubig and Cohen, 1980; Walker *et al.,* 1981, 1982; Magleby and Pallotta, 1981). Recovery of the desensitized AChRs to its resting state after removal of the agonist is a slow process occurring over minutes (Boyd and Cohen, 1980a; Sakmann *et al.,* 1980).

In the absence of agonist, most of the AChRs are in the resting low-affinity state, but a significant fraction of the unliganded AChR preexist in the high-affinity desensitized conformation. This equilibrium is described by the allosteric constant M (Fig. 7). Reported values for M for *Torpedo californica* range between 4.5 (Weiland and Taylor, 1979) and 10 (Barrantes, 1978; Heidmann and Changeux, 1979; Boyd and Cohen, 1980a). Since AChR activation and desensitization processes can go to completion within the incubation period of the assay (~10 sec), as described in the preceding section, it is clear that measurements of integrated $^{22}Na^{+}$ or $^{86}Rb^{+}$ uptake are not sensitive to the intermediate kinetics that lead to the accumulation of the measured amount of radioisotope inside the vesicles prior to the stabilization of a relatively long-lived desensitized state. However, important qualitative information about the functional characteristics of reconstituted AChRs can still be obtained, for example, through manipulation of the AChR

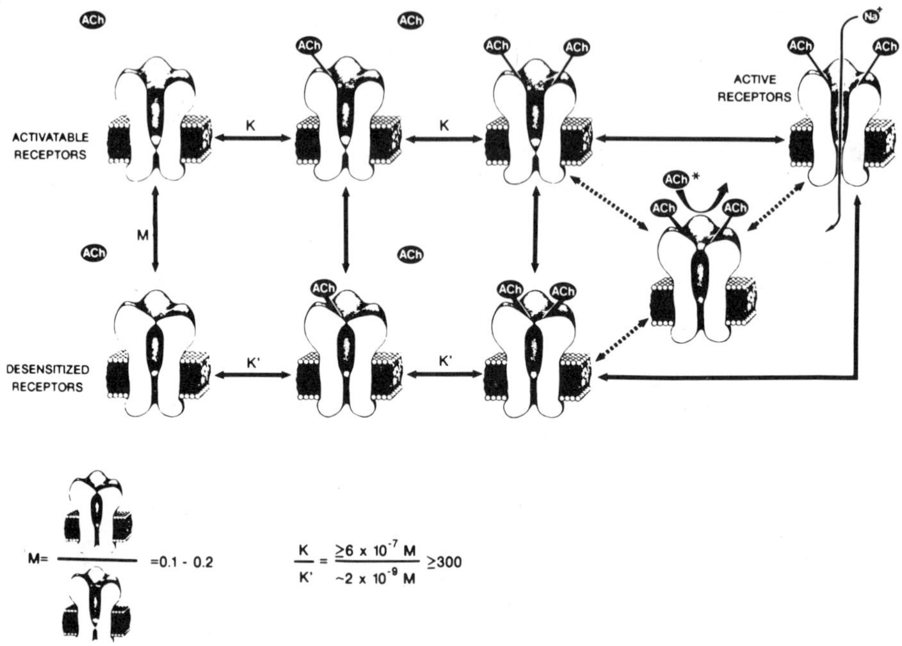

Figure 7. Schematic representation of the major conformational transitions that underlie AChR function. K and K' indicate the intrinsic equilibrium dissociation constants for the agonist of the resting and desensitized state of the AChR, respectively. M is the allosteric constant describing the isomerization equilibrium of the interconversion between unliganded activatable and desensitized AChRs. The scheme indicates that the doubly liganded AChR is able to open a cation-selective channel. Desensitization of the AChR is accompanied by the closing of the channel and an increase in binding affinity for the agonist (indicated in the figure by thickening and shortening of the bars connecting ACh with the AChR). Desensitization is a two-step process that may involve a transient intermediate state in which ACh binds to a low-affinity site, as indicated in the figure (Heidmann and Changeux, 1980; Neubig and Cohen, 1980; Walker *et al.*, 1981). The nature of the various closed states of the channel as indicated in the figure are merely illustrative and are, in reality, unknown. For precise values for *Torpedo californica* of the constants K, K', and M, see Weiland and Taylor (1979) or Boyd and Cohen (1980a). *: Binding of acetylcholine (ACh) to a site of low affinity. (From Anholt *et al.*, 1984.)

concentration. The dose–response characteristics of AChR activation by agonists depend on the concentration of AChR. When the AChR concentration is high with respect to the dissociation constant for the agonist, bound agonist will constitute a significant fraction of the total agonist concentration. The midpoint of the dose–response curve will, as a result, be displaced as a function of AChR concentration and will asymptotically approach a constant value at infinite dilution of AChR (Fig. 8) (Anholt *et al.*, 1982). This displacement effect is twice as large for ACh as for carbamylcholine (CCh), reflecting the higher binding affinity for ACh of both resting and desensitized states of the AChR. The effect of AChR concentration on the midpoint of the dose dependence of AChR desensitization

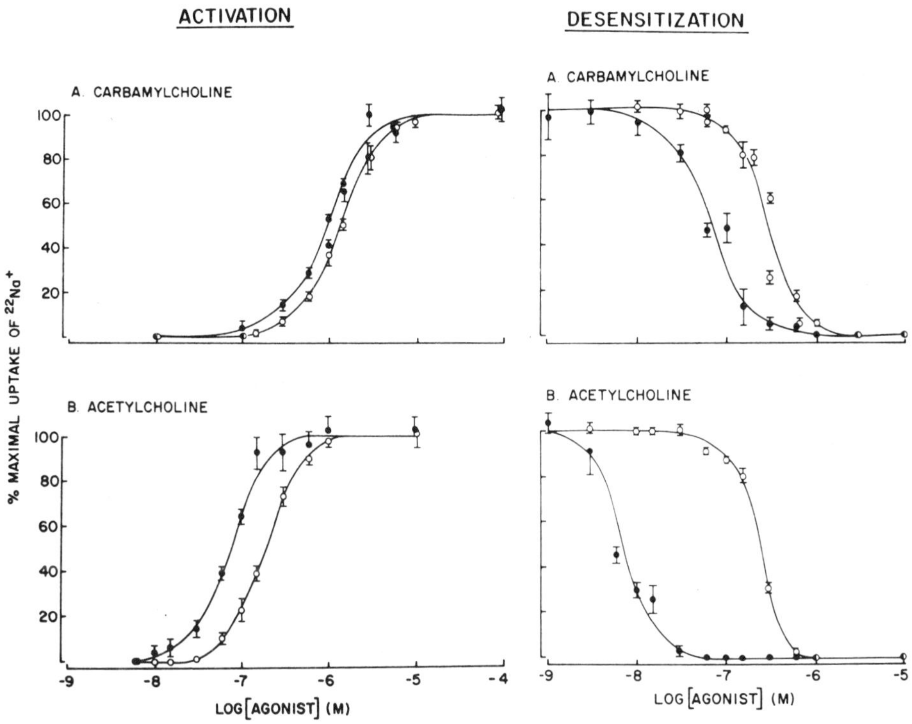

Figure 8. Dose dependence for activation and desensitization of reconstituted AChRs. Equilibrium measurements were made using a $^{22}Na^{+}$ uptake assay. Counts at the maximal response were approximately tenfold higher than the background control. The dose dependence for AChR desensitization was assayed by addition of 10^{-4} M CCh after a preincubation period of at least 30 min with the indicated concentration of agonist in the absence of radioisotope. Assays were performed at 48 nM (●) and 800 nM (○) ligand binding sites. The dose–response characteristics for activation and desensitization will depend on the AChR concentration when it is high with respect to the dissociation constant for the agonist, since the AChR will, under these conditions, deplete the concentration of free ligand by binding a significant fraction of the total agonist. As a result, the dose–response curve will be displaced depending on the affinity of the receptor for the agonist. The larger displacement of the dose–response curves for ACh as compared to those for CCh reflects the expected higher affinity for ACh. The fact that the dose–response curves for desensitization for each agonist are more sensitive to the receptor concentration than the corresponding dose–response curves for activation indicates that desensitization of the receptor is accompanied by an increase in binding affinity for the receptor. It should be noted that the midpoint of the dose–response curve for receptor activation at low receptor concentrations is at least an order of magnitude lower when cumulative responses are measured at equilibrium than when initial flux rates are measured. (Adapted from Anholt *et al.*, 1984.)

is even more dramatic (Fig. 8), since the desensitized AChR has higher affinity (by two orders of magnitude) than the resting AChR for both agonists. Thus, it appears from functional measurements that reconstituted AChRs distinguish between ligands and, at least qualitatively, undergo state transitions in a way similar to AChRs in the native membrane.

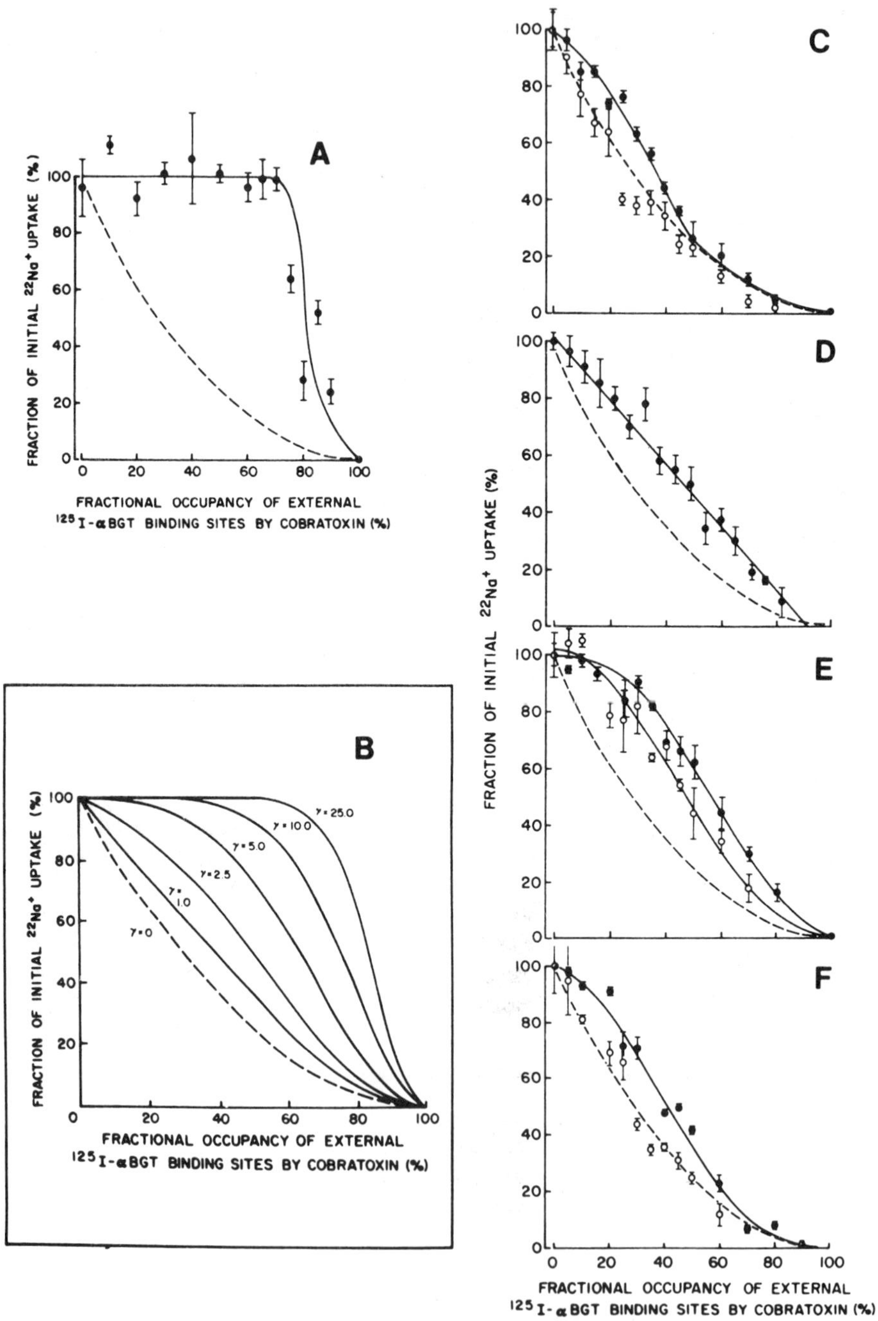

Figure 9. Titration of the CCh-induced cation uptake response by inhibition with cobratoxin. Native electric organ membranes or reconstituted vesicles were incubated for 2 hr at room temperature with increasing amounts of *Naja naja siamensis* toxins III (Karlsson *et al.*, 1971) and assayed for uptake of $^{22}Na^{+}$ at 10^{-4} (●) or 2×10^{-6} M (○) CCh. Data are compiled from triplicate measurements. The

2.3.2. Inhibition by Cobratoxin: A Tool for the Assessment of Intravesicular Volume Limitations on the Radioisotope Uptake Assay

Inhibition of AChR function by cobratoxin gives information about the extent of intravesicular volume limitations on the measured radioisotope uptake after activation by agonist. Since the probability of channel opening is greatest when two agonist molecules are bound per AChR, the concentration dependence of inhibition of AChR function by cobratoxin is expected to be parabolic (Sine and Taylor, 1980). In native membranes, the dense packing of AChRs (Fig. 9A) and the concomitant limitation of the internal volume per AChR ($\sim 2 \times 10^{-20}$ liter/AChR) lead to equilibration of the intravesicular volume with the external concentration of the radioisotope before termination of the assay. Hence, a large fraction of AChRs have to be occupied by cobratoxin before a reduction in the agonist-induced uptake of cations is observed (Fig. 9A) (Moore *et al.*, 1979; Lindstrom *et al.*, 1980; Anholt *et al.*, 1982).

A mathematical model of the cobratoxin titration experiment that considers different extents of volume limitation on the cation uptake response was developed (Anholt *et al.*, 1982). This generates a family of curves that approach the predicted parabolic inhibition curve as restraints on the available internal volume per AChR are relieved (Fig. 9B) (Anholt *et al.*, 1982). At relatively low extents of volume limitations per AChR (indicated by the parameter γ as defined in the legend to Fig. 9B), shallow sigmoids may, within experimental error, appear linear. This apparent linearity is enhanced if one assumes heterogeneity in vesicle sizes. Figures 9C–E show the range of observed toxin titration behaviors of regular reconstituted vesicles after freezing and thawing. The inhibition curves vary slightly, depending on the extent to which fusion has occurred as a result of the freeze–thaw cycle. The $^{22}Na^+$ uptake induced by a lower agonist concentration generates toxin titration curves shifted in the direction of the predicted parabola (Fig.9C, E). Cholesterol-supplemented vesicles, after freezing and thawing, have relatively large internal volumes per AChR ($\sim 25 \times 10^{-20}$ liter/AChR) and show only slight deviations from this parabolic curve. Deviations are small at saturating concentrations of CCh, and the predicted titration curve is followed closely at 2 μM CCh, a concentration that elicits between one-third and one-half of the maximal response (Fig. 9F).

---

dotted line in each panel indicates the predicted parabolic decline according to the equation $f(y) = (1 - y)^2$ for a model of doubly liganded control of channel gating in the absence of volume limitations in which $f(—y)$ is the fraction of AChRs that remain active having both ligand binding sites unoccupied by cobratoxin and y represents the fraction of all binding sites occupied by cobratoxin. (A) Native electric organ membranes; (B) a theoretical qualitative description of the effect of limited internal volume on the cobratoxin titration curve. The curves are described by the equation $f(y) = 1 - \exp[-\gamma(1 - y)^2]/[1 - \exp(-\gamma)]$ in which $f(y)$ is the remaining fraction of AChR-mediated $^{22}Na^+$ uptake, y the fractional occupancy of [^{125}I]α-BGT binding sites by cobratoxin, and γ a parameter that relates the integrated $^{22}Na^+$ uptake per AChR channel to the available internal volume per AChR. Volume restrictions become more severe as the value of γ increases. When $\gamma \rightarrow 0$, the above equation obtains the form $f(y) = (1 - y)^2$ (for derivation, see Anholt *et al.*, 1982). (C–E) Variable behavior of the regular reconstituted vesicles after a freeze–thaw cycle. (F) Typical behavior of cholesterol-supplemented reconstituted vesicles after a freeze–thaw cycle. (From Anholt *et al.*, 1982.)

The relative contributions of desensitization and volume limitations in determining the amplitude of the $^{22}Na^{+}$ uptake response vary with the agonist concentration. At low agonist concentrations, when few AChRs are activated, desensitization is expected to be the primary limitation on the $^{22}Na^{+}$ or $^{86}Rb^{+}$ uptake. At higher agonist concentrations, volume limitations will become important in an increasing fraction of the vesicles. In these vesicles, equilibration of the intravesicular volume with the external radioisotope concentration will occur at submaximal agonist concentrations. Accordingly, the midpoint of the dose–response curve for AChR activation shifts to lower agonist concentrations. Limitations on the internal volume per AChR may complicate investigation of AChR function but are much less severe in reconstituted vesicles ($\sim 8 \times 10^{-20}$ liter/AChR) than in the native membrane ($\sim 2 \times 10^{-20}$ liter/AChR). They can be entirely prevented at low agonist concentrations through the use of cholesterol-supplemented liposomes and a freeze–thaw cycle to promote the formation of large vesicles ($\sim 25 \times 10^{-20}$ liter/AChR).

2.4. Identification of Membrane Components Essential for Postsynaptic Signal Transduction

2.4.1. AChR Monomer as the Functional Entity

A unique advantage of a reconstituted system is that it allows the fractionation of the native membrane into its individual components and the subsequent determination of their respective importance in the recovery of specific functions after reassembly.

The realization that the presence of supplementary lipids prevents inactivation of the ion channel in cholate solution during the purification of AChRs by affinity chromatography allowed the isolation and reconstitution of functionally intact AChRs (Huganir *et al.*, 1979; Changeux *et al.*, 1979; Lindstrom *et al.*, 1980; Popot *et al.*, 1981). These studies established that the $\alpha_2\beta\gamma\delta$ complex by itself is sufficient to express the transduction of ligand binding into the opening of a cation-specific channel. Moreover, with few exceptions (Schindler *et al.*, 1984), it is generally agreed that AChR monomers and dimers are functionally equivalent (Popot *et al.*, 1981; Anholt *et al.*, 1980; Wu and Raftery, 1981; Wu *et al.*, 1981). This notion is based on the observation that purified monomers and dimers, reconstituted separately in lipid vesicles, are indistinguishable with respect to agonist-induced activation and desensitization (Anholt *et al.*, 1980) and that functional activity of reconstituted AChRs is unaffected after mild treatment with reducing agents that convert dimers into monomers (Anholt *et al.*, 1980; Popot *et al.*, 1981). Furthermore, the observation that AChRs remain functional after extensive trypsinization (Huganir and Racker, 1980) or after depletion of nonreceptor peripheral proteins (Criado and Barrantes, 1984), treatments that also convert dimers into monomers, constitutes additional evidence in support of the notion that the $\alpha_2\beta\gamma\delta$ monomeric form of the AChR represents its functional unit.

2.4.2. Subfractionation of the AChR Monomer

Reconstitution studies aimed at the complete dissociation of the five component subunits and their *in vitro* reassembly into a functional complex have thus far not been successful. This is not surprising in light of the fact that the sequential assembly of AChR subunits during *in vivo* synthesis is a complex process in which maturation of the toxin binding site requires a 15- to 30-min posttranslational period (Merlie and Sebbane, 1981; Merlie *et al.*, 1982). Moreover, after assembly, the subunits of the mature AChR can only be dissociated under denaturing conditions. However, partial renaturation of the α subunits isolated by preparative gel electrophoresis in sodium dodecylfulfate was reported; it was expressed either as recovery of low-affinity α-BGT binding displaceable by cholinergic ligands after removal of the detergent (Haggerty and Froehner, 1981; Gershoni *et al.*, 1982) or as high-affinity binding of the toxin that was not displaceable by agonists (Tzartos and Changeux, 1983).

The most elegant study on the formation of the AChR assemblies consisting of different complements of the five subunits was conducted by Mishina *et al.*, (1984). They coupled cloned cDNAs of all four AChR subunits to a simian virus 40 vector, amplified the messages in COS monkey cells, and injected resulting mRNAs into *Xenopus oocytes*. Translation of the mixture of injected mRNAs encoding all four subunits of the AChR resulted in the synthesis of functional AChR in the oocytes. From studies in which mRNAs coding for some of the subunits were selectively deleted during injection, they deduced that all four subunits are required for the assembly of a functional AChR, although only the α subunit appears indispensable for α-BGT binding.

2.4.3. Effect of Membrane Lipid Composition on AChR Function

Several laboratories have studied the effect of the lipid environment on AChR function (e.g., McNamee and Ochoa, 1982). Early studies with pure lipids established a mixture of phosphatidylethanolamine and phosphatidylserine (3:1 w/w) supplemented with cholesterol or other neutral lipis such as α-tocopherol or certain quinones as the optimal lipid environment (Kilian *et al.*, 1980; Dalziel *et al.*, 1980; McNamee *et al.*, 1982; Ochoa *et al.*, 1983; Criado *et al.*, 1982). However, as discussed above, changes in the lipid composition of the reconstitution mixture can alter the size and hence the volume of the resulting vesicles as well as the orientation of AChR in the liposomes (Anholt *et al.*, 1982; Ochoa *et al.*, 1983; Criado *et al.*, 1982, 1984). Therefore, evaluation of AChR function must be approached with caution in such studies.

The best-documented report on the effect of the lipid environment on AChR function is the systematic study by Criado *et al.* (1984) in which the alterations in vesicle structure are analyzed in parallel with alterations in AChR function. It was found that unsaturated phosphatidylethanolamines in combination with 28–35 mol% cholesterol hemisuccinate permit a maximal ion-translocation response of the reconstituted AChR. The cumulative results of several studies using pure lipid mixtures indicate that AChR function is optimally retained in a fluid, unsaturated, cholesterol-supplemented membrane environment. This is consistent with the fact that the native membrane of *Torpedo* electric organ is a fairly unsaturated membrane with a high content of cholesterol (Gonzalez-Ros *et al.*, 1982). A more extended discussion of the effects of the lipid environment on the function of the reconstituted AChR is presented in Chapter 10 by McNamee and colleagues.

2.5. Limitations of Equilibrium Measurements of AChR Function in Reconstituted Vesicles

Equilibrium measurements in reconstituted vesicles, by virtue of their technical simplicity, are useful as a convenient and rapid screening system for macroscopic effects on AChR function (e.g., effects of monoclonal antibodies, as reported by Lindstrom, Chapter 9). They are, however, complicated by the following limiting factors: (1) restrictions on the available volume per AChR for the accumulation of radiosotope during the assay time; and (2) measurement of a cumulative cation uptake response rather than measurement of initial flux rates. Through the use of cholesterol-supplemented vesicles, volume restrictions can be minimized or, at low agonist concentrations, eliminated (Fig. 9F). Thus, the major limitations of the system remains the measurement of a cumulative cation uptake response, which is insensitive to the intermediate kinetic steps involved in AChR function (e.g., during the assay time, activation as well as desensitization processes take place which lead to displacements of the midpoint of the dose–response curve for AChR activation toward a lower agonist concentration). Methods to study the kinetics of neurotransmitter-regulated cation fluxes through AChR receptor channels in reconstituted systems consist of rapid stop-flow measurements in reconstituted vesicles, as discussed in Chapter 10 by McNamee and colleagues (see also Hess *et al.*, 1983), and of electrical measurements in planar lipid bilayers. The transformation of reconstituted vesicles containing AChR into planar lipid bilayer membranes is discussed in the following section.

3. Reconstitution of AChR in Planar Lipid Bilayers

3.1. Rationale

The ultimate criterion of the success of the AChR channel reconstitution resides in the detailed comparison of the channel properties of the purified AChR with those of the native system. Single-channel recordings from the purified AChR reconstituted in planar lipid bilayers provide an assay that resolves, in real time, molecular transitions proceeding in a single AChR molecule. Knowledge of the kinetic, pharmacological, and ion-conduction characteristics of the purified AChR is the prerequisite necessary for the investigation of how manipulations in structure modify functional properties on a molecular level (see Section 4).

Planar lipid bilayers are assembled via apposition of two monolayers (Montal and Mueller, 1972; Montal, 1974; Montal *et al.*, 1981). The bilayers can be formed across apertures in Teflon® septi (Montal and Mueller, 1972; Montal, 1974; Montal *et al.*, 1981) or at the tip of patch pipets (Wilmsen *et al.*, 1983; Schuerholz and Schindler, 1983; Suarez-Isla *et al.*, 1983; Coronado and Latorre, 1983). In both modalities, the monolayers are derived from the reconstituted AChR vesicles (Schindler, 1979, 1980; Schindler and Quast, 1980; Schindler *et al.*, 1984; Nelson *et al.*, 1980; Labarca *et al.*, 1984a,b, 1985a,b; Montal *et al.*, 1984), described in detail in the previous section.

3.2. Transformation of Reconstituted Vesicles into Monolayers at the Air–Water Interface

A necessary condition for the assembly of planar lipid bilayers from vesicle-derived monolayers is that the surface pressure, π, of the monolayer be $\geq$25 dyne/cm (Schindler, 1979, 1980). Surface pressure is extremely sensitive to the composition of the aqueous subphase. Schindler and collaborators studied AChRs from *T. marmorata* and *californica* in planar lipid bilayers (Schindler and Quast, 1980; Schindler, 1982; Schindler *et al.*, 1984). The channel properties were found to depend on the surface properties of the monolayers and on the lipid composition of the reconstituted vesicles. Two requirements for the functional integrity of the reconstituted AChR channel were identified: (1) cholesterol, which appears to stabilize the channel; and (2) a monolayer cohesive pressure of ~30 dyne/cm, which appears to reduce the aggregation of AChRs in the membrane.

We routinely use the following procedure to derive monolayers from purified AChR reconstituted vesicles (see Section 2) with surface pressures sufficient to support the assembly of bilayers. A 50μl of sample of reconstituted vesicles (the AChR concentration ranges between 0.2 and 1.0 μM; soybean lipid is 32 mg/ml supplemented with 8 mg/ml cholesterol) is diluted with 25mM HEPES (N-2-hydroxypiperazine-N′-2-ethanesulfonic acid)–tris, pH 7.4, to a final volume of 250 μl and centrifuged at top speed in a Beckman Airfuge (122,000 × *g*) for 20 min. The pellet is resuspended in 0.5 M NaCl buffered with 2.5 mM tricine [N-tris-(hydroxymethyl)-methylglycine] or HEPES, pH 7.4, and $CaCl_2$ (in the range between 0.5 to 5 mM) in a final volume of 0.3 ml, sonicated for 15 sec, and incubated for 20 min at 20°C. Such a suspension is introduced into the bilayer chamber by gently delivering 20-μl drops (see Section 3.3). The surface pressure of monolayers at the air–water interface can be monitored by the method of Wilhelmy in a monolayer trough with two movable barriers and several compartments (Fromherz, 1975). Figure 10 illustrates the time course of change in surface pressure caused by the formation of a monolayer at the air–water interface from a suspension of AChR reconstituted vesicles (Labarca *et al.*, 1984a).

The actual transfer of purified AChRs from the reconstituted vesicles to monolayers can be determined using [^{125}I]α-BGT-labeled AChRs according to Schindler and Quast (1980; Labarca *et al.*, 1984a). An aliquot of α-BGT-labeled reconstituted vesicles is deposited in one chamber of the multicompartment surface trough. Monolayers at surface pressures of 14–18 dyne/cm are transferred to a

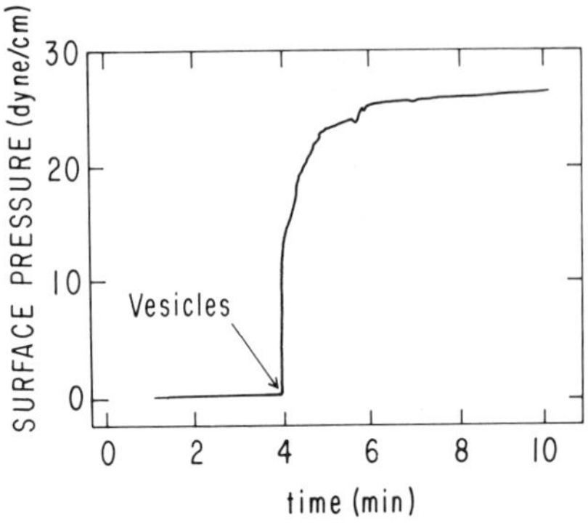

Figure 10. Time course of the change in surface pressure caused by the formation of vesicle-derived monolayers at an air–water interface. (From Labarca *et al.*, 1984a.)

vesicle-free compartment, compressed to a π of 30 dyne/cm, and collected for analysis. The radiolabled α-BGT measured in the monolayer provides an estimate of the density of AChR (AChR molecules/cm^2) in the monolayer. Under the conditions described for bilayer reconstitution studies, the estimated AChR density is $\sim 10^7/cm^2$ (see Labarca *et al.*, 1984a).

3.3. Assessment of the Pharmacological Integrity of the Reconstituted AChR Channel

The reconstituted AChR in planar lipid bilayers displays the macroscopic phenomena associated with the AChR in the native membrane, namely activation and desensitization in the presence of cholinergic agonists. The agonist-activated membrane conductance increases with agonist concentration and is inhibited by cholinergic antagonists (*d*-tubocurarine and hexamethonium) and by α-BGT (Nelson *et al.*, 1980; Labarca *et al.*, 1984a; Montal *et al.*, 1984). At high current resolution, the opening and closing of single AChR channels is clearly discerned (Nelson *et al.*, 1980; Schindler *et al.*, 1984; Boheim *et al.*, 1981; Labarca *et al.*, 1984a,b; Montal *et al.*, 1984). Single-channel currents are activated by ACh, CCh, and suberyldicholine (SubCh). Channels are not observed when the reconstituted vesicles are preincubated with an excess of α-BGT, which specifically inhibits the agonist-activated flux response in vesicles (see Section 2.4.2).

3.3.1. Comparison of Single-Channel Recordings Obtained from Planar Lipid Bilayers across Teflon® Apertures or at the Tip of Patch Pipets

Thus far, similar channel properties have been recorded from reconstituted AChRs with both modalities of bilayer support, namely bilayers formed across apertures in Teflon® septi or at the tip of patch pipets (Montal *et al.*, 1984; Suarez-Isla *et al.*, 1983; Labarca *et al.*, 1984a,b, 1985a,b). Therefore, in the description of AChR channel properties that follows, no distinction is made with respect to the specific modality used. However, the technique of forming a lipid bilayer from

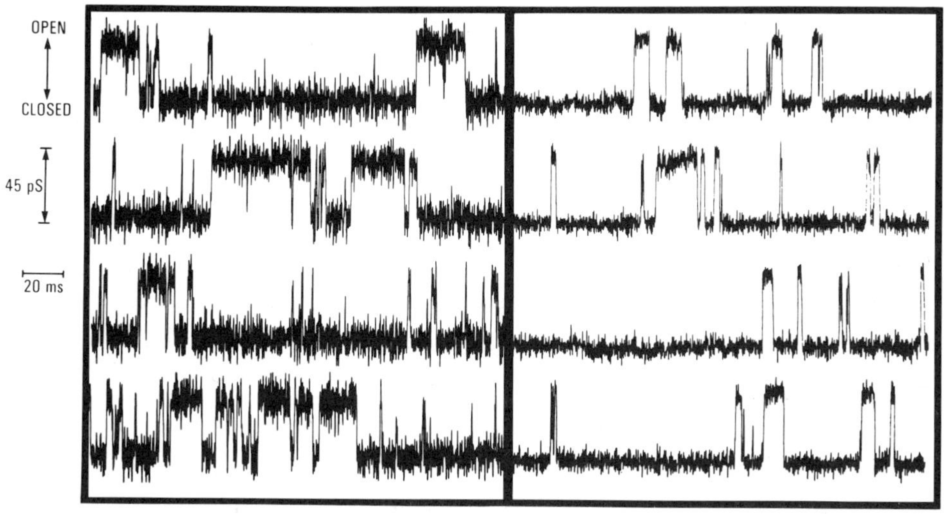

Figure 11. Single AChR channel currents activated by suberyldicholine (0.1 μM, left-hand panel) and acetylcholine (1 μM, right-hand panel) at an applied voltage of 100 mV. The recordings were obtained from a bilayer formed either across a Teflon® aperture (left record) or at the tip of a patch pipet (right record). The records were low-pass filtered at 2 kHz (8-pole Bessel filter). As indicated, an upward deflection corresponds to a channel-opening event. Bilayers were formed in 0.44 M NaCl (left) or 0.5 M NaCl (right), 0.5 mM $CaCl_2$, 2.5 mM tricine, pH 7.4. The single-channel conductances were 42 pS and 47 pS, respectively.

two monolayers at the tip of a patch pipet allows recordings of single-channel currents to be obtained at a higher time resolution and with better signal-to-noise ratio than from bilayers formed across Teflon® apertures (Suarez-Isla *et al.*, 1983).

Figure 11 shows single AChR channel currents recorded from planar lipid bilayers formed either across an aperture in Teflon® (left-hand panel) or at the tip of a patch pipet (right-hand panel). Four stretches from a continuous record are illustrated. In both panels, two distinct channel states are clearly discerned, corresponding to the closed and open states of the channel. The single-channel conductance is estimated from the conductance histogram illustrated in Fig. 12. Two Gaussian curves are identified, corresponding to closed and open channel states. The two curves (noisy data) are well fitted by the sum of two Gaussians (smooth curve). The first, with a mean peak value at zero current, corresponds to the channel-closed state. The second is associated with the channel-open state. At 100 mV, its peak value is centered around 4.2 pA and 4.7 pA, respectively, for the two bilayer types. The single-channel conductance, γ, is the current difference between the two peaks divided by the applied voltage: in these recordings, γ = 42 pS (0.44 M NaCl, left-hand panel) and 47 pS (0.5 NaCl, right-hand panel). The two Gaussians are well separated, but their widths are larger for the records from bilayers in Teflon® apertures, indicating a lower signal-to-noise ratio. Note that the width of the Gaussians is larger for the open state than for the closed

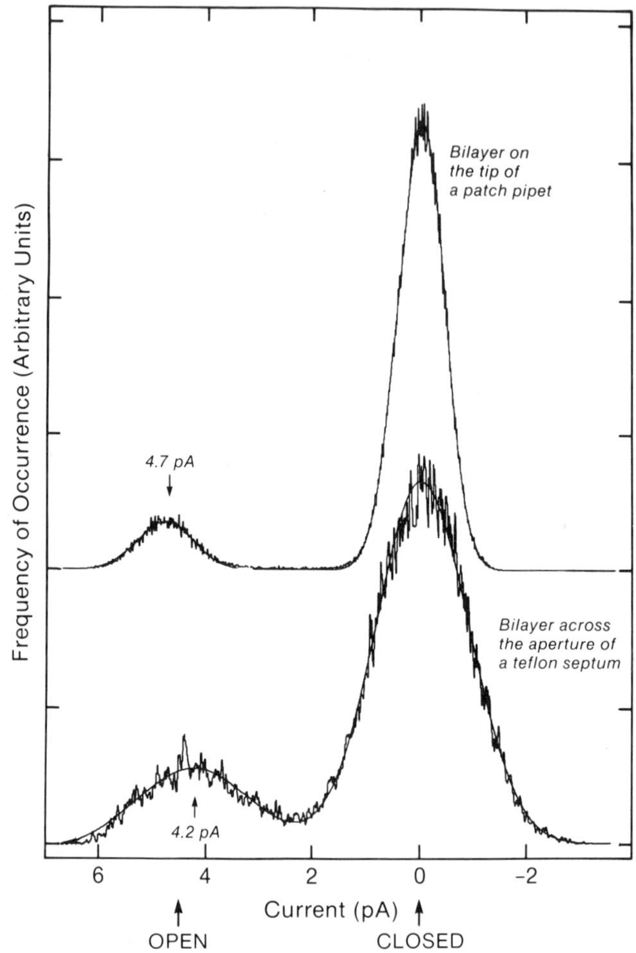

Figure 12. Single-channel conductance histograms of the recordings illustrated in Fig. 11. Two Gaussian distributions are clearly discerned, corresponding to the channel-closed state (peaks at zero current) and the open state (peaks at 4.2 and 4.7 pA). At the applied voltage of 100 mV, γ = 42 pS and 47 pS. The relative areas under the closed and open curves reflect the probability of each state at 100 mV, indicating that the channel is preferentially closed. The width parameter or standard deviation of the Gaussian distribution is significantly larger for the open state (>10%).

state, a feature especially conspicuous in the record from patch pipets. This is a measure of the current noise and indicates that the open-channel noise is >10% larger than the noise of the closed channel. This result agrees with data obtained by Sigworth (1982) in native membranes and appears to reflect the increased current introduced by the fluctuations of the protein domains forming the channel as the ions traverse through it.

The properties measured in both systems are therefore practically identical, but the records obtained from patch pipets are clearly superior.

3.3.2. Ion Conduction through the Reconstituted AChR Channel: Saturation, Selectivity, and Block

The single AChR channel current–voltage relationship is linear in the range of -100 mV $\leq V \leq$ 100 mV. The single-channel conductance is ohmic and independent of the type or concentration of the agonist for ACh, CCh, and SubCh (Labarca *et al.*, 1984a; Montal *et al.*, 1984). Single-channel currents increase with Na^+ activity to a maximum conductance of 95 pS and show a half-saturation point of 395 mM. The apparent ion selectivity sequence, derived from single-channel current recordings, is: $NH_4^+ > Cs^+ > Rb^+ > Na^+ \gg Cl^-, F^-, SO_4^{2-}$ (Labarca *et al.*, 1984a). The selectivity and saturation characteristics of the open AChR channel in planar lipid bilayers are in agreement with those determined in muscle cells (Adams *et al.*, 1980, 1981; Dwyer *et al.*, 1980; Hamill and Sakmann, 1981; Horn and Patlak, 1980; Dani and Eiseman, 1984). In addition, a quarternary ammonium derivative of the local anesthetic lidocaine (QX-222) is known to block ion conduction through open AChR channels in muscle (Neher and Steinbach, 1978). A similar action is exerted on the purified *Torpedo* AChR reconstituted in lipid bilayers as evidenced from the appearance of fast "flickering" between open and closed states and a drastic reduction in open-channel lifetimes (Labarca *et al.*, 1984a). This blocking action is voltage dependent, being prominent when the *cis* side of the bilayer is positive. In contrast, no detectable block is measured when QX-222 is present on the *trans* side of the membrane, suggesting that the drug is effective only on the normally extracellular-facing side of the channel, in agreement with data on native AChR (Horn *et al.*, 1980; Ruff, 1982; Ogden *et al.*, 1981). Thus, it is evident from single-channel studies that all tested aspects of ion conduction through the reconstituted AChR channel are functionally retained in the planar lipid bilayer.

3.3.3. Channel-Gating Kinetics

3.3.3a. The AChR Channel has Two Open States. Precise information about the number of open and closed states in which the AChR channel can exist, the pathways to enter into or exit from such states, and the rates of transition between the various states is essential for a rigorous understanding of the AChR channel at the molecular level.

Single-channel recordings are especially amenable for detailed statistical analysis. The major analytical tool is the computation of distributions of dwell times (lifetimes) in the open and closed states of the channel. Single-channel recordings obtained from cells by the patch-clamp technique (Neher and Sakmann, 1976; Colquhoun and Sakmann, 1981) and from the purified AChR reconstituted in planar lipid bilayers (Labarca *et al.*, 1984b, 1985a,b; Montal *et al.*, 1984) have led to a major revision of the kinetic schemes used to describe the function of the AChR channel. The conventional model of the AChR in which channels are open or closed (Fig. 7) and the distribution of open times is well described by a single exponential with the predicted time constant, cannot account for the single-channel results.

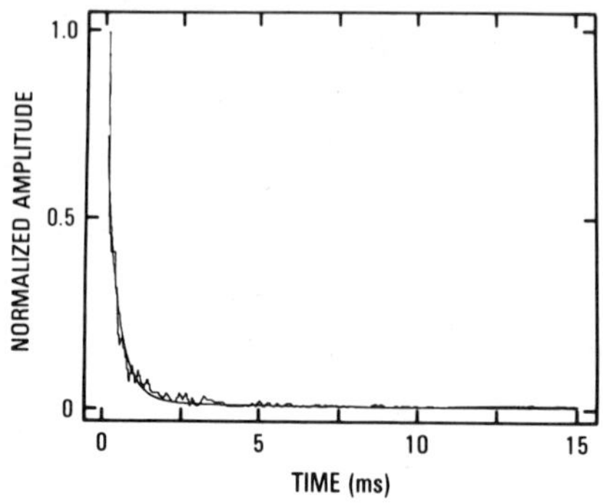

Figure 13. Probability density analysis of dwell times in the open state of the AChR channel activated by 0.1 μM ACh and recorded at 100 mV. The data are well fitted by a sum of two exponentials, with $p = 0.99$. The fitted curve (smooth curve) is superimposed on the histogram of the actual data (noisy curve). The parameters of the fitted curve are: $A_S = 0.92$, $\tau_S = 0.42$ msec; $A_L = 0.08$, $\tau_L = 4.3$ msec. The total number of openings, N, analyzed in this particular histogram is 967.

A major conclusion derived from the analysis of single-channel current records from the reconstituted AChR channel is that the distribution of channel open times is well fit by a sum of two exponentials, reflecting the existence of at least two open states (Labarca *et al.*, 1984b, 1985a,b). This is illustrated in Fig. 13A. Records in which only one channel is open at any given time are analyzed. The fitted curve (smooth curve) is superimposed on the histogram of the actual data (noisy curve). The area under the probability density function (A) and time constants for the short (S) and the long (L) components have been calculated. For ACh at an applied voltage of 100 mV, the typical values of the time constants of the short and long openings are 0.5 msec and ~4 msec (Labarca *et al.*, 1985a,b). The time constants of these two states depend on the choice of agonist, following the sequence SubCh > ACh > CCh in agreement with single-channel data from native AChR in cells (Jackson *et al.*, 1982; Trautman and Feltz, 1980). These two kinetically distinct open states, however, have the same single-channel conductance (Labarca *et al.*, 1984a,b; Montal *et al.*, 1984; see also Suarez-Isla *et al.*, 1983; Tank *et al.*, 1983).

3.3.3b. The Occurrence of the Long-Lived State Increases with ACh Concentration. The high signal-to-noise ratio of the records obtained from patch pipets (Figs. 11, 12) enables a clear identification of states and, therefore, a more accurate and reliable estimation of the kinetic parameters involved in channel gating. Because of the high resolution of these measurements, a detailed correlation between the frequency of occurrence of short and long openings and the concentration of ACh could be obtained (Labarca *et al.*, 1985b).

The probability density of open dwell times is well described by a sum of two exponentials at agonist concentrations in the range 0.01 μM ≤ ACh ≤ 0.1 mM. This study indicates that the area ratio of the two components (A_L/A_S ratio) increases with the concentration of ACh, whereas the corresponding time constants, τ_S and τ_L, of the short and long kinetic processes involved in channel closure are not altered systematically by agonist concentration. Similar results are obtained with CCh (Labarca *et al.*, 1984a). Thus, the frequency of occurrence of the long openings increases with the concentration of agonist (Labarca *et al.*, 1985b). Simmilar results have been obtained from single AChR channel recordings in adult muscle cells (D. Colquhoun and B. Sakmann, personal communication).

3.3.3c. Inferences about Molecular Mechanisms of Channel Gating. The results of the kinetic analysis described above allow one to reject a set of kinetic schemes that predict either that the A_L/A_S ratio is a constant or that the time constants vary with agonist concentration (Jackson *et al.*, 1983). A minimum kinetic scheme that is consistent with the experimental data is as follows:

$$\begin{array}{ccccc} C + 2L & \underset{k_{21}}{\overset{ck_{12}}{\rightleftharpoons}} & CL + L & \underset{k_{32}}{\overset{ck_{23}}{\rightleftharpoons}} & CL_2 \\ & & k_{42} \updownarrow k_{24} & & k_{53} \updownarrow k_{35} \\ & & OL & & OL_2 \end{array} \qquad (1)$$

where C represents closed-channel states and O open states and L denotes the ligand. This scheme is consistent with the independence of both τ_S and τ_L of agonist concentration, implying that liganding is to closed states. Such a model predicts a linear dependence with unitary slope of the A_L/A_S ratio on agonist concentration when plotted as the log (A_L/A_S) versus log[ACh]. The function, determined experimentally, indeed shows a unitary slope within the range of ACh concentration between 0.01 μM and 0.1 μM. At higher ACh concentrations, the A_L/A_S ratio approaches a constant value. Such behavior does not appear to arise from channel blocking by the agonist (Colquhoun and Sakmann, 1981), since the τ_S calculated from the distribution of open times are independent of ACh concentration. Formally, this deviation could be accounted for by introducing an additional, short-lived, doubly liganded open state to the right of CL_2 in scheme 1. Models equivalent to 1 consisting of multiple open states with higher extents of ligand occupancy or with a more complex closed-state structure could also account for the experimental results. The simplest inference derived from our observations is that both singly and doubly liganded states of the AChR molecule undergo transitions to the open conformation.

Note that the dependence on ACh concentration of the permeability response, assayed in the same reconstituted vesicles from which the planar bilayers are derived, shows that the fraction of desensitized AChRs increases from ~50% at 0.01 μM ACh to 100% at 0.1 μM ACh (Fig. 8). This is the range of ACh concentrations at which the A_L/A_S ratio is most sensitive to the concentration of ACh.

3.3.3d. Autocorrelation Analysis of AChR Channel-Gating Kinetics. A different approach to the analysis of single channels that, in addition to the conventional computation of dwell-time distributions, takes advantage of autocorrelation and autocovariance functions was developed. This method allows the extraction of information about the pathways connecting the open and closed states that is contained in such autocorrelation functions. It is only applicable to stationary processes in membranes containing a single channel. Extended accounts of both the theoretical formalism (Fredkin *et al.*, 1985) and the experimental application to the AChR channel have appeared (Labarca *et al.*, 1985a).

The autocorrelation analysis for the open state of the AChR channel activated

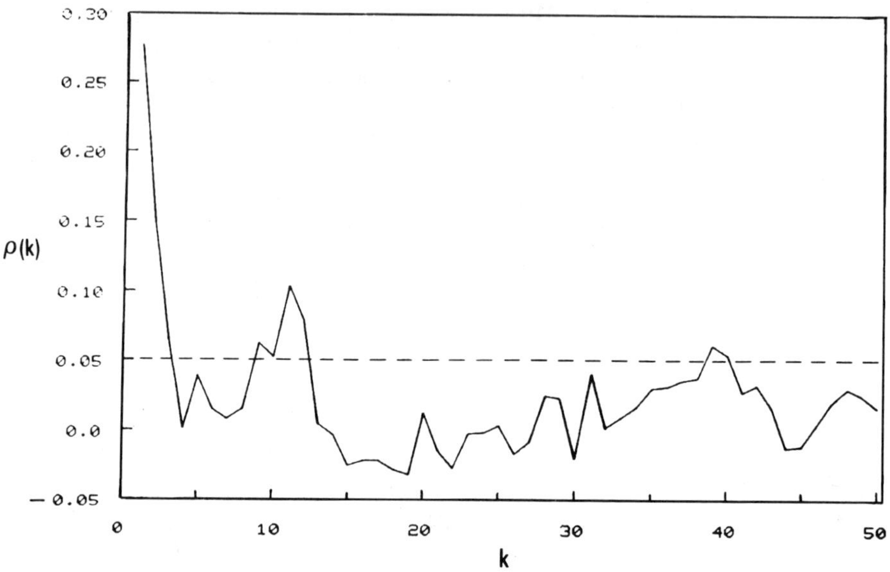

Figure 14. Autocorrelation function for the open-state aggregate of the reconstituted AChR channel. The analysis considers a sequence of open dwell times T_i, T_{i+1}, . . . and computes the covariance function:

$$\hat{\Gamma}(k) = \frac{1}{n-k} \sum_{i=1}^{n-k} (T_i - \bar{T})(T_{i+k} - \bar{T}),$$

where

$$\bar{T} = \frac{1}{n} \sum_{i=1}^{n} T_i \text{ and } k = 0,1,2, \ldots$$

The correlation function:

$$\hat{\rho}(k) = \frac{\hat{\Gamma}(k)}{\hat{\Gamma}(0)}$$

is illustrated. The dashed horizontal lines are at plus and minus twice the approximate standard error of the estimate in the case of white noise. The recordings were obtained from bilayers formed at the tip of patch pipets. The ACh concentration was 50 μM, and the applied voltage was 100 mV. The total number of openings analyzed was 1600. (From Labarca *et al.*, 1985a.)

by ACh is shown in Fig. 14. The theory described by Fredkin *et al.*, (1985) states that the correlation function decays geometrically. Since a correlation is significant only if it is larger in absolute value than twice the approximate standard error of the estimate in the case of white noise (horizontal dashed lines), it is

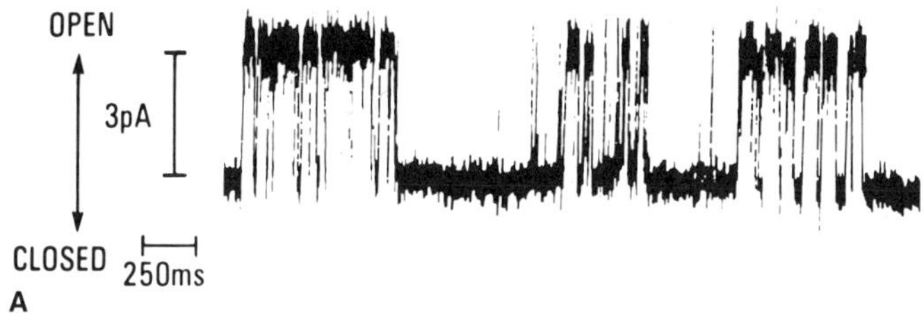

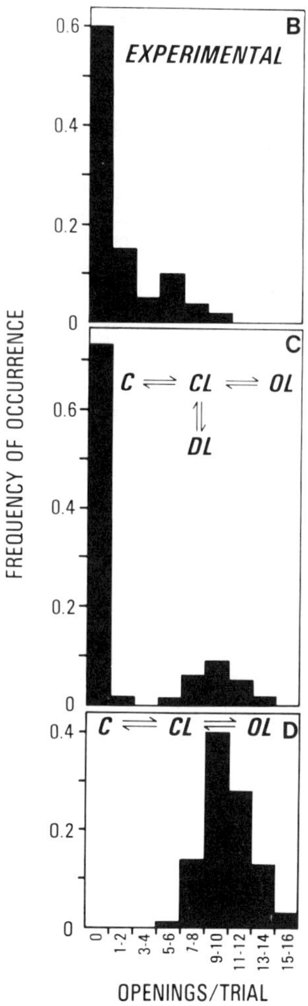

Figure 15. (A) Single AChR channel currents activated by desensitizing concentrations of agonist. This record, displayed at low time resolution, shows paroxysms of single AChR channels separated by quiescent periods. The channels were activated by 1 μM SubCh and recorded at V = 100 mV. (B–D) Frequency of channel opening in the presence of desensitizing concentrations of CCh. Records of single AChR channel currents activated with 1 mM CCh and recorded at 70 mV. (B) Experimental data; (C) Monte Carlo simulation with the four-state model:

$$\begin{array}{c} C \rightleftharpoons CL \rightleftharpoons OL \\ \phantom{C \rightleftharpoons{}} \upharpoonleft\downharpoonright \\ \phantom{C \rightleftharpoons{}} DL \end{array}$$

which includes a desensitized state (DL); (D) Monte Carlo simulation with the three state model:

$$C \rightleftharpoons CL \rightleftharpoons OL$$

(From Montal *et al.*, 1984.)

evident that substantial correlation is present near zero. This result indicates that there are at least two entry/exit states through which the open and closed aggregates communicate (Labarca *et al.*, 1985a).

The conditional probability analysis of single-channel currents of the purified AChR (Labarca *et al.*, 1984b) provided evidence of two kinetically distinct open

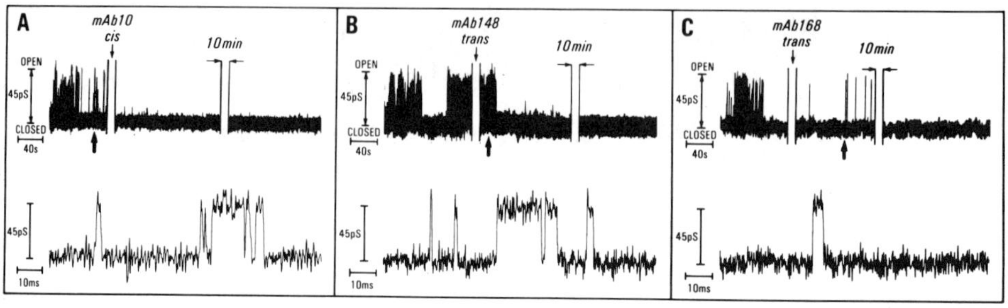

Figure 16. Monoclonal antibodies specific to the β and γ subunits block the channel activity of the reconstituted AChR. Single AChR channel currents were activated by 10 μM ACh and recorded at 100 mV. The concentration of AChR was 0.05 μM. The recordings were obtained from bilayers formed at the tips of patch pipets. In A, MAb 10 inhibits single AChR channel activity. At the arrow, mAb 10 (0.1 μM) was added to the *cis* chamber. The seal resistance was 5 GΩ. In B, MAb 148 inhibits single AChR channel activity. At the arrow, 3.8 μM mAb 148 was added to the *trans* chamber. The seal resistance was 6 GΩ. In C, MAb 168 inhibits single AChR channel activity. At the arrow, 1.3 μM mAb 168 was added to the *trans* chamber. The seal resistance was 7 GΩ. In A, B, and C, the lower panel shows the section of the record indicated by the arrow at higher time resolution (note the change in time calibration). (From Blatt *et al.*, 1986.)

states. The autocorrelation analysis further supports that notion and, in addition, indicates that these two open states can be entered or exited via at least two pathways, in other words, can be reached from at least two closed states. This new information extracted from the autocorrelation analysis is an independent demonstration of the existence of at least two open states and supports the kinetic scheme outlined above (Labarca *et al.*, 1985b). Jackson *et al.* (1983) applied a different correlation test based on the analysis of pairs of openings to AChR channel in cultured cells and reached a similar conclusion.

3.3.3e. Desensitization. Single-channel currents occur in paroxysms of channel activity followed by quiescent periods during which little or no activity is observed (Sakmann *et al.*, 1980; Montal *et al.*, 1984; Labarca *et al.*, 1984a). This pattern is more pronounced as the agonist concentration increases and is reflected in histograms of channel-opening frequencies (Montal *et al.*, 1984). Long single-channel records are divided into 100-msec sections (trials), and the histograms show how many 100-msec trials have 0, 1, 2, etc. openings. Figure 15 shows a typical histogram for 1 mM CCh (upper panel). The peak at zero openings corresponds to the quiescent periods. The peak at ~6 openings/100-msec trial corresponds to the number of openings in a typical paroxysm. Monte Carlo simulations with the simplest three-state model consisting of two closed (unliganded, C, and liganded, CL) and one open state (OL), do not resemble the recorded pattern of channel activity, especially at high agonist concentration. Inclusion of a desensitized liganded state (DL) reproduces the qualitative features of channel recordings (Montal *et al.*, 1984). The occurrence of paroxysms of channel activity thus seems to result from the transit of AChR through its active conformation, from which it can open several times before desensitizing (see also Katz and Thesleff, 1957; Sakmann *et al.*, 1980).

3.3.3f. Monoclonal Antibodies Specific to the β and γ Subunits Affect Channel Gating. A library of monoclonal antibodies (mAbs) against the AChR has been used to identify regions of the AChR involved in channel function (see Lindstrom, Chapter 9). Monoclonal antibodies directed against determinants on specific subunits have been evaluated with regard to their ability to alter the single-channel properties of the reconstituted AChR (Lindstrom *et al.*, 1981a,b, 1983; Tzartos and Lindstrom, 1980, 1981; Blatt *et al.*, 1984, 1986.)

Monoclonal antibodies that bind to the main immunogenic region on the extracellular domain of the α subunit and away from the ACh binding site (Fig. 1) and to a cytoplasmic determinant on the δ subunit altered neither the single-channel conductance nor the channel open times. In contrast, three mAbs directed against determinants on the β and γ subunits inhibited channel activity (Lindstrom *et al.*, 1983; Blatt *et al.*, 1984, 1986). This is illustrated in Fig. 16. Monoclonal antibody 10, which binds to an extracellular domain of the β subunit and also cross reacts with the α subunit (Fig. 1), caused inhibition of the AChR channel when added to the same compartment as ACh (Fig. 16A). Monoclonal antibody 148, which recognizes a cytoplasmic domain of the β subunit, and mAb 168, directed against a cytoplasmic domain of the γ subunit, inhibited single AChR channel activity only when added to the compartment opposite to that containing ACh, in agreement with their specificity for determinants on the cytoplasmic surface of the AChR (see Sargent *et al.*, 1984; Wan and Lindstrom, 1985) (Fig. 16B, C). The inhibition of channel activity can be accounted for by a reduction in the number of active channels and not by an alteration in single-channel characteristics. In agreement, mAb 10 was previously shown to inhibit the CCh-activated influx of $^{22}Na^+$ into reconstituted vesicles when added after their formation (Lindstrom *et al.*, 1983). In contrast, mAbs 148 and 168 inhibited the flux response only if present during the vesicle reassembly process, implying that inhibition was conditional to binding to a cytoplasmic AChR domain (Wan and Lindstrom, 1985). Interestingly, the monovalent Fab fragments of these mAbs do not affect channel activity.

4. Conclusions and Future Prospects

The molecular physiology of postsynaptic signal transduction has been reduced to studies on the transmembrane events proceeding in the purified AChR molecule. The concepts and techniques required for the resolution and reconstitution of the system are now well established. The full cycle of solubilization, purification, and reconstitution of the AChR can be achieved without impairment of channel function. Therefore, the stage is set to investigate how both chemical and genetic modifications in the structure of the AChR affect its channel gating.

One interesting protein modification is phosphorylation, which is widely recognized as a regulatory mechanism in signal transduction (see Nestler *et al.*, 1984). The *Torpedo* AChR is a substrate for three distinct protein kinases, namely a cAMP-dependent protein kinase (Huganir and Greengard, 1983), protein kinase

C, and an endogenous tyrosine-specific protein kinase (Huganir *et al.*, 1984). The latter phosphorylates a single intracellular site of the β, γ, and δ subunits. The effect of AChR phosphorylation on channel function can be evaluated using reconstituted AChR in planar lipid bilayers. The foreseeable trends are likely to encompass the use of monoclonal antibodies to probe AChR function (Blatt *et al.*, 1984, 1986; Lindstrom *et al.*, 1983; Gonzalez-Ros *et al.*, 1984; Goldberg *et al.*, 1983; Donnelly *et al.*, 1984) in conjunction with site-directed antibodies (Lindstrom *et al.*, 1984; Neumann *et al.*, 1984). The successful application of site-specific oligonucleotide metagenesis to the AChR (Mishina *et al.*, 1985) will give impetus to expand the use of molecular genetics to channel proteins. Individually, the results of these approaches are open to several interpretations. However, it is anticipated that the integration of the information obtained using site-directed mutagenesis, peptide-specific antibodies, and chemical modification to known sites on the AChR subunits and the single-channel recordings in reconstituted receptors will eventually lead to the clarification of structure–function relationships in the AChR.

ACKNOWLEDGMENTS. We are indebted to Jon Lindstrom, Myrta Montal, and Yoav Blatt for their collaboration and to Laura Castrejon for her invaluable assistance. This work was supported by grants from the National Institutes of Health (EY-02084) and the Department of the Army Medical Research (17–82–C2221).

A1. Appendix 1: Methods for the Solubilization, Purification, and Reconstitution of Functional AChRs

A1.1. Preparation of AChR-Rich Electric Organ Membranes

The procedure described here is based on the method of Elliott *et al.* (1980), according to which AChR-rich membrane vesicles are purified on the basis of their relatively high density because of their high concentration of AChR. It is essential to use protease inhibitors during the initial homogenization step to prevent proteolytic degradation of the AChR. Frozen *Torpedo californica* electric organ (Pacific Biomarine, Venice, CA), typically 400 g, is broken into chunks, mixed with 1 volume of room-temperature 10 mM Na-phosphate buffer, pH 7.5, 10 mM NaN_3, 5mM EDTA, and agitated occasionally for about 15 min to allow the tissue to partially thaw. All subsequent steps should be performed at 4°C. Iodoacetamide and phenylmethanesulfonylfluoride are added to 5 mM and 1 mM, respectively. The mixture is homogenized in a 1-gallon Waring blender for 2 min, and aliquots of about 200 ml are rehomogenized for 2 min in a small 1-quart Waring blender at high speed. The homogenate is then centrifuged at low speed for 10 min at 5,000 rpm in a Beckman JA-21 rotor. The supernatant is strained through cheesecloth to remove large fragments of tissue. To increase the yield of AChR, the pellet can be rehomogenized in an equal volume of buffer and centrifuged as before, and the strained supernatant can be pooled with the first one. The low-speed pellet is discarded, and the supernatant is centrifuged in a Beckman type 19 rotor at 16,000 rpm for 90 min. This crude membrane pellet can serve as the starting material for solubilization of AChRs to be further purified on toxin-agarose (see below).

In order to prepare membranes more enriched in AChR, the high-speed pellet is resuspended in 3 volumes of 10 mM Na-phosphate buffer, pH 7.5, 10 mM NaN_3, 5 mM EDTA, in a Virtis homogenizer 4 × 15 sec at medium speed. After a brief spin in a serologic centrifuge to drain the foam, 60% sucrose is added to the suspension to a final concentration of 32% (w/w) as measured with a refractometer. This suspension is layered at 12 ml/tube over 14 ml of 36% (w/w) sucrose in 10 mM Na-phosphate buffer, pH 7.5, 10 mM NaN_3, 5 mM EDTA. After centrifugation for 60 min at 45,000 rpm in a Beckman Ti-50.2 rotor, the floating fat pad and pellet are discarded. The band of AChR-rich membranes is collected, diluted with an equal volume of distilled water, and then centrifuged for 30 min at 20,000 rpm in the Ti-50.2 rotor. The resulting membrane pellet is suspended in a small volume of 10 mM Na-phosphate buffer, pH 7.5, 100 mM NaCl, 10 mM NaN_3, using an 18-guage needle and syringe. Specific activities of these membranes range between 1.0 and 2.0 μmole [^{125}I]α-BGT binding sites per gram protein, and major contaminants, as revealed by polyacrylamide gel electrophoresis in sodium dodecylsulfate, are bands at 43,000 molecular weight and several high-molecular-weight bands at about M_r 100,000.

A1.2. Alkaline Extraction of AChR-Rich Electric Organ Membranes

This method, used to further purify the above AChR-rich membranes, is based on the premise that extraction of AChR-rich electric organ membranes with low-ionic-strength pH 11 buffer removes peripheral membrane proteins, in this case the 43,000 protein (Neubig *et al.*, 1979; Moore *et al.*, 1979; Elliott *et al.*, 1979; Barrantes *et al.*, 1980). Electric organ membranes enriched in AChR are diluted with cold distilled water, and the pH of the suspension is raised to 11.2 by slowly adding 0.2 M NaOH while stirring. The alkali-treated membranes are pelleted by centrifugation for 30 min at 20,000 rpm in a Beckman Ti-50.2 rotor. The glassy pellets are suspended in 10 mM Na-phosphate buffer, pH 7.5, 100 mM NaCl, 10 mM NaN_3. The pH is measured and adjusted to 7.5 using tenfold concentrated Na-phosphate buffer. The specific activity of the alkali-washed membranes is routinely 3.2 ± 0.2 μmole [^{125}I]α-BGT binding sites per gram protein. These membranes should be devoid of the 43,000 protein, but high-molecular-weight contaminants are usually still present. The membrane suspensions can be fozen in aliquots in a methanol–dry ice bath and stored at −20°C without loss of activity.

A1.3. Solubilization of AChR-Rich Membranes

The AChR-rich membranes are dissolved in 2% Na-cholate [recrystallized according to Kagawa and Racker (1971) if obtained from Sigma or used straight if obtained from Interchem, Montluçon, France] in the presence of the desired concentration of crude soybean lipids (L-α-phosphatidylcholine type IIS, Sigma, St. Louis, MO; routinely 5 mg/ml) in 10 mM Na-phosphate buffer, pH 7.5, 100 mM NaCl. Crude soybean lipids are prepared as a 150 mg/ml stock dispersed in distilled water by sonication under argon or nitrogen in a Bransonic® bath sonicator. The solubilized membranes are gently shaken for 1 hr at 4°C. Undissolved material can be removed either by centrifugation at 165,000 × *g* for 30 min or, for small volumes, by centrifugation for 15 min in an Eppendorf microfuge at 4°C. both procedures give comparable AChR concentrations in the supernatants.

A1.4. Purification of Solubilized AChRs on Toxin-Agarose

The extract of electric organ membranes is added to toxin-agarose [0.5 mg of *Naja naja siamensis* toxin III (Karlsson *et al.*, 1971) per milliliter of Sepharose C14B] and is shaken gently for 1 hr. Application of about 4 nmol AChR/ml gel leads routinely to binding of approximate two-thirds of the AChRs in the extract to the resin. The resin is poured into a column and washed with 250 to 300 volumes of cholate/lipid buffer (2% Na-cholate, 5 mg/ml of soybean lipid, 100 mM NaCl, 10 mM Na-phosphate buffer, pH 7.5). Conventionally, AChRs eluted from the affinity column are collected on a DEAE gel, from which they can be eluted at high salt concentrations. However, the presence of lipid in the solution prevents the use of DEAE, since the lipid adsorbs to and coats the ion-exchange column. Instead, AChRs can be collected on a column of Con A-agarose (one-tenth the volume of the toxin-agarose column) connected to the toxin-agarose column. The AChR can be eluted from the toxin agarose by 1 mM benzoquinonium chloride (Stirling-Winthrop Pharmaceuticals), a competitive antagonist, or 1 M CCh in cholate/lipid buffer, recirculating through the two columns overnight by means of a peristaltic pump. Benzoquinonium is preferable as eluting agent since it is effective at a 1000-fold lower concentration than CCh and its presence can be visually monitored by the pink color of the solution. The same benzoquinonium solution can be used repeatedly for several preparations.

As the AChR is eluted from the toxin-agarose, it is bound to the Con A-agarose, which can then be washed free of the eluting ligand. Incubation at room temperature rather than in the cold is essential for optimal elution from the Con A-agarose. The AChR can be eluted using 1 column volume of 1 M α-methyl-D-mannoside, 5 mM NaN_3, 1 mM EDTA in cholate/lipid buffer for 2 hr at room temperature. The elution yield from the Con A column amounts to ~25% of the AChR originally bound to the toxin-agarose. The specific activity of the eluate is routinely ~8 μmol [^{125}I]α-BGT binding sites per gram protein. The total yield of toxin-agarose-purified AChR, compared to the AChR content in the initial homogenate, ranges between 5% and 10%. Polyacrylamide gel electrophoresis in SDS usually reveals only the four AChR polypeptides, free of contaminants, after staining with Coomassie brilliant blue or using the more sensitive ammoniacal silver stain procedure (Oakley *et al.*, 1980).

A1.5. Measurement of AChR Concentration

The AChR concentration is measured by immune precipitation as the concentration of [^{125}I]α-BGT binding sites (Lindstrom *et al.*, 1981a). α-Bungarotoxin can be purified from the crude venom of *Bungarus multicinctus*, obtained from Miami Serpentarium, according to Ravdin and Berg (1979) and iodinated (Lindstrom *et al.*, 1976). Aliquots (0.5 ml) of vesicles diluted to about 10^{-8} M α-BGT binding sites in 0.5% Triton X-100, 100 mM NaCl, 10 mM Na-phosphate buffer, pH 7.5, are incubated with 3×10^{-8} [^{125}I]α-BGT for 15 min at room temperature. Then, 5 μl of a threefold concentration of a 40% ammonium sulfate cut of rat anti-*Torpedo* AChR serum is added for at least 1 hr. Next, goat antirat IgG is added for 30 min. The immune precipitate is pelleted by 2 min of centrifugation in an Eppendorf microfuge. The washed pellets are counted in a γ counter. Counts of ^{125}I pellets from reaction mixtures without AChR are subtracted from all samples. All assays should be performed in triplicate. To measure protein, the method of Lowry *et al.*, (1951) can be used with bovine serum albumin as standard and using cholate/lipid buffer as the blank.

A1.6. Incorporation of AChRs into Liposomes

The total lipid concentration in the membrane extract or the column eluate is adjusted to 25 mg/ml using a sonicated dispersion of soybean lipids in distilled water. Supplementary cholesterol can be included when desired by mixing a 100 mg/ml solution of soybean lipids in chloroform with a 100 mg/ml solution of cholesterol (Sigma, St. Louis, MO) in chloroform in the desired proportions. After evaporation of the chloroform under nitrogen or argon, the dry lipid film is suspended in a 2% Na-cholate solution in 100 mM NaCl, 10 mM Na-phosphate buffer, pH 7.5, by sonication. It is essential to remove all traces of chloroform such that no chloroform odor is perceptable prior to addition of AChR. Recommended final concentration of the components in the reconstitution mixture are 25 mg/ml total lipid, 2% Na-cholate, 100 mM NaCl, 10 mM Na-phosphate buffer, pH 7.5, and 1–2 μM AChR. Liposomes containing AChR are formed through removal of the detergent by dialysis for 16–18 hr against 500 volumes of 10 mM Na-phosphate buffer, pH 7.5, 100 mM NaCl, 10 mM NaN_3, followed by dialysis for the same period against 500 volumes of 145 mM sucrose, 10 mM NaN_3, 10 mM Na-phosphate buffer, pH 7.5. When desired, the resulting reconstituted preparation can be subjected to a freeze–thaw cycle by placing it in a freezer at −20°C or in liquid nitrogen followed by thawing at ambient temperature. This results in sealing of leaky vesicles and leads to vesicle fusion (Anholt *et al.*, 1982).

A1.7. Measurement of AChR Orientation

The fraction of AChRs in native or reconstituted vesicles accessible to [^{125}I]α-BGT, operationally defined as the "sidedness" of the membranes, is measured by comparing the total AChR concentration observed after solubilization in Triton X-100 (as described above) with the AChR concentration observed with intact vesicles. The assay is conducted in 100 mM NaCl, 10 mM Na-phosphate buffer, pH 7.5 (Lindstrom *et al.*, 1980).

A1.8. Assay of AChR Function

The AChR function in the reconstituted vesicles is assayed at room temperature as the amplitude of the cumulative uptake of $^{22}Na^+$ or $^{86}Rb^+$ (New England Nuclear) during a 10-sec incubation period in the presence of agonist. Entrapped $^{22}Na^+$ or $^{86}Rb^+$ can be separated from the external radioisotope by passage through a Dowex cation-exchange column according to the method of Gasko *et al.* (1976).

Dowex 50W-8X100 (Sigma, St. Louis, MO) is washed with distilled water by decantation until the supernatant is clear. 2-Amino-2-hydroxymethyl-1,3-propanediol (Trizma®) base (Sigma, MO) is then added to equal volumes of resin and water at the ratio of 440 g/kg of resin. After stirring for an hour, additional Trizma® is added if necessary to bring the pH to 9.8. Then the tris is decanted, and the beads are washed with many changes of water until the pH approaches that of the water. Disposable columns (about 0.5 × 8 cm) are poured in short Pasteur pipets loosely plugged with glass wool. A small plastic funnel is attached to the top of each column with Tygon® tubing. The capacity of these columns is sufficiently large to allow repeated use for up to three assays. Immediately before the first use, each column is washed with 3 ml of 170 mM sucrose, 3.3 mg/ml of bovine serum albumin. When $^{22}Na^+$ is used, the assay should be conducted behind lead shielding.

$^{22}Na^+$ or $^{86}Rb^+$ (5 μl of 0.2 mCi/ml) and 5 μl of water or a tenfold concentrated solution of the agonist are mixed together in Eppendorf microfuge tubes. The assay is initiated by rapidly injecting 40 μl of vesicles in 10 mM Na-phosphate buffer, pH 7.5, 145 mM sucrose, 10 mM NaN_3, with a Pipetman™ pipetter. After mixing with the pipetter, the mixture is transferred onto a Dowex column in a total elapsed time of about 10 sec. Sucrose solution (3 ml of 175 mM) is carefully added to elute the vesicles. Eluates are counted in a γ counter for $^{22}Na^+$ or in a scintillation counter for $^{86}Rb^+$). For each sample, triplicate CCh (Sigma, MO) responses and triplicate water blanks should be measured. The agonist-activated cation uptake is defined as the difference between the activity in the water samples and those containing agonist. Standard errors are generally within 10% of the mean. When ACh (Sigma, St. Louis, MO) is used as agonist, the assay should be preceded by a 30-min preincubation period of the vesicles with 1 μM echothiophate iodide (Ayerst Laboratories, Inc., New York, NY) or a similar effective ACh esterase inhibitor.

The apparent internal volume of the vesicles can be measured by mixing 40 μl of vesicles, 5 μl of $^{22}Na^+$ or $^{86}Rb^+$, and 5 μl of H_2O as above, in triplicate, but incubating at 40°C for 48 hr prior to passage through Dowex®.

A2. Appendix 2: Methods for the Reconstitution of Functional AChRs in Planar Lipid Bilayers

A2.1. Bilayers Across Apertures in Teflon® Septi

A2.1.1. The Chamber

The chamber consists of two compartments fabricated in black Teflon®, which is chosen for its flexibility. Reconstitution studies also in progress in our laboratory that utilize this system include light-activated proteins (e.g., rhodopsin, bacteriorhodopsin, and reaction centers from photosynthetic bacteria) (for review, see Montal *et al.,* 1981; Montal, 1985). (Black Teflon® TFE is available from Fluorocarbon, P.O. Box 3339, 1754 South Clementine Street, Anaheim, CA 92083.) Figure 17A illustrates the chamber routinely used. The aperture in the front provides access to the bilayer for light-activation experiments. There are two holes (1.5 mm) in each compartment, through which sections of polyethylene tubing are inserted to connect the chamber with the syringe used to raise or lower the water level (Fig. 17B). These holes can also be used for perfusion experiments to replace the aqueous phase in each compartment. The chamber capacity is 1.0 ml. The chamber interior should be smooth, avoiding steps that hinder the continuous flow of the interface when the water levels are displaced. The cleaning of the chamber is of utmost importance. The routine procedure is to rinse thoroughly with water and then with chloroform, ethanol, ether, and acetone in this sequence. Thereafter, the chambers are immersed in boiling 1 M HCl (15 min), then in 1 M NaOH (15 min), and finally in water (30 min).

A2.1.2. Fabrication of the Septum and the Aperture

The two compartments of the chamber are separated by a Teflon® film pierced in the center with an aperture (~200 μm diameter). The Teflon® film can be obtained from Saunders Corp., Los Angeles, CA (No. S-25, thickness 12.5 μm) or from Yellow Spring Instruments, Yellow Springs, OH (Membrane Kit 5937 or 5776, thickness 12.5 μm; this is the high-sensitivity membrane used in oxygen electrodes). The septum is fabricated, following the modification of Schindler and Feher (1976) (see Montal, 1974), as a "sandwich" of the thin film (12.5 μm) between two thick Teflon® O-rings (50 μm). The aperture is cut with an electrically sharpened syringe needle (27 gauge, Hamilton or B-D Yale) (see also Chapter 13 by Hartshorne *et al.*). Sharpening is conducted in 5 M HCl solution applying 1.5 V (with the positive pole connected to the needle) and by repeated immersion and removal of the needle tip. The Teflon® film is supported on a new plexiglass plate. The needle is mounted on a syringe and inserted in a device that guides the needle while gentle pressure is applied against the Teflon® film. The hole is cut by a single turn around the axis. The quality of the hole is checked under a mircoscope, but the ultimate test is the ease of forming membranes, their resistance, and their stability. The hole should be coated with 2 μl of 0.5% (v/v) hexadecane in hexane or in chloroform: methanol (2:1).

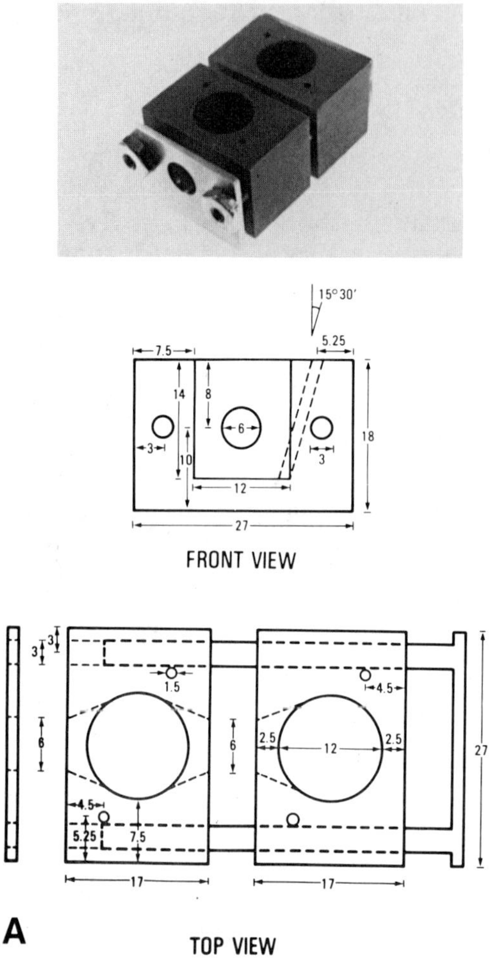

Figure 17. Experimental system for the formation of planar lipid bilayers across apertures in Teflon® septi and for recording single AChR channel currents. (A) The chamber. Top: Photograph of the chamber. Middle and Bottom: Diagram of the chamber with construction specifications, illustrating front (middle) and top (bottom) views. Dimensions in millimeters. (B) A complete setup. The chamber is located inside an aluminum box (a Faraday cage). The chamber holder is fitted with water inlets and outlets to allow studies under constant temperature and at different temperatures. The syringes (disposable) used to displace the water levels are located to the right and also are enclosed in a metallic compartment. The syringes are connected to the chamber via polyethylene tubing. Two Ag/AgCl electrodes are immersed in Teflon® reservoirs containing 1 M KCl and connected on one end to the chamber compartments via salt bridges (1 M KCl in 2% w/v agar) and on the other to the amplifier (rectangular box at the left) via high-quality chasis BNC connector. The experimental setup is mounted on a vibration-isolation platform.

A2.1.3. Routine Procedure to Reconstitute AChR in Planar Lipid Bilayers

The chambers are filled with 0.5 ml of the indicated buffer prior to the addition of the vesicles. For routine studies, the medium is 0.5 M NaCl, 0.5 to 5 mM $CaCl_2$, 2.5 mM tricine or HEPES buffer, pH 7.4, and 50 μM 4,4′-diisothiocyanatostilbene-2,2′-disulfonic acid (DIIDS). At this DIIDS concentration, the *Torpedo electroplax* chloride channel is blocked (Miller and White, 1984). The composition of the electrolyte solution in the two chambers is the same.

If symmetric bilayers are formed, the same protocol is applied to both monolayer chambers. Alternatively, when asymmetric bilayers are formed, the AChR-containing monolayer compartment (*cis*) is prepared as previously described, whereas the monolayer compartment, which contains only lipid (*trans*), is derived from liposomes devoid of AChR. The liposomes, from partially purified soybean phospholipids (Kagawa and Racker, 1971) at a concentration of 20 mg/ml in 80

Figure 17. *(continued)*

mM tricine buffer, pH 8.0, are prepared by a 10-min sonication under argon in a BRANSONIC, (Heat Systems Ultrasonic, Plainview, New York) or Laboratory Supplies Co., Inc. (Hicksville, New York, Model G112SPIT) water-bath sonicator. These liposomes, 50 μl samples, are then subjected to the same incubation conditions as the reconstituted AChR vesicles described above. After a few minutes, each monolayer achieves a surface pressure of ~30 dyne/cm. Thereafter, the level of the AChR-containing compartment is raised over the aperture in the Teflon® septum by introducing additional buffer into the chamber from a reservoir. Next, the second monolayer is apposed to the septum, and a bilayer is formed across the aperture. Membrane formation is continuously followed by monitoring the membrane capacitance. The reconstituted bilayers exhibit a capacitance of ~0.8 $\mu F/cm^2$. After formation, the membranes are left unperturbed for 10 min. The electrical conductance is monitored by applying voltages of up to ±200 mV. Bilayers with changes in conductance during this period are discarded. Thereafter, the pharmacological agent under study is introduced into the AChR compartment and mixed with a magnetic stirrer.

A2.2. Bilayers at the Tip of Patch Pipets

A2.2.1. The Chamber

The chamber is commercially available. A single well is cut out from a multiwell Disposo-tray™ available from Linbro (Flow laboratories, Inc., McLean,

Virginia 22102; catalogue No. 76-336-05). Each tray has 48 flat-bottom wells (0.8 × 0.6 cm); the well capacity is ~0.4 ml. Each well is used only once.

A2.2.2. Fabrication of the Patch Pipet

A detailed description of pipet fabrication is provided by Hamill *et al.*, (1981). Standard hematocrit capillaries made of flint glass (BLU-TIP, plain; I.D. 1.1–1.2 mm, Lancer, St. Louis, MO) are used. The capillaries are immersed in 1 N nitric acid overnight, rinsed thoroughly with deionized water, and stored in deionized water. Immediately before pulling, the capillaries are rinsed with methanol and dried by a nitrogen gas flush. The pulling procedure is performed in a vertical pipet puller (David Kopf, 700C, Tujunga, CA) and consists of two steps: first the capillary is extended (8 mm) to produce an intermediate thin section. This section is then recentered with respect to the heating element by an upward displacement of ~5 mm. The final pull produces two pipets with an opening (tip) diameter of approximately 0.8 μm with a steep taper. The tip size can be controlled by adjusting the heater current in the second pull. No further treatment such as fire polishing or coating is necessary. The tip size is adjusted to yield ~10 MΩ open pipet resistance when filled and immersed in 0.5 M NaCl, 0.5 mM $CaCl_2$, and 2.5 mM HEPES, pH 7.4. Pipets should be used within 4 hr after fabrication.

The two compartments separated by the membrane are connected to the amplifier by means of two Ag/AgCl electrodes. The active electrode, identified as the pipet electrode, is connected to a BNC connector pin of the shielded pipet holder, a component of the patch clamp system (see Section A2.3). The pipet, previously filled with buffer, is placed in the pipet holder. The holder is connected to the headstage of the patch-clamp system, which is designed to be mounted directly on a micromanipulator. The reference electrode is immersed in the chamber used for bilayer formation and is connected to the high-quality ground connection on the headstage of the patch clamp system. The pipet potential is determined with reference to this ground. Positive or negative pressures are applied to the pipet with a syringe.

A2.2.3. Routine Procedure to Reconstitute AChRs in Planar Lipid Bilayers

The process of bilayer formation from monolayers at the air–water interface at the tip of patch pipets is fundamentally analogous to that of lipid bilayer formation from two monolayers at the air–water interface across an aperture in a Teflon® partition that separates two aqueous compartments (Montal and Mueller, 1972; Wilmsen *et al.*, 1983; Suarez-Isla *et al.*, 1983; Schuerholz and Schindler, 1983; Coronado and Latorre, 1983). The method is as follows:

1. Lipid protein monolayers are derived from the reconstituted AChR vesicles. A patch pipet is introduced into the solution under positive pressure in order to avoid its occlusion.
2. The pipet is removed from the solution. The lipid head groups of the monolayers are attached to the polar glass pipet while the hydrocarbon tails contact air.

3. The pipet is reimmersed into the solution. This leads to the apposition of the hydrocarbon tails of the attached monolayer to those of the original monolayer, thereby forming a bilayer at the tip of the patch pipet.

It should be noted that the interaction of the lipid monolayer with the bilayer support is different in the two modalities: whereas the lipid hydrocarbon tails interact with the hydrophobic Teflon® septum, the lipid polar head groups interact with the polar glass pipet.

The reconstituted bilayers exhibit membrane resistances in the range of $5\ \mathrm{G\Omega} \leq R \leq 25\ \mathrm{G\Omega}$. All the planar bilayer experiments are performed at 22 ± 2°C.

A2.3. Electrical Recordings and Data Processing

The current recordings from bilayers in Teflon® septi are obtained with a current amplifier designed for high speed and high resolution. The critical strategy is the minimization of the effective feedback capacitance on the current-to-voltage converter (Analogue Devices 528, Burr-Brown 3523J, or National Semiconductor LF357AH). This allows measurements of signals at a gain of 10^{-10} A/V with 1 pA of noise (peak-to-peak amplifier time constant, τ = 250 μsec) and at a gain of 10^{-9} A/V with 2 pA of noise (amplifier τ = 120 μsec) (Montal and Mueller, 1972; Labarca *et al.*, 1984a). For the current recordings from patch pipets, a commercially available extracellular patch-clamp system is used (List L/M EPC-5 and EPC-7, List Electronics, Darmstadt, Federal Republic of Germany, and Medical Systems Corp., Greenvale, NY) (Suarez-Isla *et al.*, 1983; Labarca *et al.*, 1985a,b; Hamamoto and Montal, 1985). Voltage is applied and current is measured with Ag/AgCl electrodes (pellet electrodes E 255 and E 206 from *In Vivo* Metric Systems, Healdsburg, CA). Constant voltage is supplied by a DC source (Omnical 2001, WPI Instruments, New Haven, CT), and the *trans* side of the membrane is defined as zero voltage. Accordingly, a negative applied voltage corresponds to a depolarization in the electrophysiological convention.

Membrane current is recorded directly on a storage oscilloscope or amplified and stored on a RACAL 4DS tape recorder (frequency response DC to 10kHz, Hythe, Southampton, England) for subsequent analysis. After filtering, the recordings are digitized at a sampling interval of 100 μsec in a PDP 11/34 computer (Digital Equipment Corporation). The digitized records are inspected on a computer-generated display and compared to a matching routine in which the computer generated a virtual reproduction of the actual recording: the computer-generated rectangular pulses are checked for fidelity in matching the channel openings and closings. The duration of the generated rectangular pulses corresponds to the open times, and the intervals between the pulses are the closed times. The parameters of the generated rectangular pulses, position and width, are stored in the computer. The single-channel conductance is calculated from conductance histograms fitted by the sum of two Gaussian functions corresponding to the closed- and open-channel states. Histograms of dwell times in the open and closed states of the generated pulses are automatically produced. Histograms of dwell times, normalized to have a total area equal to 1, are fitted by a probability density function of the form:

$$f(t) = \sum_{i=1}^{N} A_i e^{-t/\tau_i}$$

The amplitudes A_i and the time constants τ_i of the exponential fit obtained by a χ^2 minimization algorithm are derived (Labarca *et al.*, 1984a,b, 1985a).

A2.4. Efficiency and Reliability of the Reconstituted System

Efficiency is operationally used to describe the fraction of the total bilayers formed, under given conditions, that exhibit functional activity. Efficiency is stringently dependent on the quality of the AChR preparation: the higher the specific activity, the higher the reconstitution efficiency. With high-quality preparations (see Section 2), the efficiency should be better than 60%. It should be stressed that the technique of forming a bilayer samples discrete areas of monolayer, and, therefore, nonrandom distributions of AChR in the monolayer will affect the reconstitution efficiency.

References

Adams, D. J., Dwyer, T. M., and Hille, B., 1980, The permeability of endplate channels to monovalent and divalent metal cations, *J. Gen. Physiol.* **75**:493–510.

Adams, D. J., Nonner, W., Dwyer, T. W., and Hille, B., 1981, Block of end-plate channels by permeant cations in frog skeletal muscle, *J. Gen. Physiol.* **78**:593–615.

Anderson, D. J., Blobel, G., Tzartos, S., Gullick, W., and Lindstrom, J., 1983, Transmembrane orientation of an early biosynthetic form of acetylcholine receptor δ-subunit determined by proteolytic dissection in conjunction with monoclonal antibodies, *J. Neurosci.* **3**:1773–1784.

Anholt, R., 1981, Reconstitution of acetylcholine receptors in model membranes, *Trends Biochem. Sci.* **6**:288–291.

Anholt, R., Lindstrom, J., and Montal, M., 1980, Functional equivalence of monomeric and dimeric forms of purified acetylcholine receptor from *Torpedo californica* in reconstituted lipid vesicles, *Eur. J. Biochem.* **109**:481–487.

Anholt, R., Lindstrom, J., and Montal, M., 1981, Stabilization of acetylcholine receptor channels by lipids in cholate solution and during reconstitution in vesicles, *J. Biol. Chem.* **256**:4377–4387.

Anholt, R., Fredkin, D. R., Deerinck, T., Ellisman, M., Montal, M., and Lindstrom, J., 1982, Incorporation of acetylcholine receptors into liposomes: Vesicle structure and acetylcholine receptor function, *J. Biol. Chem.* **257**:7122–7134.

Anholt, R., Montal., M., and Lindstrom, J., 1983, Incorporation of acetylcholine receptors in model membranes: An approach aimed at studies of the molecular basis of neurotransmission, *Peptide Protein Rev.* **1**:95–137.

Anholt, R., Lindstrom, J., and Montal, M., 1984, The molecular basis of neurotransmission: Structure and function of the nicotinic acetylcholine receptor, in: *The Enzymes of Biological Membranes*, ed. 2 (A. Martonosi, ed.), Plenum Press, New York. pp. 335–401.

Barrantes, F. J., 1978, Agonist-mediated changes of the acetylholine receptor in its membrane environment, *J. Mol. Biol.* **124**:1–26.

Barrantes, F. J., Neugebauer, D.-C., and Zingsheim, H. P., 1980, Peptide extraction by alkaline treatment is accompanied by reangement of the membrane bound acetylcholine receptor from *Torpedo marmorata*, *FEBS Lett.* **112:**73–78.

Blatt, Y., Montal, M. S., Lindstrom, J., and Montal, M. 1984, Effect of antireceptor monoclonal antibodies on single channel currents of purified acetylcholine receptor reconstituted in lipid bilayers, *Biophys. J.* **45:**311a.

Blatt, Y., Montal, M. S., Lindstrom, J., and Montal, M., 1986, Monoclonal antibodies specific to the β and γ subunits of the *Torpedo* acetylcholine receptor inhibit channel gating, *J. Neurosci.* (in press).

Boheim, G., Hanke, W., Barrantes, F. J., Eibl, H., Sakmann, B., Fels, G., and Maelicke, A., 1981, Agonist-activated ionic channels in acetylcholine receptor reconstituted into planar lipid bilayers, *Proc. Natl. Acad. Sci. U.S.A.* **78:**3586–3590.

Boyd, N. D., and Cohen, J. B., 1980a, Kinetics of binding of [^{3}H]-acetylcholine and ^{3}[H]-carbamylcholine to *Torpedo* postsynaptic membranes: Slow conformational transitions of the cholinergic receptor, *Biochemistry* **19:**5344–5353.

Boyd, N. D., and Cohen, J. B., 1980b, Kinetics of binding of [^{3}H]-acetylcholine to *Torpedo* postsynaptic membranes: Association and dissocation rate constants by rapid mixing and ultrafiltration, *Biochemistry* **19:**5353–5358.

Brunner, Y., Skrabal, P., and Hauser, H., 1976, Single bilayer vesicles prepared without sonication: Physico–chemical properties, *Biochim. Biophys. Acta* **455:**322–331.

Cartaud, J., Sobel, A., Rousselet, A., Devaux, P., and Changeux, J.-P., 1981, Consequences of alkaline treatment for the ultra-structure of the acetylcholine receptor-rich membranes from *Torpedo marmorata* electric organ, *J. Cell Biol.* **90:**418–426.

Cash, D. J., Aoshima, H., and Hess, G. P., 1981, Acetylcholine-induced cation translocation across cell membranes and inactivation of the acetylcholine receptor: Chemical kinetic measurements in the millisecond time region, *Proc. Natl. Acad. Sci. U.S.A.* **78:**3318–3322.

Chang, H. W., and Bock, E., 1977, Molecular forms of the acetylcholine receptor: Effects of calcium ions and a sulfhydryl reagent on the occurrence of oligomers, *Biochemistry* **16:**4513–4520.

Changeux, J.-P., 1981, The acetylcholine receptor: An allosteric membrane protein, *Harvey Lect.* **75:**85–254.

Changeux, J.-P., Heidmann, T., Popot, J., and Sobel, A., 1979, Reconstitution of a functional acetylcholine regulator under defined conditions, *FEBS Lett.* **105:**181–187.

Changeux, J.-P., Devillers-Thiery, A., and Chemouilli, P., 1984, Acetylcholine receptor: An allosteric protein, *Science* **255:**1335–1345.

Claudio, T., Ballivet, M., Patrick, J., and Heinemann, S., 1983, Nucleotide and deduced amino acid sequences of *Torpedo californica* acetylcholine receptor γ subunit, *Proc. Natl. Acad. Sci. U.S.A.* **80:**111–115.

Colquhoun, D., and Sakmann, B., 1981, Fluctuations in the microsecond time range of the current through single acetylcholine receptor ion channels, *Nature* **294:**464–466.

Conti-Tronconi, B. M., and Raftery, M. A., 1982, The nicotinic cholinergic receptor: Correlation of molecular structure with functional properties, *Annu. Rev. Biochem.* **51:**491–530.

Coronado, R., and Latorre, R., 1983, Phospholipid bilayers made from monolayers on patch-clamp pipettes, *Biophys. J.* **43:**231–236.

Criado, M., and Barrantes, F. J., 1984, Conversion of acetylcholine receptor dimers to monomers upon depletion of non-receptor peripheral proteins, *Biochim. Biophys. Acta* **798:**374–381.

Criado, M., Eibl, H., and Barrantes, F. J., 1982, Effects of lipids on acetylcholine receptor: Essential need of cholesterol for maintenance of agonist-induced state transitions in lipid vesicles, *Biochemistry* **21:**3622–3629.

Criado, M., Eibl, H., and Barrantes, F. J., 1984, Functional properties of the acetylcholine receptor incorporated in model lipid membranes. Differential effects of chain length and head group of phospholipids on receptor affinity states and receptor-mediated ion-translocation, *J. Biol. Chem.* **259:**9188–9198.

Criado, M., Hochschwender, S., Sarin, V., Fox, L. J., and Lindstrom, J., 1985, Evidence for unpredicted transmembrane domains in acetylcholine receptor subunits, *Proc. Natl. Acad. Sci. U.S.A.* **82:**2004–2008.

Dalziel, A. W., Rollins, E. A., and McNamee, M. G., 1980, The effect of cholesterol on agonist-induced flux in reconstituted acetylcholine receptor vesicles, *FEBS Lett.* **122**:193–196.

Damle, V., McLaughlin, M., and Karlin, A., 1978, Bromoacetylcholne as an affinity label of the acetylcholine receptor from *Torpedo californica, Biochem. Biophys. Res. Commun.* **84**:845–851.

Dani, J. A., and Eisenman, G., 1984, Acetylcholine-activated channel current–voltage relations in symmetrical Na^+ solutions, *Biophys J.* **45**:10–12.

Devilliers-Thiery, A., Giraudat, J., Bentaboulet, M., and Changeux, J. P., 1983, Complete mRNA coding sequence of the acetylcholine binding α-subunit of *Torpedo marmorata* acetylcholine receptor: A model for the transmembrane organization of the polypeptide chain, *Proc. Natl. Acad. Sci. U.S.A.* **80**:2067–2071.

Donnelly, D., Milhovilovic, M., Gonzalez-Ros, J. M., Ferragut, J. A., Richman, D., and Martinez-Carrion, M., 1984, A non-cholinergic site-directed monoclonal antibody can impair agonist-induced ion flux in *Torpedo californica* acetylcholine receptor, *Proc. Natl. Acad. Sci. U.S.A.* **81**:7999–8003.

Dwyer, T. M., Adams, D., and Hille, B., 1980, The permeability of the endplate-channel to organic cations in frog muscle, *J. Gen. Physiol.* **75**:469–492.

Elliott, J., Dunn, S. M. J., Blanchard, S. G., and Raftery, M. A., 1979, Specific binding of perhydrohistrionicotoxin to *Torpedo* acetylcholine receptor, *Proc. Natl. Acad. Sci. U.S.A.* **76**:2576–2579.

Elliott, J., Blanchard, S. G., Wu, W., Miller, J., Strader, C. D., Hartig, P., Moore, H.-P., Racs, J., and Raftery, M. A., 1980, Purification of *Torpedo californica* postsynaptic membranes and fractionation of their constituent proteins, *Biochem. J.* **185**:667–677.

Epstein, M., and Racker, E., 1978, Reconstitution of carbamylcholine-dependent soidum ion flux and desensitization of the acetylcholine receptor from *Torpedo californica, J. Biol. Chem.* **253**:6660–6662.

Fairclough, R. H., Finer-Moore, J., Love, R. A., Kristofferson, D., Desmeules, P. J., and Stroud, R. M., 1983, Subunit organization and structure of an acetylcholine receptor, *Cold Spring Harbor Symp. Quant. Biol.* **48**:9–20.

Finer-Moore, J., and Stroud, R. M., 1984, Amphipathic analysis and possible formation of the ion-channel in an acetylcholine receptor, *Proc. Natl. Acad. Sci. U.S.A.* **81**:155–159.

Fredkin, D. R., Montal, M., and Rice, J. A., 1985, Identification of aggregated Markovian models: pplication to the nicotinic acetylcholine receptor in: *Proceedings of the Berkeley Conference in Honor of Jerzy Neyman and Jack Kiefer, Volume I* (L. M. LeCam and R. A. Olshen, eds.) pp. 269–289. Wadsworth, Belmont, California.

Froehner, S. C., 1981, Identification of exposed and buried determinants of the membrane-bound acetylcholine receptor from *Torpedo californica, Biochemistry* **20**:4905–4915.

Froehner, S. C., Gulbrandsen, V., Hyman, C., Jeng, A. Y., Neubig, R. R., and Cohen, J. B., 1981, Immunofluorescence localization at the mammalian neuromuscular junction of the M_r 43,000 protein of *Torpedo* postsynaptic membranes, *Proc. Natl. Acad. Sci. U.S.A.* **78**:5230–5234.

Fromherz, P., 1975, Instrumentation for handling monomolecular films at an air–water interface, *Rev. Sci. Instrum.* **46**:1380–1385.

Gasko, O. D., Knowles, A. F., Shertzer, H. G., Suolinna, E. M., and Racker, E., 1976, The use of ion-exchange resins for studying ion transport in biological systems, *Anal. Biochem.* **72**:57–65.

Gershoni, J. M., Palade, G. E., Hawrot, E., Klimowicz, D. W., and Lentz, T. L., 1982, Analysis of α-bungarotoxin binding to *Torpedo* acetylcholine receptor by electrophoretic transfer techniques, *J. Cell. Biol.* **95**:422a.

Goldberg, G., Mochly-Rosen, D., Fuchs, S., and Lass, Y., 1983, Monoclonal antibodies modify acetylcholine-induced ionic channel properties in cultured myoballs, *J. Membr. Biol.* **76**:123–128.

Gonzalez-Ros, J. M., Paraschos, A., and Martinez-Carrion, M., 1980, Reconstitution of functional membrane-bound acetylcholine receptor from isolated *Torpedo californica* receptor protein and electroplax lipids, *Proc. Natl. Acad. Sci. U.S.A.* **77**:1796–1800.

Gonzalez-Ros, J. M., Llanillo, M., Paraschos, A., and Martinez-Carrion, M. 1982, Lipid environment of acetycholine receptor from *Torpedo californica, Biochemistry* **21**:3467–3474.

Gonzalez-Ros, J. M., Ferragut, J. A., and Martinez-Carrion, M., 1984, Binding of anti-acetylcholine receptor antibodies inhibits the acetylcholine receptor mediated cation flux, *Biochem. Biophys. Res. Commun.* **120**:368–375.

Gullick, W. J., and Lindstrom, J. M., 1983, Mapping the binding of monoclonal antibodies to the acetylcholine receptor from *Torpedo californica, Biochemistry* **22**:3312–3320.

Gullick, W. J., Tzartos, S., and Lindstrom, J. 1981, Monoclonal antibodies as probes of acetylcholine receptor structure. I. Peptide mapping, *Biochemistry* **20**:2173–2180.

Guy, H. R., 1984, A structural model of the acetylcholine receptor channel based on partition energy and helix packing calculations, *Biophys J.* **45**:249–261.

Gysin, R., Wirth, M., and Flanagan, S. D., 1981, Structural heterogeneity and subcellular distribution of nicotinic synapse-associated proteins, *J. Biol. Chem.* **256**:11373–11376.

Haggerty, J. G., and Froehner, S. C., 1981, Restoration of ^{125}I-α-bungarotoxin binding activity to the α subunit of *Torpedo* acetylcholine receptor isolated by gel electrophoresis in sodium dodecyl sulfate, *J. Biol. Chem.* **256**:8294–8297.

Hamamoto, T., and Montal, M., 1986, Functional reconstitution of bacterial cytochrome oxidases in planar lipid bilayers, *Methods Enzymol.* **126**:123–138.

Hamill, O. P., and Sakmann, B., 1981, Multiple conductance states of single acetylcholine receptor channels in embryonic muscle cells, *Nature* **294**:462–464.

Hamill, O. P., Marty, A., Neher, E., Sakmann, B., and Sigworth, F., 1981, Improved patch-clamp techniques for high resolution current recording from cells and cell-free patches, *Pfluegers Arch.* **391**:85–100.

Hamilton, S. L., McLaughlin, M., and Karlin, A., 1979, Formation of disulfide-linked oligomers of acetylcholine receptor in membrane from *Torpedo* electric tissue, *Biochemistry* **18**:155–163.

Heidmann, T., and Changeux, J.-P., 1979, Fast kinetic studies on the interaction of a fluorescent agonist with the membrane-bound acetylcholine receptor from *Torpedo marmorata*, *Eur. J. Biochem.* **94**:255–279.

Heidmann, T., and Changeux, J.-P., 1980, Interaction of a fluorescent agonist with the membrane-bound acetylcholine receptor from *Torpedo marmorata* in the millisecond time range: Resolution of an intermediate conformational transition and evidence for positive cooperative effects, *Biochem. Biophys. Res. Commun.* **97**:880–896.

Heidmann, T., Sobel, A., Popot, J.-L., and Changeux, J.-P., 1980, Reconstitution of a functional acetylcholine receptor: Conservation of the conformational and allosteric transitions and recovery of the permeability response role of lipids, *Eur. J. Biochem.* **110**:35–55.

Hess, G. P., Cash, D. J., and Aoshima, H., 1983, Acetylcholine receptor-controlled ion translocation: Chemical kinetic investigations of the mechanism, *Annu. Rev. Biophys. Bioeng.* **12**:443–473.

Horn, R., and Patlak, J., 1980, Single channel currents from excised patches of muscle membrane, *Proc. Natl. Acad. Sci. U.S.A.* **77**:6930–6934.

Horn, R., Brodwick, M. S., and Dickey, W. D., 1980, Asymmetry of the acetylcholine channel revealed by quaternary anesthetics, *Science* **210**:205–207.

Huganir, R. L., and Greengard, P., 1983, C-AMP-dependent protein kinase phosphorylates the nicotinic acetylcholine receptor, *Proc. Natl. Acad. Sci. U.S.A.* **80**:1130–1134.

Huganir, R. L., and Racker, E., 1980, Endogenous and exogenous proteolysis of the acetylcholine receptor from *Torpedo californica, J. Supramol. Struct.* **14**:215–221.

Huganir, R. L., and Racker, E., 1982, Properties of proteoliposomes reconstituted with acetylcholine receptor from *Torpedo californica, J. Biol. Chem.* **257**:9372–9378.

Huganir, R. L., Schell, M. A., and Racker, E., 1979, Reconstitution of the purified acetylcholine receptor from *Torpedo californica, FEBS Lett.* **108**:155–160.

Huganir, R. L., Miles, K., and Greengard, P. 1984, Phosphorylation of the nicotinic acetylcholine receptor by an endogenous tyrosine-specific protein kinase, *Proc. Natl. Acad. Sci. U.S.A.* **81**:6968–6972.

Jackson, M. B., Lecar, H., Askanas, V., and Engel, W. K., 1982, Single cholinergic receptor channel currents in cultured human muscle, *J. Neurosci.* **2**:1465–1473.

Jackson, M. B., Wong, B. S., Morris, C. E., Lecar, H., and Christian, C. N., 1983, Successive openings of the same acetylcholine receptor-channel are correlated in their open times, *Biophys. J.* **42**:109–114.

Kagawa, Y., and Racker, E. 1971, Partial resolution of the enzymes catalyzing oxidative phosphorylation: XXV Reconstitution of vesicles catalyzing ^{32}Pi–ATP exchange, *J. Biol. Chem.* **246:**5477–5487.

Kao, P. W., Dwork, A. J., Kaldany, R. R.-J., Silver, M. L., Wideman, J., Stein, S., and Karlin, A., 1984, Identification of the α subunit half-cystine specifically labeled by an affinity reagent for the acetylcholine receptor binding site, *J. Biol. Chem.* **259:**11662–11665.

Karlin, A., 1980, Molecular properties of nicotinic acetylcholine receptors, in: *The Cell Surface and Neuronal Function* (G. Poste, G. Nicholson, and C. Cotman, eds.), pp. 191–260, Elsevier/North-Holland Biomedical Press, New York.

Karlin, A., Weill, C. L., McNamee, M. G., and Valderrama, R., 1975, Facets of the structures of acetylcholine receptors from electrophorus and *Torpedo, Cold Spring Harbor Symp. Quant. Biol.* **40:**203–213.

Karlin, A., Cox, R., Kaldany, R.-R., Lobel, P., and Holtzman, E., 1983, The arrangement and functions of the chains of the acetylcholine receptor of *Torpedo* electric tissue, *Cold Spring Harbor Symp. Quant. Biol.* **48:**1–8.

Karlsson, E., Arnberg, H., and Eaker, D., 1971, Isolation of the principal neurotoxins of two *Naja naja* subspecies, *Eur. J. Biochem.* **21:**1–16.

Katz, B., and Thesleff, S., 1957, A study of the "desensitization" produced by acetylcholine at the motor end-plate, *J. Physiol.* (*Lond.*) **138:**63–80.

Kilian, P. L., Dunlap, C. R., Mueller, P., Schell, M. A., Huganir, R. L., and Racker, E., 1980, Reconstitution of acetylcholine receptor from *Torpedo californica* with highly purified phospholipids: Effects of α-tocopherol, phylloquinone and other terpenoid quinones, *Biochem. Biophys. Res. Commun.* **93:**409–414.

Kistler, J., Stroud, R. M., Klymkowsky, M. W., Lalancette, R. A., and Fairclough, R. H., 1982, Structure and function of an acetylcholine receptor, *Biophys. J.* **37:**371–383.

Labarca, P., Lindstrom, J., and Montal, M., 1984a, Acetylcholine receptor in planar lipid bilayers: Characterization of the channel properties of the purified nicotinic acetylcholine receptor from *Torpedo californica* reconstituted in planar lipid bilayers, *J. Gen. Physiol.* **83:**473–496.

Labarca, P., Lindstrom, J., and Montal, M., 1984b, The acetylcholine receptor channel from *Torpedo californica* has two open states, *J. Neurosci.* **4:**502–507.

Labarca, P., Rice, J., Fredkin, D., and Montal, M., 1985a, Kinetic analysis of channel gating: Application to the cholinergic receptor channel and to the chloride channel from *Torpedo californica, Biophys. J.* **47:**469–478.

Labarca, P., Montal, M. S., Lindstrom, J., and Montal, M., 1985b, The occurrence of long openings in the purified cholinergic receptor channel increases with acetylcholine concentration, *J. Neurosci.* **5:**3409–3413.

Lee, T., Witzemann, V., Schimerlik, M., and Raftery, M. A., 1977, Cholinergic ligand induced affinity changes in *Torpedo californica* acetylcholine receptor, *Arch. Biochem. Biophys.* **183:**57–63.

Lindstrom, J., Lennon, V., Seybold, M., and Whittingham, S., 1976, Experimental autoimmune myasthenia gravis and myasthenia gravis: Biochemical and immunochemical aspects, *Ann. N.Y. Acad. Sci.* **274:**254–274.

Lindstrom, J., Merlie, J., and Yogeeswaran, G., 1979, Biochemical properties of acetylcholine receptor subunits from *Torpedo californica, Biochemistry* **18:**4465–4470.

Lindstrom, J., Anholt, R., Einarson, B., Engel, A., Osame, M., and Montal, M., 1980, Purification of acetylcholine receptors, reconstitution into lipid vesicles, and study of agonist-induced cation channel regulation, *J. Biol. Chem.* **255:**8340–8350.

Lindstrom, J., Einarson, B., and Tzartos, S., 1981a, Production and assay of antibodies to acetylcholine receptors, *Methods Enzymol.* **74:**432–460.

Lindstrom, J. M., Tzartos, S. J., and Gullick, W. J., 1981b, Structure and function of acetylcholine receptors studies using monoclonal antibodies, *Ann. N.Y. Acad. Sci.* **377:**1–19.

Lindstrom, J., Tzartos, S., Gullick, W., Hochschwender, S., Swanson, L., Sargent, P., Jacob, M., and Montal, M., 1983, Use of monoclonal antibodies to study acetylcholine receptors from electric organs, muscle, and brain and the autoimmune response to receptor in myasthenia gravis, *Cold Spring Harbor Symp. Quant. Biol.* **48:**89–99.

Lindstrom, J., Criado, M., Hochschwender, S., Fox, J. L., and Sarin, V., 1984, Immunochemical tests of acetylcholine receptor subunit models, *Nature* **311:**573–575.

Lowry, O. H., Rosebrough, N. Y., Farr, A. L., and Randall, R. Y., 1951, Protein measurement with the Folin phenol reagent, *J. Biol. Chem.* **193**:265–275.

Magleby, K. L., and Pallotta, B. S., 1981, A study of desensitization of acetylcholine receptors using nerve-released transmitter in the frog, *J. Physiol. (Lond.)* **316**:225–250.

McNamee, M. G., and Ochoa, E. L. M., 1982, Reconstitution of acetylcholine receptor function in model membranes, *Neuroscience* **7**:2305–2319.

McNamee, M. G., Ellena, J. F., and Dalziel, A. W., 1982, Lipid–protein interactions in membranes containing the acetylcholine receptor, *Biophys. J.* **37**:103–104.

Merlie, J. P., and Sebbane, R., 1981, Acetylcholine receptor subunits transit a precursor pool before acquiring α bungarotoxin binding activity, *J. Biol. Chem.* **256**:3605–3608.

Merlie, J. P., Sebbane, R., Tzartos, S., and Lindstrom, J., 1982, Inhibition of glycosylation with tunicamycin blocks assembly of newly synthesized acetylcholine receptor subunits in muscle cells, *J. Biol. Chem* **257**:2694–2701.

Miller, C., and White, M. M., 1984, Dimeric structure of single chloride channels from *Torpedo electroplax, Proc. Natl. Acad. Sci. U.S.A.* **81**:2772–2775.

Mishina, M., Kurosaki, T., Tibomatsu, T., Morimoto, Y., Noda, M., Yamamoto, T., Terao, M., Lindstrom, J., Takahashi, T., Kuno, M., and Numa, S., 1984, Expression of functional acetylcholine receptor from cloned cDNAs, *Nature* **307**:604–608.

Mishina, M., Tobimatsu, T., Imoto, K., Tanaka, K., Fujita, Y., Fukuda, K., Kurasaki, M., Takahashi, H., Morimoto, Y., Hirose, T., Inayama, S., Takahashi, T., Kuno, M., and Numa, S., 1985, Location of functional regions of acetylcholine receptor α-subunit by site-directed mutagenesis, *Nature* **313**:364–369.

Montal, M., 1974, Formation of bimolecular membranes from lipid monolayers, *Methods. Enzymol.* **32**:545–556.

Montal, M., 1985, Functional reconstitution of membrane proteins in planar lipid bilayer membranes, in: *Techniques for Analysis of Membrane Proteins* (C. I. Ragan and R. Cherry, eds.) Chapman and Hall, London (in press).

Montal, M., and Mueller, P., 1972, Formation of bimolecular membranes from lipid monolayers and a study of the electrical properties, *Proc. Natl. Acad. Sci. U.S.A.* **69**:3561–3566.

Montal, M., Darszon, A., and Schindler, H., 1981, Functional reassembly of membrane proteins in planar lipid bilayers, *Q. Rev. Biophys.* **14**:1–79.

Montal, M., Labarca, P., Fredkin, D. R., Suarez-Isla, B. A., and Lindstrom, J., 1984, Channel properties of the purified acetylcholine receptor from *Torpedo californica* reconstituted in planar lipid bilayer membranes, *Biophys. J.* **45**:165–174.

Moore, H.-P. H., and Raftery, M. A., 1980, Direct spectroscopic studies of cation translocation by *Torpedo* acetylcholine receptor on a time scale of physiological relevance, *Proc. Natl. Acad. Sci. U.S.A.* **77**:4509–4513.

Moore, H.-P., Hartig, P. R., and Raftery, M. A., 1979, Correlation of polypeptide composition with functional events in acetylcholine receptor-enriched membranes from *Torpedo californica, Proc. Natl. Acad. Sci. U.S.A.* **76**:6265–6269.

Neher, E., and Sakmann, B., 1976, Single channel currents recorded from membrane of denervated frog muscle fibers, *Nature* **260**:799–802.

Neher, E., and Steinbach, J. H., 1978, Local anesthetics transiently block currents through single acetylcholine receptor channels, *J. Physiol. (Lond.)* **277**:153–176.

Nelson, N., Anholt, R. Lindstrom, J., and Montal, M., 1980, Reconstitution of purified acetylcholine receptors with functional ion channels in planar lipid bilayers, *Proc. Natl. Acad. Sci. U.S.A.* **77**:3057–3061.

Nestler, E. J., Walaas, S. I., and Greengard, P., 1984, Neuronal phosphoproteins: Physiological and chemical implications, *Science* **225**:1357–1367.

Neubig, R. R., and Cohen, J. B., 1980, Permeability control by cholinergic receptors in *Torpedo* postsynaptic membranes: Agonist dose–response relations measured at second and millisecond times, *Biochemistry* **19**:2770–2779.

Neubig, R. R., Krodel, E. K., Boyd, N. D., and Cohen, J. B., 1979, Acetylcholine and local anesthetic binding to *Torpedo* nicotinic postsynaptic membranes after removal of nonreceptor peptides, *Proc. Natl. Acad. Sci. U.S.A.* **76**:690–694.

Neumann, D., Fridkin, M., and Fuchs, S., 1984, Anti-acetylcholine receptor response achieved by immunization with a synthetic peptide from the receptor sequence, *Biochem. Biophys. Res. Commun.* **121**:673–679.

Noda, M., Takahashi, H., Tanabe, T., Toyosato, M., Fusutani, Y., Hirose, T., Asai, M., Inayama, S., Miyata, T., and Numa, S., 1982, Primary structure of α-subunit precursor of *Torpedo californica* acetylcholine receptor deduced from cDNA sequence, *Nature* **299**:793–797.

Noda, M., Takahashi, H., Tanabe, T., Toyosato, M., Kikyotani, S., Hirose, T., Asai, M., Takashima, H., Inayama, S., Miyata, T., and Numa, S., 1983a, Primary structures of β and δ-subunit precursors of *Torpedo californica* acetylcholine receptor deduced from cDNA sequences, *Nature* **301**:251–255.

Noda, M., Takahashi, H., Tanabe, T., Toyosato, M., Kikotani, S., Furutani, Y., Hirose, T., Takashima, H., Inayama, S., Miyata, T., and Numa, S., 1983b, Structural homology of *Torpedo californica* acetylcholine receptor subunits, *Nature* **302**:528–532.

Oakley, B. R., Kirsch, D. R., and Morris, N. R. 1980, A simplified ultrasensitive silver stain for detecting proteins in polyacrylamide gels, *Anal. Biochem.* **105**:361–363.

Ochoa, E. L. M., Dalziel, A. W., and McNamee, M. G., 1983, Reconstitution of acetylcholine receptor function in lipid vesicles of defined composition, *Biochim. Biophys. Acta* **727**:151–162.

Ogden, D. C., Siegelbaum, S. A., and Colquhoun, D., 1981, Block of acetylcholine-activated ion channels by an uncharged local anesthetic, *Nature* **289**:596–598.

Popot, J.-L., Cartaud, J., and Changeux, J.-P, 1981, Reconstitution of a functional acetylcholine receptor: Incorporation into artificial lipid vesicles and pharmacology of the agonist-controlled permeability changes, *Eur. J. Biochem.* **118**:203–214.

Raftery, M. A., Hunkapiller, M. W., Strader, C. D., and Hood, L. E., 1980, Acetylcholine receptor: Complex of homologous subunits, *Science* **208**:1454–1457.

Raftery, M. A., Dunn, S. M. J., Conti-Tronconi, B. M., Middlemas, D. S., and Crawford, R. D., 1983, The nicotinic acetylcholine receptor: Subunit structure, functional binding sites and ion transport properties, *Cold Spring Harbor Symp. Quant. Biol.* **48**:21–33.

Ravdin, P. M., and Berg, D. K., 1979, Inhibition of neuronal acetylcholine sensitivity by α-toxins from *Bungarus multicinctus* venum, *Proc. Natl. Acad. Sci. U.S.A.* **76**:2072–2076.

Reynolds, J. A., and Karlin, A., 1978, Molecular weight in detergent solution of acetylcholine receptor from *Torpedo californica, Biochemistry* **17**:2035–2038.

Ross, M. J., Klymkowsky, M. W., Agard, D. A., and Stroud, R. M., 1977, Structural studies of a membrane-bound acetylcholine receptor from *Torpedo californica, J. Mol. Biol.* **166**:635–659.

Ruff, R. L., 1982, The kinetics of local anesthetic blockage of end-plate channels, *Biophys. J.* **37**:625–631.

St. John, P. A., Froehner, S. C., Goodenough, D. A., and Cohen, J. B., 1982, Nicotinic postsynaptic membranes from *Torpedo*: Sidedness, permeability to macromolecules, and topography of major polypeptides, *J. Cell. Biol.* **92**:333–342.

Sakmann, B., Patlak, J., and Neher, E., 1980, Single acetylcholine-activated channels show burst-kinetics in presence of desensitizing concentrations of agonist, *Nature* **286**:71–73.

Sargent, P. B., Hedges, B. E., Tsavaler, L., Clemons, L., Tzartos, S., and Lindstrom, J. M., 1984, The structure and transmembrane nature of the acetylcholine receptor in amphibian skeletal muscle as revealed by cross-reacting monoclonal antibodies, *J. Cell Biol.* **98**:609–618.

Schindler, H., 1979, Exchange and interactions between lipid layers at the surface of a liposome solution, *Biochim. Biophys. Acta* **555**:316–336.

Schindler, H., 1980, Formation of planar bilayers from artificial or native membrane vesicles, *FEBS Lett.* **122**:77–79.

Schindler, H., 1982, Concepts and techniques for membrane transport reconstitution, in: *Transport in Biomembranes: Model Systems and Reconstitution* (R. Antolini, A. Gliozzi, and A. Gorio, eds.), pp. 75–85, Raven Press, New York.

Schindler, H., and Feher, G., 1976, Branched bimolecular lipid membranes, *Biophys. J.* **16**:1109–1113.

Schindler, H., and Quast, U., 1980, Functional acetylcholine receptor from *Torpedo marmorata* in planar membranes, *Proc. Natl. Acad. Sci. U.S.A.* **77**:3052–3056.

Schindler, H., Spillecke, F., and Neumann, E., 1984, Different channel properties of *Torpedo* acetylcholine receptor monomers and dimers reconstituted in planar membranes, *Proc. Natl. Acad. Sci. U.S.A.* **81**:6222–6226.

Schuerholz, T., and Schindler, H., 1983, Formation of lipid-protein bilayers by micropipette-guided contact of two monolayers, *FEBS Lett.* **152**:187–190.

Sealock, R., 1982, Visualization at the mouse neuromuscular junction of a submembrane structure in common with *Torpedo* postsynaptic membranes, *J. Neurosci.* **2**:918–923.

Shah, D. O., and Schulman, J. M., 1967, Enzymic hydrolysis of various lecithin monolayers employing suface pressure and potential technique, *J. Colloid Interface Sci.* **25**:107–119.

Sigworth, F. J., 1982, Fluctuations in the current through open ACh-receptor channels, *Biophys. J.* **37**:309.

Sine, S., and Taylor, P., 1979, Functional consequences of agonist-mediated state transitions in the cholinergic receptor, *J. Biol. Chem.* **254**:3315–3325.

Sine, S. M., and Taylor, P., 1980, The relationship between agonist occupation and the permeability response of the cholinergic receptor revealed by bound cobra α-toxin, *J. Biol. Chem.* **255**:10144–10156.

Sobel, A., Weber, M., and Changeux, J.-P., 1977, Large-scale purification of the acetylcholine receptor protein in its membrane-bound and detergent-extracted forms from *Torpedo marmorata* electric organ, *Eur. J. Biochem.* **80**:215–224.

Strader, C. D., and Raftery, M. A., 1980, Topographic studies of *Torpedo* acetylcholine receptor subunits as a transmembrane complex, *Proc. Natl. Acad. Sci. U.S.A.* **77**:5807–5811.

Strader, C. B. D., Revel, J.-P., and Raftery, M. A., 1979, Demonstration of the transmembrane nature of the acetylcholine receptor by labeling with anti-receptor antibodies, *J. Cell Biol.* **83**:499–510.

Suarez-Isla, B. A., Wan, K., Lindstrom, J., and Montal, M., 1983, Single channel recordings from purified acetylcholine receptors reconstituted in bilayers formed at the tip of patch pipets, *Biochemistry* **22**:2319–2323.

Tank, D. W., Huganir, R. L., Greengard, P., and Webb, W. W., 1983, Patch-recorded single-channel currents of the purified and reconstituted *Torpedo* acetylcholine receptor, *Proc. Natl. Acad. Sci. U.S.A.* **80**:5129–5133.

Tarrab-Hazdai, R., Geiger, B., Fuchs, S., and Amsterdam, A., 1978, Localization of acetylcholine receptor in excitable membrane from the electric organ of *Torpedo*: Evidence for the exposure of receptor antigenic sites on both sides of the membrane, *Proc. Natl. Acad. Sci. U.S.A.* **75**:2497–2501.

Trautmann, A., and Feltz, A., 1980, Open time of channels activated by binding of two distinct agonists, *Nature* **286**:291–293.

Tzartos, S. J., and Changeux, J.-P., 1983, High affinity binding of α-bungarotoxin to the purified α-subunit and to its 27,000-dalton proteolytic peptide from *Torpedo marmorata* acetylcholine receptor: Requirement for sodium dodecyl sulfate, *EMBO J.* **2**:381–387.

Tzartos, S. J., and Lindstrom, J. M., 1980, Monoclonal antibodies used to probe acetylcholine receptor structure: Localization of the main immunogenic region and detection of similarities between subunits, *Proc. Natl. Acad. Sci. U.S.A.* **77**:755–759.

Tzartos, S., and Lindstrom, J., 1981, Production and characterization of monoclonal antibodies for use as probes of acetylcholine receptors, in: *Monoclonal Antibodies in Endocrine Research* (R. Fellows and G. Eisenbarth, eds.), pp. 69–86, Raven Press, New York.

Walker, J. W., McNamee, M. G., Pasquale, E., Cash, D. J., and Hess, G. P., 1981, Acetylcholine receptor inactivation in *Torpedo californica* electroplax membrane vesicles: Detection of two processes in the millisecond and second time regions, *Biochem. Biophys. Res. Commun.* **100**:86–90.

Walker, J. W., Takeyasu, K., and McNamee, M. G., 1982, Activation and inactivation kinetics of *Torpedo californica* acetylcholine receptor in reconstituted membranes, *Biochemistry* **21**:5384–5389.

Wan, K. K., and Lindstrom, J. M., 1985, Effects of monoclonal antibodies on the function of acetylcholine receptors purified from *Torpedo californica* and reconstituted into vesicles, *Biochemistry* **24**:1212–1221.

Webber, M., David-Pfeuty, T., and Changeux, J-P., 1975, Regulation of binding properties of the nicotinic receptor protein by cholinergic ligands in membrane fragments from *Torpedo marmorata*, *Proc. Natl. Acad. Sci. U.S.A.* **72:**3443–3447.

Weiland, G., and Taylor, P., 1979, Ligand specificity of state transitions in the cholinergic receptor: Behavior of agonists and antagonists, *Mol. Pharmacol.* **15:**197–212.

Weill, C. L., McNamee, M. G., and Karlin, A., 1974, Affinity labeling of purified acetylcholine receptor from *Torpedo californica*, *Biochem. Biophys. Res. Commun.* **61:**997–1003.

Wennogle, L. P., and Changeux, J.-P., 1980, Transmembrane orientation of proteins present in acetylcholine receptor-rich membranes from *Torpedo marmorata* studied by selective proteolysis, *Eur. J. Biochem.* **106:**381–393.

Wennogle, L. P., Oswald, R., Saitoh, T., and Changeux, J.-P., 1981, Dissection of the 66,000 dalton subunit of the acetylcholine receptor, *Biochemistry* **20:**2492–2497.

Wilmsen, V., Methfessel, C., Hanke, N., and Boheim, G., 1983, Channel current fluctuation studies with solvent-free lipid bilayers using Neher-Sakmann pipettes, in: *Physical Chemistry of Transmembrane Ion Motions* (G. Spach, ed.), pp. 479–485, Elsevier, Amsterdam.

Wise, D. S., Karlin, A., and Schoenborn, B. P., 1979, An analysis by low-angle neutron scattering of the structure of the acetylcholine receptor from *Torpedo californica* in detergent solution, *Biophys. J.* **28:**473–496.

Wu, W. C. S., and Raftery, M. A., 1979, Carbamylcholine-induced rapid cation efflux from reconstituted membrane vesicles containing purified acetylcholine receptor, *Biochem. Biophys. Res. Commun.* **89:**26–35.

Wu, W. C. S., and Raftery, M. A., 1981, Functional properties of acetylcholine receptor monomeric and dimeric forms in reconstituted membranes, *Biochem. Biophys. Res. Commun.* **99:**436–444.

Wu, W. C. S., Moore, H.-P., and Raftery, M. A., 1981, Quantitation of cation transport by reconstituted membrane vesicles containing purified acetylcholine receptors, *Proc. Natl. Acad. Sci. U.S.A.* **78:**775–779.

Young, E. F., Ralston, E., Blake, J., Ramachandran, J., Hall, Z. W., and Stroud, R. M., 1985, Topological mapping of acetylcholine receptor: Evidence for a model with five transmembrane segments and a cytoplasmic COOH-terminal peptide, *Proc. Natl. Acad. Sci. U.S.A.* **82:**626–630.

Chapter 9

IMMUNOLOGIC ANALYSIS OF THE ACETYLCHOLINE RECEPTOR

JON LINDSTROM

1. Introduction

Studies of receptors have come a long way in the past decade or so. It is no longer possible to study their pharmacological or electrophysiological properties without serious consideration of their macromolecular properties. A cascade of new techniques (e.g., affinity labeling, affinity purification, reconstitution, patch clamping, monoclonal antibody production, gene cloning, *in vitro* gene expression, antisense mRNA) is overcoming many of the old barriers to molecular characterization of receptors (e.g., small amounts of receptor, lack of biochemical probes, lack of techniques for solubilizing, reconstituting, and characterizing receptors in membranes). In this cascade of techniques, today's new technique (e.g., *in vitro* mutagenesis) frequently seems to threaten to make yesterday's new technique (e.g., reconstitution) seem passe even before the approach can be thoroughly established and utilized. In fact, all of these techniques have virtues and limitations, and many of these techniques will be required in state-of-the-art studies of receptors. It is clear, however, that a continuance of productivity in studies of receptors will require knowledge of several of these techniques. One of the technologies that has been and will continue to be useful in studies of receptors from the initial stages of identification through purification and characterization and into studies of synthesis and the molecular basis of function is the use of antibodies.

This chapter is an attempt to illustrate how antibodies can be used to study the structure and function of acetylcholine receptors. I hope to convince the reader that antibodies are very useful experimental tools.

In the case of acetylcholine receptors, the antibodies are also objects of study because autoantibodies to acetylcholine receptors are responsible for impairing neuromuscular transmission, thereby causing the muscular weakness and fatigability that characterize myasthenia gravis (MG) (Lindstrom, 1984). Autoantibodies to other receptors, for example, those for insulin (Kahn *et al.*, 1982) or thyroid-stimulating hormone (Rees Smith and Buckland, 1982), have also proven to be both pathologically significant and useful as experimental reagents.

JON LINDSTROM • The Salk Institute, San Diego, California 92138.

Atlhough antisera to purified receptors and receptor subunits have been available for many years (e.g., Lindstrom *et al.*, 1979b), monoclonal antibodies (mAbs) are proving even more valuable (e.g., Lindstrom, 1985a). The explosion of receptor subunit sequence information available from molecular genetic studies (e.g., Numa *et al.*, 1983) has greatly helped immunologic studies by permitting the generation of antibodies of predetermined specificity to synthetic peptides corresponding to sequences deduced from receptor subunit cDNAs (e.g., Lindstrom *et al.*, 1984; Ratnam and Lindstrom, 1984; Criado *et al.*, 1985).

In the coming years mAbs raised against impure immunogens should prove valuable in purifying and characterizing receptors (e.g., Fraser and Venter, 1980). These antibodies, and in some cases autoantibodies, should prove valuable in determining the antigenic structure of receptors. Antibodies are also proving valuable in selecting receptor subunit cDNA clones by hybrid-selected translation (Merlie *et al.*, 1983a, and expression cloning (Young and Davis, 1983) should be possible. These antibodies, and in some cases autoantibodies, should prove valuable in determining the antigenic structure of receptors.

When cDNA sequences are determined, antisera and mAbs to synthetic peptides should prove to be very useful, along with *in vitro* mutagenesis, as approaches to studying the structure and function of receptors. Antibodies are required to identify receptor subunits in studies involving receptor synthesis (Merlie *et al.*, 1983b), and they can also be used to study the development of receptor conformation (Merlie and Lindstrom, 1983), problems not readily amenable to molecular genetic techniques alone.

This chapter begins with a brief description of the biochemical and antigenic structure of the acetylcholine receptor and how antibodies can be used in these studies. I do not discuss reconstitution of receptors into vesicles because this has been dealt with in detail in the preceding chapter by my collaborators Mauricio Montal, Robert Anholt, and Pedro Labarca. I do, however, briefly describe some of the studies done in my laboratory and in Mauricio Montal's laboratory on the effects of mAbs on the function of purified receptors reconstituted in these vesicles and indicate some of the directions of our future studies.

In general, I describe the uses of antibodies conceptually, to show what antibodies can be used for. I do not provide methodological detail but point out how the experiments were done in order to suggest approaches that might be taken in other systems. Compendia of many relevant techniques are available, and, of course, details are given in the relevant papers. However, in the final section I discuss methodological aspects of these experiments, both listing where to go for step-by-step methods and pointing out some of the virtues or pitfalls that may be relevant when applying these techniques to other systems.

2. Structure of the Receptor

2.1. Biochemical Structure

Acetylcholine receptors from fish electric organs are composed of four kinds of subunits in the mole ratio $\alpha_2\beta\gamma\delta$ (Reynolds and Karlin, 1978; Lindstrom *et al.*,

1979a; Raftery *et al.*, 1980). The acetylcholine binding sites are on the extracellular surface of α subunits within 1 nm of cysteine α192 [because this residue is affinity labeled by cholinergic analogues developed by Karlin and coworkers (Kao *et al.*, 1984)].

Acetylcholine receptors from *Electrophorus* muscle are composed of subunits identical to those in receptors from *Electrophorus* electric organ except, perhaps, for small differences in glycosylation (Lindstrom *et al.*, 1983b; Gullick and Lindstrom, 1983a). The α subunits of receptors from bovine muscle exhibit 81% sequence homology with α subunits of receptors from *Torpedo* electric organ, whereas β and γ subunits of bovine receptors exhibit only 59% and 56% homology with their counterparts in *Torpedo* (Noda *et al.*, 1983a; Takai *et al.*, 1984; Tanabe *et al.*, 1984).

Each receptor is actually a small gene family. Sequence homologies between the subunits suggest that all four subunits evolved from a common precursor (Raftery *et al.*, 1980). In the case of *Torpedo californica*, the subunits exhibit 36–41% sequence homology with one another (Noda *et al.*, 1983b). The sequence homologies among α, β, γ, and δ subunits suggest that their basic structure is similar (Noda *et al.*, 1983b).

Stroud and coworkers have proposed that the subunits are organized like barrel staves around a central cation channel through the molecule (Kistler *et al.*, 1982). This rationalizes several observations. One, the cation channel is a component of purified receptor monomers (Anholt *et al.*, 1980). Two, there is a negatively staining pit in the center of the receptor monomer (Bon *et al.*, 1984). Three, noncompetitive blockers that appear to bind to a site in the channel can covalently label any of the subunits depending on the amino acid reactivity of the reactive group on a particular blocker (Heidmann and Changeux, 1984). And four, the structural homologies of the subunits make it unlikely that one forms a transmembrane channel whereas the others do not. Which parts of the amino acid sequences of the subunits are involved in transmembrane domains and which of these domains are involved in lining the channel is unclear, as is discussed below.

It is clear that the overall shape of the receptor viewed from above is an 85-Å-diameter doughnut with about a 10-Å-diameter hole (Kistler *et al.*, 1982; Bon *et al.*, 1984). Viewed from the side, it is about 110 Å long with about 55 Å on the extracellular surface and only about 15 Å on the cytoplasmic surface of the membrane (Kistler *et al.*, 1982). The organization of the subunits around the central channel is controversial, but it is generally agreed that the two α subunits are not adjacent (Kistler *et al.*, 1982; Karlin *et al.*, 1983; Bon *et al.*, 1984).

All of the subunits contain carbohydrate on their extracellular surfaces (Lindstrom *et al.*, 1979a; Anderson and Blobel, 1981), at least some of the subunits have covalently attached lipid (Olson *et al.*, 1984), and at least some are phosphorylated (Vandlen *et al.*, 1979; Gordon *et al.*, 1983). The significance of all of these posttranslational modifications is unclear. However, it is clear that some or all are responsible for the multiple differences in properties of receptors noted at various stages of development and regeneration (Dennis *et al.*, 1981). For example, a serum from an MG patient demonstrated to be monospecific for carbohydrate was used to show that junctional and extrajunctional receptors differ in glycosylation near one of the two acetylcholine binding sites (Hall *et al.*, 1983).

Acetylcholine receptors at mature neuromuscular junctions are localized at the tips of folds in the postsynaptic membrane (Salpeter and Harris, 1983) in intimate associations with specific components of the basal lamina on the extracellular surface (Nitkin *et al.*, 1983) and cytoskeletal elements on the cytoplasmic surface (Burden *et al.*, 1983). Details of these interactions or their functional effects are unclear, but interesting results abound. For example, an extrinsic membrane protein of 43,000 apparent molecular weight is associated with the cytoplasmic surface of the receptor (Sealock *et al.*, 1984), probably through β subunits (Burden *et al.*, 1983), and this protein binds actin (Walker *et al.*, 1984), indicating that it may function as a link between receptors and cytoskeletal elements.

2.2. Antigenic Structure

All parts of a native protein are not equally immunogenic (Atassi, 1984), although by using synthetic peptides as immunogens it is possible to make antibodies to almost any part of a protein (Wilson *et al.*, 1984). In general, only antibodies to sequences exposed on the surface can bind to native proteins (Atassi, 1984). Antibodies bind to structures (''determinants'' or ''epitopes'') formed by five to seven amino acids (Atassi, 1984). These epitopes can be ''continuous'' or ''discontinuous'' (Atassi, 1984). Continuous epitopes are formed by a contiguous sequence of amino acids, and antibodies to such epitopes will bind to the corresponding free peptide, though perhaps with lower affinity than to the peptide in the conformation it assumes in the native protein. Discontinuous eiptopes are formed by amino acids that are adjacent in the intact protein but not immediately contiguous in the primary structure of the protein. Antibodies to discontinuous epitopes are absolutely conformation dependent and cannot bind to the denatured protein. Although epitopes are frequently formed by prominent structural features such as ends, loops, or protuberances of polypeptide chains (Atassi, 1984), and highly hydrophilic sequences are frequently antigenic (Hopp and Woods, 1981), it is not yet possible to predict the antigenic structure of a protein from even its crystal structure, much less its primary structure.

In the case of the acetylcholine receptor, the antigenic structure of the protein is of special interest because of its role as autoantigen. For most purposes antibodies are just practical tools, and it is simply important to understand a bit about how the tools work and to select the right one for the job.

Several fundamental aspect of the antigenic structure of acetylcholine receptors are known. When native acetylcholine receptors are used as immunogen, more than half of the antibodies formed are directed at the ''main immunogenic region'' (MIR) (Tzartos and Lindstrom, 1980; Tzartos *et al.*, 1981). More than half of MG patient antireceptor antibodies are also directed at the MIR (Tzartos *et al.*, 1982). It is a group of closely spaced, highly conformation-dependent continuous epitopes (Lindstrom, 1984). Large peptide fragments of α that contain the MIR have been identified (Gullick *et al.*, 1981), and small synthetic peptides of α have been shown not to be the MIR (Juillerat *et al.*, 1984; Lindstrom *et al.*, 1984), but the primary sequence of the MIR has not yet been determined. The MIR is not formed by the most hydrophilic sequence of α (Juillerat *et al.*, 1984; Lindstrom *et al.*, 1984). Although the acetylcholine binding sites are not usually a dominant immunogenic determinant, mAbs can be made that bind at or near these sites (James *et al.*, 1980; Gomez *et al.*, Mochley-Rosen and Fuchs, 1981; Watters and Maelicke, 1983). Monoclonal antibodies to many parts of the receptor can be obtained by immunizing with intact receptor or its SDS-denatured subunits (Tzartos and Lindstrom, 1980; Tzartos *et al.*, 1981, 1982, 1983; Gullick and Lindstrom, 1983b); SDS-denatured subunits are much less immunogenic than is native receptor (Lindstrom *et al.*, 1978), and the antibodies that are formed are directed preferentially at the cytoplasmic surface of the receptor (Froehner *et al.*, 1983; Sargent *et al.*, 1983). The N-terminal and C-terminal sequences of each subunit are negligible as immunogens in the intact receptor (Lindstrom *et al.*, 1984), but antibodies can be made to these synthetic peptides (Neumann *et al*,. 1984; Ratnam and Lindstrom, 1984). Antibodies to the N termini can bind to denatured subunits but not to intact receptor. Antibodies to the C termini can bind to both denatured subunits and intact receptors.

2.3. Use of Antibodies to Receptor and Synthetic Receptor Peptides to Study Receptor Structure

Antibodies have been useful throughout the course of studies of acetylcholine receptor structure. For example, when receptor was first purified and before it was possible to test whether the purified component was the physiologically significant receptor by reconstitution, demonstration that immunization with the purified component caused an autoimmune response to receptor and provoked antibodies that blocked receptor function showed that the purified component was at least part of the real receptor (Patrick and Lindstrom, 1973). The observation that antisera to each of the subunits could bind to all of the receptors showed that each subunit was a component of the receptor and helped to resolve arguments over the subunit composition of receptors that had raged for years as a result of a proteolytic artifact that could make receptors appear to consist only of α subunits (Lindstrom *et al.*, 1980b). Similar applications for antibodies will no doubt be found in the study of other receptors.

The development of mAb technology has been especially valuable (Kohler and Milstein, 1975). This is highly labor intensive but has many virtues. One of its greatest virtues has been irrelevant to the study of nicotinic receptors: it is possible to start with a crude immunogen and, with suitable assays, select a mAb specific for a receptor (Fraser and Venter, 1980). This approach will see increasing use in many systems but was irrelevant to nicotinic receptors, which were purified before this technology developed. Another approach that has been applied to acetylcholine receptors as a model system and which may prove valuable in other systems is the use of antiidiotype mAbs. This method permits making antireceptor antibodies using a receptor ligand. Erlanger and coworkers (Cleveland *et al.*,

1983) have demonstrated that antireceptor mAbs can be made by immunizing with a conformationally rigid cholinergic analogue (which provokes antianalogue antibodies with binding sites, "idiotypes," some of which may resemble receptor binding sites) and then selecting for antireceptor mAbs (which are made as regulatory antiidiotype antibodies, some of which also bind to the ligand-binding sites of receptor).

Monoclonal antibodies that cross reacted with more than one subunit (Tzartos and Lindstrom, 1980) gave the first indication of structural homologies between the subunits, which were later confirmed by sequencing studies (Raftery *et al.*, 1980). Monoclonal antibodies have also provided subunit-specific reagents in relatively unlimited amounts for studies of receptor structure, synthesis, and function. They are also proving useful for providing large amounts of highly concentrated antibodies to synthetic peptides that are poor immunogens.

Antibodies are also proving very useful in contemporary studies of receptor structure. For example, analysis of the distribution of hydrophobic amino acids in the primary structure of receptor subunits led three groups to propose that four stretches of at least 20 hydrophobic amino acids formed four transmembrane α-helical domains (Claudio *et al.*, 1983; Devillers-Thiery, 1983; Noda *et al.*, 1983b) (see Fig. 1A). Two other groups (Guy, 1983; Finer-Moore and Stroud, 1984) proposed that the cation channel through the membrane should have a hydrophilic lining to account for its electrophysiological properties (Lewis and Stevens, 1983), so they proposed a fifth amphipathic transmembrane domain formed by a sequence in the form of an α helix in which all of the hydrophilic residues occurred on one side of the helix, which would form the "barrel stave" contributed by each subunit to the lining of the channel (see Fig. 1B).

The N terminus of each subunit is expected to be on the extracellular surface, because each of the subunits has a leader sequence that is cleaved as it is synthesized (Noda *et al.*, 1983b). Further, there is direct evidence that the N termini are on the extracellular surface (Anderson *et al.*, 1982). Therefore, any model with an even number of transmembrane domains would have the C terminus on the extracellular surface, whereas any model with an odd number of transmembrane domains would end up with the C terminus on the cytoplasmic surface.

To find out where the C termini of the subunits were, we (Lindstrom *et al.*, 1984) had ten amino acid peptides corresponding to the C termini of the subunit cDNAs (Noda *et al.*, 1983b) synthesized. A tyrosine was added to the N-terminus of each of these peptides so that they could be labeled with ^{125}I. Then our library of subunit-specific mAbs was screened for their ability to bind to these peptides by indirect immune precipitation assays. Only in the case of β subunits did we find mAbs that bound to a C-terminal peptide. However, antisera to β and γ subunits also bound the C-terminal peptides of their respective subunits. Addition of detergent-solubilized receptor inhibited binding of the labeled peptides by these antibodies, showing that the C termini were exposed on the surface of the receptor. Receptor-rich membrane vesicles did not compete for antibody binding, showing that the C termini were not accessible on the extracellular surface of these vesicles, which are known to seal right-side out (Hartig and Raftery, 1979). Further, if the vesicles were permeabilized with saponin, they could compete for antibody binding. This worked even better if the vesicles were extracted at pH 11 or extracted with lithium diiodosalycilate, conditions that are known to extract the 43,000-molecular-weight protein from the cytoplasmic surface. These experiments showed that the C termini were exposed on the cytoplasmic surface and that the four-transmembrane-domain model must be wrong. Although consistent with the five domain model, they only prove that there is an odd number of domains.

Later, antisera (Ratnam and Lindstrom, 1984) and mAbs (unpublished) were made to the C-terminal peptides, and similar experiments showed that the C termini of all of the subunits were exposed on the cytoplasmic surface. Recently, Young *et al.* (1985) have confirmed this observation using an electron microscopic technique. Further, Young *et al.* (1985) showed that residues 350–358 of the β subunit were on the cytoplasmic surface, and La Rochelle *et al.* (1985) showed that residues 360–377 of the γ and δ subunits were on the cytoplasmic surface. These results further support the idea of a homologous arrangement of the polypeptide chains in the subunits and indicate that there must be an odd number of transmembrane domains before this cytoplasmic domain and an even number after it.

Parenthetically, antisera to the N-terminal peptides were also made (Neumann *et al.*, 1984; Ratnam and Lindstrom, 1984). These bound to denatured subunits but not to intact receptors. The symmetric lack of accessibility of the N termini of all four subunits and accessibility of the C termini of all four subunits support the idea that all four subunits have fundamentally homologous structures.

These experiments also answered another question about receptor structure. The α subunits have an apparent molecular weight of 40,000 or less based on their electrophoretic behavior, yet α cDNA indicates a protein molecular weight of 50,116 (Noda *et al.*, 1982). The N-terminal sequence of α subunits has been determined (Raftery *et al.*, 1980) and is identical to the N terminus predicted by the cDNA, but it was possible that proteolysis near the C terminus could produce α subunits with a real molecular weight of 40,000. However, the observation that antibodies to the C-terminal ten amino acids of α predicted by its cDNA bind to the mature α subunit shows that the C terminal is not proteolyzed (Ratnam and Lindstrom, 1984).

These experiments illustrate several fundamental points. They show how difficult it is to predict protein structure from primary sequence data and how easy it is to test some predictions of structural models by immunologic techniques using synthetic peptides based on cDNA sequences. Undoubtedly, the continuation of such studies will contribute significantly to determining the structure of receptor subunits and should be generally useful.

These experiments also illustrate several technical points. They show how one can use preexisting antisera that contain antibodies to a synthetic peptide to get a lot of information from that peptide without taking the time to make antibodies to it. They also show how immune precipitation and immune adsorbtion techniques can be used to provide information about the transmembrane orientation of antigenic determinants far more quickly and easily than is possible by immunohistochemical localization.

Subunit-specific mAbs and antipeptide sera have been used in light and

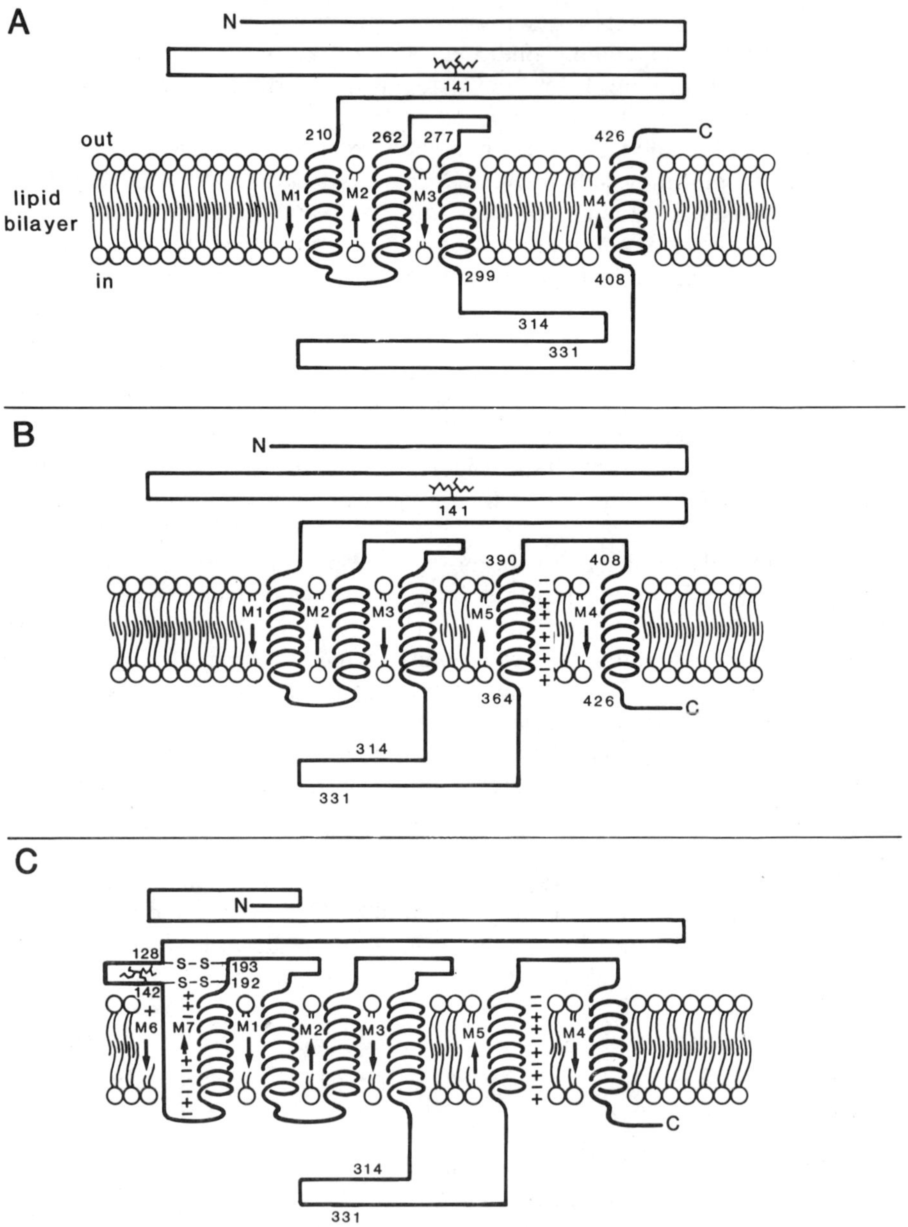

Figure 1. Alternative models for the transmembrane orientation of the polypeptide chain in acetylcholine receptor subunits. An α subunit is shown. It is presumed that β, γ, and δ subunits would have similar structures. (A) A model by three groups (Claudio *et al.*, 1983; Noda *et al.*, 1983b; Develliers-Thiery *et al.*, 1983) that proposes four hydrophobic α-helical transmembrane domains. This model is based on analysis of the distribution of hydrophobic amino acids in the subunit sequence. (B) A model proposed by two other groups (Guy, 1983; Finer-Moore and Stroud, 1984) that proposed a fifth amphipathic α-helical transmembrane domain. This model is also based on a theoretical analysis of amino acid distribution in the sequence. (C) A model (Criado *et al.*, 1985) that incorporates all five of the previously suggested domains and adds two others. All of the previously suggested domains are in-

electron microscopy experiments to determine the transmembrane orientation of antigenic determinants on receptors (Sargent *et al.*, 1983; La Rochelle *et al.*, 1985; Young *et al.*, 1985). Antireceptor mAbs have been localized histochemically by enzyme or gold-labeled autoantibodies (Sargent *et al.*, 1983; La Rochelle *et al.*, 1985) or by direct labeling with peroxidase (Jacob *et al.*, 1984). This is time-consuming and tedious work. When saponin is used for permeabilization, it wreaks havoc with morphology, and these experiments cannot be done with antisubunit sera or even low-affinity or low-titer antipeptide sera or mAbs. The most efficient and elegant of these techniques have been developed by Sealock and coworkers (La Rochelle *et al.*, 1985), who proved that the sequence 360–377 of α subunits was exposed on the cytoplasmic surface of *Torpedo* receptor-rich membrane fragments by labeling the extracellular surface of membrane fragments with 15-nm colloidal gold particles (attached to protein A, bound to antitoxin antibodies, bound to α-bungarotoxin) and then showing that the opposite side of the membrane was labeled by mAbs to the peptide (visualized by antimouse antibodies followed by protein A labeled with 5-nm colloidal gold particles).

Another example of the use of synthetic peptides concerns attempts to localize the MIR. Noda *et al.* (1982) attempted to identify the MIR by analyzing the sequence for the cDNA of α, which they had determined. They proposed that the MIR should be the most hydrophilic sequence in the extracellular domain of their model, α161–166. By using synthetic peptides including this sequence and techniques similar to those described above, we could show that this sequence was not the MIR (Lindstrom *et al.*, 1984), was not a significant antigenic determinant, and was not on the extracellular surface (Criado *et al.*, 1985). Antibodies to the MIR did not bind to the peptides α159–169 or to 152–167 (Lindstrom *et al.*, 1984). Juillerat *et al.* (1984) had similar results. We made mAbs to α152–167 and by competitive binding with other peptides mapped them to α152–159 (Criado *et al.*, 1985). Not only did these mAbs not compete for binding to the native receptor with mAbs to the MIR, they bound to the cytoplasmic surface.

This observation was very interesting, because α152–159 had been proposed to be extracellular in all previous models (Fig. 1A,B), but our observation that it was cytoplasmic meant that there must be two previously unsuspected transmembrane domains (Fig. 1C). The α subunits are contranslationally glycosylated at a single site on the extracellular surface (Merlie *et al.*, 1983a). There is only a

---

cluded simply to provide an odd total number of transmembrane domains, which is necessary to account for the observation that the C terminus is on the cytoplasmic surface (Lindstrom *et al.*, 1984; Ratnam and Lindstrom, 1984). The two additional domains are included to account for the observations that α152–159 is on the cytoplasmic surface (Criado *et al.*, 1985) and that α192 is on the extracellular surface (Kao *et al.*, 1984). A loop between α128 and 142 formed by disulfide bonds between α128 and α192 and α142 and α193 is depicted, which several groups (Noda *et al.*, 1982; Luyten *et al.*, 1984; McCormick and Atassi, 1984) have proposed to be the acetylcholine binding site. We have (unpublished) evidence to the contrary. Which cysteine is bound to which is not known. Since cysteines 192 and 193 are unique to α, it is assumed that the analogues of cysteines 128 and 142 would be linked by disulfide bonds in β, γ, and δ subunits. The N terminus is depicted as tucked under to account for the observation that it is not accessible to antibodies in the native receptor (Ratnam and Lindstrom, 1984; Neumann *et al.*, 1984). (Reproduced from Criado *et al.*, 1985.)

single sequence typical of N-glycosylation sites in α (Noda *et al.*, 1982), which makes it likely that α is glycosylated at asparagine α141. Therefore, it is likely that α141 is extracellular. The ten-amino-acid sequence between α142 and α152 is too short a stretch to form an α helix. The stretch between α159 and α192 (the cysteine near the acetylcholine binding site on the extracellular surface; Kao *et al.*, 1984) is long enough to form an α helix. Most of the hydrophilic amino acids would occur on one side of this helix.

It is interesting to consider that this transmembrane domain might be an amphipathic domain lining the cation channel. This is especially appealing because α192, which is adjacent to the acetylcholine binding site, is near the extracellular rim of this domain. Adjacent cysteines at α192 and α193 are unique to α subunits (Noda *et al.*, 1983b). It is known that there is a disulfide bond involving α192 that is adjacent to the acetylcholine binding site (Kao *et al.*, 1984). It has been suggested that the adjacent cysteines, α192 and 193, are not disulfide bonded to one another because in all other proteins known, adjacent cysteins are not disulfide bonded (Noda *et al.*, 1982; Kao *et al.*, 1984). In α subunits there are cysteines at α128 and α142, and there are comparable cysteines in all of the other subunits (Noda *et al.*, 1983b). It is possible that in α subunits α128 and 142 are disulfide bonded to α192 and 193, whereas in β, γ, and δ, which lack analogues of 192 and 193, the analogues of α128 and α142 are disulfide bonded to one another. If this were so, then the non-α-helical transmembrane domain α142–152 would exist as a loop with the amphipathic transmembrane domain $\alpha > 159$–$\alpha < 192$ covalently linked at its top by a disulfide bond adjacent to the acetylcholine binding site. This might suggest that the acetylcholine binding site regulated channel opening through conformation changes affecting these two transmembrane domains.

Isn't model building seductive? It is also useful, because it leads to testable hypotheses. And immunologic techniques can be useful in testing the predictions of these hypotheses.

As a final example of the usefulness of immunologic techniques in testing structural models of the acetylcholine receptor, consider proposals for the location of the acetylcholine binding site. Several groups have proposed that a disulfide-linked loop formed by α128–142 might form the acetylcholine binding site (Noda *et al.*, 1982; Plumer *et al.*, 1984; McCormick and Atassi, 1984; Luyten *et al.*, 1984). The basic rationale is that this sequence should be extracellular and near a disulfide bond. One group modeled this with CPK atomic models and showed that it could bind CPK models of cholinergic ligands (Luyten *et al.*, 1984). McCormick and Atassi (1984) reported that the sequence α125–147 efficiently bound [^{3}H]acetylcholine. However, Plumer *et al.* (1984) reported that antibodies to the peptide α127–132 did not compete with cholinergic ligands for binding to receptor.

We made a synthetic peptide α127–143 joined into a loop by a disulfide bond between α128 and α142 and labeled with ^{125}I at tyrosine α127 (J. Lindstrom, M. Criado, V. Sarin, and J. Fox, unpublished data). By indirect immune precipitation, this peptide bound to antibodies in antisera to native receptor and to α subunits, but antisera to β, γ, or δ subunits did not bind it, nor did any of 34 mAbs to α subunits. This sequence is highly homologous in all subunits. In this loop, α subunits differ from β, γ, and δ at only α129, 131, 134, and 139 (Noda *et al.*, 1983b). Therefore, it is very likely that the α-specific antibodies to this peptide bind to an epitope formed by the six amino acids α129–134. If this loop forms the acetylcholine binding site, some or all of these amino acids must also form the acetylcholine binding site.

It is possible to make mAbs to the acetylcholine binding site whose binding is inhibited by both α-bungarotoxin and small cholinergic ligands (James *et al.*, 1980; Gomez *et al.*, 1981; Mochley-Rosen and Fuchs, 1981; Watters and Maelicke, 1983). However, the antibodies to $[^{125}I]$α127–143 were not inhibited by receptors in vesicles whether or not the vesicles were permeabilized. Thus, unlike antisite antibodies, these antibodies could not bind to membrane-bound receptor. The antibodies could bind to detergent-solubilized receptors, but this was not inhibited by α-bungarotoxin. In fact, the antibodies bound best to receptors denatured by SDS. These results clearly show that α128–142 does not form the acetylcholine binding site. One possible explanation for the observation of McCormick and Atassi (1984) is that their peptide chelated $[^3H]$acetate (released by hydrolysis of acetylcholine) between lysine 125 and lysine 145, which are opposite to one another in the disulfide-linked loop. If this were true, carbamylcholine or curare would not be expected to reduce binding of ^3H to the peptide, whereas unlabeled acetylcholine, which is readily hydrolyzed, would inhibit ^3H binding.

These experiments illustrate the techniques one could use with another peptide to test whether it did form the acetylcholine binding site. Our data suggest that α128–142 is probably located near the protein–lipid interface and not readily accessible in membrane-bound receptors.

This is consistent with the model in Fig. 1C. However, this model should not be taken as gospel truth; it simply incorporates all of the current data and helps to make predictions that can be tested by methods like those described to help further refine the model.

2.4. Use of Antibodies to Study Receptor Synthesis

Receptor subunit synthesis has been studied most extensively in three systems: dog pancreas endoplasmic reticulum using *Torpedo* electric organ mRNA (Anderson and Blobel, 1981; Anderson *et al.*, 1982, 1983), the mouse BC3H-1 muscle cell line (Merlie *et al.*, 1983a,b; Merlie and Lindstrom, 1983), and *Xenopus* oocytes injected with mRNAs expressed in COS cells from cloned cDNAs (Mishina *et al.*, 1984). In all of these systems, the use of subunit-specific sera or mAbs has been critical for identifying the subunits. Further, in some cases the antibodies could be used to study conformational maturation and assembly of the subunits (Merlie and Lindstrom, 1983). Finally, in one case, the synthetic system was used to determine the transmembrane orientation of the antigenic determinants of the mAbs (Anderson *et al.*, 1983).

Receptor subunits have N-terminal leader sequences of about 20 amino acids (Noda *et al.*, 1983b), which interact with a signal recognition protein (Anderson *et al.*, 1982) in the endoplasmic reticulum membrane and are cleaved off after the sequence has been transported across the membrane, leaving the mature N terminus on what will become the extracellular surface of the membrane (Anderson *et al.*, 1983). The subunits are glycosylated on their extracellular surfaces as they are synthesized (Merlie *et al.*, 1982), but this carbohydrate is probably modified and added to later in the Golgi apparatus (Fambrough and Devreotes, 1978; Pumplin and Fambrough, 1982).

Merlie and coworkers (1983a) have shown that less than 1 min is required to synthesize an α subunit in a BC3H-1 cell, whereas another 30 min are required for conformational maturation and transport preceeding assembly, 60 min further are required for assembly of the subunits into intact receptors, and, finally, 60 more minutes are required for their transport to the surface. Newly synthesized subunits are detected by indirect immune precipitation of [^{35}S]methionine pulse-labeled subunits with antisubunit mAbs. Three or four times more α subunits are synthesized than are assembled, and the remainder are rapidly destroyed. Newly synthesized α subunits cannot bind α-bungarotoxin or mAb 35, a highly conformation-sensitive mAb to the MIR (Merlie and Lindstrom, 1983). Conformational maturation is detected by the ability of α subunits to bind α-bungarotoxin and mAb 35, which is acquired before assembly. Assembly is detected by the ability of an anti-α mAb to indirectly immune precipitate β, γ, and δ subunits. (Indirect immune precipitation means using antiantibody or fixed *S. aureas* to precipitate the mAb and the antigen to which it is bound.)

With an α-subunit-specific mAb, subunit synthesis can be detected *in vitro*. This provides the assay system for doing hybrid-selected translation and was used to identify an α cDNA clone from BC3H-1 (Merlie *et al.*, 1983a). In future experiments with other channel proteins, it seems very likely that mAbs raised using crude immunogens will frequently be used to screen by hybrid selected translation or, better, in expression vector systems (Young and Davis, 1983) to screen for cDNAs. This cDNA provided an assay for α mRNA, and the mAb was used to measure α subunits and show that synthesis is regulated at the levels of both transcription and translation (Olson *et al.*, 1983a,b) and regulated by denervation (Merlie *et al.*, 1984).

Numa and his coworkers (Mishina *et al.*, 1984) have shown that cloned cDNAs for the subunits of receptor from *Torpedo* can be used in appropriate vectors to produce large amounts of subunit mRNAs in COS monkey cells. Then they injected subunit mRNAs in various combinations into *Xenopus* oocytes and determined what subunits were synthesized by indirect immune precipitation with mAbs and determined electrophysiologically whether functional receptors were formed. They found that the amount of subunits formed depended on the combination of mRNAs injected. Functional receptors were efficiently formed only when all four subunit mRNAs were injected, although at a low (3–4%) level of efficiency γ and δ subunits appeared interconvertable. Clearly, this study is a precursor to much more sophisticated studies using *in vitro* mutagenesis to alter the sequence of the mRNAs and then the same system to assay their expression and function. This will be a very powerful approach.

Mishina *et al.* (1985) have recently published the first follow-up paper on site-directed mutagenesis of α subunits, which is remarkable both for the large number of mutations studied (26) and the very small number of clear conclusions that can be drawn from them. The mutations consisted of 20 deletions of 5–35 amino acids and six single amino acid replacements. It would have been more pleasing to see replacement of the deleted amino acids by, say, glycine to minimize distortions in adjacent areas, but apparently this is technically more difficult. It was observed, for example, that deletion of α224–237 (part of the putative transmembrane domain M1) abolished toxin binding and acetylcholine sensitivity. But it also abolished assembly with the δ subunit. Deletion of α279–284 also abolished toxin binding and function but did not effect δ assembly. Changing cysteine α128 or 142 to serine abolished toxin binding, function, and assembly of δ, whereas changing cysteine α192 or 193 to serine preserved some toxin binding, assembly of δ, but no function. Some small deletions within the putative amphipathic domain M5 retained some toxin binding and assembly with δ but prevented function, whereas others abolished function; and deletion of the whole stretch α355–389 retained some function. Conversion of asparagine α141 to aspartic acid, which might prevent glycosylation, also abolished toxin binding, function, and assembly with δ. Merlie *et al.* (1982) had previously shown that prevention of glycosylation of receptor in BC3H-1 muscle cell cultures impaired assembly of subunits and caused rapid destruction of the unassembled subunits. Clearly, site-directed mutagenesis experiments can provide a lot of information but also have their limitations and will have to proceed in conjunction with a number of other approaches to finally determine what the structure of acetylcholine receptors is, how it is assembled and metabolically regulated, and how it functions.

3. Effects of Antibodies on Receptor Function

Most antibodies to acetylcholine receptors do not affect receptor function (Lindstrom *et al.*, 1981b). This is a mixed blessing. It means that finding mAbs that affect function is very difficult, but it means that when a mAb does affect function, one can be reasonably sure that this indicates that the site to which it is bound is functionally important. Prior to the sequencing of subunit cDNAs, the libraries of mAbs that we generated reflected the relative antigenicity of various domains on the receptor. Now that subunit sequences are available and we can begin building and testing detailed models, we can use synthetic peptides to generate antisera and mAbs of predetermined specificity to probe the functional roles of sequences that are not normally immunogenic. Structural models can be used to predict interesting peptides. This approach is in some ways inferior to *in vitro* mutagenesis. However, this approach also has its virtues. As I have tried to indicate above, antibodies can be useful in studying many aspects of a receptor, so that a mAb that does not affect carbamylcholine-induced $^{22}Na^{+}$ flux might be very useful for studying receptor synthesis or structure and might also be used in other tissues and other species and might also be relevant as a model autoantibody.

A number of groups have made mAbs that bind at the acetylcholine binding site and are inhibited from binding by small cholinergic ligands (James *et al.*, 1980;

Gomez *et al.*, 1981; Mochley-Rosen and Fuchs, 1981; Watters and Maelicke, 1983). Antisite mAbs are, of course, effective at blocking receptor function (e.g., Gomez and Richman, 1983). Also, mAbs have been made that bind near the acetylcholine binding site and, therefore, are inhibited by α-bungarotoxin (molecular weight 8000) but not by small cholinergic ligands (e.g., Mihovilovic and Richman, 1984). A number of these mAbs and some antireceptor sera have the property that they inhibit binding of α-bungarotoxin to only one of the two binding sites. An MG patient autoantibody specific for N-acetylglucosamine also has this property (Hall *et al.*, 1983), which suggests that the two α subunits differ in glycosylation. This autoantibody can block receptor function, presumably by blocking access of agonist to one of the two sites. It is not clear in all cases how large an area of the antigen is covered by a bound antibody, but X-ray crystallography suggests that in some cases the contact area can be 20 × 25 Å (Amit *et al.*, 1985). However, in other cases bound antibodies can make very fine discriminations; for example, Watters and Maelicke (1983) report mAbs all of which are inhibited from binding by α-bungarotoxin and carbamylcholine but some of which are not inhibited by curare or decamethonium.

Antiidiotype mAbs to antibodies to the agonist bisQ have been made that bind to acetylcholine binding sites (Cleveland *et al.*, 1983). Such mAbs, not surprisingly, block activation of receptors by agonists (B. Erlanger, K. Wan, and J. Lindstrom, unpublished data). Most of these mAbs are directed at antigens that are absolutely conformationally dependent.

For antibodies to be useful in modern studies of how the structure of the receptor produces its function, it is critical that the antibodies recognize continuous antigenic determinants. If the antibodies are raised against native receptor rather than denatured receptors or peptides, and if they do not bind to denatured subunits so that their binding sites can be mapped, such antibodies are no more useful than the abundance of reversible ligands in existance and less useful than affinity-labeling reagents already in existence.

The greatest use for antibodies in studies of receptor function is as probes for the function of domains outside the acetylcholine binding site that may be involved in the structure of the channel, structures regulating its opening, closing, or desensitization, or structures involved in interactions with basal lamina or cytoskeletal components.

In addition, the functional effects of antibodies to acetylcholine receptor domains that are important autoimmunogens in EAMG and MG are pathologically interesting. In this case, observation of no functional effect of an antibody is as important as observation of an inhibitory effect. Parenthetically, antireceptor autoantibodies can also be agonists, as occurs, for example, in the case of autoantibodies to TSH receptors in Graves' disease (Rees-Smith and Buckland, 1982).

A very interesting example of a mAb that inhibits acetylcholine receptor function has been reported by Donnelly *et al.* (1984). They describe an anti-α mAb that inhibits receptor function noncompetitively and suggest that it may be a probe for the cation channel or a related domain. Clearly, more precise mapping of the binding site for this mAb should prove quite instructive.

In the remainder of this section, I review some of the observations we have made on the effects of mAbs on receptor function. I end by discussing some of the technical problems we have encountered in such studies.

Monoclonal antibodies to the MIR do not block carbamylcholine-induced $^{22}Na^+$ influx into vesicles of reconstituted *Torpedo* acetylcholine receptor (Lindstrom *et al.*, 1981b). Furthermore, these mAbs alter neither the duration nor the conductance of single-channel openings induced by carbamylcholine or suberyldicholine in reconstituted *Torpedo* AChRs in planar bilayers as monitored electrophysiologically (Lindstrom *et al.*, 1983a; Blatt *et al.*, 1985). Neither do they affect receptor function in rat muscle cells (Scheutze *et al.*, unpublished data) or frog muscle cells (A. Grinelle, unpublished data). This shows that the most frequent specificity of antibody to native receptor and most frequent specificity of autoantibody do not alter receptor activation even though it binds to the extracellular surface of α subunits.

Monoclonal antibody 13 is specific for α subunits and acts from the extracellular surface to inhibit carbamylcholine-induced $^{22}Na^+$ flux in vesicles of reconstituted receptor (Lindstrom *et al.*, 1981b).

Monoclonal antibody 10 was made against β subunits but cross reacts at lower affinity with a homologous sequence of α subunits (Tzartos and Lindstrom, 1980). It probably cross links these subunits within a receptor because, unlike some other subunit-specific mAbs (Conti-Tronconi *et al.*, 1981), it cannot cross link receptor monomers (unpublished data), which implies that both arms of the antibody are probably bound to a single receptor. Monoclonal antibody 10 binds to the extracellular surface and noncompetitively blocks carbamylcholine-induced $^{22}Na^+$ flux in reconstituted vesicles (Lindstrom *et al.*, 1981b). It also completely blocks single receptor channel openings of receptors reconstituted in planar bilayers (Lindstrom *et al.*, 1983a,b; Blatt *et al.*, 1985).

Monoclonal antibody 148 binds to β subunits, and mAb 168 binds to γ subunits (Gullick and Lindstrom, 1983b). Both mAbs block carbamylcholine-induced $^{22}Na^+$ influx in reconstituted vesicles (Wan and Lindstrom, 1985). Further, both mAbs also completely block the opening of channels in receptors reconstituted in planar bilayers (Blatt *et al.*, 1985). These mAbs also block receptor function when injected into *Xenopus* muscle cells (Y. Tamaki and Y. Kiddokoro, unpublished data). These mAbs are capable of cross linking receptors (Wan and Lindstrom, 1985). Their monovalent Fab fragments bind to receptors but have little or no effect. Therefore, their effect must result from the conformational deformation induced by cross linking homologous subunits in adjacent receptors. Antibodies to the MIR are very effective at cross linking α subunits in adjacent receptors (Conti-Tronconi *et al.*, 1981), yet they have no effect on receptor function (Lindstrom *et al.*, 1983a). Therefore, the sites to which mAbs 148 and 168 bind must be functionally important. Both of these mAbs slow the rate of α-bungarotoxin binding to receptors in reconstituted vesicles (Wan and Lindstrom, 1985). Therefore, they must create some global change in receptor conformation that is propagated from their binding sites on the cytoplasmic surface of β and γ subunits to the extracellular surface of α subunits.

When the binding sites of mAbs 13, 10, 148, 168, and the MIR are mapped on the primary structures of subunits, which is, of course, in progress, then it should be possible to interpret their blocking effects more precisely. This should help to reveal the structural basis of receptor function.

Monoclonal antibody 237, which binds to the sequence 152–159 on the cytoplasmic surface of α (Criado *et al.*, 1985), has no effect on receptor function (M. Criado and J. Lindstrom, unpublished data). Studies in the near future of mAbs to other synthetic peptides and their effects on receptor function should help to localize the acetylcholine binding sites, the channel, and other functionally interesting domains.

There can be irritating technical technical problems in studying the effects of mAbs on receptor function. Monoclonal antibodies that bind to the extracellular surface are easiest to work with because they can be used on intact cells, native-receptor-rich vesicles, and preformed reconstituted vesicles; mAbs to cytoplasmic domains can present a problem in the case of reconstituted vesicles, the mAbs are usually added to the receptor–detergent–lipid mixture before dialysis to remove detergent. Spurious blockage can be obtained by mAbs altering the interaction of receptor with lipid so that the receptors reconstitute inside out (Lindstrom *et al.*, 1981b). Because high receptor concentrations (micromolar) are used during reconstitution (Lindstrom *et al.*, 1980a,b; Anholt *et al.*, 1981, 1982), if the mAb preparation is an ammonium sulfate cut of hybridoma supernatant, then a large excess of serum proteins is present and can nonspecifically inhibit vesicle formation. If affinity chromatography on antiantibody agarose is used to purify the antibody, the denaturing conditions required for eluting the mAb from the column frequently also inactivate the mAb. It is possible to avoid this by growing mAbs in low-serum medium and purifying them by ion exchange (Hochschwender *et al.*, 1985).

The low signal-to-noise ratio of carbamylcholine-induced $^{22}Na^+$ flux in reconstituted vesicles (2/1 to 5/1) and the possibility of small artifactual interference in the reconstitution process make it difficult to screen many mAbs for small effects on receptor function. Thus far, all the mAbs that we have found to block receptor function block function completely. However, the observation that the mean amplitude and duration of open acetylcholine receptor channels in muscle cells treated with antireceptor serum is reduced (Heinemann *et al.*, 1977) suggests that some mAbs may be found that selectively alter these properties. We would especially like to find mAbs that alter the duration or conductance of open channels without blocking them completely. This would permit us to interpret most precisely the functional roles of the structures to which they bind.

4. Techniques

4.1. Purifying Receptors and Receptor Subunits

I have recently described the general strategies involved in the purification and characterization of receptors (Lindstrom, 1985a). Detailed methods that we have used for purifying acetylcholine receptors by affinity chromatography are compiled along with techniques for raising and assaying antireceptor antibodies in an article in *Methods in Enzymology* (Lindstrom *et al.*, 1981a). Technical compilations by other authors are also available (for example, Karlin *et al.*, 1976). Also, books are available that contain technical compilations for the purification of several kinds of receptors (for example, Venter and Harrison, 1984).

"Western" blotting techniques using antibodies to detect proteins transferred to paper after acrylamide gel electrophoresis have become very important and have recently been reviewed (Gershoni and Palade, 1983; Towbin and Gordon, 1984). These techniques frequently use nitrocellulose, but we frequently use DPT paper (Seed, 1982), which binds the transferred protein covalently. This is especially valuable because a single sample of precious protein can be transferred and exposed to an antibody, the antibody can then be removed with SDS, and the protein can then be reused with other antibodies (e.g., Gullick and Lindstrom, 1983a).

We have purified receptor subunits by SDS denaturation and preparative electrophoresis (Lindstrom *et al.*, 1979a,b). When this was initially done, our motive was, in part, to make antisubunit sera. Now, with the availability of mAb technology, it would probably be more efficient to immunize with whole SDS-denatured receptor and simply select for subunit-specific mAbs.

4.2. Immunizing with Receptors, Subunits, and Synthetic Peptides

Antibodies can be found with a wide range of properties; therefore, it is important to decide from the outset what you want the mAbs for and then immunize and select so that you get the mAbs you want. For example, if you want to determine the antigenic structure of the receptor, you will want to immunize with native receptor and see what sort of antibodies you get. If you want mAbs that will recognize nascent subunits or work on Western blots, you will want to immunize with SDS-denatured receptor or subunits. If you want mAbs that block function, your initial selective assay should be a functional assay. If you want mAbs that block function and can be mapped to the primary structure, then you should immunize with denatured receptor and select for mAbs that inhibit function, and so forth.

Although only a few parts of a protein are normally immunogenic, it has been observed that antibodies can be raised against synthetic peptides corresponding to almost any part of a protein (Wilson *et al.*, 1984). Custom synthesis from commercial sources is available (for example, Bachem, Los Angeles). If antibodies that can bind to the surface of the intact protein are wanted, then hydrophilic sequences of at least ten amino acids are a good bet, and peptides corresponding to the N or C termini may be better immunogens (Palfreyman *et al.*, 1984). Frequently, the primary structure will suggest domains that may be functionally interesting. Peptides as short as six or seven amino acids may be used (Plumer *et al.*, 1984). In our hands (Lindstrom *et al.*, 1984; Ratnam and Lindstrom, 1984; Criado *et al.*, 1985; M. Criado, M. Ratnam, V. Sarin, J. Fox, and J. Lindstrom, unpublished data), peptides of 15 to 20 amino acids have proven more immunogenic than shorter peptides, perhaps both because more antigenic determinants are available and because they are more likely to assume the conformation they have in the native protein.

It is our strategy to simultaneously make overlapping shorter peptides, which can later be used to map mAb binding sites within the larger peptide by competitive inhibition. It is also our strategy to add a terminal tyrosine to an unnatural end of the peptide so that it can be labeled by ^{125}I and coupled to a carrier by bis-diazobenzidine (BDB) (Lindstrom *et al.*, 1984; Ratnam and Lindstrom, 1984). It is typical to attach small peptides to a highly immunogenic carrier like keyhole limpet hemocyanin (KLH) or bovine serum albumin (BSA) by BDB or glutaraldehyde (Palfreyman *et al.*, 1984). We prefer terminal coupling rather than coupling through tyrosines or lysines in the middle of the peptide, which might destroy antigenic determinants. Because coupling to carrier is not always necessary or beneficial (Young *et al.*, 1983), we inject both the coupled and uncoupled peptide. We have found that at a dose rate of 10–100 μg peptide/rat in 200 μl of complete Freund's adjuvant or 100–1000 μg peptide/rabbit in 1000 μl of adjuvant every 2 weeks, months may be required to achieve high titers. In many cases, despite changes in carrier, coupling procedure, and dose, high titers were not achieved. Therefore, we have made it a standard practice to select mAbs. This not only provides large amounts of high-titer mAb from an animal with a low titer but avoids having to affinity purify the antipeptide antibody; but this takes time and effort even though fusions are very likely to be successful and are easy to screen.

4.3. Assays for Quantitating Antibodies and Screening for Monoclonal Antibodies

Immunoprecipitation assays using [^{125}I]-labeled antigens and antiantibody are extremely useful. Detailed techniques used with acetylcholine receptors have been described (Lindstrom *et al.*, 1981a). These assays are highly sensitive and easily quantitated over a wide range in terms of moles of antigen bound per liter of antibody solution. As illustrated in Section 2.3 (or Lindstrom *et al.*, 1984), antisera to whole receptor or subunits can be used with a labeled peptide, if it is immunogenic, to provide antibodies of predetermined specificity without having to make antipeptide antibodies.

Enzyme-linked immunosorbent assays (ELISA) or other assays in which the antigen is fixed and bound antibody is measured using antiantibody labeled with enzymes or ^{125}I have mixed virtues. These assays can be sensitive and with suitable reading devices can be very rapidly read in large numbers. The range of linearity is much less than with radioimmune assays, quantitation in moles of antigen bound is not so easy, and attachment of the antigen to the plate inevitably obscures some antigenic determinants, in some cases selectively. Several ELISA assays for antireceptor antibodies have been described (Dwyer *et al.*, 1983; Kobayashi *et al.*, 1984). We described an assay for antisubunit antibodies using purified peptides on plastic wells (Hochschwender *et al.*, 1985) and for antipeptide antibodies using intact receptor on plastic wells (Lindstrom *et al.*, 1984; Criado *et al.*, 1985). Assays using antigen fixed to plastic wells or Millititer® plates are especially useful in initially screening for monoclonal antibodies because so many assays can be done quickly. Monoclonal antibodies have been selected using histochemical screening and functional assays of ligand-binding inhibition, but this is awfully tedious.

4.4. Making Monoclonal Antibodies

We have made all of our mAbs by immunizing rats and fusing their spleen cells with mouse myelomas (Tzartos and Lindstrom, 1980; Tzartos *et al.*, 1981, 1983; Lindstrom *et al.*, 1984). This is because we wanted to use rat mAbs as model autoantibodies, because we knew that mice were more resistant to EAMG than rats, and because we had characterized rat EAMG in great detail. We have developed techniques for producing large amounts of mAbs in culture using low-serum medium (Hochschwender *et al.*, 1985). Three chapters describe our methods in detail (Lindstrom *et al.*, 1981a; Hochschwender *et al.*, 1985; Fraser and Lindstrom, 1985).

The "industry standard" for making mAbs is the use of immunized mice. For most purposes this is a good idea because the methods are most common, the mAbs can be grown as ascites as well as in culture, and most reagents available for purifying and characterizing the mAbs are adapted to mice. Several handbooks with detailed techniques for mAb production in mice are available (e.g., EMBO–SKMB, 1980; Fraser and Lindstrom, 1985). Also, a book on the use of mAbs to study receptors is available (Venter *et al.*, 1984).

To date, methods for immunization, fusion, selection, growth, and purification of mAbs have not been optimized, since most workers are interested in using mAbs as tools rather than an end in themselves and continuing with a technique that works rather than optimizing every parameter. Making mAbs is expensive and labor intensive, but they are very useful tools. mAbs of a wide range of binding specificities and immunoglobulin subclasses are available, so it is important to first decide the purpose mAbs and then design immunization and selection techniques to obtain the mAb wanted.

Making mAbs can be a major commitment, and for many purposes one may be better off collaborating with someone already in the business. However, making mAbs provides tools for so many approaches, from receptor identification to characterization, cloning synthesis, and function, that it can be worth a major commitment.

References

Amit, A., Mariuzza, R., Phillips, S., and Poljak, R., 1985, Three dimensional structure of an antigen–antibody complex at 6Å resolution, *Nature* **313**:156–158.

Anderson, D., and Blobel, G., 1981, *In vitro* synthesis, glycosylation, and membrane insertion of the four subunits of *Torpedo* acetylcholine receptor, *Proc. Natl. Acad. Sci. U.S.A.* **78**:5598–5602.

Anderson, D., Walter, P., and Blobel, G., 1982, Signal recognition protein is required for the integration of acetylcholine receptor delta subunit, a transmembrane glycoprotein, into the endoplasmic reticulum membrane, *J. Cell Biol.* **93:**501–506.

Anderson, D., Blobel, G., Tzartos, S., Gullick, W., and Lindstrom, J., 1983, Transmembrane orientation of an early biosynthetic form of acetylcholine receptor delta subunit determined by proteolytic dissection in conjunction with monoclonal antibodies, *J. Neurosci.* **3:**1773–1784.

Anholt, R., Lindstrom, J., and Montal, M., 1980, Functional equivalence of monomeric and dimeric forms of purified acetylcholine receptor from *Torpedo californica* in reconstituted lipid vesicles, *Eur. J. Biochem.* **109:**481–187.

Anholt, R., Lindstrom, J., and Montal, M., 1981, Stabilization of acetylcholine receptor channels by lipids in cholate solution and during reconstitution in vesicles, *J. Biol. Chem.* **256:**4377–4387.

Anholt, R., Fredkin, D., Deerinck, T., Ellisman, M., Montal, M., and Lindstrom, J., 1982, Incorporation of acetylcholine receptors into liposomes: Vesicle structure and acetylcholine receptor function, *J. Biol. Chem.* **25:**7122–7134.

Atassi, M. Z., 1984, Antigenic structure of proteins, *Eur. J. Biochem.* **145:**1–20.

Blatt, Y., Montal, M., Lindstrom, J., and Montal, M., 1985, Monoclonal antibodies directed against epitopes in the beta and gamma subunits of the *Torpedo* cholinergic receptor affect channel gating, *J. Neurosci.* (in press).

Bon, F., Lebrun, E., Gomel, J., Van Rapenbusch, R., Cartand, J., Popot, J.-L., and Changeux, J.-P., 1984, Image analysis of the heavy form of the acetylcholine receptor from *Torpedo marmorata*, *J. Mol. Biol.* **176:**205–237.

Burden, S., DePalma, R., and Gottesman, G., 1983, Crosslinking of proteins in acetylcholine receptor-rich membranes: Association between the beta subunit and the 43kd subunit protein, *Cell* **35:**687–692.

Claudio, T., Ballivet, M., Patrick, J., and Heinemann, S., 1983, *Torpedo californica* acetylcholine receptor 60,000 dalton subunit: Nucleotide sequence of cloned cDNA deduced amino acid sequence, subunit structural predictions, *Proc. Natl. Acad. Sci. U.S.A.* **80:**1111–1115.

Cleveland, W., Wasserman, N., Sarangarajon, R., Penn, A., and Erlanger, B., 1983, Monoclonal antibodies to the acetylcholine receptor by a normally functioning auto anti-idiotypic mechanism, *Nature* **305:**56–57.

Conti-Tronconi, B., Tzartos, S., and Lindstrom, J., 1981, Monoclonal antibodies as probes of acetylcholine receptor structure. II: Binding to native receptor, *Biochemistry* **20:**2181–2191.

Criado, M., Hochschwender, S., Sarin, V., Fox, J., and Lindstrom, J., 1985, Evidence for additional transmembranous domains in acetylcholine receptor subunits, *Proc. Natl. Acad. Sci. U.S.A.* **82:**2004–2008.

Dennis, M., Ziskind-Conhaim, L., and Harris, A., 1981, Development of neuromuscular junctions in rat embryos, *Dev. Biol.* **81:**266–279.

Devillers-Thiery, A., Giraudat, J., Bentaboulet, M., and Changeux, J.-P., 1983, Complete mRNA coding sequence of the acetylcholine binding alpha subunit of *Torpedo marmorata* acetylcholine receptor: A model for the transmembrane organization of the polypeptide chain, *Proc. Natl. Acad. Sci. U.S.A.* **80:**2067–2071.

Donnelly, D., Mihovilovic, M., Gozalez-Ros, J., Ferragut, J., Richman, D., and Martinez-Carrion, M., 1984, A noncholinergic site-directed monoclonal antibody can impair agonist-induced ion flux in *Torpedo californica* acetylcholine receptor, *Proc. Natl. Acad. Sci. U.S.A.* **81:**7999–8003.

Dwyer, D., Bradley, R., Urquhart, C., and Kearney, J., 1983, An enzyme-linked immunosorbent assay for measuring antibodies muscle acetylcholine receptor, *J. Immunol. Methods* **57:**111–119.

EMBO–SKMB, 1980, *Hybridoma Techniques*, Cold Spring Harbor Laboratories, New York.

Fambrough, D., and Devreotes, P., 1978, Newly synthesized acetylcholine receptors are located in the Golgi apparatus, *J. Cell Biol.* **76:**237–244.

Finer-Moore, J., and Stroud, R., 1984, Amphipathic analysis and possible formation of the ion channel in an acetylcholine receptor, *Proc. Natl. Acad. Sci. U.S.A.* **81:**155–159.

Fraser, C., and Veuter, J., 1980, Monoclonal antibodies to β adrenergic receptors: Use in purification and molecular characterization of β receptors, *Proc. Natl. Acad. Sci. U.S.A.* **77:**7034–7038.

Fraser, C., and Lindstrom, J., 1985, The use of monoclonal antibodies in receptor characterization and purification, in: *Receptor Biochemistry and Methodology*, Vol. 3 (C. Venter and L. Harrison, eds.), Alan R. Liss, New York (in press).

Froehner, S., Douville, K., Klink, S., and Culp, W., 1983, Monoclonal antibodies to cytoplasmic domains of the acetylcholine receptor, *J. Biol. Chem.* **258:**7112–7120.

Gershoni, J., and Palade, G., 1983, Protein blotting: Principles and applications, *Anal. Biochem.* **131:**1–15.

Gomez, C., and Richman, D., 1983, Anti-acetylcholine receptor antibodies directed against the alpha bungarotoxin binding site induce a unique form of experimental myasthenia, *Proc. Natl. Acad. Sci. U.S.A.* **80:**4089–4093.

Gomez, C., Richman, D., Burres, S., and Arnason, B., 1981, Monoclonal hybridoma anti-acetylcholine receptor antibodies: Antibody specificity and effect of passive transfer, *Ann. N.Y. Acad. Sci.* **377:**97–109.

Gordon, A., Milfay, D., and Diamond, I., 1983, Identification of a molecular weight 43,000 protein kinase in acetylcholine receptor-enriched membranes, *Proc. Natl. Acad. Sci. U.S.A.* **80:**5862–5865.

Gullick, W., and Lindstrom, J., 1983a, Comparison of the subunit structure of acetylcholine receptors from muscle and electric organ of *Electrophorus electricus*, *Biochemistry* **22:**3801–3807.

Gullick, W., and Lindstrom, J., 1983b, Mapping the binding of monoclonal antibodies to the acetylcholine receptor from *Torpedo californica, Biochemistry* **22:**3312–3320.

Gullick, W., Tzartos, S., and Lindstrom, J., 1981, Monoclonal antibodies as probes of acetylcholine receptor structure. I: Peptide mapping, *Biochemistry* **20:**2173–2180.

Guy, R., 1983, A structural model of the acetylcholine receptor channel based on partition energy and helix packing calculations, *Biophys. J.* **45:**249–261.

Hall, Z., Roisin, M., Gu, Y., and Gorin, P., 1983, A developmental change in the immunological properties of acetylcholine receptors at the rat neuromuscular junction, *Cold Spring Harbor Symp. Quant. Biol.* **48:**101–108.

Hartig, P., and Raftery, M., 1979, Preparation of right side out, acetylcholine receptor enriched intact vesicles from *Torpedo californica* electroplaque membranes, *Biochemistry* **18:**1146–1150.

Heidmann, T., and Changeux, J.-P., 1984, Time resolved photolabeling by the noncompetitive blocker chlorpromazine of the acetylcholine receptor in its transiently open and closed ion channel conformations, *Proc. Natl. Acad. Sci. U.S.A.* **81:**1897–1901.

Heinemann, S., Bevan, S., Kullberg, R., Lindstrom, J., and Rice, J., 1977, Modulation of the acetylcholine receptor by anti-receptor antibody, *Proc. Natl. Acad. Sci. U.S.A.* **74:**3090–3094.

Hochschwender, S., Langeberg, L., Schneider, D., and Lindstrom, J., 1985, Exploring the structure of the acetylcholine receptor, in: *Hybridoma Technology in the Biosciences and Medicine* (T. Springer, ed.), pp. 223–238, Plenum Press, New York.

Hopp, T., and Woods, K., 1981, Prediction of protein antigenic determinants from amino acid sequences, *Proc. Natl. Acad. Sci. U.S.A.* **78:**3824–3828.

Jacob, M. Berg, D., and Lindstrom, J., 1984, A shared antigenic determinant between the *Electrophorus* acetylcholine receptor and a synaptic component on chick ciliary ganglion neurons, *Proc. Natl. Acad. Sci. U.S.A..* **81:**3223–3227.

James, R., Kato, A., Rey, M., and Fulpius, B., 1980, Monoclonal antibodies directed against the neurotransmitter binding site of nicotinic acetylcholine receptor, *FEBS Lett.* **120:**145–148.

Juillerat, M., Barkas, T., and Tzartos, S., 1984, Antigenic sites of the nicotinic acetylcholine receptor cannot be predicted from the hydrophotocity profile, *FEBS Lett.* **168:**143–148.

Kahn, C., Kasuga, M., King, G., and Grunfeld, C., 1982, Autoantibodies to insulin receptors in man: Immunological determinants and mechanisms of action, *Ciba Found. Symp.* **90:**91–104.

Kao, P., Swork, A., Kaldany, R., Silver, M., Wideman, J., Stein, S., and Karlin, A., 1984, Identification of the alpha subunit half cystine specifically labeled by an affinity reagent for the acetylcholine receptor binding site, *J. Biol. Chem.* **259:**11662–11665.

Karlin, A., McNamee, M., Weill, C., and Valderrama, R., 1976, Methods of isolation and characterization of the acetylcholine receptor, in: *Methods in Receptor Research* (M. Blecher, ed.), pp. 1–35, Marcel Dekker, New York.

Karlin, A., Cox, R., Kaldany, R.-R., Lobel, P., and Holtzman, E., 1983, The arrangement and functions of the chains of the acetylcholine receptor of *Torpedo* electric tissue, *Cold Spring Harbor Symp. Quant. Biol.* **48:**1–8.

Kistler, J., Stroud, R., Klymkowsky, M., Lalancette, R., and Fairclough, R., 1982, Structure and function of an acetylcholine receptor, *Biophys. J.* **37:**371–383.

Kobayashi, N., Sugita, H., Terada, E., Ghoda, A., Okudaira, H., Ogita, T., and Miyamoto, T., 1984, A solid phase enzyme immunoassay for anti-acetylcholine receptor antibody in myasthenia gravis patients, *J. Immunol. Meth.* **73:**267–272.

Kohler, G., and Milstein, C, 1975, Continuous cultures of fused cells secreting antibody of predefined specificity, *Nature* **256:**495–497.

La Rochelle, W., Wray, B., Sealock, R., and Froehner, S. 1985, Immunochemical demonstration that amino acids 360–377 of the acetylcholine receptor gamma subunit are cytoplasmic, *J. Cell. Biol.* **100:**684–691.

Lewis, C., and Stevens, C., 1983, Acetylcholine receptor channel ionic selectivity: Ions experience an aqueous environment, *Proc. Natl. Acad. Sci. U.S.A.* **80:**6110–6113.

Lindstrom, J., 1984, Use of monoclonal antibodies in the study of myasthenia gravis, in: *Monoclonal Antibodies: Probes for the Study of Autoimmunity and Immunodeficiency* (G. Eisenbarth and B. Haynes, eds.), pp. 259–296, Academic Press, New York.

Lindstrom, J., 1985a, Nicotinic acetylcholine receptors: Use of monoclonal antibodies to study synthesis, structure, function, and autoimmune response, in: *Receptor Biochemistry and Methodology*. Vol. IV (J. Venter, C. Fraser, and J. Lindstrom, eds.), pp. 21–57, Alan R. Liss, New York.

Lindstrom, J., 1985b, Techniques for studying the biochemistry and cell biology of receptors, in: *Neurotransmitter Receptor Binding*, (H. Yamamura, S. Enna, and M. Kuhar, eds.), pp. 123–152, Raven Press, New York.

Lindstrom, J., Einarson, B., and Merlie, J., 1978, Immunization of rats with polypeptide chains from Torpedo acetylcholine receptor causes an autoimmune response to receptors in rat muscle, *Proc. Natl. Acad. Sci. U.S.A.* **75:**769–773.

Lindstrom, J., Merlie, J., and Yogeeswaran, B., 1979a, Biochemical properties of acetylcholine receptor subunits from *Torpedo californica*, *Biochemistry* **18:**4465–4470.

Lindstrom, J., Walter, B., and Einarson, B., 1979b, Immunochemical similarities between subunits of acetylcholine receptors from *Torpedo, Electrophorus*, and mammalian muscle, *Biochemistry* **18:**4470–4480.

Lindstrom, J., Anholt, R., Einarson, B., Engel, A., Osame, M., and Montal, M., 1980a, Purification of acetylcholine receptors with functional cation channels and reconstitution into lipid vesicles, *J. Biol. Chem.* **255:**8340–8350.

Lindstrom, J., Gullick, W., Conti-Tronconi, B. and Ellisman, M., 1980b, Proteolytic nicking of the acetylcholine receptor, *Biochemistry* **19:**4791–4795.

Lindstrom, J., Einarson, B., and Tzartos, S., 1981a, Production and assay of antibodies to acetylcholine receptors, *Methods. Enzymol.* **74:**432–460.

Lindstrom, J., Tzartos, S., and Gullick, B., 1981b, Structure and function of acetylcholine receptors studied using monoclonal antibodies, *Ann. N.Y. Acad. Sci.* **377:**1–19.

Lindstrom, J., Tzartos, S., Gullick, W., Hochschwender, S., Swanson, L., Sargent, P., Jacob, M., and Montal, M., 1983a, Use of monoclonal antibodies to study acetylcholine receptors from electric organs, muscle, and brain and the autoimmune response to receptor in myasthenia gravis, *Cold Spring Harbor Symp. Quant. Biol.* **48:**89–99.

Lindstrom, J., Cooper, J., and Swanson, L., 1983b, Purification of acetylcholine receptors from the muscles of *Electrophorus electricus*, *Biochemistry* **22:**3796–3800.

Lindstrom, J., Criado, M., Hochschwender, S., Fox, L., and Sarin, V., 1984, Immunochemical tests of acetylcholine receptor subunit models, *Nature* **311:**573–575.

Luyten, W., Kallaris, K., Kyte, J., Heinemann, S., and Patrick, J., 1984, A model for the acetylcholine binding site of the acetylcholine receptor, *Neurosci. Soc. Abstr.* **212:**10.

McCormick, D., and Atassi, M., 1984, Localization and synthesis of the acetylcholine binding site in the alpha chain of the *Torpedo californica* acetylcholine receptor, *Biochem. J.* **224:**995–1000.

Merlie, P., and Lindstrom, J., 1983, Assembly *in vivo* of mouse muscle acetylcholine receptor: Identification of an alpha subunit species which may be an assembly intermediate, *Cell* **34:**747–757.

Merlie, J., Sebbane, R., Tzartos, S., and Lindstrom, J., 1982, Inhibition of glycosylation with tunicamycin blocks assembly of newly synthesized acetylcholine receptor subunits in muscle cells, *J. Biol. Chem.* **257**:2694–2701.

Merlie, J., Sebbane, R., Gardner, S., and Lindstrom, J., 1983a, Regulation of acetylcholine receptor gene expression: Molecular cloning of a cDNA specific for alpha subunit of the receptor from the mouse muscle cell line BC3H-1, *Proc. Natl. Acad. Sci. U.S.A.* **80**:3845–3849.

Merlie, J., Sebbane, R., Gardner, S., Olson, E., and Lindstrom, J., 1983b, The regulation of acetylcholine receptor expression in mammalian muscle, *Cold Spring Harbor Symp. Quant. Biol.* **48**:135–146.

Merlie, J., Isenberg, K., Russell, S., and Sanes, J., 1984, Denervation supersensitivity in skeletal muscle: Analysis with a cloned cDNA probe, *J. Cell Biol.* **99**:332–335.

Mihovilovic, M., and Richman, D., 1984, Modification of alpha bungarotoxin and cholinergic ligand binding properties of *Torpedo* acetylcholine receptor by a monoclonal anti-acetylcholine receptor antibody, *J. Biol Chem.* **259**:15051–15059.

Mishina, M., Kurosaki, T., Tobimatsu, T., Morimoto, Y., Noda, M., Yamamoto, T., Terao, M., Lindstrom, J., Takahashi, T., Kuno, M., and Numa, S., 1984, Expression of functional acetylcholine receptor from cloned cDNAs, *Nature* **307**:604–608.

Mishina, M., Tobimatsu, T., Imoto, K., Tanaka, K., Fujita, Y., Fukuda K., Kurasaki, M., Takahashi, H., Morimoto, Y., Hirose, T., Inayama, S., Takahashi, T., Kuno, M., and Numa, S., 1985, Location of functional regions of acetylcholine receptor alpha subunit by site-directed mutagenesis, *Nature* **313**:364–369.

Mochley-Rosen, C., and Fuchs, S., 1981, Monoclonal anti-acetylcholine receptor antibodies directed against the cholinergic binding site, *Biochemistry* **20**:5920–5924.

Neumann, D., Fridkin, M., and Fuchs, S., 1984, Anti-acetylcholine receptor response achieved by immunization with a synthetic peptide from the receptor sequence, *Biochim. Biophys. Res. Commun.* **121**:673–679.

Nitkin, R., Wallace, B., Spira, M., Godfrey, E., and McMahan, V., 1983, Molecular components of the synaptic basal lamina that direct differentiation of regenerating neuromuscular junctions, *Cold Spring Harbor Symp. Quant. Biol.* **48**:653–666.

Noda, M., Takahashi, H., Tanabe, T., Toyosato, M., Furutani, Y., Hirose, T., Asai, M., Inayama, S., Miyata, T., and Numa, S., 1982, Primary structure of alpha subunit precursor of *Torpedo californica* acetylcholine receptor deduced from cDNA sequence, *Nature* **299**:793–797.

Noda, M., Furutani, Y., Takahashi, H., Toyosato, M., Tanabe, T., Shimizu, S., Kikyotani, S., Kanayo, T., Hirose, T., Inayama, S., and Numa, S., 1983a, Cloning and sequence analysis of calf cDNA and human genomic DNA encoding alpha subunit precursor of muscle acetylcholine receptor, *Nature* **305**:818–823.

Noda, M., Takahashi, H., Tanabe, T., Toyosato, M., Kikyotani, S., Furutani, Y., Hirose, T., Takashima, H., Inayama, S., Miyata, T., and Numa, S., 1983b, Structural homology of *Torpedo californica* acetylcholine receptor subunits, *Nature* **302**:528–532.

Numa, S., Noda, M. Takahashi, H., Tanabe, T., Toyosato, M., Furutani, Y., and Kykyotani, S., 1983, Molecular structure of the nicotinic acetylcholine receptor, *Cold Spring Harbor Symp. Quant. Biol.* **48**:57–71.

Olson, E., Glaser, L., Merlie, J., Sebbane, R., and Lindstrom, J., 1983a, Regulation of surface expression of acetylcholine receptors in response to serum and cell growth in the BC3H1 muscle cell line, *J. Biol. Chem.* **258**:13946–13953.

Olson, E., Glaser, L., Merlie, J., and Lindstrom, J., 1983b, Expression of acetylcholine receptor alpha subunit mRNA during differentiation of the BC3H1 muscle cell lines, *J. Biol. Chem.* **259**:3330–3336.

Olson, E., Glaser, L., and Merlie, J., 1984, Alpha and beta subunits of the nicotinic acetylcholine receptor contain covalently bound lipid, *J. Biol. Chem.* **259**:5364–5367.

Palfreyman, J., Aitcheson, T., and Taylor, P., 1984, Guidelines for the production of polypeptide specific antisera using small synthetic oligopeptides as immunogens, *J. Immunol. Methods.* **75**:383–393.

Patrick, J., and Lindstrom, J., 1973, Autoimmune response to acetylcholine receptor, *Science* **180**:871–872.

Plumer, R., Fels, G., and Maelicke, A., 1984, Antibodies against preselected peptide to map functional sites on the acetylcholine receptor, *FEBS Lett.* **178:**204–208.

Pumplin, D., and Fambrough, D., 1982, Turnover of acetylcholine receptors in skeletal muscle, *Annu. Rev. Physiol.* **44:**319–335.

Raftery, M., Hunkapillar, M., Strader, D., and Hood, L., 1980, Acetylcholine receptor: Complex of homologous subunits, *Science* **208:**1454–1457.

Ratnam, M., and Lindstrom, J., 1984, Structural features of the nicotinic acetylcholine receptor revealed by antibodies to synthetic peptides, *Biochem. Biophys. Res. Commun.* **122:**1225–1233.

Rees-Smith, B., and Buckland, P., 1982, Structure function relations of the thyrotropin receptor, *Ciba Found. Symp.* **90:**114–125.

Reynolds, J., and Karlin, A., 1978, Molecular weight in detergent solution of acetylcholine receptor from *Torpedo californica, Biochemistry* **17:**2035–2038.

Salpeter, M., and Harris, M., 1983, Distribution and turnover rate of acetylcholine receptors throughout the junction folds at a vertebrate neuromuscular junction, *J. Cell Biol.* **96:**1781–1785.

Sargent, P., Hedges, B., Tsavaler, L., Clemmons, L., Tzartos, S., and Lindstrom, J., 1983, The structure and transmembrane nature of the acetylcholine receptor in amphibian skeletal muscles revealed by crossreacting monoclonal antibodies, *J. Cell Biol.* **98:**609–618.

Sealock, R., Wray, B., and Froehner, S., 1984, Ultrastructural localization of the M_r 43,000 protein and the acetylcholine receptor in *Torpedo* postsynaptic membranes using monoclonal antibodies, *J. Cell Biol.* **98:**2239–2244.

Seed, B., 1982, Diazotizable arylamine cellulose papers for the coupling and hybridization of nucleic acids, *Nucleic Acids. Res.* **10:**1799–1810.

Takai, T., Noda, M,. Furutani, Y., Takahashi, H., Notake, M., Shimizu, S., Kayano, T., Tanabe, T., Tanaka, K., Hirose, T., Inayama, S., and Numa, S., 1984, Primary structure of gamma subunit precursor of calf-muscle acetylcholine receptor deduced from the cDNA sequence, *Eur. J. Biochem.* **143:**109–115.

Tanabe, T., Noda, M., Furutani, Y., Takai, T., Takahashi, H., Tanaka, K., Hirose, T., Inayama, S., and Numa, S., 1984, Primary structure of beta subunit precursor of calf muscle acetylcholine receptor deduced from cDNA sequence, *Eur. J. Biochem.* **144:**11–17.

Towbin, H., and Gordon, J., 1984, Immunoblotting and dot immunobinding: Current status and outlook, *J. Immunol. Methods* **72:**313–340.

Tzartos, S., and Lindstrom, J., 1980, Monoclonal antibodies used to probe acetylcholine receptor structure: Localization of the main immunogenic region and detection of similarities between subunits, *Proc. Natl. Acad. Sci. U.S.A.* **77:**755–759.

Tzartos, S., Rand, D., Einarson, B., and Lindstrom, J., 1981, Mapping of surface structures on *Electrophorus* acetylcholine receptor using monoclonal antibodies, *J. Biol. Chem.* **256:**8635–8645.

Tzartos, S., Seybold, M., and Lindstrom, J., 1982, Specificity of antibodies to acetylcholine receptors in sera from myasthenia gravis patients measured by monoclonal antibodies, *Proc. Natl, Acad. Sci. U.S.A.* **79:**188–192.

Tzartos, S., Langeberg, L., Hochschwender, S., and Lindstrom, J., 1983, Demonstration of a main immunogenic region on acetylcholine receptors from human muscle using monoclonal antibodies to human receptor, *FEBS Lett.* **158:**116–118.

Vandlen, R., Wilson, C., Eisenach, J., and Raftery, M., 1979, Studies of the composition of purified *Torpedo californica* acetylcholine receptor and its subunits, *Biochemistry* **18:**1845–1854.

Venter, C., and Harrison, L. (eds.), 1984, *Receptor Biochemistry and Methodology*, Vol. 2, *Receptor Purification Procedures*, Alan R. Liss, New York.

Venter, C., Fraser, C., and Lindstrom, J. (eds.), 1984, *Receptor Biochemistry and Methodology*, Vol. 4, *Monoclonal and Anti-idiotypic Antibodies: Probes for Receptor Structure and Function*, Alan R. Liss, New York.

Walker, J. Boustead, C., and Witzemann, V., 1984, The 43K protein, V_1, associated with acetylcholine receptor containing membrane fragments is an actin-binding protein, *EMBO J.* **3:**2287–2290.

Wan, K., and Lindstrom, J., 1985, Effects of monoclonal antibodies on the function of acetylcholine receptors purified from *Torpedo californica* reconstituted into vesicles, *Biochemistry*, **24:**1212–1221.

Watters, D., and Maelicke, A., 1983, Organization of ligand binding sites at the acetylcholine receptor: A study with monoclonal antibodies, *Biochemistry* **22:**1811–1819.

Wilson, I., Niman, H., Houghton, R., Cherenson, A., Connolly, M., and Lerner, R., 1984, The structure of an antigenic determinant of a protein, *Cell* **37:**767–778.

Young, R., and Davis, R., 1983, Efficient isolation of genes by using antibody probes, *Proc. Natl. Acad. Sci. U.S.A.* **80:**1194–1198.

Young, C., Schmitz, H., and Atassi, M., 1983, Antibodies with preselected specificities to protein regions evoked by immunization with free synthetic peptides, *Immunol. Commun.* **12:**419–428.

Young, E., Ralston, E., Blake, J., Ramachandran, J., Hall, Z., and Stroud, R., 1985, Topological mapping of acetylcholine receptor: Evidence for a model with five transmembrane segments and a cytoplasmic COOH-terminal peptide, *Proc. Natl. Acad. Sci. U.S.A.* **82:**626–630.

Chapter 10

FUNCTION OF ACETYLCHOLINE RECEPTORS IN RECONSTITUTED LIPOSOMES

MARK G. McNAMEE, OWEN T. JONES, and TUNG MING FONG

1. Introduction

The nicotinic acetylcholine receptor (AChR) from the electric organ of various *Torpedo* species is the best-characterized neurotransmitter receptor and one of the best-characterized integral membrane proteins. The relative ease with which AChR can be isolated and purified from *Torpedo californica* electroplax in large quantities, coupled with the similarity between *Torpedo* AChR and mammalian skeletal muscle AChR, makes the *Torpedo* receptor an ideal model protein for studies of receptors, ion channels, and membrane proteins in general. The function of the AChR can be divided into four subfunctions, each of which reveals an intriguing facet of the receptor protein: (1) ligand recognition and binding; (2) coupling of ligand binding to a cation-specific ion channel; (3) cation permeation; and (4) desensitization, a complex process by which the ion channel is reversibly inactivated in the prolonged presence of activating ligands. Under normal physiological conditions, the acetylcholine-induced opening of AChR ion channels leads to a transient increase in cation permeability that results in membrane depolarization. The depolarization can ultimately lead to muscle contraction, signal conduction, or voltage changes depending on the tissues involved.

A unique feature of current AChR research is the feasibility of studying AChR function with a broad range of techniques including biochemical, cell biological, electrophysiological, ultrastructural, genetic, and biophysical approaches. Recent reviews provide a comprehensive summary of progress on many aspects of acetylcholine receptor research (Changeux *et al.*, 1984; McNamee and Ochoa, 1982; Popot and Changeux, 1984), and Table I provides a summary of some of the molecular properties of the receptor protein. The emphasis in this chapter is on recent experiments designed to characterize the function of purified receptors in reconstituted lipid vesicles (liposomes) of defined composition. The reconstituted

MARK G. McNAMEE, OWEN T. JONES, and TUNG MING FONG • Department of Biochemistry and Biophysics, University of California, Davis, California 95616.

Table I. Summary of Biochemical Properties of *Torpedo californica* Acetylcholine Receptor

	Monomer	Dimer
Molecular weight	250,000	500,000
Sedimentation coefficient	9S	13S
Stokes radius	7 nm	9 nm
Radius of gyration (neutron diffraction)	4.6 nm	—
Isoelectric point	4.9	4.9
Subunit composition	$\alpha_2\beta\gamma\delta$	$\alpha_2\beta\gamma\delta$-S-S-$\delta\gamma\beta\alpha_2$
Subunit molecular weights		
α 40,000[a]; 53,649[b]		
β 50,000[a]; 56,060[b]		
γ 60,000[a]; 58,053[b]		
δ 65,000[a]; 59,792[b]		
α-Neurotoxin binding	8 nmol/mg	8 nmol/mg
Specific activity	2 mol/monomer	4 mol/dimer
Acetylcholine binding (specific activity)	8 nmol/mg	8 nmol/mg
Carbohydrate composition	4–7%, all subunits Glucosamine, mannose, glucose, galactose, sialic acid	
Phosphorus content	10 mol/monomer (includes phosphoserine)	

[a] By SDS-PAGE.
[b] By amino acid sequence analysis.

vesicles are ideal for quantitative rapid kinetics analyses of transmembrane ion fluxes, analyses of lipid effects on AChR molecular properties, studies of chemical modification effects on receptors, and characterization of lipid–protein interactions. Furthermore, the AChR serves as a good protein for basic studies of general problems in reconstitution.

2. Incorporation of Purified Acetylcholine Receptor into Liposomes

2.1. Solubilization and Purification of AChR

2.1.1. Isolation of AChR-Containing Membranes from Electric Tissue

The electric organs from *Torpedo californica* marine rays are easily removed from the animal by simple dissection techniques, and the isolated organs can be used immediately to prepare AChR or can be rapidly frozen in liquid nitrogen and stored indefinitely in liquid nitrogen or at −80°C. Organs from adult *Torpedo californica* yield 600–1200 g of usable tissue. Another commonly used species,

Torpedo marmorata, is much smaller in size and yields 100–200 g of tissue per fish.

Receptor purification usually begins with the preparation of a partially purified membrane fraction enriched in AChR. Most procedures involve vigorous homogenization of minced tissue, low-speed centrifugation to remove connective tissue and large fragments, and high-speed centrifugation to pellet the membranes. All steps are carried out rapidly at 0–4°C, and it is essential to include a chelating agent such as EDTA in the homogenization buffer to prevent proteolysis. Some investigators include additional antiprotease agents (such as PMSF and leupeptin) as a precautionary measure. Sulfhydryl-blocking agents such as iodoacetamide or N-ethylmaleimide are usually added to prevent an internal sulfhydryl–disulfide interchange that converts the naturally occuring dimer form of the *Torpedo* receptor to a mixture of monomers and dimers.

More sophisticated membrane fractionation procedures are available to purify AChR-rich membranes if biochemical or biophysical studies are to be conducted directly on the native membrane fractions. Among the most common procedures used are sucrose gradient centrifugation and alkaline extraction. The alkaline extraction removes many peripheral membrane proteins, including a set of proteins of apparent M_r = 43,000. Purified native membranes containing almost 50% by weight of the total protein as AChR have been isolated. Such membranes have been studied extensively, and they provide an appropriate system against which to compare the properties of reconstituted membranes. However, it is generally not necessary to use highly purified membranes as the starting material for receptor purification.

2.1.2. Solubilization of Acetylcholine Receptor

The acetylcholine receptor is an integral membrane protein and can be effectively solubilized by many different ionic and nonionic detergents. However, the choice of the detergent and the solubilization conditions clearly play a critical role in determining whether or not the AChR can be successfully reincorporated into lipid vesicles with complete recovery of both ligand-binding and ion-channel properties. Many of the early studies of AChR isolation, purification, and characterization were carried out in Triton X-100, an effective detergent that preserves ligand binding but leads to essentially irreversible loss of ion-channel properties. Most successful reconstitution procedures use sodium cholate, an anionic detergent that is compatible with the preservation of both AChR-binding and channel functions. The use of cholate was strongly influenced by the work of Racker and co-workers, who used cholate for many years in successful reconstitutions of many different membrane proteins (see Miller and Racker, 1979). In fact, Epstein and Racker (1978) first convincingly demonstrated that *Torpedo* AChR solubilized from native membrane could be reconstituted in Asolectin® membranes with recovery of ion-flux activity. (Asolectin® is a trade name for a partially purified lipid extract prepared from soybeans. It is available from Associated Concentrates, Woodside, NY; a similar preparation is sold by Sigma.) The role of specific detergents and lipids on the functional properties of reconstituted membranes is more fully discussed in Section 2.2.

In a typical solubilization experiment, AChR-containing membranes are extracted with 1% cholate. The supernatant after centrifugation contains 80% of the AChR and approximately 50% of the total protein, thus giving a modest increase in purity. Higher levels of extraction can be achieved using higher cholate concentrations (Anholt *et al.*, 1981), but it then becomes necessary to add additional lipid to prevent irreversible denaturation of the AChR ion channel.

2.1.3. Affinity Column Purification of AChR

The detergent-solubilized AChR from *Torpedo californica* can be purified to homogeneity in a single step using affinity chromatography. Two general types of affinity columns are commonly used: toxin columns and small ligand columns.

Columns prepared by covalently linking snake α-neurotoxin proteins to a gel matrix offer the advantage of high specificity and tight binding. However, they suffer the disadvantage of slow and incomplete elution. It is usually necessary to elute the column overnight with large volumes of a competitive ligand and trap the released AChR on a second, less specific column such as an ion-exchange or lectin column. The AChR can then be eluted from the second column at a reasonable protein concentration (Lindstrom *et al.*, 1980).

The small ligand columns are likely to be less specific and less efficient, but they can be eluted rapidly with minimal dilution of the AChR. An affinity column prepared from Affi-Gel 401 (a Bio-Rad Co. product with a free sulfhydryl as the reactive species) and bromoacetylcholine (an AChR agonist) has been very useful in purifying *Torpedo* AChR (see Ochoa *et al.*, 1983). The wash and elution buffers are always supplemented with phospholipids to prevent displacement of essential lipids from the receptor surface by the detergent. The most commonly used lipid is dioleoylphosphatidylcholine (DOPC), a synthetic lipid available from Avanti Co. (Birmingham, AL). Failure to include lipids results in irreversible loss of ion-channel properties, and the addition of lipids at this stage provides one point in the procedure where lipid alterations can be attempted.

2.1.4. Characterization of AChR Purity

Since the purified, solubilized AChR cannot be directly assayed for ion-channel properties, assessment of AChR purity rests on equilibrium binding assays and on examination of SDS-polyacrylamide gels. The SDS gels are useful because extensive proteolysis appears to have minimal adverse effects on ligand-binding or ion-channel properties. Ultimately, the assays using reconstituted membranes for ion-channel function and ligand-binding state interconversions provide the best indicator of functional AChR purity.

2.1.4a. Equilibrium Toxin Binding. The simplest assay for AChR binding sites takes advantage of the high specificity and nearly irreversible nature of α-bungarotoxin binding. Iodinated α-bungarotoxin ([^{125}I]BTX) can be readily prepared by several procedures, or it can be purchased from most major radionuclide suppliers. Excess toxin is allowed to interact with a solution containing AChR in detergent solution for 30–60 min at room temperature. The long reaction time insures complete equilibrium, since the rate of toxin binding is highly dependent on reaction conditions. The bound and unbound toxins can then be separated by ion-exchange techniques taking advantage of the fact that the toxin is positively charged at neutral pH values and the AChR–toxin complex retains a net negative charge because of the relatively large size of the negatively charged protein. A filter method using DEAE paper was first introduced by Schmidt and Raftery (1973), and variations on the method are still widely used. Typical concentrations of toxin and AChR sites are 10 nM and 2 nM, respectively, in assay volumes of 50–100 μl.

In the $[^{125}I]$BTX binding assay, the purified AChR contains 8 nmol binding sites/mg total protein. The result is consistent with one binding site per 125,000 daltons and is consistent with the known subunit structure and native structure of the protein.

2.1.4b. SDS Polyacrylamide Gel Electrophoresis. The SDS gel electrophoresis system originally described by Laemmli (1970) works effectively to separate the four bands characteristic of pure *Torpedo* AChR. The ratio of Coomassie blue staining intensities for the α, β, γ, and δ subunits is 2:1:1:1, and the apparent relative molecular weights are M_r = 40,000, 50,000, 60,000, and 65,000. Proteolysis of AChR during purification gives rise to additional lower-molecular-weight bands and initially causes an apparent increase in the intensity of the M_r = 40,000 band as the higher-molecular-weight subunits are partially degraded. The four AChR bands are readily detected even in partially purified membranes, and the relative intensities correlate well with the results of toxin-binding assays.

2.2. Incorporation of AChR into Liposomes

When a membrane is disrupted by a detergent, the micelles formed are in dynamic equilibrium. A formalism for the addition and removal of detergents from a membrane is given by Reynolds (1982):

$$\begin{array}{ccccc} \text{Pr-PL}_{(\text{Bilayer})} & \overset{\text{D}}{\rightleftharpoons} & \text{Pr-PL-D}_{(\text{Bilayer})} & \overset{\text{D}}{\rightleftharpoons} & \text{Pr-PL-D}_{(\text{Fragment})} \\ & & & & \upharpoonleft\downharpoonright \text{D} \\ \text{D} + \text{PL-D}_{(\text{Micelle})} & + & \text{Pr-D}_{(\text{Micelle})} & \overset{\text{D}}{\rightleftharpoons} & \text{Pr-PL-D}_{(\text{Mixed micelle})} \end{array}$$

where D is the detergent and PL and Pr refer to the phospholipid and protein components of the membrane, respectively.

In principle all the steps are reversible, and attempts at reconstitution are based on the premise that by reducing the detergent to below its critical micelle concentration (CMC), the limited aqueous solubility of the lipid and protein will cause aggregation and reformation of the lipid–protein bilayer.

2.2.1. A Practical Procedure for AChR Reconstitution

In this section a brief outline of a reconstitution procedure that can be reliably used to prepare membrane vesicles containing *Torpedo californica* AChR is pre-

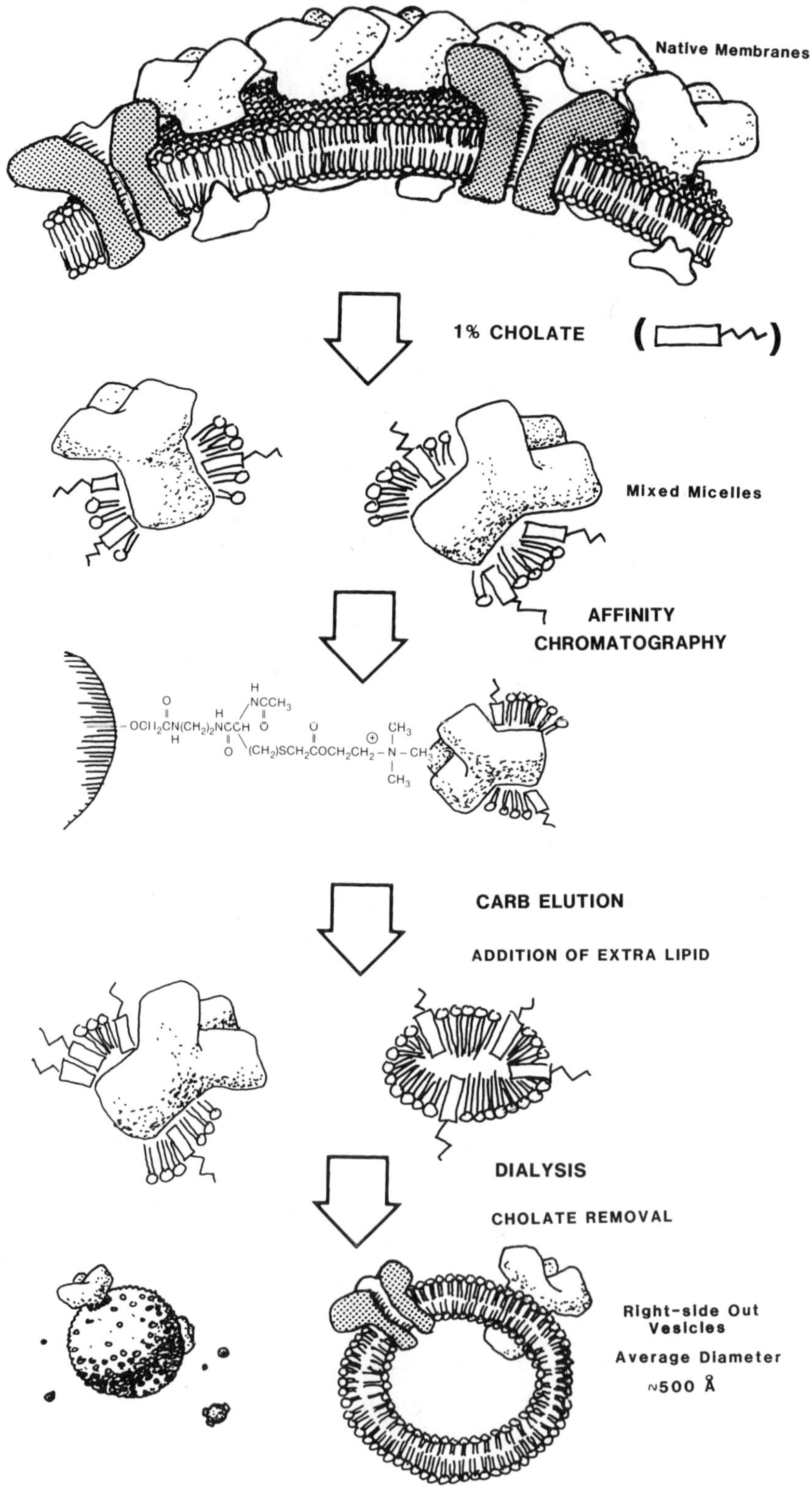
Native Membranes
1% CHOLATE
Mixed Micelles
AFFINITY
CHROMATOGRAPHY
CARB ELUTION
ADDITION OF EXTRA LIPID
DIALYSIS
CHOLATE REMOVAL
Right-side Out
Vesicles
Average Diameter
∼500 Å

sented. Figure 1 schematically outlines the overall procedure. Later sections discuss in more detail the rationale for the procedure and the options available for variations and improvements.

Purified acetylcholine receptor in 1% (W/V) cholate is used as the starting material for reconstitution. As outlined in Section 2.1, the AChR is eluted from the affinity column with carbamylcholine, and a typical preparation of pooled fractions from a 600-g preparation will contain 30–40 ml of AChR at a protein concentration of 1–1.5 mg/ml in a buffer (buffer A) containing 100 mM NaCl, 10 mM MOPS, 0.1 mM EDTA, 0.02% NaN_3 (pH 7.4), and 1 mg/ml DOPC as the lipid.

The AChR-containing solution is adjusted to 1 mg/ml protein using buffer A containing 1% cholate and 1 mg/ml DOPC, and then a solution containing 60 mg/ml Asolectin dispersed in buffer A containing 4% cholate is added to give a final cholate concentration of 2% and a total lipid concentration of 20 mg/ml.

The clear solution (which is slightly yellow because of the Asolectin®) is then dialyzed against 4 liters of buffer A. The buffer is changed at least three times over a 48-hr period, and during the dialysis period the solution becomes slightly opalescent. The dialysis allows for vesicle formation as the cholate is removed and also removes the carbamylcholine from the mixture.

When dialysis is complete, the vesicles can be stored at 0–4°C for up to 3 weeks with no apparent loss of binding or ion-flux activity. The vesicles can also be frozen in lipid nitrogen and kept indefinitely. However, as noted in Section 3.3, freeze–thaw cycles alter vesicle size, and frozen and unfrozen samples cannot be directly compared. Freezing at −20°C appears to lead to rapid loss of AChR activity.

2.2.2. Selection of Detergents

In general, the selection of a particular detergent for reconstitution is based more on precedent and empirical factors than on well-understood scientific principles. There are, of course, certain conditions that must be met if a detergent is to be useful. The detergent must be readily available in pure form, it must solubilize but not denature the protein, and it must be removable. The critical micelle concentration (CMC) of a detergent is the most useful parameter for predicting ease of removal (at least by dialysis procedures), and a relatively high CMC is usually desirable. The charge on a detergent can also be important, particularly if ion-exchange steps are necessary during purification of the protein. Several reviews of the properties of detergents and their use in membrane studies provide a more complete discussion of relevant issues (Helenius *et al.*, 1979).

In the case of AChR, a wide range of detergents have been used for solubilization and purification. As indicated in Section 2.1.2, sodium cholate is an effective detergent for all aspects of reconstitution. Some of the most widely used

←

Figure 1. Purification and reconstitution of AChR by cholate solubilization, affinity chromatography, and dialysis methods.

Name	Structure	CMC Wt %	CMC mM	Success in reconstitution
Sodium cholate (Anionic)	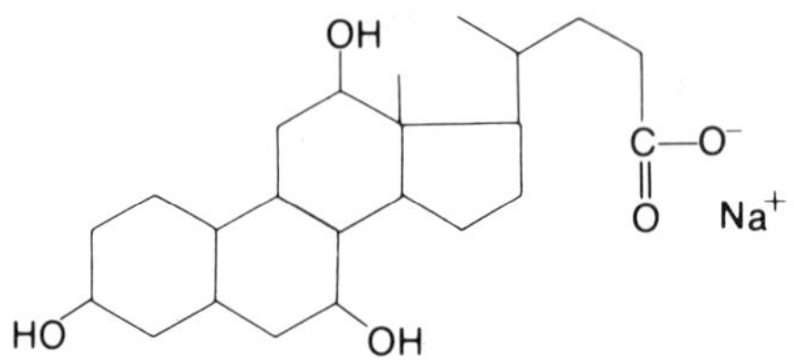	0.60	14	+ + + +
Octylglucoside (neutral)	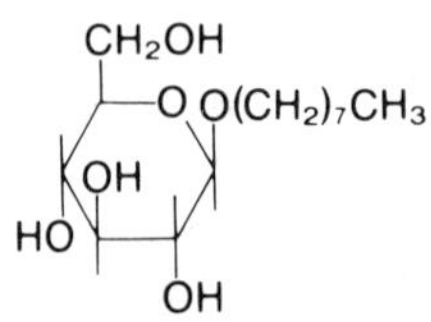	0.65	25	+ −
Triton X-100 (Neutral)		0.01	0.2	−
Sodium dodecyl sulfate (SDS) (anionic)	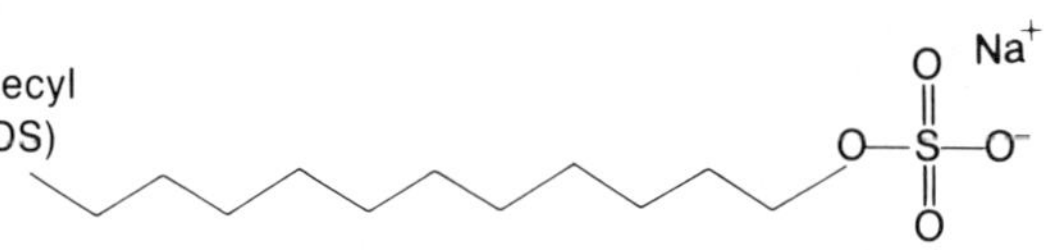	0.24	8.2	−

Figure 2. Structures of detergents commonly used in acetylcholine receptor studies.

detergents and their properties are outlined in Fig. 2. Sodium dodecylsulfate (SDS) is the only detergent that dissociates the AChR into its five subunits. Although a reconstitution of AChR from the individual subunits would be a useful procedure, it appears that subunit dissociation always leads to irreversible denaturation. There is, however, some evidence for partial recovery of binding activity for the α subunit after SDS treatment (Haggerty and Froehner, 1981; Tzartos and Changeux, 1984).

There is some controversy over the use of octylglucoside as a detergent, since one group reports good activity for AChR after reconstitution from octylglucoside (Gonzalez-Ros *et al.*, 1980, 1981) and other groups show complete inactivation of AChR (Anholt *et al.*, 1981; Huganir and Racker, 1982). Some of the different detergent effects may result from complex interactions among the lipids, the protein, and the detergents.

2.2.3. Selection of Lipids

The choice of a particular lipid to use in reconstitution must be made carefully. Experience has shown that particular lipids and lipid mixtures can give rise to many different types of membranes with variations in lamellar structure, leakiness, or heterogeneity in size and composition. Additionally, the role of lipids in preservation of receptor function must be distinguished from indirect effects on the reconstitution process. Lipid selection depends initially on particular experimental requirements. For example, if receptor-containing samples are required simply to measure functional properties of the protein, then excellent methods for reconstitution are available using either native lipids or Asolectin®. In contrast, the study of specific lipid effects on receptor function requires the use of synthetic lipids in order to provide highly defined lipid environments for reconstitution. The common sources and characteristics of the major lipids employed in receptor reconstitution are given below.

2.2.3a. Native Lipids. The lipid composition of native *Torpedo* membranes (Schiebler and Hucho, 1978; Popot *et al.*, 1978; Gonzalez-Ros *et al.*, 1982) is typical of that derived from neural tissues (Deutsch and Kelly, 1981). Approximately 50 mol% of the total lipids are phospholipids, mainly phosphatidylethanolamine, phosphatidylcholine, and phosphatidylserine, with smaller amounts of sphingomyelin and phosphatidylinositol. The fatty acids contain a range of chain lengths and degrees of unsaturation. Particularly prominent are the high levels (25 mol%) of decosahexanoyl (22:6) acyl chains, mainly associated with the ethanolamine phosphatides (Popot *et al.*, 1978; Gonzalez-Ros *et al.*, 1982). The remaining lipid fraction is composed almost entirely of unesterified cholesterol with low amounts of triglycerides and other neutral lipids.

An important consequence of the complex lipid composition of the native lipids is the predominance of components whose physical and chemical properties are poorly understood (see, for example, Lee, 1983).

2.2.3b. Asolectin®. Asolectin® (soybean lecithin) is a crude mixture of phospholipids and neutral lipids (Letters, 1964). Attractive features of Asolectin® are its well-characterized ability to support reconstitution of functional receptors coupled with its widespread availability at low cost. The major disadvantage of Asolectin® is that, like native lipids, it is highly heterogeneous. Removal of neutral lipids from Asolectin® by acetone extraction has been reported to diminish the activity of receptors reconstituted into the remaining phospholipids (Kilian *et al.*, 1980).

2.2.3c. Synthetic Lipids. In order to obtain highly defined lipid environments suitable for biophysical studies, several investigators have attempted to reconstitute acetylcholine receptors into mixtures of controlled composition using synthetic lipids (Ochoa *et al.*, 1983; Criado *et al.*, 1984). Although phosphatidylcholines are readily available, lipids with other headgroups are difficult to synthesize and are consequently expensive. One convenient route to lipids of different headgroups is to take advantage of the simple hydrolysis and transphosphatidylation reactions catalyzed by the enzyme phospholipase D (Comfurius and Zwaal, 1977). The ease with which a diverse range of lipids can be prepared from the corresponding phosphatidylcholines makes this the method of choice.

Although many synthetic lipids do not support ion-flux activity, mixtures of dioleoylphosphatidylcholine/dioleoylphosphatidic acid/cholesterol in the molar ratio of 56 : 19 : 25 appear to give good reconstitution of functional receptors (Ochoa *et al.*, 1983).

2.2.4. Techniques for Detergent Removal

The essential step in achieving successful reconstitution of functional AChR-containing vesicles is to remove the detergent from the mixtures of lipid, protein, and detergent in order to permit the spontaneous formation of bilayer vesicles.

2.2.4a. Dialysis Techniques. Dialysis was originally introduced by Racker for reconstitution of oxidative phosphorylation enzymes (Kagawa and Racker, 1971) and has proven to be the most widely used and useful method for detergent removal. The number of changes of buffer required during dialysis can be reduced by using detergent-absorbing beads, for example, Amberlite XAD-2 or Biobeads SM-2 (Holloway, 1973). The use of such beads also enhances the applicability of using detergents with low CMCs. Using a continuous dialysis apparatus can be advantageous as the rate of detergent removal can be carefully controlled (Milsmann *et al.*, 1978), and it can dramatically reduce the dialysis time. In some cases, too-rapid removal of detergent can inactivate membrane proteins through inadequate mixing of lipid and protein components. In some instances, very slow removal of detergent is required, for example, in preparing ordered samples for X-ray or EM studies. Slow dialysis is achieved by including decreasing amounts of detergents in the dialysis buffer.

Detergent removal can be assessed qualitatively by the increase in absorbance resulting from light scattering by liposomes and quantitatively by using radiolabeled detergents. Radiolabels such as [^{14}C]DPPC and [^{125}I]BTX are useful for determining the levels of lipid and AChR during dialysis. In most studies such losses are negligible (Ochoa *et al.*, 1983).

2.2.4b. Gel Filtration Techniques. Detergent removal is also possible using gel filtration techniques. The detergent is removed simply by passing the reconstitution mixture over a gel exclusion column. As the mixture passes through the resin, the protein is integrated into liposomes, which separate and elute from the column before the detergent. Efficient reconstitution of AChR into Asolectin® has been achieved using Sephadex G-25 (Huganir and Racker, 1982) and Sephadex G-50 (Sobel *et al.*, 1980). The resulting AChR pool apparently contains lower levels of residual cholate than achieved by dialysis. Detergent removal by the gel filtration method gives a broader size distribution of vesicles than found by dialysis (Popot *et al.*, 1981; Nozaki *et al.*, 1982). In addition, agonist-induced ion-flux responses for receptors reconstituted by this method are claimed to be better than by cholate dialysis (Huganir and Racker, 1982). The most significant advantage of this technique is its rapidity, although it is not convenient for large numbers of samples.

2.2.4c. Dilution Techniques. Cholate dilution was first employed by Racker in reconstitution of cytochrome oxidase (Racker, 1972). Cholate dilution has the advantage of being rapid, and it is particularly useful in minimizing the time a protein spends in detergent solution. In the cholate dilution technique, a mixed micellar solution of receptor at a slightly lower concentration of detergent than that used for reconstitution by other methods is diluted 20-fold into a suitable buffer. The dilution process reduces the detergent concentration to below its CMC and allows formation of receptor-containing liposomes. The resulting vesicles can be concentrated by centrifugation and freeze-thawed to increase the agonist-induced ion permeability response. The dilution technique has been employed with some success in two different laboratories (Sobel *et al.*, 1980; Huganir *et al.*, 1979).

2.2.5. Importance of Lipid-to-Protein Ratio in Reconstitution

For many years attempts to reconstitute AChR ion flux were unsuccessful. A major breakthrough in reconstitution strategy was the observation by Epstein and Racker (1978) that removal of lipids during solubilization could cause irreversible inactivation of the receptor. Subsequently, lipids were shown to be essential for preservation of both receptor channel function and ligand affinity-state transitions characteristic of the desensitization process (see McNamee and Ochoa, 1982).

Additional lipids were also shown to be essential for protection of the AChR channels during incorporation into vesicles. For example, lipid-to-protein ratios of 16:1 (w/w) during cholate dialysis have been found to give optimum yields of vesicles in which agonist-induced ion-permeability responses can be measured (Anholt *et al.*, 1981). At this ratio the surface density of the receptors is about five times less than in native *Torpedo* vesicles. For a higher total lipid-to-protein ratio, the lipid-to-protein ratio of the receptor-containing vesicles was found to remain constant at 16:1 (w/w). The excess lipids appear to form a second population of small vesicles devoid of receptors. Receptor reconstitution by cholate dialysis at suboptimal lipid-to-protein ratios appears to irreversibly inactivate receptor ion-channel function (Anholt *et al.*, 1981).

2.2.6. Methods for Controlling Lipid Composition

2.2.6a. Exchange Procedures. Lipid exchange is most readily accomplished on the same affinity column as used to purify the receptor (Ochoa *et al.*, 1983; Ellena *et al.*, 1983). Complete exchange of the desired lipid for receptor-associated lipids is achieved by washing the column twice with a micellar solution of detergent and test lipid [typically 2 mg/ml lipid in 1% (w/v) cholate]. The lipid pools are allowed to equilibrate for several hours, after which a final wash step is employed. The choice of lipids for equilibration is usually identical to that required for reconstitution. Most often dioleoylphosphatidylcholine can serve as a general-purpose lipid for preparations in which lipids such as Asolectin® are added after removal of AChR from the column.

2.2.6b. Equilibrium and Dialysis. In principle, equilibrattion of micellar solutions of receptor and test lipids allows thorough mixing of lipid pools. Following detergent removal, the lipid composition of the membranes reflects that of the equilibration mixture.

Several factors affect the equilibration process, including temperature, pH, and detergent concentration. An often overlooked factor, however, is the ionic strength. Increases in ionic strength tend to reduce electrostatic interactions including those involved in the action of both ionic (Helenius *et al.*, 1979) and, more surprisingly, nonionic detergents (Ueno *et al.*, 1984). An increase in ionic strength has been found to cause a decreased incorporation of AChR into Asolectin® liposomes (Popot *et al.*, 1981). Particularly important factors in determining how well equilibration occurs are the physical and chemical characteristics of the lipids to be equilibrated. To prevent aggregation and formation of multilamellar structures, it is necessary to add lipids in a highly dispersed form. This is best accomplished by hydrating a thin film of dried lipids with the detergent solution and sonicating the mixture under nitrogen at a temperature above the phase-transition temperature of the lipids. In preparing mixtures of phospholipids, it is best to mix the lipids in organic solvent before drying.

2.2.6c. Lipid Substitution. The lipid substitution method was initially developed for (Ca^{2+} + Mg^{2+})-ATPase of rabbit muscle sarcoplasmic reticulum (Warren *et al.*, 1974) and subsequently extended to other membrane proteins including (Na^{+} + K^{+})-ATPase (Hilden and Hokin, 1976) and hydroxybutyrate dehydrogenase (Houslay *et al.*, 1975). Additionally, recent experiments suggest the applicability of this method to the acetylcholine receptor (O. T. Jones, J. H. Eubanks, and M. G. McNamee, unpublished observations).

Briefly, the lipids surrounding the membrane protein are allowed to equilibrate with exogenous test lipids dispersed in a suitable detergent (usually cholate). The lipid–protein complexes are then separated from excess lipid and detergent by centrifugation into a detergent-free sucrose density gradient. In one step about 90% of the original lipid can be replaced by test lipid. If the process is repeated, the level of substitution of the original lipid by test lipid is greater than 98%. The lipid substitution technique is rapid compared to column exchange procedures.

3. Physical Characterization of Reconstituted Liposomes

3.1. Size Distribution of Liposomes

Two major factors in receptor reconstitution that limit the magnitude of the ion-flux response are the densitization processes intrinsic to the AChR protein and the rate at which ions equilibrate between the inside and outside of the vesicles. For AChR-containing membranes, the size of the vesicles has a central role in the latter process because the flux rate is proportional to the moles of AChR per liter of internal volume and the magnitude of the response depends directly on vesicle size. Vesicles with a small internal volume rapidly come to equilibrium with the external fluid, and this complicates the measurement of initial rates of ion flux (Hopfer, 1978).

Methods that measure size distribution indirectly, for example, light scattering, density gradient centrifugation, or measurement of internal volume, can cause artifactual results. This is particularly true when the vesicles are leaky or the membranes exist as sheets (Mimms *et al.*, 1981). Direct measurements of vesicle size using electron microscopy and/or gel exclusion chromatography should be considered essential requirements of any serious reconstitution study.

3.1.1. Electron Microscopy

Electron microscopy (EM) has proved useful in characterizing the effect of lipids and reconstitution conditions on the size distributions and gross morphologies of liposomes. It is very useful in discerning changes in size and volume distributions, for example, before and after freeze–thaw cycles and in the presence and absence of different lipids. Electron microscopy of negatively stained reconstituted preparations has been employed by several laboratories to characterize liposome size and heterogeneity in both Asolectin® (Anholt *et al.*, 1982) and synthetic lipid mixtures (Criado *et al.*, 1984).

Sizing of vesicles by EM is not without problems. For very large vesicles (>5000 nm), the usefulness of negative-stain EM is limited, as vesicle disruption during sample preparation can arise. Sizing problems also occur with multilamellar structures (Olson *et al.*, 1979). The advantages and disadvantages of EM in sizing liposomes have been discussed by Bangham *et al.* (1974).

3.1.2. Gel Filtration Techniques

An additional method to characterize liposome size is gel-exclusion chromatography. Gel chromatography of liposomes on large-pore agarose was initially introduced as a method for purifying small unilamellar vesicles formed on sonication of aqueous dispersions of phospholipids (Huang, 1969). More recently, gel chromatography has been used as a method for determining the size distribution of AChR-containing liposomes. Because of the large size of the liposomes, gels with extremely large pore sizes are required. The most useful gel materials have been Bio-Gel-A-150m (100–200 mesh, Bio-Rad Inc.) with a fractionation range of 1–150 million (Anholt *et al.*, 1982) and Sephacryl S-1000. Comparisons of size distributions for liposomes determined by EM and gel chromatography on Sephacryl have shown good agreement (Reynolds *et al.*, 1983).

One particular problem with sizing columns is that the gels may alter the size of the vesicles or cause preferential adsorption onto the column. These problems can be alleviated by thorough equilibration of the column with lipid solutions prior to use (Reynolds *et al.*, 1983). The size distribution of liposomes can be determined by calibrating the column using latex beads of known sizes. The elution of liposomes from the columns can be monitored by light scattering (absorbance at 400 nm), by phosphate analysis, or by incorporating radioactive lipids into the vesicles. The AChR-containing liposomes can be monitored by standard protein determination procedures or by [^{125}I]BTX binding.

3.2. Methods for Separating Receptor-Containing Liposomes

The observation of significant heterogeneity in reconstituted vesicles has important consequences for studying receptor function, since the initial rates of ion translocation are highly dependent on the internal volumes of the vesicles. In heterogeneous vesicle populations additional problems arise, especially if the AChR-containing vesicles only represent a fraction of the total internal volume. In these samples, the apparent flux response will be grossly underestimated.

The simplest method for estimating the percentage of liposomes associated with receptor was introduced by Lindstrom's group (Anholt *et al.*, 1982). Antibodies raised against the receptor were shown to be able to immunoprecipitate AChR-containing liposomes from vesicle suspensions. Immunoprecipitation of all the AChR in a reconstituted sample by antireceptor serum followed by antiantibody, left 62% of the equilibrium volume in solution. Only 26% of the lipid was precipitated, providing evidence that reconstitution of receptor into Asolectin® by dialysis gave both receptor-free and receptor-containing vesicles.

Another method based on a similar prinicple has been attempted (M. G. McNamee and P. A. Badger, unpublished observations). The acetylcholine receptor is a glycoprotein that can bind with high affinity to lectins (Lindstrom *et al.*, 1980). Passage of reconstituted liposomes through an affinity column of immobilized lectin was tested for retention of receptor-containing vesicles. Preliminary studies showed the method to be partially successful. A possible improvement would be to use gels with larger beads.

One feature of the above methods is that they are only likely to work well if the orientation of the AChR is outward facing, i.e., similar to that found *in vivo*. It is an intriguing aspect of AChR reconstitution that in apparently all studies most of the receptors are in this orientation (Anholt *et al.*, 1981; Huganir and Racker, 1982; Ochoa *et al.*, 1983; Criado *et al.*, 1984). Receptor orientation is determined by measuring [^{125}I]BTX binding without detergents and then with detergents to disrupt the vesicles and expose internal sites.

An attempt to obtain a homogeneous population of kinetically active vesicles has been described for receptor-containing vesicles prepared from eel membranes (Hess *et al.*, 1983). These investigators equilibrated the entire vesicle population with a high-density cation (cesium). They then altered the buoyant density of the active vesicles by allowing rapid equilibration with a less dense cation (sodium) in the presence of agonist. The active vesicle fraction then floated to lower densities on centrifugation in a sucrose/cesium chloride gradient.

3.3. Use of Freeze–Thaw Techniques to Alter Liposome Size and Receptor Distribution

Reconstituted liposomes prepared by the dialysis method for detergent removal are often heterogeneous in size and protein distribution, in part because the equilibrium constants for any of the reaction steps in solubilization are not known, and it is difficult to obtain mixed protein–lipid–detergent micelles exclusively (Reynolds, 1982). However, many experimental designs, especially quantitative measurements of ion influx and efflux, require reconstituted liposomes with homogeneous size and channel distribution (Bernhardt and Neumann, 1980). It is also desirable to increase the liposome size to improve the amplitude in any

Table II. Effects of Freeze–Thaw Procedure on the Internal Volume and Ion-Flux Response of Reconstituted Liposomes Containing AChR[a]

Lipid composition	Internal volume		Influx response	
	Nonfrozen	Frozen	Nonfrozen	Frozen
DOPC	87	2445	0	0.91
DOPC-PA-CH 56:19:25	398	1259	0.51	3.39
Asolectin®	418	1826	16.0	2.26

[a] AChR-containing liposomes were prepared by the cholate dialysis method. Lipid-to-AChR (w/w) ratio were 50:1, and the lipid concentration was 33 mM. Internal volumes are expressed as nanoliters of solution trapped within the vesicles per micromole of total added lipid after 48 hr equilibration at 4°C as measured by the amount of $^{86}Rb^+$ that elutes with the vesicle fraction after ion-exchange chromatography. Influx response is expressed as [cpm(+carb)-cpm(−carb)]/cpm(−carb) after correction for background counts in the absence of membranes.

flux analysis. The freeze–thaw method is a very simple technique to increase liposome size (Anholt *et al.*, 1982) and randomize protein distribution. Freeze–thaw basically involves several cycles of freezing the membrane samples in liquid nitrogen and thawing them at 4°C or room temperature. This method has been used to increase liposome size by a factor of 2 to 28 in internal volume depending on the lipid composition (Table II; see also Ochoa *et al.*, 1983). Such freeze–thaw cycles appear to promote fusion among liposomes and accelerate the process of reaching the most stable equilibrium state of liposome structure.

4. Functional Characterization of Reconstituted Liposomes

4.1. Techniques for Measuring Receptor-Mediated Cation Fluxes

4.1.1. Manual Assays

The function of AChR in reconstituted liposomes can most easily be assessed by measuring the influx of tracer ions into liposomes in response to agonist. The basic principle of a manual ion influx assay is shown in Fig. 3, in which $^{86}Rb^+$ is the tracer ion and carbamylcholine (Carb) is the agonist. After AChR-containing liposomes are incubated with $^{86}Rb^+$ and Carb for 30 sec, the reaction mixture is placed onto a disposable column containing Dowex 50W resin, and the column is washed with a salt-free sucrose solution to remove external Rb^+ (Walker *et al.*, 1982). In addition to the ion-exchange method, a filtration technique of retaining membranes has been used to remove external tracer ions from native electroplax membrane preparations by Millipore® filters (Delegeane and McNamee, 1980) or glass fiber filters (Neubig and Cohen, 1980). The ion-exchange method is not very suitable for native membranes, probably because of the difference in the electrostatic properties of membranes. For a reaction time as long as 30 sec, all AChR will usually be desensitized, so that it is not necessary to add quencher or channel blocker to prevent further influx or efflux. During the course of ion influx, no membrane potential will be developed, since the movement of each Rb^+ ion is

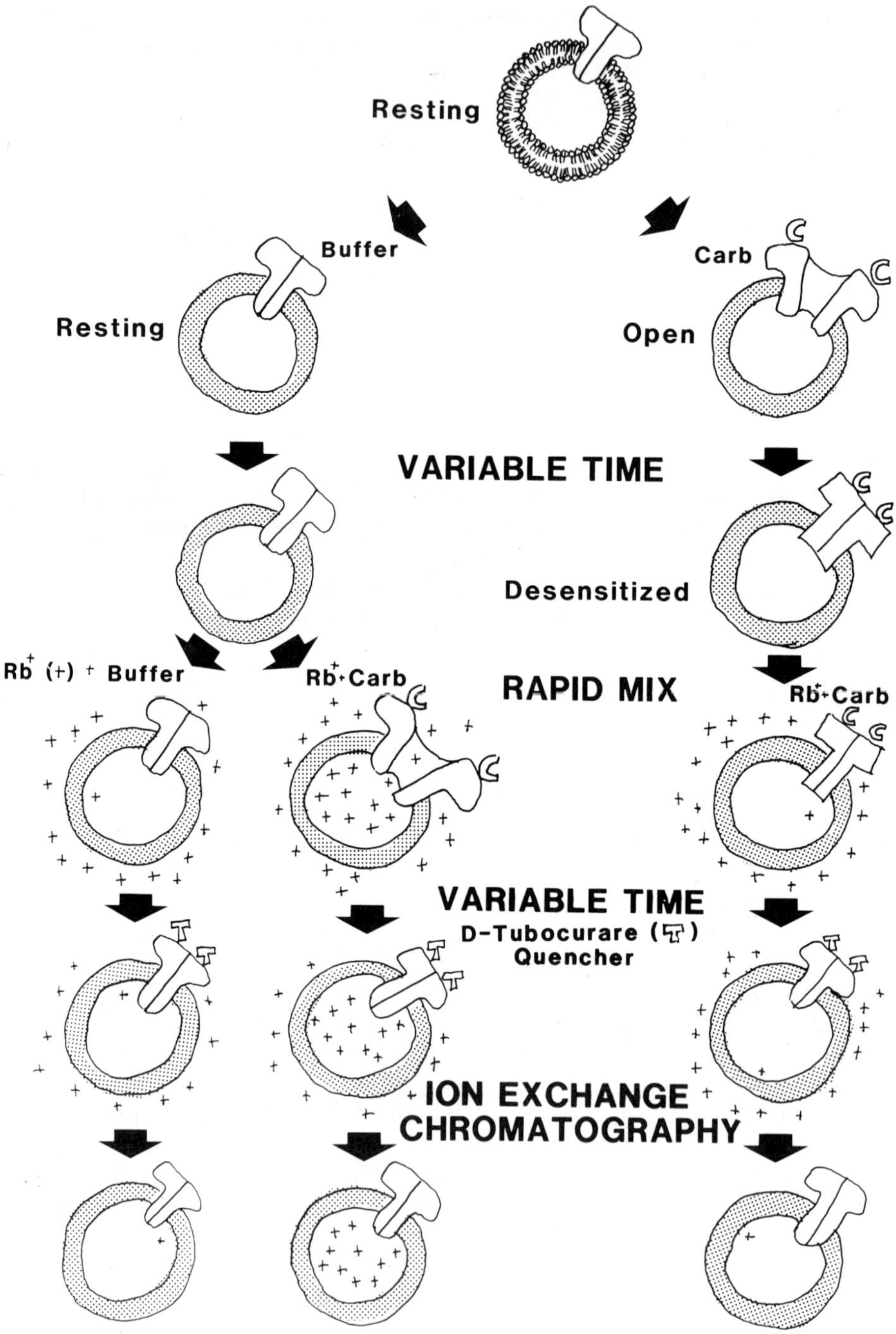

Figure 3. Rapid kinetic measurements of AChR activation and desensitization by quench-flow techniques.

accompanied by the movement of Na^+ ion in the opposite direction. This influx procedure will determine the integrated influx response, which represents the internal volume of reconstituted liposomes containing functional AChR when high concentrations of agonist, such as 1 mM Carb, are used. The total internal volume of the entire reconstituted liposome preparation can be determined by allowing the influx reaction to proceed for a very long time (e.g., 48 hr) to assure complete equilibration. From the internal volume measurements and the known lipid concentration, it is possible to estimate liposome dimensions by assuming a homogeneous size distribution (Lichtenberg *et al.*, 1981).

The manual assay is moderately sensitive to changes in AChR activity provided the vesicles are large enough to avoid instantaneous equilibration and the concentration of active AChR is low enough to give a relatively slow influx rate. The self-quenching associated with AChR desensitization effectively limits the real influx time to a few seconds. Hence, it is best to regard the manual assay as a qualitative indicator of AChR activity. The assay is simple and is ideally suited for "all-or-none" measurements of AChR activity.

The manual influx assay has been used to study the lipid composition requirement for AChR function (Ochoa *et al.*, 1983; Fong and McNamee, 1985). As shown in Table II, the ion-gating activity of AChR requires the presence of cholesterol and negatively charged phospholipids in the membranes in addition to phosphatidylcholine. The effects of chemical modifications on ion permeability control have also been studied by the manual flux assay (Walker *et al.*, 1981, 1984). It was found that alkylation of AChR with maleimide derivatives after dithiothreitol reduction results in complete inhibition of the flux response, which is attributable to the labeling of the reduced disulfide groups at the agonist binding site and other sites. When liposomes containing functional AChR are prepared, this procedure can be modified to demonstrate receptor desensitization on binding of agonist (see Fig. 3 for an analogous procedure in the millisecond time domain).

4.1.2. Quench-Flow–Rapid Kinetics Techniques

4.1.2a. Basic Principles. As mentioned in Section 4.1.1, the manual ion flux assay only determines the integrated flux response. However, AChR forms a cation channel with a mean open time of several milliseconds. It is therefore necessary to measure the ion influx in the millisecond time domain in order to obtain more detailed information about the ion-gating event. Fast kinetic techniques utilizing flow systems, such as stopped-flow, quenched-flow, etc., have been developed to investigate rapid biochemical reactions (for review, see Sha'afi and Fernandez, 1983). When radioactive tracer ions are used to study the ion-permeability control properties of AChR, a quenched-flow system can be employed to obtain kinetic parameters for certain models. The quench-flow technique mainly employs a rapid mixing device for starting the reaction and stopping it at a certain time, usually in the range of milliseconds to seconds. In the case of AChR, competitive antagonists such as *d*-tubocurarine or noncompetitive blockers such as dibucaine are frequently used to block the ion channel rapidly at the end of the influx reaction.

As shown in Fig. 3, two kinds of experiments are designed to study the receptor activation and inactivation (or desensitization), respectively. The left part of the diagram schematizes the activation experiment in which the reaction is started by rapidly mixing AChR-containing liposomes with Rb^+ and Carb. After a certain period of time (see Section 4.1.2b for the mechanical design), the reaction is stopped by rapidly mixing the reaction mixture with *d*-tubocurarine. The quenched reaction mixture is then passed through an ion-exchange column to remove external tracer ions, and the amount of internal ions is quantitated by scintillation counting (Walker *et al.*, 1982). The right part of Fig. 3 shows the inactivation experiments in which a preincubation step is carried out before the influx reaction starts. The influx reaction step in the inactivation experiments is carried out by mixing the preincubation mixture with a fixed high concentration of Carb for a fixed period of time (to obtain the initial influx rate rather than rate constant) so that the remaining activity of AChR after preincubation is determined. Therefore, only the preincubation step is carried out with variable time and variable concentrations of Carb.

4.1.2b. Design of the Apparatus. Both home-built quench-flow systems (Cash and Hess, 1981; Walker *et al.*, 1982) and commercial quench-flow systems (Neubig and Cohen, 1980; Heidmann *et al.*, 1983a) have been used for investigations of AChR-mediated transmembrane ion flux. The Cash and Hess device combines the conventional quench-flow (Lymn and Taylor, 1970) with the pulsed quench-flow (Fersht and Jakes, 1975) in a single integrated system consisting of two or more pneumatically driven piston units (also called towers, as in Fig. 4). Each piston unit may, in turn, consist of two or more syringes that contain liposome solution, agonist plus tracer ion, and quenching solution, respectively. A 100-msec stroke time for the piston displacement is usually used to minimize effects of laminar flow and turbulent flow, which may cause errors in the reaction times. For reaction times less than 100 msec, conventional quench-flow (or continuous mode as in Fig. 4) is employed, in which the reaction time is determined by the length of the tubing between two mixers. The influx reaction proceeds along with the flow of solution until the solution is mixed with quencher in the second mixer. For reaction times longer than 100 msec, pulsed mode is used, in which the reaction time is determined by the time delay between two pistons. The flux reaction proceeds inside the tubing until the second piston is actuated. In the case of inactivation experiments, the connections of syringes are modified as shown in Fig. 5. Perincubation is achieved by using the pulse mode, whereas influx reaction is achieved by the continuous mode with a fixed reaction time.

4.1.2c. Quenching Efficiency. The experimental design of a quench-flow system demands that the reaction be stopped within the mixing dead time. This requires that a highly efficient channel blocker must be used for quenching. Such quenchers and their corresponding effective concentrations can be screened and selected based on whether or not they can inhibit the flux response instantaneously by the manual ion flux assay. By mixing the AChR-containing liposomes with Rb^+ and quencher candidate simultaneously, with or without agonist, one can select a candidate at a proper concentration. Some cautions should be exercised in selecting quenchers. First, because of the amphiphilic characteristics of many

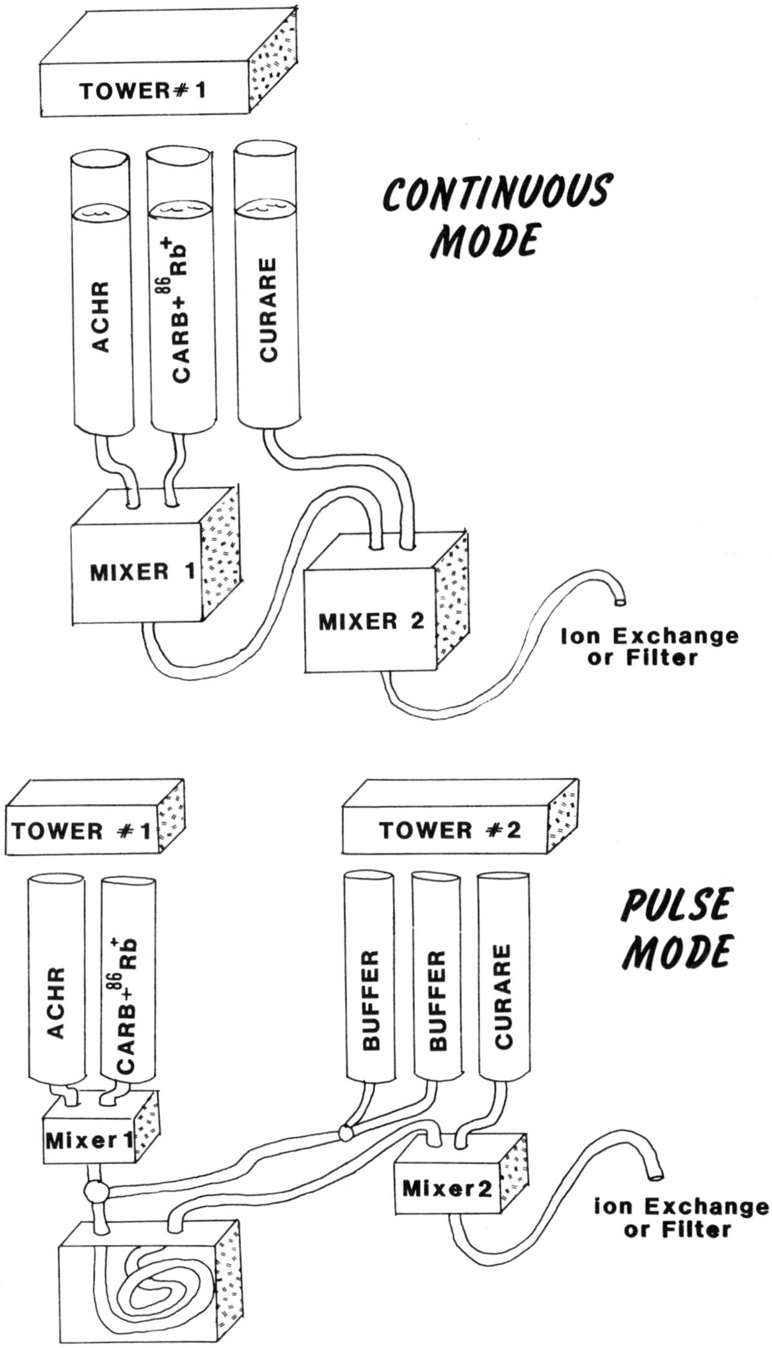

Figure 4. Schematic representation of the quench-flow system for measuring AChR activation.

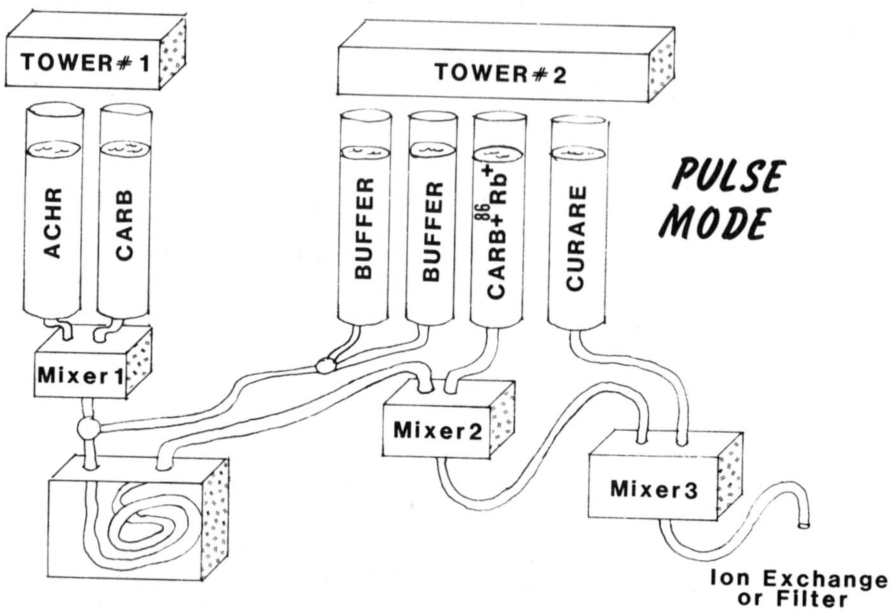

Figure 5. Schematic representation of the quench-flow system for measuring AChR densensitization.

noncompetitive blockers, a combination of four manual assays, i.e., with or without agonist and with or without quencher, should be done to assure that the quencher does not cause any artifact such as vesicle leakage. Secondly, these quencher molecules should not cause any macroscopic or microscopic aggregation of liposomes. For example, 15 mM *d*-tubocurarine is a suitable quencher for blocking AChR-containing Asolectin® liposomes, but it causes aggregation of PC–PA–CH liposomes, whereas 0.1 mM dibucaine is efficient for blocking the relatively slow ion flux into AChR-containing PC–PA–CH liposomes but ineffective for the fast ion flux into AChR-containing Asolectin® liposomes.

4.1.2d. Data Acquisition and Analysis. In order to determine the amount of tracer ions trapped inside the reconstituted liposomes, the external tracer ions must be removed by ion-exchange chromatography. The eluate is collected in scintillation vials and counted without fluor. Calibration is needed to transform the raw counting-rate data into values that reflect the total Rb^+ trapped inside the vesicles. The fraction of authentic reaction mixture in the final collected solution is dependent on the reaction time and the mode of operation when the variable-length reaction tube is changed for different reaction times. It is necessary to calculate the fraction of authentic reaction mixture in the final collected solution, and thus the total counting rates in the reaction mixtures are used in data analysis. After correcting for control levels, the data set can be analyzed by least-squares regression. The ion influx process can be described by the following equation:

$$dM/dt = J^*(M_0 - M)[R^*L_2] \qquad (1)$$

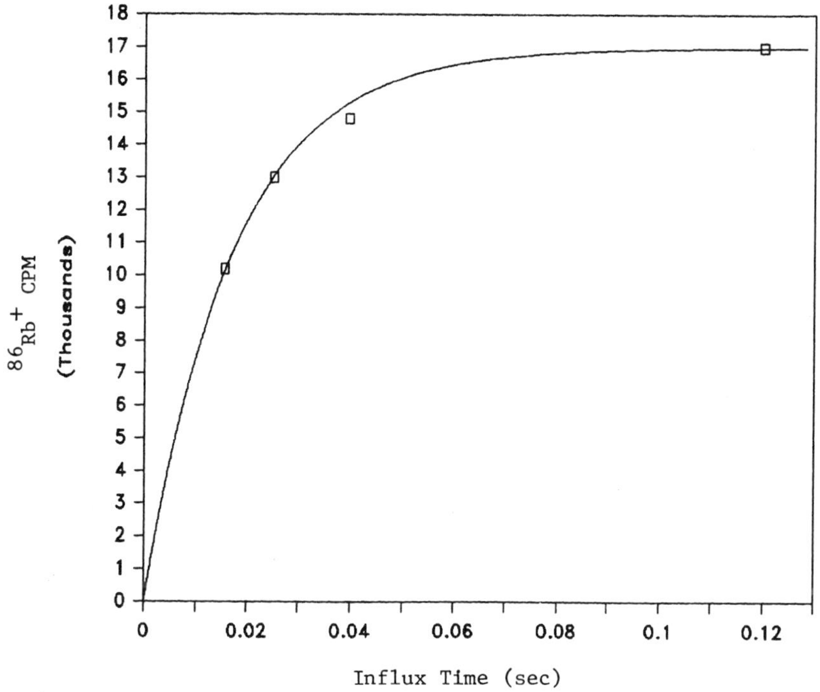

Figure 6. Time course of $^{86}Rb^+$ influx into AChR–Asolectin® vesicles at 0.3 mM carbamylcholine. The data are fit by the equation CPM = $CPM_\infty[1-\exp(-J_a t)]$, where J_a = 58 sec^{-1}.

in which M and M_0 represent the tracer ion concentration inside and outside the liposomes, respectively. J^* is the intrinsic rate constant of ion translocation through the ion channel, and $[R^*L_2]$ is the concentration of open channels in moles per liter of liposome internal volume.

For activation experiments in which the influx process is measured within 300 msec, the data can be analyzed by the first-order rate equation. One typical influx reaction curve is shown in Fig. 6, in which J_a is the first-order rate constant obtained from linear regression analysis of the data. Since activation experiments were carried out within a short time relative to the desensitization process, the following mechanism has been proposed:

$$R \rightleftharpoons RL \rightleftharpoons RL_2 \rightleftharpoons R^*L_2(\text{open})$$

With the rapid equilibrium assumption for agonist binding, it can be shown from equation 1 that

$$J_a = J^*R_0/[1 + \Phi(1 + K_1/L)^2] \tag{2}$$

in which J_a is the apparent first-order rate constant of ion influx, R_0 is the total AChR concentration in moles per liter of vesicle internal volume, Φ is the equi-

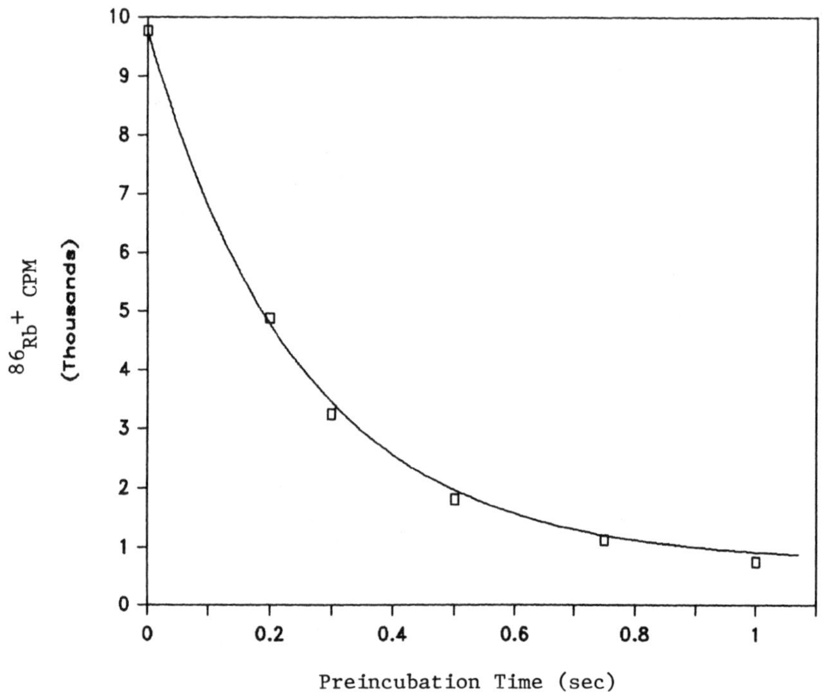

Figure 7. Time course of fast inactivation of AChR-mediated ion influx. AChR–Asolectin® vesicles were preincubated with 1 mM carbamylcholine for times given on the abscissa followed by a 18-msec $^{86}Rb^+$ influx measurement. The data are fit by the equation $CPM = (CPM_0 - CPM_{pe})e^{-\alpha t} + CPM_{pe}$, where $\alpha = 4\ sec^{-1}$.

librium constant of channel closing, K_1 is the dissociation equilibrium constant of agonist binding to the resting state of AChR, and L is the agonist concentration. When J_a is measured as a function of agonist concentration, nonlinear least-squares curve fitting of the data will give the value of $J_{a(max)}$, which is equal to $J^*R_0/(1 + \Phi)$. Since R_0 can be determined independently from manual ion flux assays and protein assays, one can calculate the value of $J^*/(1 + \Phi)$ from the above parameters obtained from curve fitting. Unfortunately, it is not possible to obtain the individual values of J^* and Φ from macroscopic experimental data.

For inactivation experiments, the remaining activity of AChR as a function of preincubation time is observed to be a first-order decay if the preincubation times are no longer than 3 sec. Figure 7 represents a typical set of data from fast inactivation experiments. This fast inactivation process can be approximated by an exponential function:

$$[R^*L_2] = [R^*L_2]_0 e^{-\alpha t} \tag{3}$$

where α is the rate constant of fast inactivation and t is the preincubation time (not to be confused with the influx reaction time), whose values are not much larger than that of $1/\alpha$. Even at the end of fast inactivation (several seconds), however, a residual influx is still observed. Therefore, the fast inactivation process should be described by the following equation:

$$[R^*L_2] = ([R^*L_2]_0 - (R^*L_2]_{pe})e^{-\alpha t} + [R^*L_2]_{pe} \quad (4)$$

in which the subscripts 0 and "pe" represent $t = 0$ and pseudoequilibrium, respectively. If we define $J_{a(0)} = J^*[R^*L_2]_0$, $J_i = J^*[R^*L_2]_{pe}$, and $J_a(t) = J^*[R^*L_2]$, equation 4 will be modified as:

$$J_a(t) = (J_{a(0)} - J_i)e^{-\alpha t} + J_i \quad (5)$$

in which $J_{a(0)}$ is the same constant as J_a that is obtained from activation experiment at a certain agonist concentration. To simplify the data analysis, $J_a(t)$ is not measured directly. Rather, the influx reaction is carried out at a fixed time with a high concentration of agonist in order to obtain the initial influx rate. In this case, the relative values of initial influx rate will approximate the relative values of $J_a(t)$.

Equation 5 does not predict the complete desensitization of AChR. Walker *et al.* (1982) found, however, that the inactivation process is in fact a biphasic process. Equation 5 is thus modified as:

$$J_a(t) = (J_{a(0)} - J_i)e^{-\alpha t} + J_i e^{-\beta t} \quad (6)$$

in which β is the rate constant of the slow inactivation. In the slow inactivation experiments, the preincubation times are chosen under conditions of $t \gg 1/\alpha$. Therefore, equation (6) is simplified to:

$$J_a(t) = J_i e^{-\beta t} \quad (7)$$

The same experimental strategy as that in fast inactivation experiments is adopted to obtain the value of the slow inactivation constant β from the measurement of influx amplitude as a function of preincubation time.

If, however, the influx rate constants and inactivation constants are to be directly measured after preincubation, or the influx process is monitored continuously over a long period of time, the influx process is no longer a simple first-order process. It will be complex with contributions from desensitization as well as activation. Hence, the data should be analyzed by an integrated rate equation describing such complex process:

$$M = M_0[1 - \exp(-\{J_{a(0)}[1 - \exp(-\alpha t)]/\alpha + J_i[1 - \exp(-\beta t)]/\beta\})] \quad (8)$$

which can be derived from equations 1 and 6.

4.1.3. Stopped-Flow–Fluorescence-Quenching Techniques

4.1.3a. Basic Principles. Many heavy atoms possess static fluorescence-quenching properties, which arise from complex formation with fluorophores. Some heavy-atom cations have ionic radii similar to that of K^+, rendering them a good tracer ion for measuring the ion translocation through AChR. The fluorescence-quenching technique was first introduced by Moore and Raftery (1980) to measure ion translocation through AChR in the millisecond time scale. In order to monitor the amount of tracer ions inside liposomes, fluorescent molecules must be present only inside the liposomes so that a decrease in fluorescence intensity will reflect the influx process of quenching ions. This method utilizes the same rapid mixing apparatus that is used in quenched-flow systems to start the reaction, but a different mechanical design is chosen so that the flow of the reaction mixture is stopped where the reaction is monitored spectrophotometrically, from which comes the name "stopped flow."

The basic procedure of the fluorescence-quenching technique is shown in Fig. 8. Water-soluble fluorescent dye is loaded into reconstituted liposomes by freeze–thaw cycles. The external fluorophores are then removed by gel chromatography. The influx reaction is started by rapidly mixing the liposomes with heavy-atom cations such as Tl^+ or Cs^+ in the presence of agonist, and the fluorescence decrease is monitored continuously. A major advantage of stopped-flow techniques over quenched-flow techniques is the ease of data acquisition and the small amount of samples required to obtain the same amount of information as obtained from quench-flow experiments, since each fluorescence decay recorded represents a complete time course of an ion-influx process, which is equivalent to one set of multiple data points obtained by the discontinuous quench-flow method.

4.1.3b. Design of the Apparatus. The mechanical design and electrical triggering in a stopped-flow system are very similar to that of quenched flow. The only difference is that after mixing the reaction mixture is collected and ready for chemical analysis in the quenched-flow experiments, whereas in stopped-flow experiments the reaction mixture is retained inside the thermostated cuvette for spectrophotometric analysis as shown in Fig. 9. To measure the activation of AChR by agonist, only single mixing (analogous to the continuous mode of quenched-flow system) is required, and the shortest measurable time is only limited by the mixing dead time. For inactivation experiments, a sequential mixing system (analogous to the pulse mode of the quench-flow system) can be used in which the preincubation with agonist is achieved by the first mixer and the influx reaction is started by the second mixer (Karpen *et al.*, 1983).

4.1.3c. Selection of Fluorescent Dye and Quenching Cation. An effective fluorescent dye useful in the stopped-flow experiments must satisfy several conditions. Its fluorescence emission spectrum should be reasonably separate from the protein emission spectrum. It should be polar enough so that it does not partition into membranes. Finally, it must not affect the functions of AChR. The fluorescence probe 8-amino-1,3,6-naphthalenetrisulfonate (ANTS) was first introduced by Moore and Raftery (1980) to monitor the rapid ion flux. The Tl^+ ion is a very

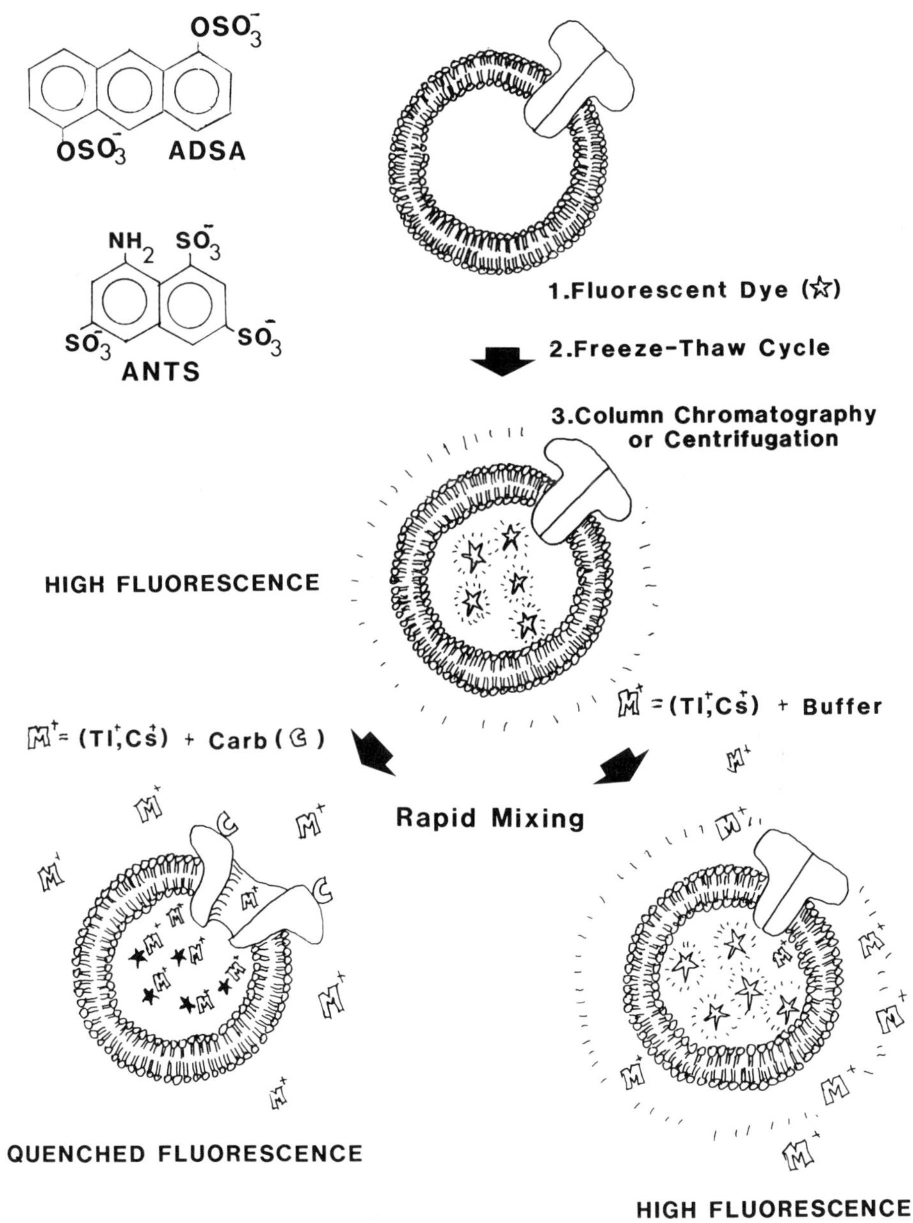

Figure 8. Rapid kinetic measurements of AChR activation and desensitization by fluorescence quenching techniques.

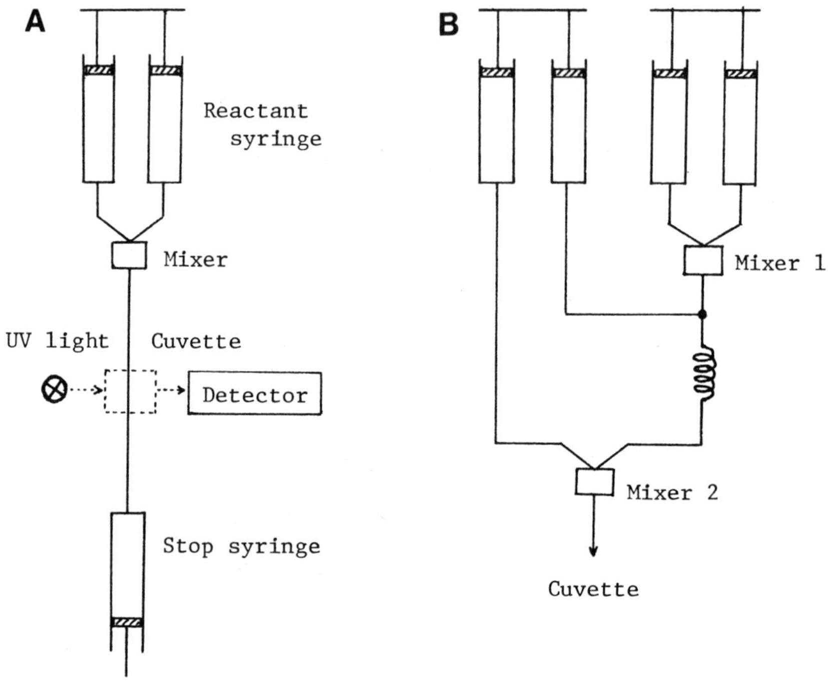

Figure 9. Schematic representation of stopped-flow systems for measuring AChR activation and desensitization. (A) Single mixing system; (B) sequential mixing system.

efficient fluorescence quencher with ANTS and anthracene-1,5-disulfonic acid (ADSA). However, the use of $T1^+$ ion also suffers from several disadvantages: T1C1 has a very low solubility, resulting in the use of NO_3^- in the place of $C1^-$, the physiological anion. Furthermore, the passive leakage rate of $T1^+$ through membrane bilayers is much higher than that of Rb^+ (Wu *et al.*, 1981; Rabon and Sachs, 1981).

Karpen *et al.* (1983) introduced the use of ADSA as the fluorescent probe that is quenched by Cs^+ ions. CsCl is soluble in aqueous solution, and its passive leakage rate through membranes is very low. Therefore, the use of Ca^+ represents an improvement in the fluorescent-quenching techniques, although Cs^+ has a lower quenching efficiency than $T1^+$.

4.1.3d. Data Acquisition and Analysis. The design of stopped-flow systems makes it possible to automate data acquisition, another advantage over quenched-flow system. The fluorescence spectrometer in the stopped-flow system can be interfaced with an analog–digital converter, and thus the digitalized fluorescence decay arising from the influx process can be analyzed immediately. The rapid influx of cations into AChR-containing liposomes is a first-order process for activation experiments carried out within 300 msec. By combining the first-order rate equation with the Stern–Volmer equation of fluorescence quenching, one can calculate the rate constant of ion influx from the fluorescence decay.

The ion influx process as measured by Cs^+ influx has been compared with the results obtained from $^{86}Rb^+$ influx by the quenched-flow method (Hess *et al.*, 1983). In addition, K^+, Na^+, and Li^+ influxes have been studied by quenched-flow methods and flame emission spectroscopy (Ramseyer *et al.*, 1981; Heidmann *et al.*, 1983a). All these methods produced consistent kinetic parameters describing the activation of AChR from *Electophorus electricus*.

4.2. Techniques for Measuring Agonist and Antagonist Binding

4.2.1. Equilibrium Binding Assays

Equilibrium binding assays were used in early attempts to characterize AChR properties and relate the agonist binding to receptor functions, and they led to a great controversy over the question of how agonists interact with AChR. The equilibrium dissociation constant for an agonist such as acetylcholine was found to be on the order of 10^{-8} M (for review, see Taylor *et al.*, 1983), which is about four orders of magnitude smaller than the value deduced from voltage-clamp experiments and chemical kinetic experiments (Dionne *et al.*, 1978; Cash and Hess, 1980; Walker *et al.*, 1982). This paradox was partially resolved by studying the change in receptor binding properties as a function of time after adding agonist. As explained in Section 4.2.2, at least two conformers of AChR coexist in the membranes, one with a low agonist binding affinity and the other with a high agonist binding affinity. The low-affinity state of AChR, whose agonist-binding affinity is still one or two orders of magnitude smaller than the value obtained from electrophysiological and chemical kinetic experiments, is favored in the absence of agonist, but the high-affinity state is favored in the presence of agonist. Therefore, mainly the high-affinity state was detected under equilibrium binding conditions. It has been proposed that the low-affinity state of AChR is an intermediate state of desensitized AChR and that the resting state has an ultralow affinity for agonists (Changeux *et al.*, 1984).

In addition to radioactive agonist, fluorescent analogues of agonists have been used in equilibrium binding studies (Prinz and Maelicke, 1983a). The binding of such fluorescent agonists to AChR results in a decrease in fluorescence intensity, which can be related to the fraction of bound ligand. The advantage of this method is that there is no need to physically separate the bound ligands from the unbound ligands.

4.2.2. Detection of Low- and High-Affinity Agonist-Binding States

4.2.2a. Toxin Binding Kinetics Technique. One of the characteristics of AChR is the desensitization phenomenon in which prolonged binding of agonist leads to channel closing (Katz and Thesleff, 1957). The onset of receptor desensitization is relatively slow (in the second to minute time domain), whereas channel activation occurs in microsecond to millisecond time domains. Katz and Thesleff

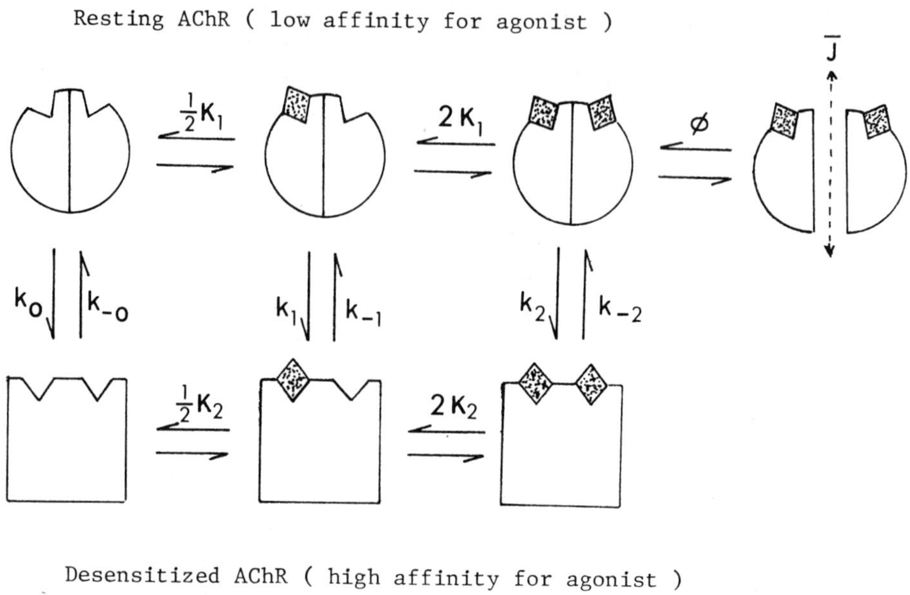

Figure 10. Minimum mechanism of AChR in which $k_c = k_0/k_0$, $K_{c1} = k_{-1}/k_1$, and $K_{c2} = k_{-2}/k_2$.

analyzed several models and found that the cyclic model fit their experimental data well if the affinity of A for R was much lower than its affinity for R′.

$$\begin{array}{ccc} A + R & \rightleftharpoons & AR \\ \upharpoonleft\downharpoonright & & \upharpoonleft\downharpoonright \\ A + R' & \rightleftharpoons & AR' \end{array}$$

Biochemical studies provided direct evidence for the two conformers coexisting in the membranes with different affinities (Weber *et al.*, 1975; Weiland *et al.*, 1977; Quast *et al.*, 1978). A major breakthrough was the introduction of toxin binding kinetics to monitor the transition from low- to high-affinity state (Weber and Changeux, 1974). Snake toxins (such as α-neurotoxin and α-bungarotoxin) and agonists binding to AChR in a mutually exclusive manner. Whereas the agonist association with AChR is virtually diffusion controlled, the toxin association rate is intermediate. Therefore, rapid equilibriumn for agonist binding can be assumed when the toxin association kinetics are measured in the presence of agonist; and during the course of toxin binding (measured within 2 min), the interconversion between two states can be ignored if [R] is low and [R] ≪ [Toxin].

Figure 10 schematizes the model that is consistent with experimental data. It is apparent from this model that in the presence of agonist the apparent toxin-binding rate constant will be lowered because of the occupation of some sites by agonists. If AChR is preincubated with agonist first, the equilibrium between low-affinity and high-affinity states will be perturbed and reestablished, resulting in more high-affinity conformers and fewer low-affinity conformers. Therefore, after the preincubation (i.e., at a new equilibrium state), the toxin-binding rate constant will be lowered further because of the increased amounts of high-affinity receptor and the decreased number of binding sites available to toxin.

This technique has been used to measure the Carb affinity for AChR in different lipid environments (Fong and McNamee, 1985). The Carb-binding equilibrium constants for both low- and high-affinity states are virtually unaltered by different lipid environments, indicating that the lipid environment has little effect on the agonist binding itself. However, what is important is that the transition from low- to high-affinity state is highly dependent on the lipid environment in an all-or-none manner.

4.2.2b. Fluorescence Techniques. An alternate method for studying agonist binding has been introduced with the application of stopped-flow techniques (Heidmann and Changeux, 1979; Prinz and Maelicke, 1983b). When a fluorescent agonist analogue (dansyl derivative of acetylcholine or nitrobenzo-2-oxa-1,3-diazole derivative of acetylcholine) binds to AChR, its fluorescent intensity may increase or decrease depending on the particular intermolecular interaction involved. Heidmann and Changeux (1979) demonstrated that three relaxation processes can be observed when AChR binds the fluorescent agonist. A rapid increase in fluorescence intensity in the millisecond range is correlated to the association at preexisting high affinity sites for agonists. An intermediate relaxation process in the subsecond range corresponds to the association at low-affinity sites. A slow relaxation process in the second range is analyzed in terms of isomerization from low-affinity receptor state to high-affinity receptor state. The equilibrium constants of agonist binding calculated from the corresponding rate constants are consistent with those obtained from equilibrium binding assays and toxin-binding kinetics in the presence of agonists.

Dunn and Raftery (1982) studied the agonist-binding kinetics by a different approach and discovered a very surprising phenomenon: AChR was labeled by the fluorescent probe aminoethyl iodoacetate ester derivative of a nitrobenzo-2-oxa-1,3-diazole group, and it was found that agonist enhanced its fluorescence intensity. This effect was agonist concentration dependent and was inhibited by α-bungarotoxin but not by *d*-tubocurarine nor histrionicotoxin. The equilibrium constant of agonist binding evaluated by this method is in excellent agreement with the value obtained from electrophysiological studies. A model with more than two agonist-binding sites was proposed, and the results were interpreted as implying that separate binding sites may be responsible for activation and desensitization, respectively. Although the detailed nature of the covalent labeling by this fluorescence probe has not been studied, these results clearly pose a new controversial issue concerning the site at which agonist binding causes the channel opening.

4.2.3. Binding of Noncompetitive Blockers

In addition to competitive antagonists such as *d*-tubocurarine, a heterogeneous group of compounds known as noncompetitive blockers also inhibit the ion-gating activity of AChR in both native and reconstituted membranes. This class of potent pharmacological agents includes amine local anesthetics, sedatives, antimalarial drugs, hallucinogens, histrionicotoxin, the lipophilic cation triphenylmethylphosphonium, and some other amphiphilic compounds. It has been suggested that they may interfere directly or indirectly (or both) with the ion channel of AChR, although detailed mechanisms for their action are not fully understood.

4.2.3a. Histrionicotoxin and Phenycyclidine. Histrionicotoxin (HTX) is a spiropiperidine alkaloid isolated from the skin extracts of a Colombian frog, and its binding to AChR cannot be inhibited by α-bungarotoxin or agonist (Elliott and Raftery, 1979). Equilibrium binding assays of $[^3H]H_{12}$-HTX indicated that there is one HTX site per two ACh binding sites, and the K_D is 3×10^{-7} M (Eldefrawi *et al.*, 1978). Agonist-induced ion flux response is also inhibited by HTX.

Similar properties were found for phencyclidine (PCP) (Oswald *et al.*, 1983). Ultraviolet-light-induced covalent labeling revealed that both PCP and HTX label all four subunits to various extents (Oswald and Changeux, 1981), which is consistent with the notion that the binding site is located at the central part of AChR (the ion channel portion). Therefore, PCP and HTX are considered specific channel blockers, and they can be used to distinguish two categories of effects exerted by other noncompetitive blockers.

4.2.3b. Local Anesthetics. The effects of local anesthetics and other compounds on the conformational transitions of AChR have been studied by fluorescence stopped-flow techniques using a fluorescent agonist (Heidmann *et al.*, 1983b). All the compounds tested were found to stabilize the desensitized state of AChR exhibiting a high affinity for agonists. Two classes of binding sites on AChR for these noncompetitive blockers were found: (1) a single high-affinity, histrionicotoxin-sensitive site, which is postulated to be located at the ion channel; (2) a population (10–20) of low-affinity, histrionicotoxin-insensitive sites, which are postulated to be located at the lipid–protein interface. The association of noncompetitive blockers at the high-affinity site is agonist dependent, since such rapid association is not observed with competitive antagonists (Heidmann and Changeux, 1984). Ultraviolet-light-induced labeling of AChR by $[^3H]$chlorpromazine revealed that the association in the presence of agonist takes place at a site common to all four subunits, which is in good agreement with the binding behavior of HTX. Therefore, the actions of local anesthetics can be viewed in two different mechanisms: blocking the ion channel directly and accelerating the desensitization process.

5. Mechanisms of Receptor Activation and Inactivation Deduced from Rapid Kinetics Analyses of Reconstituted Membranes

The rates of ion influx or efflux measured with either quench-flow or stopped-flow measurements become most useful when evaluated in the context of specific models of receptor activation and inactivation. Ideally, the directly measured rates will lead to the formulation of unique mechanisms that can be tested and compared with other biochemical or physiological measurements.

5.1. Receptor Activation

The initial rates of cation influx into reconstituted *Torpedo* membranes have been measured both by quench-flow techniques (Walker *et al.*, 1982, 1984; see also Section 4.1) and by fluorescence-quenching techniques (Wu and Raftery, 1981). The methods used for the reconstituted membranes are basically the same as those used for detailed analyses of ion fluxes in native membranes from electric eel (see Hess *et al.*, 1983) and from various *Torpedo* species (Hess *et al.*, 1982; Moore and Raftery, 1980; Neubig and Cohen, 1980; Heidmann *et al.*, 1983b; Martinez-Carrion *et al.*, 1984).

The magnitude of the maximum J_a value (100 sec^{-1}) and the agonist concentration dependence of the initial rates of $^{86}Rb^+$ influx are consistent with a model of AChR activation that involves agonist binding to two identical sites and an efficient coupling of ligand binding to channel opening. The Hill coefficient for the agonist concentration dependence of initial rates of influx was 1.8, and the apparent K_D for carbamylcholine binding to two sites was 1.5×10^{-4} M. From the fluorescence data obtained by Moore and Raftery (1980) for native membranes and by Wu and Raftery (1981) for reconstituted membranes, it was not possible to clearly distinguish between models that require two or one bound ligands prior to channel opening.

An important parameter that can be extracted (with difficulty) from the kinetic analyses is the number of ions that move through a single channel when it is open. The single-channel conductance is readily measured in patch-clamp recordings, and it is useful to compare the results deduced from the macroscopic measurements in vesicles with the true conductances. Since the measured flux response represents the rate of approach of the ion concentration on each side of the vesicles to equilibrium, factors such as the number of functional ion channels per vesicle, the size of the vesicles, the probability of channel opening, the absolute ion concentrations, and the membrane potential must be considered in calculating the single-channel conductance. The most thorough attempt at comparing rapid kinetic measurements with patch-clamp measurements has been made by Hess *et al.* (1984) using *Electrophorus electricus* electric organ cells and membrane. The key result is that the specific rate constant for ion flux in the vesicles (3×10^7 M^{-1} sec^{-1}) agress to within a factor of 2 with the value derived from single-channel conductance measurements (5×10^7 M^{-1} sec^{-1}). Such complete analyses have not yet been attempted for *Torpedo* receptors, but calculations by Moore and Raftery (1980) and Walker *et al.* (1982) are consistent with single-channel conductances within an order of magnitude of the expected value of about 10^7 M^{-1} sec^{-1}.

5.2. Receptor Inactivation (Desensitization)

The desentization phenomenon represents a time-dependent decrease in the average number of functional ion channels in the prolonged presence of activating ligands. The time-dependent inactivation of ion channels in reconstituted membranes containing AChR is easily measured using the pulse-quench experimental design described in Section 4.1.2. Membranes are preincubated with various concentrations of carbamylcholine for various times and then mixed with $^{86}Rb^{+}$ and a high concentration of carbamylcholine to initiate influx. Influx is allowed to proceed for a short time (e.g., 15 msec), and then the entire mixture is quenched and analyzed. It is presumed that the short influx at high ligand concentrations provides a measure of the number of remaining functional channels, since an initial, maximal rate is being measured. Results of such inactivation analysis show that the inactivation is first order and ligand concentration dependent. Replots of the inactivation rate versus ligand concentration are hyperbolic, suggesting that the inactivation process is noncooperative, unlike the situation with channel activation. The maximum rate constant for channel inactivation in reconstituted membranes is 5 sec^{-1}, corresponding to a half-time for inactivation of 140 msec, a value similar to the values measured in native *Torpedo* membranes and in eel membranes.

An interesting aspect of the inactivation measurements described above was first discovered in native membranes (Walker *et al.*, 1981) and later confirmed in reconstituted membranes (Walker *et al.*, 1982). If influx was measured for much longer times than the 15-msec times used to determine the number of active channels, a second, slower phase of inactivation was detected. The slower phase was also ligand concentration dependent, displayed first-order kinetics, and reached a maximum value of 0.10 sec^{-1} ($t_{1/2}$ = 6.9 sec). The slow inactivation process was not apparent in ccl membranes, which showed a low-level but steady influx rate after the fast inactivation process was complete. The concentration dependence of the slow inactivation process ($K_D = 2.5 \times 10^{-4}$ M) was different from that of the fast process, suggesting that the fast and slow inactivation events might be proceeding by independent pathways. In fact, the apparent K_D for the fast process (1.3×10^{-3} M) was significantly higher than the K_D for ligand activation, suggesting that the fast process in *Torpedo* might be regulated by a separate ligand-binding site (Walker *et al.*, 1982).

The two-phase inactivation of AChR function has been independently observed by single-channel analysis of frog muscle in time domains comparable to those described here (Sakmann *et al.*, 1980). Recently, Aoshima (1984) reported that a very slow phase of AChR inactivation could be detected in eel membranes ($t_{1/2}$ = 15 min). A comprehensive model that incorporates the two stages of inactivation has not yet been developed.

5.3. Relationship between Ligand Binding and Ion Flux

The rapid kinetics measurements described above do not provide direct information about the ligand-binding sites or the coupling of the binding to ion-channel activation and inactivation. Since the initial binding events are undoubtedly very fast and the initial channel opening is also fast (microsecond time domain), it is unlikely that chemical kinetics measurements will be widely available to study the processes. However, the rate constants and the concentration dependences do permit the calculation of apparent binding constants within the context of a minimal mechanism. Chemical kinetics measurements are also ideally suited for measuring the slow interconversions among AChR binding states that are likely to be involved in the inactivation pathways. The few studies that have attempted to compare directly the relationship between ligand binding and flux responses have used highly purified native membranes from *Torpedo*. However, the reconstituted membranes can also be used and offer the possibility that different aspects of the binding and flux can be selectively altered.

Using dansylcholine as a ligand to monitor the AChR binding states and a rapid kinetics assay for ion flux, Heidmann *et al.* (1983a) quantitatively established a close correlation between binding to the two normal acetylcholine binding sites and the activation and inactivation of ion-flux responses. Following the initial suggestion of Neubig and Cohen (1980), they inferred the existence of a very low-affinity binding state that corresponds to the resting, activatable state of the AChR.

6. Interactions of Lipids with the Receptor in Reconstituted Membranes

6.1. Electron Spin Resonance (Spin-Labeling) Techniques

Electron spin resonance (ESR) techniques have been applied to the study of membrane structures for nearly two decades (for review, see Marsh, 1981; Devaux, 1983). It is a common practice in membrane biophysics to employ a nitroxide radical either as a free molecular probe or as a reporter group attached to lipid and protein molecules. The major contribution of spin-labeling techniques has been to study molecular motions, since the ESR spectra of these nitroxide groups are very sensitive to their motions including both rate and amplitude. The typical range of motion that can be studied by ESR methods is 10^{-7}–10^{-11} sec. However, with the application of the saturation transfer method, molecular motions with correlation times of 10^{-3}–10^{-7} sec can also be studied. A disadvantage is that the spin-labeling technique involves the introduction of a reporter group, and thus perturbation of native biological structures is always possible.

6.1.1. Membrane Fluidity

Membrane fluidity is a convenient but imprecise term encompassing all molecular motions throughout the membrane bilayer. It includes such phenomena as lateral and rotational motions, *trans–gauche* isomerization of C–C bonds, and anisotropic motion of phospholipid fatty acyl chains. Unfortunately, we cannot define their relative contributions affecting membrane protein functions, resulting in the usage of a general term "fluidity." In spite of these complications, some of these features can still be determined by spin-labeling techniques.

First of all, the degree of *trans–gauche* isomerization of C–C bonds and the amplitude of swing motion of fatty acyl chains are characterized by a physical quantity known as the order parameter. When the motion is fast on the ESR timescale ($\tau_c < 10^{-9}$ sec), the line positions are determined by a time-average version of the spin Hamiltonian in which the hyperfine splittings are determined solely by the time-average motional amplitude (θ), which is the angle between the membrane normal and the long axis of the acetyl chain. The order parameter is defined as $S = 0.5\ (3\langle\cos^2\theta\rangle - 1)$. It can be shown that

$$S = \frac{A_{\|} - A_{\perp}}{A_{zz} - 0.5\ (A_{xx} + A_{yy})} \qquad (9)$$

(Gaffney and Chen, 1977). A_{xx}, A_{yy}, and A_{zz} are the principal elements of the hyperfine tensor, which is equal to the spectral splitting in the single-crystal spectrum when the magnetic field is parallel to the *x*, *y*, and *x* axes of the molecular coordinate system, respectively. $A_{\|}$ and $A_{\perp}$ are the experimental hyperfine splittings when the magnetic field is parallel ($\|$) or perpendicular ($\perp$) to the membrane normal.

The interconversion of the two conformations of AChR has been studied as a function of membrane fluidity as measured by the order parameter of spin-labeled fatty acid (Fong and McNamee, 1985). A very interesting trend was observed in which membrane fluidity in a narrow range of the order parameter scale was necessary to enable AChR to undergo conformational changes.

6.1.2. Boundary Lipids and Specifity of AChR–Lipid Interactions

Many experiments have shown that spin-labeled lipids are motionally restricted when they are in contact with membrane proteins (Jost *et al.*, 1973, 1977), with a change in either the rate or the amplitude of motion. As discussed in Section 6.1.1, such perturbation from protein molecules will cause changes in line shape and line positions compared to the nonperturbed lipid molecules. Since the exchange rate of lipid molecules between the protein-perturbed lipid domains (sometimes called boundary lipid domains) and the protein-free bilayer lipid domains is slow on the EPR time scale ($k < 10^7$ sec^{-1}), these two populations of lipid molecules will be distinguishable by ESR techniques, resulting in the observation of composite spectra consisting of two components: a boundary component and a mobile component.

These two components can be resolved by computer spectral subtraction methods (Ellena *et al.*, 1983), as shown in Fig. 11. Double integration of the composite spectrum (spectrum A) and the mobile spectrum (spectrum E) will give the percentage of spin-labeled lipids that are not in contact with AChR. By assuming a multiple-binding-equilibrium model for the lipid–protein interaction in membranes (Brotherus *et al.*, 1981), the relative affinity of spin-labeled lipids for AChR (K_{av}) can be evaluated, where K_{av} is the average association equilibrium constant of the spin-labeled lipid relative to that of solvent lipid, e.g., DOPC.

$$\begin{aligned} L_b + L_f^* &\rightleftharpoons L_b^* + L_f \\ K_{av} &= (L_b^* L_f)/(L_f^* L_b) \end{aligned} \qquad (10)$$

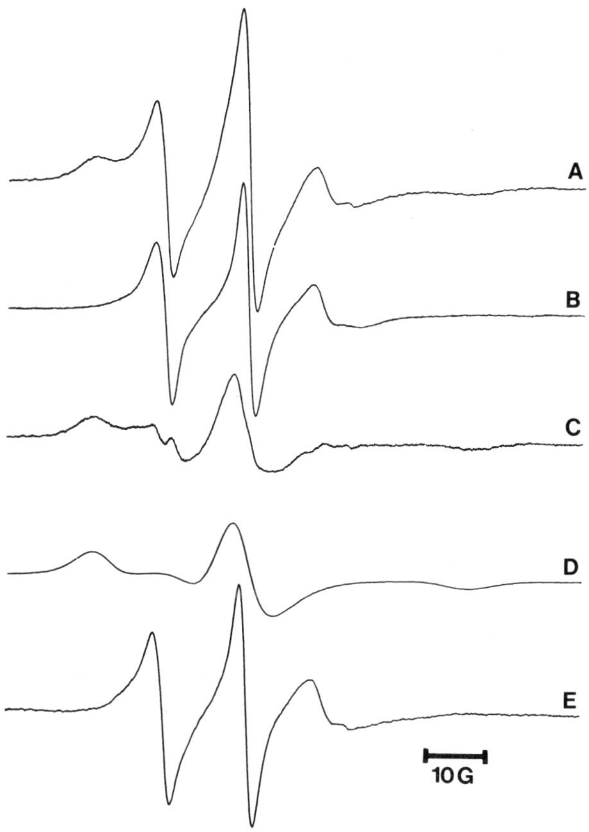

Figure 11. The EPR spectral component subtraction method. Spectra were recorded from 16-doxylstearic acid in (A) AChR–DOPC membranes at 0°C; (B) DOPC liposomes at −5°C served as the mobile component, and (D) ether–pentane–ethanol (1:1:3) at −122°C served as the protein-perturbed component. (C) Difference spectrum between spectra A and B, which represents 60% protein-perturbed component. (E) Difference spectrum between spectra A and D, representing 40% mobile component.

where L_b is DOPC immediately adjacent to protein, L_f^* is labeled lipid not adjacent to protein, L_b^* is labeled lipid immediately adjacent to protein, and L_f is DOPC not adjacent to protein. All molecules referred to are intramembranous. A series of spin-labeled lipids have been tested in this analysis (Ellena *et al.*, 1983). The results demonstrated that different kinds of lipids interact differentially with AChR, corresponding to different binding affinity. Sterols, phosphatidic acid, and fatty acids have a relatively high binding affinity for AChR compared to other kinds of lipids.

In parallel studies, it was found that cholesterol and negatively charged phospholipids were required to support the ion-gating activity of AChR (Ochoa *et al.*, 1983; Fong and McNamee, 1985), whereas fatty acids block the ion-flux response (Andreasen and McNamee, 1980). The latter is interpreted as perturbation of the functionally significant interaction between AChR and cholesterol or negatively charged phospholipids.

6.1.3. Interaction of Spin-Labeled Local Anesthetics with AChR

The actions of local anesthetics on biological membranes are not well understood. Both direct anesthetic–protein interactions and general perturbation of membrane structure have been suggested. As discussed in Section 4.2.3b, many local anesthetics are found to bind to AChR at multiple binding sites. Spin-labeled methyl analogues of intracaine derivatives have been synthesized, and complementary results were obtained. A protein-perturbed component (moderately immobilized) is observed only if the anesthetic is positively charged (Earnest *et al.*, 1984), and the cationic form corresponds to a higher potency in inhibiting the ion-flux response of AChR-containing liposomes than the neutral form (Blickenstaff and Wang, 1985). The existence of this protein-perturbed component is consistent with the binding of noncompetitive blockers at the low-affinity sites (Heidmann *et al.*, 1983b). In the presence of agonist, a highly immobilized component appears, which is correlated to the binding at a high-affinity allosteric site for the anesthetic (Palma *et al.*, 1985). These results suggest multiple mechanisms for local anesthetic actions.

6.2. Fluorescence-Quenching Techniques

Fluorescence spectroscopy has been a powerful tool in studying membrane phenomena (Lee, 1982; Lakowicz, 1983). The techniques available fall into two main categories: those relying on the intrinsic fluorescence of proteins and those employing direct labeling of the protein molecules. In this section, only those techniques that show promise in studying lipid–protein interactions in reconstituted AChR systems are discussed.

It is now well established that many membrane proteins are fluorescent. This is also true of the AChR in native (Bonner *et al.*, 1976; Barrantes, 1978), solubilized (Kaneda *et al.*, 1982), and reconstituted (O. T. Jones, unpublished observations) lipid environments. The fluorescence characteristics are typical of those found in integral membrane proteins ($\lambda_{ex} = 290$ nm, $\lambda_{em} = 336$ nm), presumably arising from the many tryptophan and, to a lesser extent, tyrosine residues in the primary amino acid sequence of AChR (Noda *et al.*, 1983).

It has been shown that the intrinsic fluorescence of membrane proteins can be reduced (i.e., quenched) by a variety of compounds present in the lipid bilayer (Marsh and Barrantes, 1978). Such compounds include nitroxide-labeled fatty acids and cholesterol analogues and brominated phospholipids. The quenching phenomenon makes it possible to quantitate lipid–protein interactions using recently developed fluorescence-quenching techniques (London and Feigenson, 1981; East and Lee, 1982). Reconstitution of AChR with spin-labeled or brominated lipids may lead to a reduction of the intrinsic fluorescence as a result of the static quenching of the fluorescent amino acid residues in the intramembranous regions of the protein. The extent of quenching is proportional to the mole fraction of quenching lipids in the membranes, which is given by the equation:

$$F/F_o = F_{MIN} + (F_o - F_{MIN})(1 - X_Q)^n \qquad (11)$$

where F_o is the fluorescence intensity in the absence of quenching lipids, and F is the fluorescence intensity in a mixture containing a mole fraction (X_Q) of quenching lipids. The index n describes the average number of lipid sites close enough to an average fluorophore (e.g., tryptophan) to cause quenching. Considering the lipid environment of AChR by an exchange-equilibrium model (see Section 6.1.2), binding constants for particular lipids relative to DOPC can be determined from the relationship:

$$K_{A/DOPC} = X_A(1 - X_{DOPC})/[(1 - X_A)X_{DOPC}] \qquad (12)$$

where X_A and X_{DOPC} are the mole fractions of lipid A and DOPC that cause identical levels of fluorescence quenching.

The fluorescence-quenching method can also be used to test the displacement of boundary lipid by other hydrophobic molecules such as local anesthetics, fatty acids, and cholesterol (Lee, 1982). With this method, binding at sites other than the lipid–protein interface may also be examined (Simmonds *et al.*, 1982).

Direct labeling of AChR by a fluorescence probe has been extensively used to study ligand-binding behaviors (see, for example, Dunn and Raftery, 1982). However, few studies have been done specifically to examine the effects of lipids on the fluorescence properties of labeled AChR molecules. The feasibility of this approach is demonstrated by the recently reported photolabeling studies of AChR-rich membranes (Gonzalez-Ros *et al.*, 1983). It has been shown previously that AChR can be covalently labeled by a pyrene derivative in the intramembranous regions of the β and γ subunits (Sator *et al.*, 1979). By fluorescence-quenching methods, agonists that bind at the extramembranous portions of AChR were shown to affect the conformation of those regions of the receptor accessible to lipids. Such long-range effects presumably reflect large conformational changes within the protein.

6.3. Other Techniques

Few other physical techniques have been applied to study AChR–lipid interactions. One example is the use of Raman spectroscopy to determine the secondary structure of AChR in reconstituted dielaidoylphosphatidylcholine membranes (Yager *et al.*, 1984). The AChR is composed of 38% α helix and 33% β sheet based on the analysis of the amide I vibrational region. This is not unacceptably different from the theoretical values of 44% α helix and 27% β sheet deduced from the Fourier analysis of the primary amino acid sequence (Finer-Moore and Stroud, 1984).

Another potential technique is ^2H NMR spectroscopy, which can be used to study the motion of amino acid residues accessible to membrane lipids. With the application of cDNA cloning and *in vitro* expression techniques, it is now possible to obtain expression of functional AChR from its cDNA (Mishina *et al.*, 1984), and thus AChR with a single [^{2}H]amino acid enrichment can be ultimately used in the characterization of protein dynamics.

7. Summary and Conclusions

This chapter has outlined some of the unique features of reconstituted lipid vesicles that provide good experimental opportunities for studying the structural and functional properties of the acetylcholine receptor. The ability to directly correlate biochemical and biophysical parameters with initial rates of ion flux is the most striking advantage of the vesicle system. The reconstituted vesicles highlight two important features of the acetylcholine receptor: (1) its intrinsic importance as an ion channel; and (2) its role as a model protein for fundamental studies of lipid–protein interactions and reconstitution.

Future efforts will be directed at attempts to define more precisely the causal relationships between lipid–protein interactions and AChR function and to develop more fully a kinetic scheme that incorporates the two stages of AChR inactivation. The vesicles might be particularly useful in obtaining a complete understanding of the interactions of noncompetitive antagonists, such as local anesthetics, with both the membrane and the receptor itself.

References

Andreasen, T. J., and McNamee, M. G., 1980, Inhibition of ion permeability control of acetylcholine receptor from *Torpedo californica* by long-chain fatty acids, *Biochemistry* **19**:4719–4726.

Anholt, R., Lindstrom, J., and Montal, M., 1981, Stabilization of acetylcholine receptor channels by lipids in cholate solution and during reconstitution in vesicles, *J. Biol. Chem.* **256**:4377–4387.

Anholt, R., Fredkin, D. R., Deerinck, T., Ellisman, M., Montal, M., and Lindstrom, J., 1982, Incorporation of acetylcholine receptor into liposomes: Vesicle structure and acetylcholine receptor function, *J. Biol. Chem.* **257**:7122–7134.

Aoshima, H., 1984, A second, slow inactivation process in acetylcholine receptor-rich membrane vesicles prepared from *Electrophorus electricus, Arch. Biochem. Biophys.* **235**:312–318.

Bangham, A., Hill, M. W., and Miller, N. G. A., 1974, Preparation and use of liposomes as models of biological membranes, *Methods Membrane Biol.* **1**:1–68.

Barrantes, F. J., 1978, Agonist-mediated changes of the acetylcholine receptor in its membrane environment, *J. Mol. Biol.* **124**:1–26.

Bernhardt, J., and Neumann, E., 1980, Analysis of gated flux from or into sealed membrane fragments, *J. Theor. Biol.* **86**:649–661.

Blickenstaff, G. D., and Wang, H. H., 1985, The effects of spin labeled local anesthetics on acetylcholine receptor-mediated ion flux, *Biophys. J.* **47**:261a.

Bonner, R., Barrantes, F. J., and Jovin, T. M., 1976, Kinetics of agonist-induced intrinsic fluorescence changes in membrane-bound acetylcholine receptor, *Nature* **263**:429–431.

Brotherus, J. R., Griffith, O. H., Brotherus, M. O., Jost, P. C., Silvius, J. R., and Hokin, L., 1981, Lipid–protein multiple binding equilibria in membranes, *Biochemistry* **20**:5261–5267.

Cash, D. J., and Hess, G. P., 1980, Molecular mechanism of acetylcholine receptor-controlled ion translocation across cell membranes, *Proc. Natl. Acad. Sci. U.S.A.* **77**:842–846.

Cash, D. J., and Hess, G. P., 1981, Quenched flow technique with plasma membrane vesicles: Acetylcholine receptor mediated transmembrane ion flux, *Anal. Biochem.* **112**:39–51.

Changeux, J.-P., Devillers, T., and Chemovilli, P., 1984, Acetylcholine receptor: An allosteric protein, *Science* **225**:1335–1345.

Comfurius, P., and Zwaal, R. F. A., 1977, The enzymatic synthesis of phosphatidylserine and purification by CM-cellulose column chromatography, *Biochim. Biophys. Acta* **488**:36–42.

Criado, M., Eibl, H., and Barrantes, F. J., 1982, Effects of lipids on acetylcholine receptor. Essential need of cholesterol for maintenance of agonist-induced state transitions in lipid vesicles, *Biochemistry* **21**:3622–3629.

Criado, M., Eibl, H., and Barrantes, F. J., 1984, Functional properties of the acetylcholine receptor incorporated into model lipid membranes, *J. Biol. Chem.* **259**:9188–9198.

Delegeane, A. M., and McNamee, M. G., 1980, Independent activation of the acetylcholine receptor from *Torpedo californica* at two sites, *Biochemistry* **19**:890–895.

Deutsch, J. M., and Kelly, R. B., 1981, Lipids of synaptic vesicles: Relevance to the mechanism of membrane fusion, *Biochemistry* **20**:378–385.

Devaux, P. H., 1983, ESR and NMR studies of lipid–protein interactions in membranes, in: *Biological Magnetic Resonance*, Vol. 5 (L. J. Berliner and J. Reubin, eds.), pp. 183–299, Plenum Press, New York.

Dionne, V. E., Steinbach, J. K., and Stevens, C. F., 1978, An analysis of the dose–response relationship at voltage-clamped frog neuromuscular junctions, *J. Physiol.* (*Lond.*) **281**:421–444.

Dunn, S. M. J., and Raftery, M. A., 1982, Activation and desensitization of *Torpedo* acetylcholine receptor: Evidence for separate binding sites, *Proc. Natl. Acad. Sci. U.S.A.* **79**:6757–6761.

Earnest, J. P., Wang, H. H., and McNamee, M. G., 1984, Multiple binding sites for local anesthetics on reconstituted acetylcholine receptor membranes, *Biochem. Biophys. Res. Commun.* **123**:862–868.

East, J. M., and Lee, A. G., 1982, Lipid selectivity of the calcium and magnesium ion dependent adenosinetriphosphatase, studied with fluorescence quenching by a brominated phospholipid, *Biochemistry* **21**:4144–4151.

Eldefrawi, M. E., Eldefrawi, A. T., Mansour, N. A., Daly, J. W., Witkop, B., and Albuquerque, E. X., 1978, Acetylcholine receptor and ionic channel of *Torpedo electroplax*: Binding of perhydrohistrionicotoxin to membrane and solubilized preparation, *Biochemistry* **17**:5474–5484.

Ellena, J. F., Blazing, M. A., and McNamee, M. G., 1983, Lipid–protein interactions in reconstituted membranes containing acetylcholine receptor, *Biochemistry* **22**:5523–5535.

Elliott, J., and Raftery, M. A., 1979, Binding of perhydrohistrionicotoxin to intact and detergent-solubilized membranes enriched in nicotinic acetylcholine receptor, *Biochemistry* **18**:1868–1874.

Epstein, M., and Racker, E., 1978, Reconstitution of carbamylcholine-dependent sodium ion flux and desensitization of the acetylcholine receptor from *Torpedo californica, J. Biol. Chem.* **253**:6660–6662.

Fersht, A. R., and Jakes, R., 1975, Determination of two reaction pathways for the aminoacylation of tRNA: Application of the pulsed quenched flow technique, *Biochemistry* **14**:3350–3356.

Finer-Moore, J., and Stroud, R. M., 1984, Amphipathic analysis and possible formation of the ion channel in an acetylcholine receptor, *Proc. Natl. Acad. Sci. U.S.A.* **81**:155–159.

Fong, T. M., and McNamee, M. G., 1985, Correlation between acetylcholine receptor function and structural properties of membranes, *Biochemistry* (in press).

Gaffney, B. J., and Chen, S. C., 1977, Spin-label studies of membranes, in: *Methods in Membrane Biology*, Vol. 8 (E. D. Korn, ed.), pp. 291–358, Plenum Press, New York.

Gonzalez-Ros, J. M., Paraschos, A., and Martinez-Carrion, M., 1980, Reconstitution of functional membrane-bound acetylcholine receptor from isolated *Torpedo californica* receptor protein and electroplax lipids, *Proc. Natl. Acad. Sci. U.S.A.* **77**:1796–1800.

Gonzalez-Ros, J. M., Paraschos, A., Farach, M. C., and Martinez-Carrion, M., 1981, Characterization of acetylcholine receptor isolated from *Torpedo californica* elecroplax through the use of an easily removable detergent, B-D-octylglucopyranoside, *Biochim. Biophys. Acta* **643**:407–420.

Gonzalez-Ros, J. M., Llanillo, M., Paraschos, A., and Martinez-Carrion, M., 1982, Lipid environment of acetylcholine receptor from *Torpedo californica, Biochemistry* **21**:3467–3473.

Gonzalez-Ros, J. M., Farach, M., and Martinez-Carrion, M., 1983, Ligand-induced effects at regions of acetylcholine receptor accessible to membrane lipids, *Biochemistry* **22**:3807–3811.

Haggerty, J. G., and Froehner, S. C., 1981, Restoration of ^{125}I-α-bungarotoxin binding activity to the α-subunit of *Torpedo* acetylcholine receptor isolated by gel electrophoresis in sodium dodecyl sulfate, *J. Biol. Chem.* **256**:8296–8297.

Heidmann, T., and Changeux, J.-P., 1979, Fast kinetic studies on the interaction of a fluorescent agonist with the membrane-bound acetylcholine receptor from *Torpedo marmorata, Eur. J. Biochem.* **94**:255–279.

Heidmann, T., and Changeux, J.-P., 1984, Time resolved photolabeling by the noncompetitive blocker chlorpromazine of the acetylcholine receptor in its transiently open and closed ion channel conformations, *Proc. Natl. Acad. Sci. U.S.A.* **81**:1897–1901.

Heidmann, T., Bernhardt, J., Neumann, E., and Changeux, J.-P., 1983a, Rapid kinetics of agonist binding and permeability response analyzed in parallel on acetylcholine receptor rich membranes from *Torpedo marmorata, Biochemistry* **22**:5452–5459.

Heidmann, T., Oswald, R. E., and Changeux, J.-P., 1983b, Multiple sites of action for noncompetitive blockers on acetylcholine receptor rich membrane fragments from *Torpedo marmorata, Biochemistry* **22**:3112–3127.

Helenius, A., McCaslin, D. R., Fries, E., and Tanford, C., 1979, Properties of detergents, *Methods Enzymol.* **56**:734–749.

Hess, G. P., Pasquale, E. B., Walker, J. W., and McNamee, M. G., 1982, Comparison of acetylcholine receptor-controlled cation flux in membrane vesicles from *Torpedo californica* and *Electrophorus electricus*: Chemical kinetic measurement in the milliseconds region, *Proc. Natl. Acad. Sci. U.S.A.* **79**:963–967.

Hess, G. P., Cash, D. J., and Aoshima, H., 1983, Acetylcholine receptor controlled ion translocation: Chemical kinetic investigations of the mechanism, *Annu. Rev. Biophys. Bioeng.* **12**:443–473.

Hess, G. P., Kolb, H.-A., Lauger, P., Schoffeniels, E., and Schwarze, W., 1984, Acetylcholine receptor (from *E. electricus*): A comparison single channel current recordings and chemical kinetics measurements, *Proc. Natl. Acad. Sci. U.S.A.* **81**:5281–5285.

Hilden, S., and Hokin, L., 1976, Coupled Na^+–K^+ transport in vesicles containing a purified (Na K)ÅTPase and only phosphatidylcholine, *Biochem. Biophys. Res. Commun.* **69**:521–527.

Holloway, P. W., 1973, A simple procedure for removal of Triton-X-100 from protein samples, *Anal. Biochem.* **53**:304–308.

Hopfer, U., 1978, Isolated membrane vesicles as tools for analysis of epithelial transport, *Am. J. Physiol.* **233**:E445–E449.

Houslay, M. D., Warren, G. B., Birdsall, N. J. M., and Metcalfe, J. C., 1975, Lipid phase transitions control β-hydroxybutyrate dehydrogenase activity in defined-lipid protein complexes, *FEBS Lett.* **51**:146–150.

Huang, C.-H., 1969, Studies on phosphatidylcholine vesicles. Formation and physical characteristics, *Biochemistry* **8**:344–352.

Huganir, R. L., and Racker, E., 1982, Properties of proteoliposomes reconstituted with acetylcholine receptor from *Torpedo californica, J. Biol. Chem.* **257**:9372–9378.

Huganir, R. L., Schnell, M. A., and Racker, E., 1979, Reconstitution of the purified acetylcholine receptor from *Torpedo californica, FEBS Lett.* **108**:155–160.

Jost, P. C., Griffith, O. H., Capaldi, R. A., and Vanderkooi, G., 1973, Evidence for boundary lipid in membranes, *Proc. Natl. Acad. Sci. U.S.A.* **70**:480–484.

Jost, P. C., Nadakavukaren, K. K., and Griffith, O. H., 1977, Phosphatidylcholine exchange between the boundary lipid and bilayer lipid domains in cytochrome oxidase containing membranes, *Biochemistry* **16**:3110–3114.

Kagawa, Y., and Racker, E., 1971, Partial resolution of the enzymes catalyzing oxidative phosphorylation, *J. Biol. Chem.* **246**:5477–5487.

Kaneda, N., Tanaka, F., Kohno, M., Hayashi, K., and Yagi, K., 1982, Change in the intrinsic fluorescence of acetylcholine receptor from *Narke japonica* upon binding with cholinergic ligands, *Arch. Biochem. Biophys.* **218**:376–383.

Karpen, J. W., Sachs, A. B., Cash, D. J., Pasquale, E. B., and Hess, G. P., 1983, Direct spectrophotometric detection of cation flux in membrane vesicles: Stopped-flow measurements of acetylcholine receptor-mediated ion flux, *Anal. Biochem.* **135:**83–94.

Katz, B., and Thesleff, S., 1957, A study of the desensitization produced by acetylcholine at the motor end plate, *J. Physiol.* (*Lond.*) **138:**63–80.

Kilian, P., Dunlap, C., Mueller, P., Schell, M. A., Huganir, R., and Racker, E., 1980, Reconstitution of acetylcholine receptor from *Torpedo californica* with highly purified phospholipids: Effect of α-tocopherol, phylloquinone and other terpenoid quinones, *Biochem. Biophys. Res. Commun.* **93:**409–414.

Laemmli, U. K., 1970, Cleavage of structural proteins during assembly of the head of bacteriophage T4, *Nature* **277:**680–682.

Lakowicz, J. R., 1983, *Principles of Fluorescence Spectroscopy*, Plenum Press, New York.

Lee, A. G., 1982, Membrane studies using fluorescence spectroscopy, in: *Techniques in the Life Sciences—Lipid and Membrane Biochemistry, B422,* (T. R. Hesketh, H. L. Kornberg, J. C. Metcalfe, C. H. Northcote, C. I. Pogson, and K. F. Tipton, eds.), pp. 1–49, Elsevier, Amsterdam.

Lee, A. G., 1983, Lipid phase transitions and mixtures, in: *Membrane Fluidity in Biology*, Vol. 2, *General Principles* (R. C. Aloia, ed.), pp. 43–88, Academic Press, New York.

Letters, R., 1964, The application of a two dimensional paper chromatographic technique to the analysis of phospholipids, *Biochem. J.* **93:**313–316.

Lichtenberg, D., Freire, E., Schmidt, C. F., Barenholz, Y., Felgner, P. L., and Thompson, T. E., 1981, Effect of surface curvature on solubility, thermodynamic behavior, and osmotic activity of dipalmitoylphosphatidylcholine single lamellar vesicles, *Biochemistry* **20:**3462–3467.

Lindstrom, J., Anholt, R., Einarson, B., Engle, A., Osame, M., and Montal, M., 1980, Purification of acetylcholine receptor and reconstitution into lipid vesicles and study of agonist-induced cation channel regulation, *J. Biol. Chem.* **255:**8340–8350.

London, E., and Feigenson, G. W., 1981, Fluorescence quenching in model membranes. 2. Determination of the local lipid environment of the calcium adenosine triphosphatase from sarcoplasmic reticulum, *Biochemistry* **20:**1939–1948.

Lymn, R. W., and Taylor, E. W., 1970, Transient state phosphate production in the hydrolysis of nucleotide triphosphates by myosin, *Biochemistry* **9:**2975–2983.

Marsh, D., 1981, Electron spin resonance: Spin labels, in: *Membrane Spectroscopy* (E. Grell, ed.), pp. 51–142, Springer-Verlag, Berlin.

Marsh, D., and Barrantes, F. J., 1978, Immobilized lipid in acetylcholine receptor rich membranes from *Torpedo marmorata, Proc. Natl. Acad. Sci. U.S.A.* **75:**4329–4333.

Martinez-Carrion, M., Gonzalez-Ros, J. M., Mattingly, J. R., Ferragut, J. A., Farach, M. C., and Donneley, D., 1984, Fluorescence probes for the studies of acetylcholine receptor function, *Biophys. J.* **45:**141–143.

McNamee, M. G., and Ochoa, E. L. M., 1982, Reconstitution of acetylcholine receptor function in model membranes, *Neuroscience* **7:**2305–2319.

Miller, C., and Racker, E., 1979, Reconstitution of membrane transport functions, in: *The Receptors*, Vol. 1 (R. D. O'Brien, ed.), pp. 1–31, Plenum Press, New York.

Milsmann, M. H. W., Schwendener, R. A., and Weder, H.-G., 1978, The preparation of large single bilayer liposomes by a fast and controlled dialysis, *Biochim. Biophys. Acta* **512:**147–155.

Mimms, L., Zampighi, G., Nozaki, Y., Tanford, C., and Reynolds, J. A., 1981, Phospholipid vesicle formation and transmembrane protein incorporation using octyl glucoside, *Biochemistry* **20:**833–840.

Mishina, M., Kurosaki, T., Tobimatsu, T., Morimoto, Y., Noda, M., Yamamoto, T., Terao, M., Lindstrom, J., Takahashi, T., Kuno, M., and Numa, S., 1984, Expression of functional acetylcholine receptor from cloned cDNAs, *Nature* **307:**604–608.

Moore, H.-P. H., and Raftery, M. A., 1980, Direct spectroscopic studies of cation translocation by *Torpedo* acetylcholine receptor on a time scale of physiological relevance, *Proc. Natl. Acad. Sci. U.S.A.* **77:**4509–4513.

Neubig, R. R., and Cohen, J. B., 1980, Permeability control by cholingeric receptors in *Torpedo* postsynaptic membranes: Agonist dose response relations measured at second and millisecond times, *Biochemistry* **19:**2770–2779.

Noda, M., Takahashi, H., Tanabe, T., Toyosato, M., Kikyotani, S., Furutani, Y., Hirose, T., Takashima, H., Inayama, S., Miyata, T., and Numa, S., 1983, Structural homology of *Torpedo californica* acetylcholine receptor subunits, *Nature* **302**:528–531.

Nozaki, Y., Lasic, D. D., Tanford, C., and Reynolds, J. A., 1982, Size analysis of phospholipid vesicle preparations, *Science* **217**:366–376.

Ochoa, E. L. M., Dalziel, A. W., and McNamee, M. G., 1983, Reconstitution of acetylcholine receptor function in lipid vesicles of defined composition, *Biochim. Biophys. Acta* **727**:151–162.

Olson, F., Hunt, C. A., Szoka, F. C., Vail, W. J., and Papahadjopoulos, D., 1979, Preparation of liposomes of defined size distribution by extrusion through polycarbonate membranes, *Biochim. Biophys. Acta* **557**:9–23.

Oswald, R., and Changeux, J.-P., 1981, Ultraviolet light induced labelling by noncompetitive blockers of the acetylcholine receptor from *Torpedo marmorata, Proc. Natl. Acad. Sci. U.S.A.* **78**:3925–3929.

Oswald, R., Heidmann, T., and Changeux, J.-P., 1983, Multiple affinity states for noncompetitive blockers revealed by [^{3}H]phencyclidine binding to acetylcholine receptor rich membrane fragments from *Torpedo marmorata, Biochemistry* **22**:3128–3136.

Palma, A., Herz, J., Wang, H. H., and Taylor, P., 1985, The binding of a spin-labeled local anesthetic to acetylcholine receptor at an allosteric site, *Biophys. J.* **47**(2):258a.

Popot, J.-L., and Changeux, J.-P., 1984, Nicotinic receptor of acetylcholine: Structure of an oligomeric integral membrane protein, *Physiol. Rev.* **64**:1162–1239.

Popot, J.-L., Demel, R. A., Sobel, A., van Deenan, L. L. M., and Changeux, J.-P., 1978, Interaction of the acetylcholine (nicotinic) receptor protein from *Torpedo marmorata* electric organ with monolayers of pure lipids, *Eur. J. Biochem.* **85**:27–42.

Popot, J.-L., Cartaud, J., and Changeux, J.-P., 1981, Reconstitution of a functional acetylcholine receptor. Incorporation into artificial lipid vesicles and pharmacology of the agonist-controlled permeability changes, *Eur. J. Biochem.* **118**:203–214.

Prinz, H., and Maelicke, A., 1983a, Interaction of cholinergic ligands with purified acetylcholine receptor protein. I. Equilibrium binding studies, *J. Biol. Chem.* **258**:10263–10271.

Prinz, H., and Maelicke, A., 1983b, Interaction of cholinergic ligands with the purified acetylcholine receptor protein. II. Kinetic studies, *J. Biol. Chem.* **258**:10273–10182.

Quast, U., Schimerlik, M., Lee, T., Witzemann, V., Blanchard, S., and Raftery, M. A., 1978, Ligand-induced conformation changes in *Torpedo californica* membrane-bound acetylcholine receptor, *Biochemistry* **17**:2405–2414.

Rabon, E. C., and Sachs, G., 1981, Thallium interaction with the gastric (K,H)-ATPase, *J. Membr. Biol.* **62**:19–27.

Racker, E., 1972, Reconstitution of cytochrome oxidase vesicles and conferral of sensitivity to energy transfer inhibitors, *J. Membr. Biol.* **10**:221–235.

Ramseyer, G. O., Morrison, G. H., Aoshima, H., and Hess, G. P., 1981, Vidicon flame emission spectroscopy of Li^+, Na^+, and K^+ fluxes mediated by acetylcholine receptor in *Electrophorus electricus* membrane vesicles, *Anal. Biochem.* **115**:34–41.

Reynolds, J. A., 1982, Interactions between protein and amphiliphiles, in: *Lipid–Protein Interactions*, vol. 2 (P. C. Jost and O. H. Griffith, eds.), pp. 193–224, John Wiley & Sons, New York.

Reynolds, J. A., Nozaki, Y., and Tanford, C., 1983, Gel exclusion chromatography on S-1000 Sephacryl: Application to phospholipid vesicles, *Anal. Biochem.* **130**:471–474.

Sakmann, B., Patlak, J., and Neher, E., 1980, Single acetylcholine activated channels show burst kinetics in the presence of desensitizing concentrations of agonist, *Nature* **286**:71–73.

Sandermann, H., 1982, Lipid-dependent membrane enzymes: A kinetic model for cooperative activation in the absence of cooperativity in lipid binding, *Eur. J. Biochem.* **127**:123–128.

Sator, V., Gonzalez-Ros, J. M., Calvo-Fernandez, P., and Martinez-Carrion, M., 1979, Pyrenesulfonyl azide: A marker of acetylcholine receptor subunits in contact with the membrane hydrophobic environment, *Biochemistry* **18**:1200–1206.

Schiebler, W., and Hucho, F., 1978, Membranes rich in acetylcholine receptor: Characterization and reconstitution to excitable membranes from exogenous lipids, *Eur. J. Biochem.* **85**:55–63.

Schmidt, J., and Raftery, M. A., 1973, A simple assay for the study of solubilized acetylcholine receptor, *Anal. Biochem.* **52**:349–354.

Sha'afi, R. I., and Fernandez, S. M., (eds.), 1983, *Fast Methods in Physical Biochemistry and Cell Biology*, Elsevier, Amsterdam.

Simmonds, A. C., East, J. M., Jones, O. T., Rooney, E. K., McWhirter, J., and Lee, A. G., 1982, Annular and non-annular binding sites on the (Ca^{2+}, Mg^{2+})-ATPase, *Biochim. Biophys. Acta* **693:**398–406.

Sobel, A., Heidmann, T., Cartaud, J., and Changeux, J.-P., 1980, Reconstitution of functional acetylcholine receptor. Polypeptide chains, ultrastructure and binding sites for acetylcholine and local anesthetics, *Eur. J. Biochem.* **110:**13–33.

Taylor, P., Brown, R. D., and Johnson, D. A., 1983, The linkage between ligand occupation and response of the nicotinic acetylcholine receptor, in: *Current Topics in Membranes and Transport*, Vol. 18, *Membrane Receptors* (A. Kleingeller and B. R. Martin, ed.), pp. 407–444, Academic Press, New York.

Tzartos, S. J., and Changeux, J.-P., 1984, Lipid-dependent recovery of α-bungarotoxin and monoclonal antibody binding to the purified α-subunit from *Torpedo marmorata* acetylcholine receptor, *J. Biol. Chem.* **259:**11512–11519.

Ueno, M., Tanford, C., and Reynolds, J. A., 1984, Phospholipid vesicle formation using non-ionic detergents with low monomer solubility. Kinetic factors determine vesicle size and permeability, *Biochemistry* **23:**3070–3076.

Walker, J. W., Lukas, R. J., and McNamee, M. G., 1981, Effects of thio-group modifications on the ion permeability control and ligand binding properties of *Torpedo californica* acetylcholine receptor, *Biochemistry* **20:**2191–2199.

Walker, J. W., Takeyasu, K., and McNamee, M. G., 1982, Activation and desensitization kinetics of *Torpedo californica* acetylcholine receptor in reconstituted membranes, *Biochemistry* **21:**5384–5389.

Walker, J. W., Richardson, C. A., and McNamee, M. G., 1984, Effects of thio-group modifications of *Torpedo californica* acetylcholine receptor on ion flux activation and inactivation kinetics, *Biochemistry* **23:**2329–2338.

Warren, G. B., Birdsall, N. J. M., Lee, A. G., and Metcalfe, J. C., 1974, Lipid substitution: The investigation of functional complexes of single species of phospholipid and a purified calcium transport protein, in: *Membrane Proteins in Transport and Phosphorylation* (G. F. Azzione, M. E. Klingenberg, E. Quagliariello, and N. Siliprandi, eds.), pp. 1–12, North-Holland, Amsterdam.

Weber, M., and Changeux, J.-P., 1974, Binding of *Naja nigricollis* [^{3}H]alpha-toxin to membrane fragments from *Electrophorus* and *Torpedo* organs: Effect of cholinergic agonists and antagonists on the binding of tritiated alpha-neurotoxin, *Mol. Pharmacol.* **10:**15–34.

Weber, M., David-Pfeuty, T., and Changeux, J.-P., 1975, Regulation of binding properties of the nicotinic receptor protein by cholinergic ligands in membrane fragments from *Torpedo marmorata, Proc. Natl. Acad. Sci. U.S.A.* **72:**3443–3447.

Weiland, G., Georgia, B., Lappi, S., Schnell, C. F., and Taylor, P., 1977, Kinetics of agonist-mediated transitions in state of the cholinergic receptor, *J. Biol. Chem.* **252:**7648–7656.

Wu, W. C.-S., and Raftery, M. A., 1981, Reconstitution of acetylcholine receptor function using purified receptor protein, *Biochemistry* **20:**694–701.

Wu, W. C.-S., Moore, H.-P. H., and Raftery, M. A., 1981, Quantitation of cation transport by reconstituted membrane vesicles containing acetylcholine receptor, *Proc. Natl. Acad. Sci. U.S.A.* **78:**775–779.

Yager, P., Chang, E. L., Williams, R. W., and Dalziel, A. W., 1984, The secondary structure of acetylcholine receptor reconstituted in a single lipid component as determined by Raman spectroscopy, *Biophys. J.* **45:**26–28.

Part III

SODIUM CHANNEL

Chapter 11

SKELETAL MUSCLE SODIUM CHANNELS

Isolation and Reconstitution

J. C. TANAKA, R. E. FURMAN, and R. L. BARCHI

1. Introduction

In mammalian muscle, a voltage-sensitive sodium channel controls the transient increase in membrane conductance that produces an action potential in the sarcolemma and T-tubular membranes. The time- and voltage-dependent characteristics of the currents regulated by this channel have been studied extensively over the past few decades, first using the traditional approach of voltage clamp (Adrian *et al.*, 1970) and more recently at the single-channel level using patch-clamp technology (Sigworth and Neher, 1980).

Rapid progress has been made in the past 10 years in the biochemistry of sodium-channel proteins, and the muscle sodium channel has been no exception in this regard. Building on early work with the binding of radiolabeled toxins to the sodium channel *in situ*, the sodium channel protein has been solubilized and purified from rat and rabbit skeletal muscle. The purified protein was reconstituted into artificial lipid vesicles and characterized by isotopic flux measurements and single-channel electrical recordings. Although the purification of these proteins on the basis of their toxin-binding capacity represents an important step forward, it is their functional reconstitution that provides the necessary proof of their integrity as a membrane ion channel.

This chapter deals with the purification, reconstitution, and functional assessment of the sodium channel from mammalian skeletal muscle. Since the biochemical aspects of the purification and characterization of this channel have been the subject of other reviews (Barchi, 1984; Barchi *et al.*, 1984), those aspects of the work are considered only briefly here. We concentrate instead on the functional reconstitution of the purified sodium channel and on approaches to its study in a defined lipid environment.

Although reconstitution as a biochemical technique has been applied to a variety of membrane proteins, the tricks needed for successful reconstitution of

J. C. TANAKA, R. E. FURMAN, and R. L. BARCHI • David Mahoney Institute of Neurological Sciences and Department of Neurology, University of Pennsylvania, Philadelphia, Pennsylvania 19104.

a particular protein are usually not discussed and often are not obvious or logical. We take this opportunity to discuss some of our tricks for reconstitution of the sodium channel as well as some of the approaches that have not been useful with our reconstitution systems.

2. Isolation of Muscle Sodium Channels

2.1. Purification and Physical Properties

We have purified the voltage-dependent sodium channel from rat (Barchi *et al.*, 1980; Barchi, 1983) and rabbit (Kraner *et al.*, 1985) skeletal muscle and obtained partial purification of this protein from human skeletal muscle (R. Roberts and R. L. Barchi, unpublished data). The membrane isolation procedure used with rat muscle (Barchi *et al.*, 1979) produces membranes that are predominantly sarcolemmal in origin, whereas the procedure used with rabbit muscle (Rosemblatt *et al.*, 1981) generates predominantly T-tubular membranes. Comparable channel proteins have also been isolated from eel electric organ (Agnew *et al.*, 1978; Miller *et al.*, 1983; Norman *et al.*, 1983) and rat brain (Hartshorne and Catterall, 1981; Hartshorne *et al.*, 1982; Barhanin *et al.*, 1983) and are considered elsewhere in this volume (see Chapters 8–10).

In each of the skeletal muscle preparations, the sodium channel could be solubilized by a variety of nonionic detergents (Barchi and Murphy, 1980); for most of our work NP-40 was used. When care was taken to insure optimal ratios of detergent to phospholipid, the solubilized skeletal muscle sodium channel exhibited stable high-affinity saxitoxin (STX) binding at 0–2°C for at least 24 hr. For biochemical manipulations and purification procedures, exogenous phospholipid such as soybean phosphatides (Asolectin®) must be added to all detergent-containing buffers in a 1:5 molar ratio to the detergent in order to stabilize this high-affinity STX binding (Agnew and Raftery, 1979).

The physical properties of the solubilized skeletal muscle sodium channel can be estimated by a combination of gel filtration and density-gradient centrifugation (Barchi *et al.*, 1980). All three of the channels that we have examined in this way appear remarkably similar (Table I). Each exhibits an apparent Stokes

Table I. Physical Properties of Mammalian Skeletal Muscle Sodium Channels

	Stokes radius (Å)	S_{H_2O}[b]	S_{D_2O}[c]	$S_{w,20}$[d]	v	Apparent M_r
Rat	86	9.1	6.1	9.8	0.83	314,000
Rabbit	86	8.6	5.9	8.8	0.82	295,000
Human[a]	85	8.3	5.4	9.1	0.83	287,000

[a] Values derived from fresh human muscle homogenized directly in NP-40 (R. Roberts and R. L. Barchi, unpublished observations).
[b] Apparent S value as compared to standard globular proteins on sucrose gradients prepared in H_2O.
[c] Apparent S values on sucrose gradients prepared in D_2O.
[d] Calculated $S_{20,w}$.

radius of about 8.6 nm in its mixed micelle with detergent and phospholipid. The partial specific volume of the mixed micelle for each is about 0.83 ml/g. Using an estimate for the actual partial specific volume of the protein of 0.73 ml/g, we calculated the molecular weight of the channel protein in the absence of detergent and lipid to be about 300,000.

The rat sarcolemmal and rabbit T-tubular channels have been purified to >90% theoretical homogeneity by chromatography on a weak anion-exchange resin followed by affinity chromatography on a column of immobilized wheat germ agglutinin and centrifugation on a sucrose density gradient (Barchi *et al.*, 1980; Barchi, 1983; Kraner *et al.*, 1985). Membranes are typically solubilized at 2–3 mg protein/ml in 1% NP-40. The initial ion-exchange chromatography is performed on a Sepharose resin containing 20-atom hydrophilic spacer arms terminated with a guanidinium residue; elution is accomplished with a gradient of choline chloride. The saxitoxin binding peak (~550 mM choline chloride) is transferred directly to the WGA-Sepharose column, washed, and eluted with 50 mM N-acetylglucosamine. This column has the advantage of simultaneously desalting and concentrating the peak while providing an additional 10- to 20-fold purification of the channel. The channel protein eluted from this column may then be fractionated on a 3–14% sucrose gradient by centrifugation at 175,000 × *g* for 12 hr.

Peak fractions from these final sucrose gradients have exhibited saxitoxin (STX) binding activity in excess of 3000 pmol/mg protein. Assuming an average molecular weight for the total protein from physical measurements of 300,000, the theoretical maximal specific activity of the purified protein would be ~3300 pmol/mg, suggesting that these final preparations can easily exceed 90% purity.

2.2. Subunit Composition

All three of the skeletal muscle sodium channels characterized in our laboratory contain a large glycoprotein subunit of ~260 kDa that runs diffusely and with anomalous migratory characteristics on SDS-PAGE (Fig. 1). This subunit appears to be very susceptible to proteolytic nicking during the initial homogenization of the skeletal muscle and the subsequent preparation of membrane fractions; the product of this proteolytic activity is a diffuse band centered about 150 kDa and possibly one smaller component of 45 kDa (Barchi, 1983; Casadei *et al.*, 1984). In order to inhibit proteolysis, protease inhibitors are present throughout the membrane isolation as well as the channel purification. For work with skeletal muscle, we typically use the protease inhibitors phenylmethyl sulfonyl fluoride, pepstatin, iodoacetamide, and EGTA in solutions for the muscle homogenization and membrane preparation; EGTA, phenylmethyl sulfonyl fluoride, and pepstatin are present throughout the channel purification after membrane solubilization. *o*-Phenanthroline, leupeptin, and iodoacetamide have also been added to the channel purification at various times, but their presence appears to have no additional protective effect.

The purest preparations of each of these mammalian skeletal muscle channels also contain a small 38-kDa component (Barchi, 1983; Kraner *et al.*, 1985). This

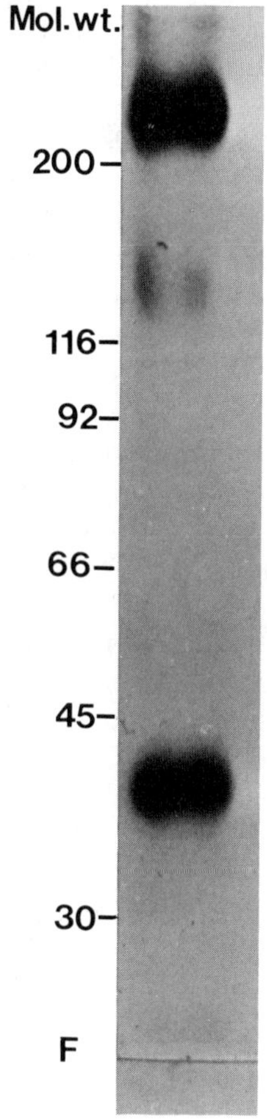

Figure 1. Subunit composition of the purified sodium channel from rabbit skeletal muscle. The purified channel contains a large glycoprotein with an apparent mol. wt. of about 260,000 and a smaller component of 38,000 mol. wt. Proteins were labeled with Bolton–Hunter reagent prior to electrophoresis, separated on a 7–20% gradient SDS-polyacrylamide gel, and visualized by autoradiography.

subunit comigrates with the 260-kDa subunit as well as with the peak of functional channel activity on all columns and gradients. A similar 38-kDa component is seen in the sodium channel purified from rat brain (Hartshorne and Catterall, 1981) but is not found in the channel isolated from eel electroplax (Agnew *et al.*, 1978; Miller *et al.*, 1983).

Immunoaffinity columns constructed with either polyclonal or monoclonal antibodies generated against the purified rat sodium channel can be used to purify the channel protein from muscle that has been homogenized directly in detergent and protease inhibitors so that proteolysis can be minimized (Casadei *et al.*, 1985). These columns isolate a 260-kDa protein from fresh rat muscle. The 38-kDa component is also present, although occasionally in amounts that do not appear to be stoichiometrically related to the 260-kDa protein. Since buffers of high ionic strength are used to wash contaminant protein from these immunoaffinity columns prior to channel elution, it is possible that the 38-kDa component is only loosely associated with the 260-kDa protein in its native state and becomes dissociated during isolation procedure.

The role of the 38-kDa component remains to be resolved. In the case of the eel channel, its presence does not seem necessary for functional reconstitution of the channel (Rosenberg *et al.*, 1984), and in mammals the major polypeptide toxin-binding site and the saxitoxin binding site appear to be located on the 260-kDa component (Beneski and Catterall, 1980; Sharkey *et al.*, 1983; Barhanin *et al.*, 1984a). The 38-kDa component may indeed represent an integral part of the mammalian channel; alternatively, the peptide could be an associated regulatory protein, a cytoskeletal element linked to the channel, or an unrelated but persistent contaminant. In our experience, however, all rat or rabbit muscle channel preparations that exhibited functional activity after reconstitution contained the 38-kDa component (Tanaka *et al.*, 1983; Kraner *et al.*, 1985).

The sodium channels purified from rat sarcolemma and rabbit T-tubular membranes are closely related. The migratory properties of the large glycoprotein subunit and 38-kDa component of each are indistinguishable on SDS-PAGE. A number of monoclonal antibodies raised against the rat sarcolemmal sodium channel cross react with the rabbit channel in a solid-phase radioimmunoassay (Casadei *et al.*, 1984; Kraner *et al.*, 1985). On Western blots, these antibodies react specifically with the 260-kDa glycoprotein in both rat and rabbit muscle preparations. As is discussed below, the rat and rabbit reconstituted sodium channels exhibit similar functional profiles in regard to toxin activation and block, cation selectivity, and single-channel conductance.

3. Reconstitution of the Purified Channel

After purification of the saxitoxin-binding protein was accomplished, the major question was whether the purified protein could modulate membrane sodium permeability. The first step toward evaluating the function of an ion channel is the reinsertion of the protein into a lipid bilayer under conditions that allow ionic flux or current to be measured. The remainder of this chapter deals with various aspects of the functional reconstitution of rat and rabbit muscle sodium channels first into lipid vesicles for flux measurements and subsequently into planar bilayers for single-channel studies.

The reconstitution of purified membrane proteins into artificial lipid vesicles is becoming a standard procedure in the biochemist's repertoire. The ability to measure an activated ion influx into these small sealed vesicles depends on both the properties of the vesicles and the conductance of the reconstituted channel. For flux measurements, the vesicles should be large and unilamellar in order to enclose as large a volume as possible; they should also be nonaggregating and should have a low resting permeability to monovalent cations. Unfortunately, there are no recipes that can be used to form vesicles fulfilling all these requirements; furthermore, the insertion of a protein into a vesicle often alters its characteristics dramatically. We have found that many factors influence the physical characteristics of the reconstituted vesicles; some of the factors are predictable, but many are not.

3.1. Techniques for Vesicle Formation

For flux measurements and planar bilayer studies, we reconstituted the purified muscle sodium channel into phospholipid liposomes by using Biobeads SM2 to remove the nonionic detergent (Weigele and Barchi, 1982). For standard reconstitutions, 250 mg of egg phosphatidylcholine was dried under vacuum for at least 3 hr to insure the complete removal of all solvents. The dry lipid was solubilized with 10% NP-40 to give a final concentration of 50 mg lipid/ml. This solution was then added to the purified sodium channel to yield a final concentration of 5 mg/ml of lipid and 1% detergent. Typically, fractions from a two-step purification were used that bound 900–2800 pmol STX/mg protein. Washed Biobeads (Holloway, 1973) were added to the solution at 0.3 mg/ml, and the vesicles were formed during a 3-hr rotation at 10°C. Vesicles were separated from the Biobeads by filtration through glass wool.

For patch recordings on reconstituted vesicles, nonionic detergents introduced serious electrical artifacts that obscured true channel activity (see Section 5.1). For these studies, we carried out channel purification and reconstitution in the zwitterionic detergent CHAPS (Hjelmeland *et al.*, 1983). Purification followed the previously described protocol (Kraner *et al.*, 1985), but all buffers contained 10 mM CHAPS without added lipid instead of the nonionic detergent NP-40. For reconstitution, purified channel protein in 10 mM CHAPS buffer was added to the dried lipid mixture, usually phosphatidylethanolamine/phosphatidylserine (PE/PS) in 70:30 ratio, to give a final concentration of 5 mg/ml. After a few minutes of vortexing, the lipid was usually solubilized; the solution (1.2 ml) was then applied to a 0.9 × 18 cm Sephadex G-50 column equilibrated in 100 mM NaCl, 10 mM HEPES, pH 7.4, and 0.5 mM EGTA. Vesicles were collected in the column void volume.

3.2. Vesicle Characteristics

One goal of our reconstitution experiments was to estimate the fraction of saxitoxin-binding sites that showed functional channel activity following reconstitution. Functional activity was estimated using a sodium channel neurotoxin to activate the channel (see below) and stimulate $^{22}Na^{+}$ influx into vesicles containing purified channels. For experiments of this sort, the characteristics of the reconstituted vesicles themselves can limit the resolution of the assay. The two most important of these properties are the intravesicular space and the monovalent cation leakage.

3.2.1. Characterization of Vesicle Size and Intravesicular Space

Many approaches to determining vesicle size are available, such as light scattering, included volume measurements, gel filtration, and electron microscopy with negative staining or freeze–fracture replicas. We have examined egg phosphatidylcholine vesicles containing the purified sarcolemmal channel by both freeze–fracture EM and column chromatography (Weigele and Barchi, 1982). Reconstitutions were carried out under conditions designed to incorporate about one channel per ten vesicles.

Gel filtration of vesicles containing the reconstituted sodium channel provided information on the distribution of STX binding sites as a function of vesicle size. A 0.9 × 57 cm Sepharose 2B-CL column was calibrated with latex beads of various diameters, and the recovery of [^{14}C]phosphatidylcholine and [^{3}H]STX binding was followed in a typical population of channel-containing vesicles. Sixty percent of the total lipid and [^{3}H]STX binding associated with the void volume of the column contained vesicles >1000 Å diameter, whereas the remainder eluted with smaller vesicles averaging ~300 Å diameter (Fig. 2).

For freeze–fracture electron microscopy, the vesicles were concentrated by centrifugation, fixed with 1% gluteraldehyde, infiltrated in 50% glycerol, and freeze fractured. Vesicles reconstituted with ~20 μg protein/ml again exhibited a bimodal size distribution, with about half of the vesicles having diameters 1000 Å to 5000 Å. Many of these vesicles contained discrete intramembranous particles not seen in vesicles prepared without the sodium channel protein (Fig. 2). The remainder of the vesicles appeared quite small and more uniform, with diameters of ~300 Å.

Several methods are available for measuring the intravesicular space, including gramicidin permeation, ^{22}Na^{+} or [^{14}C]glucose equilibration (Rhoden and Goldin, 1979), and calcein equilibration (Oku *et al.*, 1982). The method chosen should be accurate, inexpensive, and should not interfere with vesicle formation or subsequent flux measurements. The method we use for routine determinations of the trapped space is gramicidin-stimulated ^{22}Na^{+} uptake. Gramicidin forms monovalent-cation-selective pores permitting rapid equilibration of the total trapped space. Measurements can then be done at the time of the experiment on the same vesicles used for toxin activation.

The presence of protein in the reconstitution buffer had a marked effect on the intravesicular space (Table II). For example, at 0.038 mg protein/ml, the space was <10% that of protein-free vesicles made under identical conditions, although the molar ratio of lipid to protein was at least 51,000:1. To reduce this effect of the sodium channel protein on vesicle formation, we attempted to keep our protein concentrations below 20 μg/ml in the reconstitution. The properties of the phospholipid also influenced the size of the vesicles formed as well as their morphology. Generally, we have been able to form unilamellar vesicles only with uncharged lipids such as PC or with PC/PE mixtures. Vesicles made with charged lipids were multilamellar and leaky, obscuring the batrachotoxin (BTX) activation in all but the best reconstitutions.

3.2.2. Channel Orientation and Distribution

Saxitoxin binding was measured on reconstituted fractions before and after the addition of a small amount of detergent to disrupt the vesicles (Weigele and

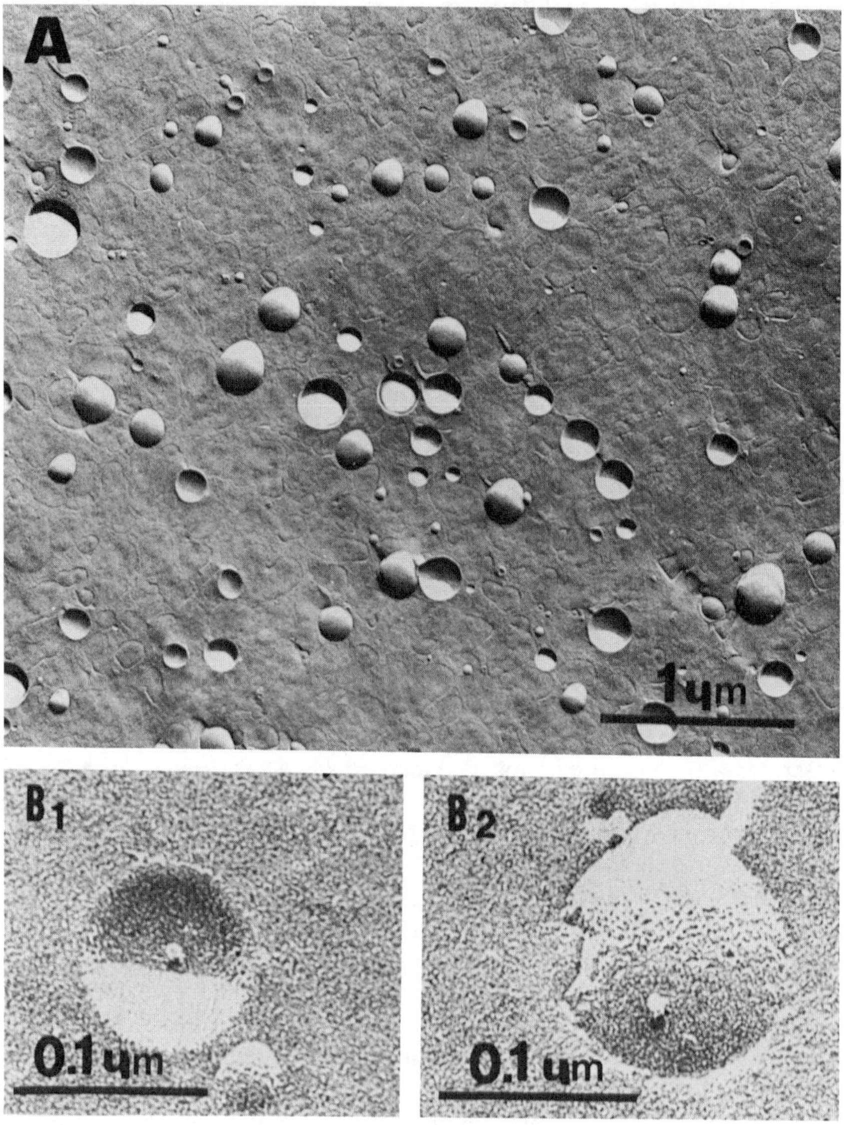

Figure 2. Size distribution of reconstituted vesicles. (A) Freeze-fracture electron microscopy of reconstituted vesicles formed from phosphatidylcholine alone demonstrates unilamellar profiles with smooth and featureless fracture planes. (B) Vesicles formed with the purified sodium channel have the same appearance and size characteristics, but they contain intramembranous particles (~100 Å diameter) in both the E (outer half) face (B_1) and the P (inner half) face (B_2) of the bilayer (Bars = 0.1 μm). (C) Reconstituted vesicles with the purified sodium channel chromatographed on Sepharose 2B-CL exhibit two size populations, one averaging 300 Å and the other >1000 Å. (D) Specific [^{3}H]STX binding activity (●, total; □, nonspecific) is uniformly distributed in both size groups. (From Weigele and Barchi, 1982.)

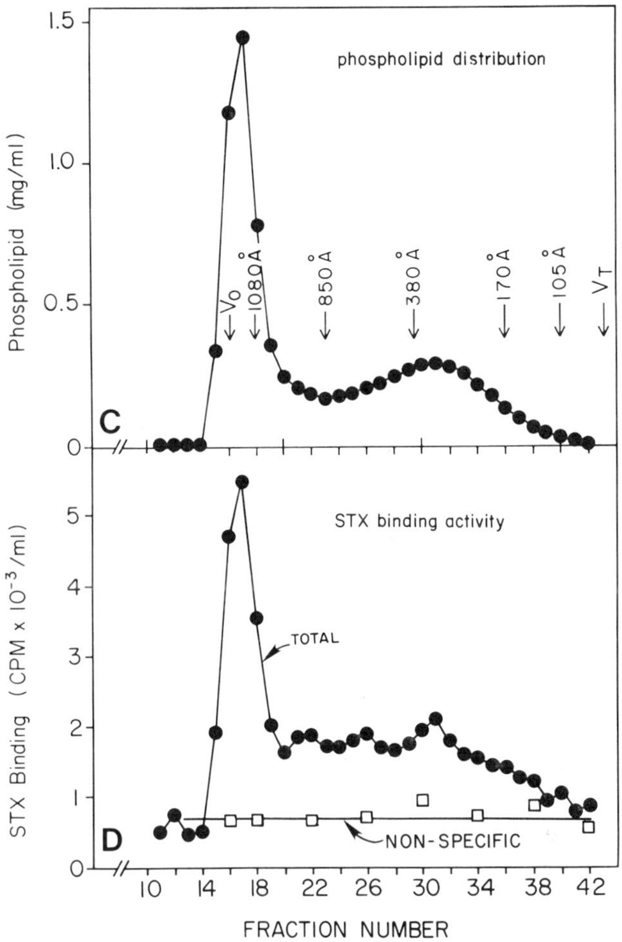

Figure 2. (*continued*)

Barchi, 1982; Kraner *et al.*, 1985). Only outward-facing channels will be detected in the intact vesicles, whereas the detergent-disrupted vesicles will show the total binding to both inward- and outward-facing channels. In our standard PC reconstitutions, both the rat and rabbit sodium channels orient with 50% of the channels facing outward, suggesting random channel insertion during vesicle formation. Further evidence for random insertion is obtained from STX block of activated cation flux (see below).

The distribution of [^{3}H]STX binding in the reconstituted vesicles parallels the [^{14}C]lipid distribution with ~60% of the binding eluting in the void volume of the 2B-CL Sepharose fractionation (see Fig. 2). Under similar reconstitution conditions, then, we would expect to see a correlation between the number of binding sites and the activated flux. In general, we find little correlation between the

Table II. Properties of Skeletal Muscle Sodium Channel Reconstituted Vesicles[a]

	Saxitoxin binding (pmol/ml)	Gramicidin space (μl/mg PC)		Activation ratio	Leak influx (cpm/150 μl)	Protein conc. (mg/ml)
A. Pooled peak	20	1.2		4.1	1563	0.012
1:2	10	1.8		2.2	2054	0.006
1:5	4	1.8		2.4	625	0.002
1:10	2	1.5		2.3	909	0.001
		Before F/T	After F/T			
B. Pooled peak	20.8	0.96	0.27	1.10	4052	0.025
1:2	10.4	1.19	0.36	1.04	2053	0.012
1:4	5.2	1.17	0.45	1.03	8367	0.006
1:10	2.1	1.21	1.15	1.30	7039	0.003
C. Pooled fraction	45.9	0.29		1.3	3400	0.038
Pooled fraction	43.6	0.30		1.5	2696	0.064
Buffer		2.55				—

[a] Three reconstitutions were examined: (A) purified rabbit sodium channel reconstituted into egg PC; (B) purified rat sodium channel reconstituted in a 70:30 mixture of PE/PS; (C) rat channel reconstitued in PC. In A and B, serial dilutions of the pooled peak were reconstituted separately. In C, two peak fractions were reconstituted with a twofold difference in specific activity. The reconstitution protocol was described in the text; each fraction had a final detergent concentration of 1% and 5 mg/ml lipid. The gramicidin space was measured as described; PE/PS vesicles were measured before and after one freeze–thaw cycle.

biochemical parameters of the purification and the activation ratio. The extremes of our rabbit reconstitutions dramatically illustrate this point; one preparation had a low specific saxitoxin binding activity of 956 pmol/mg and a high flux ratio of 7.4; a similar purification had a high specific activity of 2440 pmol/mg but produced a flux ratio of only 2.0. We did find that the flux ratios were significantly better with purified rabbit channels (6.2 ± 3.7, n = 24) than rat (2.3 ± 1.1, n = 63), even though there was only a small increase in the average specific activity of the purified rabbit fractions used (1540 ± 672 pmol/mg) as compared to the rat preparations (1174 ± 631 pmol/mg).

We have considered several possible causes for the low correlation between saxitoxin binding and activated flux ratios. In view of the dramatic effect of very low concentrations of protein on the intravesicular space (see Table II), the most likely possibilities are channel aggregation within the vesicles or partition of channels into bulk lipid aggregates that do not form vesicles; either of these would result in a low activation ratio but not a loss of binding sites. Another possibility is, of course, that a low activation ratio reflects damaged, nonactivatable channels that still bind STX. At first glance this explanation may seem particularly applicable to the muscle reconstitution, since that channel is apparently more sensitive to proteolysis than either the rat brain or electric eel channels. A closer look, however, suggests that such an interpretation is overly simplified at best, since we see no correlation between the activation ratio and the amount of 150-kDa protein on SDS-PAGE, which is thought to be proteolyzed sodium channel protein.

4. Flux Measurements with Reconstituted Sodium Channels

The voltage-sensitive sodium channel in muscle has a number of biophysical and biochemical properties that uniquely define it among membrane ion channels (see Hille, 1984). Saxitoxin binding is one such characteristic, and that property is used to identify the channel during its biochemical isolation; the binding does not, however provide evidence of channel function. From electrophysiological and pharmacological studies we can derive a list of the properties that the reconstituted protein should possess. This list includes demonstration of appropriate toxin pharmacology, the ability to select for sodium among similar monovalent cations, a characteristic maximal rate of ion flow or single-channel conductance, and characteristic voltage-dependent channel transitions between conducting and nonconducting states. Proof that the purified STX-binding protein is the voltage-dependent sodium channel demands evidence for as many of these channel properties as possible in the reconstituted vesicles.

In order to measure sodium-channel gating in reconstituted vesicles, neurotoxins have been used to activate the channel in the absence of a membrane potential (Narahashi, 1974; Catterall, 1980). In our system, the uptake of radiolabeled cations was measured following the removal of extravesicular cations on a Dowex ion-exchange column (Weigele and Barchi, 1982). Typically, vesicles were made in a buffer containing 100 mM NaCl. Following the incubation with an activating neurotoxin, a small amount of $^{22}Na^+$ was added to initiate the influx. At timed intervals, aliquots were filtered through 2.0-ml ion-exchange columns, and the isotope within the vesicles was determined by scintillation counting.

A limit to the resolution of the ion transport in these experiments is the presence of a nonspecific background leak in the vesicles. Since vesicles without protein have a very small leak, this larger leak could reflect a fraction of damaged channels that are no longer toxin sensitive or could result directly from the insertion of a large protein into the ordered bilayer. Since the magnitude of this leak can become the limiting factor in flux studies with the sodium channel, the characteristics of this nonspecific uptake are considered first.

4.1. Characterization of the Leakage Flux

Phosphatidylcholine vesicles formed in identical buffer solutions to those made with purified channels have a negligible cation permeability, equivalent to 25×10^{-15} mol $^{22}Na^+$/mg PC per min at an extravesicular concentration of $^{22}Na^+$ of 0.63 μM; PC vesicles with 15–50 μg/ml of protein have a 15- to 100-fold greater leak. Vesicles made with the same concentration of protein in a lipid mixture containing PE/PS (70:30) have a leak 500 times larger than PC vesicles (Table II). Increases in the protein concentration in the range of 15–50 μg/ml lead to a striking decrease in the trapped space as well as to increases in the background influx. Apparently the sodium channel protein disturbs the vesicle formation at very dilute channel concentrations (20–50 μg/ml). We attempted to reduce this effect by varying the buffer (HEPES versus tris and phosphate), divalent cations (Mg^{2+}, Ca^{2+}, or chelators), and concentration and species of monovalent cations. With the exception of the protein concentration, these other factors exert little influence on the resultant leakage influx.

The difficulty that we experienced in correlating the activation ratio with the number of saxitoxin binding sites in the purified fraction led us to consider whether the leak represented conducting sodium channels that still bound STX but were damaged in some other way. We therefore measured the temperature dependence of the leak influx, its cation selectivity, and its susceptibility to STX block. The Q^{10} for the leak in the range 5–37°C was ~1.2; no reduction in the leak was seen with STX, and no selectivity was observed among monovalent cations. We concluded that if the leak were through damaged, aggregated channels, they retained none of the selectivity or pharmacology appropriate to the sodium channel.

4.2. Functional Activity of the Reconstituted Sodium Channel

The alkaloid toxins batrachotoxin and veratridine act from either side of the membrane to shift the voltage dependence of sodium channel activation and prevent or delay channel inactivation (Ulbricht, 1969; Khodorov, 1978; Quandt and Narahashi, 1982; Huang *et al.*, 1982). Vesicles containing the purified reconsti-

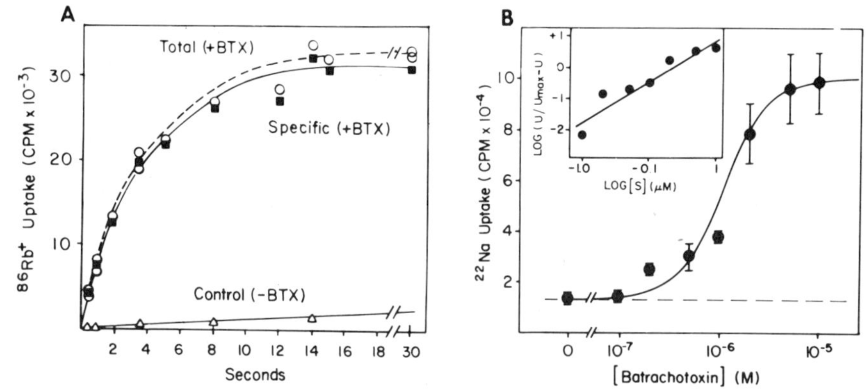

Figure 3. (A) Influx of $^{86}Rb^+$ into vesicles containing the purified sodium channel. The purified channel was reconstituted into egg phosphatidylcholine vesicles and incubated for 45 min at 36°C with 5 μM batrachotoxin (○) or with ethanol alone (△). Specific batrachotoxin-activated influx (■), defined as total minus control influx, occurred with a half-time of about 3 sec in this example. Fluxes were measured after equilibration of the vesicles at 22°C. (B) Dose-dependent activation of the purified, reconstituted channel by batrachotoxin. Vesicles containing the purified channel were incubated for 45 min at 36°C with the indicated concentrations of batrachotoxin or with an equivalent amount of ethanol alone. Influx of $^{22}Na^+$ was then measured after 5 sec, a period sufficient for all vesicles containing open channels to equilibrate with this cation. (Insert) Hill plot of the same data indicating an apparent K_d of 2 μM and a Hill coefficient of 1.2.

tuted channel demonstrated a rapid influx of $^{22}Na^+$ in response to batrachotoxin activation that was essentially complete within the first few seconds of the assay. The addition of STX to the assay medium inhibited ~50% of this activated flux, as expected if the activated channels were randomly oriented and only channels facing outward were blocked by this polar toxin. When vesicles were made in the presence of STX and the extravesicular STX was removed by gel filtration, the BTX-activated flux was again ~50% of the total activated flux, suggesting block of the complementary inward-facing channel population (Fig. 3). Subsequent addition of STX to the outside of these vesicles eliminated all BTX-activated flux. Thus, all channels capable of activation by BTX also contained a functional binding site for STX.

Veratridine also activated a $^{22}Na^+$ influx that was blocked with saxitoxin. Dose–response curves for batrachotoxin (Fig. 3) and veratridine activation of both rat and rabbit reconstituted sodium channel indicated K_ds of 5μM and 35 μM, respectively, in good agreement with measurements on native channels. The K_i for STX block (measured with $^{86}Rb^+$ uptake at 5 sec) was 5 nM in BTX-activated vesicles. Taken as a whole, these results demonstrate a pharmacological profile for the reconstituted STX-binding protein that is comparable to that of the native sodium channel (Catterall, 1977) and confirm that the reconstituted proteins modulate $^{22}Na^+$ membrane permeability in response to alkaloid activation, providing strong evidence for channel function.

4.3. Ion Selectivity

4.3.1. Batrachotoxin- and Veratridine-Modified Channels

The time course of BTX-activated cation influx into vesicles containing the purified channel was measured with a quenched-flow apparatus designed by J. Eccleston (see Gutfreund, 1969; Fersht and Jakes, 1975). This system has a mixing time of 4 msec and can be used for sample volumes as small as 20 μl. Motor-driven syringes rapidly mixed BTX-activated vesicles with labeled cations and ejected them through a variable-length delay tube into a Dowex slurry (Tanaka *et al.*, 1983). The length of the ejection tube controlled the incubation time for relatively fast measurements; for longer times, the mixed vesicles were held in the tube a set time before expulsion.

This quenched-flow device allowed clear resolution of $^{42}K^+$, $^{86}Rb^+$, and $^{137}Cs^+$ equilibration rates into vesicles containing the purified rat or rabbit sodium channel. The early linear phase of $^{42}K^+$ uptake was used to calibrate the resolution of the system; extrapolation to zero uptake indicated a 90-msec dead time, during which uptake could not be measured accurately. For both the rat and rabbit channels, $^{22}Na^+$ influx was complete prior to 90 msec, but an upper limit of 50 msec could be placed on the half-time for $^{22}Na^+$ equilibration (Fig. 4). From this estimate, selectivity ratios were calculated for the other cations; these values are summarized in Table III. Although there are reports that BTX reduces the selectivity of the sodium channel (Khodorov, 1978; Huang *et al.*, 1979; Frelin *et al.*, 1981), the reconstituted channel clearly retains much of its native selectivity (Hille, 1972; Pappone, 1980). The absolute half-times for equilibration of these cations and their relative selectivities were virtually identical with rat sarcolemmal or rabbit T-tubular channel protein in the vesicles, again underscoring the similarity between these channels.

Cation uptake into veratridine-activated vesicles occurred several orders of magnitude more slowly than that seen in the same vesicles after batrachotoxin activation. The selectivity of the veratridine-modified channel was again determined by comparing the half-times for $^{42}K^+$, $^{86}Rb^+$, and $^{137}Cs^+$ to the half-time for $^{22}Na^+$ uptake. The sequence was the same as that of native or BTX-modified channels, but relative selectivity among cations appeared greatly reduced.

The slow equilibration rates seen with veratridine may reflect the relatively infrequent channel openings produced by this toxin as compared to batrachotoxin, and part of the apparent loss in selectivity may result from the longer duration of these opening events when they do occur. Thus, for veratridine-activated channels, channel opening rather than ion movement through the open channel may be the rate-limiting step. Since the vesicles fill rapidly when the channel is open, we might not be able to distinguish the filling of $^{22}Na^+$ from $^{42}K^+$ in these vesicles because one channel opening could permit equilibration of both ions.

4.3.2. Temperature Dependence

Further evidence for differences in the rate-limiting step for flux measurements through channels activated by batrachotoxin or veratridine was provided

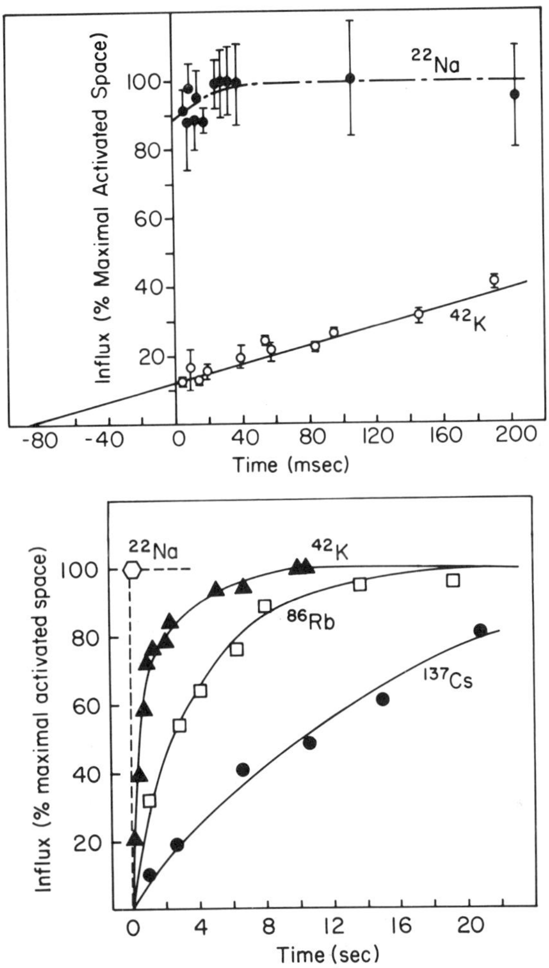

Figure 4. (A) Quenched-flow measurements of $^{22}Na^{+}$ and $^{42}K^{+}$ influx in batrachotoxin-stimulated vesicles at short incubation periods. Data points represent the specific batrachotoxin-stimulated uptake corrected for control uptake. Each point represents the mean ± S.D. of three or more separate measurements after correction for control influx. The effective quenching time for the apparatus was determined by extrapolation of the initial linear uptake rate for $^{42}K^{+}$ into batrachotoxin-activated vesicles (○), yielding a value of 90 msec. The $^{22}Na^{+}$ influx on this time scale (●) was very rapid, and an initial rate could not be resolved; an upper limit of 50 msec for the half-time for $^{22}Na^{+}$ could be estimated based on the measured quenching time of 90 msec for the system. (B) The time course of $^{42}K^{+}$, $^{86}Rb^{+}$, and $^{137}Cs^{+}$ uptake into batrachotoxin-activated vesicles as determined using the quenched-flow apparatus. Specific uptake data from several preparations, corrected for control uptake and plotted as percent maximal activated space, are shown for $^{42}K^{+}$ (▲), $^{86}Rb^{+}$ (□), and $^{137}Cs^{+}$ (●). The maximal space accessible through activated sodium channels was the same for all cations in any given preparation. Maximal $^{22}Na^{+}$ uptake was already observed at the shortest interval shown on this graph (⬡). All measurements shown were made at 18°C. (From Tanaka *et al.*, 1983.)

Table III. Cation Selectivity in the Reconstituted Sodium Channel[a]

	Batrachotoxin							Veratridine						
A. Rabbit T-tubular sodium channel														
Sequence	Na^+	>	K^+	>	Rb^+	>	Cs^+	Na^+	≃	K^+	>	Rb^+	>	Cs^+
$T^{1/2}$	<50 msec		390 msec		2.6 sec		6.5 sec	1.1 min		1.2 min		2.7 min		3.6 min
Ratio	1	:	0.13	:	0.019	:	0.008	1	:	0.92	:	0.41	:	0.31
B. Rat sarcolemmal sodium channel														
Sequence	Na+	>	K^+	>	Rb^+	>	Cs^+	Na^+	>	K^+	>	Rb^+	>	Cs^+
$T^{1/2}$	<50 msec		350 msec		2.5 sec		10 sec	1.0 min		1.2 min		2.0 min		2.6 min
Ratio	1	:	0.14	:	0.02	:	0.005	1	:	0.83	:	0.50	:	0.38

[a] Adapted from Tanaka *et al.* (1983) and Kraner *et al.* (1985).

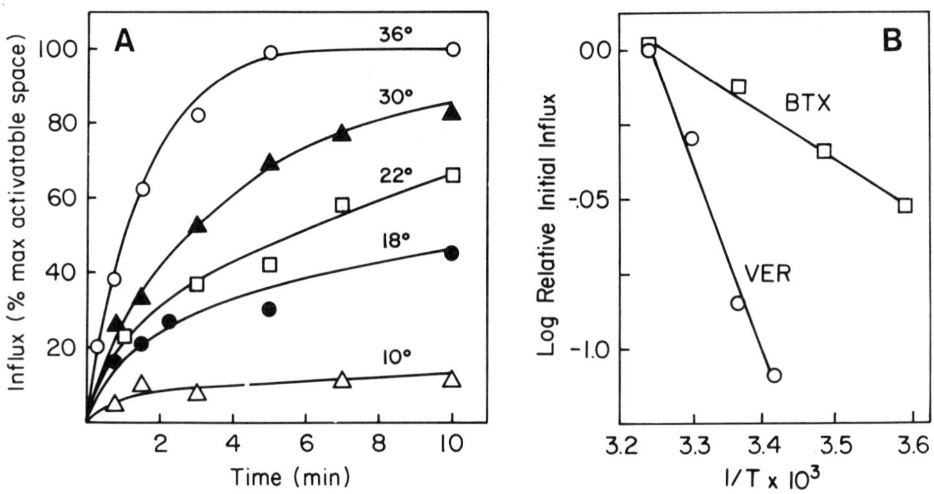

Figure 5. Temperature dependence of veratridine-stimulated $^{22}Na^+$ influx and Arrhenius plot of $^{86}Rb^+$ uptake in batrachotoxin- and veratridine-activated vesicles. In A, vesicles were activated with 5×10^{-4} M veratridine and equilibrated at the experimental temperature prior to addition of isotope. Influx was measured at 36 (○), 30 (▲), 22 (□), 18 (●), and 10°C (△). The corrected specific uptake data are plotted as percent maximum stimulated uptake. In B, initial uptake rates were measured for $^{86}Rb^+$ influx in two parallel populations of vesicles stimulated with either batrachotoxin (□), or veratridine (○) at temperatures between 5°C and 36°C. In each case, activation was carried out at 36°C for 45 min with the appropriate toxin, and vesicles were preequilibrated to the assay temperature prior to initiation of cation uptake. For ease of comparison, the stimulated uptake rates were normalized to the rate obtained in each preparation at 36°C. Linear regressions of the results of the Arrhenius plots were used to calculate activation energies. (From Tanaka *et al.*, 1983.)

by determining the activation energies of ion influx through reconstituted channels opened by each toxin. Vesicles were activated with batrachotoxin or veratridine at 36°C for 45 min, and the temperature dependence of subsequent cation influx was then measured (Tanaka *et al.*, 1983). Cation influx through veratridine-activated channels was very temperature dependent, with a Q_{10} of ~3.0 between 10°C and 37°C. The temperature dependence of batrachotoxin-activated influx was much lower, with a Q_{10} of ~1.5 over the same temperature range (Fig. 5). The activation energy calculated for veratridine influx is ~30 kcal/mol, similar to values reported for the sodium-channel activation process (Schauf, 1973), supporting the concept that some step leading to the channel opening may be the rate-limiting step for cation influx in veratridine-activated vesicles. The Q_{10} of 1.5 observed for cation influx in BTX-activated vesicles, on the other hand, suggests an activation energy of ~6 kcal/mol and approaches the temperature dependence of ion movement in solution, supporting the notion that BTX channels can be considered essentially open during the flux measurement.

This interpretation of the difference between the rates of vesicle filling with batrachotoxin and veratridine is consistent with results from ion-flux measurements on cultured cells (Catterall, 1975) and with electrophysiological measurements (Quandt and Narahashi, 1982), which suggest that BTX essentially locks the channel open at 0 mV. Veratridine is a partial agonist, however, opening the channel much less frequently. Although the measured half-time for $^{22}Na^+$ influx in veratridine-activated vesicles was about 1 min, we estimate that the half-time for the filling of a 1500-Å vesicle through an open channel having a conductance of 20 pS would be no more than 20–40 msec (Miller and Racker, 1979). Measurements of $^{22}Na^+$ uptake into our small vesicles after veratridine activation may well reflect the probability of first channel openings during that interval rather than the rate of cation movement through a sodium channel held open by veratridine.

4.4. Voltage-Dependent Toxin Activation

Depolarization of a sodium channel *in situ* normally leads to a brief increase in the membrane sodium conductance that lasts about 1 msec before the channel inactivates (Sigworth and Neher, 1980; Horn *et al.*, 1981). This inactivation process then prevents the channel from reopening unless the membrane potential is hyperpolarized. If reconstituted sodium channels could be activated in vesicles, this 1 to 2 msec open time would be too brief to allow sufficient isotope influx to be measured accurately. Electrophysiological studies have shown that sodium channels activated by batrachotoxin or veratridine retain their voltage-dependent activation, although the relationship between activation and voltage is shifted about 40 mV in the hyperpolarizing direction (Huang *et al.*, 1982; Leibowitz *et al.*, 1985). The other major action of these toxins is to slow or eliminate the inactivation process so that channels remain opened with prolonged depolarization. We reasoned, therefore, that voltage-dependent channel gating could be demonstrated in vesicles containing BTX-activated channels if sufficiently hyperpolarizing potentials were used.

In order to establish a membrane potential in the reconstituted vesicles, a potassium concentration gradient was created across the vesicle membrane, and the vesicles were then rendered permselective to K^+ by the addition of valinomycin. Vesicles were formed in KCl buffer with 2 μM STX trapped internally to block inward-facing channels and create a functionally uniform population of outward-oriented channels. External saxitoxin and KCl were then removed and replaced with choline chloride by gel filtration. Following the addition of 10^{-7} M valinomycin (Andreoli *et al.*, 1967; Mueller and Rudin, 1967), the hyperpolarized vesicles were incubated with batrachotoxin. The transmembrane potential was controlled during influx measurements by mixing the vesicles with buffer containing an appropriate external K^+ concentration calculated from the Nernst equation.

The voltage dependence of these BTX-activated channels was determined by measuring $^{22}Na^+$ uptake at various membrane potentials set by altering external $[K^+]$. We found that the specific $^{22}Na^+$ flux was maximal at low membrane potentials ($K_o = K_i$) and completely inhibited at large hyperpolarizing potentials (K_o = O) (Fig. 6). The flux increased sharply over a range of external potassium

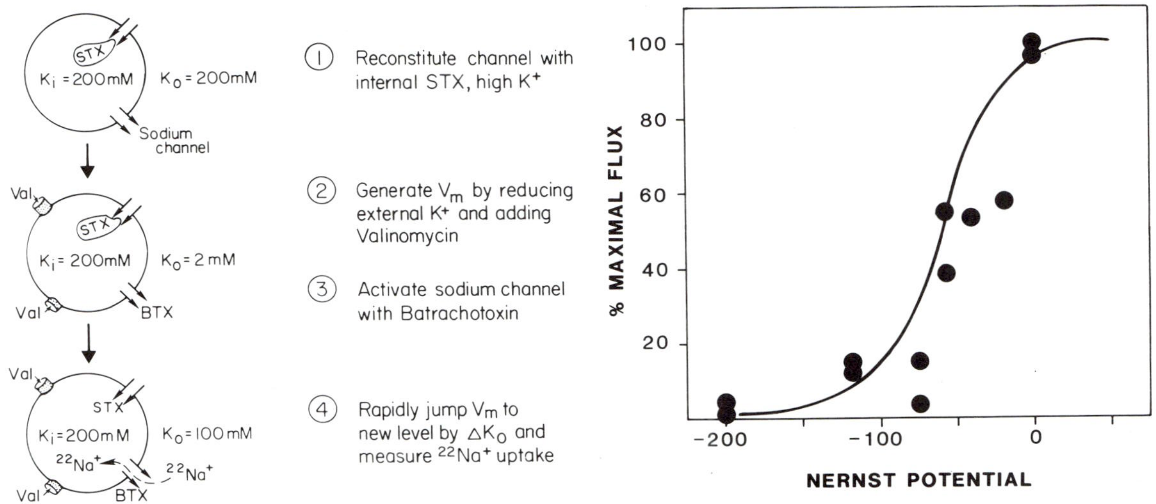

Figure 6. The reconstituted rabbit muscle sodium channel retains voltage-dependent gating after batrachotoxin activation. The $^{22}Na^+$ influx is plotted as a function of the Nernst potential calculated from the potassium gradient across the vesicles. The inhibition of batrachotoxin-activated influx with hyperpolarization demonstrates that all the channels retain the voltage-dependent properties observed with single-channel measurements (see Fig. 8). The experimental paradigm, shown on the left, is detailed in the text.

Table IV. Properties of Reconstituted Sodium Channels

	Rat muscle	Rabbit muscle
A. Pharmacological properties	$K_{0.5}$	$K_{0.5}$
Batrachotoxin	1.5 μM	1.5 μM
Veratridine	35 μM[a]	25 μM
Saxitoxin[f]	5 nM[b]	
Lidocaine[f]	0.4 mM[c]	
Benzocaine[f]	0.6 mM[c]	
Anemonia sulcata[f]		Not detected
Leiurus quinquestriatus[f]		Not detected
B. Selectivity		
Batrachotoxin	Na > K > Cs > Rb	Na > K > Cs > Rb
Veratridine	Na > K > Cs > Rb	Na > K > Cs > Rb
C. Single-channel properties		
Conductance	14 pS[d]	22 pS[e]
Midpoint of batrachotoxin activation	−100 mV	−95 mV

[a] Weigele and Barchi (1982).
[b] Tanaka *et al.* (1983).
[c] Barchi and Tanaka (1984).
[d] 100 mM NaCl/100 mM NaCl.
[e] 200 mM NaCl (*trans*)/500 mM NaCl (*cis*).
[f] Determined in the presence of 5 μM BTX

concentrations corresponding to predicted Nernst potentials between −100 and −40 mV, the general range observed for batrachotoxin-modifed channels *in situ*.

These experiments clearly demonstrate that all the batrachotoxin-activated and saxitoxin-inhibitable sodium channels present in the reconstituted vesicle population retained at least one of the voltage-sensitive properties characteristic of this channel and that the general relationship between membrane potential and channel activation was comparable to that of BTX-treated channels in their native membrane. Although this approach lacks the temporal resolution of single-channel recording and cannot provide information on channel kinetics, it yields insight into the response of the entire channel population that is not available from patch clamping and planar bilayer measurements.

Preliminary experiments show that veratridine activation is also blocked in vesicles hyperpolarized with KCl, although this experiment is more difficult because the slower ion equilibration times result in a larger background influx.

5. Single-Channel Measurements

The functional characteristics of single sodium channels place stringent demands on electrical recording techniques used for reconstitution studies. A sodium channel *in situ* typically opens and closes within a millisecond in response to depolarizing changes in potential. The peak current through a channel is ~1 pA, which represents the transport of $\sim 10^3$ ions through the open channel during these brief openings (Sigworth and Neher, 1980). Both patch-clamp and planar bilayer recording can be used to study purified sodium channels, although the latter approach lacks the frequency response needed for resolution of channel kinetics in the absence of modifying toxins (Krueger *et al.*, 1983). For single-channel measurements, we initially attempted to patch clamp vesicles containing the reconstituted channel. Because of our low success rate with single-channel patch recordings and only modest success with the revised channel purification in CHAPS, we also examined channel events after vesicle fusion into planar bilayers. The frequency of success has not been high with this method either, but the planar bilayer allows us to study the same reconstituted vesicles used for flux studies.

5.1. Patch-Clamp Recording

Successful patch clamping of large proteoliposomes was first described by Tank *et al.*, (1982). We modified his procedure to form large liposomes from PE/PS (70:30) at a total lipid concentration of 5 mg/ml in 1% NP-40 after detergent removal with Biobeads and a freeze–thaw step. High-resistance inside-out patches were easily pulled from these liposomes, but we found that the patches were electrically noisy and often produced single-channel records resembling those expected from an ion channel. These artifactual detergent channels were present in PE/PS vesicles made from several batches, brands, and types of nonionic detergents, including Triton X-100, Lubrol WX and PX, octylglucoside, and OPE-10, a purified constituent of Triton X-100 (Fig. 7).

Vesicles formed with sodium cholate (Tank *et al.*, 1982) were electrically silent, but the negative charge on cholate interfered with the ion-exchange step of the sodium-channel purification. Using the zwitterionic detergent CHAPS, however, we were able to produce electrically silent large PE/PS liposomes, which could then be used for patch-clamp experiments with the purified sodium channel. Reliable formation of large liposomes after a freeze–thaw cycle required several tricks. In the absence of protein, vesicles were most readily formed in buffer containing 1 mM $CaCl_2$. However, the sodium-channel protein in the buffer led to flocculation of the lipid if calcium was also present in the buffers; 0.5 mM EGTA was therefore added before detergent removal by gel filtration through a Sephadex G-50 column. Column fractions were frozen and thawed once to enlarge the vesicles. Typically, only one or two fractions produced vesicles sufficiently large for patch clamping; the remaining fractions produced vesicles barely visible under light microscopy. Better results were obtained by removing the detergent in either alkali metal cations other than sodium ($Rb \cong K > Na \cong Cs \gg Li$) or other anions ($SO_4 \gg Cl >$ methylsulfate). Fluxes could not be measured in these vesicles because of their multilamellar morphology.

For patch-clamp studies, the experimental chamber was either a 35-mm plastic Petri dish or was fabricated from a glass coverslip sealed with Kronig's cement to a milled piece of Plexiglas®. Small aliquots (1 to 10 μl) of liposomes were placed onto the floor of the experimental chamber, and the reconstitution buffer was

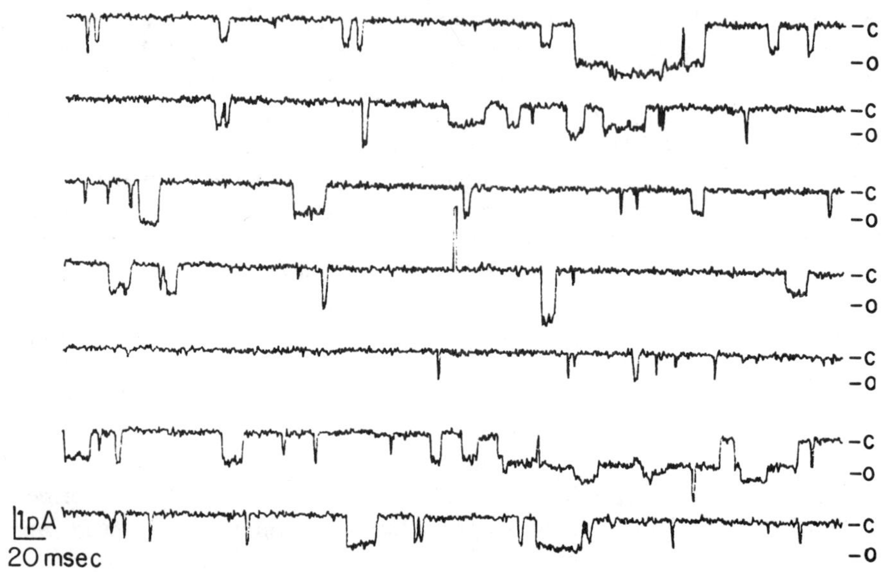

Figure 7. Channellike conductances recorded from liposomes made with a nonionic detergent. These single "channels" were recorded from an inside-out patch pulled from a PE/PS (70:30) liposome formed by solubilizing the lipids in the nonionic detergent OPE-10, removing the detergent with BioBeads, and freeze-thawing. This patch shows two discrete channels, the larger of which has a conductance of 20 pS. Gating in these channels does not appear to be voltage dependent. Patch held at −100 mV.

carefully layered over the aliquot. Sufficient numbers of patchable liposomes usually remained in the unstirred layer at the bottom of the chamber. Pretreatment of the glass coverslips with 0.5 mg/ml of poly-L-lysine was sometimes used to enhance adhesion of liposomes to the coverslip. Conventional patch microelectrodes fabricated from borosilicate capillaries were used in preference to soda lime capillaries, since the latter generally produced poor seals. After obtaining a gigaseal on the liposome, inside-out patches were formed by quickly withdrawing the electrode and passing the tip through the air–water interface. Currents across the patch were recorded with the electronics described by Hamill *et al.* (1981).

Voltage-dependent and pharmacological behavior consistent with native sodium channels was observed with several inside-out patches from CHAPS-purified sodium channels (Barchi *et al.*, 1984). Single channel openings were recorded as a function of transmembrane potential in a BTX-activated inside-out patch. Hyperpolarization shifted the channel from a mostly open to mostly closed state. The curve of percentage open time versus membrane potential was shifted in the hyperpolarizing direction, as observed with *in situ* BTX-activated sodium channels (see Huang *et al.*, 1982). The slope of the curve was shallower than that observed with native channels, possibly reflecting some alteration of the protein during the purification or reconstitution. Some patches showed spontaneous, infrequent burst openings with depolarization before BTX activation. Immediately after BTX activation, the opening bursts converted to continuous channel openings and closings. Before BTX activation, these bursts were completely suppressed at potentials less than −100 mV.

The number of sodium channels seen with the patch clamp was very small, however, despite excellent signals in the flux assays. To increase the likelihood of recording sodium channel transitions, we tried fusing vesicles containing CHAPS-purified channels into preformed bilayers at the tip of a patch pipette, equilibrating protein–detergent–lipid micelles under a monolayer at an air–water interface, forming patches with proteoliposomes spread at an air–water interface, and fusing vesicles into planar bilayers. None of these techniques increased our success at recording single-channel events in the CHAPS-purified rat sarcolemmal sodium channel.

5.2. Planar Bilayer Reconstitution

More recently, we have used the approach of Hartshorne *et al.*, (1985) with the NP-40-purified sodium channel from rabbit membranes. Planar bilayers were formed across a 350-μm orifice in a Teflon® cup using 1-palmitoyl, 2-oleoyl PE and 1-palmitoyl, 2-oleoyl PC (80:20) in decane (50 mg lipid/ml). The solution on the *cis* side of the membrane contained 500 mM NaCl, 10 mM HEPES, pH 7.4, 0.15 mM $CaCl_2$, 0.1 mM $MgCl_2$, and 0.05 mM EGTA, whereas the *trans* side substituted 200 mM NaCl in the solution. Potentials were applied across the membrane, and currents were recorded by the headstage of a Dagan Model 8900 patch clamp in the *trans* chamber; the *cis* side was grounded through a Ag–AgCl wire.

Sodium-channel protein was purified from T-tubules using conventional techniques after solubilization in NP-40. The purified protein was reconstituted in egg PC as described above, and isotopic flux studies were carried out to document the reconstitution of functional channels. Vesicles containing these channels were activated with BTX for 45 min at 37°C and then drawn up into a small right-angled capillary tube that was positioned over the bilayer. Small aliquots (0.5 μl) were eluted onto the bilayer surface every few minutes until a channel event was observed with ±70-mV potential steps. Withdrawal of the tube prevented further fusions.

With this approach, single-channel activity was observed with a higher-frequency than with the liposome–patch-clamp method. Sodium channels generally incorporated into the bilayer with their extracellular TTX site facing the *trans* chamber. The single-channel conductance averaged 22 pS, and the current–voltage relationship was linear over the range examined. The channel activation was steeply voltage dependent, with the 50% opening point at −95 mV (Fig. 8). A shift of 5 mV about the midpoint dramatically altered the ratio of times spent in the open and closed states. Application of TTX to the *cis* chamber (intracellular side of the channel) had no effect on channel activity, but applying the toxin to the *trans* (extracellular) side closed the channel for greater than 99% of the time (Fig. 9). Strong depolarization decreased the percentage of time the channel spent in the blocked state, as described by Krueger *et al.* (1983).

The question that persists using the planar bilayer fusion for single-channel

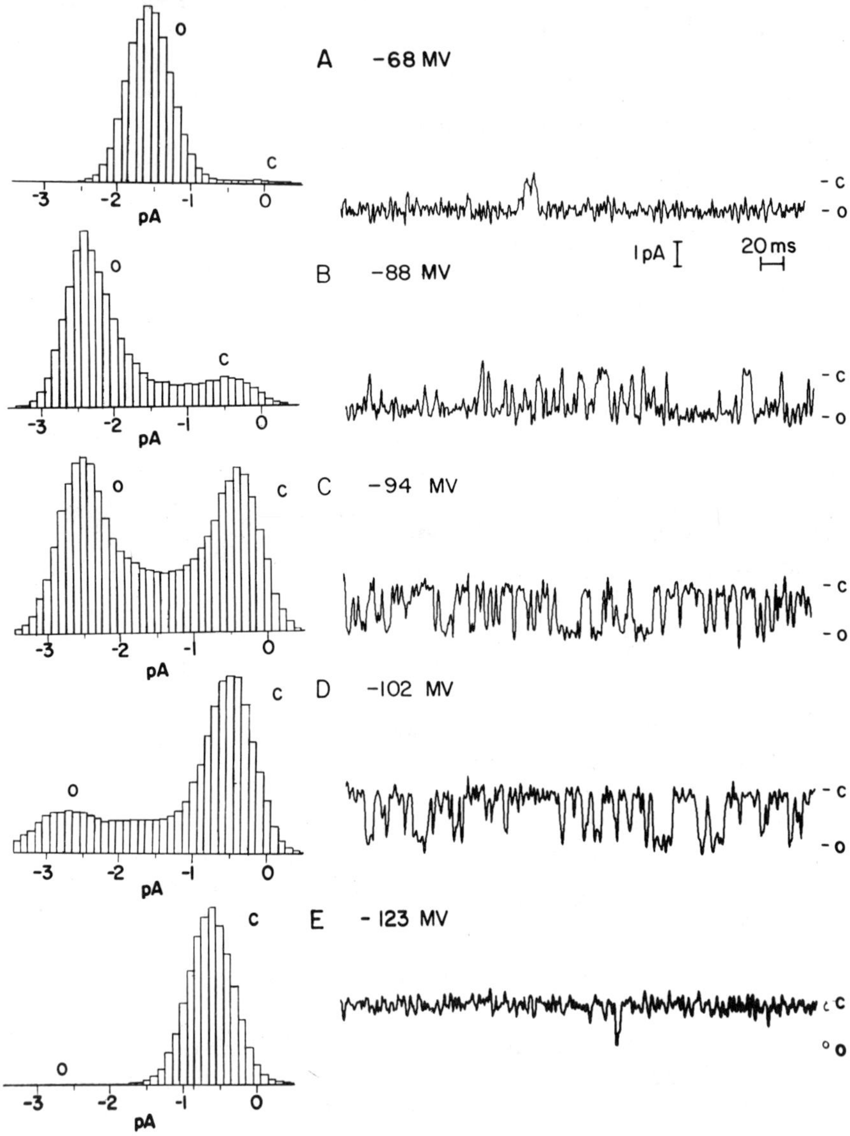

Figure 8. Amplitude histograms and corresponding current records from a purified rabbit sodium channel reconstituted into a planar bilayer, illustrating the voltage dependence of channel gating. Hyperpolarization (A to E) increasingly shifts the channel from the open (O) to the closed state (C). The point at which the batrachotoxin-activated channel was open 50% of the time was about −95 mV (physiological convention; extracellular side of channel was *trans*). Single-channel conductance was 22 pS. Current records are filtered at 400 Hz. Bilayers were formed as described in the text. *Cis* solution: 500 mM NaCl, 10 mM HEPES, 0.15 mM $CaCl_2$, 0.1 mM $MgCl_2$, 0.05 mM EGTA, pH 7.4. *Trans* solution was the same except for 200 mM NaCl and 1×10^{-6} M batrachotoxin. Temperature was 22°C.

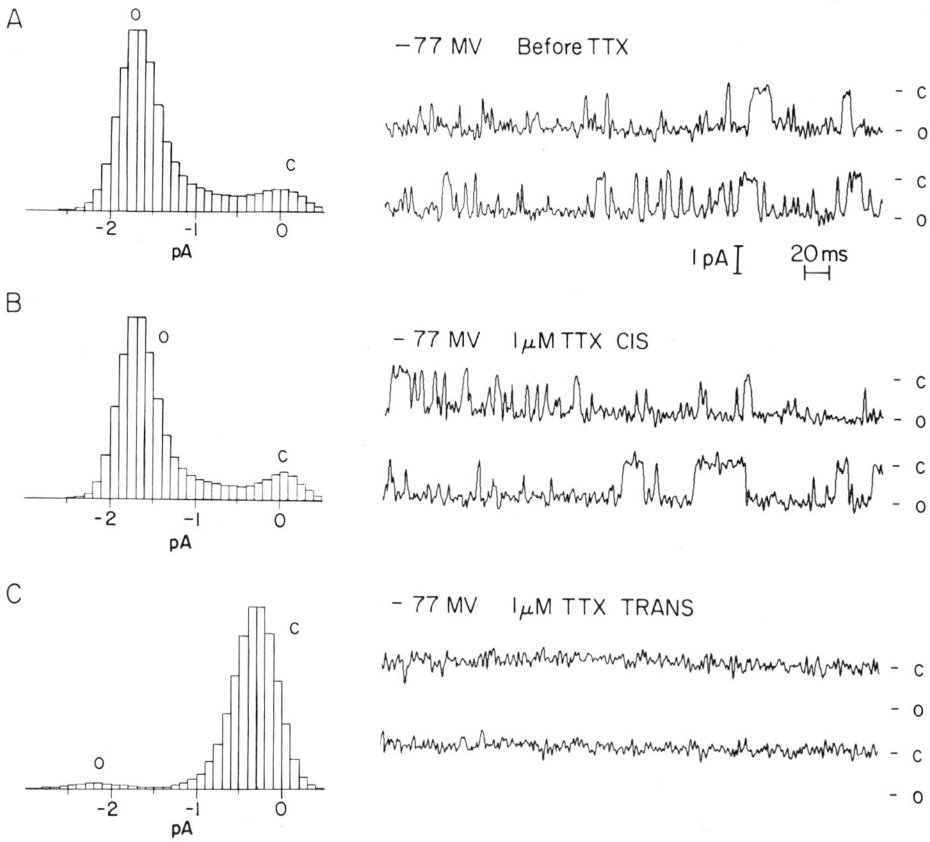

Figure 9. Reconstituted sodium channels are blocked by tetrodotoxin. Amplitude histograms and representative current records are shown for (A) a single channel activated by 1 μM batrachotoxin and held at −77 mV, (B) the same channel following addition of 1 μM tetrodotoxin to the *cis* chamber (no change is seen), and (C) following addition of 1 μM tetrotodoxin to the *trans* chamber, which blocks the channel more than 95% of the time. Recording conditions and solutions are the same as Fig. 8.

analysis is whether the channels one sees are a subset of the population or are truly representative of the reconstitution. The single-channel conductance, voltage dependence, and sensitivity to TTX seen with these purified single channels in planar bilayers closely resemble those previously reported for native sodium channels. However, channels that do not fuse with the bilayer or are not activated contribute no information in these studies. Analysis of the reconstitution requires, then, evidence of both macroscopic and microscopic channel function.

6. Summary

On the basis of reconstitution studies that allow direct measurement of cation fluxes and single-channel properties, we can state with certainty that the protein originally purified from skeletal muscle as a saxitoxin-binding component is in fact the voltage-sensitive sodium ion channel. In its purified form, this channel retains its capacity to gate cation currents in response to pharmacological activation and block, to select sodium ions among other cations of similar charge, and to respond dramatically to changes in membrane potential. Its single-channel properties are comparable to those of a native channel, and flux measurements confirm that the voltage-dependent gating seen with single channels extends to all the channels that are capable of BTX activation and STX block.

Several important aspects of the reconstitution system remain to be clarified. For example, the correlation between the specific STX-binding activity of a particular preparation and the activated flux seen after reconstitution is not good, suggesting that other as yet undefined variables are present in the reconstitution.

Reconstitution has so far provided no insight into the role of the 38-kDa component of the mammalian sodium channel, since this band was present in all our rat and rabbit purified preparations. Our reconstitution studies have also been unable to provide evidence for or against skeletal muscle sodium channel subtypes. Although recent reports, including *Tityus* toxin binding to surface but not T-tubular channels (Jaimovich *et al.*, 1983; Barhanin *et al.*, 1984b), varying TTX binding affinities in cultured muscle cells (Frelin *et al.*, 1983; Sherman *et al.*, 1983) and denervated muscle (Harris and Thesleff, 1971; Redfern and Thesleff, 1971), differences in voltage-clamp properties of TTX-resistant channels (Pappone, 1980), and differential staining in skeletal muscle immunohistochemistry using sodium-channel antibodies (B. Haimovich and R. L. Barchi, unpublished data), have provided evidence for muscle channel subtypes, clear differences have not yet been resolved in the biochemistry of the purified muscle channel or its function after reconstitution.

These and other issues will certainly be addressed in the near future. The recent elucidation of the primary sequence of the eel sodium channel (Noda *et al.*, 1984) and the active work now under way to clone both the mammalian muscle and brain channels point the way toward the next major chapter in the molecular study of this ion channel, the detailed correlation of structure with function.

ACKNOWLEDGMENTS. Work described in this chapter was supported in part by NIH grants NS-08075 and NS-18013 and by a grant from the Muscular Dystrophy Association. Dr. Furman is the recipient of an NIH Teacher Investigator Development Award. The authors would like to thank Dr. Paul Mueller for advice and assistance in our reconstitution studies and acknowledge the technical assistance of Ms. Lois Murphy and Ms. Sheri Irons and the help of Nancy Goodman in preparing this manuscript. Other investigators who contributed significantly to various aspects of the work reviewed include Dr. J. Weigele, Dr. S. Cohen, and Ms. S. Kraner.

References

Adrian, R. H., Chandler, W. K., and Hodgkin, A. L., 1970, Voltage clamp experiments in striated muscle fibers, *J. Physiol.* (*Lond.*) **208:**607–644.

Agnew, W. S., and Raftery, M. A., 1979, Solubilized tetrodotoxin binding component from the electroplax of *Electrophorus electricus*. Stability as a function of mixed lipid–detergent micelle composition, *Biochemistry* **18:**1912–1919.

Agnew, W. S., Levinson, S. R., Brabson, J. S., and Raftery, M. A., 1978, Purification of the tetrodotoxin-binding component associated with the voltage-sensitive sodium channel from *Electrophorus electricus* electroplax membranes, *Proc. Natl. Acad. Sci. U.S.A.* **75:**2606–2610.

Andreoli, T. E., Tieffenberg, M., and Tosteson, D. C., 1967, The effect of valinomycin on the ionic permeability of thin lipid membranes, *J. Gen. Physiol.* **50:**2527–2545.

Barchi, R. L., 1983, Protein components of the purified sodium channel from rat skeletal muscle sarcolemma, *J. Neurochem.* **40:**1377–1385.

Barchi, R. L., 1984, Voltage-sensitive Na^+ ion channels: Molecular properties and functional reconstitution, *Trends. Biochem. Sci.* **9:**358–361.

Barchi, R. L., and Murphy, L. E., 1980, Size characteristics of the solubilized sodium channel saxitoxin binding site from mammalian sarcolemma, *Biochim. Biophys. Acta* **597:**391–398.

Barchi, R. L., and Tanaka, J. C., 1984, Cation gating and selectivity in a purified, reconstituted, voltage-dependent sodium channel, *Biophys. J.* **45:**35–37.

Barchi, R. L., Weigele, J., Chalikian, D., and Murphy, L., 1979, Muscle surface membranes. Preparative methods affect apparent chemical properties and neurotoxin binding, *Biochim. Biophys. Acta* **550:**59–76.

Barchi, R. L., Cohen, S. A., and Murphy, L. E., 1980, Purification from rat sarcolemma of the saxitoxin-binding component of the excitable membrane sodium channel, *Proc. Natl. Acad. Sci. U.S.A.* **77:**1306–1310.

Barchi, R. L., Tanaka, J. C., and Furman, R. F., 1984, Molecular characteristics and functional reconstitution of muscle voltage-sensitive sodium channels, *J. Cell. Biochem* **26:**135–146.

Barhanin, J., Pauron, D., Lombet, A., Norman, R. I., Vijverberg, P. M., Giglio, J. R., and Lazdunski, M., 1983, Electrophysiological characterization, solubilization and purification of the *Tityus* toxin receptor associated with the gating component of the Na^+ channel from rat brain, *EMBO J.* **2:**915–920.

Barhanin, J., Schmidt, A., Lombet, A., Wheeler, K. P., Lazdunski, M., and Ellory, J. C., 1984a, Molecular size of different neurotoxin receptors on the voltage-sensitive Na^+ channel, *J. Biol. Chem.* **258:**700–702.

Barhanin, J., Ildefonse, M., Rougier, O., Sampaio, S. V., Giglio, J. R., and Lazdunski, M., 1984b, *Tityus* γ toxin, a high affinity effector of the Na^+ channel in muscle, with a selectivity for channels in the surface membrane, *Pflugers Arch.* **400:**22–27.

Beneski, D. A., and Catterall, W. A., 1980, Covalent labeling of protein components of the sodium channel with a photoactivable derivative of scorpion toxin. *Proc. Natl. Acad. Sci. U.S.A.* **77:**639–643.

Casadei, J. M., Gordon, R. D., Lampson, L. A., Shotland, D. L., and Barchi, R. L., 1984, Monoclonal antibodies against the voltage-sensitive Na^+ channel from mammalian skeletal muscle, *Proc. Natl. Acad. Sci. U.S.A.* **81:**6227–6231.

Casadei, J. M., Gordon, R. D., and Barchi, R. L., 1985, Immunoaffinity purification of the voltage dependent sodium channel from mammalian skeletal muscle, *J. Biol. Chem.* (in press).

Catterall, W. A., 1975, Activation of the action potential Na^+ ionophore of cultured neuroblastoma cells by veratridine and batrachotoxin, *J. Biol. Chem.* **250:**4053–4059.

Catterall, W. A., 1977, Activation of the action potential Na^+ ionophore by neurotoxins, *J. Biol. Chem.* **252:**8669–8676.

Catterall, W. A., 1980, Neurotoxins that act on voltage-sensitive sodium channels in excitable membranes, *Annu. Rev. Pharmacol. Toxicol.* **20:**15–43.

Fersht, A., and Jakes, R., 1975, Demonstration of two reaction pathways for the aminoacylation of tRNA. Application of the pulsed quenched flow technique, *Biochemistry* **14:**3350–3356.

Frelin, C., Vigne, P., and Lazdunski, M., 1981, The specificity of the sodium channel for monovalent cations, *Eur. J. Biochem.* **119:**437–442.

Frelin, D., Vigne, P., and Lazdunski, M., 1983, Na^+ channels with high and low affinity tetrodotoxin binding sites in mammalian skeletal muscle cells, *J. Biol. Chem.* **258:**7256–7259.

Gutfreund, H., 1969, Rapid mixing: Continuous flow, *Methods Enzymol.* **16:**229–249.

Hamill, O. P., Marty, A., Neher, E., Sakmann, B., and Sigworth, F. J., 1981, Improved patch-clamp techniques for high-resolution current recording from cells and cell free membrane patches. *Pflugers. Arch.* **391**:85–100.

Harris, J. B., and Thesleff, S., 1971, Studies on tetrodotoxin resistant action potentials in denervated skeletal muscle, *Acta Physiol. Scand.* **83**:382–388.

Hartshorne, R. P., and Catterall, W. A., 1981, Purification of the saxitoxin receptor of the sodium channel from rat brain, *Proc. Natl. Acad. Sci. U.S.A.* **78**:4620–4624.

Hartshorne, R. P., Messner, D. J., Coppersmith, J. C., and Catterall, W. A., 1982, The saxitoxin receptor of the sodium channel from rat brain. Evidence for two nonidentical β subunits, *J. Biol. Chem.* **257**:13888–13891.

Hartshorne, R., Keeler, B. U., Talvenheimo, J. A., Catterall, W. A., and Montal, M., 1985, Functional reconstitution of the purified brain sodium channel in planar lipid bilayers, *Proc. Natl. Acad. Sci. U.S.A.* **82**:240–244.

Hille, B., 1972, The permeability of the sodium channel to metal catins in myelinated nerve, *J. Gen. Physiol.* **59**:637–658.

Hille, B., 1984, *Ionic Channels of Excitable Membranes*, Sinauer, Sunderland, MA.

Hjelemeland, L. M., Nebert, D. W., and Osborne, J. C., Jr., 1983, Sulfobetaine derivatives of bile acids: Nondenaturing surfactants for membrane biochemistry, *Anal. Biochem.* **130**:72–82.

Holloway, F. W., 1973, A simple procedure for removal of Triton X-100 from protein samples, *Anal. Biochem* **53**:301–308.

Horn, R., Patlak, J., and Stevens, C. F., 1981, Sodium channels need not open before they inactivate, *Nature* **291**:426–427.

Huang, L. M., Catterall, W. A., and Ehrenstein, G., 1979, Comparison of ionic selectivity of batrachotoxin-activated channels with different tetrodotoxin dissociation constants, *J. Gen. Physiol.* **73**:839–854.

Huang, L. M., Moran, N., and Ehrenstein, G., 1982, Batrachotoxin modifies the gating kinetics of sodium channels in internally perfused neuroblastoma cells, *Proc. Natl. Acad. Sci. U.S.A.* **79**:2082–2085.

Jaimovich, E., Chicheportiche, R., Lombet, A., Lazdunski, M., Ildefonse, M., and Rougier, O., 1983, Differences in the properties of Na^+ channels in muscle surface and t-tubular membranes revealed by tetrodotoxin derivatives, *Pflugers. Arch.* **397**:1–5.

Khodorov, B. I., 1978, Chemicals as tools to study nerve fiber sodium channels; effects of batrachotoxin and some local anesthetics, in: *Membrane Transport Processes*, Vol. 2 (D. C. Tosteson, A. Y. Ovchennikov, and R. Latorre, eds.), pp. 153–174. Raven Press, New York.

Kraner, S. D., Tanaka, J. C., Matesic, D. R., and Barchi, R. L., 1985, Purification and functional reconstitution of the voltage-sensitive sodium channel from rabbit T-tubular membranes, *J. Biol. Chem.* **260**:6341–6347.

Krueger, B. K., Worley, J. F., and French, R. J., 1983, Single sodium channels from rat brain incorporated into planar lipid bilayer membranes, *Nature* **303**:172–175.

Leibowitz, M. D., Sutro, J. B., and Hille, B., 1985, Four lipid-soluble toxins modify sodium channel gating, *Biophys. J.* **47**:32a.

Miller, C., and Racker, E., 1979, Reconstitution of membrane transport functions, in: *The Receptors*, Vol. 1 (R. D. O'Brien, ed.), p. 16, Plenum Press, New York.

Miller, J. A., Agnew, W. S., and Levinson, S. R., 1983, Principal glycopeptide of the tetrodotoxin saxitoxin binding protein from *Electrophorus electricus*: Isolation and partial chemical and physical characterization, *Biochemistry* **22**:462–470.

Mueller, P., and Rudin, D. O., 1967, Development of K^+–Na^+ discrimination in experimental bimolecular lipid membranes by macrocyclic antibiotics, *Biochem. Biophys. Res. Commun.* **26**:398–404.

Narahashi, T., 1974, Chemicals as tools in the study of excitable membranes, *Physiol. Rev.* **54**:814–889.

Noda, M., Shimizu, S., Tsutomu, T, Takai, T., Kayano, T., Ikeda, T., Takahashi, H., Nakayama, H., Kanaoka, U., Minamino, N., Kangawa, K., Matsuo, H., Raftery, M. A., Hirose, T., Inayama, S., Hayashida, H., Miyata, T., and Numa, S., 1984, Primary structure of *Electrophorus electricus* sodium channel deduced from cDNA sequence, *Nature* **312**:121–127.

Norman, R. I., Schmidt, A., Lombet, A., Barhanin, J., and Lazdunski, M., 1983, Purification of binding protein from *Tityus* toxin identified with the gating component of the voltage-sensitive Na^+ channel, *Proc. Natl. Acad. Sci. U.S.A.* **80:**4164–4168.

Oku, N., Kendall, D. A., and Macdonald, R. C., 1982, Measurement of trapped space by calcein quenching and comparison with gramicidin measured spaces, *Biochim. Biophys. Acta* **691:**332–340.

Pappone, P., 1980, Voltage-clamp experiments in normal and denervated mammalian skeletal muscle fibers, *J. Physiol.(Lond.)* **306:**377–410.

Quandt, F. N., and Narahashi, T., 1982, Modification of single Na^+ channels by batrachotoxin, *Proc. Natl. Acad. Sci. U.S.A.* **79:**6732–6736.

Redfern, P., and Thesleff, S., 1971, Action potential generations in denervated rat skeletal muscle, *Acta Physiol. Scand.* **81:**557–564.

Rhoden, V., and Goldin, S. M., 1979, Formation of unilamellar lipid vesicles of controllable dimensions by detergent dialysis, *Biochemistry* **18:**4172–4176.

Rosemblatt, M., Hidalgo, C., Vergara, C., and Ikemoto, N., 1981, Immunological and biochemical properties of transverse tubule membranes isolated from rabbit skeletal muscle, *J. Biol. Chem.* **256:**8140–8148.

Rosenberg, R. L., Tomiko, S. A., and Agnew, W. S., 1984, Reconstitution of neurotoxin-modulated ion transport by the voltage-regulated sodium channel isolated from the electroplax of *Electrophorus electricus*, *Proc. Natl. Acad. Sci. U.S.A.* **81:**1239–1243.

Schauf, C. L., 1973, Temperature dependence of the ionic current kinetics of *Myxicola* giant axons, *J. Physiol. (Lond.)* **235:**197–205.

Sharkey, R. S., Beneski, D. A., and Catterall, W. A., 1983, Specific labeling of the α and β, subunits of the sodium channel by photoreactive derivatives of scorpion toxin, *Biochemistry* **23:**6078–6086.

Sherman, S. J., Lawrence, J. C., Messner, D. J., Jacoby, K., and Catterall, W. A., 1983, Tetrodotoxin-sensitive sodium channels in rat muscle cells developing *in vitro*, *J. Biol. Chem.* **258:**2488–2495.

Sigworth, F. J., and Neher, E., 1980, Single Na^+ channel currents observed in cultured rat muscle cells, *Nature* **287:**447–449.

Tanaka, J. C., Eccleston, J. F., and Barchi, R. L., 1983, Cation selectivity characteristics of the reconstituted voltage-dependent sodium channel purified from rat skeletal muscle sarcolemma, *J. Biol. Chem.* **258:**7519–7526.

Tank, D. W., Miller, C., and Webb, W. W., 1982, Isolated-patch recording from liposomes containing functionally reconstituted chloride channels from *Torpedo electroplax, Proc. Natl. Acad. Sci. U.S.A.* **79:**7749–7753.

Ulbricht, W., 1969, The effect of veratridine on excitable membranes of nerve and muscle, *Ergeb. Physiol. Biol. Chem. Exp. Pharmacol.* **61:**18–71.

Weigele, J. B., and Barchi, R. L., 1982, Functional reconstitution of the purified sodium channel protein from rat sarcolemma, *Proc. Natl. Acad. Sci. U.S.A.* **79:**3651–3655.

Yoshi, M., Scruggs, V., and Narahashi, T., 1984, Effect of veratridine on single sodium channel currents, *Soc. Neurosci. Abstr.* **9:**674.

Chapter 12

RECONSTITUTION OF THE SODIUM CHANNEL FROM *ELECTROPHORUS ELECTRICUS*

W. S. AGNEW, R. L. ROSENBERG, and S. A. TOMIKO

1. Introduction

The sodium (Na^+) channel that transiently depolarizes nerve and muscle membranes in the initial phase of the action potential (Hodgkin and Huxley, 1952; Cahalan, 1980) is perhaps the archetypical voltage-gated channel. Advances in the biochemical isolation, characterization, and functional reconstitution of Na^+-channel proteins from the electroplax of electric fish (Miller *et al.*, 1983; Norman *et al.*, 1983; Rosenberg *et al.*, 1984a,b), mammalian skeletal muscle (Barchi, 1983; Weigele and Barchi, 1982; Tanaka *et al.*, 1983), and mammalian brain (Hartshorne and Catterall, 1984; Talvenheimo *et al.*, 1982; Tamkun *et al.*, 1984) have recently been made (for review, see Agnew, 1984). In addition, the cDNA for the principal peptide component of the electroplax channel has been cloned and sequenced, providing the first extensive information about the primary structure and allowing deductions of the possible secondary and tertiary structure of the protein (Noda *et al.*, 1984). These biochemical, biophysical, and molecular biological approaches, including specific protein chemical modifications, site-specific mutagenesis, electron microscopic structural analysis, and functional reconstitution will provide important insights into the structures and mechanisms of the channel. This chapter concentrates on reconstitution studies with the purified electroplax Na^+ channel.

The tasks attending reconstitution include the isolation of the protein in a well-preserved state and its subsequent reincorporation into a satisfactory membrane bilayer. In addition, however, it is essential to be able to measure conveniently and precisely the various functional properties of the protein, including ion transport, interaction with various drugs, neurotoxins, and anesthetics, and, of course, the changes in conformational state in response to rapid alteration of membrane potential. Unlike enzymes that are relatively slow, channels pass millions of ions per second. The Na^+ channel normally responds not merely to

W. S. AGNEW, R. L. ROSENBERG and S. A. TOMIKO • Department of Physiology, Yale University School of Medicine, New Haven, Connecticut 06510.

membrane potential but to rapid changes in membrane potential, and the activity cycle of the channel is completed within a few milliseconds. Thus, rather novel techniques will be required to create the signals that are detected and propagated by these channels and to follow channel activity.

In this chapter we describe (1) the isolation of the Na^+ channel protein from electroplax in a form suitable for functional reconstitution; (2) procedures for incorporating the protein into membrane vesicles; (3) the assay of channel-mediated ion transport by radiotracer flux measurements; (4) a rapid, extremely sensitive spectrofluorometric assay of transport; and (5) patch-clamp techniques used to record purely voltage-regulated single-channel activity. It should be evident that a great deal remains to be learned about optimizing conditions for functional reconstitution and for fully exploiting these preparations.

2. Purification of the Tetrodotoxin-Binding Protein

The most convenient biochemical assay for the Na^+ channel in membrane fragments or detergent-solubilized preparations is specific binding of [^{3}H]tetrodotoxin (TTX) or [^{3}H]-saxitoxin (STX) derivatives (Agnew *et al.*, 1978; Levinson *et al.*, 1979). These toxins are stoichiometric, completely selective, high-affinity (K_d ~10^{-8}–10^{-9} M) markers for the channel, and binding of these compounds is not lost on solubilization with detergent.

The TTX-binding protein from electroplax has been purified to homogeneity (Miller *et al.*, 1983) and identified as a large, heavily glycosylated polypeptide of M_r ~270,000. In comparison, proteins from mammalian skeletal muscle (Barchi, 1983) and brain (Hartshorne and Catterall, 1984) contain a large peptide together with one or two smaller species of 37,000–39,000 daltons. The procedures for purification of the TTX-binding protein, which were described by Miller *et al.* (1983), have been modified at several steps for reconstitution studies. For routine studies, we have used material subjected to only two steps of fractionation, which is ~80% of homogeneity.

2.1. Methods

2.1.1. The Electric Organ

The electric organs of *Electrophorus electricus* represent perhaps the richest preparative source of the Na^+ channel. The main organs of a 1-m animal weigh 0.5–1.0 kg and are composed of a single type of cell, the electroplax, together with myelinated axons from cholinergic electromotor neurons (Keynes and Martins-Ferriera, 1953). Only the caudal, innervated face is excitable. Approximately 1–2% of the surface is subsynaptic, containing the acetylcholine receptor. The remainder of this face contains Na^+ channels at an estimated surface density of ~500 μm^{-2} (Levinson, 1975). This suggests that ~10% of the intrinsic membrane protein is Na^+ channel. One 300-g organ contains 15–18 nmol (4.5–5.4 mg) of sodium channel protein, in comparison with 12–14 nmol from 60 rat brains (Hartshorne and Catterall, 1984), or 0.6 nmol from 300 g of rat skeletal muscle (calculated from Barchi, 1983). The dissected organs can be stored frozen at −70°C for several months with no detectable loss of neurotoxin-binding activity.

2.1.2. Preparation of Starting Membranes

Starting membranes, washed free of soluble proteins, are prepared in two steps of homogenization and centrifugation, as described in detail by Miller *et al.* (1983). These can be used immediately or may be stored frozen at −70°C for several months without losing activity.

The "crudeness" of these membranes should be underscored: the final homogenate contains large amounts of extracellular connective material, primarily collagen and acid mucopolysaccharides. Only about 7–15 mg of membrane phospholipid is present per gram of packed pellet (Agnew and Raftery, 1979). A brief, low-speed centrifugation of the first homogenate can remove much of this material, but large amounts of the membranes are also removed. The connective tissue does not interfere with solubilization and is easily removed by centrifugation after addition of detergent. However, the crude membranes are not "microsomes" readily suited to binding, transport studies, or fusion with planar lipid bilayers.

2.1.3. Solubilization

All nonionic detergents tested, including Triton X-100, Nonidet P-40, Brij-56, and Lubrol PX, are effective in solubilizing TTX-binding sites from the crude membranes. Lubrol PX yields the most efficient solubilization (~90%), and the extracts are comparatively stable (Moore *et al.*, 1982). Cholate and deoxycholate are less effective in solubilizing the electroplax membranes, and the extracts are less stable (Agnew *et al.*, 1983). However, because these detergents form small micelles, resolution during gel filtration is greatly enhanced. In recent studies, CHAPS {3-[(3-cholamidopropyl)dimethylammonio]-1-propane sulfonate, a zwitterionic detergent derived from cholate; Hjelmeland, 1980}, was found to be very effective, with solubilization efficiencies and stability characteristics similar to Lubrol PX. Fractionation may be as favorable as with cholate. Furthermore, CHAPS can be removed from a solution by dialysis or gel filtration, facilitating subsequent reconstitution steps. It should be emphasized that the protein solubilized with all detergents must be stabilized with low molar ratios of membrane phospholipid (Agnew and Raftery, 1979). Generally, during fractionation, egg phosphatidylcholine (PC) at a ratio of 1 mol per 7 mol of detergent (never less than 1 per 15) is used.

Starting membrane suspensions are solubilized at 0°C in 1% (w/v) Lubrol PX in the presence of a mixture of protease inhibitors (50 μM *o*-phenanthroline, 50 μM 1-tosylamido-2-phenylethylchloromethylketone, 100 μM phenylmethylsulfonylfluoride, 1 μg/ml pepstatin A). An average preparation begins with 80–100 ml of membrane suspension (2 ml buffer/g packed membranes). The suspension containing detergent is stirred for 10–30 min; we have found that vigorous homogenization does not improve solubilization efficiency.

The suspension is centrifuged (100,000 × g × 1 hr), and the supernatant is carefully removed from the pellet. This extract typically contains 150 pmol TTX-binding sites per milliliter and 5–7 mg protein per milliliter. Solid KC1 is added to bring the final concentration to 200 mM.

2.1.4. Fractionation

Purification essentially involves two stages, one of ion-exchange chromatography and one of gel-exclusion chromatography. Throughout, care is taken to keep the temperature close to 0°C, the protease inhibitors listed above are included, and the lipid/detergent ratio is maintained at 1/7.

The ion-exchange step capitalizes on the acidic nature of the protein (Agnew *et al.*, 1978; Miller *et al.*, 1983), resulting in avid binding to the anion-exchange resin diethylaminoethyl (DEAE) Sephadex A-25, and the presence in the extract of an acid mucopolysaccharide material, which binds to the resin irreversibly, preventing uptake of nearly all other protein species. Before use, the resin is washed in 0.5 N NaOH, 0.5 N HC1, equilibrated in 0.5 M Na phosphate (pH 7.4), and then in 50 mM Na phosphate (pH 7.4) as recommended by the manufacturers. It is then equilibrated in buffer containing 0.1% (w/v) Lubrol PX, 0.183 mg/ml phosphatidylcholine (PC), 200 mM KC1, 1 mM EGTA, 50 mM Na phosphate (pH 7.4), and the protease inhibitors described above. Moist DEAE resin is added to the detergent extract (1.1 g resin/ml), stirred for 30 min at 0°C, and then washed with 5 volumes of the same buffer in a suction filtration funnel. This results in the adsorption of more than 90% of the TTX-binding sites but less than 2% of the total protein. The washed resin is loaded into a 2.6 × 30 cm column, and the TTX-binding protein is eluted stepwise in the same buffer containing 600 mM KC1. This step desorption serves to elute the protein in a minimal volume. The TTX-binding activity appears at an elution volume of 0.67 of the bed volume. Fractions containing >80 pmol TTX binding sites per milliliter are usually combined; this pool typically contains 5–7 nmol of sites (45–55% of initial) at 400 pmol/ml with specific activity of 350–600 pmol/mg protein. The DEAE resin may not be reused even after acid and base recycling because of nearly irreversible adsorption of the acid mucopolysaccharides.

The second step of purification capitalizes on the large Stokes radius of the protein (~90 Å; Agnew *et al.*, 1978; W. S. Agnew, H. F. Rudnick, and M. C. Emerick, unpublished data). This is larger even than the acetylcholine receptor (Karlin, 1980) and enables an efficient fractionation by gel filtration. Unlike the procedures described by Agnew *et al.* (1978) or Miller *et al.* (1983), TTX is not used to stabilize the protein. A Sepharose 6B column of 1.6 × 90 cm is used. Before delivery to the column, the DEAE desorption pool is concentrated to <2 ml by ultrafiltration in a 43-mm Amicon pressure cell equipped with an XM-50 membrane. The Sepharose column is equilibrated and eluted with 100 mM Na phosphate buffer, pH 7.4, containing 0.1% Lubrol PX, 0.183 mg/ml PC, 1 mM EGTA, 0.02% Na azide, and protease inhibitors. In part to compensate for the

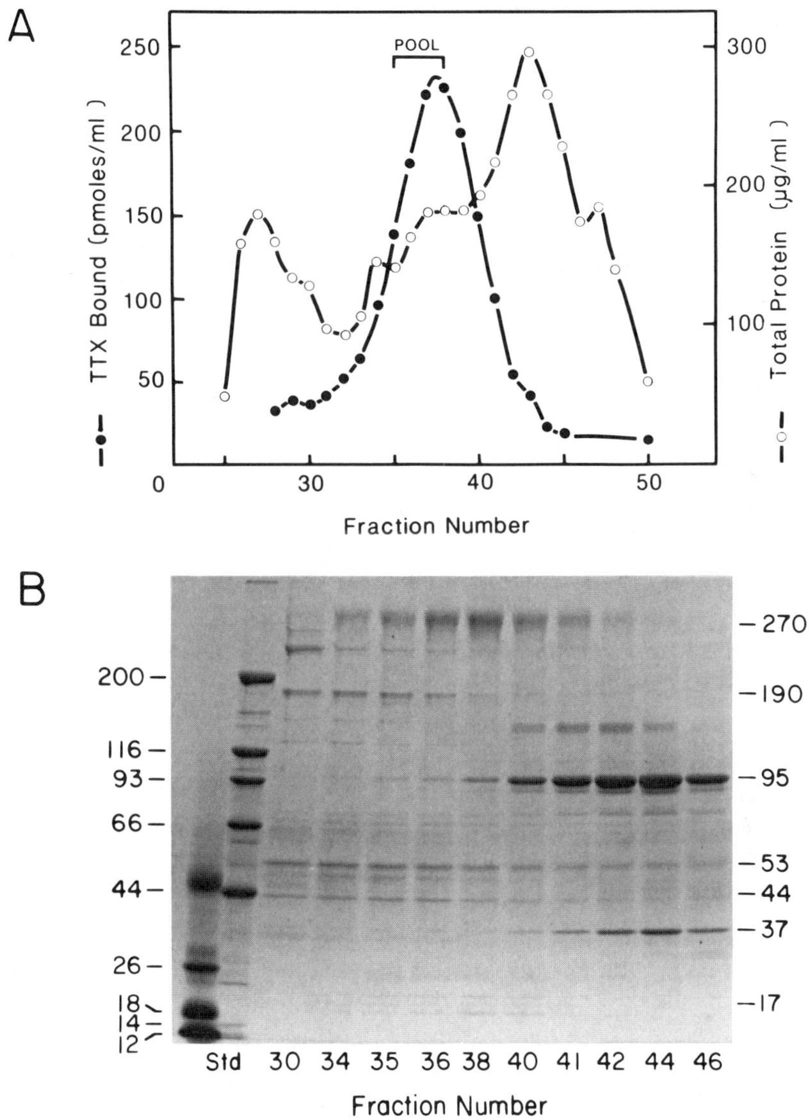

Figure 1. Sepharose 6B column elution profile. A: (●) Tetrodotoxin-binding activity and (○) protein were determined as described in Rosenberg *et al.* (1984a). Fraction size was 2.5 ml. B: Sodium dodecylsulfate polyacrylamide gel of fractions from the Sepharose 6B column. Acrylamide concentration varied linearly from 5% to 12%. Each lane contained 10–30 µg protein, and peptides were stained by Coomassie Brilliant Blue. Standard proteins were (in daltons): cytochrome *c* (12,300), lysozyme (14,300), β-lactoglobulin (18,400), α-chymotrypsinogen (25,700), ovalbumin (44,000), bovine serum albumin (66,200), phosphorylase *b* (92,500), β-galactosidase (116,250), and myosin (200,000). Molecular weights of the peptides in the experimental samples were determined from a plot of log M_r versus relative migration for the standard peptides.

omission of stabilizing TTX, the column and fractions are maintained at 0–1°C throughout the run. Fractions are assayed for protein, [^{3}H]-TTX binding, and polypeptide composition by SDS-PAGE.

2.2. The Protein

The results of a typical fractionation on the Sepharose column are shown in Fig. 1. In part A, the elution of the TTX-binding protein and of total protein is shown. The TTX-binding protein emerges as a single peak ~0.55 × bed volume and appears as a shoulder on the profile of total protein. The "leading edge" fractions containing the highest specific activity of TTX binding are pooled for further study. A typical pool contains 200 pmol TTX-binding sites per milliliter at a specific activity of 1100 pmol/mg protein, and the recovery is ~30%. These are 40–50% of the yields and specific activities obtained when TTX is present for stabilization (Miller *et al.*, 1983).

The peptide compositions of the fractions from the Sepharose column elution are illustrated by the SDS-acrylamide gradient gels shown in Fig. 1B. The elution of the large peptide of M_r ~270,000 peaks at fraction 38, corresponding to the peak of TTX-binding activity. Smaller peptides are present in variable quantities in all fractions analyzed. Peptides of 20,000–40,000 daltons are not found in the peak fractions, although smaller species of 15,000–20,000 daltons are sometimes observed. Densitometer scans of the gels show that approximately 80% of the total protein in pooled fractions is accounted for by the 270,000-dalton peptide.

3. Reconstititution Strategies and Methods

3.1. General Considerations

Detergent solubilization represents disassembly of the membrane and the dispersal of lipid and protein constituents in an extensive micellar "phase." Reconstitution is an attempt to reverse the solubilization. By selectively removing the detergent, the micellar phase is reduced, and membrane constituents are increasingly likely to share the same mixed micelle. As detergent is fully stripped away, membrane vesicles spontaneously reform, with the protein incorporated. These vesicles may then be manipulated as necessary to alter membrane geometry for a particular experimental protocol. In the case of the Na^+ channel, many of the expected pharmacological and biophysical properties of the channel survive these manipulations.

3.2. Reconstitution Methods

In studies that preceded our investigations, Weigele and Barchi (1982) and Talvenheimo *et al.* (1982) successfully used a polystyrene resin (Bio-Beads® SM-

2; Holloway, 1973) to remove Nonidet P-40 and Triton X-100 from STX-binding proteins purified from mammalian muscle and brain membranes. In their procedures, the purified proteins were supplemented with detergent-solubilized lipid before detergent removal. In an earlier study of the D-glucose transporter from erythrocyte membranes, Kasahara and Hinkle (1977) also used Bio-Beads® to remove detergent. These investigators adsorbed detergent in the absence of exogenous lipid and then reinserted the aggregated, delipidated protein into preformed sonicated liposomes in a freeze–thaw–sonication (FTS) cycle. Also, Villegas *et al.* (1977) used a FTS cycle to incorporate unsolubilized Na^+ channels into sonicated liposomes.

In studies of the purified electroplax protein (Agnew *et al.*, 1983; Rosenberg *et al.*, 1984a,b), we found that procedures similar to those of Weigele and Barchi (1982) and Talvenheimo *et al.* (1982) could be used. However, we obtained larger and more consistent flux signals with a modified procedure including the FTS described by Kasahara and Hinkle (1977) and Villegas *et al.* (1977).

The reconstitution scheme is illustrated in Fig. 2. The purified protein, in detergent suspension with the small amount of lipid used for stabilization, is stripped of detergent by addition of Bio-Beads®. This leaves the protein in association with lipid, probably in small vesicles, as reflected in enhanced stability at elevated temperature (Agnew *et al.*, 1983; Rosenberg *et al.*, 1984a). The material is then supplemented with exogenous pure lipid in the form of sonicated liposomes. Freezing and thawing produces multilamellar liposomes (Tank and Miller, 1983; Pick, 1981) containing the incorporated protein. As described in Sections 4 and 6, these may be used for patch-recording experiments, or may be briefly sonicated to produce intermediate sized, mostly unilamellar vesicles suitable for radiotracer or fluorescence transport assays. Because we were particularly interested in avoiding conductance artifacts in patch-recording experiments, we avoided adding unnecessary detergent at any stage. The procedure has the advantage that toxins, drugs, or solutes can be introduced to the vesicle interior during the freeze–thaw cycle.

Specifically, the solubilized, purified TTX-binding protein (~150 μg protein in 0.1% Lubrol PX, 0.183 mg/ml PC) is mixed with 0.3 g of moist, washed Bio-Beads® (Holloway, 1973) per milliliter of solution. No lipid or detergent is added at this stage. The suspension is agitated for 5 hr at 4°C, and the beads are removed by filtering through a small plug of polyester floss. To increase the number and size of lipid vesicles, this preparation is mixed with sonicated liposomes of pure lipid (10 mg/ml final) and subjected to a freeze–thaw–sonication cycle (see Section 3.2.2.).

As an alternative to adding the liposomes after removing detergent, they may be added to the soluble preparation before the resin is added. As before, no further additions of detergent are made. The small amount of detergent present (0.1%) is apparently sufficient to disrupt but not fully solubilize the added lipid (i.e., 10 mg/ml). Bio-Beads® (0.3 g/ml) are then added to remove the detergent, and the suspension is agitated as before. The resulting reconstituted vesicles are then enlarged in a FTS cycle as before.

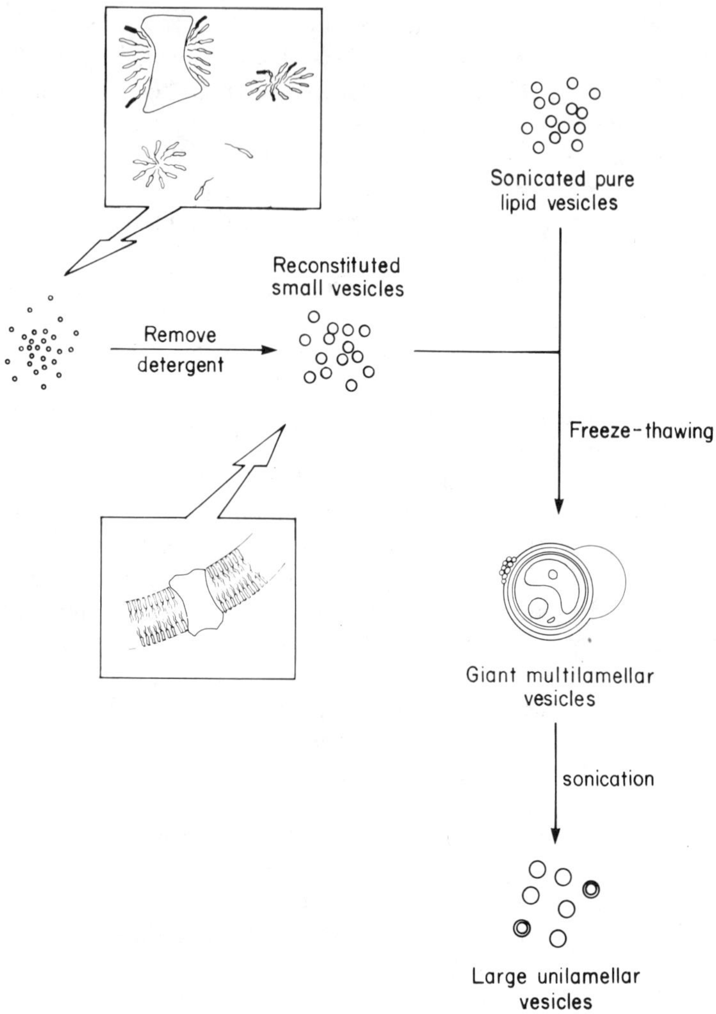

Figure 2. Scheme for Na^+ channel reconstitution.

3.2.1. Lipids

Our initial observations indicate that a mixture of pure lipids containing phosphatidylethanolamine (PE), phosphatidylserine (PS), and phosphatidylcholine (PC) is especially suitable for both patch-recording experiments and flux assays. A mixture of PE : PS : PC (5 : 4 : 1 molar ratio) yields large multilamellar vesicles in the freeze–thaw step with smooth surfaces suitable for forming high-resistance seals with patch electrodes. When sonicated briefly, the resulting vesicles give large signal-to-background ratios in the flux assays, probably because of their high volume-to-surface ratio. Pure PC was successfully used in some early radiotracer flux experiments, but signals were smaller than those with the lipid mixture. Furthermore, freezing and thawing PC produces clusters of aggregated smaller vesicles, which cannot be used for patch-recording. Asolectin, unfractionated lipids from soybean, formed large vesicles after freeze–thawing (Tank and Miller, 1983) but resulted in the loss of TTX-binding activity as well as poor ion-flux signals.

In most of our experiments we have used 10 mg of sonicated lipid mixture per milliliter, added either before or after the Bio-Bead® treatment. The concentration may be reduced to 0.2 mg/ml in efforts to increase the protein-to-lipid ratio, with no apparent negative effects on vesicle fusion in the freeze–thaw step or final vesicle morphology (M. C. Emerick, unpublished data). Care is taken not to expose the protein to solubilized lipid containing more than 50 mol% phosphatidylserine, as severe loss of activity may result (Agnew and Raftery, 1979).

3.2.2. Freeze–Thaw–Sonication

The FTS steps vary depending on the sample size or experiment to be performed. For tracer flux studies in which relatively large volumes (>1 ml) are prepared, the Bio-Bead®-treated protein and lipid suspension is frozen in a Pyrex® tube by immersion in dry-ice–acetone for 1 min, thawed at room temperature for 5–10 min, and sonicated for 10 sec. For smaller volumes (0.1–0.3 ml), the sonication is abbreviated to 2–4 sec. Small volumes (~0.1 ml) of freeze–thaw vesicles for patch-clamp experiments are prepared by placing the suspension (in a small Pyrex® tube) in contact with solid dry ice for 1 min and thawing at room temperature for 1 min.

The FT and FTS methods produce heterogeneous vesicle populations. After extensive sonication, the liposomes are fairly uniform, about 300–500 Å in diameter (by gel filtration). After the freeze–thaw cycle, the particles are larger than 6000 Å in diameter, with many particles 10–50 μm observed under the microscope. After brief sonication, the average vesicle appears to be intermediate between these extremes; some of these may be multilamellar (W. E. Bollinger, unpublished data). Vesicle size and heterogeneity are important variables for optimizing reconstitution procedures. In flux assays, the signal and kinetics will be greatly affected by the size and heterogeneity. Large vesicles will have greater probabilities of having active channels, but vesicles with larger volumes and smaller surface-to-volume ratios will exhibit slower equilibration kinetics. As yet, we cannot readily vary the channel surface density or control the vectorial orientation of the proteins.

3.2.3. Aqueous Environment

Ionic and osmotic compositions of the internal and extravesicular solutions are important experimental variables. We find that sulfate is a better counterion than phosphate for studies involving either radiotracer or fluorometric assays. The soluble preparation is routinely dialyzed against 84 mM Na_2SO_4, 1 mM EGTA, 10 mM HEPES–NaOH, pH 7.4, before reconstitution. The osmotic strength of all solutions that contact the vesicles is adjusted to within 2% of the internal osmolarity.

4. Channel-Mediated Radiotracer Influx

The first test for channel reconstitution is a qualitative determination of whether essential functional mechanisms are present, with little detailed information as to how "healthy" the protein may be. Radiotracer flux assays can demonstrate ion transport, presumably reflecting presence of the channel itself. There is an extensive arsenal of drugs, neurotoxins, and anesthetic compounds that act on Na^+ channels (Catterall, 1980). Toxins such as veratridine, batrachotoxin (BTX), grayanotoxin, aconitine, and others greatly perturb the channel, causing loss of inactivation, shifting the voltage dependence of activation, modifying activation kinetics, and altering channel conductance, ion selectivity, and interactions with other neurotoxins. These may be used to chronically activate the channel. Tetrodotoxin, STX, and local anesthetics are effective at blocking Na^+ currents through toxin-activated channels. Hence, with these and other compounds, we can manipulate the channel, permitting a biochemical/pharmacological screening of the preparations.

4.1. The $^{22}Na^+$ Influx Assay

The assay employs Dowex 50, a strong cation-exchange resin, to separate internalized radiotracer Na^+ from external ions (Gasko *et al.*, 1976; Weigele and Barchi, 1982; Talvenheimo *et al.*, 1982). Neurotoxin-treated or control vesicles are diluted into a buffer containing $^{22}NaC1$, incubated for specified times, and quickly percolated through small columns of Dowex 50 in the Tris form [Prerinsed with isotonic sucrose containing 3.3 mg/ml bovine serum albumin (BSA); Gasko *et al.*, 1976]. The Dowex binds all the external $^{22}Na^+$, but internalized tracer is eluted along with the vesicles and quantified by liquid scintillation counting. Because the columns exchange permeant for impermeant cations, loss of internalized $^{22}Na^+$ during the elution through the column is retarded by a large inside-negative diffusion potential, which develops after efflux of only a few cations per vesicle.

Vesicles are preincubated with or without TTX or local anesthetics 15 min at 0°C. They are then incubated at 30°C with veratridine or veratridine vehicle for 5 min or with BTX or BTX vehicle for 45 min. The $^{22}Na^+$ influx is initiated by 1 : 4 dilution of the prepared sample into isotonic Tris sulfate (175 mM, pH 7.4) containing the same concentration of toxins and ~10 μCi/ml $^{22}NaC1$. The sample is maintained at 30°C for defined intervals. To stop influx, 0.2 ml of the sample is applied to a Dowex column and eluted with 2 ml of isotonic 220 mM sucrose plus 1 mg/ml BSA, 0°C.

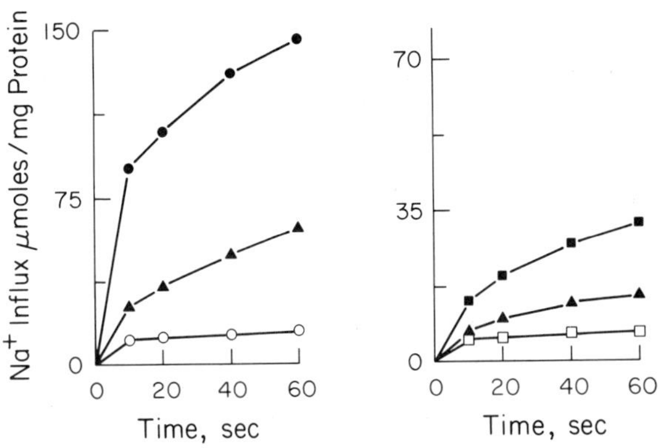

Figure 3. Neurotoxin-stimulated $^{22}Na^+$ uptake. Samples were preequilibrated in 84 mM Na_2SO_4, 1 mM EGTA, and 10 mM HEPES–NaOH, pH 7.4. Left panel: Vesicles were treated with (●) 5 μM BTX in ethanol (1% final, v/v) or (▲) 5 μM BTX + 1 μM TTX external, or (○) ethanol vehicle alone (1% final, v/v), and influx was measured as described in the text ($n = 1$). Right panel: Vesicles were treated with (■) 100 μM veratridine or (▲) 100 μM veratridine + 1 μM TTX external or (□) the veratridine vehicle alone (170 mM Tris-SO_4, pH 7.4), and influx was measured ($n = 1$).

4.2. Results

4.2.1. Alkaloid Neurotoxins Including Batrachotoxin and Veratridine Stimulate $^{22}Na^+$ Influx

Several important Na^+ channel functions are immediately apparent from experiments such as those shown in Fig. 3. The influx of radiotracer Na^+ was stimulated tenfold by 5 μM BTX and fivefold by 100 μM veratridine. Half-maximal stimulation by veratridine occurred at a concentration of 18 μM (Rosenberg *et al.*, 1984a), in agreement with values from studies of intact electroplax (Bartels and Rosenberry, 1973) as well as other reconstituted preparations (Weigele and Barchi, 1982; Talvenheimo *et al.*, 1982). In fluorescence-quench experiments (see Section 5), the $K_{\frac{1}{2}}$ for BTX was 169nM ± 44(SD), also in the normal pharmacological range.

4.2.2. Conductance Blockade by TTX and Local Anesthetics

Tetrodotoxin added to the vesicle exterior blocked approximately 65% of the neurotoxin-activated influx (Fig. 3). This partial blockade may be explained by the randomized vectorial orientation of the reconstituted channel; external TTX can bind only to channels facing outside-out. Consistent with this, when TTX was equilibrated with the vesicle interior during the disruptive FTS step, influx was blocked completely (Rosenberg *et al.*, 1984a). Also, direct [^{3}H]-TTX-binding experiments have indicated that 60–70% of the sites are accessible from the vesicle exterior.

Lipid-soluble local anesthetics, including tetracaine and dibucaine, blocked influx completely when added to the outside of reconstituted vesicles (Rosenberg *et al.*, 1984a). QX-222, a positively charged derivative of lidocaine that is poorly permeable, blocked only a fraction of the signal, presumably that residing in inside-out-oriented channels (Rosenberg *et al.*, 1984a). The effects of external TTX and QX-222 were approximately additive.

Parallel measurements of [^{3}H]-TTX binding and blockade of $^{22}Na^+$ influx (Rosenberg *et al.*, 1984a) indicated that binding correlated closely with inhibition. The dissociation constant for TTX reported was 33 nM. (More recent studies based on Tl^+ transport inhibition have yielded dissociation constants of 4.3 nM ± 2.1(SE), although some variation from preparation to preparation is seen; Tomiko *et al.*, 1986.)

Thus, these experiments reveal the presence of the ion-channel and receptor sites for the alkaloid neurotoxins, the guanidinium channel blockers, and local anesthetics. Furthermore, the assay can be used to measure the ion selectivity of the channel and may be adapted to quench-flow apparatus with time resolutions of less than 100 msec (Tanaka *et al.*, 1983).

4.2.3. Use of Diffusion Potentials to Amplify Transport

It is possible to increase radiotracer influx markedly through the use of asymmetric permeant ion concentration gradients as described by Garty *et al.* (1983). Before reconstitution, the sample is dialyzed against 80–100 mM Na_2SO_4. After forming the vesicles, the samples are equilibrated with the relevant neurotoxins. Instead of being diluted into buffer containing $^{22}Na^+$, the samples are eluted through small columns of Dowex 50 equilibrated in 175 mM Tris sulfate. This removes all external Na^+, replacing it with impermeant Tris ions. The efflux of Na^+ through activated Na^+ channels causes a retarding diffusion potential to develop in those vesicles. When the vesicles are added to buffer containing tracer $^{22}Na^+$, there is still little net efflux but relatively free exchange of high-specific-activity external Na^+ ions for the unlabeled ions within the vesicles.

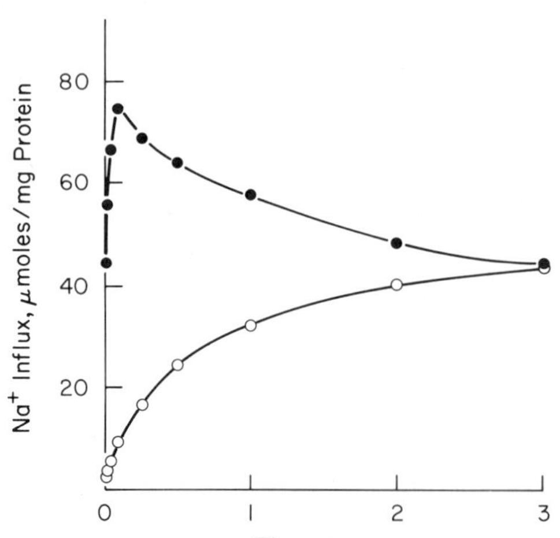

Figure 4. Batrachotoxin-stimulated $^{22}Na^+$ influx amplified by diffusion potential. Vesicles prepared in Na_2SO_4 solution as in Fig. 3 were treated with (●) 5 μM BTX or (○) ethanol vehicle alone. Each sample was applied to a Dowex 50 column and rinsed with isotonic Tris-SO_4 to exchange $Tris^+$ for external Na^+. Samples were supplemented with BTX (to 5 μM) or vehicle and with Na_2SO_4 to 8.4 mM final ($[Na^+]_{in}/[Na^+]_{out} \sim 10$). Assay was initiated by addition of $^{22}Na^+$ (10 μCi/ml), and influx was measured at the times indicated.

Initially, then, there are two important thermodynamic disequilibria for Na^+ ions: the entropic disequilibrium caused by differences in specific activity of $^{22}Na^+$ on either side of the membrane and the actual Na^+ ion concentration gradient across the membrane. The open Na^+-selective channels permit the label to equilibrate rapidly but do not allow the Na^+ concentration gradient to dissipate. This results in transient trapping of label in the vesicle interior, as illustrated in Fig. 4. There is a much greater percentage uptake of the total radiotracer than in the earlier experiments, and the ''overshoot'' is transient. As the diffusion potential gradually dissipates, net efflux is observed. This experiment demonstrates that diffusion potentials may be generated that last for 10–30 min; in some cases they last up to several hours.

5. Channel-Mediated Fluorescence-Quench Experiments

We have recently established a sensitive fluorescence assay for neurotoxin-modulated ion transport by the reconstituted Na^+ channel (Tomiko *et al.*, 1984, 1986). The basis of the assay, as first applied to the nicotinic acetylcholine receptor by Moore and Raftery (1980), is collisional quenching of a highly fluorescent chromophore by a heavy metal cation. A membrane-impermeant fluorophore, 8-amino-1,3,6-naphthalenetrisulfonate (ANTS), is trapped in the liposome interior during the freeze–thaw cycle, and external chromophore is removed by gel filtration. The four charges on the molecule prevent adsorption to the lipid and leakage from the vesicles. Extravesicular Tl^+ ions can pass through activated sodium channels (Hille, 1972; Khodorov, 1978) and quench the fluorescence of the internalized dye.

5.1. General Considerations

Quenching of ANTS fluorescence is a nonlinear function of $[Tl^+]$ that is described by the Stern–Volmer relationship (equation 1). The quenching signal is independent of the chromophore concentration over a wide range. In Fig. 5, the quench curve for 12 mM ANTS in solution is compared with that for ANTS trapped in lipid vesicles: the intravesicular concentration is 30 mM; the overall concentration in solution is 9 μM. In the inset, the data are replotted according to the equation

$$F_0/F = 1 + K_{sv}\cdot[Tl^+] \quad (1)$$

where F_0 is the initial fluorescence and F is the quenched fluorescence. Stern–Volmer constants (K_{sv}) are 51.8 M^{-1} and 49.7 M^{-1}, respectively.

To enhance the fluorescence signal, the conditions are chosen to concentrate Tl^+ selectively in vesicles containing activated Na^+ channels, as described above for the amplified radiotracer flux. After loading of the vesicles with high concentrations of Na_2SO_4, external Na^+ is exchanged for $Tris^+$; subsequent addition of

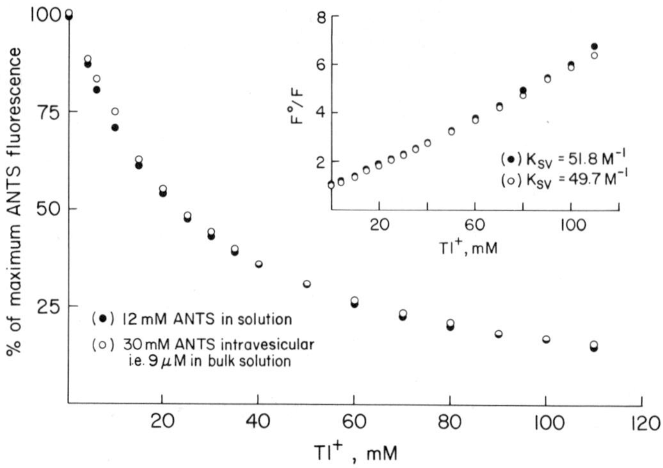

Figure 5. Thallium quench of ANTS fluorescence (●) in solution and (○) internalized in vesicles is equivalent. (●) Fluorescence of 12 mM Na_3ANTS (pH 7.4 in 8 mM Na_2SO_4) was measured in the presence of increasing amounts of $TlNO_3$. Ionic strength was held constant by substituting $TlNO_3$ for $NaNO_3$ in solutions buffered with 10 mM HEPES, pH 7.4. (○) Reconstituted vesicles were equilibrated with 30 mM ANTS (pH 7.4) during FTS, and external ANTS was removed by gel filtration. Each sample was diluted in appropriate $TlNO_3/NaNO_3$ solution; overall ANTS concentration was 9 μM (i.e., intravesicular volume was 0.03% of the suspension). Fluorescence was measured 21 hr later. Inset: Data plotted according to equation 1. Slope of line = Stern–Volmer constant (K_{sv}).

low concentrations of external Tl^+ results in transient accumulation beyond the external concentration. In vesicles that are nonspecifically leaky to ions, Tl^+ only equilibrates with the external concentration, yielding low levels of background quench. The result is a marked enhancement of the transport signal and an improvement of the signal-to-background ratio.

5.2. Methods

5.2.1. Equipment

The assay system described here resolves the signal in the seconds time scale, making it comparable to the radiotracer assay. The background leak into reconstituted liposomes is sufficiently slow to permit easy detection of the channel-mediated signal. The assay can easily be adapted to include a stopped-flow apparatus and can be used to evaluate Tl^+ influx in the millisecond time scale (Moore and Raftery, 1980; Garcia and Miller, 1984).

In the experiments described here, fluorescence is monitored by an FS950 Fluoromat (Kratos Analytical Instruments) fitted with standard 28-μl quartz flow cell (illuminated volume 20 μl). A blue-filtered excitation lamp with an emissions peak at 365 nm, a 365-nm band excitation filter, and a 470-nm cutoff emission filter provide optimal conditions for detecting ANTS fluorescence while excluding wavelengths that harm Na^+ channels (~280 nm; Fox, 1974) or that excite Tl^+ fluorescence (excitation ~214 nm, emmission 310–450 nm; Udenfriend, 1962). Fluorescent output is monitored on a multimeter in parallel with a variable-speed chart recorder. Sample solutions of ANTS-loaded vesicles (see Section 5.2.2) and Tl^+ are loaded into two 1-ml disposable syringes. Tygon® tubing connects the sample syringes to a glass capillary Y connector (with beaded glass irregularities inside to create turbulent mixing); Teflon® tubing (0.3 mm I.D.) leads to the flow-cell cuvette. For sample mixing, a rigid bar placed over the syringe plungers is pushed to deliver equal volumes from both syringes to the mixing chamber and flow cell.

5.2.2. Sample Preparation

Trisodium ANTS may be prepared as a 120 mM stock solution in 84 mM Na_2SO_4, 20 mM HEPES–NaOH (pH 7.4); ANTS (3 or 30 mM) is internalized into the vesicles during the disruptive FTS cycle. Sonication is generally limited to 2 sec because of the small sample volumes (260 μl). To remove external ANTS, 0.2 ml of the FTS preparation is centrifuged through 1.6-ml columns of Sephadex G-50 equilibrated with isotonic Na_2SO_4, 10 mM HEPES–NaOH, pH 7.4 (cf. Levinson *et al.*, 1979). As described by Tomiko *et al.* (1986), the eluate from this step contained more than 90% of the vesicles but less than 0.1% of the total ANTS. The sample was then diluted to 2–5 ml with the Na_2SO_4 buffer and stored at 0°C.

5.2.3. Experimental Protocols

In experiments illustrated here, pretreatment with inhibitory drugs or toxins was for 12 min at 0°C. Incubation with or without veratridine was for 6 min at 30°C, with or without BTX for 36 min at 30°C. After toxin incubation, 200 μl of sample was applied to a Dowex column and eluted with 1.2 ml isotonic Tris-SO_4 to remove external Na^+. (Alternatively, the sample was diluted with Na_2SO_4 solution.) A 0.6-ml aliquot, supplemented with toxin(s), was drawn into the sample syringe and connected to the assay system. The other syringe was preloaded with $TlNO_3$ (or $NaNO_3$). Three 1-min time courses were recorded after depressing the plunger to deliver 75–100 μl from each syringe though the mixing chamber.

5.3. Results

5.3.1. Neurotoxin-Modulated Transport Measured Fluorometrically

The neurotoxin-specific signals reported by Tl^+ quench accurately reproduced signals measured with radiotracer assays. In Fig. 6, the results of treatment with BTX and veratridine, with or without TTX, are illustrated for comparison with Fig. 3. The signal with BTX was always larger than that obtained with veratridine. External TTX blocked 30–65% of both signals, consistent with the radiotracer flux experiments. As before, when TTX was present inside and outside,

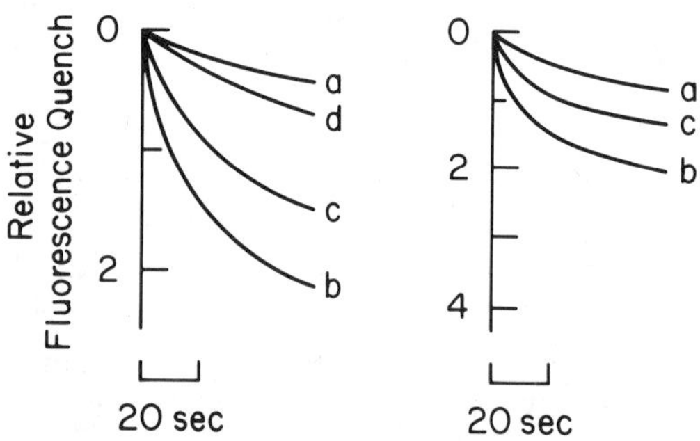

Figure 6. Neurotoxin-stimulated fluorescence quenching. Left panel: Vesicles containing 3 mM ANTS were treated with (a) ethanol vehicle, (b) 5 μM BTX, (c) 5 μM BTX + 1 μM TTX external, or (d) 5 μM BTX + 1 μM TTX added before FTS, i.e., internal and external. Right panel: Vesicles containing 30 mM ANTS were treated with (a) tris-SO_4 vehicle, (b) 100 μM veratridine, or (c) 100 μM veratridine + 1 μM TTX external. In both experiments, toxin-treated samples were eluted through Dowex, and fluorescence was measured in 40 mM $TlNO_3$. Curves represent chart recorder tracings from superimposable triplicate time courses. The BTX consistently produced greater quench signals than veratridine, and external plus internal TTX blocked these signals more completely than external TTX alone.

the inhibition was very nearly complete. Dibucaine routinely blocked 70–100% of the stimulated flux.

5.3.2. Signal Amplification

Records from standard fluorescence-quench experiments are shown in Fig. 7. Traces in the left column are from samples exposed to increasing concentrations of Tl^+ in the absence of an imposed diffusion potential. Even at the lowest concentration of Tl^+, a BTX-activated signal was detected, although at the limits of the assay sensitivity. As the concentration of Tl^+ increased, both the background and the BTX-stimulated signal were increased. The percentage stimulation by BTX actually declined, but the difference between the two traces was more easily discerned.

The amplification of the signal by replacing external Na^+ with $Tris^+$ is clear from the second column of traces. The signal in the presence of BTX was greatly enhanced, whereas background quench remained the same or decreased. That the enhanced signal was caused by the "overshoot" of internal Tl^+ is shown in Fig. 8. As observed in the radiotracer experiments (Fig. 4), the concentration of internal Tl^+ peaked transiently. As the diffusion potential dissipated by nonspecific leakage, the enhanced signal declined. When $[Na^+]_{in}/[Tl^+]_{out} = 10$, maximal quench in BTX-treated vesicles occurred within 10–20 min ($n = 5$), followed by a gradual return to a lower level of quenching. The shape of the curve resembles the radiotracer overshoot experiment, dampened somewhat because of the nonlinear relationship of $[Tl^+]$ to the quench signal.

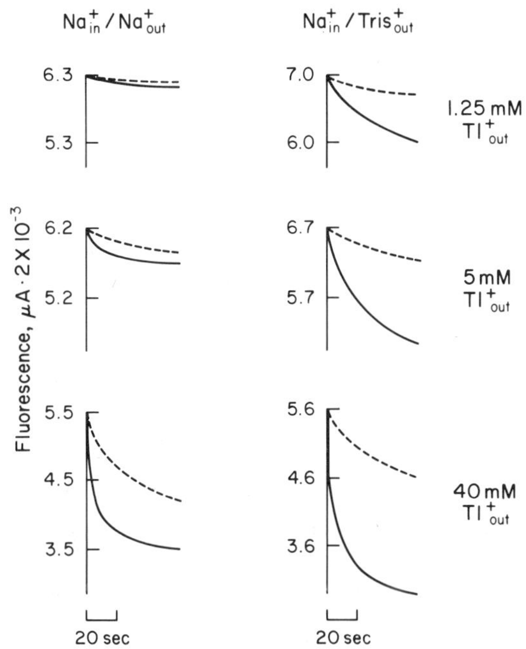

Figure 7. Amplification of alkaloid-stimulated quench by diffusion potential. Vesicles containing 30 mM ANTS were incubated in 5 μM BTX in ethanol (solid lines) or ethanol alone (dotted lines) for 36 min at 30°C. Fluorescence of vesiculated ANTS was measured during 1 min in 1.25, 5.0, or 40 mM $TlNO_3$. Left column: Before assay, samples were diluted into 6 parts of isotonic Na_2SO_4/10 mM HEPES, pH 7.4. Right column: Before assay, samples were applied to Dowex columns and rinsed with 6 parts isotonic tris-SO_4, pH 7.4, to exchange $Tris^+$ for external Na^+.

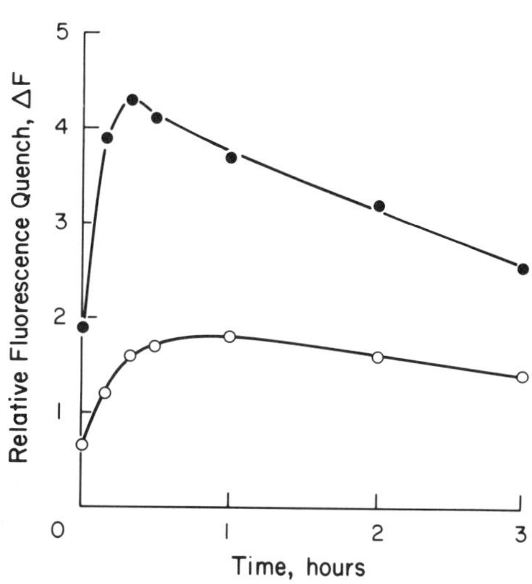

Figure 8. Fluorescence quench signal attains transient maximum in presence of diffusion potential. Vesicles (preequilibrated in 13.5 mM Na_2SO_4, 163 mM tris-SO_4, 1 mM EGTA, pH 7.4) containing 3 mM ANTS were incubated in (●) 5 μM BTX or (○) ethanol vehicle. $Tris^+$ was exchanged for external Na^+ by passage through Dowex. Samples were mixed with isotonic tris-SO_4 containing 8 mM $TlNO_3$, so that $[Na^+]_{in}$/$[Tl^+]_{out}$ ~ 10. Aliquots were withdrawn, and fluorescence was measured at the times indicated. The decline in fluorescence relative to t = 0 is plotted on the ordinate. A pronounced overshoot was observed in the BTX-stimulated quench signal as $[Tl^+]_{in}$ transiently exceeded $[Tl^+]_{out}$.

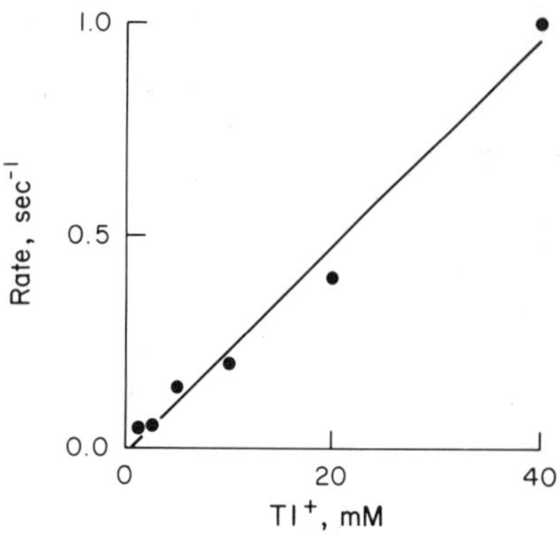

Figure 9. Quenching rates are proportional to external Tl^+ concentration. Vesicles containing 3 mM ANTS were incubated ±5 μM BTX, a diffusion potential was created, and fluorescence was measured in triplicate in six Tl^+ concentrations. The quench observed in 1.25 mM Tl^+ was measured at 20 sec. The times at which this extent of stimulated quenching was reached at each $[Tl^+]$ were then determined, and their reciprocals plotted against $[Tl^+]$. Calculations based on extents of reaction in 1.25 mM Tl^+ at 10, 40, and 60 sec likewise yielded linear relations.

5.3.3. Fluorescence-Quench Reflects the Rate of $T1^+$ Influx

The signals observed by this procedure have time courses of many seconds. The kinetics could in principle derive from the rate of transport through the activated channels or from the kinetics of channel activation, with each vesicle equilibrating essentially instantaneously after channel opening. That the former is likely to be the case is indicated by the dependence of transport rates on the external $T1^+$ concentration. The maximal level of quenching in the presence of a diffusion potential was set by the initial concentration of Na^+ inside. Vesicles exposed to lower concentrations of external $T1^+$ required more time to equilibrate. This is illustrated by plotting initial rates as a function of external $[T1^+]$. In Fig. 9, results are presented from an experiment in which $[Na^+]_{in} = 185$ mM and $[T1^+]_{out}$ varied from 1.25 to 40 mM. The quench signal obtained at 1.25 mM $T1^+$ at 20 sec was measured, and the time required to reach this level of toxin-stimulated quench at higher $T1^+$ concentrations was measured. A linear rate of 0.024 sec^{-1} mM^{-1} was observed. Because a single, chronically activated channel should be able to equilibrate a small (~500–1000 Å diameter) vesicle in a few milliseconds, these observations suggest that we observe quenching in populations of relatively large (~1 μm diameter) vesicles.

5.3.4. Effects of $T1^+$ on the Reconstituted Na^+ Channel

The assay depends on two key assumptions. First, $T1^+$ must be able to pass through open Na^+ channels. In native membranes, the permeability ratio $P_{T1}{}^+/P_{Na}{}^+$ is approximately 0.33 (Hille, 1972); after BTX modification, the ratio is ~1.3 (Khodorov, 1978; Huang, *et al.*, 1979). The $T1^+$ transport may saturate at lower concentrations than Na^+ or Li^+ (Hille, 1972). Secondly, the assay requires that $T1^+$ not alter or damage the permeability characteristics of the channel. Although Hille (1972) has stated that $T1^+$ directly destroys Na^+ channels in frog node of Ranvier in less than 1 min, Landowne (1975) reported that squid axons were stable to lower concentrations of $T1^+$ for extended times.

Several types of control experiments provide evidence that the reconstituted protein is not adversely affected by exposure to high concentrations of $T1^+$ (Tomiko *et al.*, 1986). In radiotracer flux experiments, BTX-treated vesicles were stripped of external Na^+, exposed to 20 mM external Tl^+ for 15 min, then assayed for BTX-stimulated $^{22}Na^+$ uptake. The extended exposure to elevated internal and external Tl^+ had no effect on the percentage stimulation of $^{22}Na^+$ influx. Exposure to elevated [Tl^+] also failed to destabilize the detergent-solubilized TTX-binding site. The loss of sodium channels observed in intact cells may stem from the more complex intracellular solutions, perhaps resulting in formation of precipitates in or around the channel.

5.4. Special Considerations

There are several potential sources of artifact with this assay. For example, interfering fluorescence from drugs must be considered; dibucaine contributes a background fluorescence (not quenched by Tl^+) that must be taken into account. Another potential problem stems from precipitation of ANTS with Tl^+. We have not determined the exact composition, but the salt is a dense, white crystal that can form in the concentration ranges of interest. Although 100 mM $TlNO_3$ + 30 mm ANTS precipitates, 50 mM Tl^+ + 15 mM ANTS is soluble. Thus, for prolonged experiments in which Tl^+ concentrates transiently in the vesicle interior, 3 mM ANTS is used.

The greatest source of variability stems from the sonication step. The freeze–thaw cycle is used to introduce ANTS into the vesicle interior after reconstitution and produces the larger vesicles required if the time course is to be followed in the seconds time scale. A 5-sec sonication of the small sample volume often results in complete loss of activity; peak activity was found with a 2 to 3-sec sonication. Extended sonication reduces the vesicle size so that flux through an open channel may equilibrate the interior in tens of milliseconds rather than tens of seconds. A 10-pS channel will equilibrate a 1000-Å-diameter vesicle in 30 msec, but equilibration of a 10,000-Å vesicle will require approximately 30 sec. The energy intensity of sonication cannot be regulated precisely and varies with sample volume and tube geometry.

5.5. Sensitivity and Reliability

The assay is extremely sensitive. Because the signal is caused by a reporter group in the vesicle interior, dilute suspensions may be used. Thus, a routine time course requires 1–10 fmol of TTX-binding protein, allowing thousands of assays to be performed with a single reconstituted preparation of 300–600 pmol. Each preparation, stored in ice, is stable for more than 3 weeks. The assay is extremely reproducible, and replicate time courses are nearly always superimposable. The procedure avoids the use of expensive and potentially harmful energetic isotopes. It yields a continuous record of transport, and one person can produce triplicate time courses in 6 min. We think this procedure will be of great help in optimizing reconstitution conditions, in the screening of drugs, toxins, and anesthetic compounds for action on the channel, and in studies involving defined chemical modifications of the protein.

6. Single-Channel Experiments

Radiotracer influx and fluorescence flux assays provide important tests of Na^+ channel reconstitution. However, experiments using these techniques have so far required the use of neurotoxins to open the Na^+ channel, as have reconstitution studies involving planar lipid bilayers (Krueger *et al.*, 1983; Moczydlowski *et al.*, 1984; Hartshorne *et al.*, 1985). Patch-clamp techniques for recording single-channel currents (Hamill *et al.*, 1981), applied to reconstituted preparations (Tank and Miller, 1983), permit precise voltage control and measurement of the picoampere currents of millisecond duration that are mediated by the unmodified channel. Recent studies of the electroplax channel with this approach indicate that the protein can be studied in this way (Rosenberg *et al.*, 1984b; Rosenberg, 1985).

6.1. Methods

A small aliquot (<1 μl) of freeze–thaw reconstituted vesicles, prepared as described above, is pipetted into a small bath (200 μl) on a microscope stage. The bath contains 80 mM NaC1 or KC1, 10 mM HEPES, pH 7.4, and 5 mM $MgC1_2$ (or 1 mM $MgC1_2$ plus 1 mM $CaC1_2$). The vesicles settle into an unstirred layer at the bottom of the bath and easily can be visualized under Nomarski or Hoffman optics at 400×. Figure 10 shows a photograph and a cartoon representation of a typical preparation. Large vesicles of ~10 μm diameter can be seen, often associated with clusters of smaller vesicles. Thin, ethereal membrane "blebs" protrude from some of the vesicles (cf. Tank and Miller, 1983). In our experience, these blebs, although difficult to see, provide the best membrane surface for making stable, high-resistance seals.

The electrodes are pulled, coated, and polished as described by Corey and Stevens (1983). Because the vesicles have a very fluid membrane, the best results are obtained with electrodes pulled and polished to a very small tip diameter (approximately 10 MΩ in 100 mM KC1; ~0.6 μm tip diameter based on the calculations and determinations of Sakmann and Neher, 1983). Seals between membrane blebs and the electrode surface form readily, and only very slight suction is required; even moderate suction usually causes the entire vesicle to be drawn inside the electrode. Seal resistances are usually 10–50 GΩ. Membrane patches are detached from the vesicle by rapidly withdrawing the electrode and briefly passing the tip through the air–water interface (Hamill *et al.*, 1981; Tank and Miller, 1983).

The electronics used for data acquisition (designed by D. P. Corey) have been described in detail (cf. Sigworth, 1983; Hamill *et al.*, 1981). Briefly, the membrane

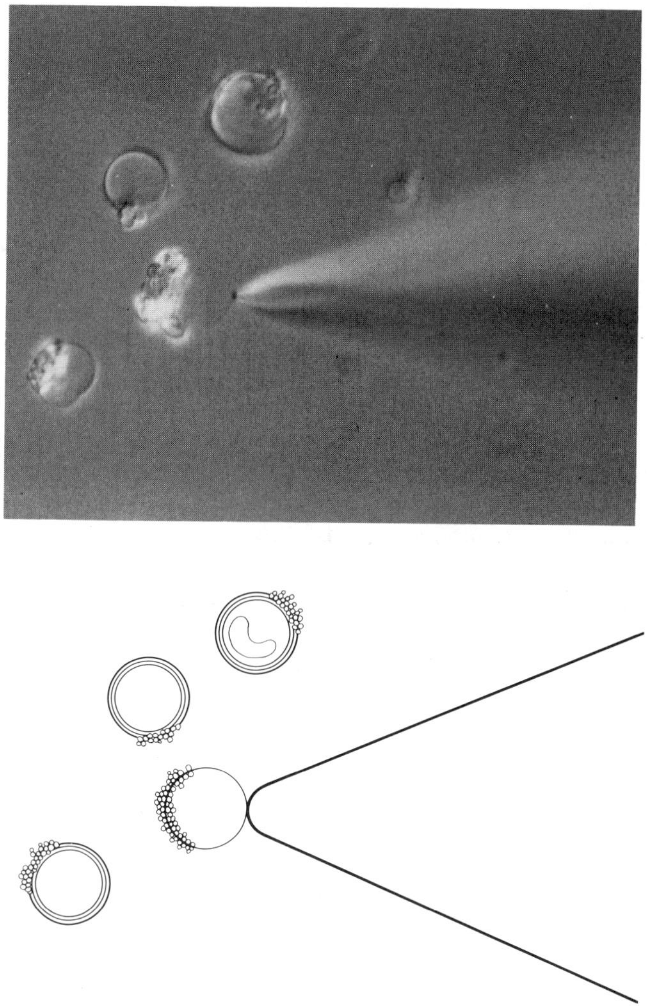

Figure 10. Large multilamellar vesicles used for patch-clamp studies. Top panel: Photomicrograph of vesicles (10 mg PE:PS:PC, 5:4:1, and 60 μg protein per ml) diluted 100-fold in 90 mM NaCl, 5 mM $MgCl_2$, 5 mM HEPES–NaOH, pH 7.4. Vesicles settled to the bottom of the chamber and were photographed at ~400× with Nomarski optics. Bottom panel: Schematic representation of the vesicle structures observed, including a patch electrode in contact with one of the membrane "blebs" described in the text.

patches are voltage clamped with a patch-clamp amplifier interfaced with a PDP 11/23 computer. A computer-driven 12-bit D/A converter supplies signals to the command input, and the membrane was clamped to predetermined holding and sampling voltages for preset durations. Membrane currents were filtered at 1000 Hz (eight-pole tunable Bessel low pass) and digitized at a sample rate of >0.2

msec. The voltages reported follow the "cellular" convention for inside-out patches; voltages are referenced to the vesicle exterior (now facing the electrode interior), and currents towards the vesicle interior (i.e., towards the bath) are represented by downward transitions in the recordings. Experimental current recordings are corrected by analog compensation of capacity transients (Sigworth, 1983) and by digital and analogue subtraction of leak and offset currents.

There are a few difficulties with this method. (1) Long-lived, high-resistance seals are difficult to achieve, and a considerable time may be invested for each satisfactory membrane patch. (2) Each successfully detached patch may not contain an active Na^+ channel. In our experience, the number of patches containing active channels is lower than a rough calculation of the density of TTX-binding sites would imply. Very slow relaxation from an inactivated state may be responsible. Or, the ethereal membrane blebs most satisfactory for forming seals may have particularly low channel surface densities. Also, a fraction of the membrane patches may be oligolamellar. (3) The randomized orientation of the channels in the membrane means that "depolarization" may be in either voltage direction, and holding the membrane at -100 mV may, in fact, be causing inactivation. One of our primary concerns at present is to resolve the problems which complicate the routine use of patch-clamp technology with these reconstituted membrane vesicles.

6.2. Results

6.2.1. Single-Channel Currents Recorded from Reconstituted Na^+ Channels

Figure 11A shows the single-channel currents recorded from a voltage-clamped patch of membrane containing purified Na^+ channels. The membrane was held at -100 mV and was depolarized for 30 msec to -25 mV. Approximately half of the depolarization epochs showed distinct inward currents, and the channel events were clustered towards the beginning of each depolarization. This patch apparently contained two channels as evidenced by one event of the current reaching twice the single-channel current level. In general, simultaneous opening events were not observed, although multiple openings per epoch were common. Figure 11B shows the scaled average of the 62 records. Following the step of depolarization, there was a rapid increase in the net inward current, which peaked at 2 msec and then declined rapidly to the zero-current level. The falling phase of the averaged currents could be approximately fit by a single exponential decline with a time constant of 6–7 msec. This time course is very similar to those observed in native membranes of other tissues (Sigworth and Neher, 1980; Horn *et al.*, 1981; Aldrich *et al.*, 1983; Horn *et al.*, 1984), and it closely resembles the macroscopic Na^+ currents in voltage-clamped nerve membranes (Hodgkin and Huxley, 1952).

The ensemble of records may be analyzed for the duration of the single-channel events. The mean channel open time was 1.9 msec, in agreement with

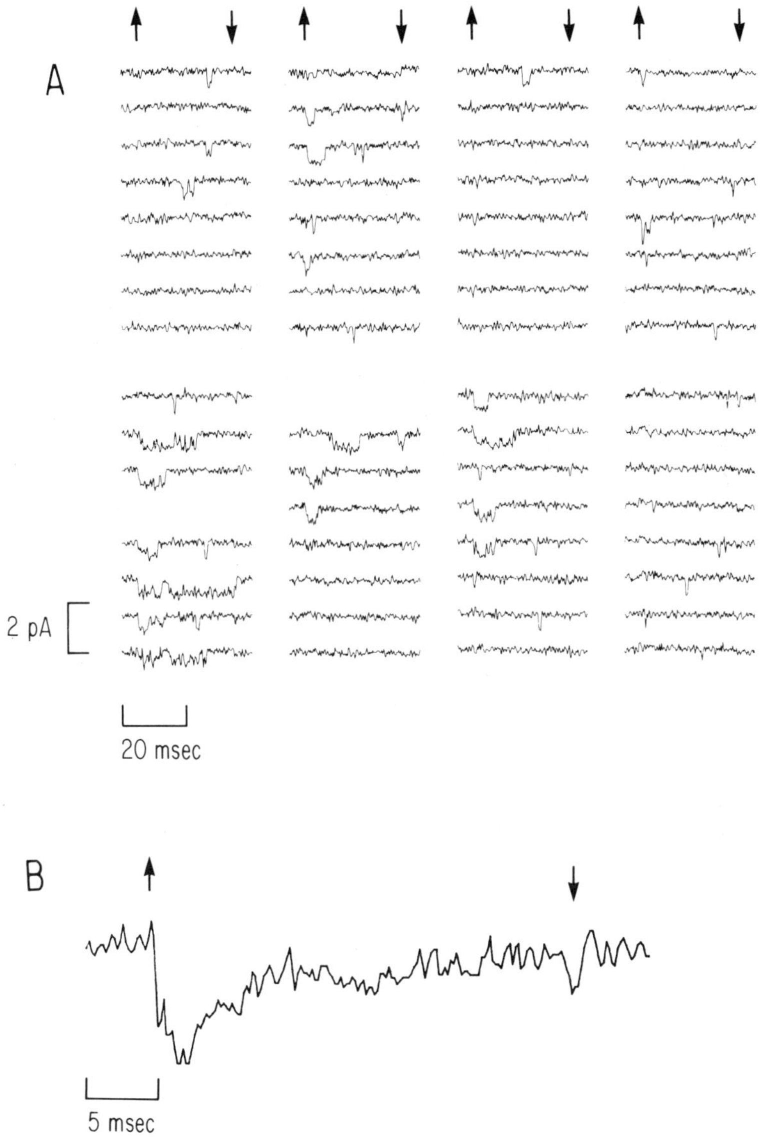

Figure 11. Single-channel currents of an excised patch containing the reconstituted purified electroplax Na^+ channel. The patch was held at -100 mV, and depolarizations to -25 mV lasting 30 msec were made at 0.75-sec intervals. The bath solution was 10 mM NaCl/80 mM KCl/5 mM $MgCl_2$/10 mM HEPES–KOH, pH 7.4, and the electrode solution was 90 mM NaCl/5 mM $MgCl_2$/10 mM HEPES–NaOH, pH 7.4. A: A collection of depolarization epochs are presented above in the sequence in which they were obtained. Channel opening is a downward deflection. B: Summated currents, scaled arbitrarily, to show the statistical probability of channels being opened during the depolarization epoch.

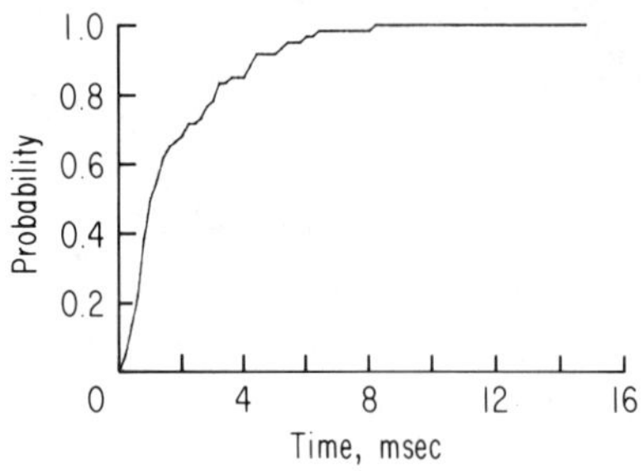

Figure 12. Channel open-time distribution. Depolarization was from -100 mV to -25 mV. Ionic conditions were as described in Fig. 11. A cumulative histogram of open durations was calculated and plotted as the probability of channel closing before time t conditional on channel opening at $t = 0$. Discrimination of opening and closing transitions was made at 0.5 of the open-channel current.

values of 0.5–3.5 msec for Na^+ channels in native membranes (Sigworth and Neher, 1980; Aldrich *et al.*, 1983; Horn *et al.*, 1984). A cumulative histogram of channel open times, plotted as the probability that the channel would have closed at some interval after opening, is shown in Fig. 12. The open times were exponentially distributed, suggesting a single open state and a single closing process (Rosenberg *et al.*, 1984b).

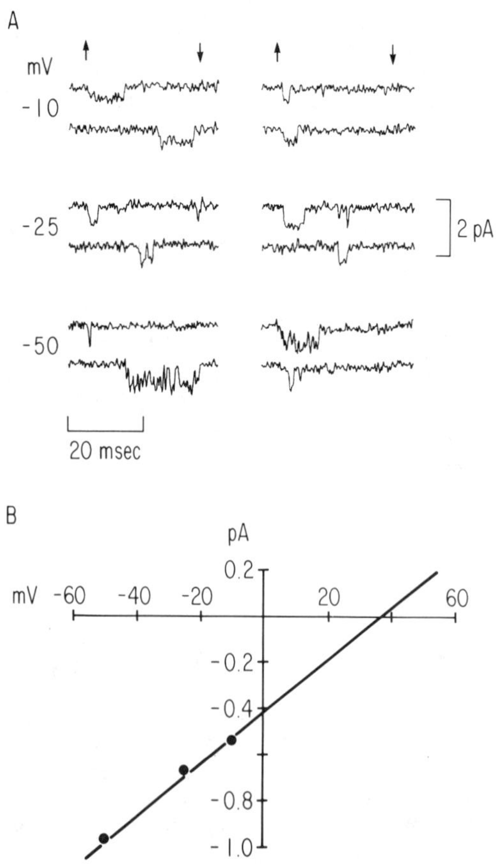

Figure 13. Selectivity of the purified channel for Na^+ over K^+. A: Representative currents recorded during 30-msec pulses from -100 mV to the voltages indicated. The pulse was applied in the period between the arrows. Ionic conditions were as described in Fig. 11 B: Mean channel current is plotted versus membrane voltage. The line drawn is a linear least-squares fit of the data, yielding a reversal potential of 37 mV. The slope conductance of the channel thus computed was 11.2 pS.

6.2.2. The Reconstituted Na^+ Channel is Selective for Na^+ over K^+

The permeation selectivity of the purified electroplax Na^+ channel was tested by measuring the single-channel currents at different command potentials in the asymmetric salt solutions described in the legend to Fig. 11. Representative records are shown in Fig. 13A, and in Fig. 13B the single-least-squares fit of the data yields an extrapolated reversal potential of 37 mV. The equation

$$E_{rev} = (RT/F) \ln \frac{Na_o + (P_K/P_{Na})K_o}{Na_i + (P_K/P_{Na})K_i} \tag{2}$$

was used to calculate a permeability ratio P_{Na}/P_K of 7. In perfused nerve, the permeability ratio is approximately 12 (Chandler and Meves, 1965; Hille, 1972), but it is subject to substantial variation. A possible source for the variability is the internal K^+ concentration; in squid axon, the selectivity ratio declines from 12.5 to 3.5 as the internal K^+ concentration is reduced below normal (Cahalan and Begenisich, 1976; Ebert and Goldman, 1976). In the intact electroplax, the internal K^+ concentration is 170 mM (Karlin, 1967); the selectivity ratio of 7 reported here is in the range expected, considering that 80 mM K^+ "internal" was used.

6.2.3. The Single-Channel Conductance of the Purified Na^+ Channel

The conductance of the channels depicted in Figs. 11 and 13 was 11.2 pS. This was in close agreement with the values of 11.1 ± 0.9 pS and 11.2 ± 1.0 pS (standard deviations) obtained from two other patches in symmetrical solutions of 90 mM NaC1, 5mM $MgC1_2$, 10 mM HEPES–NaOH, pH 7.4 (Rosenberg *et al.*, 1984b). These values were each determined on ten or more well-resolved openings from more than 30 opening events detected in each patch. These single-channel conductances stand comfortably among the values of 4–18 pS reported by others from current-fluctuation or single-channel analysis of native nerve and muscle membranes (Cahalan, 1980; Sigworth and Neher, 1980; Horn *et al.*, 1981).

6.2.4. Batrachotoxin Modifies the Single-Channel Properties of the Reconstituted Na^+ Channel

We have not characterized the single-channel properties of Na^+ channels altered with alkaloid neurotoxins extensively, although it appears somewhat easier to find active patches when the neurotoxins are present. Such modified channels should have markedly altered behavior; alkaloid neurotoxins such as BTX may alter inactivation, activation, single-channel conductances, ion selectivity, and interaction with other neurotoxins (Khodorov, 1978; Catterall, 1980; Huang *et al.*, 1979, 1984). Consistent with this, we have found that in the presence of 1 μM BTX, the channels observed exhibited properties significantly different from those described above. After exposure to BTX for 30 min, the channels opened repeatedly during a long pulse of 240 msec duration and tended to open later in the depolarization and with 15-fold longer durations (~30 msec). Summed currents showed no evidence of inactivation and did not decline during the depolarization. Two types of conductances were observed in these experiments, of 10–15 pS or 20–25 pS. In native membranes, BTX-modified Na^+ channels have been reported with a conductance of about 10 pS (Huang *et al.*, 1982, 1984), but in planar bilayers, the conductances have been reported to be in the range of 20–30 pS (Krueger *et al.*, 1983; Moczydlowski *et al.*, 1984; Hartshorne *et al.*, 1985). In several of the experiments in which 0.1 μM STX was added to the bath after seal formation, all channel events ceased, consistent with the blockade of those channels oriented outside-out. However, we have not yet studied the dose–response or reversibility of STX blockade.

7. Conclusions

The electroplax Na^+ channel is a large-molecular-weight (~300,000 daltons) protein formed by a large, heavily glycosylated polypeptide, apparently not containing smaller peptide subunits. The primary structure of the principal peptide constituent deduced from the cDNA sequence suggests an internal organization of four pseudosubunit domains. It now further seems apparent that the purified protein can be reconstituted with recovery of most, and perhaps all, of its functional properties intact. Methods described here and elsewhere in this volume, together with their descendants, should permit a systematic analysis of the molecular mechanisms for transport, permeation selectivity, and gating as well as mechanisms of drug, neurotoxin, and anesthetic effects.

Note Added in Proof

A detailed analysis of quench kinetics, accounting for the overall timecourse in terms of fractional channel activity, surface abundance, and vesicle geometry is presented by Tomiko, *et al.*, 1986.

ACKNOWLEDGMENTS. We are grateful to Dr. John Daly for generous gifts of BTX, to Dr. Y. Kishi for TTX, and to Drs. Stephen Smith, Charles Stevens, Richard Aldrich, and Toshinori Hoshi for helpful advice and encouragement. This work was supported by National Institutes of Health grants NS17928 and GM07527 and a grant from the Multiple Sclerosis Society. S.A.T. is an M.D.A. postdoctoral fellow.

References

Agnew, W. S., 1984, Voltage-regulated sodium channel molecules, *Annu. Rev. Physiol.* **46:**517–530.
Agnew, W. S., and Raftery, M. A., 1979, Solubilized tetrodotoxin binding component from the elec-

troplax of *Electrophorus electricus*. Stability as a function of mixed lipid–detergent micelle composition, *Biochemistry* **10:**1912–1919.

Agnew, W. S., Levinson, S. R., Brabson, J. S., and Raftery, M. A., 1978, Purification of the tetrodotoxin-binding component associated with the voltage-sensitive sodium channel from *Electrophorus electricus* electroplax membranes, *Proc. Natl. Acad. Sci. U.S.A.* **75:**2606–2610.

Agnew, W. S., Miller, J. A., Ellisman, M. H., Rosenberg, R. L., Tomiko, S. A., and Levinson, S. R., 1983, The voltage-regulated sodium channel from the electroplax of *Electrophorus electricus, Cold Spring Harbor Symp. Quant. Biol.* **48:**165–179.

Aldrich, R. W., Corey, D. P., and Stevens, C. F., 1983, A reinterpretation of mammalian sodium channel gating based on single channel recording, *Nature* **306:**436–441.

Barchi, R. L., 1983, Protein components of the purified sodium channel from rat skeletal muscle sarcolemma, *J. Neurochem.* **40:**1377–1385.

Bartels, E., and Rosenberry, T. L., 1973, Modification of electroplax excitability by veratridine, *Biochim. Biophys. Acta* **298:**973–985.

Cahalan, M. D., 1980, Molecular properties of sodium channels in excitable membranes, in: *The Cell Surface and Neuronal Function* (C. W. Cotman, G. Poste, and G. L. Nicolson, eds.), pp. 1–47, Elsevier/North-Holland Biomedical Press, New York.

Cahalan, M. D., and Begenisich, T., 1976, Sodium channel selectivity. Dependence on internal permeant ion concentrations, *J. Gen. Physiol.* **68:**111–125.

Catterall, W. A., 1980, Neurotoxins that act on voltage-sensitive sodium channels in excitable membranes, *Annu. Rev. Pharmacol. Toxicol.* **20:**15–43.

Chandler, W. K., and Meves, H., 1965, Voltage-clamp experiments on internally perfused giant axons, *J. Physiol.* (*Lond.*) **180:**788–820.

Corey, D. P., and Stevens, C. F., 1983, Science and technology of patch recording electrodes, in: *Single-Channel Recording* (B. Sakmann and E. Neher, eds.) pp. 53–68, Plenum Press, New York.

Ebert, G. A., and Goldman, L., 1976, The permeability of the sodium channel in *Myxicola* to the alkali cations, *J. Gen. Physiol.* **68:**327–340.

Fox, J. M., 1974, Selective blocking of the nodal sodium channels by ultraviolet radiation. I. Phenomenology of the radiation effect, *Pfluegers Arch.* **351:**287–301.

Garcia, A. M., and Miller, C., 1984, Channel mediated monovalent cation fluxes in isolated sarcoplasmic reticulum vesicles, *J. Gen. Physiol.* **83:**819–839.

Garty, H., Rudy, B., and Karlish, S. J. D., 1983, A simple and sensitive procedure for measuring isotope fluxes through ion-specific channels in heterogeneous populations of membrane vesicles, *J. Biol. Chem.* **258:**13094–13099.

Gasko, O. D., Knowles, A. F., Shertzer, H. G., Suolinna, E.-M., and Racker, E., 1976, The use of ion-exchange resins for studying ion transport in biological systems, *Anal. Biochem.* **72:**57–65.

Hamill, O. P., Marty, A., Neher, E., Sakmann, B., and Sigworth, F. J., 1981, Improved patch-clamp techniques for high-resolution current recording from cells and cell-free membrane patches, *Pfleugers Arch.* **391:**85–100.

Hartshorne, R. P., and Catterall, W. A., 1984, The sodium channel from rat brain. Purification and subunit composition, *J. Biol. Chem.* **259:**1667–1675.

Hartshorne, R. P., Keller, B. U., Talvenheimo, J. A., Catterall, W. A., and Montal, M., 1985, Functional reconstitution of the purified brain sodium channel in planar lipid bilayers, *Proc. Natl. Acad. Sci. U.S.A.* **82:**240–244.

Hille, B., 1972, The permeability of the sodium channel to metal cations in myelinated nerve, *J. Gen. Physiol.* **59:**637–658.

Hjelmeland, L. H., 1980, A non-denaturing zwitterionic detergent for membrane biochemistry: Design and synthesis, *Proc. Natl. Acad. Sci. U.S.A.* **77:**6368–6370.

Hodgkin, A. L., and Huxley, A. F., 1952, A quantitative description of membrane current and its application to conduction and excitation in nerve, *J. Physiol.* (*Lond.*) **117:**500–544.

Holloway, P. W., 1973, A simple procedure for removal of Triton X-100 from protein samples, *Anal. Biochem.* **53:**304–308.

Horn, R., Patlak, J., and Stevens, C. F., 1981, The effect of tetramethylammonium on single sodium channel currents, *Biophys J .* **36:**321–327.

Horn, R., Vandenberg, C. A., and Lange, K., 1984, Statistical analysis of single sodium channels. Effects of N-bromoacetamide, *Biophys. J.* **45:**323–335.

Huang, L. M., Catterall, W. A., and Ehrenstein, G., 1979, Comparison of ionic selectivity of batrachotoxin-activated channels with different tetrodotoxin dissociation constants, *J. Gen. Physiol.* **73:**839–854.

Huang, L. M., Moran, N., and Ehrenstein, G., 1982, Batrachotoxin modifies the gating kinetics of sodium channels in internally perfused neuroblastoma cells, *Proc. Natl. Acad. Sci. U.S.A.* **79:**2082–2085.

Huang, L. M., Moran, N., and Ehrenstein, G., 1984, Gating kinetics of batrachotoxin-modified sodium channels in neuroblastoma cells determined from single-channel measurements, *Biophys. J.* **45:**313–322.

Karlin, A., 1967, Permeability and internal concentration of ions during depolarization of the electroplax, *Proc. Natl. Acad. Sci. U.S.A.* **58:**1162–1167.

Karlin, A., 1980, Molecular properties of nicotinic acetylcholine receptors, in: *The Cell Surface and Neuronal Function* (C. W. Cotman, G. Poste, and G. L. Nicholson, eds.), pp. 191–260, Elsevier/North-Holland Biomedical Press.

Kasahara, M., and Hinkle, P. C., 1977, Reconstitution and purification of the D-glucose transporter from human erythrocytes, *J. Biol. Chem.* **252:**7384–7390.

Keynes, R. D., and Martins-Ferriera, H., 1953, Membrane potentials in the electroplates of the electric eel, *J. Physiol.* (*Lond.*) **119:**315–351.

Khodorov, B. I., 1978, Chemicals as tools to study nerve fiber sodium channels. Effects of batrachotoxin and some local anesthetics, in: *Membrane Transport Processes*, Vol. 2 (D. C. Tosteson, Y. A. Ovchinnikov, and R. Latorre, eds.), pp. 153–174, Raven Press, New York.

Krueger, B. K., Worley, J. F., and French, R. J., 1983, Single sodium channels from rat brain incorporated into planar lipid bilayers, *Nature* **303:**172–175.

Landowne, D., 1975, A comparison of radioactive thallium and potassium fluxes in the giant axon of the squid, *J. Physiol.* (*Lond.*) **252:**79–96.

Levinson, S. R., 1975, Studies on excitable membrane proteins, Ph.D. Thesis, University of Cambridge, Cambridge.

Levinson, S. R., Curatalo, C. J., Reed, J., and Raftery, M. A., 1979, A rapid and precise assay for tetrodotoxin binding to detergent extracts of excitable tissues, *Anal. Biochem.* **99:**72–84.

Miller, J. A., Agnew, W. S., and Levinson, S. R., 1983, Principal glycopeptide of the tetrodotoxin/saxitoxin binding protein from *Electrophorus electricus*: Isolation and partial chemical and physical characterization, *Biochemistry* **22:**462–470.

Moczydlowski, E., Garber, S. S., and Miller, C., 1984, Batrachotoxin-activated Na^+ channels in planar lipid bilayers: Competition of tetrodotoxin block by Na^+, *J. Gen. Physiol.* **84:**665–686.

Moore, H.-P. H., and Raftery, M. A., 1980, Direct spectroscopic studies of cation translocation by *Torpedo* acetylcholine receptor on a time scale of physiological relevance, *Proc. Natl. Acad. Sci. U.S.A.* **77:**4509–4513.

Moore, A. C., Agnew, W. S., and Raftery, M. A., 1982, Biochemical characterization of the tetrodotoxin binding protein from *Electrophorus electricus, Biochemistry* **24:**6212–6220.

Noda, M., Shimizu, S., Tanabe, T., Takai, T., Kayano, T., Ikeda, T., Takahashi, H., Nakayama, H., Kanaoka, Y., Minamino, N., Kangawa, K., Matsuo, H., Raftery, M. A., Hirose, T., Inayama, S., Hayashida, H., Miyata, T., and Numa, S., 1984, Primary structure of *Electrophorus electricus* sodium channel deduced from cDNA sequence, *Nature* **312:**121–127.

Norman, R. I., Schmid, A., Lombet, A., Barhanin, J., and Lazdunski, M., 1983, Purification of binding protein for *Tityus* γ-toxin identified with the gating component of the voltage-sensitive Na^+ channel, *Proc. Natl. Acad. Sci. U.S.A.* **80:**4164–4168.

Pick, U., 1981, Liposomes with a large trapping capacity prepared by freezing and thawing sonicated phospholipid mixtures, *Arch. Biochem. Biophys.* **212:**186–194.

Rosenberg, R. L., 1985, Functional reconstitution of the voltage-regulated sodium channel purified from the electroplax of *Electrophorus electricus*, Ph.D. dissertation, Yale University, New Haven.

Rosenberg, R. L., Tomiko, S. A., and Agnew, W. S., 1984a, Reconstitution of neurotoxin-modulated ion transport by the voltage-regulated sodium channel isolated from the electroplax of *Electrophorus electricus, Proc. Natl. Acad. Sci. U.S.A.* **81:**1239–1243.

Rosenberg, R. L., Tomiko, S. A., and Agnew, W. S., 1984b, Single-channel properties of the reconstituted voltage-regulated Na channel isolated from the electroplax of *Electrophorus electricus*, *Proc. Natl. Acad. Sci. U.S.A.* **81**:5594–5598.

Sakmann, B., and Neher, E., 1983, Geometric parameters of pipettes and membrane patches, in: *Single-Channel Recording* (B. Sakmann and E. Neher, eds.), pp. 37–51, Plenum Press, New York.

Sigworth, F. J., 1983, Electronic design of the patch clamp, in: *Single-Channel Recording* (B. Sakmann and E. Neher, eds.), pp. 3–35, Plenum Press, New York.

Sigworth, F. J., and Neher, E., 1980, Single Na^+ channel currents observed in cultured rat muscle cells, *Nature* **287**:447–449.

Talvenheimo, J. A., Tamkun, M. M., and Catterall, W. S., 1982, Reconstitution of neurotoxin-stimulated sodium transport by the voltage-sensitive sodium channel purified from rat brain, *J. Biol. Chem.* **257**:11868–11871.

Tamkun, M. M., Talvenheimo, J. A., and Catterall, W. A., 1984, The sodium channel from rat brain. Reconstitution of neurotoxin-activated ion flux and scorpion toxin binding from purified components, *J. Biol. Chem.* **259**:1676–1688.

Tanaka, J. C., Eccleston, J. F., and Barchi, R. L., 1983, Cation selectivity characteristics of the reconstituted voltage-dependent sodium channel purified from rat skeletal muscle sarcolemma, *J. Biol. Chem.* **258**:7519–7526.

Tank, D. W., and Miller, C., 1983, Patch-clamped liposomes: Recording reconstituted ion channels, in: *Single-Channel Recording* (B. Sakmann and E. Neher, eds.), pp. 91–105, Plenum Press, New York.

Tomiko, S. A., Rosenberg, R. L., and Agnew, W. S., 1984, A fluorescence assay for cation flux into liposomes containing sodium channels purified from *Electrophorus electricus, Soc. Neurosci. Abstr.* **10**:864.

Tomiko, S. A., Rosenberg, R. L., Emerick, M. C., and Agnew, W. S., 1986, A fluorescence assay for neurotoxin-modulated ion transport by the reconstituted voltage-activated Na channel isolated from eel electric organ, *Biochemistry* (in press).

Udenfriend, S., 1962, *Fluorescence Assay in Biology and Medicine*, pp. 498–501, Academic Press, New York.

Villegas, R., Villegas, G. M., Barnola, F. V., and Racker, E., 1977, Incorporation of the sodium channel of lobster nerve into artificial liposomes, *Biochem. Biophys. Res. Commun.* **79**:210–217.

Weigele, J. B., and Barchi, R. L., 1982, Functional reconstitution of the purified sodium channel protein from rat sarcolemma, *Proc. Natl. Acad. Sci. U.S.A.* **79**:3651–3655.

Chapter 13

THE RECONSTITUTED SODIUM CHANNEL FROM BRAIN

ROBERT HARTSHORNE, MICHAEL TAMKUN, and MAURICIO MONTAL

1. Introduction

The voltage-sensitive sodium channel mediates the inward sodium current during the depolarizing phase of an action potential. Because of its central role in impulse conduction, it has been the subject of extensive investigations by electrophysiological and biochemical techniques to understand its structure and function. Sodium channels have been solubilized in detergents and purified from eel electroplax (Agnew *et al.*, 1978; Miller *et al.*, 1983), rat muscle (Barchi *et al.*, 1980; Barchi, 1983), chicken heart (Lombet and Lazdunski, 1984), and rat brain (Hartshorne and Catterall, 1981, 1984; Barhanin *et al.*, 1983; Grishin *et al.*, 1984). The only functional property of sodium channels that is measurable in detergent solution is the binding of certain sodium-channel-specific neurotoxins. Therefore, it is necessary to reconstitute the purified channel protein to establish that it is functionally intact. The reconstitution of the sodium channel purified from eel electroplax (Rosenberg *et al.*, 1984a,b) and rat muscle (Weigele and Barchi, 1982; Tanaka *et al.*, 1983) is described in the two preceding chapters. Reconstitution of purified sodium channels from rat brain has been reported by Hanke *et al.* (1984). This chapter describes the reconstitution into lipid vesicles (Talvenheimo *et al.*, 1982; Tamkun *et al.*, 1984) and planar lipid bilayers (Hartshorne *et al.*, 1985) of sodium channels purified from rat brain and the measurement of sodium channel function in these two systems.

1.1. Sodium-Channel-Specific Neurotoxins

Several types of sodium-channel-specific neurotoxins (reviewed by Catterall, 1980) have been extremely useful in these studies. The first class consists of the

ROBERT HARTSHORNE and MAURICIO MONTAL • Departments of Biology and Physics, University of California San Diego, La Jolla, California 92093. MICHAEL TAMKUN • Carnegie Institute of Washington, Baltimore, Maryland 21210. Present Addresses: R. H., Department of Pharmacology, Oregon Health Sciences University, Portland, Oregon 97201, M. M., Department of Neurosciences. Roche Institute of Molecular Biology, Nutley, New Jersey, 07110.

sodium channel blockers saxitoxin (STX) and tetrodotoxin (TTX). These are small, positively charged, water-soluble, heterocyclic guanidines that bind with nanomolar K_Ds to a site on the extracellular-facing side of the sodium channel. [^{3}H]saxitoxin binds with similar affinity to both detergent-solubilized and mem brane-bound sodium channels, making it an excellent marker for the sodium channel protein during its purification (Catterall *et al.*, 1979). Binding of these toxins to the channel protein results in an inhibition of ion flux. Since these compounds cannot permeate lipid bilayers, they have no effect on the channel when they have access only to the normally intracellular-facing side (Narahashi *et al.*, 1966). Thus, STX and TTX are also valuable for determining the orientation of reconstituted sodium channels.

The second group of toxins is represented by the sodium channel activators batrahotoxin (BTX) and veratridine (Ver) (Albuquerque and Daly, 1976). These are small (mol wt. 538 and 674, respectively) lipid-soluble molecules that can interact with sodium channels from either side of the bilayer. These toxins shift the voltage dependence of activation by ~50 mV in the hyperpolarized direction so that the voltage at which 50% of the channels are activated (V_{50}) is near -90 mV (Khodorov and Revenko, 1979). In addition, they eliminate both fast and slow (BTX only) inactivation, resulting in persistently activated channels over a wide voltage range (Huang *et al.*, 1982). The ionic selectivity and single-channel conductance of sodium channels are also decreased by these toxins (Huang *et al.*, 1984; Quandt and Narahashi, 1982).

The third type of toxin is a scorpion toxin (LqTX) purified from the venom of the North African scorpion *Leiurus quinquestriatus*. This toxin is a 6700-mol.-wt. positively charged polypeptide that binds to a site on the extracellular-facing side of the sodium channel in a highly voltage-dependent manner (Catterall, 1977). The K_D at -55 mV is 2 nM, whereas at 0 mV it is 1000 nM. The LqTX acts to slow the rate of inactivation of the sodium channel (Koppenhöfer and Schmidt, 1968).

1.2. Sodium Channel Purification and Subunit Composition

The sodium channel was purified from a Triton X-100 solution of rat brain membranes as outlined in Table I by selecting the fractions that contained high concentrations of [^{3}H]STX binding sites (Hartshorne and Catterall, 1984). The purity at each step can be assessed by comparing its specific activity to the specific activity of 3200 pmol/mg calculated for pure sodium channel from its molecular weight. On this basis, the channel protein is 90% pure after the sucrose gradient step.

The sodium channel protein purified from rat brain consists of three subunits, α ($M_r \approx 260{,}000$), $\beta1$($M_r \approx 39{,}000$), and $\beta2$($M_r \approx 37{,}000$) (Hartshorne *et al.*, 1982; Hartshorne and Catterall, 1984), as shown in silver-stained SDS gels (Fig. 1) (Messner and Catterall, 1985). Several lines of evidence support this conclusion. First, in sucrose density gradients, all three polypeptides cosediment as a stoichiometric complex whose distribution exactly coincides with the distribution of

Table I. Saxitoxin Receptor Purification

Step	Saxitoxin receptor		Protein		Sp. act. (pmol/mg)	Relative purification
	pmol[a]	Percent	mg[a]	Percent		
Membrane extract	6608 ± 726	100.0	1145.0 ± 62.9	100	5.8	1.0
DEAE-sephadex	3857 ± 398	58.3	42.8 ± 8.6	3.7	90.1	15.6
Hydroxylapatite	1486 ± 276	22.4	6.8 ± 0.3	0.59	218.0	37.8
WGA sepharose	926 ± 194	14.0	0.97 ± 0.2	0.085	955.0	165.0
Sucrose gradient	513 ± 90	7.8	0.17 ± 0.03	0.015	2910.0	505.0

[a] ± SE, $n = 3$.

[^{3}H]STX binding sites in the gradient (Hartshorne and Catterall, 1984). Second, with native sodium channels in synaptosomes and in primary neuronal cultures from rat brain, the α subunit and the β1 subunit are specifically covalently labeled by photoactivated derivatives of scorpion toxin (Beneski and Catterall, 1980; Darbon *et al.*, 1983; Sharkey *et al.*, 1984). Third, the α subunit and the β2 subunit are covalently attached by at least one disulfide bond (Hartshorne *et al.*, 1982).

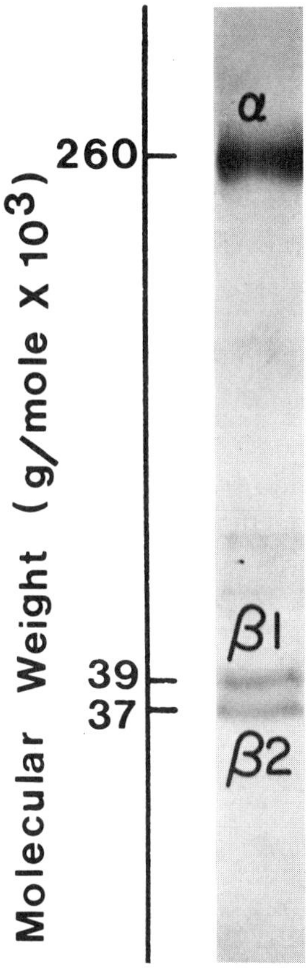

Figure 1. Subunit composition of purified sodium channels from rat brain. Purified sodium channels were dissociated by boiling in 2% sodium dodecylsulfate and 20 mM 2-mercaptoethanol and then analyzed by electrophoresis in polyacrylamide gel (Messner and Catterall, 1985). Silver staining (Oakley *et al.*, 1981) shows the three subunits of the purified protein, α, β1, and β2.

And fourth, the β subunits are not proteolytically derived from α or a larger precursor during the purification, because the amino termini of all three subunits are blocked (Grishin *et al.*, 1984).

1.3. Why Reconstitute?

Since the only functional property measurable with purified sodium channels in detergent solution is the binding of STX and TTX, a number of obvious questions about the structural and functional integrity of the purified protein remained to be answered. Has the entire structure of the sodium channel been purified, or only the portion of it that contains the STX/TTX binding site? Which other functional elements of the sodium channel survive the purification process, and how closely does their function resemble that of native sodium channels? To answer these questions, purified sodium channels are reconstituted into lipid vesicles and into planar lipid bilayers. These systems nicely complement each other. Purified sodium channels reconstituted in lipid vesicles provide a convenient system for studying the average properties of a large population of channels over a time scale that is slow compared to the opening and closing rates of the channel. This system is well suited for quantitative measurement of (1) ligand binding; (2) the fraction of functional ion channels reconstituted and their orientation; and (3) neurotoxin-activated ion flux, provided that care is taken to insure initial-rate conditions throughout the time course of the assay. The planar lipid bilayer system excels in those areas in which the vesicle system is weakest: measurement of the current through one or a small number of channels with excellent time resolution under conditions in which the transbilayer voltage is well controlled. This system allows detailed measurements of the conductance, voltage dependence, kinetic constants, and ion selectivity of single sodium channels. Although neurotoxin binding measurement is impractical, neurotoxin effects can be readily demonstrated and their concentration dependence determined if the off and on rates are slow enough to be resolved and fast enough to permit a representative sampling of the data in a reasonable time. Note that the reconstituted vesicles are the source of sodium channels for the planar lipid bilayer experiments. Therefore, successful reconstitution in planar lipid bilayers is strictly dependent on the functional integrity of the sodium channels in the reconstituted vesicles. We turn now to describe the methods used to reconstitute purified sodium channels first into lipid vesicles and then in planar lipid bilayers and to review the properties of the reconstituted sodium channel.

2. Reconstitution of Purified Sodium Channels into Phospholipid Vesicles

2.1. Methods

2.1.1. Vesicle Preparation

Reconstitution of solubilized integral membrane proteins requires the slow removal of detergent in the presence of phospholipid. Detergents commonly used for reconstitution have relatively high critical micelle concentrations and are removable by dialysis. However, the sodium channel was solubilized and purified using Triton X-100, which cannot be dialyzed because of its low critical micelle concentration. This detergent can be removed by porous polystyrene beads (Bio-Beads® SM-2) (Holloway, 1973), and this is the basis of our reconstitution procedure.

Purified rat brain sodium channels were reconstituted into egg phosphatidylcholine (PC) (Sigma) vesicles according to the following protocol (Talvenheimo *et al.*, 1982). Triton X-100 (10% w/v) in Na_2SO_4 reconstitution medium (67.5 mM Na_2SO_4, 25 mM HEPES, pH adjusted to 7.4 with tris base, 0.5 mM $MgSO_4$, and 150 mM sucrose) containing 5% w/v PC was added to sucrose gradient samples containing 30–40 pmol of purified sodium channels to yield a final concentration of 1.75% Triton X-100 and 0.85% PC. Small amounts of [^{14}C]PC were added to serve as a tracer vesicle phospholipid. Detergent was removed with Bio-Beads® (0.2 ml/ml sample) during an overnight incubation at 4°C. The Bio-Beads® were then replaced with an identical volume of fresh Bio-Beads®. The sample was incubated for an additional 2 hr at 4°C, and the Bio-Beads® were removed by filtration. Under these conditions, Bio-Beads® removed more than 98% of the Triton X-100 (determined using [^{3}H]Triton X-100). The Bio-Beads® also removed 10–20% of the protein and lipid, which is one drawback of this procedure.

Sodium channels were also reconstituted into vesicles composed of 40% egg PC and 60% purified rat brain lipid (w/w). A solution of 10% Triton X-100, 1% brain lipid, 0.25% PC in Na_2SO_4 reconstitution medium was added to purified sodium channels to give a final lipid concentration of 0.06%. Vesicles were formed during an overnight incubation with 0.3 ml Bio-Beads® per milliliter of solution as described above.

2.1.2. Characterization of the Reconstituted Vesicles

Internal volume was measured by including radioactive markers such as [^{14}C]glucose or $^{22}Na^+$ prior to reconstitution by detergent removal. The nonspecific release of these markers is slow. The efflux of $^{22}Na^+$ from these vesicles occurred with a $t_{1/2}$ of 4 hr at 4°C. The reconstituted vesicles were passed over a glass fiber filter (Whatman GF/F) or through a Dowex ion exchange column to separate internal and external [^{14}C]glucose or $^{22}Na^+$, respectively (Gasko *et al.*, 1976). Total internal volume was calculated from the initial isotope concentration. Little or no [^{14}C]glucose or $^{22}Na^+$ was bound to vesicle phospholipid, which was determined by treating the vesicles with low concentrations of Triton X-100 prior to measuring internal isotope levels.

Vesicle size was measured by both gel filtration and freeze–fracture electron microscopy (Tamkun *et al.*, 1984). We used a BioGel A150 column (1.0 × 25 cm) calibrated with uniform-size latex beads obtained from Dow Chemical Company. Vesicle diameters up to 2000 Å were accurately determined with such a column. Comigration of [^{3}H]saxitoxin ([^{3}H]STX) binding activity and vesicle phospholipid through the column demonstrated quantitative reconstitution. Vesicle aggregation can artificially create large particles, and vesicles with diameters larger than 2000

Å cannot be resolved. These problems can be avoided in freeze–fracture electron microscopy, which was the method of choice to assess vesicle size (Szoka and Papahadjopoulos, 1980).

2.1.3. Assay of Neurotoxin Binding to the Reconstituted Sodium Channel

Saxitoxin binding was measured by incubating the reconstituted vesicles with 2–10 nM [^{3}H]STX in choline chloride binding medium as previously described in detail (Tamkun and Catterall, 1981b). Separation of bound and free [^{3}H]STX was by filtration through a Whatman GF/F filter. Since the vesicles contained a trace amount of [^{14}C]PC, the filters were analyzed for both ^{3}H and ^{14}C radioactivity to quantitate [^{3}H]STX binding and vesicle retention, respectively. Approximately 20% of the vesicles were not retained by the filter. Nonspecific binding was determined in the presence of 1 μM TTX.

To determine the orientation of the reconstituted channel across the vesicle membrane, the binding of 10 nM [^{3}H]STX to vesicles in the presence of increasing concentrations of Triton X-100 was measured (Tamkun *et al.*, 1984). Following a 30-min incubation at 4°C, free and bound [^{3}H]STX were separated by centrifugation through 1.5-ml columns of Sephadex G-50 as described for binding to the solubilized STX receptor (Catterall *et al.*, 1979). A concentration of 0.2% Triton X-100 allowed [^{3}H]STX access to internally oriented, inside-out, STX binding sites without solubilizing protein. At higher detergent concentrations, the binding activity declined as a result of solubilization and denaturation of the reconstituted STX receptor. The maximum percentage increase in the [^{3}H]STX binding observed with increasing detergent was taken as an indication of the fraction of inside-out STX receptor sites.

Scorpion toxin was purified from the venom of *Leiurus quinquestriatus* and labeled with ^{125}I as described by Catterall (1976, 1977). The binding of this radiolabeled polypeptide to the reconstituted channel was measured by incubating vesicles in reconstitution medium with 0.1 nM [^{125}I]scorpion toxin ([^{125}I]LqTX) at 36°C for 15 min. The mixture was then diluted with ice-cold wash medium and filtered through a Whatman GF/F filter to separate bound and free toxin. Prior to use, the filters were incubated in wash medium containing 1 mg/ml bovine serum albumin (BSA) for 30–45 min. This preincubation was necessary to reduce [^{125}I]LqTX binding to the filter. Incubation times longer than 1 hr or with more than 1 mg/ml BSA further reduced [^{125}I]LqTX binding but also decreased vesicle retention by the filter. The LqTX binding in the presence of a membrane potential was measured as described elsewhere (Tamkun *et al.*, 1984).

Analysis of nonspecific [^{125}I]LqTX binding was more complicated than measuring binding in the presence of an excess of unlabeled LqTX. Both glass fiber filters and protein-free phospholipid vesicles demonstrated low-affinity but saturable [^{125}I]LqTX binding. To correct for this, vesicles were formed with heat-denatured sodium channels (15 min at 36°C) under conditions identical to those used to form vesicles with active channels. The [^{125}I]LqTX binding to these vesicles, which contained no [^{3}H]STX binding activity, was taken as nonspecific binding. Routinely, at concentrations of [^{125}I]LqTX below 0.2 nM, nonspecific binding comprised 25% to 35% of total binding (see Fig. 3). Of this fraction, ~40%

represented binding to GF/F filters, and the remainder nonspecific binding to reconstituted vesicles. Binding to phospholipid vesicles was characterized by high capacity and low affinity (K_D = 5 μM). In contrast, the binding to reconstituted sodium channels had a 100-fold lower K_D, 60 nM (Tamkun *et al.*, 1984).

2.1.4. Assay of Na^+ Transport by the Reconstituted Sodium Channel

The small diameter of reconstituted vesicles complicates the quantitative measurement of initial rates of ion transport. Equilibrium is likely to be reached in less than a second when rates of net ion influx are measured. The small internal volume of the vesicles also results in a low signal-to-noise ratio, especially if the measured ions also bind to vesicle phospholipid. Tanaka and co-workers (1983) used a stop-flow approach to measure initial rates of ion transport by reconstituted sodium channels purified from skeletal muscle. We used an alternative procedure (Talvenheimo *et al.*, 1982) based on an assay system developed by Epstein and Racker (1978) to measure $^{22}Na^+$ uptake by reconstituted vesicles. Vesicles were first incubated with neurotoxins to activate the channel. To initiate uptake, a 20-μl aliquot of vesicles formed in the presence of Na_2SO_4 reconstitution medium was diluted tenfold with 180 μl of tris-SO_4 medium (105 mM tris-SO_4, pH 7.4, 0.5 mM $MgSO_4$, 25 mM HEPES adjusted to pH 7.4 with tris base, and the sucrose concentration required to maintain osmolarity with the vesicle medium) containing 10μCi/ml $^{22}NaCl$ and the appropriate neurotoxins. The diluted sample was incubated at 36°C for 20 sec unless indicated otherwise and then applied to a Dowex AG50W-X8 (tris form) column (10 × 0.5 cm) in a Pasteur pipette and immediately eluted with 1 ml of 0.38 M sucrose containing 1 mg/ml of BSA. Recovery of the vesicles in the column eluate, as determined by [^{14}C]PC, approached 100%. The $^{22}Na^+$ uptake terminated within 3 sec as the cations were adsorbed to the resin.

Dilution into Na^+-free medium generates a tenfold outward-oriented Na^+ electrochemical gradient. Vesicles containing neurotoxin-activated sodium channels are selectively permeable to Na^+ ions and, therefore, should generate an inside-negative membrane potential of ~62 mV as predicted by the Nernst equation (Tamkun *et al.*, 1984). The observed $^{22}Na^+$ uptake represents isotopic equilibration rather than net influx and, therefore, occurs over a slower time course than would otherwise be expected. Since the ratio $[Na^+]_{in}/[Na^+]_{out} = 10$, $^{22}Na^+$ is concentrated within those vesicles containing active channels. This effect greatly enhances the signal-to-noise ratio of the assay system. Sodium must be the only ion capable of crossing the vesicle bilayer if the gradient and membrane potential are to be maintained. The lack of permeant counterions, coupled with the inside-negative membrane potential, also prevents isotope loss during passage over the Dowex resin.

2.2. Assessment of the Functional Activity of the Reconstituted Sodium Channel

2.2.1. Description of Reconstituted Vesicles

Figure 2 is a freeze–fracture electron micrograph of reconstituted sodium channels is egg PC vesicles. The arrow points to a vesicle with an intramem-

Figure 2. Electron micrographs of freeze–fracture replicas of purified sodium channels reconstituted in egg PC vesicles. The bar indicates 1000 Å. The arrow indicates a vesicle containing an intramembranous particle likely to be a sodium channel.

branous particle with a diameter of ~100 Å, presumably a sodium channel. The size of the vesicles is distributed in two populations, one with a mean diameter of 850 Å ± 35 Å (SE) and the other significantly larger, with a mean diameter of 2300 Å ± 50 Å (SE). Intramembranous particles are present in both populations, so there is no evidence that sodium channels are preferentially incorporated into vesicles of one size group. The average number of STX receptors per vesicle (0.75 to 2.1 from six different batches of vesicles) is kept low to facilitate ion-flux measurements (Tamkun *et al.*, 1984).

2.2.2. [^{3}H]Saxitoxin Binding and Receptor Orientation

Scatchard analysis of [^{3}H]STX binding to purified reconstituted sodium channels shows a single class of binding sites with a K_D of 2 nM (Tamkun *et al.*, 1984). This value agrees well with the 2.3 nM K_D of [^{3}H]STX binding to sodium channels in synaptosomes (Catterall *et al.*, 1979). The orientation of the STX receptors was determined as described in Section 2.1.3. on six different batches of reconstituted sodium channels. The percentage of externally oriented STX receptors ranged from 40% to 60% with a mean of 50% ± 4% (SE).

2.2.3. Scorpion Toxin Binding by the Reconstituted Sodium Channel

Scorpion toxin binding activity is an extremely sensitive biochemical indicator of the functional state of the sodium channel. Binding of this polypeptide toxin to native channels is influenced by both membrane potential and occupancy of the alkaloid toxin receptor site (Catterall, 1980). Scorpion toxin binding is also affected by the hydrophobic environment of the sodium channel. Concentrations of Triton X-100 that are tenfold lower than those that solubilize the membrane completely inhibit toxin binding (Catterall *et al.*, 1979).

Sodium channels reconstituted into egg PC vesicles do not bind [^{125}I]LqTX. Solubilization *per se* does not irreversibly denature the LqTX receptor, since Triton-X-100-solubilized channels can be reconstituted prior to purification with recovery of LqTX binding activity (Tamkun and Catterall, 1981a). Three possible reasons can account for the lack of [^{125}I]LqTX binding following reconstitution into PC vesicles: (1) the LqTX receptor may be irreversibly denatured during the purification; (2) a protein component essential for LqTX binding may separate from the sodium channel during purification; or (3) a membrane constituent necessary for LqTX binding, other than sodium channel protein itself, may be removed during purification. Endogenous lipid is a possible candidate within this third category.

To examine the role of endogenous membrane lipid, whole rat brains (obtained from Pel Freez Biologicals) were subjected to a modified Folch extraction, and the resulting lipid was purified by silicic acid chromatography (Tamkun *et al.*, 1984). This lipid fraction was a mixture of PC, phosphatidylethanolamine (PE), phosphatidylserine (PS), and cholesterol and was protein-free.

Binding of [^{125}I]LqTX to reconstituted sodium channels in rat brain lipid/PC vesicles is illustrated in Fig. 3. In the absence of unlabeled LqTX, reconstituted vesicles containing untreated sodium channels bound 1.95 fmol of [^{125}I]LqTX, whereas vesicles containing heat-denatured sodium channels bound 0.45 fmol (23% of total binding). The difference between these two values represents specific binding. Specifically bound [^{125}I]LqTX was competitively displaced by unlabeled LqTX with a K_D of 100 nM. In contrast, 100 nM unlabeled LqTX had little effect on nonspecific binding to vesicles with denatured sodium channels. Concentrations of unlabeled LqTX ≥ 100 nM displace nonspecifically bound [^{125}I]LqTX. This nonspecific binding component with a high binding capacity and a K_D of 5–10 μM was not characterized further.

To verify that the observed binding to vesicles was to the reconstituted sodium channel protein and not the vesicle lipid, a thermal inactivation protocol was performed. Purified sodium channels were incubated at 36°C for the time indicated in Fig. 4. The heated sodium channel preparation was then reconstituted using rat brain lipid and PC, and both [^{3}H]STX and [^{125}I]LqTX binding were measured. The STX-binding activity of the solubilized sodium channel is very sensitive to thermal inactivation, whereas the reconstituted material shows ther-

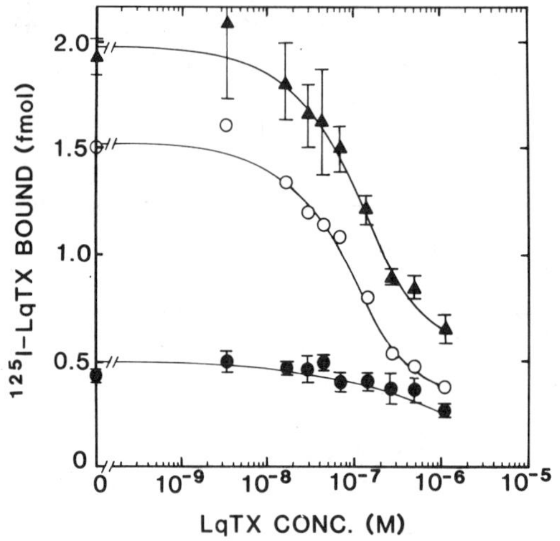

Figure 3. Specific and nonspecific binding of [^{125}I]LqTX to reconstituted vesicles. Purified sodium channels were reconstituted in vesicles composed of PC supplemented with purified rat brain lipids. Prior to reconstitution, sodium channels were either incubated at 36°C for 15 min (●) or untreated (▲). Binding was measured in the presence of 0.1 nM[^{125}I]LqTX and increasing concentrations of unlabeled LqTX as indicated. Specific binding (○) is defined as the difference between binding to vesicles containing untreated sodium channels and binding to comparable vesicles containing heat-denatured channels.

mal stability (Talvenheimo *et al.*, 1982). As shown in Fig. 4, both [^{3}H]STX and [^{125}I]LqTX binding decay in parallel. Thus, the specific LqTX binding to purified sodium channels reconstituted with brain lipids represents binding to the sodium channel protein.

The LqTX binding recovered after reconstitution with endogenous lipid was not affected by either membrane potential or occupancy of the alkaloid toxin receptor site. Either the channel has yet to be reconstituted in the proper lipid environment or the residual Triton X-100 is a problem. Recent results of Feller and co-workers (1984) indicate that bovine brain PC and PE can effectively substitute for the purified rat brain lipid, allowing reconstitution of LqTX binding. This approach is likely to lead to the recovery of voltage-dependent, alkaloid-toxin-sensitive scorpion toxin binding. (See Feller *et al.*, 1985).

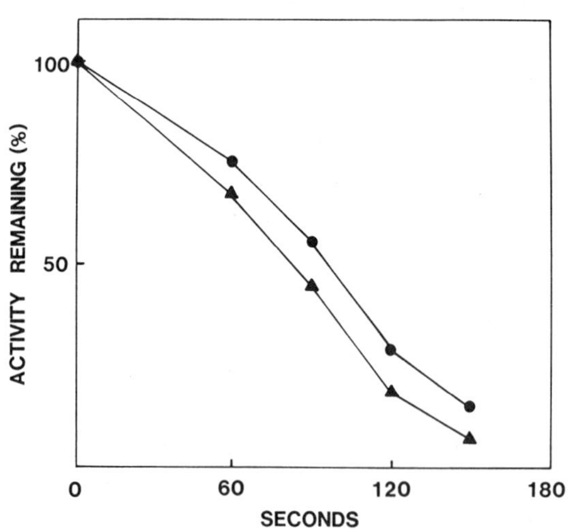

Figure 4. Thermal inactivation of the [^{3}H]STX and [^{125}I]LqTX binding activities of the purified sodium channel. Sodium channels were incubated at 36°C for the indicated times. Channels were reconstituted in vesicles composed of PC and purified rat brain lipid, then [^{3}H]STX (●) and [^{125}I]LqTX (▲) binding activities were measured. The [^{3}H]STX binding was determined in the presence of 10 nM [^{3}H]STX incubated for 15 min at 4°C. The [^{125}I]LqTX binding was measured in the presence of 0.1 nM [^{125}I]LqTX incubated for 15 min at 36°C.

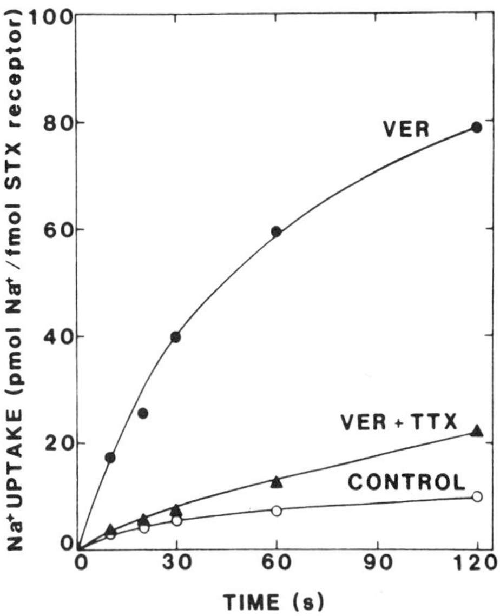

Figure 5. Time course for $^{22}Na^{+}$ uptake by purified sodium channels reconstituted in PC vesicles. The uptake measured in the presence of 100 μM veratridine (VER) (●), 100 μM veratridine together with 1 μM TTX on both sides (▲), and no added toxins (○) is illustrated.

2.2.4. Sodium Transport by the Reconstituted Sodium Channel

To demonstrate that the three polypeptides copurifying with saxitoxin-binding activity also contain components necessary for ion transport, neurotoxin-activated Na^+ uptake by the reconsituted vesicles was examined. The time course of Na^+ uptake by the reconstituted vesicles is illustrated in Fig. 5. In the absence of neurotoxins, the uptake is slow. With veratridine (100 μM), the extent of Na^+ uptake was stimulated eightfold, with half-maximal stimulation at 28 μM. This value is similar to that measured for veratridine-activated Na^+ transport in synaptosomes (Tamkun and Catterall, 1981b). If 1 μM TTX was included both inside and outside the vesicles, the extent of the veratridine-stimulated Na^+ uptake was reduced to the control level at the early time points. At longer times there was a small measurable uptake even in the presence of TTX. This uptake can be accounted for by reversible dissociation of TTX from a fraction of the activated channels, allowing Na^+ uptake at long incubation times.

2.2.5. Concentration Dependence of TTX Inhibition of Na^+ Transport by the Reconstituted Sodium Channel

The extent of veratridine-stimulated Na^+ uptake detected in the presence of increasing TTX concentrations is shown in Fig. 6. In these vesicles, ~40% of the reconstituted TTX binding sites faced the vesicle exterior (determined as described in Section 2.1.3). Tetrodotoxin, which was present only on the extravesicular compartment, maximally inhibited 40% of the veratridine-activated in Na^+ uptake, as expected. The maximal percentage of inhibition of Na^+ uptake observed with externally added TTX was consistently equal to the fraction of right-

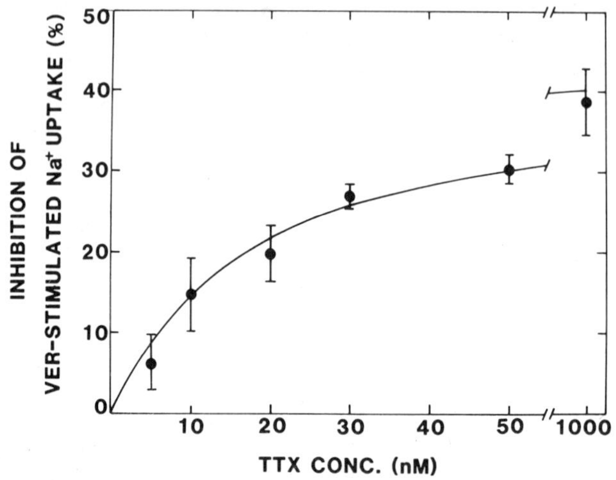

Figure 6. Concentration dependence of tetrodotoxin inhibition of veratridine-activated $^{22}Na^+$ uptake by purified sodium channels reconstituted in PC vesicles. Vesicles were incubated with 100 μM veratridine and the indicated concentration of TTX for 2 min at 37°C prior to dilution into tris-SO_4 medium containing the same toxin concentration. Each point is the mean ± SD (n = 6).

side-out oriented TTX receptors. In this experiment, the K_I was 18 nM. The average value of the K_I of TTX was 14 nM ± 6 nM (SD; n = 3), similar to the 10 nM K_D for TTX binding to native channels in rat brain membranes (Catterall *et al.*, 1979).

2.2.6. Ionic Selectivity of the Reconstituted Sodium Channel

A qualitative assessment of the ionic selectivity of the reconstituted channel can be obtained from the flux assay described (Tamkun *et al.*, 1984). The veratridine-activated sodium channels also transport Rb^+ and Cs^+ with an apparent selectivity ratio with respect to Na^+ of 1.0:0.25:0.10.

2.2.7. The Majority of the Reconstituted Sodium Channels are Functional

To verify that only the three STX receptor subunits are sufficient for Na^+ transport in the reconstituted system, it is valuable to estimate the fraction of reconstituted STX receptors that bind scorpion toxin and transport Na^+. If only 1% of the STX receptors transported Na^+, for example, it is conceivable that an "impurity" in the preparation representing 1% of total protein could be a sodium channel component required for ion transport.

The fraction of reconstituted STX receptor sites that bind scorpion toxin is readily calculated. Scatchard analysis of the binding data indicate that the 76% ± 6% (S.E.M.) of the reconstituted channels that bind STX also bind scorpion toxin. Determination of the fraction of STX binding sites that transport Na^+ can be estimated in two ways. First, the initial rates of $^{22}Na^+$ influx through veratridine-activated sodium channels in synaptosomes and in reconstituted vesicles are compared under similar conditions. Sodium channels in rat brain synaptosomes transport Na^+ at a rate of 1.4×10^6 ions/min per STX binding site (Tamkun and Catterall, 1981b) as compared to 1.0×10^6 ion/min per STX binding site for the reconstituted channel. Accordingly, 70% of the reconstituted channels are active and transport sodium at comparable rates to sodium channels *in situ*. The fraction of STX binding sites associated with functional ion channels can also be calculated from the fraction of vesicular internal volume accessible to veratridine-induced Na^+ efflux and the distributions of STX binding sites in the vesicles as described in detail by Tamkun *et al.* (1984). By this method, the fraction of STX binding sites representing functional sodium channels would range between 30% and 73% of the total STX receptor sites. Thus, the two independent estimates of sodium channel ion-transport activity indicate that a large fraction, up to 70%, of the reconstituted STX receptors contribute to the measured ion flux. This value is in agreement with the fraction of reconstituted saxitoxin binding sites that bind scorpion toxin (76%). Since the sodium channel used for these reconstitution studies was ~90% pure by SDS gel electrophoresis, the complex of the α, β1, and β2 subunits is sufficient for reconstitution of both neurotoxin-stimulated ion transport and scorpion toxin binding.

3. Planar Lipid Bilayers

3.1. Methods

The basic procedure for forming planar lipid bilayers from *n*-decane solutions of phospholipids across an aperture in a partition separating two aqueous compartments is described by Mueller *et al.* (1963) and Mueller and Rudin (1969). The procedure for fusing vesicles with planar lipid bilayers was first described by Drachev *et al.* (1974), applied to vesicles containing ion channels by Miller and Racker (1976), elaborated on by Cohen *et al.* (1982, 1984), and used to incorporate sodium channels into bilayers by Krueger *et al.* (1983). These techniques are discussed in detail in other chapters of this book. To avoid redundancy, we concentrate in this chapter on the specific modifications to these procedures that we have found valuable.

3.1.1. Aperture Fabrication

The aperture across which the bilayer is formed is a critical component of the system. Its area controls the major source of noise—the capacitance of the bilayer. Smooth, circular apertures 30–300 μm in diameter are readily produced in a 25-μm-thick Teflon® film (Yellow Springs Instruments #5775, Yellow Springs, OH) when an electrical spark generated by an automobile ignition coil (1967 Rambler, Eddie's Easy Auto, Chula Vista, CA) is passed through the Teflon®. The ignition coil, with its associated capacitor connected across its low-voltage terminals, is connected to a 12- or 22-V battery through a potentiometer and a toggle switch. An automotive spark plug with its ground electrode removed makes a convenient high-voltage electrode. It is necessary first to create a small defect in the Teflon® film by poking it with a needle or the tip of a pulled glass rod. Then support the Teflon® so that it is approximately centered in a 2- to 5-mm gap between the high-voltage and ground electrodes and discharge a spark through it by closing and opening the circuit between the battery and the coil. The size of the aperture is determined by the battery voltage and the number of sparks passed through the hole. A 12-V battery producs a ≈90-μm aperture with a single spark, and both higher and lower voltages yield smaller holes. The spark melts the Teflon® to create a smooth aperture that is an excellent support for stable, long-lasting bilayers.

3.1.2. Vesicle Formation

Purified sodium channels are reconstituted into lipid vesicles composed of 35% bovine brain PE (Sigma) and 65% bovine brain PC (Sigma) by the procedure described in Section 2.1.1. using 0.25–0.4 ml of Bio-Beads® per milliliter of reconstitution mixture. The resulting vesicles are composed of 2 mg/ml PC, 1.05 mg/ml PE, and 5–15 pmol/ml STX receptor in 50 mM NaCl, 10 mM HEPES/tris, pH 7.4, 0.5 mM $MgSO_4$, and 400 mM sucrose. In some experiments additional sucrose (1.1 M final concentration) is added to the solution just prior to reconstitution.

3.1.3. Experimental Setup

The Teflon® sheet containing the aperture is attached with silicone grease to a partition containing a 2-mm hole that separates two 450-μl chambers that have been milled from a block of Teflon®. The small aperture reduces both the capacitance of the system and its microphonic noise pickup. The Teflon® block rests on a battery-powered magnetic stirrer enclosed in a Faraday cage mounted on a vibration-free table. The *cis* chamber contains 0.5 M NaCl and 1–5 μl of vesicles containing reconstituted purified sodium channels (5–75 fmol) in medium A (10 mM HEPES/Na^+, pH 7.4, 0.15 mM $CaCl_2$, 0.1 mM $MgCl_2$, and 0.05 mM EGTA). The *trans* chamber initially contains 0.2–0.4 M NaCl in medium A plus 1 μM BTX. The chambers are connected by Ag/AgCl pellet electrodes (In Vivo Metric Systems, Healdsburg, CA) to a List Medical Electronics EPC-7 patch-clamp amplifier in voltage-clamp mode. The EPC-5 amplifier does not work because of the high capacitance of the bilayer. The *trans* electrode is set to the command voltage relative to the *cis* electrode, which is held at ground. The EPC-7 output is filtered at 3 kHz, monitored on an oscilloscope, and recorded on a RACAL 4DS FM tape recorder. All experiments are performed at 21°C ± 2°C.

3.1.4. Bilayer Formation

To form stable bilayers, the area of Teflon® around the aperture is first thoroughly cleaned in hexane then precoated on both sides with a thin film of 5%

lipid in *n*-decane (Sigma) and dried for 3–5 min under a gentle stream of nitrogen. The precoating step is important. Not enough yields bilayers that spontaneously break in a few minutes. Excess coating yields bilayers whose areas diminish over the course of several minutes, turning into thick (≥1000 Å) films. After the chambers are filled with their aqueous solutions, a small amount of the 5% lipid solution is spread across the *cis* side of the aperture with a clean sable paint brush (#2) from which most of the bristles have been removed. Voltage steps 0.1–1 mV in amplitude and 5 msec in duration are applied to the electrodes to monitor bilayer formation by capacitance change (Mueller and Rudin, 1969). A clean brush is then used to repeatedly wipe away the excess lipid solution until a bilayer is formed as indicated by a large increase in capacitance. A few more brush strokes may be needed to increase the specific capacitance of the bilayer and torus to 0.25–0.35 $\mu F/cm^2$. An alternative method to form bilayers that works well for small apertures starts by spreading on a small amount of lipid solution on the *cis* side of the aperture with a brush and then using a clean brush to break the high-resistance seal across the aperture. Carefully lowering the aqueous solution in the cis chamber with a pipet until the aperture is exposed to air and then slowly raising the solution to its original level forms a bilayer across the aperture. This will require several tries.

The lipid composition of the bilayer is also important. We have used a mixture of 40 mg/ml 1-palmitoyl-2-oleoylphosphatidylethanolamine (POPE) and 10 mg/ml 1-palmitoyl-2-oleoyl phosphatidylcholine (POPC) (Avanti Polar Lipids) in *n*-decane as suggested by Larry Weiss (Weiss *et al.*, 1984) to form stable, long-lasting bilayers. We have also used a nonoxidizable lipid, diphytanoylphosphatidylcholine (Avanti Polar Lipids) (Redwood *et al.*, 1971), as a 50 mg/ml solution in *n*-decane. This lipid works better for small apertures (<100 μm diameter) because it thins to form the bilayer configuration more readily and vesicles are more likely to fuse with it. Bilayers of both types typically have high resistances (>300 GΩ), are electrically stable between − 150 mV and + 150 mV and can last several hours.

3.1.5. Channel Incorporation into Lipid Bilayers

The fusion approach is used to incorporate lipid vesicles containing purified sodium channels into planar lipid bilayers. After formation of a planar bilayer, vesicles containing reconstituted sodium channels are added to the *cis* chamber, the solutions in the *cis* and *trans* chambers are gently stirred for 3 min while an applied voltage (V) of 70 mV is maintained across the bilayer. The current through the bilayer is then examined at V = 70 mV and −70 mV to determine if any channels have been incorporated. If not, the bilayer capacitance is rechecked, and the stirring step is repeated until one or more channels are seen (Hartshorne *et al.*, 1985). Sodium channels are usually observed within 30 min when 2 μl of vesicles, a 0.5 M_{cis}, 0.2 M_{trans} NaC1 gradient, and a POPE/POPC bilayer are used in a 220-μm aperture. Channel incorporation is stopped by eliminating the osmotic gradient across the bilayer with the addition of 4 M NaC1 to the *trans* chamber to yield a final concentration of 0.5 M. In approximately 30% of the bilayers, under the conditions above, channel incorporation could be stopped with only one sodium channel in the bilayer. Decreasing the size of the bilayer, the transbilayer osmotic gradient, the number of sodium channels per vesicle, or the vesicle concentration will increase the percentage of single-channel bilayers but will also prolong the time required before channel incorporation occurs. Preliminary experiments with vesicles containing 1.1 M sucrose suggest that this increases the rate of channel incorporation.

3.1.6. Voltage Convention

Since sodium channels are oriented randomly in the vesicles (Section 2.2.2), their orientation in the bilayer is necessarily random. This could cause confusion in comparing a voltage-dependent property of channels oriented in opposite directions. To avoid this ambiguity, the voltage sensed by the channel is defined according to the electrophysiological convention ($V = V_{\text{``intracellular'' side}} - V_{\text{``extracellular'' side}}$). The "extracellular" side of a sodium channel can readily be determined by the sidedness of TTX block at the end of an experiment. Alternatively, the polarity of the applied voltage sensed as a hyperpolarization that closes a channel will also indicate its orientation. When both methods have been employed to determine the orientation of a channel, the results have always agreed ($n = 25$).

3.2. Properties of Purified Sodium Channels in Planar Lipid Bilayers

3.2.1. Tetrodotoxin Sensitivity

The most specific test of whether an ion channel in a planar lipid bilayer is a sodium channel is its susceptibility to block by TTX or STX. Figure 7 demonstrates that the ionic current through purified sodium channels is reversibly blocked by TTX in an orientation-dependent manner (Hartshorne *et al.*, 1985). The bilayer in Fig. 7 contains four active channels. Prior examination of the polarity of the voltage that closes the channels by hyperpolarization indicated that there are three *cis*-facing channels and one *trans*-facing channel. Following addition of TTX to the *trans* chamber in Fig. 7A, the one *trans*-facing sodium channel is blocked and remains blocked for the rest of the experiment while the three *cis*-facing channels are unaffected. Addition of TTX to the *cis* side of the bilayer in Fig. 7B reduces the current in three discrete steps to the value it had prior to channel incorporation as the three *cis*-facing channels are blocked. The initial segment of Fig. 7C, on a tenfold slower time scale, shows that the channels remain blocked as long as TTX is present. Following perfusion of the *cis* chamber to remove TTX, the three *cis*-facing channels reappear, demonstrating that TTX block is reversible. The TTX sensitivity of these single-channel events is unequivocal evidence that they are mediated by the sodium channel protein.

The concentration and voltage dependence of TTX block of purified BTX-activated sodium channels are presented in Table II. The concentration dependence of block is well described by a simple one-to-one binding curve. At −50

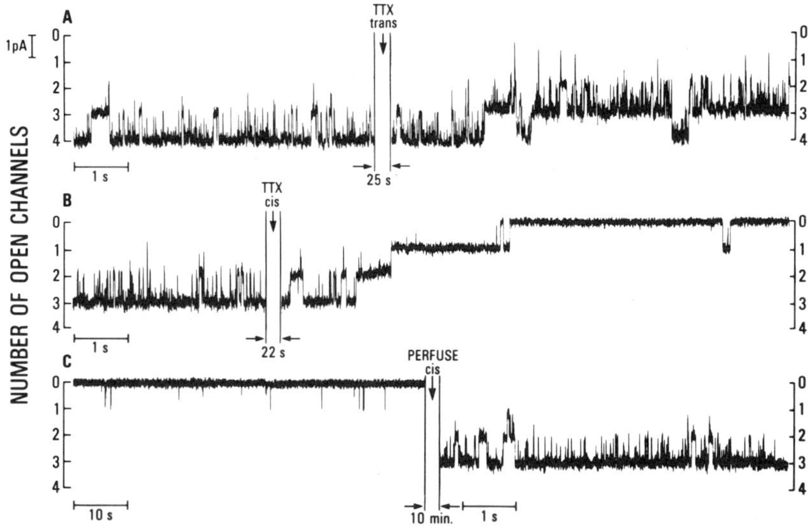

Figure 7. Tetrodotoxin block of reconstituted sodium channels in planar lipid bilayers. This POPE/POPC membrane contained one *trans*-facing and three *cis*-facing sodium channels in symmetric 0.5 M NaCl medium A with 1 μM BTX *trans*. *V* was −50 mV for the *cis*-facing channels and +50 mV for the *trans*-facing channel. The records were filtered at 500 Hz. A: Orientation dependence of TTX block. The initial current record shows four open channels interrupted by occasional brief closures. At the arrow, 40 μM TTX (final concentration) was added to the *trans* chamber. The 25-sec delay between the addition of TTX and channel block was the result of slow mixing of TTX with the solution. This was intentionally done to allow the channel-blocking event to be recorded. B: Continuation of current record in A 3 min later, following complete mixing of the *trans* chamber. Addition of 2 μM TTX to the *cis* chamber blocks the remaining channels. C: Reversibility of TTX block. With TTX on both sides, all channels remain blocked with occasional brief openings. On removal of TTX from the *cis* chamber by perfusion (exchange of 14 volumes with stirring), the membrane current returns to the pre-TTX value. The *trans*-facing channel remains blocked.

mV, the K_I for TTX block is 8.3 nM, whereas at 70 mV the K_I is 135 nM (Hartshorne *et al.*, 1985). The voltage dependence of this shift in K_I corresponds to an e-fold increase in K_I for each 43-mV depolarization. Similar values for the K_I of TTX block and its voltage dependence were found with synaptosomal sodium channels incorporated into planar bilayers by B. K. Krueger and R. J. French (personal communication) (Table II).

3.2.2. Channel Opening is Voltage Dependent

The next property of purified sodium channels evaluated in the planar bilayer system is the voltage dependence of gating in the presence of BTX. Batrachotoxin should shift the voltage dependence of sodium-channel activation in the hyperpolarized direction and eliminate inactivation. Figure 8 shows the single-channel current through a BTX-modified purified sodium channel at the holding potentials indicated. The excellent time resolution and low noise (0.3 pA RMS noise at 3

Table II. Comparison of Functional Properties of Purified and Native Sodium Channels from Rat Brain

Property	Purified	Native
BTX $K_{0.5}$	2 μM[a]	0.5 μM[b]
Veratridine $K_{0.5}$	30 μM[a]	14 μM[b]
Scorpion toxin K_D	1.9 nM[c]	2 nM[b]
STX K_D	0.4 nM[a]	0.2 nM[d]
TTX K_I at $V = -50$ mV	8.3 nM[e]	7–16 nM[f]
mV/e-fold shift in K_I (TTX)	43 mV[e]	35–41 mV[f]
V_{50} for channel opening	−91 mV[e]	−93 mV[g]
Apparent gating charge (q)	3.8[e]	4–6[g]
γ (0.5 M NaCl)	25 pS[e]	30 pS[h]
Permeability relative to Na^+		
P_{K^+}	0.14[e]	0.07[h], 0.17[i]
P_{Rb^+}	0.05[e]	0.08[i]

[a] Tamkun *et al.* (1984).
[b] Tamkun and Catterall (1981b).
[c] Feller *et al* (1985).
[d] Catterall *et al.* (1979).
[e] Hartshorne *et al.* (1985).
[f] R. J. French and B. K. Krueger, personal communication.
[g] French *et al.* (1984).
[h] Krueger *et al.* (1983).
[i] Data from N18 neuoblastoma cells; Huang *et al.* (1979).

kHz) of these records approach that of patch recording and are an example of what can be achieved using small membrane areas. In the first current record at −135 mV, the channel is closed nearly all of the time except for a few brief openings indicated by downward deflections. At more positive voltages, the channel spends an increasing fraction of time open. The voltage at which the channel is open 50% of the time (V_{50}) is −95 mV. At −55 mV and more positive voltages, the channel is open most of the time with occasional closures. As expected for a BTX-modified channel, it does not inactivate.

For purified sodium channels, the average value of V_{50} is −91 mV ± 17 mV (SE, n = 22) (Hartshorne *et al.*, 1985). This is in good agreement with the V_{50} of −93 mV for BTX-modified synaptosomal sodium channels in planar bilayers (French *et al.*, 1984), −94 mV for BTX-modified rat sarcolemmal sodium channels in planar bilayers (Moczydlowski *et al.*, 1984), −90 mV for BTX-modified sodium channels in frog node of Ranvier (Khodorov and Revenko, 1979), and −85 mV for BTX-modified sodium channels in NG-108-15 neuroblastoma cells (Huang *et al.*, 1982). The V_{50} of purified sodium channels varies from channel to channel. For most of the channels, V_{50} is between −80 mV and −105 mV, with the extreme values being −130 mV and −64 mV. Figure 9A shows the voltage dependence of channel opening for three different channels selected because of their widely differing V_{50} values. Channel opening is steeply voltage dependent over a limited voltage range for each channel. The curves are parallel to each other and shifted along the voltage axis.

Figure 9B shows a linear transformation (French *et al.*, 1984) of the curves whose slope is proportional to the apparent gating charge (q) of the sodium chan-

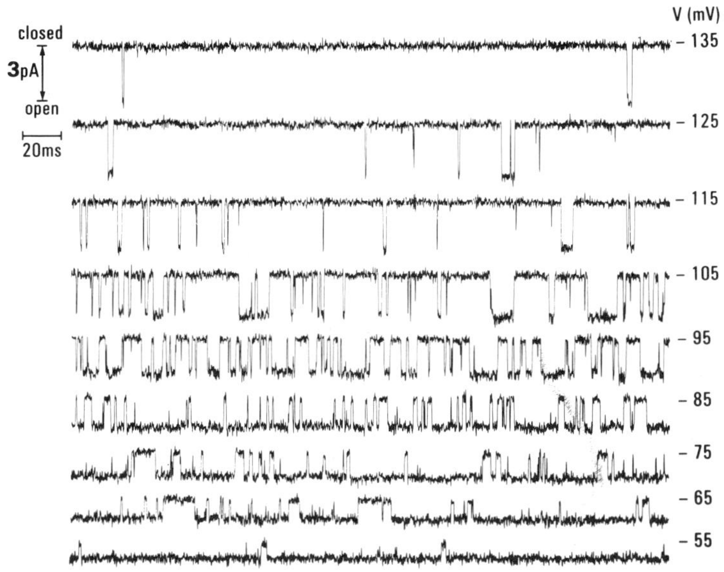

Figure 8. Single-sodium-channel current records. A *trans*-facing sodium channel was incorporated into a diphytanoyl -PC bilayer in a 70-μm aperture bathed by 0.5 M NaCl medium A *cis* and 0.2 M NaCl medium A plus 1 μM BTX *trans*. The voltages indicated were applied to the channel for 1-min intervals. Representative current records, filtered at 1 kHz, are shown.

nel. Despite the differences in V_{50}, the apparent gating charge for each of these three channels is essentially the same. The apparent gating charge for purified BTX-modified sodium channels is 3.8 ± 0.3 (SE, $n = 5$) electronic charges (Hartshorne *et al.*, 1985). Similar values of q were found for BTX-modified sodium channels from frog node of Ranvier (3.5: Khodorov and Revenko, 1979), rat brain (4–6: French *et al.*, 1984), and NG-108-15 neuroblastoma cells (4–5: Huang *et al.*, 1984). The agreement for the values of q and the average V_{50} between purified reconstituted sodium channels and native sodium channels is evidence that the voltage-sensing and gating mechanisms are present and functionally intact in the purified protein. However, the variation of V_{50} needs to be addressed.

What is the origin of the variation in V_{50} from channel to channel? The trivial explanation of an uncompensated electrode offset potential can be ruled out by experiments showing that the reversal potential for the Na^+ current is always within 1 or 2 mV of the calculated Nernst potential for Na^+. The fact that the voltage dependence curves are parallel (Fig. 9) rules out the possibility that the channels are incorporated into the bilayer in such a way that they sense only a fraction of the applied electric field. Lipid surface charge effects are unlikely since these experiments are conducted with symmetric bilayers of uncharged lipids in solutions with symmetric ionic composition. Is the variability caused by alteration of the channel protein during its purification? Probably not, because similar results were obtained with BTX-modified sodium channels incorporated into planar bi-

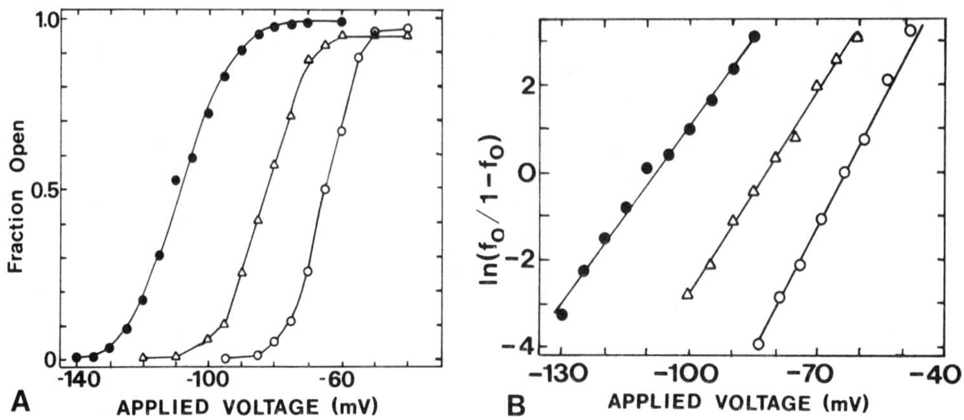

Figure 9. A: Voltage dependence of sodium channel opening. The fraction of time a channel was open was determined by integrating conductance histograms derived from 25 sec of current record at each voltage as described in Hartshorne *et al.* (1985). The voltage dependence for three different channels, selected because of their widely differing V_{50} values, is illustrated. The channels are in POPE/POPC bilayers bathed in symmetric 0.5 M NaCl medium A plus 1 μM BTX *trans*. B: Apparent gating charge. For each of the channels in Fig. 9A, the fraction of time it is open (f_0) is transformed to a linear function of voltage by plotting ln ($f_0/1 - f_0$). The lines are least-squares fit to a linearized form of the Boltzmann equation, where the slope is proportional to the apparent gating charge.

layers from native membranes (Weiss *et al.*, 1984; Moczydlowski *et al.*, 1984) and with patch-clamped BTX-modified sodium channels in neuroblastoma cells (Huang *et al.*, 1984), although in this case the variation of V_{50} from channel to channel was less extensive. At present, we favor the possibility that the variation in V_{50} reflects a naturally occuring heterogeneity in the sodium channel population, perhaps arising from variations in fixed charge on or near the voltage-sensing mechanism. The planar bilayer system is ideally suited for investigations of possible biochemical mechanisms modifying the voltage dependence of sodium channels in a well-defined environment.

3.2.3. Progress towards Voltage-Dependent Activation

A major objective of this research is the examination of voltage-dependent activation and inactivation of purified sodium channels in the absence of BTX. The high capacitance associated with bilayers in the 220-μm aperture in which the initial results were obtained precluded the observation of channel opening after a depolarizing voltage step. A large slowly decaying capacitive current saturates the amplifier for 5 msec after the change of voltage. Reducing this capacitive transient has been the impetus behind examining bilayers in small apertures. With diphytanoyl-PC/decane bilayers in 70-μm apertures, 30-mV steps result in the loss of only the first 0.3 msec of data when the capacitance cancellation circuitry in the EPC-7 is used. Digital subtraction techniques can further reduce this loss.

Since the electrical noise of these bilayers is low (see Fig. 8), the current can be filtered at 3 kHz, which would allow channel openings as short as 300 μsec to be clearly resolved from the noise.

Channel activity in the absence of BTX has not yet been observed. Either nonfunctional channels are being incorporated into the bilayer or, in the absence of BTX, sodium channels are not incorporated into the bilayer under the conditions used. The following experiment favors the second possibility. Under conditions chosen to promote fusion (5 μl of vesicles, a 0.5 M_{cis}–0.2 M_{trans} NaCl gradient, and a diphytanoyl PC bilayer in a 220-μm aperture), we attempted to incorporate sodium channels into the bilayer using the procedure described in Section 3.1.5 in the absence of BTX for 1 hr (0.5 μl of ethanol was added in place of the solution of BTX in ethanol). To stop any additional fusion, concentrated NaCl was then added to the *trans* chamber, yielding a final concentration of 0.5 M. To determine if any sodium channels had been incorporated during the 1-hr incubation, BTX (1 μM final *trans*) was added, and the current through the bilayer was monitored for 30 min. No sodium channels were seen. The contents of the *trans* chamber were then diluted with distilled water to a final concentration of 0.3 M NaCl. In each of four separate experiments, sodium channels (one to three) were then seen within 10 min. These results clearly indicate that BTX is required for vesicle fusion under these conditions. Whether BTX promotes fusion by increasing the osmotic gradient across the vesicle bilayer by increasing its sodium permeability or by causing a hydrophobic surface on the sodium channel to become exposed is not known. We are currently examining other methods of promoting channel incorporation into the bilayer to allow us to study the purified sodium channel in the absence of BTX.

3.2.4. Single-Channel Conductance and Ionic Selectivity

Purified sodium channels are selectively permeable to Na^+. This is evident from the single-channel current–voltage curves for Na^+, K^+, and Rb^+ in Fig. 10 (Hartshorne *et al.*, 1985). The average single-channel conductance (γ) in 0.5 M solutions of these ions is γ_{Na^+} = 25.2 pS ± 0.5 pS (SE, n = 10), γ_{K^+} = 3.5 pS ± 0.2 pS (SE, n = 3), and γ_{Rb^+} = 1.2 pS ± 0.1 pS (SE, n = 3). The single-channel conductance for Na^+ of 25.2 pS is a little lower than the 30 pS found by Krueger *et al.* (1983) and a bit higher than the 24 pS found by Moczydlowski *et al.* (1984) for native rat brain sodium channels incorporated into planar bilayers and activated by BTX in 0.5 M NaCl. The permeability ratios of K^+ and Rb^+ relative to Na^+ are 0.14 ± 0.02 and 0.05 ± 0.01, respectively. These permeability ratios agree well with the permeability ratios directly measured from ion-flux determinations with BTX-activated sodium channels in N18 neuroblastoma cells (P_{K^+} = 0.17 and P_{Rb^+} = 0.08, Huang *et al.*, 1979) and sodium channels purified from rat muscle (P_{K^+} = 0.14 and P_{Rb^+} = 0.02, Tanaka *et al.*, 1983). The agreement is not as good with the relative permeability of K^+ (0.07, Krueger *et al.*, 1983) determined from reversal-potential measurements on BTX-activated rat brain sodium channels in planar bilayers. The relative permeability of sodium channels calculated from reversal-potential measurements varies with the ionic

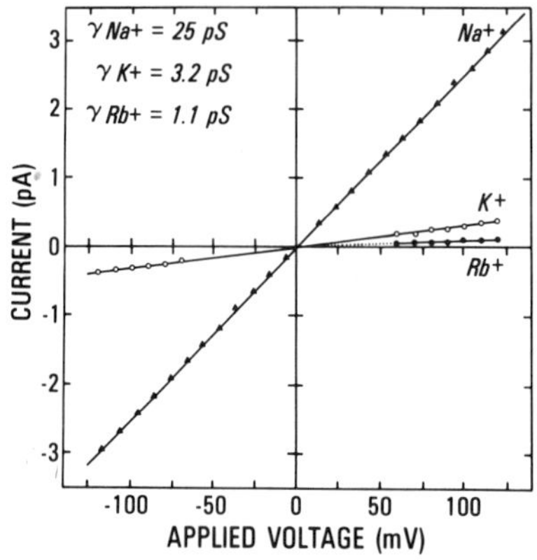

Figure 10. Current–voltage relationship of single sodium channels for Na^+, K^+ and Rb^+. The amplitude of single-channel currents was determined in 0.5 M symmetric solutions of NaCl (▲), KCl (○), and RbCl (●) in 10 mM HEPES/tris, pH 7.4, 0.15 mM $CaCl_2$, 0.1 mM $MgCl_2$, and 0.05 mM EGTA with 1 μM BTX *trans*. To accurately determine the small single-channel current in RbCl, 200 nM TTX was added to the appropriate chamber to produce long closed periods, and the current records were filtered at 50 Hz to resolve the open and blocked channels from the noise. For this reason, only the positive branch of the curve for Rb^+ is shown. The continuous lines are the least-square fits to the data points. The slope of a line is the single-channel conductance for the ion: γ_{Na^+} = 25 pS, γ_{K^+} = 3.2 pS, and γ_{Rb^+} = 1.1 pS in this experiment. The uncorrected zero-current potentials were 1.4 mV and 1.5 mV for Na^+ and K^+, respectively.

composition of the medium (Cahalan and Begenisich, 1976), which may account for this difference.

4. Conclusions

The functional properties of purified sodium channels from rat brain reconstituted into lipid vesicles and planar lipid bilayers are compared with those of native rat brain sodium channels in Table II. The functional characteristics of the purified sodium channel are comparable to those of the native sodium channel in nearly every aspect tested. Of particular importance is the finding that a large fraction of the STX binding sites are associated with functional ion channels (Tamkun *et al.*, 1984). This rules out the possibility that trace amounts of other polypeptides are involved in sodium-channel function. In addition, it indicates that the sodium channels sampled in the planar bilayer studies are likely to be representative of the whole population. The conclusion from the reconstitution studies in both lipid vesicles and planar bilayers is clear. The complex of α, β1, and β2 subunits that make up the purified saxitoxin-binding protein contains functional binding sites for TTX, LqTX, BTX, and veratridine and forms a functional ion-conducting pore that exhibits both the ionic selectivity and the voltage-dependent gating of the native sodium channel in the presence of BTX. Thus, the α, β1, and β2 subunits are sufficient to account for the entire structure of the sodium channel from rat brain.

The initial goal of reconstitution, to establish if all the functional elements of the sodium channel are in the purified protein, has now been achieved. The immediate objectives are to determine which subunits and domains of subunits are involved in opening and closing the channel, forming its structure, and governing its voltage regulation. Also, reconstitution techniques provide an especially sensitive assay to investigate the influences of the lipid environment on sodium channel function. It is anticipated that the combination of recombinant DNA technology to modify the gene(s) encoding the sodium channel (Noda *et al.*, 1984), biochemical alterations of purified sodium channels, and reconstitution studies as described here will prove valuable the pursuit of understanding of how voltage controls the gating of this channel protein.

ACKNOWLEDGMENTS. We would like to acknowledge the contributions of the following people: Dr. Jane Talvenheimo, who performed the ion-flux measurements in reconstituted vesicles and provided the vesicles for the planar bilayer studies; Dr. William Catterall, who was an instrumental collaborator in all of this work; Mr. Bernhard Keller, who did the initial bilayer experiments with sodium channels in Dr. Montal's lab and performed the single-channel conductance and ion-selectivity measurements in planar bilayers; Mr. Donald Messner, who supplied the gel shown in Fig. 1; Dr. Clara Riddle, who performed the freeze–fracture electron microscopy; Dr. John Daly, for generously supplying the BTX used in these experiments; Dr. Larry Weiss, who suggested the use of POPC/POPE bilayers; Dr. Peter Schoch, who suggested using electrical discharge for aperture fabrication; Mr. Tony Lee and Dr. Richard Greenblatt, who did the computer programming; R. H. would like to acknowledge the contributions of his new daughter Elizabeth Hartshorne with whom he had many late night discussions on the contents of this chapter. This work was supported by The National Multiple Sclerosis Society, NIH #EY02084, and DAMR 17-82-C221.

References

Agnew, W. S., Levinson, S. R., Brabson, J. S., and Raftery, M. A., 1978, Purification of the tetrodotoxin-binding component associated with the voltage-sensitive sodium channel from *Electrophorus electricus* electroplax membranes, *Proc. Natl. Acad. Sci. U.S.A.* **75:**2606–2610.

Albuquerque, E. X., and Daly, J. W., 1976, Batrachotoxin: A selective probe for channels modulating sodium conductances in electrogenic membranes, in: *The Specificity and Action of Animal, Bacterial, and Plant Toxins* (P. Cuatrecasas, ed.), pp. 298–338, Chapman and Hall, London.

Barchi, R. L., 1983, Protein components of the purified sodium channel from rat skeletal muscle sarcolemma, *J. Neurochem.* **40:**1377–1385.

Barchi, R. L., Cohen, S. A., and Murphy, L. E., 1980, Purification from rat sarcolemma of the saxitoxin-binding component of the excitable membrane sodium channel, *Proc. Natl. Acad. Sci. U.S.A.* **77:**1306–1310.

Barhanin, J., Pauron, D., Lombet, A., Norman, R., Vijverberg, H., Giglio, J. R., and Lazdunski, M., 1983, Electrophysiological characterization, solubilization, and purification of the *Tityus* γ toxin receptor associated with the gating component of the Na^+ channel from rat brain, *EMBO J.* **2:**915–920.

Beneski, D. A., and Catterall, W. A., 1980, Covalent labeling of protein components of the sodium channel with a photoactivable derivative of scorpion toxin, *Proc. Natl. Acad. Sci. U.S.A.* **77:**639–643.

Cahalan, M., and Begenisich, T., 1976, Sodium channel selectivity: Dependence on internal permeant ion concentration, *J. Gen. Physiol.* **68:**111–125.

Catterall, W. A., 1976, Purification of a toxic protein from scorpion venom which activates the action potential Na^+ ionophore, *J. Biol. Chem.* **251:**5528–5536.

Catterall, W. A., 1977, Membrane potential dependent binding of scorpion toxin to the action potential Na^+ ionophore, *J. Biol. Chem.* **252:**8660–8668.

Catterall, W. A., 1980, Neurotoxins that act on voltage sensitive sodium channels in excitable membranes, *Annu. Rev. Pharmacol. Toxicol.* **20:**15–43.

Catterall, W. A., Morrow, C. S., and Hartshorne, R. P., 1979, Neurotoxin binding to receptor sites associated with voltage-sensitive sodium channels in intact, lysed, and detergent solubilized brain membranes, *J. Biol Chem.* **254:**11379–11387.

Cohen, F. S., Akabas, M. H., and Finkelstein, A., 1982, Osmotic swelling of phospholipid vesicles causes them to fuse with a planar phospholipid bilayer membrane, *Science* **217:**458–460.

Cohen, F. S., Akabas, M. H., Zimmerberg, J., and Finkelstein, A., 1984, Parameters affecting the fusion of unilamellar phospholipid vesicles with planar bilayer membranes, *J. Cell. Biol.* **98:**1054–1062.

Darbon, H., Jover, E., Couraud, F., and Rochat, H., 1983, Photoaffinity labeling of the α and β-scorpion toxin receptors associated with rat brain sodium channel, *Biochem. Biophys. Res. Commun.* **115:**415–422.

Drachev, L. A., Jasaitis, A. A., Koulen, A. D., Kondrashin, H. A. Liberman, E. A., Hayrecek, I. B., Ostroumov, S. A., Semenov, A. Y., and Skulachev, V. P., 1974, Direct measurement of electric current generation by cytochrome oxidase, H^+-ATPase, and bacteriorhodopsin, *Nature* **249:**321–323.

Epstein, M., and Racker, E., 1978, Reconstitution of carbamylcholine-dependent sodium ion flux and desensization of the acetylcholine receptor from *Torpedo californica, J. Biol. Chem.* **253:**6660–6662.

Feller, D. J., Talvenheimo, J. A., and Catterall, W. A., 1984, Restoration of voltage dependent state change in the purified Na^+ channel from rat brain, *Soc. Neurosci. Abstr.* **10:**863.

Feller, D. J., Talvenheimo, J. A., and Catterall, W. A., 1985, The sodium channel from rat brain: Reconstitution of scorpion toxin binding site in vesicles of defined lipid composition, *J. Biol. Chem.* **260:**11542–11547.

French, R. J., Worley, J. F., and Krueger, B. K., 1984, Voltage-dependent block by saxitoxin of sodium channels incorporated into planar lipid bilayers, *Biophys. J.* **45:**301–310.

Gasko, O. D., Knowles, A. F. Shertzer, H. G., Snolinna, E. M., and Racker, E., 1976, The use of ion-exchange resins for studying ion transport in biological systems, *Anal. Biochem.* **72:**57–65.

Grishin, E. V., Kovalenko, V. A., Pashkov, V. N., and Shamotienko, O. G., 1984, Isolation and characterization of sodium channel components, *Membr. Biochem.* (*U.S.S.R.*) **1:**858–867.

Hanke, W., Boheim, G., Barhanin, J., Pauron, D., and Lazdunski, M., 1984, Reconstitution of highly purified saxitoxin-sensitive sodium channels into planar lipid bilayers, *EMBO J.* **3:**509–515.

Hartshorne, R. P., and Catterall, W. A., 1981, Purification of the saxitoxin receptor of the sodium channel from rat brain, *Proc. Natl. Acad. Sci. U.S.A.* **78:**4620–4624.

Hartshorne, R. P., and Catterall, W. A., 1984, The sodium channel from rat brain: Purification and subunit composition, *J. Biol. Chem.* **259:**1667–1675.

Hartshorne, R. P., Messner, D. J., Coppersmith, J. C., and Catterall, W. A., 1982, The saxitoxin receptor of the sodium channel from rat brain: Evidence for two nonidentical β subunits, *J. Biol. Chem.* **257:**13888–13891.

Hartshorne, R. P., Keller, B. U., Talvenheimo, J. A., Catterall, W. A., and Montal, M., 1985, Functional reconstitution of the purified brain sodium channel in planar lipid bilayers, *Proc. Natl. Acad. Sci. U.S.A.*. **82:**240–244.

Holloway, P. W., 1973, A simple procedure for removal of Triton X-100 from protein samples, *Anal. Biochem.* **53:**304–308.

Huang, L. M., Catterall, W. A., and Ehrenstein, G., 1979, Comparison of ionic selectivity of batrachotoxin-activated channels with different tetrodotoxin dissociation constants, *J. Gen. Physiol.* **73:**839–854.

Huang, L. M., Moran, N., and Ehrenstein, G., 1982, Batrachotoxin modifies the gating kinetics of sodium channels in internally perfused neuroblastoma cells, *Proc. Natl. Acad. Sci. U.S.A.* **79:**2082–2085.

Huang, L. M., Moran, N., and Ehrenstein, G., 1984, Gating kinetics of batrochotoxin modified sodium channels in neuroblastoma cells determined from single channel measurements, *Biophys, J.* **45:**313–322.

Khodorov, B. I., and Revenko, S. V., 1979, Further analysis of the mechanisms of action of batrachotoxin on the membrane of myelinated nerve, *Neuroscience* **4:**1315–1330.

Koppenhöfer, E., and Schmidt, H., 1968, Die Wirkung von Scorpiongift auf die Ionenströme des Ranvierschen Schnürrings II. Unvollständige Natrium Inaktivierung, *Pflüegers Arch.* **303:**150–161.

Kreuger, B. K., Worley, J. F., and French, R. J., 1983, Single sodium channels from rat brain incorporated into planar lipid bilayer membranes, *Nature* **303:**172–175.

Lombet, A., and Lazdunski, M., 1984, Characterization, solubilization, affinity labeling and purification of the cardiac Na^+ channel using *Tityus* toxin γ, *Eur. J. Biochem.* **141:**651–666.

Messner, D. J., and Catterall, W. A., 1985, The sodium channel from rat brain: Separation and characterization of subunits, *J. Biol. Chem.* **260:**10597–10604.

Miller, C., and Racker, E., 1976, Ca^{2+} induced fusion of fragmented sarcoplasmic reticulum with artificial bilayers, *J. Membr. Biol.* **30:**283–300.

Miller, J. A., Agnew, W. S., and Levinson, S. R., 1983, Principal glycopeptide of the tetrodotoxin/saxitoxin binding protein from *Electrophorus electricus*: Isolation and partial chemical and physical characterization, *Biochemistry* **22:**462–470.

Moczydlowski, E., Garber, S. S., and Miller, C., 1984, Batrachotoxin-activated Na^+ channels in planar lipid bilayers: Competition of tetrodotoxin block by Na^+, *J. Gen. Physiol.* **84:**687–704.

Mueller, P., and Rudin, D. O., 1969, Bimolecular lipid membranes: Techniques of formation, study of electrical properties and induction of gating phenomenon, in: *Laboratory Techniques of Membrane Biophysics* (M. Passow and R. Stampfli, eds.), pp. 141–156, Springer-Verlag, Berlin.

Mueller, P., Rudin, D. O., Tien, H. T., and Wescott, W. C., 1963, Methods for the formation of single bimolecular lipid membranes in aqueous solution, *J. Phys. Chem.* **67:**534–535.

Narahashi, T., Anderson, N. C., and Moore, J. W., 1966, Tetrodotoxin does not block excitation from inside the nerve membrane, *Science* **153:**765–767.

Noda, M., Shimizu, S., Tanabe, T., Takai, T., Kayano, T., Ikeda, T., Takahashi, H., Nakayama, H., Kanaoka, Y., Minamino, N., Kangawa, K., Matsuo, H. Raftery, M., Hirose, T., Inayama, S., Hayashida, H., Miyata, T., and Numa, S., 1984, Primary structure of *Electrophorus electricus* sodium channel deduced from cDNA sequence, *Nature* **312:**121–127.

Oakley, B. R., Kirsch, D. R., and Morris, N. R., 1980, A simplified ultrasensitive silver stain for detecting proteins in polyacrylamide gels, *Anal. Biochem.* **105:**361–363.

Quandt, F., and Narahashi, T., 1982, Modification of single Na^+ currents by batrachotoxin, *Proc. Natl. Acad. Sci. U.S.A.* **79:**6732–6736.

Redwood, W. R., Pfeiffer, F. R., Weisbach, J. A., and Thompson, T. E., 1971, Physical properties of bilayer membranes formed from a synthetic saturated phospholipid in n-decane, *Biochim. Biophys. Acta* **233:**1–6.

Rosenberg, R. L., Tomiko, S. A., and Agnew, W. S., 1984a, Reconstitution of neurotoxin-modulated ion transport by the voltage-regulated sodium channel isolated from the electroplax of *Electrophorus electricus*, *Proc. Natl. Acad. Sci. U.S.A.* **81:**1239–1243.

Rosenberg, R. L., Tomiko, S. A., and Agnew, W. S., 1984b, Single channel properties of the reconstituted voltage-regulated Na channel isolated from the electroplax of *Electrophorus electricus*, *Proc. Natl. Acad. Sci. U.S.A.* **81:**5594–5598.

Sharkey, R. G., Beneski, D. A., and Catterall, W. A., 1984, Differential labeling of the α and β_1 subunits of the sodium channel by photoreactive derivatives of scorpion toxin, *Biochemistry* **23:**6078–6086.

Szoka, F., Jr., and Papahadjopoulos, D., 1980, Comparative properties and methods of preparation of lipid vesicles (liposomes), *Annu. Rev. Biophys. Bioeng.* **9:**467–508.

Talvenheimo, J. A., Tamkun, M. M., and Catterall, W. A., 1982, Reconstitution of neurotoxin-stimulated sodium transport by the voltage-sensitive sodium channel purified from rat brain, *J. Biol. Chem.* **257:**11868–11871.

Tamkun, M. M., and Catterall, W. A., 1981a, Reconstitution of the voltage-sensitive sodium channel of rat brain from solubilized components, *J. Biol. Chem.* **256:**11457–11463.

Tamkun, M. M., and Catterall, W. A., 1981b, Ion flux studies of voltage-sensitive sodium channels in synaptic nerve ending particles, *Mol. Pharmacol.* **19:**78–86.

Tamkun, M. M., Talvenheimo, J. A., and Catterall, W. A., 1984, The sodium channel from rat brain:

Reconstitution of neurotoxin-activated ion flux and scorpion toxin binding from purified components, *J. Biol. Chem.* **259:**1676–1688.

Tanaka, J. C., Eccelston, J. F., and Barchi, R. L., 1983, Cation selectivity characteristics of the reconstituted voltage-dependent sodium channel purified from rat skeletal muscle sarcolemma, *J. Biol. Chem.* **258:**7519–7526.

Weigele, J. B., and Barchi, R. L., 1982, Functional reconstitution of the purified sodium channel protein from rat sarcolemma, *Proc Natl. Acad. Sci. U.S.A.* **79:**3651–3655.

Weiss, L. B., Green, W. N., and Andersen, O. S., 1984, Single channel studies on the gating of batrachotoxin modified sodium channels in lipid bilayers, *Biophys. J.* **45:**67.

Chapter 14

GATING OF BATRACHOTOXIN-ACTIVATED SODIUM CHANNELS IN LIPID BILAYERS

ROBERT J. FRENCH, JENNINGS F. WORLEY III, MARC B. BLAUSTEIN, WILLIAM O. ROMINE, Jr., KENNETH K. TAM, and BRUCE K. KRUEGER

1. Introduction

Sodium channels incorporated into planar lipid bilayers share many fundamental properties with sodium channels in general and thus provide a useful system for the study of various aspects of channel behavior. Use of the steroidal alkaloid toxin batrachotoxin (BTX) was a crucial aid in making the first electrical recordings from voltage-dependent sodium channels incorporated into artificial lipid membranes from native membrane vesicles from rat brain (Krueger *et al.*, 1983). Batrachotoxin has now been used to enable studies of sodium channels from other tissues (sarcolemma: Moczydlowski *et al.*, 1984a,b; canine brain: Green *et al.*, 1984) and of reconstituted, purified sodium channels from rat brain (Hartshorne *et al.*, 1985). Rosenberg *et al.* (1984) have recently reported patch-clamp recordings from sodium channels, purified from eel electric organ and reconstituted into liposomes, using depolarizing voltage steps to transiently open the channels in the absence of BTX. However, the use of relatively large (100–200 μm) planar bilayers formed from decane–lipid solutions is at present the easiest and most reliable approach to making electrical recordings from sodium channels in a "reconstituted" system.

In studying channels incorporated into artificial bilayers, a compromise must often be made between attaining a reasonably high probability of incorporation and the resolution of the recordings. Use of membranes having a diameter of 100–250 μm represents such a compromise. The large size, and hence large capacitance, of the bilayers offers a low-impedance pathway to high-frequency noise and thus limits the effective recording bandwidth. In addition, it is difficult to

ROBERT J. FRENCH and KENNETH K. TAM • Department of Biophysics, University of Maryland School of Medicine, Baltimore, Maryland 21201. JENNINGS F. WORLEY III, MARC B. BLAUSTEIN, WILLIAM O. ROMINE, Jr., and BRUCE K. KRUEGER • Department of Physiology, University of Maryland School of Medicine, Baltimore, Maryland 21201.

compensate for the capacitative current transient that follows a step in transmembrane voltage. In painted bilayers, this is a particular problem because of the electrostrictive thinning and the changes in membrane area that result from a change in transmembrane voltage. These processes, and the consequent voltage dependence of the membrane capacitance, give the charging current a complex form and extend its duration. Thus, in large bilayers, the channels must be kept open for fairly long periods (for example, with BTX) in order to allow good resolution of single-channel events. Batrachotoxin accomplishes this by abolishing slow and fast inactivation processes, shifting the activation curve in the hyperpolarizing direction, and slowing the kinetics of gating (e.g., Albuquerque *et al.*, 1971; Narahashi *et al.*, 1971; Khodorov, 1978; Huang *et al.*, 1982). Thus, the channels are open for relatively long periods (tens of milliseconds; see also Quandt and Narahashi, 1982) at steady holding potentials over a relatively wide range of voltages. Such steady-state recordings not only have enabled the study of channel gating but also have allowed unexpected new observations of voltage-dependent block of open sodium channels by the guanidinium toxins saxitoxin (STX) and tetrodotoxin (TTX) (Krueger *et al.*, 1983; French *et al.*, 1984; Moczydlowski *et al.*, 1984a,b; Green *et al.*, 1984).

In this chapter, we give an overview of some of the basic properties of the gating of BTX-activated sodium channels in planar bilayers and compare these observations with studies of sodium channel gating in other systems.

2. Practical Considerations

2.1. Bilayer Formation

"Painted" planar bilayers were formed by applying a solution of the desired lipid (40–50 mg/ml) in decane to a hole in a plastic partition separating two chambers that contained buffered salt solutions. The method was essentially that of Mueller *et al.* (1963).

The holes on which the membrane was formed were drilled in a thinly machined area in the side of a plastic cup, which defined one of the aqueous chambers. For ease of machining and durability, the materials of choice for these cups were Lexan® or Kel-F®, the latter being a chlorofluoropolyethylene with wetting properties similar to Teflon®. As a precaution to reduce the likelihood of an aqueous film forming over the edge of the cup and causing electrical leaks and drift in recordings, a thin line of petroleum jelly was applied around the top of the cup. In some experiments, a two-piece chamber divided by a Teflon® film partition was used. To form the holes in these partitions, an electrically heated fine platinum wire was used.

2.2. Channel Incorporation

Channels were incorporated from a suspension of membrane vesicles prepared from rat forebrain (Krueger *et al.*, 1979) using the procedures described by

Krueger *et al.* (1983) and French *et al.* (1984). Incorporation occurred directly from the experimental solutions, with the same solution on each side of the bilayer. Vesicles were added on only one side (*cis*) of the bilayer, and the orientation of incorporated channels was consistent with a fusion mechanism (Miller, 1978; Cohen *et al.*, 1980; 1982). Unless otherwise stated, solutions contained (in mM) 250 NaCl, 10 Na-HEPES, 0.15 $CaCl_2$, 0.1 $MgCl_2$, 0.1 EGTA, and were buffered to a pH of 7.0. Batrachotoxin (120–600 nM) was normally present on the *trans* side of the bilayer (i.e., the side opposite that of vesicle addition) throughout the experiments, although addition of BTX to the *cis* side gave equivalent results.

2.3. Recording Amplifiers

Since we used relatively large bilayer membranes, the dominant source of noise was due to the expression of the voltage noise at the input of the current-to-voltage converter across the membrane capacitance (see Sigworth, 1983, for a detailed account of the various sources in single-channel records). For some experiments we have used commercial patch clamps (EPC-7, List, supplied by Medical Systems Corporation, Great Neck, NY; also an instrument built by the electronics shop of the Department of Physiology, Yale University School of Medicine, New Haven, CT), with the head stage mounted in the shielded box containing the bilayer chamber.

We have also used simpler amplifiers built in our own electronics shop, which were completely housed in the box containing the preparation (cf. Chapter 5). These amplifiers consisted of the following elements: (1) a command amplifier (gain 0.1× to reduce noise) connected to one side of the bilayer; (2) a simple transient compensator consisting of an inverting variable-gain amplifier that fed the command signal through a variable capacitor into the input of the current-to-voltage converter; (3) a current-to-voltage converter, using either a single op-amp (Burr-Brown OPA102 or OPA111) or a discrete BIFET input stage, with a 1- or 10-GΩ feedback resistor; (4) a high-frequency boost; and (5) a gain stage. The cost of these "home-built" amplifiers was about one-tenth to one-fifth of the commercial patch clamps. For the experiments we describe here, each of these alternatives was equally satisfactory.

2.4. Analysis of Data

In this general description of the gating of BTX-activated sodium channels in bilayers, we have used a straightforward, "hands-on" approach to analysis of the data, in which open and closed levels and transition levels were determined by visual inspection of the records. More extensive accounts of various aspects of single-channel analysis have been given by Colquhoun and Sigworth (1983). Sachs *et al.* (1982) and Sachs (1983) describe programs for automating the procedures. For detailed discussions of the interpretation of single-channel data in terms of various kinetic models, see Colquhoun and Hawkes (1983), Horn and Lange (1983), and Horn and Vandenberg (1984).

All analysis programs were written in the interpretive language of DAOS (Data Analysis Operating System, Laboratory Software Associates, Melbourne, Australia). We run this software under RT-11 on microcomputer systems utilizing Digital Equipment Corporation (Marlboro, MA) LSI-11/2 and LSI-11/23 processors. DAOS provides virtual memory handling of data arrays and a large variety of single-command array arithmetic operations as well as input/output and display routines that facilitate programming and speed computations. For analysis, data were recorded continuously on FM tape during experiments. These data were subsequently digitized (12-bit resolution, usually at 2 kHz) in 4096-point segments using a digital oscilloscope (2090-3A, Nicolet Instrument Corporation, Madison, WI) and transcribed to floppy disk. Data were filtered, prior to digitization, with an eight-pole low-pass Bessel filter using a corner frequency (−3 dB) of 100–500 Hz. If filtering was required subsequent to digitization, this was done digitally using a Gaussian filter routine written in FORTRAN (Colquhoun and Sigworth, 1983) and linked to DAOS.

2.4.1. Measurement of Open and Closed Times

A transition threshold halfway between open and closed levels (see Colquhoun and Sigworth, 1983) was set visually using a horizontal display cursor. Threshold crossings were detected with DTRIG, an array-oriented digital trigger routine in DAOS. All records in a set were examined to insure that no significant drift occurred and to allow discarding of any abnormally noisy records. Transitions were then recorded, and dwell-time distributions constructed from the entire set of remaining records. For convenience in analysis, the records were divided into "slices" of 512 or 1024 points in DAOS, but where data were continuous across slice boundaries, slices were concatenated for the dwell-time analysis to minimize loss of data as incomplete events. Thus, data were usually analyzed in $4096 \times 0.5 = 2048$-msec continuous segments.

Both cumulative dwell-time distributions and probability density functions (see Colquhoun and Sigworth, 1983) were constructed. For this chapter, characteristic time constants for the distributions (equal to the arithmetic mean dwell time, $\langle t \rangle$, for single exponential distributions) were estimated by fitting the cumulative distributions. Fitting was accomplished using a nonlinear least-squares routine or by eye. The "by eye" approach allowed reasonable fits to be obtained to both components of double-exponential distributions even though a constant time bin width was used (see Colquhoun and Sigworth, 1983, for methods of generating computer fits to the density functions using a variable bin width).

2.4.2. Measurement of Fractional Open Time: The Activation Curve

Two alternative ways of determining the fractional open time, f_o, were used. The first made use of open and closed times measured as described in the preceding paragraph. Since the arithmetic mean dwell times in the open, $\langle t_o \rangle$, and

Figure 1. Visual determination of open and closed levels (horizontal cursors) for estimation of the fractional open time by the mean-current method (see text). Data were filtered at 500 Hz. $E = -110$ mV; unit current indicated by horizontal cursors, 2.75 pA; duration of record, 511.5 msec.

closed, $\langle t_c \rangle$, states were routinely calculated when the distributions of open and closed times were constructed, the fractional oper time could be estimated directly as in equation 1.

$$f_o = \langle t_o \rangle / (\langle t_o \rangle + \langle t_c \rangle) \quad (1)$$

The second method, computationally more economical, did not require the recognition of individual open and closed events but rather was based on the determination of the time-averaged mean current for a long stretch of record from several seconds to 1–2 min in length. There are two requirements for this approach to work well. The first is that there not be significant drift in the base line during the period being averaged, and the second is that there be enough clearly resolved events with all channels closed and with all channels open for these levels to be determined accurately. In practice, these levels were determined visually using horizontal cursors superimposed on the displayed records (see Fig. 1), and the whole set of records was examined before the program processed the set of records. Any records showing abnormal noise were omitted from the fractional open

Table I. Fractional Open-Time Determinations from Open and Closed Times and from Time-Averaged Currents

Experiment	E (mV)	$\langle t_o \rangle$ (msec)	$\langle t_c \rangle$ (msec)	f_o[a]	f_o[b]
03AU84	−60	20.3	2.8	0.88	0.81
	−90	8.6	4.0	0.68	0.67
08AU84	−60	51.7	3.6	0.94	0.96
	−95	11.0	3.8	0.75	0.82
	−100	8.7	6.5	0.57	0.58
	−110	3.0	20.8	0.12	0.16
22AU84A	−80	33.8	2.7	0.93	0.91
	−90	13.2	3.0	0.82	0.77
22AU84B	−60	112.4	5.55	0.95	0.98
	−80	25.7	3.15	0.89	0.90

[a] Calculated from open and closed dwell times.
[b] Determined from mean currents.

time determination. Beyond computational speed, a second advantage of this method is that it can be applied equally easily to records from one or several channels, as long as the current levels with all channels open, i_o, and with all channels closed, i_c, are known. Since the channels are open most of the time at moderate voltages, counting them is easy and unambiguous. The fractional open time is determined from the mean current, $\langle i \rangle$, as follows.

$$f_o = (\langle i \rangle - i_c)/(i_o - i_c) \tag{2}$$

In Table I, we compare determinations of fractional open times by these two methods. No significant systematic differences are apparent.

3. Voltage-Dependent Channel Gating

In contrast to the normal, sigmoid rise in sodium current (e.g., Hodgkin and Huxley, 1952a) following a depolarizing voltage step, macroscopic sodium currents activated by depolarizing voltage pulses after modification by BTX follow near-single-exponential time courses (Huang *et al.*, 1982). Nonetheless, analysis of single-channel closed-time distributions suggests that BTX-modified sodium channels, as do native channels, exhibit at least two kinetically distinguishable closed states (Huang *et al.*, 1984). These general conclusions are consistent with the studies in planar bilayer membranes, which we describe in Section 3.3.

Since BTX abolishes both fast and slow inactivation processes, the voltage-dependent channel-opening and -closing rates can be determined from the single-channel fluctuations during long periods in which the voltage is held at fixed levels. In general, opening and closing fluctuations seem to rapidly settle into a quasi-steady state after a voltage change. However, because of the slow relaxation in the membrane capacitance of the painted bilayers, a few seconds are allowed to elapse following a voltage step before data are used for analysis. Barring any biasing of the rate constants by slow changes induced by the long periods at each holding potential, it should be possible to predict the macroscopic and single-channel responses to conventional voltage-step protocols directly from the rate constants that we determine below (cf. Vandenberg and Horn, 1984). Such an analysis has not yet been done for BTX-modified channels in any system.

3.1. Gating Fluctuations at Constant Voltage

Characteristic records taken at four different voltages from a membrane containing a single BTX-activated sodium channel are shown in Fig. 2. At -60 mV, the channel spends almost all of the time open. This is a minor but consistent deviation from the observations of Huang *et al.* (1984) in patch-clamp studies on neuroblastoma cells. They observed that, at large depolarizations, the fractional open time reached a maximum of about 0.80. Other studies in bilayer systems

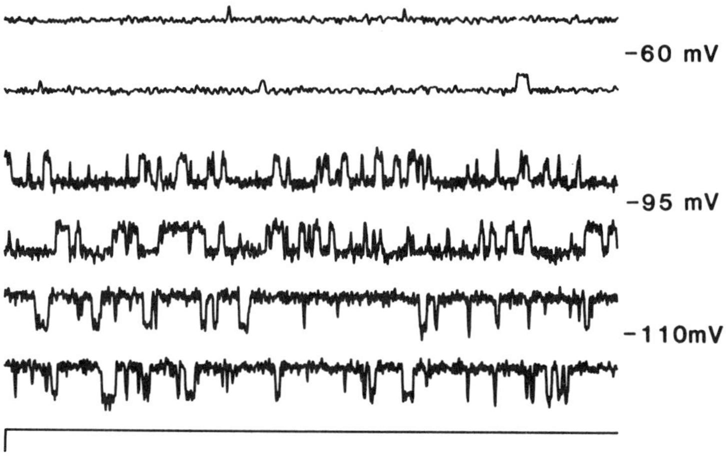

Figure 2. Single-sodium-channel current fluctuations in a painted bilayer at various voltages. Membrane PE:PS, 80:20. Bandwidth (−3 dB, eight-pole low-pass Bessel filter) −60 mV, 200 Hz; other voltages, 500 Hz. Upper level is closed in each trace. Vertical scale, 2 pA; horizontal scale, 511.5 msec.

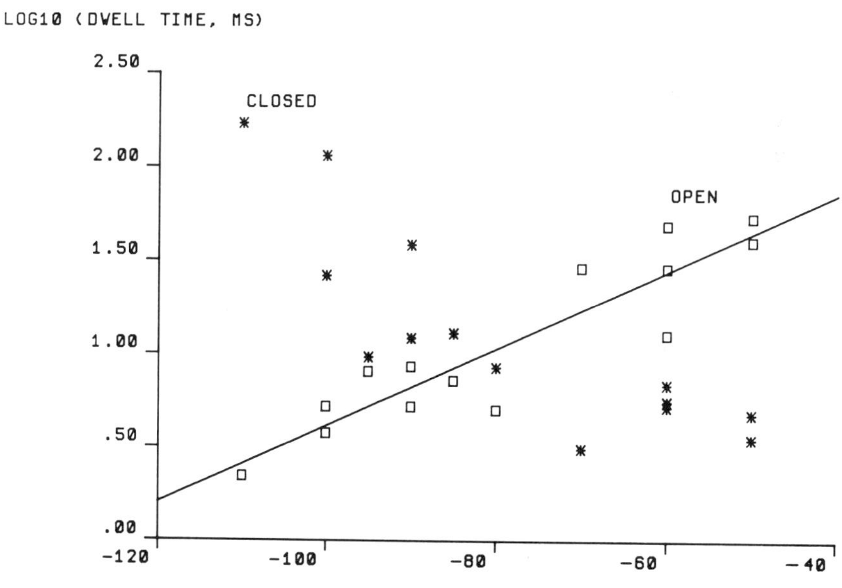

Figure 3. Voltage dependence of mean open and closed times. The straight line is the linear-regression line through the open-time data, using the equation $\ln \langle t_0 \rangle = 6.14 + 1.2\, FE/RT$. Collected data from four experiments using PE membranes.

have also shown f_o approaching limiting values between 0.9 and 1.0 (French *et al.*, 1984; Moczydlowski *et al.*, 1984a; Hartshorne *et al.*, 1985). At −60 mV, the channel spends almost all of the time open, with occasional brief closings ($\langle t_o \rangle$ = 52 msec; $\langle t_c \rangle$ = 3.6 msec). As the voltage is changed to more hyperpolarized levels, channel closings become more and more frequent, and, on the average, longer in duration. By −110 mV, the mean closed time was increased about sixfold ($\langle t_c \rangle$ = 20.8 msec), and the mean open time is much reduced ($\langle t_o \rangle$ = 3.0 msec). Thus, on depolarization, the increase in the probability of a channel being open results from both a decrease in mean closed time and an increase in mean open time (see Fig. 3).

3.2. Open-Time Distributions

In general, the open-time distributions were well approximated by single exponentials with time constants that increased with increasing depolarization. Examples of cumulative open-time distributions, determined at −60 and −110 mV, are plotted in Fig. 4, with exponential fits superimposed. In the case of the −60 mV data shown, the slight excess of long-duration events may in part be an artifact of missed brief closing events (see footnote to Table II). The size of the single-channel current at this voltage required filtering at 200 Hz to give a reasonable signal-to-noise ratio. For the distributions at more negative voltages, the filter was set at 500 Hz. Time constants for the distributions determined in one experiment are tabulated along with arithmetic mean open times in Table II. The agreement between the arithmetic mean open time and the time constants for open-time density functions is as expected for single-exponential distributions. This suggests there is a single open state of the channel and that the rate of closing from the open state is given by $k_c = 1/T_o$. It is consistent with channel closing being a time-homogeneous Markov process but does not critically test that possibility (see Horn and Vandenberg, 1984, for evidence that this is the case for closing of sodium channels in the absence of BTX). In our experiments, the mean lifetime of the open state changed by e-fold for about a 20-mV change in voltage. This result can be used to estimate the gating charge associated with channel closing (see Section 3.4.3).

3.3. Closed-Time Distributions

Closed-time distributions at voltages more positive than about −80 mV clearly required fitting by two exponential components (see Fig. 5). This suggests that, within this range of voltages, there is a significant occupancy of at least two kinetically distinct closed states. At more negative potentials, closed-time distributions were well approximated by single exponentials. This may result from the presence of two reaction steps whose lifetimes are too similar to be resolved, but additional observations would be needed to make a strong argument on this point. Our data are generally consistent with those of Huang *et al.* (1984), who suggested

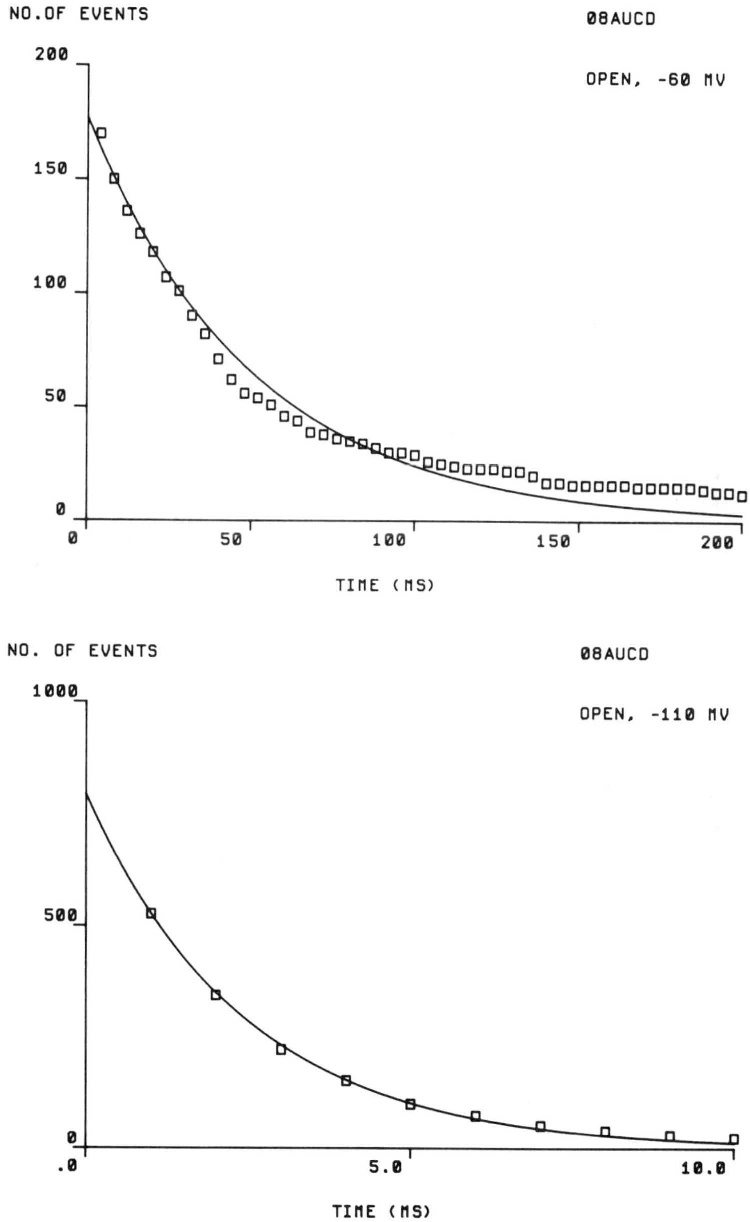

Figure 4. Cumulative open-time distributions with exponential fits (smooth curves) to the data superimposed. The ordinate represents the number of events lasting at least as long as the time on the abscissa. Time constants are 50.1 msec (−60 mV) and 2.4 msec (−110 mV). The −60 mV data were filtered at 200 Hz, and the −110 mV data were filtered at 500 Hz.

Table II. Open and Closed Times(msec) as a Function of Voltage[a]

E (mV)	T_o	$\langle t_o \rangle$	T_c	$\langle t_c \rangle$
−60	50.1	51.7	1.05, 10.5	3.6
−95	10.5	11.0	4.2	3.8
−100	9.4	8.7	7.1	6.5
−110	2.4	3.0	19.0	20.8

[a] Time constants (T_o) are determined from exponential fits to the cumulative open-time distributions; $\langle t_o \rangle$s are arithmetic mean open times. Time constants for the closed-time distributions are denoted by T_c. A double exponential was used to fit the −60 mV closed-time distribution. Mean closed times are denoted by $\langle t_c \rangle$. Data from experiment 08AU84. The finite bandwidth of the recordings can introduce significant errors into the estimates of the open and closed times. Truncation of distributions of short events caused by the inability to resolve events shorter than some dead time, T_d, leads to an overestimate of the dwell times (Neher and Steinbach, 1978). Thus, $T_{true} = T_{observed} - T_d$. We use the time constant for a first-order response at the recording bandwidth to approximate T_d (0.8 msec at 200 Hz and 0.3 msec at 500 Hz). Applying the correction, we estimate the fast component, T_{cf}, in the closed-time distribution at −60 mV to be $T_{cf}(-60)$ <0.25 msec and the open time at −110 mV to be $T_o(-110) \simeq 2.1$ msec. In addition, missed short closings cause overestimates of the complementary long open times, and missed short openings cause overestimates of the complementary closed times. For single exponential distributions, Sachs *et al.* (1982) give the correction, $T_{o,true} = T_{o,observed} \times \exp(-T_d/T_{c,true})$. Corrections for this type of error yield estimates of the "true" values of $T_o(-60)$ = 23 msec and $T_c(-110)$ = 22 msec. As might be intuitively expected, these correlations are largest at −60 mV (50–80% of the observed values), where the unit current was smallest and the records were filtered at 200 Hz. At −110 mV, where the open times become very short, corrections were 10–15% of the observed values. Errors from these sources are less than 10% at the intermediate voltages. Tabulated values are uncorrected values obtained directly from the distributions.

use of the linear scheme below to represent gating of BTX-modified channels. We do not have sufficient data to determine unambiguously the positions of the two closed states with respect to the open state.

$$C_1 \rightleftharpoons C_2 \rightleftharpoons O \qquad \text{(Scheme 1)}$$

The relationship between the rate constants in Scheme 1 and the two exponential time constants required to fit the closed-time distributions is given by Huang *et al.* (1984; see also Colquhoun and Hawkes, 1981).

3.4. Steady-State Activation Curve

The possibility that individual sodium channels gate independently of one another is suggested by the fact that the average of many single-channel records shows a time course similar to that of macroscopic currents (Sigworth and Neher, 1980). This assumption cannot be critically tested in any given preparation, however, until both single- and multichannel data are available over a wide voltage range (see Vandenberg and Horn, 1984). There is a recent suggestion that BTX-activated channels may behave as dimers in patches containing two channels (Iwasa and Ehrenstein, 1985). Among our own data, we do not see evidence of strong interaction, but we have not examined this possibility in detail.

A second consequence of independent gating of individual channels is that

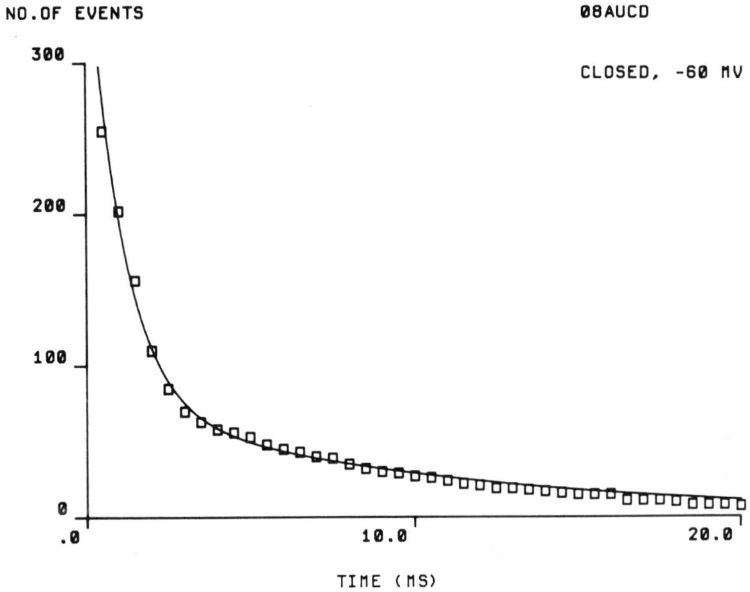

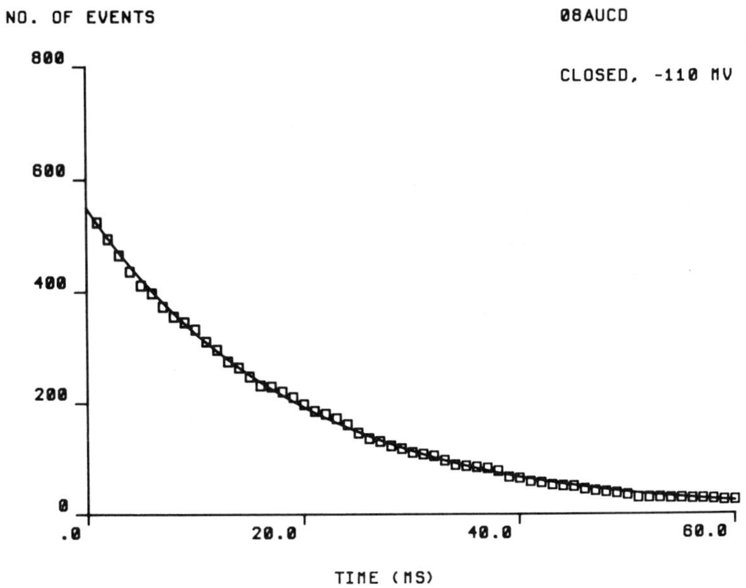

Figure 5. Cumulative closed-time distributions. Note that a double exponential was required to give a reasonable fit to the data at −60 mV. This was typical of closed-time distributions at voltages for which channels were open most of the time. Time constants for the smooth curves are 1.05 msec and 10.5 msec for short and long components, respectively, at −60 mV, and 19.0 msec at −100 mV.

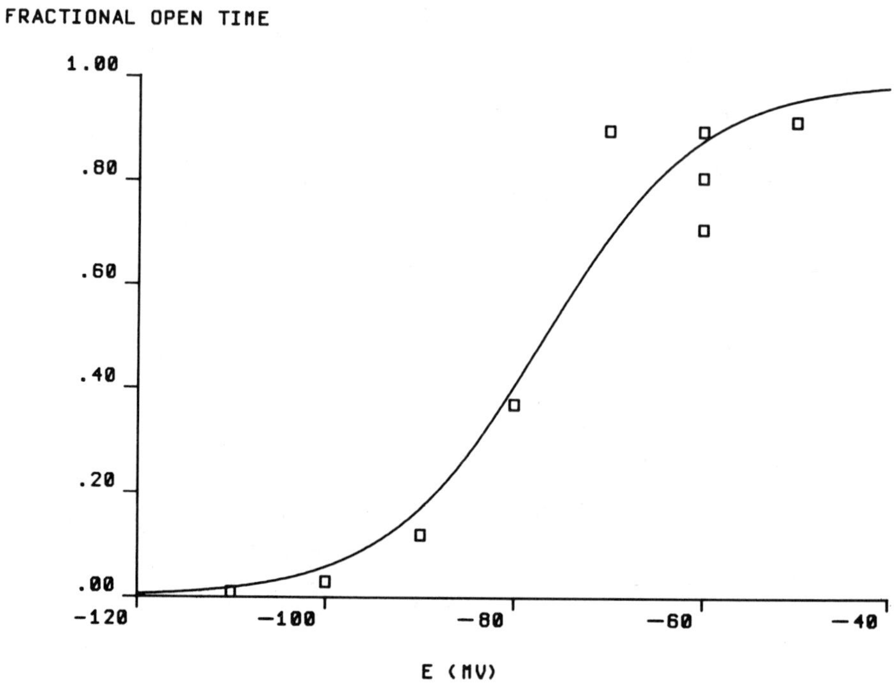

Figure 6. Fractional open time as a function of membrane voltage for single sodium channels incorporated into lipid (PE) bilayers. The smooth curve is drawn according to equation 3 of the text, with $E_{0.5} = -76$ mV and $q = 3.0$.

the form of the fractional open time versus voltage relationship should be the same as that of the macroscopic activation curve. In Fig. 6 we present an example of fractional-open-time data taken from a group of experiments in which the membrane was bathed in symmetric NaCl solutions containing 0.1 mM Ca^{2+}. The smooth curve is drawn through the data using the following expression:

$$f_o = 1/\{1 + \exp [qF(E - E_{0.5})/RT]\} \tag{3}$$

$E_{0.5}$ is the voltage at which channels are open 50% of the time and thus defines the midpoint of the curve. The apparent gating charge that would have to move through the whole transmembrane voltage during the open–closed transition is represented by q. Values of $E_{0.5}$ and q were estimated by nonlinear least-squares fits to the data obtained using a function-minimization procedure described by Colquhoun (1971). Results from experiments reported by French *et al.* (1984), together with some new data, are summarized in Table III. Our parameter values are almost identical with those determined by Moczydlowski *et al.* (1984a) for channels from rat sarcolemmal fragments and by Hartshorne *et al.* (1985) for purified channels from rat brain. Using data from 11 experiments (six determi-

Table III. Half-Open Gating Voltage ($E_{0.5}$) and Apparent Gating Charge Determined from Steady-State Activation Curves

Experiment	Lipid (PE:PS)	$[Ca^{2+}]$ (mM), ext/int	$E_{0.5}$ (mV)	q	Notes
KT	80:20 70:30	0.1/0.1	−99	3.6	Four expts, two with each lipid mixture
WOR	100:0	0.1/0.1	−76	3.0	Three expts
16NO82	72:28	0.1/0.1	−93	4.1	
83C	72:28	0.1/0.1	−83	5.3	
82J	72:28	0.1/0.1	−102	4.6	
84AAA	30:70	0.1/0.1	−96	3.0	
		10.0/0.1	−67	2.3	
3L1340	30:70	10.0/0.1	−74	3.1	
EEE640	30:70	10.0/0.1	−85	3.8	

nations, see Table III) with symmetric 0.1 mM Ca^{2+}, $E_{0.5} = 91.5 \pm 10.1$ mV and $q = 3.9 \pm 0.9\ e$ (given as means ± SD in both cases).

3.4.1. Variability in the Gating Voltage Range

The scatter in values for $E_{0.5}$, in particular, is far beyond any experimental error. This may imply that the channel protein can exhibit different gating modes (cf. observations by Hess *et al.*, 1984, on Ca^{2+} channels) or could, in principle, result from changes in the lipid environment immediately adjacent to the channel (but see Section 3.5).

In patch-clamp studies, Vandenberg and Horn (1984) observed a comparable range of values for the midpoint of the sodium channel inactivation curve in GH_3 cells. Also, various authors (Fenwick *et al.*, 1982; Marty and Neher, 1983; Vandenberg and Horn, 1984) have noted shifts of gating parameters along the voltage axis in the hyperpolarizing direction when patches were excised. Thus, such variability is not unique to the bilayer system.

3.4.2. Apparent Gating Charge

Estimates of the total charge involved in channel activation show notable agreement among different bilayer studies and are within the range of values obtained from other preparations. These estimates represent the net charge that would have to move through the total transmembrane voltage to result in the observed voltage dependence of channel opening or gating charge movement. Thus, at room temperature, 1 *e*/channel corresponds to about 25 mV/*e*-fold change. We have collected some of these results in Table IV. Apparent gating charge is about 3–5 *e*/channel, with results from the bilayer studies falling in the middle of this range.

3.4.3. Apparent Charge Movement during the Closing Step

There appears to be only a single open state, and over the range of our studies open times show an approximately exponential dependence on voltage. From the

Table IV. Apparent Gating Charge Determined in Different Preparations

Preparation and method	Modifier	Gating charge, q	Reference
Rat brain fragments, bilayer	BTX	3.9	This chapter
Rat brain, pure channels, bilayer	BTX	3.8	Hartshorne *et al.* (1985)
Rat sarcolemma, bilayer	BTX	3.8	Moczydlowski *et al.* (1984a)
Neuroblastoma, whole cell	BTX	4.6	Huang *et al.* (1982)
	None	4.6	
Frog node, g_{Na}	BTX	5.2	Dubois *et al*, (1983)
gating current	BTX	4.9	
Frog node, ionic currents	BTX	4.5–5	Khodorov (1978)
GH_3, whole-cell ionic currents	None	4.3	Vandenberg and Horn (1984)
	Trypsin	3.0	
Squid axon, ionic currents	None	6	Hodgkin and Huxley, (1952b)
	Pronase	3.6	Stimers *et al.* (1985)

semilogarithmic plot of the mean open times against voltage for the experiments in Fig. 3, a charge of 1.2 *e*/channel can be associated with the unidirectional closing step. A separate set of experiments in PE/PS membranes gave a value of 1.3 *e*/channel. If voltage affects equally the transitions into and out of the open state, this suggests conservatively that 65% of the total gating charge is associated with transitions directly into and out of the open state. The fact that we can reasonably account for only part of the total charge movement by the voltage dependence of the transitions in and out of the open state is an additional argument (see also Section 3.3) that there are at least two closed states with voltage dependent transitions between them.

Experiments of Keller *et al.* (1985) on the purified rat brain channel activated by BTX suggest a value of about 2 *e*/channel for the charge associated with the unidirectional closing step. A maximum-likelihood analysis of data from unmodified channels by Vandenberg and Horn (1984) yielded a value of 1.3 *e*/channel.

3.5. Stationarity: Do the Channels Show Different Gating Modes?

One difficulty in analyzing kinetics of BTX-activated channels is that there are occasional, abrupt, spontaneous changes in the pattern of the fluctuations even when the membrane voltage is held at a constant value (see Fig. 7). This has been observed in at least two other laboratories (Moczydlowski *et al.*, 1984a; Weiss *et al.*, 1984) and occurs in "solvent-free" Montal–Mueller (1972) membranes as well as decane–lipid painted bilayers. The latter point argues against, but does not exclude, the idea that the changes result from the channel occasionally bumping into an island or lens of lipid and solvent as it floats through the two-dimensional lipid "sea." A related idea, that there may be regions of different phase coexisting in the membrane, is also possible. Some preliminary observations in our own laboratory suggest that neither the open-channel con-

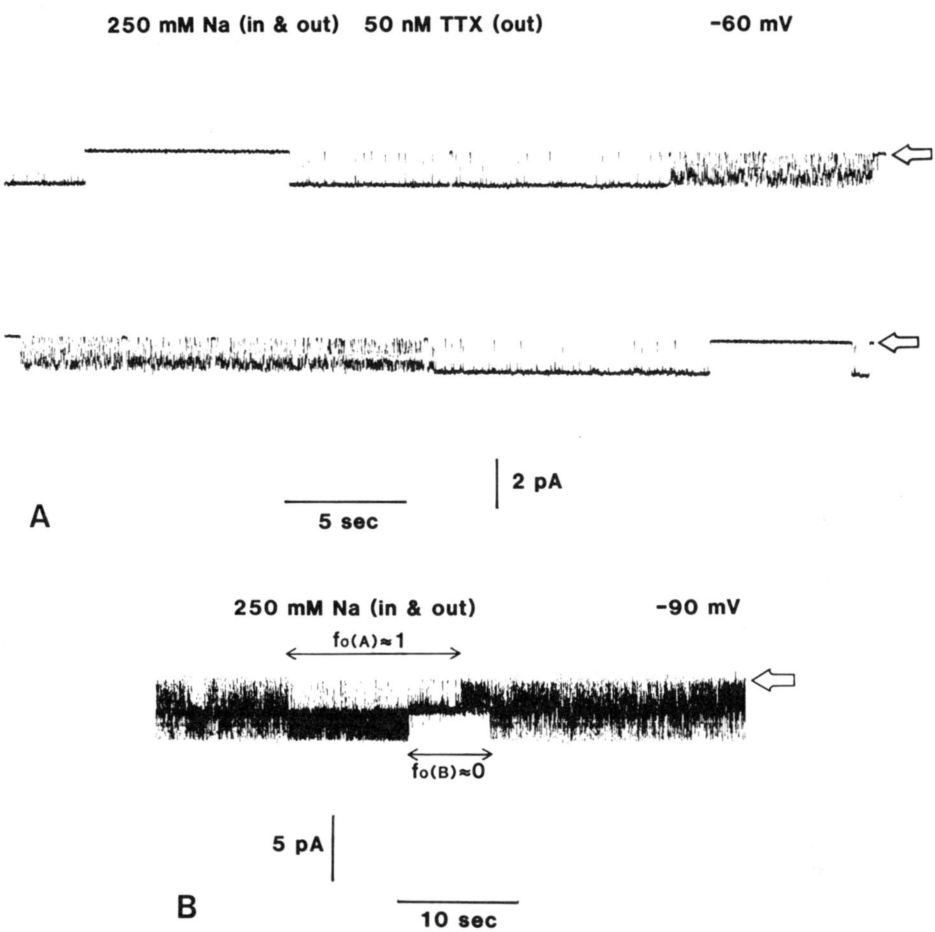

Figure 7. Records illustrating spontaneous, nonstationary behavior. Arrows indicate closed-channel currents. A: Record from a membrane containing a single sodium channel in the presence of 50 nM tetrodotoxin (TTX). Completely silent periods indicate toxin block of the channel. Note the sudden changes in the rate of gating fluctuations and the degree of "openness" during the periods in which the channel is not blocked by the toxin. Voltage, −60 mV. The unblocked periods shown in these selected sections of record are unusually long for this concentration of TTX. B: Record from a membrane containing two channels. First, it seems that one channel (A) switches to being open most of the time, then the second (B) appears to remain mostly closed for an extended period. These two unusual events overlap, an indication of possible interaction.

ductance (Worley *et al.*, 1985, and unpublished data) nor channel gating (see below) are much affected by the surface charge on the surrounding lipid. Conspicuous spontaneous shifts were so infrequent that they were excluded from data used for our analysis.

There is a large body of evidence that various channels undergo sharp changes

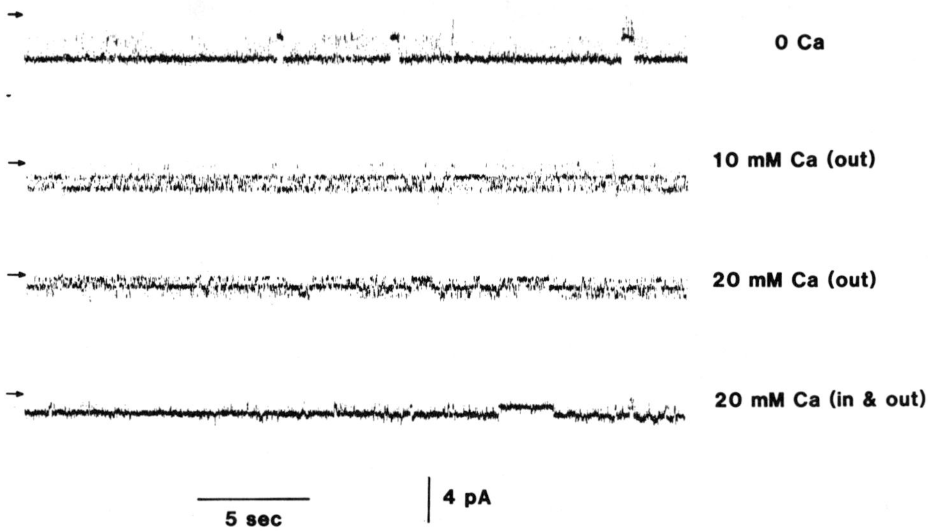

Figure 8. Effects of calcium on gating of BTX-activated sodium channels in a lipid bilayer membrane. Note that Ca^{2+} added to the external surface increases the probability of channel closing. Subsequent addition of Ca^{2+} to the opposite side again returns the probability of the channel being open to approximately the original value. The apparent single-channel conductance is reduced by Ca^{2+}, and, unlike the effect on gating, there is a further reduction in conductance when Ca^{2+} is added to the second side to make the concentrations equal on the two sides.

in gating activity on a time scale of seconds under steady repetitive stimulation or at constant voltage. Examples include calcium channels (Hess *et al.*, 1984), calcium-activated potassium channels (Moczydlowski and Latorre, 1983), glutamate-activated channels (Patlak *et al.*, 1979; Gration *et al.*, 1981), and sodium channels (Horn *et al.*, 1984; Patlak and Ortiz, 1985). Thus, it is likely that the occurrence of transitions of this type, between different gating modes or states, is a general characteristic of the behavior of channel proteins. Transitions on this time scale could have important implications for signal processing and would be difficult to study or detect without the opportunity, provided by patch-clamp and lipid-bilayer methods, to make long-term records of single-channel activity.

3.6. Effects of Calcium on Channel Gating

It has long been known that activity of excitable cells is modulated by the level of calcium ions in the bathing media. Frankenhaeuser and Hodgkin (1957) showed that raising external $[Ca^{2+}]$ causes a shift of the sodium conductance gating parameters in the depolarizing direction along the voltage axis. A similar effect may be seen at the single-channel level in the records of Fig. 8. Raising the external $[Ca^{2+}]$ to 10 mM decreases the fraction of time that the channel spends in the open state. Interestingly, this effect is reversed at least partially when Ca^{2+} is added to the internal solution to equalize the $[Ca^{2+}]$ on the two sides of the membrane. Frankenhaeuser and Hodgkin (1957), based on a suggestion of A. F. Huxley, proposed that this effect might be caused by a decrease in the effect of negative surface charge in the vicinity of the channels as a result of binding of Ca^{2+} to these charges. The view that both binding and nonspecific screening play a role is currently accepted (Hille *et al.*, 1975; McLaughlin, 1977; Hahin and Campbell, 1983).

An important question that arises is whether the fixed negative charges that fix the operating voltage range for gating are located on the membrane phospholipid or the channel molecule itself. This question is one that can be tackled directly in the bilayer system. In preliminary experiments we have observed that Ca^{2+}-induced shifts in the f_o–E relationship appear to be independent of the charge on the phospholipid (data not shown), suggesting that the gating voltage is set by charges that are an integral part of the channel molecule. Frace *et al.* (1985) reached a similar conclusion from studies on squid giant axon. We are currently completing a detailed study of this issue.

4. In Conclusion

4.1. Channels in Bilayers and Cells

The general properties of sodium channels inserted into artificial bilayers from native membrane fragments are very similar to those that have been reported in cellular preparations. Moreover, the additional biochemical violence required to purify the rat brain sodium channel does not seem to alter channel behavior significantly (Hartshorne *et al.*, 1985). A critical evaluation of the possible changes in properties of channel proteins during isolation and purification must await detailed recordings from the parent cells and from channels after insertion into bilayers. The hazards of judging the normality of properties of channels inserted into bilayers against data collected from a different species or cell type are clearly illustrated by comparing the properties of channels from two different organs of the same species, namely, skeletal muscle and brain of rat. In bilayer studies using virtually identical techniques, channels from these sources differ quantitatively in maximal conductance and apparent affinity for sodium under symmetric ionic conditions (Moczydlowski *et al.*, 1984a; Worley *et al.*, 1985) and qualitatively in their sensitivity to a peptide blocker, μ-conotoxin (Cruz *et al.*, 1985). The gating behavior of these channels from these two tissues, to the extent studied so far, appears to be very similar.

4.2. For the Future

At present, the principal technical limitations for studying sodium channels in planar bilayers are the limited recording bandwidth and the large and complex capacity transient, which is difficult to suppress and obscures events immediately following a voltage step. In large bilayers, these problems preclude study of sodium channels in the absence of BTX or another substance that eliminates slow and fast inactivation. However, the problems are not insurmountable. Both limitations are alleviated by using smaller membranes; 50-μm or smaller bilayers should provide satisfactory noise (less than 1 pA) at greater than 1 kHz and allow recording within the first millisecond after the beginning of a voltage step. Such small bilayers can easily be formed using either the "folded" bilayer technique of Montal and Mueller (1972) or the "dipping" technique described by Coronado and Latorre (1983), Suarez-Isla *et al.* (1983), and Schuerholz and Schindler (1983). Also, small areas of painted membranes may be excised on a hydrophobic, silanized glass pipet (Mueller, 1975; Andersen, 1983).

Small bilayers do, however, present a very small target area for vesicle incorporation and, for some preparations such as the brain membrane vesicles used in our studies, may preclude reliable incorporation. Refinement of channel insertion techniques, perhaps using reconstituted vesicles with a high density of purified channels to increase the probability of incorporation, would enable use of very small membrane areas for recording, improving resolution to the point that unmodified channels easily could be resolved. Use of a purified channel protein preparation is also a distinct advantage in that it reduces the chance of introducing nonspecific conductances, which cause noise and instability, into the membrane. Greater control over the incorporation process would enable insertion of channels at high density, which ultimately could enable measurements of macroscopic, single-channel, and gating currents in the same reconstituted system. It has not been possible to make these three kinds of measurements in any single preparation to date (see Vandenberg and Horn, 1984, for a discussion of the possibility in patch-clamped GH_3 cells). Regardless of where this achievement is first attained, we can expect cellular, isolated-patch, and bilayer techniques to yield complementary insights into channel-gating processes for some time.

ACKNOWLEDGMENTS. We thank Dr. John Daly for making these studies possible by providing batrachotoxin, Gus Ay for making the bilayer chambers, and Jeff Michael and Tim Sparks for construction of electronic equipment. This work was supported, in part, by grants NS-16285, NS-16775, and NS-20106 from the U.S. Department of Health and Human Services and by contract DAMD-17-C-2188 from the U.S. Army Medical Research and Development Command. K. K. Tam held fellowships from the Short Term Research Training Program of the University of Maryland, and from United Healthcare.

References

Albuquerque, E. X., Daly, J. W., and Witkop, B., 1971, Batrachotoxin: Chemistry and pharmacology, *Science* **172:**995–1002.

Andersen, O. S., 1983, Ion movement through gramicidin A channels. Single-channel measurements at very high potentials, *Biophys. J.* **41:**119–133.

Cohen, F. S., Zimmerberg, J., and Finkelstein, A., 1980, Fusion of phospholipid vesicles with planar bilayer membranes. II. Incorporation of a vesicular membrane marker into the planar membrane, *J. Gen. Physiol.* **75:**251–270.

Cohen, F. S., Akabas, M. H., and Finkelstein, A., 1982, Osmotic swelling of phospholipid vesicles causes them to fuse with a planar phospholipid bilayer membrane, *Science* **217:**458–460.

Colquhoun, D., 1971, *Lectures on Biostatistics*. Clarendon Press, Oxford.

Colquhoun, D., and Hawkes, A. G., 1981, On the stochastic properties of single ion channels, *Proc. R. Soc.* (*Lond.*) [*Biol.*] **300:**1–59.

Colquhoun, D., and Hawkes, A. G., 1983, The principles of the stochastic interpretation of ion-channel mechanisms, in: *Single-Channel Recording* (B. Sakmann and E. Neher, eds.), pp. 135–175, Plenum Press, New York.

Colquhoun, D., and Sigworth F., 1983, Fitting and statistical analysis of single channel records, in: *Single-Channel Recording* (B. Sakmann and E. Neher, eds.), pp. 191–263, Plenum Press, New York.

Coronado, R., and Latorre, R., 1983, Phospholipid bilayers made from monolayers on patch-clamp pipettes, *Biophys. J.* **43:**231–236.

Cruz, L. J., Gray, W. R., Olivera, B. M., Zeikus, R. D., Kerr, L., Yoshikami, D., and Moczydlowski, E., 1985, *Conus geographus* toxins that discriminate between neuronal and muscle sodium channels, *J. Biol. Chem.* **260:**9280–9288.

Dubois, J. M., Schneider, M. F., and Khodorov, B., 1983, Voltage dependence of intramembrane charge movement and conductance of batrachotoxin-modified sodium channels in frog node of Ranvier, *J. Gen. Physiol.* **81:**829–844.

Fenwick, E. M., Marty, A., and Neher, E., 1982, Sodium and calcium channels in bovine chromaffin cells. *J. Physiol.* (*Lond.*) **331:**559–635.

Frace, A. M., Poznansky, M., Eaton, D. C., and Brodwick, M. S., 1985, Effects of phospholipid headgroup modification in voltage clamped squid giant axon, *Biophys. J.* **47:**440a.

Frankenhaeuser, B., and Hodgkin, A. L., 1957, The action of calcium on the electrical properties of squid axons, *J. Physiol.* (*Lond.*) **137:**218–244.

French, R. J., Worley, J. F. III, and Krueger, B. K., 1984, Voltage-dependent block by saxitoxin of sodium channels incorporated into planar lipid bilayers, *Biophys. J.* **45:**301–310.

Gration, K. A. F., Lambert, J. J., Ramsey, R., and Usherwood, P. N. R., 1981, Non-random openings and concentration-dependent lifetimes of glutamate-gated channels in muscle membrane, *Nature* **291:**423–425.

Green, W. N., Weiss, L. B., and Andersen, O. S., 1984, Batrachotoxin-modified sodium channels in lipid bilayers, *Ann. N.Y. Acad. Sci.* **435:**548–550.

Hahin, R., and Campbell, D. T., 1983, Simple shifts in the voltage dependence of sodium channel gating caused by divalent ions, *J. Gen. Physiol.* **82:**785–805.

Hartshorne, R. P., Keller, B. U., Talvenheimo, J. A., Catterall, W. A., and Montal, M., 1985, Functional reconstitution of the purified brain sodium channel in planar lipid bilayers, *Proc. Natl. Acad. Sci. U.S.A.* **82:**240–244.

Hess, P., Lansman, J. B., and Tsien, R. W., 1984, Different modes of Ca channel gating behaviour favoured by dihydropyridine Ca agonists and antagonists, *Nature* **311:**538–544.

Hille, B., Woodhull, A. M., and Shapiro, B. I., 1975, Negative surface charge near sodium channels of nerve: Divalent ions, monovalent ions and pH, *Phil. Trans. R. Soc.* (*Lond.*) [*Biol.*] **270:**301–318.

Hodgkin, A. L., and Huxley, A. F., 1952a, Currents carried by sodium and potasium ions through the membrane of the giant axon of *Loligo*, *J. Physiol.* (*Lond.*) **116:**449–472.

Hodgkin, A. L., and Huxley, A. F., 1952b, A quantitative description of membrane current and its application to conduction and excitation in nerve, *J. Physiol.* (*Lond.*) **117:**500–544.

Horn, R., and Lange, K., 1983, Estimating kinetic constants from single channel data, *Biophys. J.* **43:**207–223.

Horn, R., and Vandenberg, C. A., 1984, Statistical properties of single sodium channels, *J. Gen. Physiol.* **84:**505–534.

Horn, R., Vandenberg, C. A., and Lange, K., 1984, Statistical analysis of single sodium channels. Effects of N-bromoacetamide, *Biophys. J.* **45:**323–335.

Huang, L. M., Moran, N., and Ehrenstein, G., 1982, Batrachotoxin modifies the gating kinetics of sodium channels in internally perfused neuroblastoma cells, *Proc. Natl. Acad. Sci. U.S.A.* **79:**2082–2085.

Huang, L. M., Moran, N., and Ehrenstein, G., 1984, Gating kinetics of batrachotoxin-modified sodium channels in neuroblastoma cells determined from single channel measurements, *Biophys. J.* **45:**313–322.

Iwasa, K., and Ehrenstein, G., 1985, Dimers of batrachotoxin-modified sodium channels, *Biophys. J.* **47:**191a.

Keller, B., Hartshorne, R., Talvenheimo, J., Caterall, W., and Montal, M., 1985, Channel gating kinetics of purified sodium channels modified by batrachotoxin in lipid bilayers, *Biophys. J.* **47:**439a.

Khodorov, B. I., 1978, Chemicals as tools to study nerve fiber sodium channels; Effects of batrachotoxin and some local anesthetics, in: *Membrane Transport Processes,* Vol. 2 (D. C. Tosteson, Y. A. Ovchinnikov, and R. Latorre, eds.), pp. 152–174, Raven Press, New York.

Krueger, B. K., Ratzlaff, R. W., Strichartz, G. R., and Blaustein, M. P., 1979, Saxitoxin binding to synaptosomes, membranes, and solubilized binding sites from rat brain, *J. Memb. Biol.* **50:**287–310.

Krueger, B. K., Worley, J. F. III, and French, R. J., 1983, Single sodium channels from rat brain incorporated into planar lipid bilayers, *Nature* **303:**172–175.

Marty, A., and Neher, E., 1983, Tight-seal whole-cell recording, in: *Single-Channel Recording* (B. Sakmann and E. Neher, eds.), pp. 107–122, Plenum Press, New York.

McLaughlin, S., 1977, Electrostatic potentials at membrane–solution interfaces, in: *Current Topics in Membranes and Transport,* Vol. 9 (F. Bronner and A. Kleinzeller, eds.) pp. 71–144, Academic Press, New York.

Miller, C., 1978, Voltage-gated cation channel from fragmented sarcoplasmic reticulum: Steady-state electrical properties, *J. Membr. Biol.* **40:**1–23.

Moczydlowski, E., and Latorre, R., 1983, Gating kinetics of Ca^{2+}-activated K^+ Channels from rat muscle incorporated into planar lipid bilayers. Evidence for two voltage-dependent binding reactions, *J. Gen. Physiol.* **82:**511–542.

Moczydlowski, E., Garber, S. S., and Miller, C., 1984a, Batrachotoxin-activated sodium channels in planar lipid bilayers: Competition of tetrodotoxin block by Na^+, *J. Gen. Physiol.* **84:**665–686.

Moczydlowski, E., Hall, S., Garber, S. S., Strichartz, G. R., and Miller, C., 1984b, Voltage-dependent blockade of muscle Na^+ channels by guanidinium toxins: Effect of toxin charge, *J. Gen. Physiol.* **84:**687–704.

Montal, M., and Mueller, P., 1972, Formation of bimolecular membranes from lipid monolayers and a study of their electrical properties, *Proc. Natl. Acad. Sci. U.S.A.* **69:**3561–3566.

Mueller, P., 1975, Membrane excitation through voltage-induced aggregation of channel precursors, *Ann. N.Y. Acad. Sci.* **264:**247–264.

Mueller, P., Rudin, D. O., Tien, H. T., and Wescott, W. C., 1963, Methods for formation of single bimolecular lipid membranes in aqueous solution, *J. Phys. Chem.* **67:**534–535.

Narahashi, T., Albuquerque, E. X., and Deguchi, T., 1971, Effects of batrachotoxin on the membrane potential and conductance of squid giant axons, *J. Gen. Physiol.* **58:**54–70.

Neher, E., and Steinbach, J. H., 1978, Local anaesthetics transiently block currents through single acetylcholine-receptor channels, *J. Physiol.* (*Lond.*) **277:**153–176.

Patlak, J., and Ortiz, M., 1985, Slow currents through single sodium channels of the adult rat heart, *J. Gen. Physiol.* **86:**89–104

Patlak, J. B., Gration, K. A. F., and Usherwood, P. N. R., 1979, Single glutamate-activated channels in locust muscle, *Nature* **278:**643–645.

Quandt, F. N., and Narahashi, T., 1982, Modification of single Na^+ channels by batrochotoxin, *Proc. Natl. Acad. Sci. U.S.A.* **79:**6732–6736.

Rosenberg, R. L., Tomiko, S. A., and Agnew, W. S., 1984, Single-channel properties of the reconstituted, voltage-regulated sodium channel from the electroplax of *Electrophorus electricus, Proc. Natl. Acad. Sci. U.S.A.* **81:**5594–5598.

Sachs, F., 1983, Automated analysis of single channel records, in: *Single-Channel Recording* (B. Sakmann and E. Neher, eds.), pp. 265–285, Plenum Press, New York.

Sachs, F., Neil, J., and Barkakati, N., 1982, The automated analysis of single channel data, *Pfluegers Arch.* **395:**331–340.
Schuerholz, T., and Schindler, H., 1983, Formation of lipid–protein bilayers by micropipette guided contact of two monolayers, *FEBS Lett.* **152:**187–190.
Sigworth, F., 1983, Electronic design of the patch clamp, in: *Single-Channel Recording* (B. Sakmann and E. Neher, eds.), pp. 3–35, Plenum Press, New York.
Sigworth, F., and Neher, E., 1980, Single Na^+ channel currents observed in cultured rat muscle cells, *Nature* **287:**447–449.
Stimers, J. R., Bezanilla, F., and Taylor, R. E., 1985, Sodium channel activation in the squid giant axon. Steady state properties, *J. Gen. Physiol.* **85:**65–82.
Suarez-Isla, B. A., Wan, K., Lindstrom, J., and Montal, M., 1983, Single channel recordings from purified acetylcholine receptors reconstituted in bilayers at the tip of patch pipets, *Biochemistry* **22:**2319–2323.
Vandenberg, C., and Horn, R., 1984, Inactivation viewed through single sodium channels, *J. Gen. Physiol.* **84:**535–564.
Weiss, L. B., Green, W. N., and Andersen, O. S., 1984, Single channel studies on the gating of batrachotoxin (BTX)-modified sodium channels in lipid bilayers, *Biophys. J.* **45:**67a.
Worley, J. F. III, French, R. J., and Krueger, B. K., 1985, Interactions of divalent ions with single sodium channels, *Biophys. J.* **47:**439a.

Chapter 15

ION CONDUCTION THROUGH SODIUM CHANNELS IN PLANAR LIPID BILAYERS

O. S. ANDERSEN, W. N. GREEN, and B. W. URBAN

1. Introduction

Ion-permeable channels often derive their generic names from the specialized conductance and permeability properties they possess, the voltage-dependent Na^+ channel being no exception. Apart from their use in identifying ion channels, measurements of ion selectivities and single-channel conductances also provide insight into the biophysical mechanisms underlying ion permeation through the channels. Such measurements and their interpretations may be ambiguous, even in the case of channels incorporated into the comparatively simple planar lipid bilayers. Using the voltage-dependent Na^+ channel as an example, we show in this chapter that, even though the lipid bilayer system offers many experimental advantages, the determination and the analysis of single-channel conductance and permeability data are not always simple.

2. Advantages of the Planar Bilayer System

The electrical and the chemical environments of ion-conducting channels in planar lipid bilayers can be controlled and changed with relative ease. This experimental flexibility makes the bilayer system(s) particularly useful in the study of channel-mediated ion permeability phenomena. Transmembrane potentials and the chemical composition of the electrolyte solutions on either side of the planar bilayer can be varied over a wide range, and the effects of repeated solution changes can be studied on the same ion channel. The bilayer composition can be altered if the effects of membrane environment on ion channel function are to be investigated. Conversely, if nominally identical ion channels derived from dif-

O. S. ANDERSEN, W. N. GREEN, and B. W. URBAN • Department of Physiology and Biophysics and Department of Anesthesiology, Cornell University Medical College, New York, New York 10021.

ferent biological tissues are to be compared, it is possible to do so in lipid bilayers of identical compositions. Therefore, although the channel function may be modified on transfer into the lipid bilayer (e.g., by the presence of hydrocarbons), identical ion channels should undergo the same transformations and should thus be indistinguishable in the bilayer system. Detailed comparisons are possible, particularly as it is feasible to examine a single ion channel (molecule) in a defined membrane and electrolyte environment for hours. This ability to observe individual channels in a steady state has important implications for studies on the molecular biology of ion-permeable channels.

Studies on the batrachotoxin (BTX)-modified Na^+ channel illustrate the need for a well-controlled electrochemical environment. It is generally accepted that BTX has at least three effects on Na^+ channels in biological membranes: inactivation is removed; the voltage–activation curve is shifted in the hyperpolarized direction; and the apparent ion selectivity is decreased (Khodorov and Revenko, 1979). Studies on the conductance and ion selectivity of BTX-modified Na^+ channels must, however, be interpreted in light of the BTX-induced shift in the voltage–activation curve: BTX-modified Na^+ channels will remain open at normal holding potentials (Khodorov and Revenko, 1979; Huang *et al.*, 1982; Dubois *et al.*, 1983), and ion exchange through open Na^+ channels will result (e.g., Huang *et al.*, 1979). This exchange of permeant ions through the open Na^+ channels will alter the ions' electrochemical gradients unless both the "intracellular" and the "extracellular" solutions have sufficiently large volumes relative to the number of Na^+ channels in the membrane that one can disregard the consequences of these ion movements. If the intracellular concentration of Na^+ (and other permeant ions) approaches the extracellular concentrations, the reversal potential for the Na^+ channels could eventually become indistinguishable from 0 mV irrespective of their actual K^+/Na^+ selectivity.* Unless the changing concentrations are accounted for, it is therefore not possible to use changes in the reversal potential resulting from the variation of the extracellular electrolyte composition as a measure of the ion selectivity of the Na^+ channel.

3. Measuring the Single-Channel Conductance

The permeability characteristics of ion-conducting channels can often be quantified in terms of a two-state model, where the channel is either open or closed. This behavior is illustrated for the Na^+ channel in Fig. 1. In the open state, a low-energy path is formed for ion movement across the membrane: the open channel is a catalyst for ion exchange between the two aqueous phases. In the closed state, transmembrane ion movement is negligible. This two-state de-

* The intracellular and extracellular concentrations will not, in fact, become equal to each other; they will differ by the predictions of the Gibbs–Donnan distribution (e.g., Overbeek, 1956). The reversal potential for BTX-modified Na^+ channels will thus eventually become equal to the Donnan potential. Interestingly, this will not be discernible to patch-clamp studies on cell-attached Na^+ channels, where the reversal potential will appear to be 0 mV.

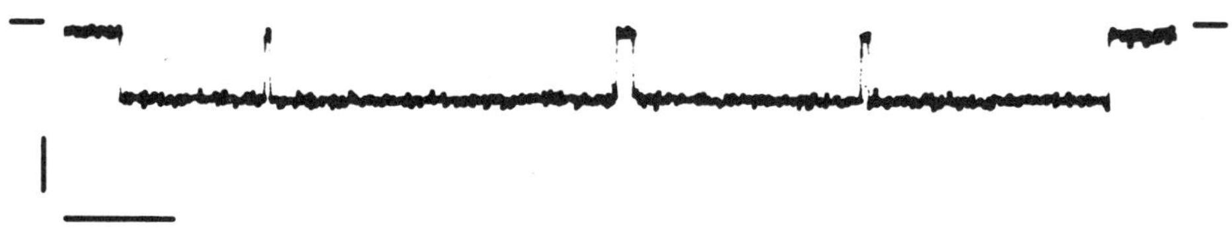

Figure 1. A single BTX-modified Na^+ channel is displayed; the channel undergoes transitions between two states, open and closed. The single-channel current transitions were obtained in a bilayer made from 1-palmitoyl-2-oleoylphosphatidylcholine and 1-palmitoyl-2-oleoylphosphatidylethanolamine (1:4) in *n*-decane (5% w/v). The aqueous phases contained 50 mM Na^+ ($+ Cl^-$) buffered to pH 7.4 with 5 mM phosphate, and 1.08 nM saxitoxin. The membrane potential was −60 mV relative to the "extracellular" phase. The calibration bars denote 2 sec (horizontally), and 1 pA (vertically); the zero-current level is denoted by the two upper markers; 22.0°C (W. N. Green, unpublished observations).

scription can readily be extended to deal with the existence of multiple conductance levels, as, for example, for alamethicin (Eisenberg *et al.*, 1973) or monazomycin (Andersen and Muller, 1982). In these cases, it is not possible to speak of a single-channel conductance.

The single-channel conductance, g, is of interest for two reasons. First, the physiological function of membrane ion channels is to regulate the flow of certain ions across the membrane. The magnitude of the single-channel conductance is thus one of several parameters that determine a channel's efficacy in controlling or modulating cell function. Second, the magnitude of the single-channel conductance varies as a function of the type and concentration of the permeant ion(s). The analysis of such conductance changes provides insight into the biophysical processes underlying ion translocation.

The single-channel conductance, g, can be calculated from I/V characteristics, either as the chord conductance, $g_c(V)$:

$$g_c(V) = i(V)/(V - V_{rev}) \quad (1)$$

or as the slope conductance, $g_s(V)$:

$$g_s(V) = \mathrm{d}i(V)/\mathrm{d}V \quad (2)$$

where $i(V)$ denotes the measured single-channel current (the difference between the total current when the channel is open and when it is closed) at the applied potential V, and V_{rev} is the reversal potential for the channel, the potential at which $i = 0$. Generally, g_c and g_s are only equal if the channel behaves as a simple ohmic resistor in the potential range of interest. (The difference between g_c and g_s is discussed in greater detail by Finkelstein and Mauro, 1977.)

3.1. Symmetrical Aqueous Phases

Notational ambiguities arising from the two different definitions of conductances are easily overcome. Much more serious are the experimental uncertainties

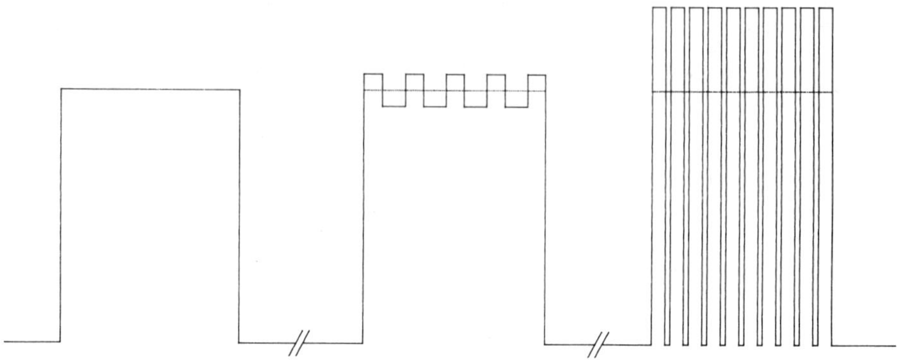

Figure 2. Schematic representation of three different channels exhibiting the same average current, but with different microstructure. The full line denotes the current that ideally should be measured. The interrupted horizontal line denotes the average current during the open state.

in the determination of single-channel conductances that can occur even in symmetrical aqueous phases, where $V_{rev} = 0$, and V is accurately known. The problem becomes obvious when the root-mean-square (r.m.s.) noise of the current through an open channel differs significantly from the noise in the base line prior to channel opening (e.g., Auerbach and Sachs, 1984, and the associated discussion). The difficulties occur because the measurement of i reflects only the time-averaged net rate of ion movement through the channel, with the time resolution being determined by the bandwidth of the recording apparatus. The magnitude of i may reflect exclusively the characteristics of the (always fully open) permeation pathway as a stable entity, or it may reflect an average of several permeation pathways, where the channel rapidly cycles through several states, each with different permeability characteristics. Figure 2 illustrates schematically three situations in which the average rate of ion movement is the same, but the underlying mechanism(s) differ.

The ability to differentiate among the three channel types displayed in Fig. 2 depends on the frequency response of the current-measuring system. The channels will be indistinguishable if the current signal is highly filtered (i.e., if the measurement of i is an average over a long time interval); they may be clearly separable (i.e., if the frequency response is sufficient to allow the individual current fluctuations to be resolved); or they may be (more or less) distinguishable if the (membrane) current in the open state is clearly noisier than that in the closed state.

These complications are illustrated by the voltage-dependent Na^+ channels (Fig. 3), where they may raise significant problems if not recognized. Failure to recognize the existence of these complications can lead to spurious estimates of the single-channel conductance.

It is apparent that the different "open" states are not equivalent. There are current transitions of large amplitude in which the r.m.s. current noise is approximately the same as when the channel is closed. Other current transitions are of smaller amplitude, and the r.m.s. current noise is visibly increased when

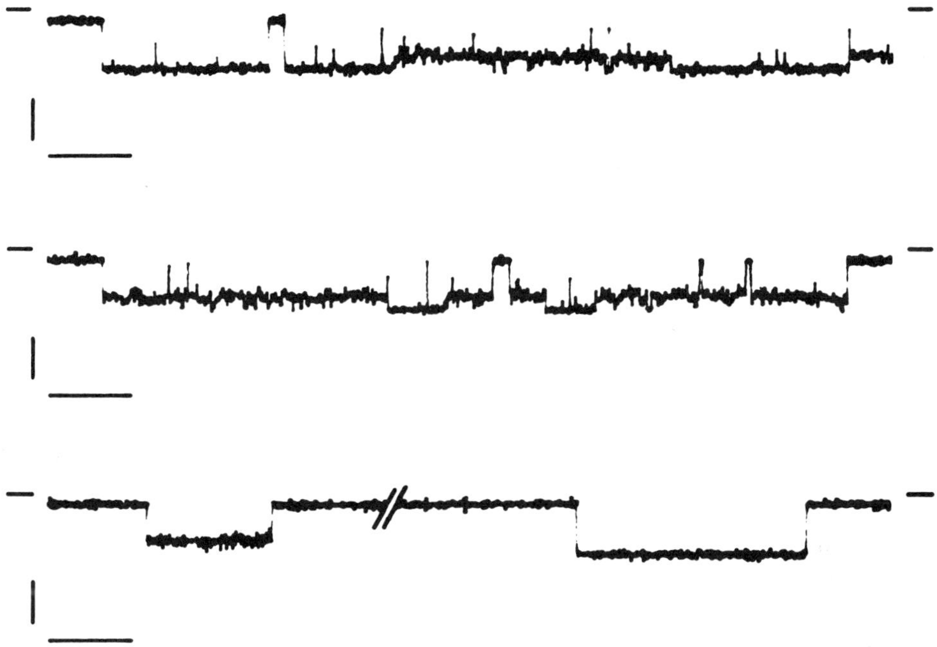

Figure 3. Examples of current traces in which the BTX-modified Na^+ channel exists in more than one type of "open" state. The Na^+ channel undergoes transitions between a well-defined open state, where there is no excess noise, and another kind of open state with a smaller current amplitude, where the open state exhibits excess noise. Note how the current transitions, initially well defined, become flickery, with short excursions to the fully open state. In the bottom trace, the channel reverts to the well-defined state. The calibration bars denote 2 sec (horizontally) and 1 pA (vertically); the zero-current level is denoted by the two upper markers; 50 mM Na^+, 24.0°C, −60 mV. Other experimental conditions as described in the legend to Fig. 1. The figure contains three 20-sec segments; there was a 17-sec gap between the top and the middle trace and a 48-sec gap between the middle and the lower (where the interruption denotes a 17-sec gap) (W. N. Green, unpublished observations). Tracings were filtered at 25 Hz.

the channel is "open." The assignment of a current value to the large transitions seems straightforward, even though the absence of excess noise could result simply from the limited bandwidth of the recording apparatus. If the increase in the r.m.s. noise is disregarded, one can calculate the magnitude of the average current for the smaller noisy transitions and thus an average conductance. This average conductance, however, does not describe the permeability characteristics of a simple two-state channel. Instead, it describes the modification of the permeability of the open state(s) by some slower processes.*

Similar complications have been described in other systems. The nicotinic acetylcholine-gated channel exhibits "subconductance" states in which the cur-

* The slow conformational changes that might be indicated by the flickery transitions in Fig. 3 would not be related to the much more rapid conformational transitions that may be associated with ion permeation *per se* (Läuger *et al.*, 1980). These latter fluctuations may also exist, but they would not be discernible in these measurements.

rent noise is substantially increased (Sachs, 1983; Auerbach and Sachs, 1984). A similar situation exists for the Ca^{2+}-activated K^+ channel, where the current noise in the open state may be increased substantially. Yellen (1984) has discussed how the existance of a large number of unresolved brief closing events may bias the estimate of single-channel currents.

The coexistence of these different current transitions leads naturally to the notions of a "well-defined current transition," where the r.m.s. noise before and after the transition is comparable, and of a "well-defined channel appearance," in which both the opening and closing transitions are well defined and the magnitude of the opening transition equals the magnitude of the closing transition (see also Andersen and Muller, 1982). The noisy current transitions or channel appearances can be denoted "flickery," similar to the notation used by Yellen (1984) to describe the Na^+-induced block of the Ca^{2+}-activated K^+ channel.

Whether an event appears to be well defined or flickery may depend on the recording bandwidth and the method of analysis. It is therefore very important always to specify as precisely as possible the criteria used to decide if a current transition or channel appearance should be included in the amplitude histogram. It cannot be overemphasized that there is no substitute for the visual inspection of the single-channel events. Once the basic pattern is understood, one may devise automated analysis algorithms (e.g., Sachs *et al.*, 1982; Andersen, 1983a). When experimental conditions change, however, it will be necessary to check again whether the algorithm still contains appropriate test criteria for the inclusion of an event into the amplitude histogram.

The different appearances of the current transitions in Fig. 3 do not seem to indicate a systematic deterioration of the Na^+ channel in this preparation, e.g., proteolysis. Experimentally the following is found: (1) a channel may start out in the flickery state and then become well defined; (2) it may start out well defined and become flickery at a later time; or (3) it may exhibit transitions between the two states. The latter situation is, in fact, illustrated in Fig. 3. Observations may last for hours without any one of the situations just described becoming dominant. Transitions between the well-defined and the flickery states could possibly result from the reversible association of some critical protein or lipid component with the Na^+ channel. This is unlikely because of the essentially infinite dilution of all extraneous components that occurs when the channel incorporates into the bilayer. It is possible that these transitions, similar to the state changes that occur in the voltage–activation curves (Green *et al.*, 1984; Moczydlowski *et al.*, 1984; Urban *et al.*, 1986), reflect conformational changes in the BTX-modified sodium channel between different conducting and possibly nonconducting states. The voltage-dependent Na^+ channel is a large protein with a complex phase space and presumably many conformations of approximately equal energy, some of which may have significantly different observable behavior.

The practical problems that may arise from the flickery channel appearances are illustrated in Fig. 4. This figure is a histogram that was obtained by including all the channel appearances in a single-channel current record. There are two peaks in the histogram. The peak representing the larger current steps contains the well-defined current transitions. The standard deviations for the two peaks

Figure 4. Amplitude histogram based on single-channel current transitions in 50 mM Na^+. The two peaks correspond to well-defined transitions (to the right) and flickery transitions (to the left). The mean ± S D (no. of observations) for the two populations were 0.955 ± 0.039 pA (46) and 1.120 ± 0.024 pA (31), respectively. The experimental conditions were as described in the legend to Fig. 1 except that the pH was maintained with 10 mM HEPES; 24.5°C, −61.5 mV (W. N. Green, unpublished observations).

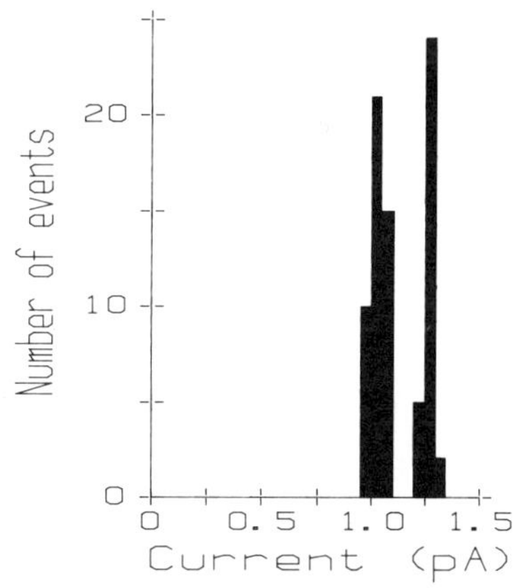

are different, being larger for the smaller peak, consistent with the larger current noise associated with the flickery channel appearances, which makes the determination of the magnitude of the current steps somewhat uncertain. These complications aside, one should note that the current steps cluster into two distinct groups, indicating that the events underlying each of the peaks are quite reproducible.

For the BTX-modified Na^+ channel, the frequency of flickery channels relative to the well-defined events varies as a function of [NaCl].* They are rare in [NaCl] ≥ 500 mM and are seen with increasing frequency as [NaCl] is decreased. In 20 mM Na^+, they seem to be the predominant conductance variant(s) (Green *et al.*, 1986a). It is therefore more difficult to obtain reliable single-channel conductances at low [Na^+], which is unfortunate in view of the critical importance of reliable conductances (and current–voltage characteristics) at low permeant ion concentrations for understanding the mechanism of ion entry into the channels (e.g., Andersen, 1983b).

Current–voltage characteristics based on well-defined current transitions are, in symmetrical solutions, effectively linear over the voltage range of immediate physiological interest (see Fig. 5). The small deviations from ideal linearity are not systematic, so $g_c = g_s$, and it is possible to use the entire current–voltage characteristic to determine the single-channel conductance by linear regression.

The single-channel conductance measurements are, apart from the appearance of the flickery conductance variant(s), quite reproducible. The average conductance determined from well-defined current transitions in 50 mM Na^+ is 19.5 ± 0.3 pS (mean ± S.E.M. for nine experiments with a total of 17 channels; data normalized to 24°C using a Q_{10} of 1.3), with a range of 18.4–20.7 pS. The conductance variation from channel to channel seems, however, to be larger than can be accounted for from the standard error of the mean for the individual am-

* It is presently unclear whether the flickery state exists for all Na^+ channels. They are clearly evident in the dog forebrain preparation, where they are seen in every batch whether or not protease inhibitors are present. They have not been reported for other preparations, however.

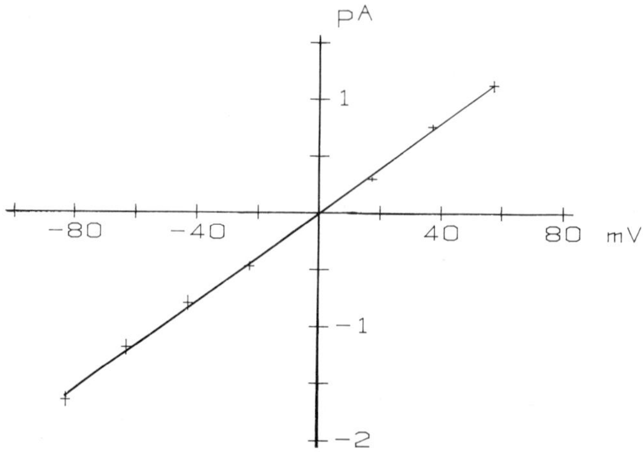

Figure 5. Current–voltage characteristics for Na^+ channels in symmetrical 50 mM Na^+. The points denote mean ± S D based on at least 22 current transitions obtained with two single-channel records obtained in a single bilayer. The conductance was estimated to be 19.6 pS based on a linear regression analysis of the data. Experimental conditions were as described in the legend to Fig. 1; 23.5°C (W. N. Green, unpublished observations).

plitude histograms, which is ~2% of the mean value. There may thus be channel-to-channel variations in the conductance. Some of these "different" Na^+ channels have altered saxitoxin affinity and binding kinetics (Green *et al.*, 1986b). The origin or significance of these variations is otherwise obscure, but it seems possible that the channels do not constitute a uniform population. Biochemical evidence for microheterogeneity of Na^+ channels has been described by Barchi (1983) and Miller *et al.* (1983). In any case, the conductance variations serve to underline the need to use data from several membranes (and channels) to establish the conductance at any given experimental condition.

3.2. Asymmetrical Aqueous Phases

The preceding discussion has focused on the technical difficulties that may be encountered in the determination of the single-channel conductance because of the existence of flickery channels. A different set of problems may arise if the ionic composition of the two electrolyte solutions bathing the channels are different and the current–voltage characteristics are nonlinear. This behavior is illustrated in Fig. 6, which is the current–voltage characteristic observed with 500 mM Na^+ at one (the "extracellular") side of the channel and 50 mM Na^+ + 450 mM K^+ at the other (the "intracellular") side.

The magnitude of the single-channel conductance will in this case depend on the potential range used for its evaluation and on whether g is calculated using equation 1 or equation 2. It should further be noted that the slope conductances

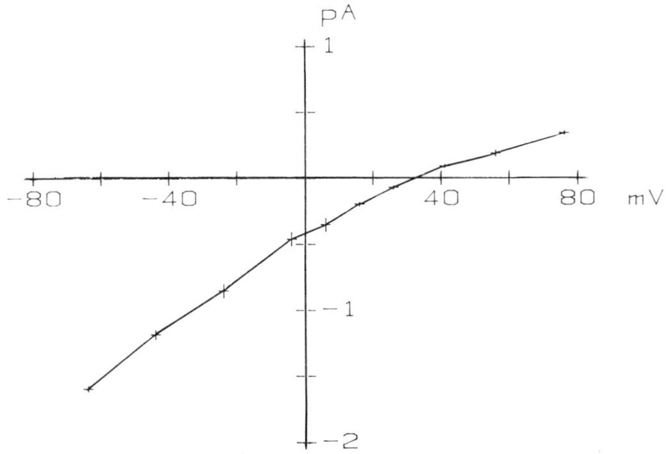

Figure 6. Current–voltage characteristics for Na^+ channels in asymmetrical solutions. The "extracellular" aqueous phase contained 500 mM Na^+; the "intracellular" aqueous phase contained 50 mM Na^+ + 450 mM K^+. Both solutions were buffered to pH 7.4 with 10 mM phosphate; 21.0°C. Experimental conditions were otherwise as described in the legend to Fig. 1. The reversal was estimated to be about 32.4 mV based on linear interpolation of the data between 26.1 and 41.1 mV. The slope conductance around V_{rev} was 10.8 pS. P_K/P_{Na} was obtained using equation 10 and was found to be 0.19 (W. N. Green, unpublished observations).

determined at both the de- and hyperpolarized voltage ranges are substantially less than the conductances determined in symmetrical solutions. These observations are straightforward consequences of the underlying permeation process. Only the slope conductances determined at very hyperpolarized or depolarized potentials allow simple comparisons with the behavior observed in symmetrical solutions. This restriction occurs because

$$j_{net} = j_i - j_o \tag{3}$$

where j_{net} denotes the net ion flux through the channel and j_i and j_o denote the unidirectional influx and efflux, respectively. The latter two fluxes can in principle be measured using tracer techniques, whereas the net flux can be measured electrically or as the difference between the two unidirectional tracer fluxes (Ussing, 1949). The unidirectional fluxes vary as a function of the extracellular and intracellular $[Na^+]$, $[Na^+]_o$ and $[Na^+]_i$, and the applied potential, V:

$$j_i/j_o = ([Na^+]_o/[Na^+]_i)^n \exp(-n \cdot e \cdot V/kT) \tag{4}$$

or

$$j_i/j_o = \exp[-n \cdot e \cdot (V - E_{Na})/KT] \tag{5}$$

where e denotes the elementary charge, k is Boltzmann's constant, T is temperature, E_{Na} is the equilibrium (Nernst) potential for Na^+, and n is the flux ratio exponent: n is equal to 1 if there are no interactions among permeating or blocking ions in the channel; n will be larger than 1 if there are single-file flux interactions among the permeating ions or if there are blocking ions present in one or both electrolyte solutions (Hodgkin and Keynes, 1955; Heckmann, 1972; Hille and Schwarz, 1978; Urban and Hladky, 1979); n may also differ from 1 if the channels interact with the permeating ion(s) through ion-induced fluctuations in the (barrier) structure of the type analyzed by Läuger *et al.* (1980).

Equations 3 and 5 can be combined to give:

$$j_{\mathrm{net}} = j_{\mathrm{i}}\cdot\{1 - \exp[n\cdot e\cdot(V - E_{\mathrm{Na}})/kT]\} \tag{6}$$

and

$$\mathrm{d}j_{\mathrm{net}}/\mathrm{d}V = \mathrm{d}j_{\mathrm{i}}/\mathrm{d}V - (\mathrm{d}j_{\mathrm{i}}/\mathrm{d}V + j_{\mathrm{i}}\cdot n\cdot e/kT)\cdot\exp[n\cdot e\cdot(V - E_{\mathrm{Na}})/kT] \tag{7}$$

Thus, j_{net} and $\mathrm{d}j_{\mathrm{net}}/\mathrm{d}V$ can be equated with j_{i} and $\mathrm{d}j_{\mathrm{i}}/\mathrm{d}V$, respectively, only when $\exp[n\cdot e\cdot(V - E_{\mathrm{Na}})/kT]$ is small, i.e., when $(V - E_{\mathrm{Na}}) < -120$ mV or so. To the extent that j_{i} at very hyperpolarized potentials does not depend on the ionic composition of the intracellular solution (an assumption that may be difficult to justify, e.g., because of screening of surface charges), one can thus at sufficiently large applied potentials equate the current and slope conductance with the current and slope conductance that would have been measured in symmetrical solutions at the same potential. Importantly, this means that measurements in asymmetric solutions ($[\mathrm{Na}^+]_{\mathrm{i}} < [\mathrm{Na}^+]_{\mathrm{o}}$) should be taken past -120 mV to permit valid comparisons with data obtained in symmetric solutions (see Fig. 7, which illustrates a particular example of a model-dependent situation.)

4. Permeability Characteristics

4.1. Single-Channel Conductances

In planar bilayers, conductance-activity relations have been studied for three different Na^+ channels: from rat skeletal muscle (Moczydlowski *et al.*, 1984), from rat forebrain (Worley *et al.*, 1985), and from canine forebrain (Green *et al.*, 1986a). The results for Na^+ channels from different tissues vary somewhat with respect to their conductances, even in the same species (cf. Moczydlowski *et al.*, 1984; Worley *et al.*, 1985). These findings suggest that differences exist between Na^+ channels from the central nervous system and from skeletal muscle, as the measurements were made in bilayers of similar composition. These differences serve to emphasize the advantages of studying channels of biological membrane origin in defined environment, where differences between nerve and muscle chan-

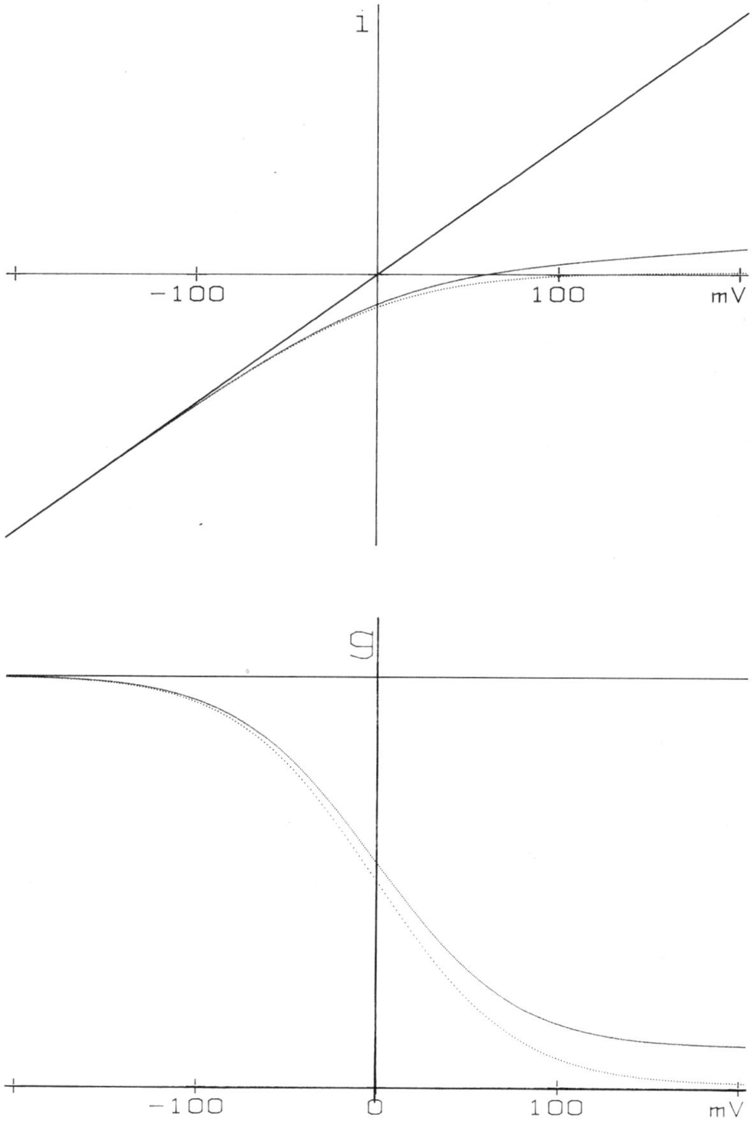

Figure 7. Top: Current–voltage characteristics for a channel subject to the constant-field approximation (Goldman, 1943). The current is expressed as $i = (e^2/kT)\cdot u\cdot V\cdot[c_O\cdot\exp(-V\cdot e/kT) - c_i]/[\exp(-V\cdot e/kT) - 1]$, where u is the ion mobility in the membrane and c_O and c_i denote the extracellular and intracellular concentrations of the permeant cation. Bottom: The corresponding slope conductance-versus-voltage characteristics. ——— denotes the behavior predicted for symmetrical aqueous phases ($c_O = c_i = 1$). ········ denotes the behavior predicted in the presence of a 1:0.1 ratio of the permeant ion between the two aqueous phases ($c_O = 1$, $c_i = 0.1$). · · · · · denotes the behavior predicted for a 1:0.01 ratio ($c_O = 1$, $c_i = 0.01$). Note how the slope conductance close to $V = 0$ in the asymmetric conditions differs appreciably from the conductance observed in symmetrical solutions.

nels, for example, are unlikely to result from different membrane environments but rather to originate from different molecular characteristics.

The Na^+ channels from rat brain and muscle show Langmuirian conductance-activity characteristics, i.e., the single-channel conductance is described by:

$$g = g_{\max} \cdot a/(a + K_g) \tag{8}$$

where $g_{\max}$ denotes the maximal conductance observed at high $[Na^+]$, a is the Na^+ activity, and K_g denotes the Na^+ activity for half-maximal conductance. Visual comparison of data with the predictions of equation 8 is sometimes difficult, and equation 8 is often transformed to yield a linear plot, e.g., the Eadie–Hofstee plot:

$$g = g_{\max} - K_g \cdot (g/a) \tag{9}$$

A plot of g versus g/a should thus yield a straight line if the conductance–activity relationship is described by equation 8. The apparent simplicity of equation 9 may be misleading, however, as the transformation tends to skew the visual impact of any deviations from the predictions of equation 8. Notwithstanding the importance of exploratory graphic analysis of the data (e.g., Wong, 1975), one should therefore critically assess whether small deviations from linearity in an Eadie–Hofstee plot have biophysical significance. The final data analysis should in any case be based on equation 8 itself or some other appropriate direct expression for the relationship between g and a.

K_g and $g_{\max}$ differ for the brain and the muscle channels from the rat. In PE:PC bilayers, one finds values of 37 mM and 32 pS, respectively, for the former and 8 mM and 21 pS, respectively, for the latter. The molecular basis for these differences is presently unknown. Importantly, however, these results were obtained with NaCl being the major electrolyte constituent in the aqueous phases, i.e., under conditions of varying ionic strength. The finding that the data follow the predictions of equation 8 argues, therefore, that there cannot be a significant (negative) surface charge density close to the channel's entrances. This is surprising in light of the effects of carboxyl-specific covalent modification of the protein on the single-channel conductance (Sigworth and Spalding, 1980; Worley *et al.*, 1985). These questions need further investigation.

In contrast to the rat channels, the conductance-activity relationship for Na^+ channels from canine brain is non-Langmuirian, as the Eadie–Hofstee plot is distinctly nonlinear (see Fig. 8). The shape of the conductance–activity relationship could in this case result from the presence of fixed negative charges close to the channel entrances, as these "surface" charges at low bulk-phase $[Na^+]$ would tend to maintain the $[Na^+]$ close to the entrance at a fixed level, independent of the bulk $[Na^+]$ (Bell and Miller, 1984). The non-Langmuirian behavior could, however, also result from other causes, e.g., multiple ion occupancy (Eisenman *et al.*, 1978; Urban and Hladky, 1979) or conformational transitions in the channel structure (Läuger *et al.*, 1980). The conductance increase seen at $[Na^+] \geq 1.0$ M may reflect multiple ion occupancy. But negative surface charges

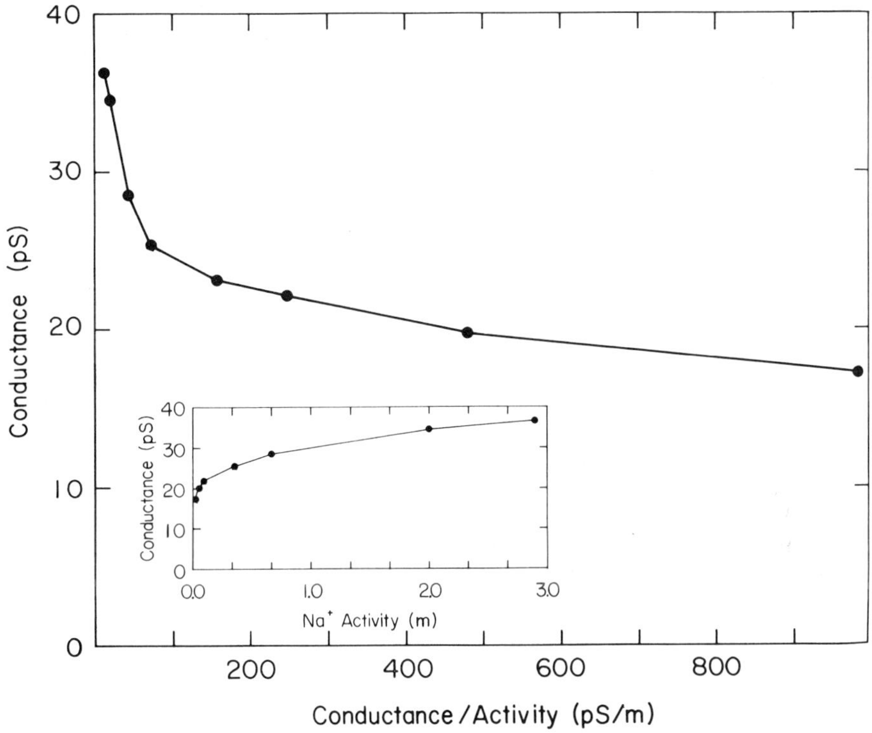

Figure 8. An Eadie–Hofstee plot of the single-channel conductance as a function of the Na^+ activity. The data are normalized to a temperature of 24.0°C, assuming a Q_{10} of 1.3. The nontransformed data are displayed in the figure inset. Experimental conditions are as described in the legend to Fig. 1 (W. N. Green, unpublished observations).

seem more likely to explain the non-Langmuirian conductance-activity behavior at the lower salt concentrations because the reversal potential in Na^+/K^+ mixtures (500 mM salt) is independent of the K^+ mole fraction and because the magnitude of the conductance varies as a function of the ionic strength at a constant [Na^+] (Green *et al.*, 1986a).

4.2. Ion Selectivity

The K^+/Na^+ selectivity can be evaluated using a variety of techniques, e.g., conductance ratios, reversal potentials, tracer fluxes. Results based on these different techniques in general will not provide the same quantitative, or even qualitative, information when there are interactions between the permeating ion and the channel (e.g., saturation) or among the permeating ions in a multiply occupied channel (Hille, 1975; Urban and Hladky, 1979). Only one of these techniques, the reversal-potential measurement, seems feasible with channels from native membrane vesicles incorporated into planar lipid bilayers because of the unavoidable presence of a variety of K^+ channels in the vesicle membrane. The presence of K^+ channels will render more difficult the identification of Na^+ channels in symmetrical K^+ solutions, and the use of K^+-channel blockers may not be practical if they also affect the magnitude of the Na^+-channel conductance. These limitations should not affect the determination of K^+ movement through Na^+ channels that have been reconstituted from purified protein (Hartshorne *et al.*, 1985), although the unavoidable presence of small amounts of other peptides may raise difficulties because of the very low yield of functional channels relative to the total protein used in the reconstitution experiments. It is therefore possible that the visible channel activity may arise from contamination by other peptides (see also Mazet *et al.*, 1984, for a discussion of this problem).

Operationally, the interpretation of reversal-potential measurements is based on the formal use of the Goldman–Hodgkin–Katz equation (Goldman, 1943; Hodgkin and Katz, 1949). The ion selectivity is quantified in terms of a permeability ratio, $P_K/P_{Na} = \alpha$, which can be obtained from:

$$V_{rev} = (kT/e) \ln\{([Na^+]_o + \alpha\cdot[K^+]_o)/([Na^+]_i + \alpha\cdot[K^+]_i)\} \qquad (10)$$

V_{rev} can be determined from single-channel current curves, as illustrated in Fig. 6. The value of P_K/P_{Na} is, however, very sensitive to even small errors in the determination of V_{rev}, and it is essential to compensate for electrode asymmetry and liquid-junction potentials. The latter can be estimated using the Henderson equation (Plettig, 1930; Salling and Siggaard-Andersen, 1971).

The reported values for P_K/P_{Na} vary by about a factor of 2 between 0.07–0.08 (Krueger *et al.*, 1983; Moczydlowski *et al.*, 1984) and ~0.18 (Green *et al.*, 1986a). The basis for these differences is not understood, although it should be noted that there also are differences in P_K/P_{Na} for Na^+ channels in biological membranes (*vide infra*). Measurements on bilayer-incorporated channels may, again, be helpful in deciding whether differences between Na^+ channels from different preparations reflect molecular differences or differences in their environment.

5. Comparison of Sodium Channels in Bilayers and Biological Membranes

The permeability characteristics of voltage-dependent Na^+ channels have been studied in a variety of preparations and, unfortunately, under quite varied ionic conditions.

Table I lists the single-channel conductances and the ionic conditions under which they were determined for 17 preparations. To facilitate the comparison among the various preparations, the measured conductances as well as the conductances normalized to 24°C have been listed.

At first glance, it appears that the conductance of BTX-modified Na^+ chan-

Table I. Single-Channel Conductances for Na^+ Channels in Different Preparations

Electrolyte composition: Extracellular Na^+ (mM)	Extracellular K^+ (mM)	Extracellular Me^{2+}[a] (mM)	Intracellular Na^+ (mM)	Intracellular K^+ (mM)	Intracellular Me^{2+}[a] (mM)	Potential range (mV)	T (°C)	Toxin	Preparation	g (pS)	g^*[b] (pS)	Reference
Lipid bilayer, unpurified membrane vesicles												
500		0.25	500		0.25	−60 to +60	22–25	BTX	Rat brain	29	29	Krueger *et al.* (1983)
200			200			−100 to +80	23–25	BTX	Rat muscle	20	20	Moczydlowski *et al.* (1984)
500			500			−60 to +60	23–24	BTX	Dog brain	25	25	Green *et al.* (1984)
Lipid bilayer, purified Na^+ channels												
140	3	1	140	3	1	−75	22–23	None	Rat brain	25	25	Hanke *et al.* (1984)
90		5	10	80	5	−60 to 0	22–24	None	Eel	11	11	Rosenberg *et al.* (1984)
90		5	10	80	5	−60 to 0	22–24	BTX	Eel	24	24	Rosenberg *et al.* (1984)
500		0.25	500		0.25	−100 to +100	19–23	BTX	Rat brain	25	26	Hartshorne *et al.* (1985)
Patch clamp on cell membrane												
140	1.4	2	(cell attached)			−70 to 0	19–22	None	Rat myoball	18	19	Sigworth and Neher (1980)
150	5	2.5		160 Cs^+		−50 to −10	*ca.* 10	None	Rat myotube	15	22	Horn *et al.* (1981)
137	5.4	2	(cell attached)			−50 to +20	16–18	None	Rat cardiac	14	17	Cachelin *et al.* (1983)
137	5.4	2	137	5.4	2	−50 to 0	16–18	None	Rat cardiac	28	33	Cachelin *et al.* (1983)
135	4	3	1	100		−40 to −20	13–14	None	Chick heart	9	11	ten Eick *et al.* (1984)
160		3	1	20 Cs^+	3	−60 to +10	9.3	None	GH_3 cells	12	18	Horn *et al.* (1984)
145	5	2.6	12	120 Cs^+		−70 to −15	8	None	N1E-115	12	18	Nagy *et al.* (1983)
125	5.5	2.6	1	100		−90 to −30	8.5	None	N1E-115	10	15	Quandt and Narahashi (1982)
125	5.5	2.6	1	100		−70 to −40	7	BTX	N1E-115	2	3	Quandt and Narahashi (1982)
130	5.4	0.8	(cell attached)			−100 to +50	20–23	BTX	NG108-15	11	11	Huang *et al.* (1984)
125	5.5	2.6	2	150 Cs^+		−60 to 0	12	Tetramethrin	N1E-115	10	14	Yamamoto *et al.* (1984)
140	2.8	3		120 Cs^+ 20 TEA^+		−40 to −10	20–22	None	Chromaffin cells	14	15	Fenwick *et al.* (1982)
600		1.5	(cell attached)				13–15	None	Tunicate eggs	10	13	Fukushima (1981)

[a] Me^{2+} denotes divalent cations irrespective of chemical species.
[b] g^* is the single-channel conductance normalized to 24°C using a Q_{10} of 1.3 (e.g., Nagy *et al.*, 1983).

nels in lipid bilayers is larger than the conductance of Na^+ channels in biological membranes. The experimental conditions are, however, sufficiently different that no quantitative comparisons can be made, particularly as all but one (Cachelin *et al.*, 1983) of the patch-clamp studies were done with asymmetric electrolyte solutions. (The only other study which approaches symmetric electrolyte solutions is that of Huang *et al.*, 1984, which, however, had 30 mM tetraethylammonium, TEA^+, in the extracellular solution.) The determination of the slope conductances is subject to the uncertainties described earlier. It should also be noted that the patch-clamp data usually were obtained under conditions in which both the "intra-" and "extracellular" solutions contain ions such as calcium and TEA^+ that may induce a voltage-dependent block of the single-channel currents (Horn *et al.*, 1981; Yamamoto *et al.*, 1984) or just screen any surface charges close to the channel entrances. It is therefore not possible to compare quantitatively single-channel conductance measurements in biological membranes with each other or with the results from bilayer-incorporated channels. The qualitative similarity among the different measurements suggests, nevertheless, that studies on the permeability characteristics of Na^+ channels in lipid bilayers should be useful for understanding the permeability properties of Na^+ channels in their native environment, particularly with regard to the gross features of the ion permeation.

The conductance measurements from the different bilayer experiments should be directly comparable, since they were performed on channels in similar bilayers with symmetrical electrolyte solutions of similar composition. The differences in single-channel conductances should thus signify differences in the channel protein(s), complementing pharmacological dissections of the channels (Moczydlowski *et al.*, Chapter 16, this volume).

Table II summarizes information on the K^+/Na^+ selectivity of Na^+ channels in bilayers and biological membranes. There are, again, quantitative variations among the different preparations, but the general features are similar for all of them. In both biological and bilayer membranes, there is an approximately twofold variation in P_K/P_{Na}, which in part at least may reflect variations in channel structure. The reported values for P_K/P_{Na} for bilayer-incorporated Na^+ channels vary, for example, between 0.07 (Krueger *et al.*, 1983; Moczydlowski *et al.*, 1984) and 0.18 (Green *et al.*, 1986a). This variation should be compared with the variation of P_K/P_{Na} among native Na^+ channels in different tissues of *Rana pipiens*, where P_K/P_{Na} varies between 0.048 (muscle) and 0.086 (nerve) (Hille, 1972; Campbell, 1976). The basis for these differences is not understood, but it seems that one would need the strength of the bilayer system to sort them out.

There are, for the reasons outlined earlier, no reliable electrophysiological data on BTX-modified Na^+ channels in biological membranes that the bilayer data can be compared with. There are, however, tracer flux studies on BTX-modified Na^+ channels in neuroblastoma cells and reconstituted Na^+ channels made under conditions in which the driving forces were well controlled (Huang *et al.*, 1979; Tanaka *et al.*, 1983), and P_K/P_{Na} was estimated to be 0.14–0.17. It is, as for the single-channel conductances, not possible to make quantitative comparisons, particularly as P_K/P_{Na} for unmodified Na^+ channels in biological membranes varies depending on the intracellular $[K^+]$ (Chandler and Meves, 1965;

Table II. K^+/Na^+ Selectivity of Voltage-Dependent Na^+ Channels

Electrolyte composition										
Extracellular			Intracellular							
Na^+ (mM)	K^+ (mM)	Me^{2+} [a] (mM)	Na^+ (mM)	K^+ (mM)	Me^{2+} [a] (mM)	Method	Toxin	Preparation	P_K/P_{Na} [b]	Reference
Lipid bilayer, unpurified membrane vesicles										
100	400	0.25	500		0.25	Rev. pot.	BTX	Rat brain	0.07	Krueger *et al.* (1983)
50	200		250			Rev. pot.	BTX	Rat muscle	0.07	Moczydlowski *et al.* (1984)
500			50	450		Rev. pot.	BTX	Dog brain	0.18	Green *et al.* (1986a)
Lipid bilayer, purified Na^+ channels										
140	3	1	140	3	1	*g* ratio[c]	None	Rat brain	ca 0.1	Hanke *et al.* (1984)
3	140	1	3	140	1					
90		5	10	80	5	Rev. pot.	None	Eel	0.14	Rosenberg *et al.* (1984)
500		0.25	500		0.25	*g* ratio	BTX	Rat brain	0.14	Hartshorne *et al.* (1985)
	500	0.25		500	0.25					
Na^+ channels in native membranes										
473		66		300		Rev. pot.	None	Squid axon	0.087	Chandler and Meves (1965)
473		66		24		Rev. pot.	None	Squid axon	0.13	Chandler and Meves (1965)
110		2		(110, 2)		d(Rev. pot.)[d]	None	Frog axon	0.086	Hille (1972)
110		2		(110, 2)		d(Rev. pot.)	None	Frog muscle	0.048	Campbell (1976)

[a] Me^{2+} denotes divalent cations irrespective of chemical species.
[b] P_K/P_{Na} is in most cases determined from reversal potential measurements using equation 10.
[c] Conductance ratio determinations are based on single-channel conductances obtained in the indicated symmetrical electrolyte solutions.
[d] d(Rev. pot.) denotes a differential reversal potential measurement in which the intracellular ion composition is unknown. The difference between the reversal potentials measured with the two indicated solutions is taken as the reversal potential, and P_K/P_{Na} is calculated according to equation 10.

Cahalan and Begenisich, 1976). There are, however, suggestions that BTX alters the intracellular entrance to the Na^+ channel, as a number of compounds that block unmodified Na^+ channels from the intracellular side are less effective on the BTX-modified Na^+ channels in the squid giant axon (Tanguy *et al.*, 1984; J. Z. Yeh, personal communication).* Despite the qualitative similarity between the permeability characteristics of native and BTX-modified Na^+ channels, one must consider seriously the possibility that significant differences may exist between native and modified Na^+ channels as well as among Na^+ channels from different tissues or species.

ACKNOWLEDGMENTS. This work was supported by NIH grant GM 21342 (O.S.A.), by the Whitaker Foundation (B.W.U.), and by the Cornell Departmental Associates Program (W.N.G.). B.W.U. is an Investigator of The New York Heart Association.

References

Andersen, O. S., 1983a, Ion movement through gramicidin A channels. Single-channel measurements at very high potentials, *Biophys. J.* **41**:119–133.

Andersen, O. S., 1983b, Ion movement through gramicidin A channels. Studies on the diffusion-controlled association step, *Biophys. J.* **41**:147–165.

Andersen, O. S., and Muller, R. U., 1982, Monazomycin-induced single channels. I. Characterization of the elementary conductance events, *J. Gen. Physiol.* **80**:403–426.

Auerbach, A., and Sachs, F., 1984, Single-channel currents from acetylcholine receptors in embryonic chick muscle, *Biophys J.* **45**:187–198.

Barchi, R. L., 1983, Protein components of the purified sodium channel from rat skeletal muscle sarcolemma, *J. Neurochem.* **40**:1377–1385.

Bell, J. E., and Miller, C., 1984, Effects of phospholipid surface charge on ion conduction in the K^+ channel of sarcoplasmic reticulum, *Biophys. J.* **45**:279–287.

Cachelin, A. B., De Peyer, J. E., Kokubun, S., and Reuter, H., 1983, Sodium channels in cultured cardiac cells, *J. Physiol.* (*Lond.*) **340**:389–401.

Cahalan, M., and Begenisich, T., 1976, Sodium channel selectivity. Dependence on internal permeant ion concentration, *J. Gen. Physiol.* **68**:111–125.

Campbell, D. T., 1976, Ionic selectivity of the sodium channel of frog skeletal muscle, *J. Gen. Physiol.* **67**:295–307.

Chandler, W. K., and Meves, H., 1965, Voltage clamp experiments on internally perfused giant axons, *J. Physiol.* (*Lond.*) **180**:788–820.

Dubois, J. M., Schneider, M. F., and Khodorov, B. I., 1983, Voltage dependence of intramembrane charge movement and conductance activation of batrachotoxin-modified sodium channels in frog node of Ranvier, *J. Gen. Physiol.* **81**:829–844.

Eisenberg, M., Hall, J. E., and Mead, C. A., 1973, The nature of the voltage-dependent conductance induced by alamethicin in black lipid membranes, *J. Membrane. Biol.* **14**:143–176.

Eisenman, G., Sandblom, J., and Neher, E., 1978, Interactions in cation permeation through the gramicidin channel. Cs, Rb, K, Na, Li, Tl, and effects of anion binding, *Biophys. J.* **22**:307–340.

* Batrachotoxin may produce significant alterations in the overall permeation pathway, as the normally completely impermeable methylammonium (Hille, 1971) seems to be able to permeate through BTX-modified Na^+ channels (Khodorov and Revenko, 1979). The significance of this finding should, however, be evaluated in light of the possible alterations of the intracellular ionic composition that could be effected by the extracellular solution changes.

Fenwich, E. M., Marty, A., and Neher, E., 1982, Sodium and calcium channels in bovine chromaffin cells, *J. Physiol. (Lond.)* **331**:599–635.

Finkelstein, A., and Mauro, A., 1977, Physical principles and formalisms of electrical excitability, in: *Handbook of Physiology,* Sect. 1: *The Nervous System,* Vol. 1, Part 1 (E. R. Kandel, ed.), pp. 161–213, American Physiological Society, Washington.

Fukushima, Y., 1981, Identification and kinetic properties of the current through a single Na^+ channel, *Proc. Natl. Acad. Sci. U.S.A.* **78**:1274–1277.

Goldman, D. E., 1943, Potential, impedance and rectification in membranes, *J. Gen. Physiol.* **27**:37–60.

Green, W. N., Weiss, L. B., and Andersen, O. S., 1984, Batrachotoxin-modified sodium channels in lipid bilayers, *Ann. N.Y. Acad. Sci.* **435**:548–550.

Green, W. N., Weiss, L. B., and Andersen, O. S., 1986a, Batrachotoxin modified sodium channels from canine forebrain in lipid bilayers. I. Ion permeation and block, *J. Gen. Physiol.* (submitted).

Green, W. N., Weiss, L. B., and Andersen, O. S., 1986b, Batrachotoxin modified sodium channels from canine forebrain in lipid bilayers. III. Characterization of the saxitoxin- and tetrodotoxin-induced block, *J. Gen. Physiol.* (submitted).

Hanke, W., Boheim, G., Barhanin, J., Pauron, D., and Lazdunski, M., 1984, Reconstitution of highly purified saxitoxin-sensitive Na^+-channels into planar lipid bilayers, *EMBO J.* **3**:509–515.

Hartshorne, R. P., Keller, B. U., Talvenheimo, J. A., Catterall, W. A., and Montal, M., 1985, Functional reconstitution of the purified brain sodium channel in planar lipid bilayers, *Proc. Natl. Acad. Sci. U.S.A.* **82**:240–244.

Heckmann, K., 1972, Single file diffusion, in: *Biomembranes,* Vol. 3 (F. Kreuzer and J.F.G. Slegers, eds.), pp. 127–153, Plenum Press, London.

Hille, B., 1971, The permeability of the sodium channel to organic cations in myelinated nerve, *J. Gen. Physiol.* **58**:599–619.

Hille, B., 1972, The permeability of the sodium channel to metal cations in myelinated nerve, *J. Gen. Physiol.* **59**:637–658.

Hille, B., 1975, Ionic selectivity, saturation and block in sodium channels: A four barrier model, *J. Gen. Physiol.* **66**:535–560.

Hille, B., and Schwarz, W., 1978, Potassium channels as multi-ion single-file pores, *J. Gen. Physiol.* **72**:409–442.

Hodgkin, A. L., and Katz, B., 1949, The effect of sodium ions on the electrical activity of the giant axon of squid, *J. Physiol. (Lond.)* **108**:37–77.

Hodgkin, A. L., and Keynes, R. D., 1955, The potassium permeability of a giant nerve fibre, *J. Physiol. (Lond.)* **128**:61–88.

Horn, R., Patlak, J., and Stevens, C. F., 1981, The effect of tetramethylammonium on single sodium channel currents, *Biophys. J.* **36**:321–327.

Horn, R., Vandenberg, C. A., and Lange, K., 1984, Statistical analysis of single sodium channels. Effects of N-bromoacetamide, *Biophys. J.* **45**:323–335.

Huang, L. Y. M., Catterall, W. A., and Ehrenstein, G., 1979, Comparison of ionic selectivity of batrachotoxin-activated channels with different tetrodotoxin dissociation constants, *J. Gen. Physiol.* **73**:839–854.

Huang, L. Y. M., Moran, N., and Ehrenstein, G., 1982, Batrachotoxin modifies the gating kinetics of sodium channels in internally perfused neuroblastoma cells, *Proc. Natl Acad. Sci. U.S.A.* **79**:2082–2085.

Huang, L. Y. M., Moran, N., and Ehrenstein, G., 1984, Gating kinetics of batrachotoxin-modified sodium channels in neuroblastoma cells determined from single-channel measurements, *Biophys. J.* **45**:313–322.

Khodorov, B. I., and Revenko, S. V., 1979, Further analysis of the mechanisms of action of batrachotoxin on the membrane of myelinated nerve, *Neuroscience* **4**:1315–1330.

Krueger, B. K., Worley, J. F., and French, R. J., 1983, Single sodium channels from rat brain incorporated into planar lipid bilayer membranes, *Nature* **303**:172–175.

Läuger, P., Stephen, W., and Frehland, E., 1980, Fluctuations of barrier structure in ionic channels, *Biochim. Biophys. Acta* **602**:167–180.

Mazet, J. L., Andersen, O. S., and Koeppe, R. E., 1984, Single-channel studies on linear gramicidins with altered amino acid sequences, *Biophys. J.* **45**:263–276.

Miller, J. A., Agnew, W. S., and Levinson, S. R., 1983, Principal glycopeptide of the tetrodotoxin/saxitoxin binding protein from electrophorus electricus: Isolation and partial chemical and physical characterization, *Biochemistry* **22**:462–470.

Moczydlowski, E., Garber, S. S., and Miller, C., 1984, Batrachotoxin-activated Na^+ channels in planar lipid bilayers: Competition of tetrodotoxin block by Na^+, *J. Gen. Physiol.* **84**:665–686.

Nagy, K., Kiss, T., and Hof, D., 1983, Single Na channels in mouse neuroblastoma cell membrane. Indications for two open states, *Pfluegers Arch.* **399**:302–308.

Overbeek, J. T. G., 1956, The Donnan equilibrium, *Prog. Biophys.* **6**:57–84.

Plettig, V., 1930, Über die Diffusionspotentiale, *Ann. Phys.* **5**:735–761.

Quandt, F. N., and Narahashi, T., 1982, Modification of single Na^+ channels by batrachotoxin, *Proc. Natl. Acad. Sci. U.S.A.* **79**:6732–6736.

Rosenberg, R. L., Tomiko, S. A., and Agnew, W. S., 1984, Single-channel properties of the reconstituted voltage-regulated Na channel isolated from the eel electroplax of *Electrophorus electricus, Proc. Natl. Acad. Sci. U.S.A.* **81**:5594–5598.

Sachs, F., 1983, Automated analysis of single-channel records, in: *Single-Channel Recording* (B. Sakmann and E. Neher, eds.), pp. 265–285. Plenum Press, New York.

Sachs, F., Neil, J., and Barkakati, N., 1982, The automated analysis of data from single ionic channels, *Pfluegers Arch.* **395**:331–340.

Salling, N., and Siggaard-Andersen, O., 1971, Liquid-junction potentials between plasma or erythrolysate and KCl solutions, *Scand. J. Clin. Lab. Invest.* **28**:33–40.

Sigworth, F. J., and Neher, E., 1980, Single Na^+ channel currents observed in cultured rat muscle cells, *Nature* **287**:447–449.

Sigworth, F. J., and Spalding, B. C., 1980, Chemical modification reduce the conductance of sodium channels in nerve, *Nature* **283**:293–295.

Tanaka, J. C., Eccleston, J. F., and Barchi, R. L., 1983, Cation selectivity characteristics of the reconstituted voltage-dependent sodium channel purified from rat skeletal muscle sarcolemma, *J. Biol. Chem.* **258**:7519–7526.

Tanguy, J., Yeh, J. Z., and Narahashi, T., 1984, Interaction of batrachotoxin with sodium channels in squid axons, *Biophys. J.* **45**:184a.

tenEick, R., Yeh, J., and Matsuki, N., 1984, Two types of voltage dependent Na channels suggested by differential sensitivity of single channels to tetrodotoxin, *Biophys. J.* **45**:70–73.

Urban, B. W., and Hladky, S. B., 1979, Ion transport in the simplest single file pore, *Biochim. Biophys. Acta* **554**:410–429.

Urban, B. W., Weiss,,L. B., Green, W. N., and Andersen, O. S., 1986, Batrachotoxin modified sodium channels from canine forebrain in lipid bilayers. II. Steady-state voltage activation, *J. Gen. Physiol.* (submitted).

Ussing, H. H., 1949, The active ion transport through the isolated frog skin in the light of tracer studies, *Acta Physiol. Scand.* **17**:1–37.

Wong, J. T. F., 1975, *Kinetics of Enzyme Mechanisms,* Academic Press, New York.

Worley, J. F., French, R. J., and Krueger, B. K., 1985, Interactions of divalent cations with single sodium channels, *Biophys. J.* **47**:439a.

Yamamoto, D., Yeh, J. Z., and Narahashi, T., 1984, Voltage-dependent calcium block of normal and tetramethrin- modified sodium channels, *Biophys. J.* **45**:337–344.

Yellen, G., 1984, Ionic permeation and blockade in Ca^{2+}-activated K^+ channels of bovine chromaffin cells, *J. Gen. Physiol.* **84**:157–186.

Chapter 16

BLOCKING PHARMACOLOGY OF BATRACHOTOXIN-ACTIVATED SODIUM CHANNELS

EDWARD MOCZYDLOWSKI, AKIRA UEHARA, and SHERWOOD HALL

1. Introduction

Biochemical progress toward the mechanism of voltage-dependent Na^+ channels has enjoyed the benefit of an extensive molecular pharmacology that includes natural specific toxins as well as synthetic chemical probes of this channel. The recently introduced method of studying chemically activated Na^+ channels in planar lipid bilayers with the use of batrachotoxin (Krueger *et al.*, 1983) makes it possible to examine functional manifestations of Na^+-channel pharmacology at the level of individual channel macromolecules. In this chapter, we summarize the results of our investigations of the blocking of batrachotoxin-activated Na^+ channels from rat skeletal muscle by three classes of pharmacological agents: specific guanidinium toxins, synthetic organic cations, and local anesthetics.

2. A Series of Naturally Occurring Saxitoxin Derivatives with Different Net Charge

Figure 1A shows the structure of saxitoxin and 11 other derivatives that have been isolated from cultures of dinoflagellates of the genus *Protogonyaulax* (Hall *et al.*, 1980). These 12 toxins comprise the complete set of possible compounds for substitutions at three positions, one of which involves a pair of stereoisomers. Saxitoxin itself has two positively charged guanidinium groups, one centered at the C-2 carbon with $pK_a = 11.4$ and the other at C-8 with $pK_a = 8.2$ (Rogers and Rapoport, 1980; Shimizu *et al.*, 1981). Substitution of a negatively charged sulfo group occurs at the N-21 position on the carbamyl tail, and a negatively charged

EDWARD MOCZYDLOWSKI • Department of Physiology and Biophysics, University of Cincinnati College of Medicine, Cincinnati, Ohio 45267. AKIRA UEHARA • Department of Physiology, Fukuoka University School of Medicine, Fukuoka 814-01, Japan. SHERWOOD HALL • Food and Drug Administration, Washington, D. C. 20204.

A

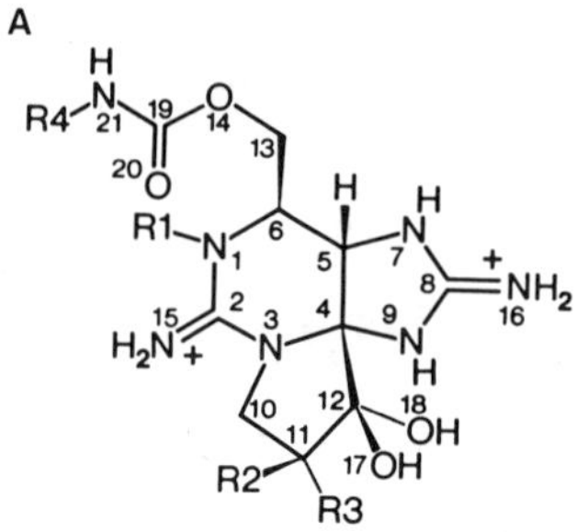

	R1	β R2	α R3	R4
STX	H	H	H	H
B1	H	H	H	SO_3^-
GTX2	H	H	OSO_3^-	H
C1	H	H	OSO_3^-	SO_3^-
GTX3	H	OSO_3^-	H	H
C2	H	OSO_3^-	H	SO_3^-
NEO	OH	H	H	H
B2	OH	H	H	SO_3^-
GTX1	OH	H	OSO_3^-	H
C3	OH	H	OSO_3^-	SO_3^-
GTX4	OH	OSO_3^-	H	H
C4	OH	OSO_3^-	H	SO_3^-

B

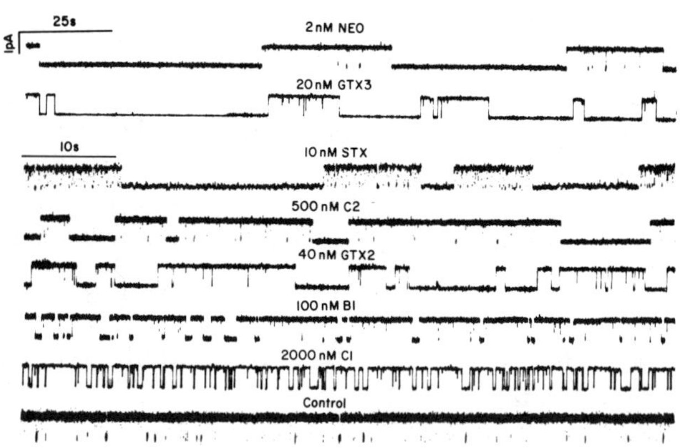

Figure 1. Block of single BTX-activated Na^+ channels by various naturally occurring saxitoxin derivatives. A: Structure of saxitoxin and 11 related toxins that have been isolated from cultures of *Protogonyaulax*. B: Single-channel current records at +50 mV in the absence (bottom trace) or presence of the indicated toxin. Channel opening is in the upward direction. (From Moczydlowski *et al.*, 1984b with permission.)

hydroxysulfate occurs at the C-11 carbon in either the β(R2) or α(R3) position. In addition, substitution of a hydroxyl group at the N-1 nitrogen of the C-2 guanidinium has been shown to introduce a third pK_a of 6.75 in neosaxitoxin (Shimizu *et al.*, 1978). A variety of chromatographic and electrophoretic evidence has shown that these molecules can be separated into three groups of net charges near 0, +1, and +2 at neutral pH (Hall, 1982). Whether compounds C3 and C4 bear a net negative charge in solution has not yet been established. In any case, these derivatives provide a unique opportunity to examine the effect of net charge and small hydrophilic substituent groups on the binding and channel-blocking behavior of a potent neurotoxin.

Figure 1B shows a series of single-channel records at +50 mV of batrachotoxin-activated Na^+ channels from rat skeletal muscle in the presence of various concentrations of seven different toxins. The bottom control record in the absence of any guanidinium toxin shows that the batrachotoxin-activated Na^+ channel is essentially open all of the time at this voltage, with only occasional brief closing events that last on the order of 10–500 msec. Each of the saxitoxin derivatives induces the appearance of discrete blocked states when added to the extracellular face of the channel. The records of Fig. 1B are shown in order of decreasing mean blocked time, with NEO having the longest-lived blocked events (mean 44 sec) and toxin C1 having the shortest (mean 0.4 sec).

The kinetics of guanidinium toxin block at the single-channel level have been shown to conform to the predictions of a reversible binding equilibrium to a single site (French *et al.*, 1984; Moczydlowski *et al.*, 1984a). The binding rate constants for the blocking reaction can readily be obtained from statistical analysis of the probability distributions of the blocked and unblocked events. One of the most striking features of guanidinium toxin block of open Na^+ channels is its strong voltage dependence, with the block being enhanced at negative voltage (Krueger *et al.*, 1983). This behavior is shown in Fig. 2, where the equilibrium dissociation constant for several toxins is plotted on a logarithmic scale against the applied voltage. To date, the blocking kinetics of 12 guanidinium toxins have been analyzed, as summarized in Table I. The results of these studies indicate that the apparent rate and equilibrium constants are well described by exponential functions of voltage in the range of +60 to −60 mV. However, as shown in Fig. 2 and Table I, the voltage dependence of the binding reaction does not appear to vary with the net charge of the toxin or the structural modifications we have examined thus far. This result contradicts the expectation of one commonly proposed model for voltage-dependent block (Woodhull, 1973), which holds that the voltage dependence arises as result of the direct entry of a charged molecule into an electric field.

The effect of varying net charge on three different binding parameters is represented in Fig. 3. The apparent net charge of the various derivatives at pH 7.4 was calculated by using the published pK_a's for saxitoxin and neosaxitoxin (Rogers and Rapoport, 1980; Shimizu *et al.*, 1978, 1981). These pK_a's may be altered in the various sulfo and hydroxysulfate derivatives and are therefore intended only as a rough estimate of the actual net charge. In any case, it is apparent from Fig. 3A that the voltage dependence of the binding equilibrium (expressed as a fraction of the electric field, $z\delta$) is essentially constant ($z\delta = 0.65 \pm 0.06$) as the net charge is varied. Figure 3B also shows that the logarithm of the dissociation rate constant is poorly correlated with net charge (correlation coefficient, $r = -0.26$), meaning that a simple ionic interaction cannot be the sole factor that determines the duration of the toxin–receptor complex. In contrast to these

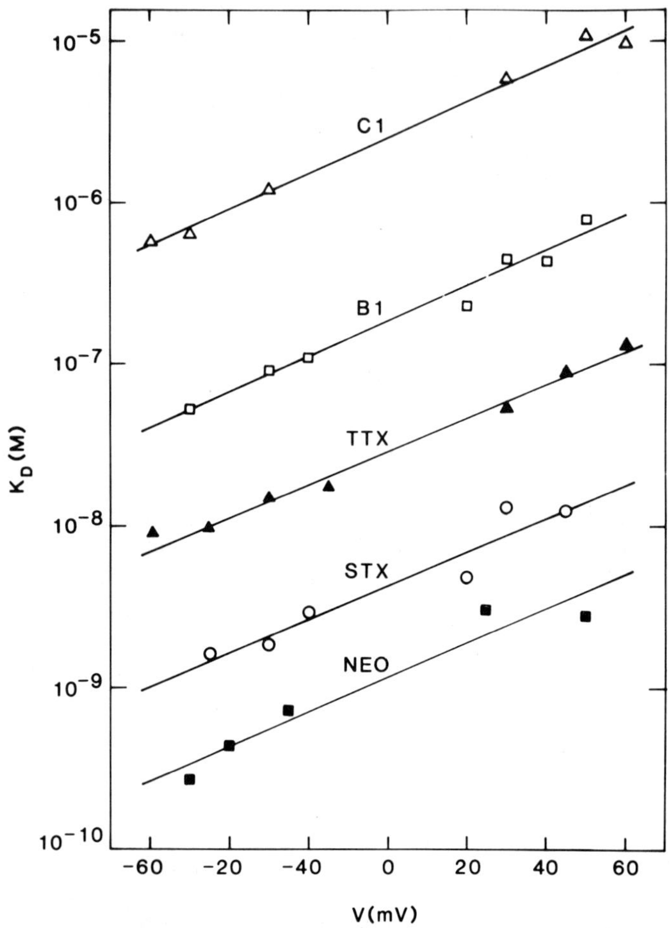

Figure 2. Voltage dependence of the equilibrium dissociation constant for various guanidinium toxins in the range of +60 to −60 mV.

results, Fig. 3C shows a very good correlation between the logarithm of the zero-voltage association rate constant and the apparent net charge ($r = 0.96$).

The latter result suggests that the bimolecular association rate of toxin is influenced by fixed negative surface charges in the vicinity of the receptor, as has been previously proposed from electrophysiological and binding studies (Henderson *et al.*, 1974; Hille *et al.*, 1975). This situation would predict a higher local concentration of positively charged toxins near the receptor as compared to net neutral toxins. If no other factors were involved, the observed association rate constant, k_{on}^*, would be expected to vary as a Boltzmann distribution with respect to the net charge, z, and the electrostatic potential at the receptor, Ψ:

$$k_{on}^* = k_{on}^0 \exp(-zF\Psi/RT) \quad (1)$$

where k_{on}^0 is the rate constant in the absence of surface potential. With this

Table I. Blocking Kinetics of BTX-activated Na^+ Channels from Rat Skeletal Muscle by Various Toxins

	Dissociation rate[a] $k_{-1}(V) = k_{-1}(0) \exp(A_{-1}V)$		Association rate[a] $k_1(V) = k_1(0) \exp(A_1 V)$		Equilibrium constant $K_D(V) = K_D(0) \exp(AV)$	
Toxin	$k_{-1}(0)$ (sec^{-1})	A_{-1} (mV^{-1})	$k_1(0)$ ($10^6\ sec^{-1}M^{-1}$)	A_1 (mV^{-1})	$K_D(0)$ (nM)	A (mV^{-1})
STX	0.061	0.015	14	−0.0092	4.3	0.024
B1	0.88	0.013	4.8	−0.012	180	0.025
GTX2	0.17	0.014	6.0	−0.011	28	0.025
C1	1.1	0.016	0.44	−0.0098	2500	0.026
GTX3	0.039	0.011	8.7	−0.013	4.5	0.024
C2	0.095	0.016	0.65	−0.014	150	0.030
NEO	0.012	0.013	10	−0.012	1.2	0.025
B2	0.023	0.012	1.2	−0.013	190	0.025
GTX1	0.036	0.021	2.4	−0.0078	15	0.028
C3	0.45	0.017	0.20	−0.0075	2300	0.025
dcSTX[b]	0.39	0.015	28	−0.014	14	0.029
TTX	0.12	0.013	4.5	−0.010	27	0.023
μ-Conotoxin GIIIA	0.007	0.026	0.06	−0.003	110	0.029

[a] Dissociation and association rate constants were measured from mean dwell times of blocked and unblocked events as described in Moczydlowski *et al.* (1984b).
[b] Decarbamoylsaxitoxin.

relationship, the slope of the line in Fig. 3C corresponds to an estimate of −44 mV for the electrostatic potential at the receptor under the ionic conditions of our experiment, 0.2 M NaCl. For the purpose of comparison, a bilayer of pure phosphatidylserine has a surface potential of −130 mV at 0.1 M monovalent salt concentration (McLaughlin, 1977).

Since these experiments were carried out in bilayers formed exclusively of neutral phospholipids (PE and PC), what could be the origin of this surface charge? One possibility is that the effect could arise from the negatively charged sialic acid groups present on the external carbohydrate portion of the Na^+ channel. For the eel electroplax Na^+ channel, the sialic acid content has been measured to be about 100 residues per 250,000-dalton glycopeptide (Miller *et al.*, 1983). Alternatively, the fixed charges of externally located acidic amino acids may also account for this surface charge.

If the results of Fig. 3C do indicate a surface charge effect, then one might also expect a similar effect on channel conductance, assuming that the toxin receptor site is in the vicinity of the outer mouth of the channel. However, we previously observed that the relationship between single-channel conductance and Na^+ concentration conforms to a simple Michaelis–Menten function with a $K_{0.5}$ of 8 mM for Na^+ (Moczydlowski *et al.*, 1984a). If there were a significant surface charge near the outer mouth of the channel, we would expect this relationship to depart severely from a rectangular hyperbola at low ionic strength, as recently shown for the sarcoplasmic reticulum K^+ channel in a bilayer containing phos-

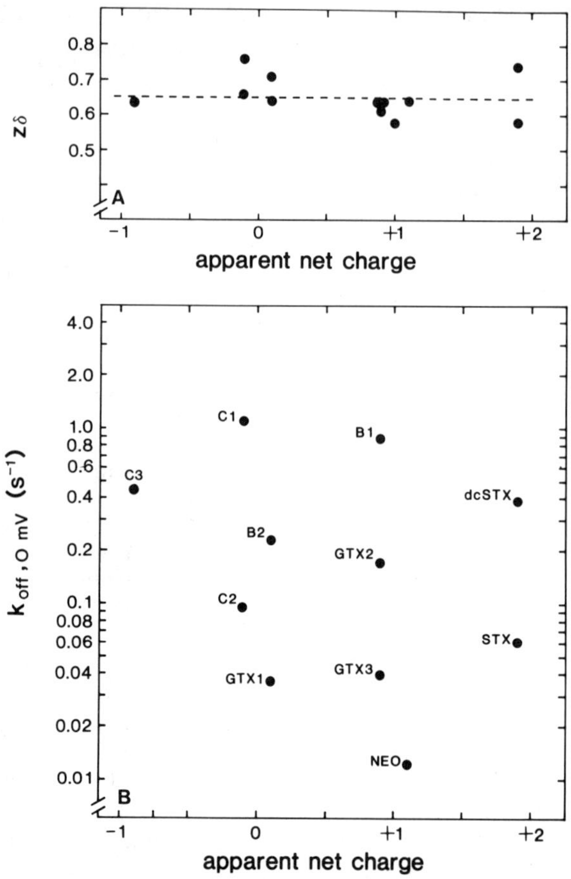

Figure 3. Correlation of blocking voltage dependence and rate constants with net charge of various toxins. A: The voltage dependence of the equilibrium dissociation constant is expressed as an effective electrical distance or effective valence, $z\delta$, as defined by $K_D(V) = K_D(0)\exp(z\delta FV/RT)$. B,C: Zero-voltage dissociation rate constants (B) or association rate constants (C) listed in Table I are plotted on a logarithmic scale against apparent net charge as defined in the text.

phatidylserine (Bell and Miller, 1984). Thus, there is an apparent discrepancy in the expectation of a surface charge effect on toxin binding and channel conductance for the muscle Na^+ channel. Further studies will be required to resolve this issue.

As a final dividend of the 12 dinoflagellate toxins, we may consider the effect of the various substituent groups on the toxin binding energy. For this purpose, the intrinsic binding energy, ΔG_i, of a chemical group can be estimated from the ratio, K_R/K_H, of equilibrium dissociation constants of a derivative containing a particular R- group to that of a molecule where the R- group is replaced by a hydrogen atom (Jencks, 1981):

$$\Delta G_i = -RT\ln(K_R/K_H) \qquad (2)$$

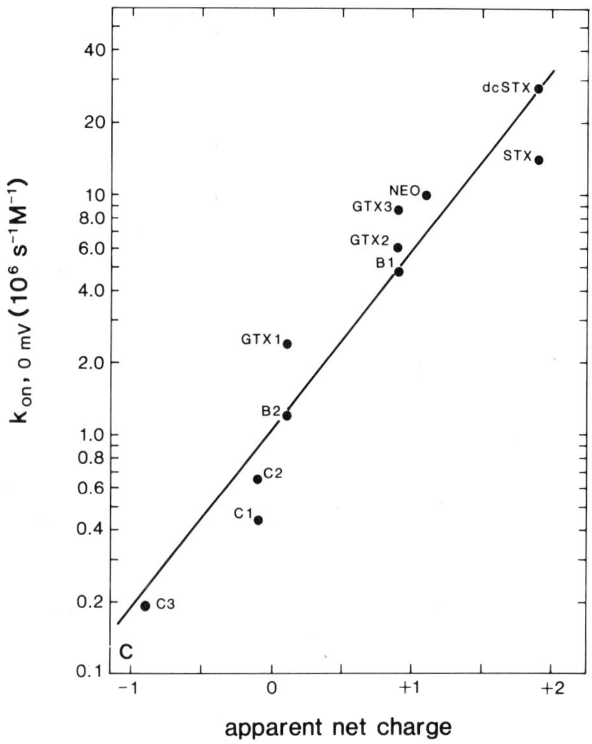

Figure 3. *(continued)*

In our case, we are able to distinguish individual contributions to the forward and reverse binding activation energies according to the similar expression:

$$\Delta G_i^\ddagger = -RT \ln(k_R/k_H) \tag{3}$$

Figure 4 summarizes the intrisic activation energies, as calculated by equation 3, for the different saxitoxin substituents, using all the possible corresponding ratios of zero-voltage rate constants that have been determined to date. A negative $\Delta G_i^\ddagger$ for the dissociation rate constant is defined as a decreased rate constant compared to the hydrogen-containing reference compound, whereas a negative $\Delta G_i^\ddagger$ for the association rate constant is defined as an increased rate constant compared to the reference. In the absence of strain, entropy losses, and conformational changes of the receptor, one would expect to observe similar values of $\Delta G_i^\ddagger$ for different pairs of toxins that vary by the same replacement of an R-group for an H atom. In fact, we do observe surprisingly similar $\Delta G_i^\ddagger$ values for the various substituents in the face of these uncertainties. Figure 4 shows that the N-21 sulfo group consistently increases k_{off} by about 0.5–1.5 kcal/mol in five different molecular pairs. A consistent increase in k_{off} is also observed for the α isomer of the C-11 hydroxysulfate, whereas consistent decreases in k_{off} are ob-

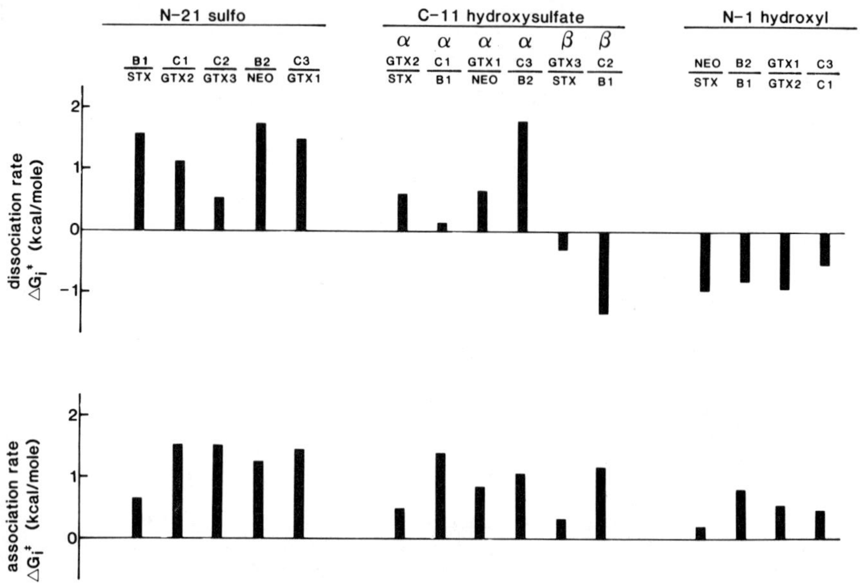

Figure 4. Intrinsic binding activation energy, $\Delta G_i^\ddagger$, associated with three different chemical substituents of saxitoxinderivatives: the N-21 sulfo group, the C-11 hydroxysulfate group, and the N-21 hydroxyl group. $\Delta G_i^\ddagger$ was calculated from the ratios of zero-voltage rate constants of various pairs of toxins according to equation 3 as described in the text.

served for both the β isomer at C-11 and the hydroxyl group at N-1. These observations can serve as the basis for testable speculations concerning the chemistry of specific bonding interactions in the receptor site (Strichartz, 1984; Moczydlowski *et al.*, 1984b). As far as the association rate is concerned, each of the three substituents causes a similar small reduction in k_{on}, corresponding to an energy of about 0.8 kcal/mol. This can conveniently be explained via the surface charge mechanism discussed previously, since each group introduces additional negative charge into the molecule.

3. Tissue-Specific Block of Muscle Na^+ Channels by μ-Conotoxin GIIIA

μ-Conotoxins are a group of homologous 22-amino-acid peptides that have been purified from venom of the marine snail *Conus geographus* (Sato *et al.*, 1983; Cruz *et al.*, 1985). Electrophysiological studies have shown that one of these purified peptides, GIIIA, reversibly blocks action potentials of mammalian muscle without affecting nerve action potentials at concentrations up to 50 μM (Cruz *et al.*, 1985). The sequence of GIIIA is extremely hydrophilic, rich in basic amino acids, and contains the unusual amino acid, hydroxyproline:

1 5 10 15
Arg·Asp·Cys·Cys·Thr·Hyp·Hyp·Lys·Lys·Cys·Lys·Asp·Arg·Gln·Cys

20
·Lys·Hyp·Gln·Arg·Cys·Cys·Ala–NH_2

The six cysteine residues of GIIIA presumably participate in intramolecular disulfide bonds, although the pairing relationships of these bonds in the native conformation have not yet been determined.

In order to investigate the basis for μ-conotoxin action at the single-channel level, we examined the effect of GIIIA on batrachotoxin-activated Na^+ channels from rat muscle and rat brain in planar bilayers. Figure 5 presents a series of single-channel experiments showing that GIIIA induces discrete blocking events in current records of muscle Na^+ channels when present on the extracellular side of the channel but has no observable effect on tetrodotoxin-sensitive Na^+ channels from rat brain under similar conditions. A statistical analysis of the blocked and unblocked events induced by GIIIA in muscle channels showed that the

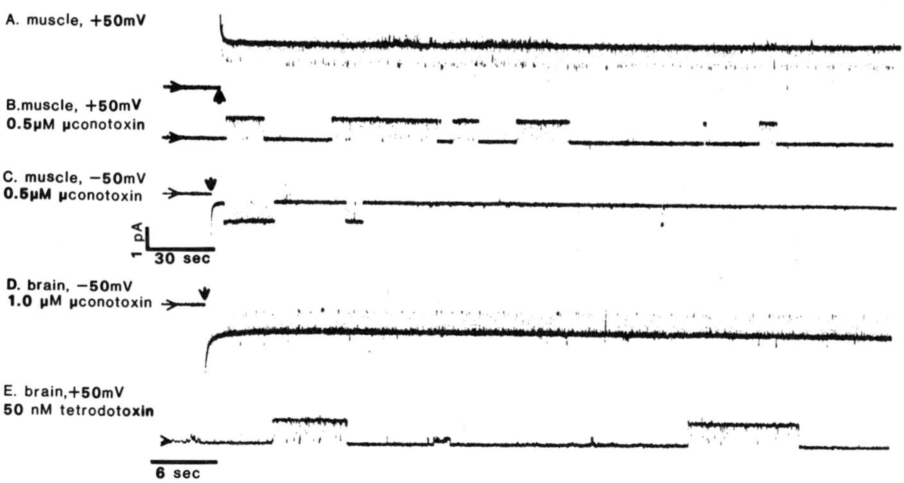

Figure 5. Effect of μ-conotoxin GIIIA on BTX-activated Na^+ channels from rat skeletal muscle or rat brain. The horizontal arrow at the left of each trace corresponds to the zero-current level. In B, C, and D, the vertical arrow marks when the holding voltage was changed from 0 mV to the indicated value as noted by the capacitative transient. In B and E, representative segments of long records are shown. A: Control behavior of a bilayer containing two muscle Na^+ channels at +50 mV; virtually identical behavior is observed at −50 mV, as previously described (Moczydlowski *et al.*, 1984a). B: A different bilayer than A containing one muscle Na^+ channel in the presence of 0.5 μM external (i.e., extracellular) GIIIA. C: The same bilayer as in B at −50 mV. D: A bilayer containing one brain Na^+ channel at −50 mV in the presence of 1 μM GIIIA on the external side. E: A record from the same bilayer as in D after addition of 50 nM tetrodotoxin to the same side as that containing GIIIA. The time scale here is five times faster than that for traces A–D. (From Cruz *et al.*, 1985 with permission.)

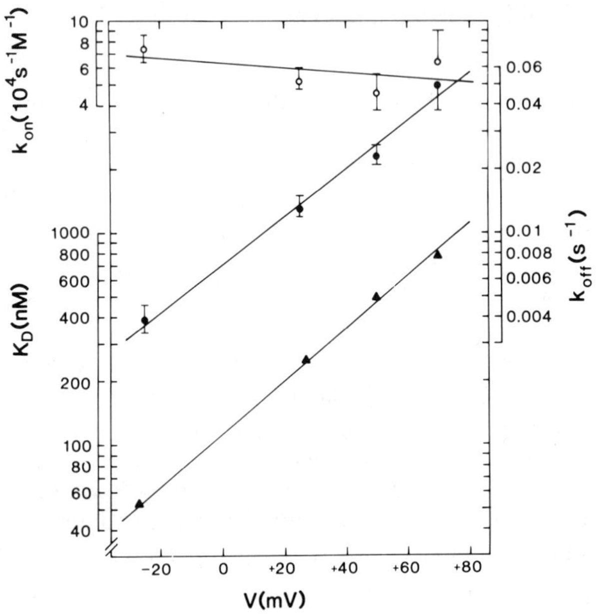

Figure 6. Voltage dependence of μ-conotoxin GIIIA blocking kinetics in BTX-activated Na^+ channels from rat muscle. Association (○) and dissociation (●) rate constants were measured from respective mean unblocked and mean blocked times in the presence of 0.5 μM GIIIA. The equilibrium dissociation constant, K_D (▲) was calculated as the ratio k_{off}/k_{on}. Fits to exponential functions of voltage are given in Table I. (From Cruz *et al.*, 1985.)

blocking kinetics conform to the expectations of a reversible binding equilibrium to a single site (Cruz *et al.*, 1985). In addition, GIIIA block is voltage dependent, with most of the voltage dependence related to the dissociation rate constant, which decreases e-fold per 43 mV of hyperpolarization (Fig. 6). Thus, the peptide toxin GIIIA blocks in a similar fashion as tetrodotoxin and saxitoxin at the single-channel level; however, GIIIA apparently recognizes a structural difference between nerve and muscle Na^+ channels. In addition, the rate constant for GIIIA binding are slower than any of the saxitoxin derivatives that have been examined to date (Table I). At 0 mV, the derived dissociation rate constant is 7×10^{-3} sec^{-1}; the association rate constant is 6×10^4 $sec^{-1}M^{-1}$, giving an equilibrium dissociation constant of 100 nM. Future binding competition studies may establish the relationship between the receptor sites for saxitoxin and GIIIA in muscle Na^+ channels.

4. Fast Block by Internal Organic Cations

Previous voltage-clamp studies have shown that many different organic cations reduce Na^+-channel currents when present on the intracellular side of the

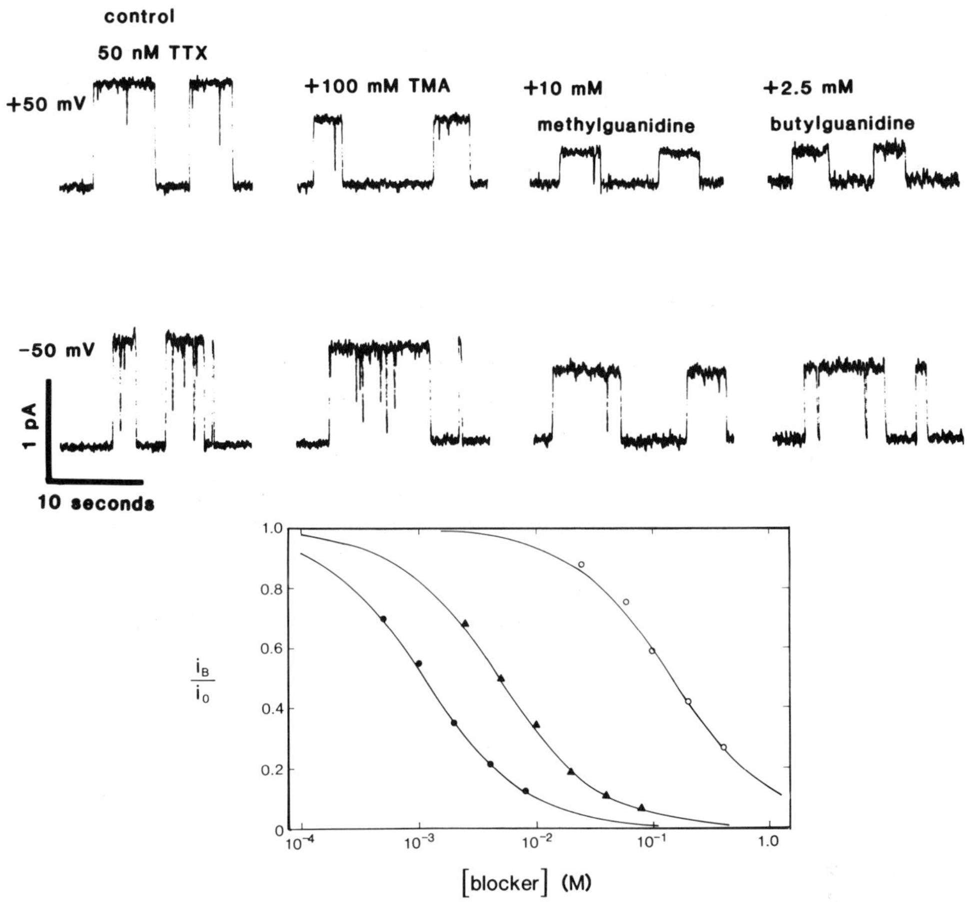

Figure 7. Fast block by three different organic cations from the internal side. The upper group of current records shows examples of single-channel events obtained in the presence of 0.2 M symmetrical NaCl, 50 nM external tetrodotoxin, and the indicated concentrations of either internal TMA, methylguanidine, or butylguanidine. The upper records were all taken at +50 mV, whereas the lower group was taken at −50 mV. Opening is upward in all cases. In the lower graph, the ratio of the single-channel current at +50 mV in the presence of the indicated concentration of internal blocker to that in the absence of blocker is plotted against blocker concentration: (●) butylguanidine, (▲) methylguanidine, and (○) TMA. The solid curves correspond to single-site binding isotherms according to $i_B/i_0 = K_B/(K_B + [B])$, with blocker dissociation constants of (●) 1.1 mM, (▲) 4.7 mM, and (○) 144 mM.

membrane (Strichartz, 1973; Cahalan and Almers, 1979; Horn *et al.*, 1981). Similarly, we find that the batrachotoxin-activated Na^+ channel exhibits a fast voltage-dependent block by organic cations from the internal side. Figure 7 shows the effect of tetramethylammonium (TMA), methylguanidine, and butylguanidine. These cations cause a decrease in the apparent conductance, as would be expected if the dwell time of the blocking ion is too short to be resolved at the frequency response of the recording (100 Hz). Figure 7A also shows that the reduction in current is greater at +50 than at −50 mV, as expected for a voltage-dependent blocker entering the electric field through the internal mouth of the channel. Figure 7B shows that the concentration dependence of blocking by the three cations is well described by a single-site binding isotherm.

A more quantitative analysis of fast block by butylguanidine is presented in Fig. 8. Figure 8A shows the current–voltage relationship for a single channel in the presence of 0, 1, and 8 mM internal butylguanidine. The ohmic behavior in the absence of blocker is transformed by butylguanidine into a rectifier that preferentially passes inward current. This behavior can quantitatively be described by a simple model of blocker binding to a single site with an equilibrium dissociation constant that varies as a Boltzmann distribution of the applied voltage. Theoretical fits to such an analysis are shown in a plot of relative current against voltage in Fig. 8B. The blocking parameters of the three organic cations of Fig. 7 are summarized in Table II, which shows that the electrical distance ($z\delta$) for these compounds is 0.5–0.6, indicating that they sense about half of the transmembrane electric field. A similar blocking effect of internal TMA was also previously reported for normal voltage-activated Na^+ channels of rat myotubes using patch-clamp recording (Horn *et al.*, 1981). We believe that the blocking action of internal organic cations results from binding within the pore and direct occlusion of Na^+ flux. This conclusion rests heavily on the work of Cahalan and Almers (1979), who showed that internal block of squid axon Na^+ channels by the quaternary ammonium compound QX-314 is relieved by raising external Na^+ concentrations. Similar experiments should be possible in the planar bilayer system. This will permit us to investigate the competitive interaction of external Na^+ and internal blocker within the ion permeation path.

Table II. Blocking Parameters for Internal Organic Cations[a]

Blocker (n)	$K_B(0)$ (mM)	$z\delta$
TMA (4)	370 ± 30	0.48 ± 0.08
Methylguanidine (5)	15 ± 1.3	0.59 ± 0.04
Butylguanidine (4)	3.1 ± 0.16	0.53 ± 0.01
QX-222 (9)	14 ± 1	0.34 ± 0.04
QX-314 (5)	5.8 ± 1.5	0.36 ± 0.03
Lidocaine (3)	2.5 ± 0.3	0.39 ± 0.01
Procaine (3)	26 ± 4	0.38 ± 0.02

[a] Current–voltage data at various blocker concentrations in experiments similar to those of Figs. 8A and 12A were analyzed according to the following equation:

$$\ln[(i_0/i_B) - 1] = z\delta FV/RT + \ln([B]/K_B(0))$$

Best fit values of $K_B(0)$ and $z\delta$ were found by linear regression. The values given are the mean and standard deviation from n experiments.

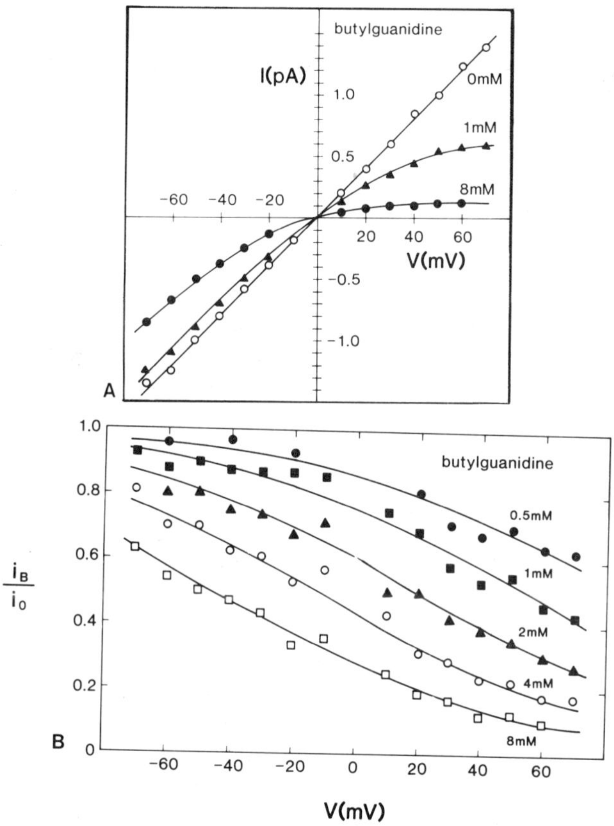

Figure 8. Voltage-dependent block by internal butylguanidine. A: Single-channel current–voltage data are plotted for the same channel in the presence of 0.2 M symmetrical NaCl and 0, 1, and 8 mM internal butylguanidine. B: The ratio, i_B/i_0, of single-channel current in the presence of various internal butylguanidine concentrations to that in the absence of blocker is plotted against voltage. Solid curves correspond to fits to the equation given in Table II with blocking parameters for butylguanidine given in that table.

5. Does the Guanidinium Group of Tetrodotoxin Bind within the Pore?

Assuming that the internal blocker methylguanidine occupies a site within the channel's passageway for Na^+, we can ask whether the binding of TTX on the external side is sensitive to occupation of the channel by a fast internal blocker. For example, binding competition could result if the guanidinium group of TTX occupied the same site reached by internal methylguanidine. This example would predict that increasing methylguanidine would lower the apparent association rate constant for TTX in a purely competitive fashion, since TTX could not bind to channel states occupied by methylguanidine. Alternatively, an enhanced TTX dissociation rate might be expected via electrostatic repulsion if the guanidinium groups of TTX and methylguanidine could simultaneously occupy adjacent sites

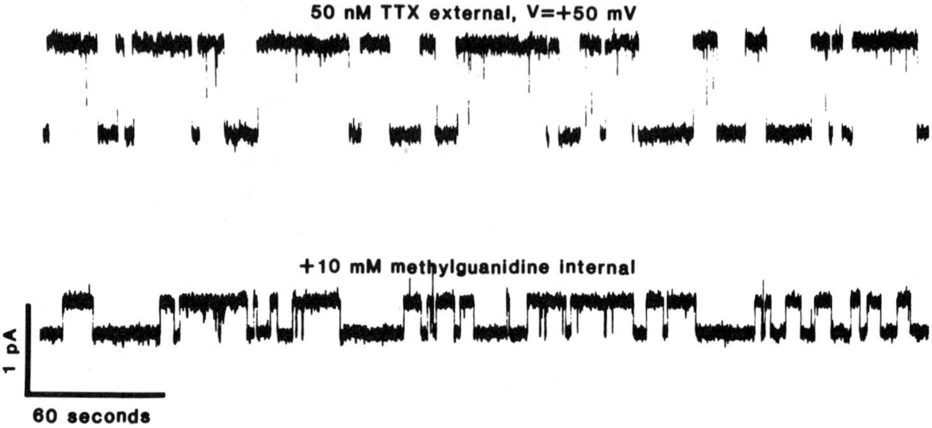

Figure 9. Effect of internal methylguanidine on TTX blocking kinetics. The top trace shows a single-channel current record (opening upward) taken in the presence of 0.2 M symmetrical NaCl and 50 nM external TTX at +50 mV. The bottom trace is a record from a similar experiment in which 10 mM methylguanidine was also present on the internal side.

within the pore. Such simple tests for binding interactions within an ion channel originated with the seminal experiments of Armstrong and his colleagues, who demonstrated that such interactions occur between K^+ and TEA derivatives in squid axon K^+ channels (Armstrong, 1975).

Figure 9 shows actual results for such an experiment on the BTX-activated Na^+ channel. A qualitative inspection of the duration of blocked and unblocked events in the presence of TTX reveals no dramatic difference in the absence or presence of 10 mM internal methylguanidine, a concentration that blocks the channel 65% of the time as measured by the decrease in conductance. A purely competitive interaction resulting from mutually exclusive binding at one site would predict that the mean duration of open events when the channel is not blocked by TTX should have increased threefold. In contrast, the mean unblocked ($\bar{\tau}_u$) and blocked ($\bar{\tau}_b$) times of TTX kinetics showed no significant difference in the absence ($\bar{\tau}_u = 11 \pm 1.6$ sec, $\bar{\tau}_b = 8.3 \pm 1.2$ sec) or presence of 10 mM internal methylguanidine ($\bar{\tau}_u = 9.1 \pm 1.1$ sec, $\bar{\tau}_b = 7.1 \pm 0.9$ sec). Similar results have been obtained in analogous experiments with the canine brain Na^+ channel using TEA as an internal blocker (W. N. Green and O. S. Andersen, personal communication).

In a previous study (Moczydlowski *et al.*, 1984a), we also found that Na^+ competes with TTX binding to single channels, but only by increasing Na^+ on the extracellular face of the channel. Increasing internal Na^+ had no significant effect on TTX binding kinetics, just as internal methylguanidine has no effect. These results, coupled with the observation that the voltage dependence of toxin binding is independent of toxin charge, suggest that the toxins do not bind deeply within the ion permeation path near sites that determine ion selectivity (Kao and Nishiyama, 1965; Hille, 1975). As an alternative hypothesis, TTX and STX may bind superficially over the outer mouth, forming a molecular lid (Kao and Walker, 1982); or they might bind away from the pore altogether and close the channel via an allosteric gatinglike mechanism.

6. The Action of Three Kinds of Local Anesthetics: Neutral, Tertiary Amine, and Quaternary Ammonium Compounds

Figure 10 shows the structures of five local anesthetics, whose effects we have begun to study on BTX-activated Na^+ channels. Included in this group are the tertiary amines procaine (pK_a = 8.9) and lidocaine (pK_a = 7.9), the quaternary ammonium derivatives of lidocaine QX-314 and QX-222, and the neutral benzocaine molecule. The complicated actions of these drugs have been studied extensively in many preparations by conventional voltage-clamp measurements of normal and BTX-modified Na^+-channel currents (for reviews see Strichartz, 1976; Strichartz and Ritchie, 1985).

In screening these compounds, we have observed two distinct effects on

Benzocaine

Procaine

Lidocaine

QX-314

QX-222

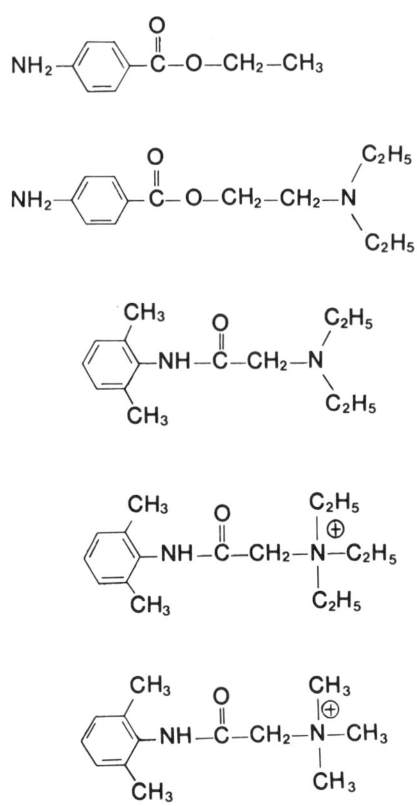

Figure 10. Structures of various local anesthetics.

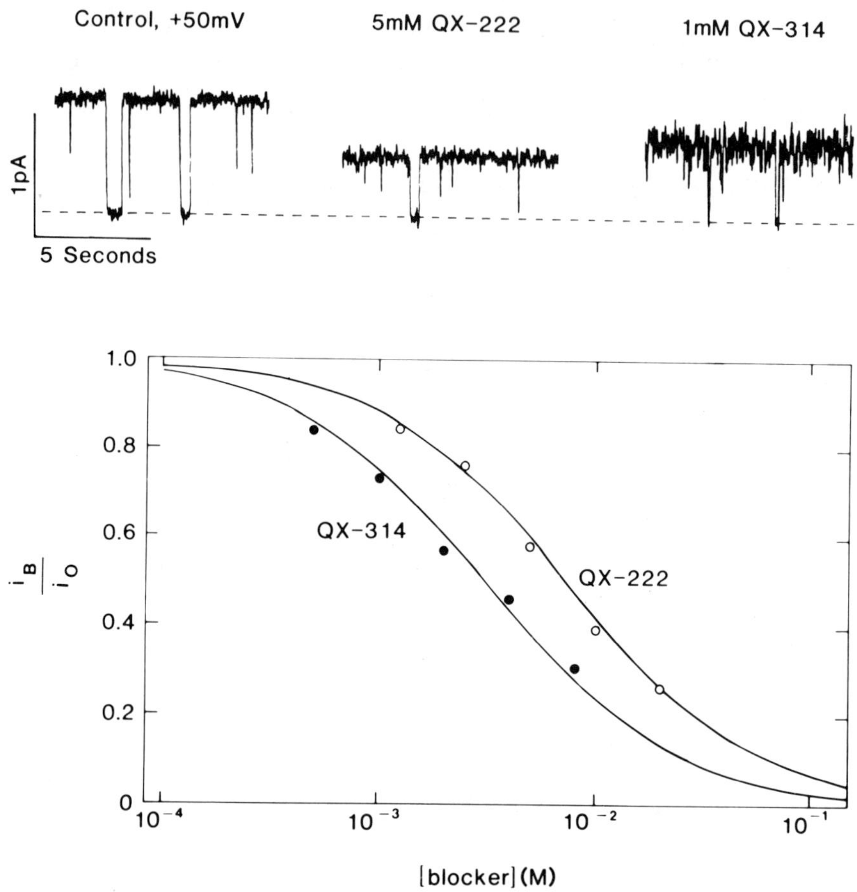

Figure 11. Effect of QX-222 and QX-314 on BTX-activated Na^+ channels. The upper traces show segments of single-channel records at +50 mV in the presence of 0.2 M symmetrical NaCl and either 5 mM internal QX-222 or 1 mM internal QX-314. The dashed line corresponds to the zero-current level. The lower graph is a plot of the ratio of the mean single channel current at +50 mV in the presence of various concentrations of QX to that in its absence. The solid curves correspond to single-site binding isotherms with blocker dissociation constants of 3.1 mM for QX-314 (●) and 7.6 mM for QX-222 (○).

single BTX-activated Na^+ channels. The first effect is exemplified by the current records of Fig. 11. These records show that the apparent single-channel conductance is reduced by the permanently charged QX compounds when present on the internal side of the channel. This effect is very similar to the action of the fast-blocking organic cations discussed in the preceding sections. Both QX-222 and QX-314 produce no observable effect when added to the external side of the channel at concentrations for which internal block is effective. The block by internal QX compounds is also voltage dependent, increasing with positive voltage, as shown in the current–voltage data of Fig. 12. One difference between the

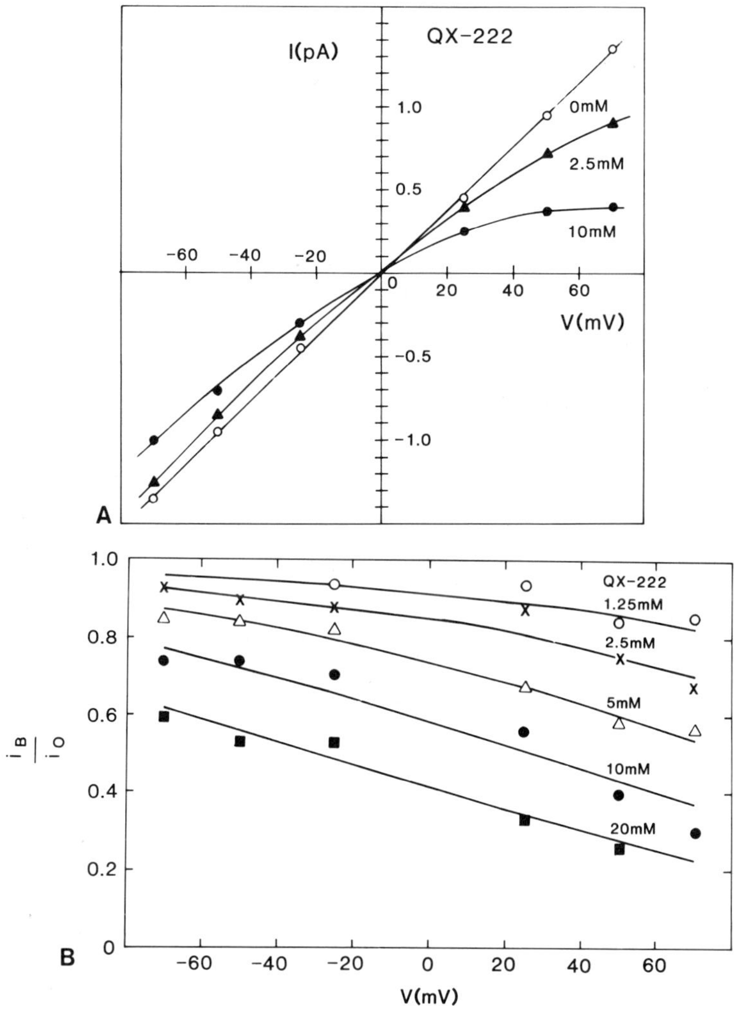

Figure 12. Voltage-dependent block by QX-222. A: Single-channel current–voltage data are plotted for the same channel in the presence of 0, 2.5, and 10 mM internal QX-222. B: The ratio, i_B/i_0, of single-channel current in the presence of various internal QX-222 concentrations to that in the absence of blocker is plotted against voltage. Solid curves correspond to fits to the equation given in Table II with blocking parameters for QX-222 given in that table.

action of the QX compounds and that of TMA or methylguanidine is that the QX derivatives induce significant noise in the open-current level, as shown in Fig. 11 for QX-314. This noise increases dramatically as the voltage is made more positive (not shown). Such an effect would be expected if the dwell time of the blocker on the channel increases with depolarization. Such behavior was recently demonstrated for internal block of normal Na^+ channels by 9-aminoacridine (Yamamoto and Yeh, 1984). The probable reason why QX-314 block appears as "noise" rather than resolved blocking events is that the dissociation rate is at the limit of the current time resolution. We have not yet attempted a quantitative analysis of this noise but instead used heavy filtering (10 Hz) to measure the mean current at various blocker concentrations and voltages. The results of this analysis (Fig. 12) show that QX block can be quantitatively explained by the same blocking model discussed above for other organic cations: voltage-dependent binding to a single site within the transmembrane electric field.

We also find that the tertiary amines, procaine and lidocaine, induced a similar fast blocking effect (not shown). Unlike the other organic cations, fast block by procaine and lidocaine is observed when these compounds are added to the chamber corresponding to either the internal or external side of the channel. However, for any given concentration, such fast blocking action is severalfold greater when the amine is added directly to the internal side. We believe that the fast blocking effect of external anesthetic is explained by the high membrane permeability of the uncharged form of these molecules. After diffusion across the bilayer from the external side, fast block is presumably mediated by the protonated species present on the internal side in the unstirred layer near the planar bilayer (Hille, 1977a,b). To support this idea, we find that the neutral benzocaine molecule does not induce a fast block effect on the conductance and that fast block by procaine and lidocaine is always enhanced by depolarization regardless of the side to which the drug is added. Comparing lidocaine to procaine, we find that procaine block is faster; i.e., procaine causes a very clean reduction in conductance, as does TMA, whereas lidocaine induces a very noisy or flickering block, as does QX-314. The fast blocking action of internal procaine and lidocaine was analyzed in the same manner as previously described for the permanently charged organic cations.

Table II summarizes the derived zero-voltage equilibrium dissociation constants and the effective electrical distances measured for seven different organic cations that exhibit fast block from the internal side of the channel. The common mode of action of simple hydrophilic organic cations and the more hydrophobic local anesthetics leads one to the conclusion that there is a common low-affinity site for many such compounds that is accessible from bulk solution on the internal side of the membrane.

In addition to this fast blocking action of local anesthetics, there is a second, completely different blocking mode that we have only observed for benzocaine and procaine. We call this effect a slow blocking mode as exemplified by the discrete long-lived blocking events shown in Figs. 13 and 14. These blocking events are induced by benzocaine and procaine at millimolar concentrations added either to the internal or external side of the channel. For procaine, the kinetics of this slow blocking phenomenon are quite different from the usual type of discrete blocking behavior that has been described at the single-channel level. Figure 13 shows that when procaine concentration is increased, we observe that the average duration of the blocked periods increases, with no clear change in the duration of open events between the slow blocks (see also Table III). Such a kinetic effect is predicted by a scheme involving preferential binding of a drug to a closed state of a channel and requiring that this closed-blocked state cannot reopen or reopens slowly:

$$\mathrm{C{\cdot}B} \overset{K_B}{\longleftrightarrow} \mathrm{C} \longleftrightarrow \mathrm{O}$$

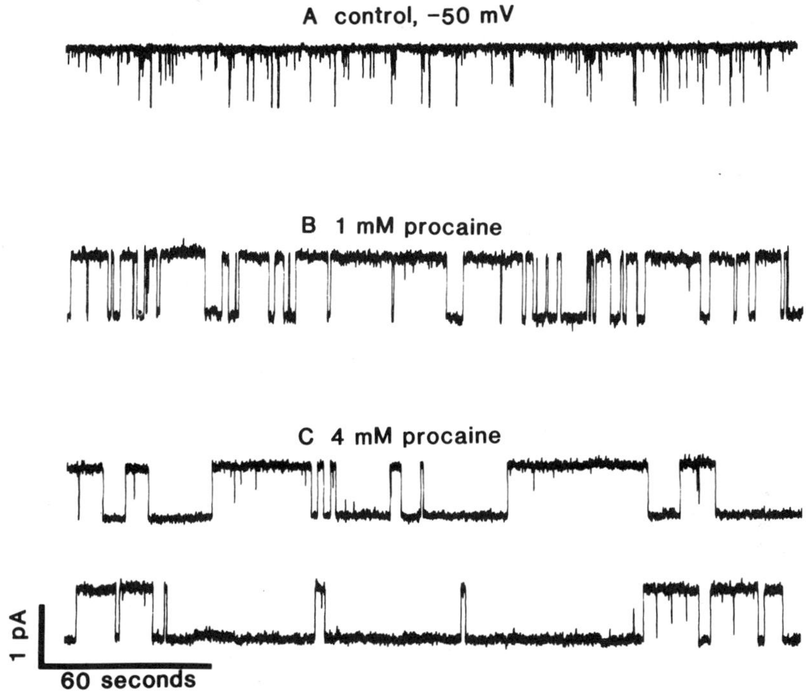

Figure 13. Effect of increasing procaine concentration on a single BTX-activated Na^+ channel. A: Control trace showing a single-channel current record at −50 mV in the presence of symmetrical 0.2 M NaCl without blockers or anesthetics. Closing events are downward in A–C. B: Record taken at −50 mV after addition of 1 mM procaine to the external side of the channel. C: Record taken at −50 mV after the 1 mM external procaine in B was increased to 4 mM external procaine.

From an intiutive consideration of the above scheme, it is apparent that an increase in blocker concentration, [B], will shift the equilibrium toward the closed–blocked state, C·B, resulting in a lengthening of the duration of closed states. Thus far, we have only been able to clearly demonstrate such an effect for procaine. Experiments with benzocaine are difficult because of its limited solubility in water.

If procaine and benzocaine do indeed bind preferentially to closed states of the BTX-activated Na^+ channel, what sort of closed states might these be? A clue to this question is present in the voltage dependence of the slow blocking phenomenon (Fig. 14). Regardless of the side to which benzocaine and procaine are added, we consistently observe that the frequency of slow blocking events increases with hyperpolarization. A compilation of mean dwell times at different voltages shows that the voltage dependence can be characterized as a decrease in the mean open or unblocked time (e-fold per ~80 mV) with increasing negative voltage (Fig. 14B). The mean blocked time is rather constant from +50 to −30 mV (Fig. 14B). This enhancement of block at negative voltage is behavior opposite that of the fast block described previously, which is enhanced by depolarization. From this point alone, it is clear that these two blocking modes are different and

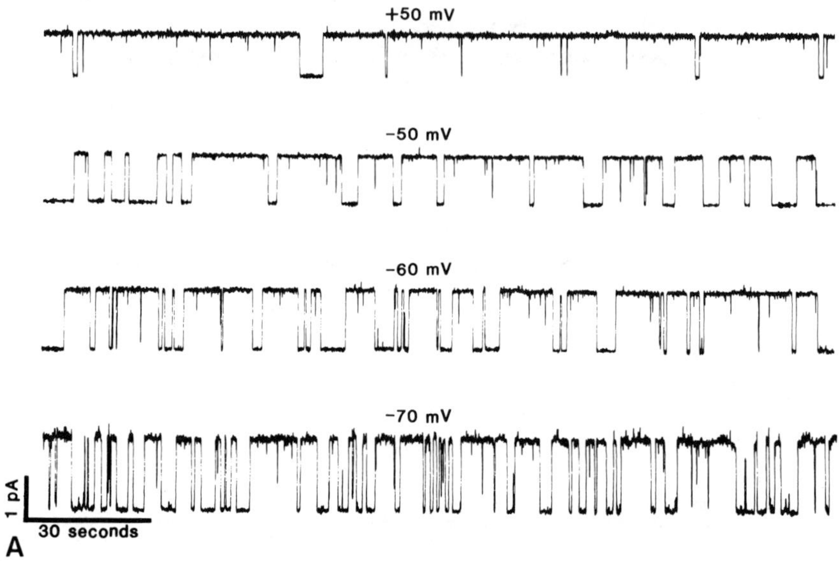

Figure 14. Effect of voltage on the kinetics of procaine- and benzocaine-induced closing events. A: Representative segments of records of a single-channel bilayer in the presence of 0.2 M symmetrical NaCl and 2 mM internal procaine are shown at four different holding voltages. Closing events are downward. A comparison with control data such as those of Fig. 13A shows that nearly all of the observed closing events longer than about 0.5 seconds are induced by procaine. B: Voltage dependence of mean closed and mean open times in the presence of procaine or benzocaine. Data such as that in A were analyzed by defining a closed event as any closing greater than or equal to 0.6 sec. Open-state events are defined as dwell times between two such closed events. Populations of such dwell times exhibited single-exponential distributions when compiled as cumulative probability histograms. The mean open and mean closed times from samples of 30–140 events for four different (●, ○, ×, ▲) single channels are plotted on a logarithmic scale against voltage: (●) 2 mM internal procaine; (○, ×) 2 mM external procaine; (▲) 0.5 mM external benzocaine.

are probably mediated via different anesthetic binding sites. The fastblocking phenomenon appears to require a cationic molecule, whereas for the slow blocking phenomenon, a neutral hydrophobic molecule like benzocaine is sufficient.

Two other peculiarities make the slow blocking phenomenon especially difficult to study quantitatively: (1) the kinetics of discrete blocking events vary markedly from channel to channel under apparently identical conditions, as illustrated by the scattered data in Fig. 14B and Table III; (2) single BTX-activated channels unpredictably "disappear" at various times after addition of local anesthetic. The kinetic heterogeneity implied by the first point may be related to heterogeneity in gating kinetics from channel to channel that we have alluded to previously (Moczydlowski *et al.*, 1984a), which has also been observed for the purified rat brain Na^+ channel (Hartshorne *et al.*, 1985). The second point is explained by the enhanced dissociation rate of batrachotoxin that is caused by local anesthetics, as documented by radioligand-binding experiments (Postma and Catterall, 1984). In our case, once batrachotoxin dissociates from the channel,

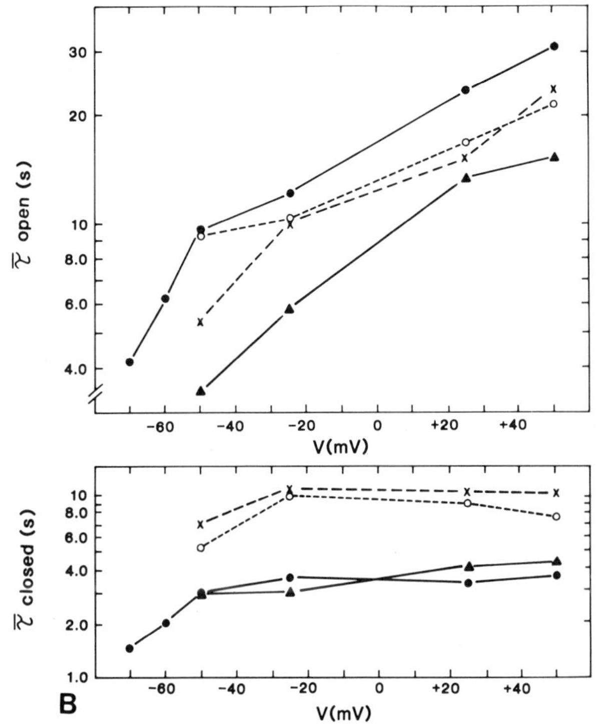

Figure 14. (*continued*)

Table III. Dependence of Closed and Open Times on Procaine Concentration[a]

Bilayer	V (mV)	[Procaine] (mM)	$\bar{\tau}_{closed}$ ± SD (n) (sec)	$\bar{\tau}_{open}$ ± SD (n) (sec)	P_0
a	−50	1	2.9 ± 2.2(37)	7.5 ± 7.3(37)	0.72
		2	7.0 ± 6.1(52)	5.3 ± 5.0(53)	0.43
b	−50	1	1.3 ± 0.8(31)	11 ± 13(30)	0.89
		4	3.3 ± 3.6(39)	9.2 ± 8.1(39)	0.74
c	−50	1	2.6 ± 2.0(25)	7.5 ± 7.1(25)	0.74
		2	14 ± 12(23)	16 ± 14(23)	0.53
		4	20 ± 18(21)	12 ± 13(21)	0.38
a	+50	1	5.9 ± 5.4(14)	41 ± 49(13)	0.87
		2	10 ± 6.7(49)	24 ± 14(46)	0.70
b	+50	1	1.8 ± 1.2(12)	26 ± 14(11)	0.93
		4	5.8 ± 6.9(22)	25 ± 28(22)	0.81

[a] The dwell times of blocking events in experiments such as that of Fig. 13 were tabulated by defining a closed event as any closing longer than or equal to 0.6 sec and an open event as periods separating such closings. Experiments are grouped individually where external procaine concentration was increased for the same channel. The mean and standard deviation for n events are listed. P_0 is the time-averaged probability of the open state.

normal voltage-dependent inactivation would be expected to resume, rendering the single-channel currents inaccessible by the low time resolution of the present bilayer system. It will be necessary to improve this resolution further in order to test this hypothesis.

7. Summary

The blocking pharmacology described in this chapter can be summarized in terms of three distinct binding sites.

The first site is a well-known specific external receptor for guanidinium toxins. Binding of such toxins to the BTX-activated open channel is voltage dependent; however, there is no apparent effect of toxin charge on this voltage dependence. In contrast, the relative association rate of various toxins does correlate with the net positive charge of the toxin, as expected for an effect involving fixed negative surface charge in the vicinity of the receptor. At present, there is no compelling evidence to indicate that such guanidinium toxins occupy part of the ion-conducting passageway. A new class of polypeptide toxins exemplified by μ-conotoxin GIIIA appear to block in a similar fashion as the classical toxins, TTX and STX. However, GIIIA may be useful in discriminating Na^+-channel subtypes as demonstrated by an affinity for skeletal muscle Na^+ channels that is about 1000-fold greater than its affinity for nerve Na^+ channels (Cruz *et al.*, 1985).

The second site is a low-affinity site accessible from the internal solution that appears to accept a variety of organic cations, including local anesthetics. Binding to this site is characterized by comparatively fast blocker dissociation rates and an enhancement of block with positive applied voltage. As concluded previously (Hille, 1977a,b), tertiary amine local anesthetics are apparently accessible to this site from the external side of the membrane after diffusion of the neutral form through the hydrophobic phase of the bilayer and the subsequent appearance of a gradient of the charged species in the internal unstirred layer.

The third site (or sites) is not well characterized but is presently defined by the discrete blocking events induced by benzocaine and procaine. This site may involve a hydrophobic domain that is exposed to anesthetic in a closed-channel conformation, since the concentration dependence of procaine blocking suggests preferential binding to closed states. The closed conformation that accepts procaine and benzocaine is favored by hyperpolarizing voltage. This closed conformation may represent certain closed states that lie in the gating activation pathway or, alternatively, may reflect a voltage-dependent change that is distinct from channel gating. The latter possibility is suggested by the observation that the voltage dependence of the slow block is not well correlated with the gating region of BTX-activated channels (−70 to −120 mV) but is similar to the shallow continuous voltage dependence of guanidinium toxin blocking in the range of +60 to −60 mV.

Our observations of the effects of local anesthetics on BTX-activated channels suggest that inhibition of the macroscopic current through such channels (Khodorov *et al.*, 1975; Albuquerque *et al.*, 1976; Creveling *et al.*, 1983; Huang and Ehrenstein, 1981) may be caused by three different microscopic events: (1) a fast block at depolarized potentials by cationic species from the internal solution; (2) slow blocking events at hyperpolarized potentials that are induced by certain anesthetics such as benzocaine and procaine; and (3) an inhibition of BTX binding caused by an increased BTX dissociation rate in the presence of local anesthetic (Postma and Catterall, 1984), marked here by the disappearance of channels. Of the local anesthetics we have tested, only procaine clearly acts by both fast and slow blocking modes simultaneously, i.e., internal procaine induces both a decreased single-channel current and the appearance of slow blocking events.

ACKNOWLEDGMENTS. Support for this work was provided by the Elma Margaret Lapp Fund of the University of Cincinnati College of Medicine, a grant from the Muscular Dystrophy Association, and an Established Investigatorship of the American Heart Association to Edward Moczydlowski. We are grateful to Dr. John Daly for supplying batrachotoxin and Drs. Chris Miller and Gary Strichartz for local anesthetics and inspiration. We thank Judy McMahan for expert assistance in preparing the manuscript.

References

Albuquerque, E. X., Brookes, N., Onur, R., and Warnick, J. E., 1976, Kinetics of interaction of batrachotoxin and tetrodotoxin on rat diaphram muscle, *Mol. Pharmacol.* **12**:82–91.

Armstrong, C. M., 1975, Potassium pores of nerve and muscle membranes, in: *Membranes: A Series of Advances* (G. Eisenman, ed.), pp. 325–358, Marcel Dekker, New York.

Bell, J. E., and Miller, C., 1984, Effects of phospholipid surface charge on ion conduction in the K^+ channel of sarcoplasmic reticulum, *Biophys. J.* **45**:279–287.

Cahalan, M. D., and Almers, W., 1979, Interactions between quaternary lidocaine, the sodium channel gates, and tetrodotoxin, *Biophys. J.* **27**:39–56.

Creveling, C. R., McNeal, E. T., Daley, J. W., and Brown, G. B., 1983, Batrachotoxin-induced depolarization and [^{3}H]bactrachotoxinin-A 20 α-benzoate binding in a vesicular preparation from guinea pig cerebral cortex: Inhibition by local anesthetics, *Mol. Pharmacol.* **23**:350–358.

Cruz, L. J., Gray, W. R., Olivera, B. M., Zeikus, R. D., Kerr, L., Yoshikami, D., and Moczydlowski, E., 1985, *Conus geographus* toxins that discriminate between neuronal and muscle sodium channels, *J. Biol. Chem.* **260**:9280–9288.

French, R. J., Worley, J. F., and Krueger, B. K., 1984, Voltage-dependent block by saxitoxin of sodium channels incorporated in planar lipid bilayers, *Biophys. J.* **45**:301–312.

Hall, S., 1982, Toxins and Toxicity of Protogonyaulax from the Northeast Pacific, Ph.D. Thesis. University of Alaska, Fairbanks.

Hall, S., Reichardt, P. B., and Neve, R. A., 1980, Toxins extracted from an Alaskan isolate of *Protogonyaulax* sp, *Biochem. Biophys. Res. Commun.* **97**:649–653.

Hartshorne, R. P., Keller, B. U., Talvenheimo, J. A., Catterall, W. A., and Montal, M., 1985, Functional reconstitution of the purified brain sodium channel in planar lipid bilayers, *Proc. Natl. Acad. Sci. U.S.A.* **82**:240–244.

Henderson, R., Ritchie, J. M., and Strichartz, G. R., 1974, Evidence that tetrodotoxin and saxitoxin act at a metal cation binding site in the sodium channel of nerve membrane, *Proc. Natl. Acad. Sci. U.S.A.* **71**:3936–3940.

Hille, B., 1975, The receptor for tetrodotoxin and saxitoxin: A structural hypothesis, *Biophys. J.* **15**:615–619.

Hille, B., 1977a, The pH-dependent rate of action of local anesthetics on the node of Ranvier, *J. Gen. Physiol.* **69**:475–496.

Hille, B., 1977b, Local anesthetics: Hydrophilic and hydrophobic pathways for the drug–receptor reaction, *J. Gen. Physiol.* **69**:497–515.

Hille, B., Ritchie, J. R., and Strichartz, G. R., 1975, The effect of surface charge on the nerve membrane in the action of tetrodotoxin and saxitoxin in frog myelinated nerve, *J. Physiol. (Lond.)* **250**:34–35P.

Horn, R., Patlak, J., and Stevens, C. F., 1981, The effect of tetramethylammonium on single sodium channel currents, *Biophys. J.* **36**:321–327.

Huang, L. M., and Ehrenstein, G., 1981, Local anesthetics QX-572 and benzocaine act at separate sites on the batrachotoxin-activated sodium channel, *J. Gen. Physiol.* **77**:137–153.

Jencks, W. P., 1981, On the attribution and additivity of binding energies, *Proc. Natl. Acad. Sci. U.S.A.* **78**:4046–4050.

Kao, C. Y., and Nishiyama, A., 1965, Action of saxitoxin on peripheral neuromuscular systems, *J. Physiol. (Lond.)* **180**:50–66.

Kao, C. Y., and Walker, S. E., 1982, Active groups of saxitoxin and tetrodotoxin as deduced from actions of saxitoxin analogues on frog muscle and squid axon, *J. Physiol. (Lond.)* **323**:619–637.

Khodorov, B. L., Peganov, E. M., Revenko, S. V., and Shishkova, L. D., 1975, Sodium currents in voltage clamped nerve fiber of frog under the combined action of batrachotoxin and procaine, *Brain Res.* **84**:541–546.

Krueger, B. K., Worley, J. F., and French, R. J., 1983, Single sodium channels from rat brain incorporated into planar lipid bilayer membranes, *Nature* **303**:172–175.

McLaughlin, S., 1977, Electrostatic potentials at membrane–solution interfaces, *Curr. Top. Membr. Transport* **9**:71–144.

Miller, J. A., Agnew, W. S., and Levinson, S. R., 1983, Principal glycopeptide of the tetrodotoxin/saxitoxin binding protein from *Electrophorus electricus:* Isolation and partial chemical and physical characterization, *Biochemistry* **22**:462–470.

Moczydlowski, E., Garber, S. S., and Miller, C., 1984a, Batrachotoxin-activated Na^+ channels in planar lipid bilayers: Competition of tetrodotoxin block by Na^+, *J. Gen. Physiol.* **84**:665–686.

Moczydlowski, E., Hall, S., Garber, S. S., Strichartz, G. S., and Miller, C., 1984b, Voltage-dependent blockade of muscle Na^+ channels by guanidinium toxins: Effect of toxin charge, *J. Gen. Physiol.* **84**:687–704.

Postma, S. W., and Catterall, W. A., 1984, Inhibition of binding of [^{3}H]batrachotoxin in A 20-α-benzoate to sodium channels by local anesthetics, *Mol. Pharmacol.* **25**:219–227.

Rogers, R. S., and Rapoport, H., 1980, The p*Ka*'s of saxitoxin, *J. Am. Chem. Soc.* **102**:7335–7339.

Sato, S., Nakamura, H., Ohizumi, Y., Kobayashi, J., and Hirata, Y., 1983, The amino acid sequences of homologous hydroxyproline-containing myotoxins from the marine snail *Conus geographus* venom, *FEBS Lett.* **155**:277–280.

Shimizu, Y., Hsu, C., Fallon, W. E., Oshima, Y., Miura, I., and Nokanishi, K., 1978, Structure of neosaxitoxin, *J. Am. Chem. Soc.* **100**:6791–6793.

Shimizu, Y., Hsu, C., and Genenah, A., 1981, Structure of saxitoxin in solution and stereochemistry of dihydrosaxitoxins, *J. Am. Chem. Soc.* **103**:605–609.

Strichartz, G. R., 1973, The inhibition of sodium currents in myelinated nerve by quaternary derivatives of lidocaine, *J. Gen. Physiol.* **62**:37–57.

Strichartz, G., 1976, Molecular mechanisms of nerve block by local anesthetics, *Anesthesiology* **45**:421–441.

Strichartz, G., 1984, Structural determinants of the affinity of saxitoxin for neuronal sodium channels: Electrophysiological studies on frog peripheral nerve, *J. Gen. Physiol.* **84**:281–305.

Strichartz, G. R., and Ritchie, J. M., 1985, The action of local anesthetics on ion channels of excitable tissues, in: *Handbook of Experimental Pharmacology: Local Anesthetics* (G. Strichartz, ed.), Springer, New York (in press).

Woodhull, A., 1973, Ionic blockage of sodium channels in nerve, *J. Gen. Physiol.* **61**:687–708.

Yamamoto, D., and Yeh, J. Z., 1984, Kinetics of 9-aminoacridine block of single Na channels, *J. Gen. Physiol.* **84**:361–377.

Part IV

OTHER CHANNELS IN MODEL MEMBRANES

Chapter 17

THE LARGE CALCIUM-ACTIVATED POTASSIUM CHANNEL

RAMON LATORRE

La ventura va guiando neustras cosas mejor de lo que acertáramos a desear; porque ves allí amigo Sancho Panza donde se descubren treinta o poco más desaforados gigantes, con quien pienso hacer batallas y quitarles a todos las vidas, con cuyos despojos comenzaremos a enriquecer. . . .

Don Quijote de la Mancha

1. Introduction

In many types of cells an increased cytoplasmic Ca^{2+} concentration triggers a K^+ conductance. This effect, first discovered in red blood cells (Gárdos, 1958), is mediated by ion channels in the membrane known as Ca^{2+}-activated K^+ channels. The patch-clamp and reconstitution techniques have revealed that several classes of Ca^{2+}-activated channels exist (for reviews, see Latorre and Miller, 1983; Petersen and Maruyama, 1984; Marty, 1983a; Latorre *et al.*, 1985; Passow, 1983). These channels are widely distributed in different cells and tissues, and they appear to play an important role in secretion, repetitive firing, action potential repolarization, and some abnormal states of the heart. All these channels are directly activated by internal Ca^{2+}, but they can be of small or large unitary conductance, highly selective or only cation selective, and voltage dependent or voltage independent.

In this chapter I discuss in detail the properties of a large-unitary-conductance Ca^{2+}-activated K^+ channel (maxi CaK channel). This channel appears to be present in the T-tubule membrane of rabbit skeletal muscle, and its characterization has been performed after its incorporation into planar lipid bilayer membranes.

Although, for the sake of clarity, I have divided this chapter into several sections, as a whole it covers three main topics. First, I describe the general characteristics of the T-tubule channel incorporated into planar bilayers and some

RAMON LATORRE • Department of Biology, Faculty of Sciences, University of Chile, and Centro de Estudios Cientificos de Santiago, Casilla 16443, Santiago 9, Chile.

reconstitution procedures. Second, I review the main characteristics of maxi CaK channels' gating kinetics, discussing some of the models proposed for this process in planar bilayers and cells. Finally, the third general section is concerned with the conduction and ion blockade properties of these channels, and I discuss some models that can reconcile the very large conductance of this channel and its high ionic selectivity.

2. Reconstitution of a Maxi CaK Channel

2.1. Ion Channel Reconstitution Always Surprises

In 1980, when Cecilia Vergara and I started working in channel reconstitution using a T-tubule preparation (Rosemblatt *et al.*, 1981), we found that channel incorporation proceeded in a way that we did not expect.

We were surprised by three characteristics of the current records shown in Fig. 1. When T-tubule vesicles were added to only one of the aqueous compartments bathing the bilayer (*cis* compartment), the membrane conductance increased in discrete steps. This was surprising, because when vesicles from sarcoplasmic reticulum containing K^+ channels (Miller, 1978) or membrane vesicles prepared from the electric organ from *Torpedo* containing Cl^- channels (White and Miller, 1979) fuse with the planar bilayers, they transfer packages of many channels to the latter. Therefore, if we equate each of the current steps shown in Fig. 1 with the incorporation of one channel into the membrane, we arrive at the conclusion that when a T-tubule membrane vesicle fuses with the planar bilayer, it transfers at most one channel, i.e., channel density is very small.

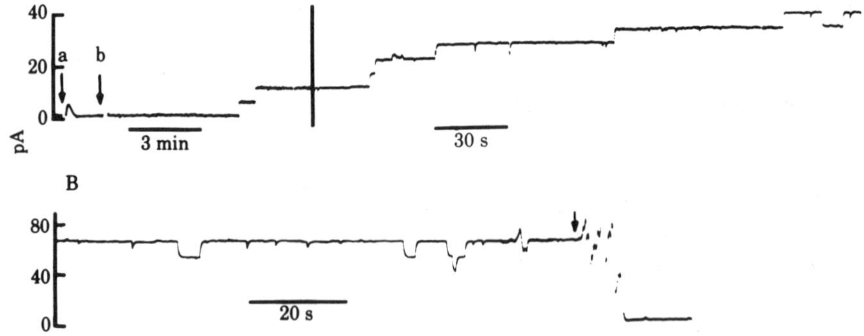

Figure 1. Addition of T-tubule vesicles to one side of a bilayer induces an increase in current in discrete steps. A: Ca^{2+} was added to a final concentration of 1 mM to the *cis* side only (a), followed by addition of T-tubule vesicles (b). Applied voltage +20 mV. B: Addition of EGTA (arrow) induces an inhibition of the current promoted by the addition of T-tubule vesicles. Applied voltage was +50 mV. Lipid bilayer made from a mixture of PE/PS = 7/3 (w/w); 100 mM K^+. (From Latorre *et al.*, 1982.)

The conductance of each step is 230 pS, much larger than any of the biological channels described at the time.

The conductance induced by the T-tubule vesicles vanished when the Ca^{2+} present on the *cis* (but not in the *trans*) side was reduced to very low levels ($<10^{-7}$ M) with EGTA. Moreover, Mg^{2+} was not able to restore this conductance.

Our hope, when we started our work, was to be able to detect in the planar bilayer some of the several channels that were supposed to exist in T-tubules (Cl^- channels, inward rectifier, etc.). Thus, we went looking for a butterfly and found a dinosaur, a "monster" channel for which we had not been searching. To make things worse, we found that the channel was very selective for K^+. Large conductance channels were not supposed to be so selective. This result is counterintuitive, since high selectivity implies strong interactions with the protein groups forming the selectivity filter, and this would tend to lower the channel conductance. Other K^+-selective channels, such as the delayed rectifier and the inward rectifier channels, have conductances ranging between 3 and 18 pS (see Latorre and Miller, 1983; Hille, 1984, for reviews).

Two reports gave us hope that the channel we detected in bilayers was actually present in the T-tubule vesicles and was not an artifact produced by the reconstitution procedure. Fink and Luttgau (1976) found that the K^+ conductance of a metabolically poisoned frog muscle fiber increases to high levels. They ascribed the increase in K^+ conductance to an increase in internal Ca^{2+} concentration as a result of fiber poisoning. On the other hand, Barrett *et al.* (1981) found a Ca^{2+}-activated slow K^+ conductance in rat skeletal myotubes.

In some cells, macroscopic current experiments indicated that the channel was activated by internal Ca^{2+} and depolarizing voltages and was blocked by external TEA (Meech, 1978). In bilayers, we found that Ca^{2+} induces channel opening only from the *cis* side and that it is blocked by very low TEA concentrations on the *trans* side. Furthermore, the fraction of time that the channel remained in the open configuration was increased by positive potentials (the *trans* side is virtual ground), which in the cell voltage convention corresponds to a depolarization (Latorre *et al.*, 1982). While our work was in progress, two reports appeared in 1981 describing Ca^{2+}-activated K^+ channels of large conductance.* One was a study on rat muscle cell culture (Pallota and Magleby, 1981), and the other was on chromaffin cells (Marty, 1981). We discuss in detail many of these characteristics of the channel later in this chapter.

2.2. Maxi CaK Channel Incorporation Hints

Our experience is that the maxi CaK channel incorporates well into decane-containing phosphatidylethanolamine (PE), phosphatidylserine (PS) membranes as well as in bilayers made from mixtures of these lipids (for a practical guide on how to make bilayers, see Chapter 5 by Alvarez). An osmotic gradient (*cis* side

* To my knowledge, this was the first case in which the same channel was detected simultaneously in artificial membranes and in cells.

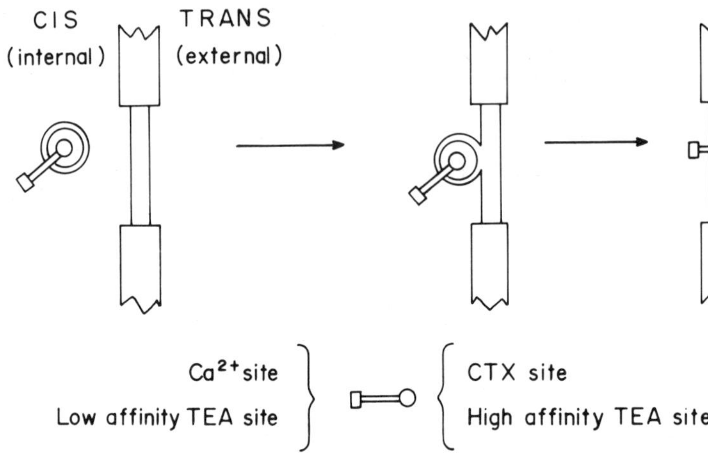

Figure 2. Fanciful representation of the fusion process of a T-tubule vesicle with a planar bilayer.

hyperosmotic) always helps in channel incorporation, but usually more than one channel is transferred to the bilayer. Most of the time, it is desirable to have only one channel in the membrane. In order to obtain single-channel membranes, two procedures can be followed: (1) symmetrical KCl solutions and patience (in this case, the time from when the vesicles are added until the first channel is obtained is extremely variable); (2) mixtures of PE and phosphatidylcholine (PC), e.g., a molar ratio of PE/PC = 4/1 and an osmotic gradient.

So far, we have failed to detect the maxi CaK channel by means of a liposome patch clamp (Tank *et al.*, 1982) or a bilayer formed on the tip of patch-clamp pipet (Coronado and Latorre, 1983). This may simply reflect a very low density of maxi CaK channels in the T-tubule preparation.

Maxi CaK channels can be incorporated into bilayers from at least four different membrane vesicle sources: rabbit or rat skeletal muscle (Latorre *et al.*, 1982; Moczydlowski and Latorre, 1983), brain synaptosomes (Krueger *et al.*, 1982), smooth muscle (Wolff *et al.*, 1985), and chromaffin cells (G. Yellen and C. Miller, personal communication). Channels transferred to the planar bilayer from the T-tubule membranes of rat or rabbit always have the same polarity; i.e., the Ca^{2+}-activating site is always at the *cis* side of the membrane. However, when we use the smooth muscle membrane preparation, channels are incorporated randomly. Apparently, this reflects a random vesicle orientation. T-tubule vesicles are mostly inside-out vesicles (Hidalgo *et al.*, 1985; a cartoon of a possible mechanism for the fusion process is given in Fig. 2), whereas a smooth muscle membrane preparation is composed of inside- and outside-out vesicles (Wibo *et al.*, 1982). So far, we have not detected differences between the maxi CaK channels from T-tubule and those from smooth muscle, but the chromaffin channel is at least one order of magnitude more sensitive to Ca^{2+}. Reconstitution of a CaK channel of smaller conductance (70–100 pS) and modulation by protein phosphorylation are discussed in Chapter 21 by Levitan.

Table I. Characteristics of Ca^{2+}-Activated K^+ Channels

	Ionic condition[b]	Conductance (pS)	Selectivity	Voltage dependence	References
Maxi CaK channels					
Rat myotubes	$140K^+$	307	$Tl^+(1.2) > K^+(1) > Rb^+(0.67) > NH_4^+(0.11) \gg Cs^+, Na^+, Li^+$	Yes	Blatz and Magleby (1984)
Rabbit T-tubule[a]	$100K^+$	260	$Tl^+(1.5) > K^+(1) > Rb^+(0.7) > NH_4^+(0.15) \gg Cs^+, Na^+, Li^+$	Yes	Vergara *et al.* (1984)
Bovine chromaffin cells	$160K^+$	265	$K^+(1) > Rb^+(0.33) \gg Na^+(0.03)$	Yes	Yellen (1984a)
Rat anterior pituitary cells	$145K^+$	208	—	Yes	Wong *et al.* (1982)
Mouse parotid acini	$145K^+$	250	$K^+ \gg Na^+$		Maruyama *et al.* (1983)
Toad smooth muscle cells	$130K^+$	250	$K^+ \gg Na^+$	Yes	Walsh and Singer (1983)
Rabbit cortical collecting tubules	$75K_o^+/15K_i^+$	90	$K^+(1) > Na^+(0.11)$	?	Hunter *et al.* 1984
Rat pancreatic B-cells	$140K^+$	244	—	Yes	Cook *et al.* (1984)
Rabbit smooth muscle[a]	$100K^+$	230	$K^+(1) > Rb^+(0.06) \gg Cs^+, Na^+, Li^+$	Yes	Wolff *et al.* (1985)
Rat axolemma[a]	$100K^+$	230	$K^+ \gg Na^+$	Yes	Krueger *et al.* (1982)
Macrophages	$145K^+$	240	$K^+ \gg Na^+$	Yes	Gallin (1984)
Small CaK channels					
Helix pomatia	$4K_o^+/400K_i^+$	19	$K^+ \gg Na^+$	Yes	Lux *et al.* (1981)
Helix aspera	$120K^+$	40–60		Yes	Ewald *et al.* (1985)
Human red cells	$150K^+$	10–40	$K^+(1) > Na^+(0.06)$	Weak	Grygorczyk and Schwarz (1983)
Frog red cells	$115K^+$	50	$K^+ \gg Cs^+, Na^+$	No	Hamill (1983)
Cation channels					
Mouse neuroblastoma	$125Na_c^+/125K_i^+$	22	$K^+ \sim Na^+ \sim Li^+ \gg Ca^{2+}$	No	Yellen (1982)
Red heart	$137K^+$	35	$K^+ \sim Na^+$	No	Colquhoun *et al.* (1981)
Rat pancreatic acini	$145K^+$	30–35		No	Maruyama and Petersen (1982)

[a] Characteristics determined in planar bilayer. For all other channels, patch-clamp technique was used.
[b] Concentration is mM. Where no designation of internal or external concentration is given, symmetrical solutions were used.

3. Types of Ca^{2+}-Activated Channels

Calcium-activated channels come in all sorts of sizes and ionic selectivities. As shown in Table I, they are found in many different kinds of cells, from neurons to macrophages. In general, it is possible to classify Ca^{2+}-activated channels according to their conductance. Thus, we have "maxi" Ca^{2+}-activated channels with conductances ranging from 100 to 260 pS and small Ca^{2+}-activated channels (17–50 pS). Small Ca^{2+}-activated channels may in turn be divided into two classes, K^+ and monovalent cation selective.

4. Models for the Maxi CaK Channel Gating Kinetics

4.1. General Characteristics of Channel Kinetics

The simplest model for channel gating is one in which a population of channels is randomly fluctuating between two conductance states (i.e., one closed state and one open state) and in which the configurational energies depend on membrane voltage, agonists concentration, or other types of stimulus. If the channel obeys Poisson statistics, or, in other words, the chance of an open channel closing or a closed channel opening is uniform in time (a system without memory), then the distribution of closed dwell times and open dwell times is exponential. However, few, if any, channels in biological membranes follow such a simple model, and maxi CaK channels are not an exception.

The maxi CaK channel from T-tubules shows a complicated gating kinetics consisting of fast fluctuations between a closed and an open state and long periods in which the channel remains quiescent (Latorre *et al.*, 1982; Moczydlowski and Latorre, 1983; Fig. 3). A closer inspection of the fast fluctuations in an expanded time scale shows that both Ca^{2+} and voltage activate the channel by increasing the fraction of time the channel spends in the open state (Fig. 3B). With a 3- to 4-kHz frequency resolution, the single-channel records reveal the presence of well-resolved closed-state events and "flickers" emanating from the open state (Fig. 1C).

4.2. An Economical Kinetic Model

The long silent periods shown in Fig. 3A are more frequent at high voltage and high Ca^{2+} concentration and have been interpreted in terms of a slow Ca^{2+} blockade (Vergara and Latorre, 1983; see below). We have analyzed channel kinetics after exclusion of the slow process and after exclusion of the flickers by filtering single-channel currents at 1 kHz (Moczydlowski and Latorre, 1983). Under these conditions, open and closed dwell times are exponentially distributed, and both mean open ($\bar{\tau}_o$) and closed ($\bar{\tau}_c$) times are affected by Ca^{2+} concentration. $\bar{\tau}_o$ is a linear function of Ca^{2+} concentration, whereas $\bar{\tau}_c$ is a linear function of

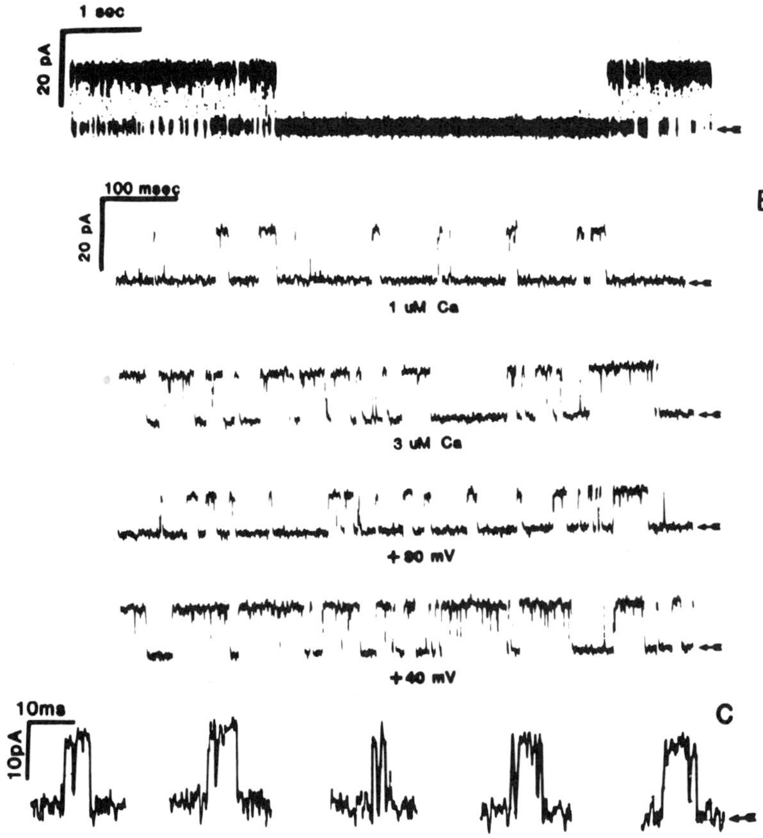

Figure 3. Different types of opening and closing kinetics of the T-tubule CaK channel. A: Slow kinetics at +50 mV. B: Effect of Ca^{2+} and voltage on the probability of opening. These records correspond to periods of channel activity between silent periods; 1 kHz filtering. C: Five examples of short channel closures (flickers); 2.5 kHz filtering, +50 mV. Phosphatidylethanolamine membrane in 0.2 M K^+ (Redrawn from Moczydlowski and Latorre, 1983.)

the reciprocal of Ca^{2+} concentration. The most economical model able to explain this paradigm of results is*:

$$\mathrm{C} \underset{}{\overset{K_1}{\rightleftharpoons}} \mathrm{CCa} \underset{\alpha}{\overset{\beta}{\rightleftharpoons}} \mathrm{OCa} \underset{}{\overset{K_2}{\rightleftharpoons}} \mathrm{Ca\text{-}O\text{-}Ca} \qquad \text{(Scheme 1)}$$

The linear kinetic scheme 1 consists of the following reaction steps: (1) binding of a Ca^{2+} ion to a closed state (C); (2) channel opening; and (3) binding of a

* Hill plots of the open probability as a function of Ca^{2+} concentration reveal that a maximum of two Ca^{2+} ions are involved in the gating kinetics (Moczydlowski and Latorre, 1983).

second Ca^{2+}, which further stabilizes the open (O) configuration of the channel. This model predicts the following relationships between $\bar{\tau}_o$ and $\bar{\tau}_c$ and Ca^{2+} concentration.

$$\bar{\tau}_o = (1/\alpha)\,(1 + [Ca^{2+}]/K_2) \qquad (1)$$

$$\bar{\tau}_c = (1/\beta)\,(1 + K_1/[Ca^{2+}]) \qquad (2)$$

Furthermore, in the limit of zero $[Ca^{2+}]$, $\bar{\tau}_o$ is voltage independent, and in the limit of infinite $[Ca^{2+}]$, $\bar{\tau}_c$ becomes voltage independent.

This observation has a profound significance in the mechanisms of channel gating. They indicate that the rate constants, β and α, for the reaction $CCa \rightleftharpoons OCa$ are voltage independent. Inasmuch as the shapes of the curves that describe the relationships between $\bar{\tau}_o$ and $[Ca^{2+}]$ and between $\bar{\tau}_c$ and $1/[Ca^{2+}]$ are voltage dependent, all the voltage dependence must reside in the equilibrium dissociation constants K_1 and K_2 (see equations 1 and 2). Actually, we found that the two dissociation constants are exponential functions of the applied voltage. As we discuss in Section 6, voltage-dependent binding constants can be explained by assuming that the binding site "senses" the electric field. Accordingly, the binding site described by K_1 senses 70–80% of the electric field, whereas the site described by K_2 senses about 95% of the field. The affinity for Ca^{2+} of the closed channel is 30-fold lower than that of the open channel. Thus, we concluded that the voltage dependence of the channel is conferred by Ca^{2+} binding to two distinct sites located within the electric field.

Very similar results were obtained with the myotube maxi CaK channel, and a model formally equivalent to the one proposed here has been put forward (Methfessel and Boheim, 1982). However, in this case it was assumed that the voltage dependence of the gating reaction resided in the open/closed conformational changes with voltage-independent Ca^{2+} binding reactions. The results shown above do not support that assumption.

It is clear that one weakness of the kinetic approach we have followed is that we do not consider the flicker events because of their very short duration. These events may in some way be related to the kinetic pathway for channel activation and should ultimately be included as part of any complete kinetic scheme. Our laboratory and others (Barrett *et al.*, 1982; Methfessel and Boheim, 1982) have noticed a substate conductance level with about 50% of the full conductance of the open state. This state shows up very rarely and appears to be independent of $[Ca^{2+}]$ and voltage. Also, we have observed shifts in the open state probability at constant applied voltage that may or may not return to the original level. Further, the $[Ca^{2+}]$ dependence varies from channel to channel by as much as 0.9 units of P_o, where P_o is the open-state probability. All these observations have also been made in a similar channel present in myotubes (Barrett *et al.*, 1982; Methfessel and Boheim, 1982). Thus, it is unlikely that they are a consequence of the channel transfer to the planar bilayer. Changes in gating behavior have also been reported for the glutamate-activated channel of locust muscle (Gration *et al.*, 1981). It has been suggested that this behavior is a consequence of inherent

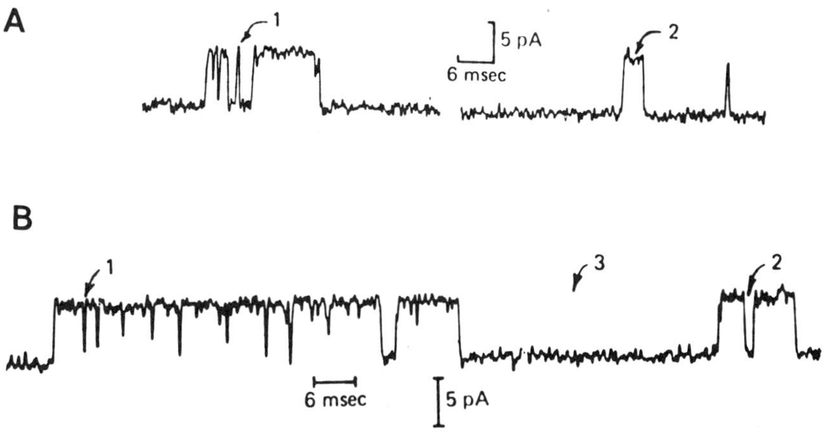

Figure 4. Opening and closing kinetics of a single maxi CaK channel from myotube. Currents were recorded from an excised patch containing a single CaK channel. A: Opening indicated by arrow 1 was probably a short opening. Arrow 2 indicates an opening with a high probability of being a long opening event. This record was taken at $[Ca^{2+}] = 0.25$ μM and +30 mV. B: Arrows 1, 2, and 3 indicate short, intermediate, and long closures, respectively. This record was taken at $[Ca^{2+}] = 0.5$ μM and +30 mV (Redrawn from Magleby and Pallota, 1983a.)

time-dependent variations in the gating equilibria, i.e., slow conformational changes. It is tempting to speculate here that such behavior may also arise from regulatory mechanisms such as protein phosphorylation (De Peyer *et al.*, 1982; see also Chapter 21 by Levitan).

4.3. More Complicated Models are Possible

A detailed analysis of the maxi CaK channel from cultured rat muscle cells has been done by Magleby's group (Barrett *et al.*, 1982; Magleby and Pallota, 1983a,b). In this channel, a third-power relationship is found between the probability of opening and the internal Ca^{2+} concentration. In their analysis, they did not exclude the flickers or events of short duration. Open dwell times are well described by the sum of two exponentials, but the distribution of closed dwell times requires the sum of three exponentials in order to be fitted adequately. These results indicate the presence of at least three closed and two open states, which are identified in Fig. 4. Magleby and Pallota (1983a) proposed the following kinetic scheme:

$$\begin{array}{lll} \text{C} \rightleftharpoons \text{CCa} \rightleftharpoons & \text{Ca-C-Ca} & \\ \quad\ \downarrow\uparrow & \ \ \downarrow\uparrow & \\ \quad\ \text{OCa} & \text{Ca-O-Ca} \xrightleftharpoons[?]{} & \text{Ca}_3\text{O} \end{array} \qquad \text{(Scheme 2)}$$

In agreement with reaction scheme 2, Magleby and Pallota (1983a) found that the frequency of openings with a larger $\bar{\tau}_o$ is a steeper function of Ca^{2+} concentration than the frequency of openings with a shorter $\bar{\tau}_o$. This result suggests that the state giving rise to the long openings requires more Ca^{2+} ions (Ca-O-Ca) than does the state giving rise to short openings (OCa). In fact, the long closed time varies as the inverse square of the Ca^{2+} concentration.* This result indicates that the long open state requires binding of two Ca^{2+} ions (i.e., scheme 2). The model predicts that mean open times should be independent of Ca^{2+} concentration. Contrary to this prediction, Magleby and Pallota (1983a) found that, although the short open intervals are independent of Ca^{2+}, the mean open times for long openings are Ca^{2+} concentration dependent. They argue that it is always possible to include a third open state with three bound Ca^{2+} to account for the concentration dependence of the mean time for long openings. A model in which three Ca^{2+} ions are involved in the opening reaction is also consistent with a Hill coefficient of 3.

However, the model of scheme 2 appears to be incomplete (McManus and Magleby, 1985; Magleby and McManus, 1985). First, McManus and Magleby found that in the absence of Mg^{2+} the Hill coefficient is 3.6, suggesting that at least four Ca^{2+} ions binding to the channel for maximum activity; second, the distribution of closed dwell times needs the sum of five exponentials to be described, whereas the open dwell-time distribution is well described by the sum of three exponentials. If these data are taken at face value, they imply the presence of at least six closed and three open states. Furthermore, on the basis of their kinetic data, Magleby and McManus (1985) concluded that open states with short mean times are adjacent to closed states with long mean times, whereas open states with short mean times are adjacent to closed states of long duration.

4.4. Summary and Conclusions

Maxi CaK channels are activated by internal (*cis*) Ca^{2+} and depolarizing voltages. The number of Ca^{2+} ions needed to maximally open the channel is different in different channels, and it varies from two in the T-tubule channel to four in myotube channels. Micromolar amounts of Ca^{2+} are needed to activate the channel. The Ca^{2+} concentration required to activate the channel to 50% of its maximum open probability to 20 mV is <1 μM in pituitary cells (Wong *et al.*, 1982) and 10–50 μM in the T-tubule channel incorporated into PE membranes (Moczydlowski and Latorre, 1983). Activation of all the other maxi CaK channels studied so far falls in this range of Ca^{2+} concentration. The probability of finding an open channel increases e-fold per 12–15 mV in muscle cells in culture (Barrett *et al.*, 1982), chromaffin cells (Marty, 1983a,b), and in the T-tubule channel and per 9 mV in a pituitary cell line (Wong *et al.*, 1982) and macrophages (calculated

* As predicted by the model, the long mean closed time that coresponds to a transition CCa→C decreases as the Ca^{2+} concentration increases. On the other hand, the intermediate mean closed time increases as Ca^{2+} concentration is raised.

Table II. Permeability Ratios P_{X^+}/P_{K^+} for Different K^+ Channels[a]

Ion	K^+ channel				
	Maxi CaK[b]	*Aplysia* CaK channel[c]	Node of Ranvier delayed rectifier[d]	Snail neuron delayed rectifier[e]	Starfish inward rectifier[f]
Tl^+	1.2–1.5	0.99	2.3	1.3	1.5
K^+	1.0	1.0	1.0	1.0	1.0
Rb^+	0.7–0.8	0.69	0.91	0.74	0.35
NH_4^+	0.11–0.15	0.11	0.13	0.15	0.035
Na^+	0.01–0.03	0.009	0.01	0.07	0.03
Li^+	0.02	0.011	0.02	0.09	—
Cs^+	0.05	0.03	0.08	0.18	0.03

[a] Permeability ratios calculated from reversal potentials.
[b] Blatz and Magleby, 1984; Yellen, 1984a; Vergara, 1983; Latorre *et al.* 1983; unpublished data.
[c] Gorman *et al.*, 1982.
[d] Hille, 1975.
[e] Reuter and Stevens, 1980.
[f] Hagiwara and Takahashi, 1974.

from Gallin, 1984). It is not clear at present how many closed and open states are necessary to fit all the kinetic data, but it is clear that very complex kinetic models will be necessary. All the data available indicate that transfer of maxi CaK channels to lipid bilayers does not alter their kinetic properties.

5. Selectivity, Conduction, and Blockade

5.1. Ion Selectivity

Most of the maxi CaK channels studied in some detail show the common property of being highly selective with an unusually large conductance (Table II). From reversal potential measurements under biological conditions, we found that both the T-tubule and the smooth muscle channels in planar bilayer have a selectivity sequence of $Tl^+ > K^+ > Rb^+ > NH_4^+$ with permeability ratios of 1.5, 1.0, 0.7, and 0.15 (Vergara, 1983; Latorre *et al.*, 1983; Wolff *et al.*, 1985; unpublished results); Cs^+, Na^+, and Li^+ are immeasurably permeant, and they all induced a voltage-dependent blockade (Latorre *et al.*, 1983; D. Naranjo, X. Cecchi, O. Alvarez, D. Wolff, and R. Latorre, unpublished results; see Section 6.2). The same selectivity sequence with almost identical permeability ratios has been found for the cultured rat muscle cell channel (Blatz and Magleby, 1984).

A similar selectivity sequence was obtained in chromaffin cells (Yellen, 1984a). An interesting finding is that although Rb^+ has almost the same permeability as K^+, its conductance is much smaller (Yellen, 1984a; Blatz and Magleby, 1984). Yellen (1984a) has suggested that Rb^+ binds to a saturable site in the channel, thus decreasing its own current.

The alkali cation selectivity sequence found for maxi CaK channels follows very closely that of other K^+ channels such as the delayed rectifier of the squid axon, the node of Ranvier (Bezanilla and Armstrong, 1972; Hille, 1973), and the Ca^{2+}-activated channel from *Aplysia* (Gorman *et al.*, 1982; see Table II). The inward rectifier selectivity sequence of starfish egg cells (Hagiwara and Takahashi, 1974) parallels that of maxi CaK channels. There are some quantitative differences, however. The inward rectifier discriminates more closely than maxi CaK channels among monovalent cations: Rb^+ and NH_4^+ are less permeant through the former channel, i.e., K^+ (1) > Rb^+ (0.3) > NH_4^+ (0.03). Quantitative and qualitative differences are found when maxi CaK channels are compared with the delayed rectifier from snail neurons (Reuter and Stevens, 1980). The latter is measurably permeant to Cs^+, Na^+, and Li^+ and shows a completely different selectivity sequence. In particular, Li^+ is more permeable than Na^+.

The selectivity filter of maxi CaK channels appears to share some common structural features with some K^+-selective channels such as the node of the squid axon delayed rectifier and the *Aplysia* CaK channel. Rubidium is the largest (r = 0.148 nm) cation able to permeate through the delayed rectifier channel; Cs^+ (r = 0.169 nm) does not pass. The selectivity filter is viewed in this case as a narrow part of the channel with a diameter of 0.296 to 0.338 nm formed by a ring of oxygens provided by the channel protein (Hille, 1973). These oxygens provide the right environment to totally or partially dehydrate the ion and closely interact with it. The selectivity filter can be deformable, but in very selective channels this flexibility cannot be high (for a more detailed discussion on selectivity filters, the reader should consult Reuter and Stevens, 1980; Latorre and Miller, 1983; Hille, 1975; Eisenman and Horn, 1983).

5.2. Ion Conduction

If we take an average conductance for maxi CaK channels of 250 pS at 200 mM K^+, from Ohm's law, a 0.1-V applied voltage will force a current of 25 pA through the channel. This current corresponds to an ion flow through the channel of 1.5×10^8 ions/sec. This rate is very close to a diffusion-limited flux. In other words, flux through maxi CaK channels is comparable to the rate of diffusion up to an aperture of about 5 nm in diameter.

In view of the high selectivity of maxi CaK channels, such a high channel conductance represents a paradox. We have discussed how high conductance and high selectivity may arise in maxi CaK channels that cannot be occupied by more than one ion at a time (Latorre and Miller, 1983). However, evidence is mounting that Maxi CaK channels are actually multi ion pores. I try to give an account of this evidence in Section 5.3 and also present some "primitive" models that can account (at least partially) for it.

5.3. Conductance–K^+ Concentration and Current–Voltage Relationships

In a previous discussion of the conductance versus concentration behavior of the maxi CaK channels in bilayers, we suggested that this channel operates by

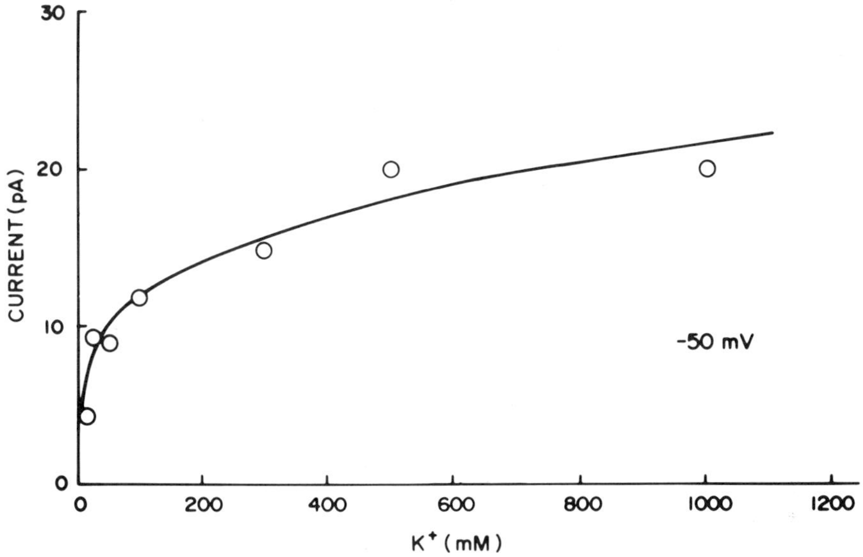

Figure 5. Current versus K^+ concentration for maxi CaK channel from smooth muscle incorporated into a PE bilayer. Solid curve is that predicted by the three-barrier, two-site barrier model of Fig. 10. For details, see text (D. Wolff, X. Cecchi, O. Alvarez, and R. Latorre, unpublished data).

a single-ion mechanism based on the Michaelis–Menten behavior of the unit conductance (Latorre and Miller, 1983). However, we have since found that these experiments were complicated by the presence of trishydroxymethylaminoethane (tris) as a buffer. Tris is a blocker of the channel and gives rise to artifactual Michaelis–Menten behavior with respect to K^+ (Vergara, 1983). When tris is absent in the solutions bathing the channel, the conductance–concentration relationship deviates markedly from a Langmuir isotherm (Vergara *et al.*, 1984; Moczydlowski *et al.*, 1985).

The same result is found for the case of smooth muscle maxi CaK channels incorporated into bilayers (Fig. 5; Wolff *et al.*, 1985). Many types of energy profiles could underlie this apparent negative cooperativity, for example, two initially different noninteracting sites or two initially identical interacting sites. This explanation demands multiple-ion occupancy of the channel. However, a single site that can relax slowly to two or more conformations can also explain this type of data. Whichever is the best model to explain the data, the conductance–concentration curve cannot be explained with a simple one-ion channel model (e.g., Läuger, 1973).*

The current–voltage relationship measured in symmetrical K^+ for the chro-

* Blatz and Magleby (1984) have reported that the conductance–activity curve for the culture rat muscle channel using asymmetrical solutions can be fitted by a single-ion model. However, they used N-methyl-D-glucamine as a buffer. This cation, like tris, may give rise to artifactual Michaelis–Menten kinetics.

maffin channel is linear at low voltages but, at high voltages (>100 mV), bends toward the voltage axis (is sublinear: Yellen, 1984a). In the T-tubule or smooth muscle channel, this sublinearity is a function of K^+ concentration. In the range of $\pm$ 100 mV, the curves show deviations from linearity at K^+ concentrations <20 mM. The sublinearity in the current–voltage relationship can arise from the intrinsic properties of the channel, and models in which the ions must pass across a sequence of energy barriers separating wells (binding sites) have commonly been used to explain this type of behavior (e.g., Hagglund *et al.*, 1979). If the conduction process could be described using this type of model, a picture of the binding sites and their locations in the channel could be obtained.

The ionic species considered in this analysis can be permeating ions or ions that can block the channel. Any conduction model for the maxi CaK channel will be based strongly on the current–voltage characteristics measured in a wide range of K^+ concentration and voltages. A model with two barriers at the channel entrances and a low central barrier (three-barrier, two-site model; see Fig. 10) predicts, as found experimentally, a sublinear current–voltage relationship at low ionic activity. However, a decrease in channel conductance as voltage is increased can also be obtained if the ion flow is diffusion limited (Läuger, 1976; Andersen and Procopio, 1980; Andersen, 1982). In this case, channel conductance at high voltages (or low concentrations) decreases because aqueous diffusion up to the channel entrance is rate limiting.

6. Probing the Channel with Blocking Ions

6.1. Some Words about Blockade: Three Different Modes of Blockade

A large variety of cations and a minor component of a scorpion venom are able to block maxi CaK channels. The basic blocking model presented below was developed by Woodhull (1973), and it assumes that once a channel is opened it becomes available for blocking according to the scheme:

$$\text{closed} \underset{\beta}{\overset{\alpha}{\rightleftharpoons}} \text{open} \underset{k_{-1}}{\overset{k_1[\text{B}]}{\rightleftharpoons}} \text{blocked} \qquad \text{(Scheme 3)}$$

where [B] is the concentration of the blocker. If we assume that the blocking reaction is in equilibrium (it is much faster) than the gating reaction, the apparent open-state conductance corresponds to a time-averaged value determined by k_1 and k_{-1}. Therefore, increasing [B] reduces this time-averaged conductance according to the expression:

$$\gamma/\gamma_o = ([\text{B}]/K_d + 1)^{-1} \qquad (3)$$

where γ_o is the conductance of the channel in the absence of the blocker and K_d is the apparent dissociation constant, which can be voltage-dependent if the site of blockade is within the electric field. If the site of blockade is located at a fractional distance, δ, down the electric field, the site "senses" a fraction, δV, of the applied voltage, V. Therefore, K_d will be voltage dependent according to the relationship (Woodhull, 1973):

$$K_d(V) = K_d(0)\exp(-z\delta FV/RT) \tag{4}$$

where z is the valence of the blocking ion. Combining equations 3 and 4, we obtain

$$\gamma/\gamma_o = \{1 + [B]/K_d(0)\exp(z\delta FV/RT)\}^{-1} \tag{5}$$

Thus, by measuring γ as a function of voltage at a constant concentration of the blocker, we can obtain the parameters $z\delta$ and $K_d(0)$. The quantity $z\delta$, which measures the voltage dependence of the block, is usually called the effective valence of the blocking reaction (Hille and Schwartz, 1978). For channels that cannot be occupied by more than one ion at a time, and when monovalent blocking ions and monovalent permeant ions are used, the value of $z\delta$ cannot exceed 1. For multiion pores, $z\delta$ may be larger than 1 despite the fact that the blocking ion is monovalent. Scheme 3 also predicts an apparent increase in the mean open time concomitant with the channel conductance decrease, inasmuch as a blocked channel must pass through the open state before it can close.

If the blocking reaction is slowed down but is still faster than the gating reaction, the open state will be interrupted by short-lived closings ("flickerings"). This flickering phenomenon was first described by Neher and Steinbach (1978) with local anesthetics acting in the acetylcholine receptor channel. From a molecular point of view, each short closure corresponds to the diffusion of the blocking molecule up to the blocking site, where it remains for a randomly distributed period of time. When the blocker leaves the site, it allows the channel to open again until the next blocker molecule enters. For a flickering blockade, by measuring mean open ($\bar{\tau}_o$) and mean blocked ($\bar{\tau}_b$) times, it is possible to obtain the rate constants for the blocking reaction:

$$\bar{\tau}_o = 1/(\alpha + k_1[B]) \tag{6}$$

$$\bar{\tau}_b = 1/k_{-1} \tag{7}$$

When $\alpha \ll k_1[B]$, $\bar{\tau}_o \sim 1/k_1[B]$. Scheme 3 predicts, therefore, that $1/\bar{\tau}_o$ should increase linearly with [B] and that $\bar{\tau}_b$ should be independent of [B].

When the residence time of the blocking molecule in the site is very long, channel kinetics will be described by periods in which the channel is active, interrupted by long periods of quiescence. In this case, the mean burst time, $\bar{\tau}_B$, is given by

$$\bar{\tau}_B = (1/k_1[B])(1 + \beta/\alpha) \tag{8}$$

and the mean block time is exactly as in Equation 7. Termination of a burst leaves the set of closed and open states of Scheme 3. The mean burst time given by

Equation 8 differs from the case in which the blocking reaction is faster than the gating in that termination of a burst in the latter implies leaving the set of states open and blocked (Neher and Steinbach, 1978). Of course, it is possible to set the experimental conditions in such a way that $\alpha \gg \beta$, and Equation 8 is reduced to $\bar{\tau}_B = 1/k_1[B]$ (Vergara and Latorre, 1983). Fast, flickering, and very slow types of blockades have been described for maxi CaK channels.

6.2. Alkali Cation Blockade

Both Li^+ and Na^+ can block the T-tubule channel from the *cis* (internal) side with $z\delta$ of 0.38 and 0.7, respectively. The $K_d(0)$ necessary to fit the voltage-dependent block induced by these two ions is 77 mM for Li^+ and 134 mM for Na^+ (D. Naranjo, D. Wolff, O. Alvarez, and R. Latorre, unpublished results). We have found very little or no Na^+ blockade from the *trans* side. Since channel blockade by Li^+ and Na^+ behaves as an ion binding to a single discrete site, it is tempting to suggest that the degree of hydration is crucial in determining how far an ion penetrates into the channel: Na^+, the least hydrated, penetrates farther than Li^+. Both Li^+ and Na^+ blockade were inferred from the channel conductance decrease, i.e., "fast" blockade (our data were filtered at 1 kHz).

Marty (1983b) found that in chromaffin cells Na^+ induces a voltage-dependent flickering type of blockade (channel kinetics recorded at 5 kHz). This blockade was relieved by external K^+, showing that the sodium-binding site is located in the permeation pathway. Together with its blocking effect, Na^+ reduces the opening probability, suggesting that Na^+ can interact with Ca^{2+}-binding sites. This effect was also relieved by external K^+. Reaction scheme 3 predicts a lengthening of the mean burst time when the blocking ion is present. Therefore, in CaK channels, the lengthening of the mean burst time induced by the blocker can be masked by interaction of the blocker with the Ca^{2+}-activating site(s). Marty's results also suggest that the conduction system and the Ca^{2+}-activating site are located in close proximity (see also Section 8)

Also using chromaffin cells, Yellen (1984a,b) confirmed and extended Marty's findings. Yellen (1984a) found that millimolar concentrations of internal Na^+ reduce the average current through the channel and produce a flickering noise in the open-channel currents, but is was not possible to resolve the flicker into discrete blocking and unblocking events (Fig. 6).

However, Yellen (1984a) applied Fitzhugh's theoretical treatment for a two-state process and a first-order filter (Fitzhugh, 1983) and obtained the blocking and unblocking rates for Na^+ blockade even though individual blocking events are not resolved by the recording system. This method, "amplitude distribution analysis," is based on the fact that for a blocker that is not fast enough or slow (where rate constants can be measured directly), the current amplitude distributions are not exactly Gaussian in shape. Unequal blocking and unblocking rates give a skewed distribution. This distribution is sensitive to changes in the blocking and unblocking rates. By amplitude distribution analysis, it is found that the rates for Na^+ blockade in the chromaffin channel are in the microsecond range. Furthermore, in agreement with the predictions of Scheme 3 (Equations 6 and 7), the

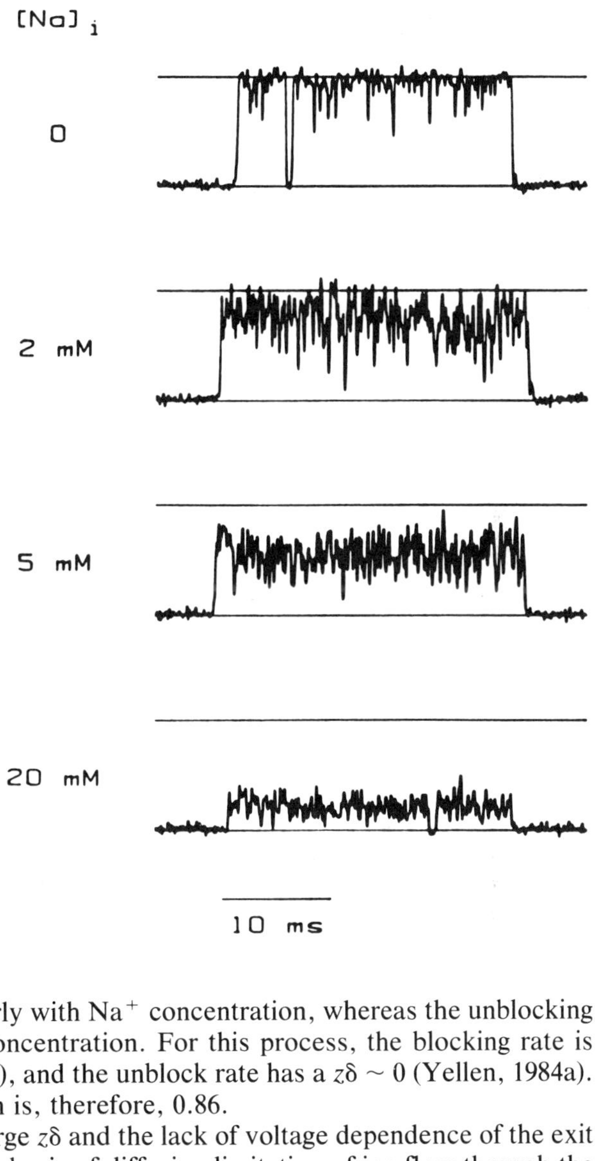

Figure 6. Flickering block induced by internal sodium. Single CaK channel from chromaffin cell. The lines indicate base line and open-channel current; the difference between the lines is 24.4 pA. $[Ca^{2+}]$ = 0.3 μM, membrane voltage +80 mV. Note that internal Na^+ induces flickering and decreases the channel conductance. (Redrawn from Yellen, 1984a.)

blockade rate increases linearly with Na^+ concentration, whereas the unblocking rate is unaffected by Na^+ concentration. For this process, the blocking rate is voltage dependent ($z\delta = 0.86$), and the unblock rate has a $z\delta \sim 0$ (Yellen, 1984a). The $z\delta$ of the overall reaction is, therefore, 0.86.

Yellen argued that the large $z\delta$ and the lack of voltage dependence of the exit steps can be explained on the basis of diffusion limitation of ion flow through the channel. Diffusion limitations in K^+ flow promote a depletion of K^+ in the region near the mouth of the channel, producing a decrease in the entry rate for this ion, which brings, as a consequence, an increase in channel occupancy by Na^+. Depletion of K^+ from the mouth also promotes an increase in the resistance of this region, and, hence, a current flowing through the channel causes a voltage drop across the diffusional access resistance. It is clear that if, for example, 20% of the applied voltage were to drop in this resistance, a $z\delta = 0.2$ will be found even in the absence of any other contribution to the voltage dependence of Na^+ entry, and the "actual" $z\delta$ would be 0.66 (i.e, 0.86 − 0.2). Therefore, changes in channel

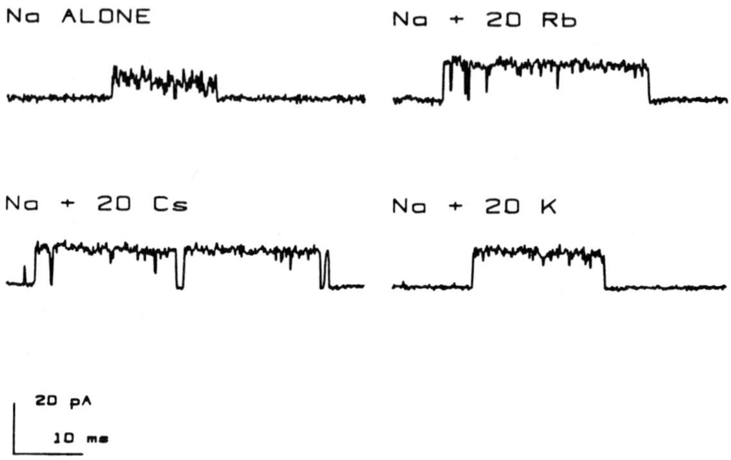

Figure 7. External Rb^+, Cs^+, and K^+ relieve internal Na^+ blockade. CaK channel from chromaffin cell. Membrane voltage +60 mV. (Redrawn from Yellen, 1984b.)

occupancy by the blocking ion and a voltage drop in the resistance may be leading factors giving rise to large $z\delta$. However, experiments that directly demonstrate the existence of a diffusion-limited rate step in the process of ion conduction are still lacking.

For single-ion channels, it is predicted that in the presence of two competing ions (e.g., Na^+ and K^+) only the Na^+ blocking rate should be affected by al-

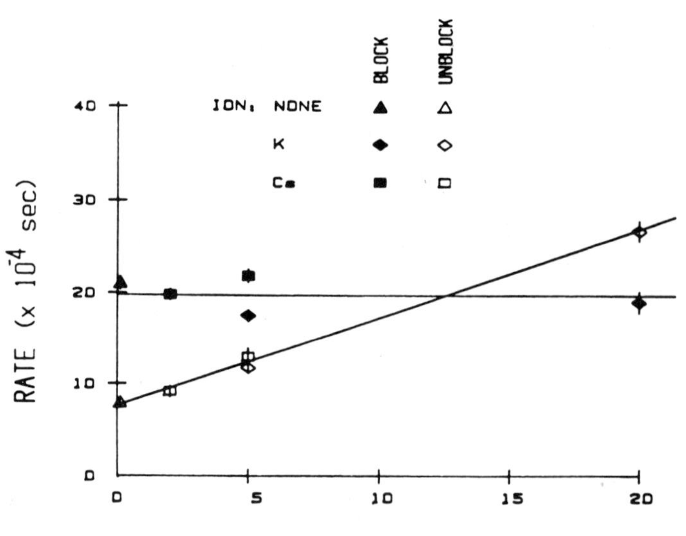

Figure 8. Rate constants for internal Na^+ blockade at different external K^+ or Cs^+ concentrations. Chromaffin cell CaK channel. Membrane voltage +60 mV. (Redrawn from Yellen, 1984b.)

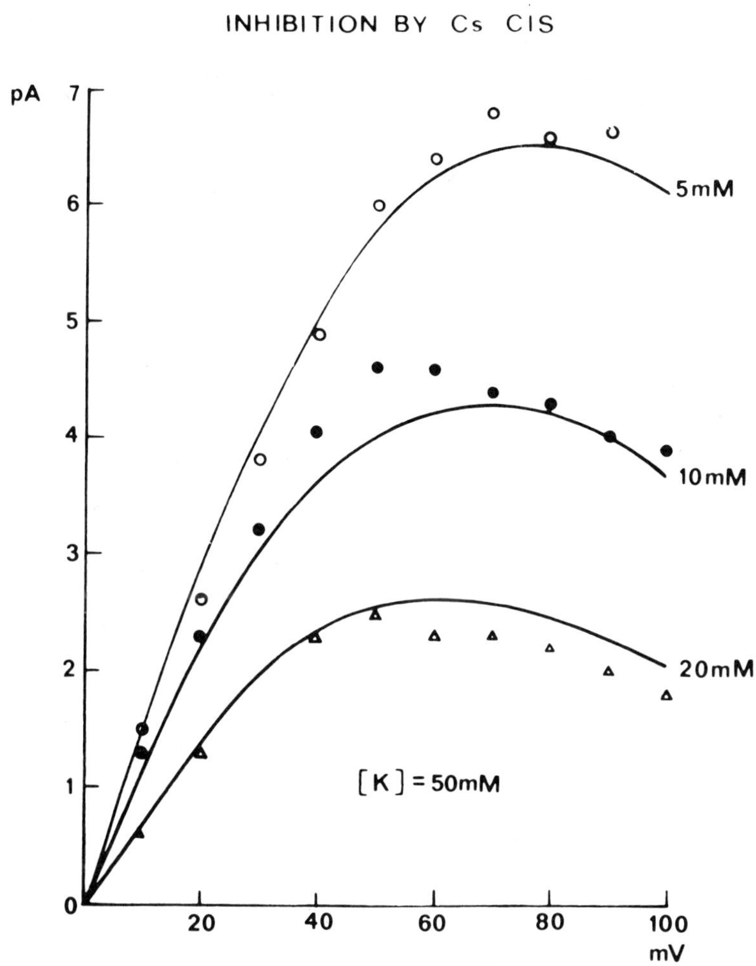

Figure 9. *Cis* (internal) Cs^+ blockade of the maxi CaK channel from T-tubule incorporated into a PE membrane. Cs^+ was added at the indicated concentrations to the *cis* side only. The K^+ concentration was 50 mM. At this concentration, the central I–V relationship (not shown) was linear in the range of 0 to +100 mV. Solid lines are those predicted by the three-barrier, two-site model shown in Fig. 10. (For more details, see text; X. Cecchi, D. Wolff, O. Alvarez, and R. Latorre, unpublished observations.)

terations in the K^+ concentration. For the chromaffin channel, it is found that Na^+ blockade can be relieved by external K^+ (Fig. 7). In disagreement with a single-ion model for the channel, K^+-induced relief is brought about exclusively by an increase in the exit rate (Yellen, 1984b; Fig. 8). Moreover, external Cs^+, which cannot carry current through the channel, and external Rb^+ relieve internal Na^+ blockade in the same manner as K^+. These observations are consistent with a model in which the channel can be occupied simultaneously by more than one ion. The second ion entering the channel (e.g., K^+, Cs^+, or Rb^+) increases the rate of exit of the blocking ion through a coulombic interaction. In other words,

Table III. Cesium Blockade Characteristics in Two CaK Channels[a]

	External (*trans*) blockade		Internal (*cis*) blockade	
Channel	$z\delta$	$K_D(0)$ (mM)	$z\delta$	$K_D(0)$ (mM)
Chromaffin	1.0	160	0.25	100
T-tubule	1.4	19	0.70	55

[a] Data for the chromaffin CaK channel were obtained from Yellen (1984a). The $z\delta$ parameter for *trans* Cs^+ blockade of the T-tubule channels is that obtained at *trans* $[Cs^+] > 1.7$ mM (see text).

occupancy of a blocked channel by a second ion increases the free energy in the place where the blocking site is located (see below). The data can also be explained with a strict knock-out mechanism (Hodgkin and Keynes, 1955; Armstrong, 1975; Yellen, 1984b).

Cesium is able to block T-tubule channels from both sides of the membrane, but it does so in strikingly different manners (X. Cecchi, D. Wolff, D. Naranjo, O. Alvarez and R. Latorre, unpublished observations). From the *trans* side, Cs^+ blockade is highly voltage dependent. The effective valence for the blocking reaction is also a steep function of Cs^+ concentration. The value of $z\delta$ rose from 0.7 at 0.1 mM Cs^+, reached 1.4 at 1.7 mM, and changed very little above this Cs^+ concentration. The characteristic of the Cs^+ *trans* blockade gives further support to the contention that maxi CaK channels are multiion pores. From the *cis* side, Cs^+ blocks the channel with a $K_d(0) = 55$ mM and a $z\delta = 0.7$ (Fig. 9). The effective valence of the value of $z\delta$ in *cis* Cs^+ blockade is independent of Cs^+ concentration. Although qualitatively the characteristics of Cs^+ blockade of the T-tubule channel are similar to those found in chromaffin cells (Yellen, 1984a), the blocking parameters are different (Table III). The differences in effective valence and affinity for Cs^+ of the ion-binding sites suggest structural differences in the conduction systems of these two channels. *Trans* Cs^+ blockade of the T-tubule channel is not relieved by increasing the K^+ concentration on the *trans* side only. On the contrary, blockade is slightly increased by this experimental manipulation. This observation is also consistent with a multiion pore model.

6.3. A Barrier Model for Maxi CaK Channels

Maxi CaK channels have several characteristics that suggest that the conduction process cannot be described using an electrodiffusion formalism. These are:

1. Single-channel conductance is a saturating function of potassium concentration.

2. Rubidium is only slightly less permeant than K^+ through the channel (see Table II), but channel conductance when the current is carried by Rb^+ is much smaller than the conductance when the current is carried by K^+. Actually, Rb^+ blocks the channel.
3. Maxi CaK channels do not obey the independence principle (Hodgkin and Huxley, 1952; Blatz and Magleby, 1984).

Given the shape of the current–voltage relationship (see above), we (O. Alvarez, X. Cecchi, D. Wolff, and R. Latorre, unpublished data) have chosen to model conduction through maxi CaK channels with a three-barrier, two-site model in which more than one ion can be in the channel at a time. The following experimental findings suggest that the channel can accept more than one ion at a time:

1. The channel conductance-versus-concentration curve cannot be described using a one-ion pore model since it cannot be fitted with a single hyperbola (Fig. 5; Vergara *et al.*, 1984; but see Blatz and Magleby, 1984).
2. *Trans* Cs^+ blockade is strongly voltage dependent. The effective valence of the blocking reaction is concentration dependent, and at "high" Cs^+ concentrations, it is larger than 1.
3. External K^+ relieves internal Na^+ blockade exclusively by increasing the Na^+ exit rate (Yellen, 1984b).

The two-binding-site model we have developed is based on standard rate theory (Hille and Schwarz, 1978) and is shown in Fig. 10. The solid line is the free-energy profile "seen" by a K^+ ion in the absence of both an applied electric field and other ions within the channel. The dotted line describes the free-energy profile "seen" by a K^+ ion when the channel is already loaded with one K^+. In order to obtain the energy profile shown in Fig. 10, we first have to define all the peak and well energies at zero voltage, the electrical distances at which the wells are located, and the increase in barrier energies when one well is loaded or the two wells are loaded. Given these parameters, the energies of wells and peaks can be calculated as a function of voltage and occupancy. Finally, having all the energies calculated, the rate constants can be obtained. Rates of transition, k, are of the form $k = Q \exp(-\Delta G)$, and for entry rates this expression is multiplied by the ion activity in the medium. The factor $Q = kT/h$, where k is the Boltzmann and h the Planck constant, has been scaled to give the ion transport in picoamperes: $Q = 10^6$ at 20°C.

Once the rate constants are obtained, it is possible to evaluate the probability of each of the states of the diagram which allows for the calculation of the

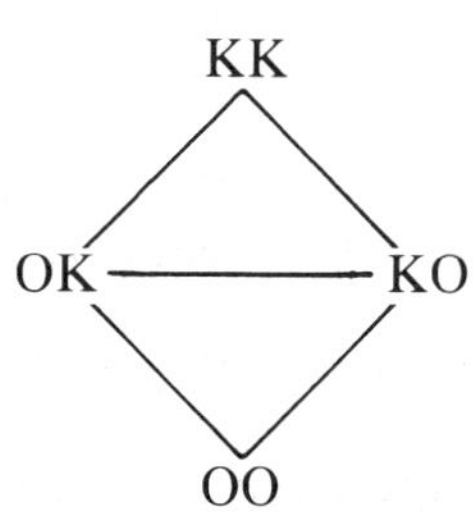

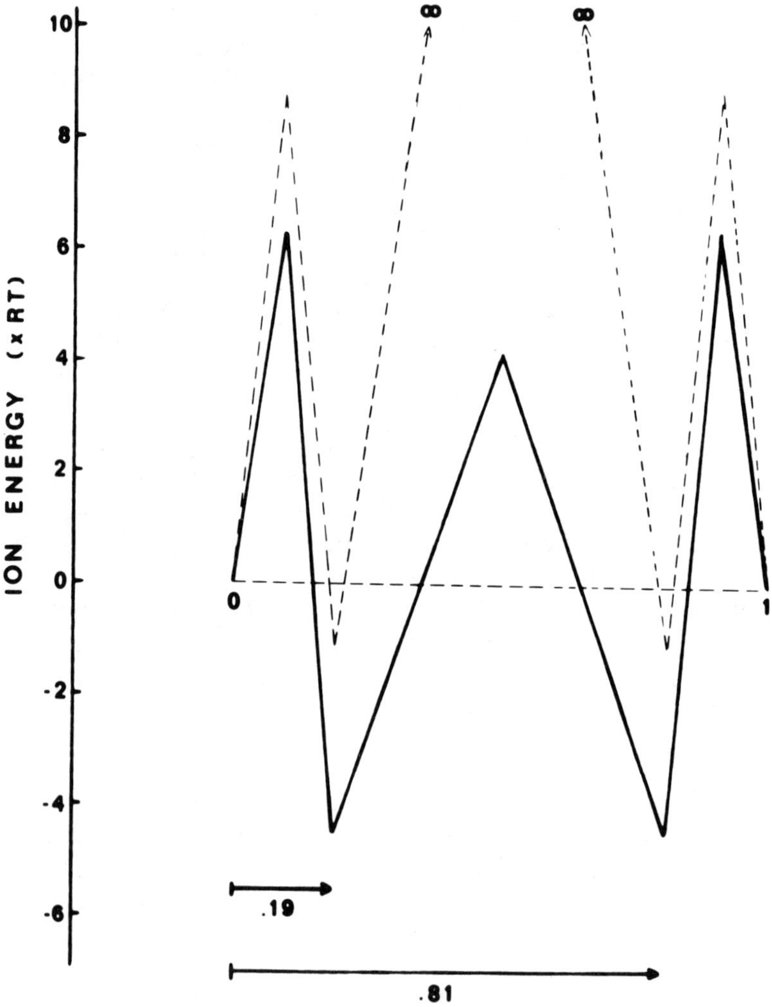

Figure 10. Three-barrier, two-site model for the maxi CaK channel of T-tubule. The solid line represents the free-energy profile for a single K^+ ion in the absence of both an applied electric field and other ions within the channel. The dashed line is the energy profile for a K^+ ion when the channel is already loaded with another K^+ ion. Fractional electrical distances to the *cis* (left) and *trans* sites are indicated by the arrows. The model predicts that the entry of a second K^+ ion is 12-fold more difficult than the entry of the first K^+ ion and that the exit of a K^+ ion when both wells are occupied is 2.5-fold faster than when there is a single ion in the channel. (For more details, see text.)

steady-state- ion currents (I), since this is simply the net transition rate across any energy barrier, e.g., the net current across the central barrier is

$$I = k_{\text{KO-OK}} P_{\text{KO}} - k_{\text{OK-KO}} P_{\text{OK}}$$

where P_{KO} and P_{OK} are, respectively, the probabilities of the states KO and OK given in the diagram. For more details, the reader should consult Hille and Schwarz (1978). A new method to calculate the steady-state values for all the states of any kinetic diagram recently has been put forward (Hagglund *et al.*, 1984).

Diffusion in this barrier model requires vacancies, unlike the strict "knock-on" theory in which an ion can obtain sufficient kinetic energy that can "knock-on" (knock out) another ion from a site (Hodgkin and Keynes, 1955).

The energy barrier profile shown in Fig. 10 gives a good fit to the channel current amplitudes-versus-voltage curve and to the channel conductance-versus-K^+ concentration relationship (solid line in Fig. 5). We note that (1) entry of a second K^+ ion is 12-fold more difficult than the entry of the first K^+ ion, and (2) exit of a K^+ ion when both wells are loaded is 2.5-fold faster. Therefore, the model predicts that double occupancy has a low probability and accounts for the large conductance of the CaK channel. We have extended this model to include *cis* Ca^{2+} (fast) and *cis* Cs^+ blockade by adding the necessary states to the diagram. The fast Ca^{2+} blockade (see Section 6.6) of the T-tubule channel is well explained by increasing the energy of the *cis* peak to $13kT$ and decreasing the *cis* energy well to $-6.3kT$. The central energy has been set so that *cis* Ca^{2+} cannot reach the *trans* well. Cesium *cis* blockade is also well described by the two-site, three-barrier model (solid line, Fig. 9). In this case, the *cis* barrier has a value of $5.5kT$, and the *cis* well of $-6.2kT$. However, the *cis* well location needs to be changed to a $\delta = 0.5$.

6.4. Conclusion

Evidence has accumulated for maxi CaK channels that ions interact strongly with the channel and with other ions diffusing across them. Many lines of evidence reveal that, like the delayed and inward rectifier channels, maxi CaK channel are multiion pores. Barrier models that consider multiion occupancy appear to describe ion conduction data (see above; Yellen, 1984a). However, these models are still very primitive, and processes that may contribute to the overall ion conduction, such as diffusion limitation and knock-on, need to be studied in detail. Unidirectional flux ratio measurements and measurements of channel conductance in the presence of mixtures of permeant ions (e.g., Cl^- and K^+) are needed. In a multiion pore, it is expected to find unidirectional flux ratios given by a high power of the electrochemical activity ratio that exhibit "anomalous mole fraction" behavior.

6.5. Quaternary Ammonium Ions

Large cations such as tetraethylammonium (TEA) and nonylmethylammonium are able to block maxi CaK channels (Vergara *et al.*, 1984; Yellen, 1984a; Blatz and Magleby, 1984). For the T-tubule channel, *trans* and *cis* blockade are well described by equation 5 with $K_d(0)$s for the blocking reaction of 0.29 mM

Table IV. Blocking Parameters for Mono- and Divalent Organic Cations[a]

	$K_D(0)$ (mM)	$z\delta$
Tetramethylammonium	54 ± 6	0.29 ± 0.7
Tetraethylammonium	41 ± 15	0.29 ± 0.02
Phenyltrimethylammonium	1.89 ± 0.1	0.32 ± 0.07
Nonyltrimethylammonium	0.008 ± 0.001	0.37 ± 0.13
Nonyltriethylammonium	0.003 ± 0.001	0.34 ± 0.10
Decamethonium	0.005 ± 0.001	0.54 ± 0.05

[a] Blocking parameters were obtained by a least-squares fit to the g/g_0 vs. V curves and according to equation 5. Blockers were added to the *cis* side only. The [Ca^{2+}] in all the experiments was 3 μM (J. Villarroel, O. Alvarez and R. Latorre, unpublished results).

and 45 mM, respectively, Thus, the *cis* affinity is 155-fold lower than the affinity of the *trans* site. Further, *trans* TEA blockade is independent of applied voltage, whereas *cis* blockade is well described by equation 5 with a $z\delta = 0.3$. These results suggest that the *cis* and *trans* TEA receptors are different and that the T-tubule channel might, at each side, contain a mouth wide enough to accommodate TEA ($r = 0.4$ nm). Similar, although not identical, results have been obtained in the CaK channels of chromaffin and in cultured rat muscle cells (Yellen, 1984a; Blatz and Magleby, 1984). We have recently extended these results to include a number of monovalent quaternary ammonium (QA) ions of different sizes and hydrophobicities (Table IV). Figure 11 shows that when these are added to the *cis* side, two types of blockers can be distinguished: ''quite'' fast blockers (e.g., TMA) and flickering blockers (e.g., nonytrimethylammonium). Table IV shows that there is a general trend for blocking potency to increase with increasing size, and we attributed this to stronger interactions of the hydrophobic residues of QA compounds with hydrophobic regions of the channel. Steady-state block by QA ions showed similar voltage dependence, and all appear to interact with a site located at 30% of the electrical potential drop measured from the *cis* side.

Two points are of interest to note here when comparing the present results with those obtained in the squid axon delayed rectifier. (1) Replacement of the ethyl groups by methyl groups in the long alkyl-chain derivatives does not render the compound inactive (cf. nonyltrimethylammonium and nonyltriethylammonium, Table IV). In contrast, octyltrimethylammonium produced no detectable inhibition of the squid axon potassium conductance (Armstrong, 1975). (2) Block by TMA has the same voltage dependence as the blockade induced by TEA. On the other hand, TMA blockade of the delayed rectifier is about twice as steeply dependent on voltage ($z\delta = 0.36$) than TEA blockade ($z\delta = 0.2$) (French and Shoukimas, 1981). Therefore, the detailed molecular structure of the TEA binding site of maxi CaK channels appears to be different from that of the delayed rectifier.

Table IV shows that decamethonium (a divalent *bis*-quaternary ammonium ion) is also able to block the T-tubule channel, but the blockade is more voltage

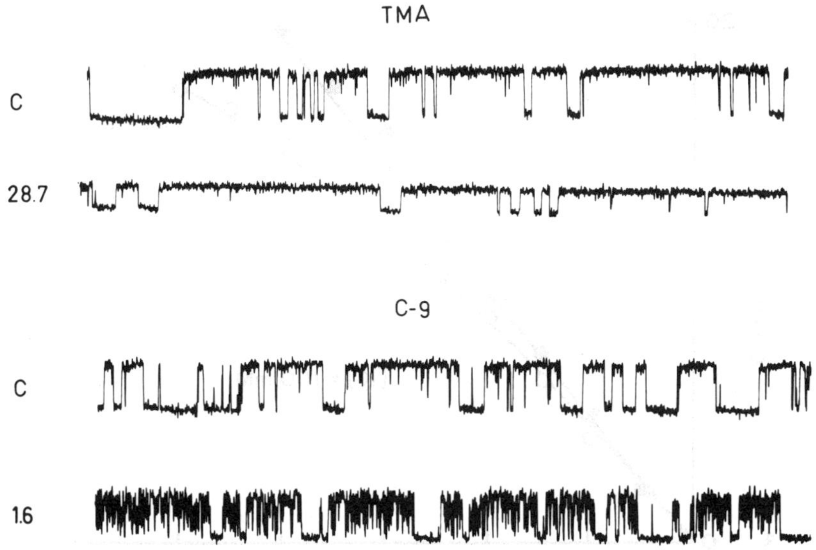

Figure 11. Fast and flickering blockage induced by quaternary ammonium (QA) ions on the T-tubule CaK channel. Records of representative channel fluctuations are shown in the absence (C) or in the presence of QA ions. Numbers to the left are the concentrations of TMA or nonyltrimethylammonium C-9 in the *cis* solution. An apparent increase in the mean open (mean burst time for C-9) times was observed for both QA ions. Membrane voltage was +40 mV. $[K^+]$ = 100 mM; $[Ca^{2+}]$ = 3 μM; PE membrane (J. Villarroel, O. Alvarez, and R. Latorre, unpublished observations).

dependent than the one found for QA ions. It is possible that the large effective valence of decamethonium (about twice that of QA ions) reflects the binding of this divalent cation in a bent-over conformation with both charges interacting with the "TEA site" (e.g., Miller, 1982).

An interesting result is that nonyltriethylammonium is not effective as a blocker of the T-tubule channel from the *trans* side even at concentrations as high as 1 mM. This suggests that the *trans* binding site for TEA is very different from a structural point of view from the *cis* side. A similar result was obtained when QA ions were applied externally to the node of Ranvier (Armstrong and Hille, 1972).

6.6. Ca^{2+} and Ba^{2+} Blockade

Cis Ca^{2+} induces a fast and a slow blockade of the T-tubule maxi CaK channel. The fast Ca^{2+} blockade is clearly present at $[Ca^{2+}] > 1$ mM (Fig. 12) and is voltage dependent ($z\delta$ = .02). The presence of this type of blockade makes it necessary to work at very low Ca^{2+} concentrations when characterizing the current–voltage relationship for this channel, especially when a wide range of voltages is to be studied. The presence of Ca^{2+} can give rise to artifactual asymmetric current–voltage relationships.

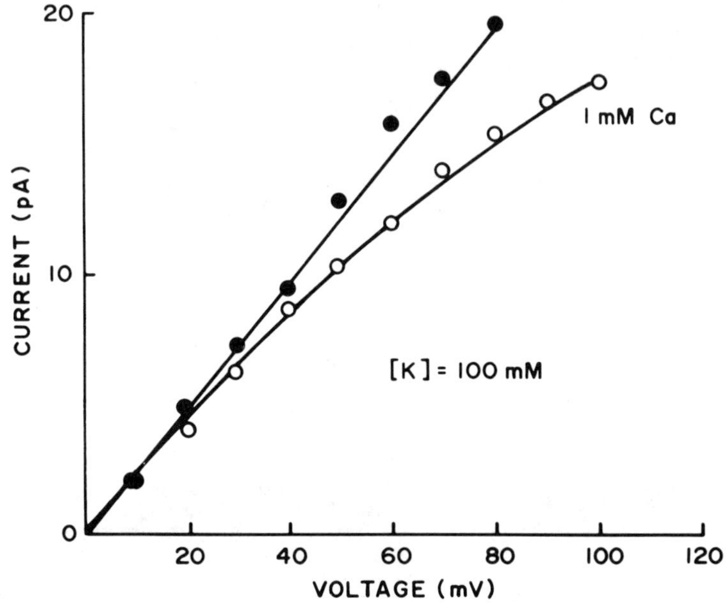

Figure 12. *Cis* "fast" Ca^{2+} blockade. The current–voltage relationships for the open CaK channel from T-tubule are shown. Solid circles show the central I–V curve taken at $[Ca^{2+}] = 3\ \mu M$. Solid lines are the predictions of the three-barrier, two-site model of Fig. 10. Phosphatidylethanolamine membrane (D. Wolff, X. Cecchi, O. Alvarez, and R. Latorre, unpublished observations).

In Section 4.1, we mentioned that the T-tubule channel shows a very slow Ca^{2+}-blocking process. This has been studied in detail (Vergara and Latorre, 1983). The top record of Fig. 13 shows that the channel activity can be viewed as bursts of activity (labeled b) and quiescent periods. We found that the duration

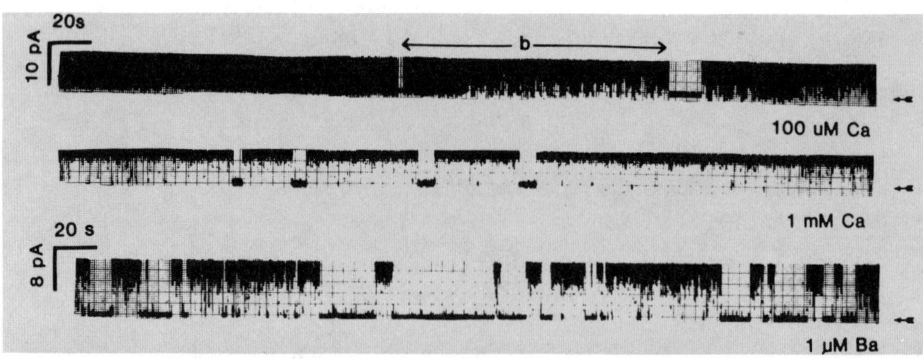

Figure 13. *Cis* Ca^{2+} and Ba^{2+} slow blockade. Membrane voltage was +40 mV. PE/PS = 7/3 in symmetrical 100 mM K^+. Ba^{2+} records were taken in the presence of 0.1 mM Ca^{2+}. (Redrawn from Vergara and Latorre, 1983.)

of the burst is decreased and the number of quiescent periods is increased if we increase either the [Ca^{2+}] or the voltage. In other words, the slow process has a [Ca^{2+}] and voltage dependence opposite to that of the fast process. We have quantified the slow process by characterizing the mean burst ($\bar{\tau}_B$) and the mean time for the slow closings ($\bar{\tau}_s$) as functions of [Ca^{2+}] and voltage. The data obtained are in agreement with reaction scheme 3, in which the gating kinetics are fast compared to those of the Ca^{2+} blockade. At high probability of opening, we found that, as the model demands, $\bar{\tau}_B$ is a linear function of 1/[Ca^{2+}] and $\bar{\tau}_S$ is independent of [Ca^{2+}]. We also found that the voltage dependence of the blockade resides almost exclusively in the forward rate constant k_1, as if Ca^{2+} must surmount a highly asymmetric barrier before reaching the blocking site. From the voltage dependence of the block, we conclude that the site senses approximately 65% of the electric field. The $K_d(0)$ is 0.29 M.

Another result that indicates that the slow Ca^{2+} blockade is governed by a reaction scheme like 3 is the fact that the number of quiescent periods increases dramatically as the [K^+] is reduced. In terms of the blocking scheme 3, these results can be interpreted as competition between Ca^{2+} and K^+ for a site located in the channel lumen. Furthermore, only the entry rate is affected by K^+, indicating that the T-tubule channel behaves in this case like a single-ion channel. It is possible that when the channel is loaded with one Ca^{2+} no other ion is allowed inside the conduction system because of its divalent nature. In other words, electrostatic repulsion, when the channel is loaded with one Ca^{2+} ion, hinders the entrance to the channel of a second ion.

Barium added to the *cis* side also induces an increase in the frequency of slow closing (Vergara and Latorre, 1983). However, the analysis of this effect is made simpler by the fact that Ba^{2+} does not activate the channel. Figure 13 shows that Ba^{2+} is able to induce slow kinetics at much lower concentrations than Ca^{2+}. Accordingly, we assume that Ba^{2+} as well as Ca^{2+} can block an open channel through a reaction scheme similar to scheme 3. Of course, the Ba^{2+} blockade must be studied under experimental conditions in which there is minimum interference from the Ca^{2+} blockade. The Ba^{2+} blockade is voltage dependent, and the Ba^{2+}-binding site senses about 80% of the field. The voltage dependence of the Ba^{2+} blockade resides mainly in the blocking rate constant. The $K(0)$ for this process is 36 μM. The following experimental evidence indicates that Ba^{2+} enters the pore when it blocks: (1) the blocking equilibrium and kinetics are strictly competitive with K^+; (2) increasing the *trans* [K^+] modifies only the Ba^{2+} entry rate (contract with Na^+ blockade). These two observations cannot be reconciled with a model in which Ba^{2+} binds to a site outside the K^+ diffusion pathway.

In conclusion, the "slow gating" appears to be a reflection of a slow blockade reaction induced by Ca^{2+}. Furthermore, we conclude that both Ca^{2+} and Ba^{2+} bind to the same site, because the voltage dependence and the general characteristics of the blockade are about the same; i.e., the divalent ions travel the same electrical distance before reaching the binding site. However, Ba^{2+} binding to the site is 10^4-fold stronger. The type of blocking promoted by these ions is different from that caused by alkali cations or TEA (see above). For those ions, the blocking rate is much faster than the channel-closing rate, whereas the reverse situation is true for Ca^{2+} and Ba^{2+}.

Barium blocks most of the K^+ channels studied until now, including the delayed rectifier and the inward rectifier (for a review, see Latorre and Miller, 1983). In these channels, the characteristics of the Ba^{2+} blockade are well explained in terms of a multiion pore (Armstrong *et al.*, 1982). On the other hand, in the absence of a knock-off mechanism and the purely competitive behavior between divalent cations and K^+, we tend to suggest that loading of the channel with a divalent cation like Ca^{2+} and Ba^{2+} does not allow other ions into the channel. A Ba^{2+} blockade similar to the one described here has recently been found in the Maxi CaK channel of rabbit cortical collecting tubules (Hunter *et al.*, 1984).

7. Maxi CaK Channels and Toxins

7.1. Charybdotoxin Blocks Maxi CaK Channels

Until now, purification of channel-forming proteins has depended on the existence of substances able to bind to the channel with high specificity. Thus, the biochemistry of the ACh receptor channel and the sodium channel was characterized mainly because bungarotoxin and tetrodotoxin, respectively, were known to block these channels in a highly specific manner and at very low concentrations. To find a suitable ligand for maxi CaK channels is a necessary task in order to perform similar studies with this protein(s).

Figure 14 shows that *Leiurus quinquestratus* venom (LQV) is able to promote dramatic changes in the kinetics of T-tubule maxi CaK channels incorporated into planar bilayers (Miller *et al.*, 1985). Addition of LQV to the *trans* side (external side of the channel) induces long periods in which the channel remains quiescent. As the LQV concentration increases, the inhibition of the channel becomes increasingly pronounced, and this can be appreciated in the shortening of active periods. It is important to mention here that LQV is only active from the *trans* side of the channel. These characteristics can be interpreted as the binding and dissociation of single toxin molecules to single CaK channels according to the reaction (Miller *et al.*, 1985).*

$$\text{active} \underset{k_{\text{off}}}{\overset{k_{\text{on}}[\text{TX}]}{\rightleftharpoons}} \text{blocked} \qquad \text{(Scheme 4)}$$

The quiescent periods are then intervals of time in which a toxin molecule is associated with a single channel. This kinetic scheme predicts that the active and

* It is not clear at present whether or not the toxin can interact with the closed state, inducing a closed–blocked state. For Ba^{2+} blockade, it has been shown that this divalent caation blocks the channel from the open state (Vergara and Latorre, 1983).

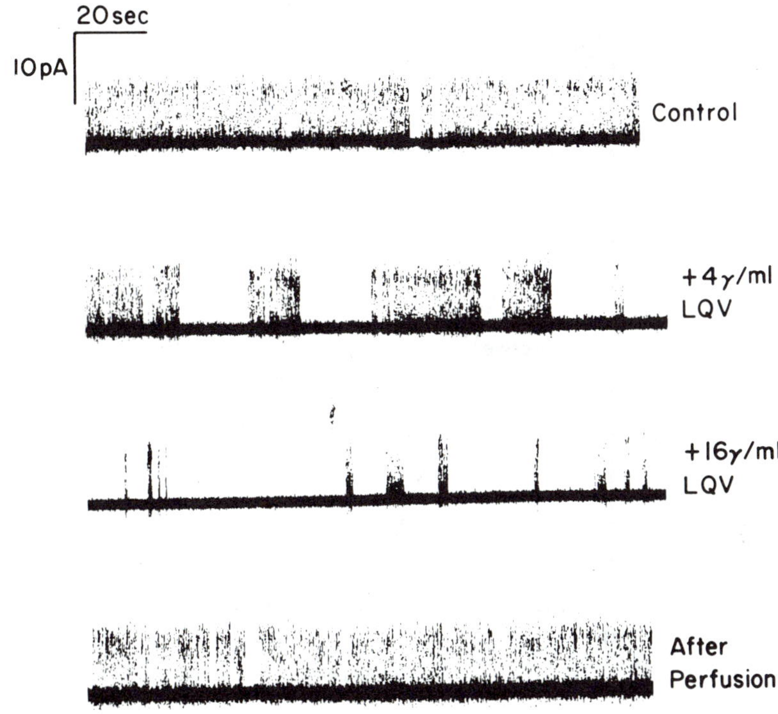

Figure 14. Concentration dependence and reversibility of the inhibition of the CaK from T-tubule induced by *Leiurus* venom. Leiurus venom was added to the *trans* (external) side only at the indicated concentrations. Bottom record was taken after extensive perfusion of the *trans* side. PE/PC = 4/1 membrane. The *cis* solution contained 200 mM K^+, 80 μM Ca^{2+}. The *trans* solution contained 90 mM Na^+, 10 mM K^+, and 3 μM Ca^{2+}. (Redrawn from Miller *et al.*, 1985.)

blocked times must be exponentially distributed and that the mean blocked time must be independent of toxin concentration.

However, the LQV is a mixture of about 50 proteins, and, therefore, the active component of the crude venom needs to be identified and purified. A two-step fractionation using cation-exchange chromatography of crude venom and Bio-gel® fractionation results in a recovery of 70% of total toxin activity with 1–2% of the protein, corresponding to a 50-fold purification. Cation-exchange chromatography produced multiple protein peaks, but toxin activity was observed only at high salt elution (>500 mM). The partially purified active component has a $M_r \cong 7000$. For this component, we have coined the name "charybdotoxin" (CTX), since we can view the Maxi CaK channels as a whirlpool (*Charybdis*, a female monster of classical mythology) against which the toxin is directed. Charybdotoxin activity is sensitive to reduction with dithiothreitol and to treatment with chymotrypsin. Thus, CTX is a small, basic, disulfide-linked protein.

When CTX is added to the *trans* (external) side only, and in agreement with the predictions of Scheme 4, we found that the dwell time in the blocked and

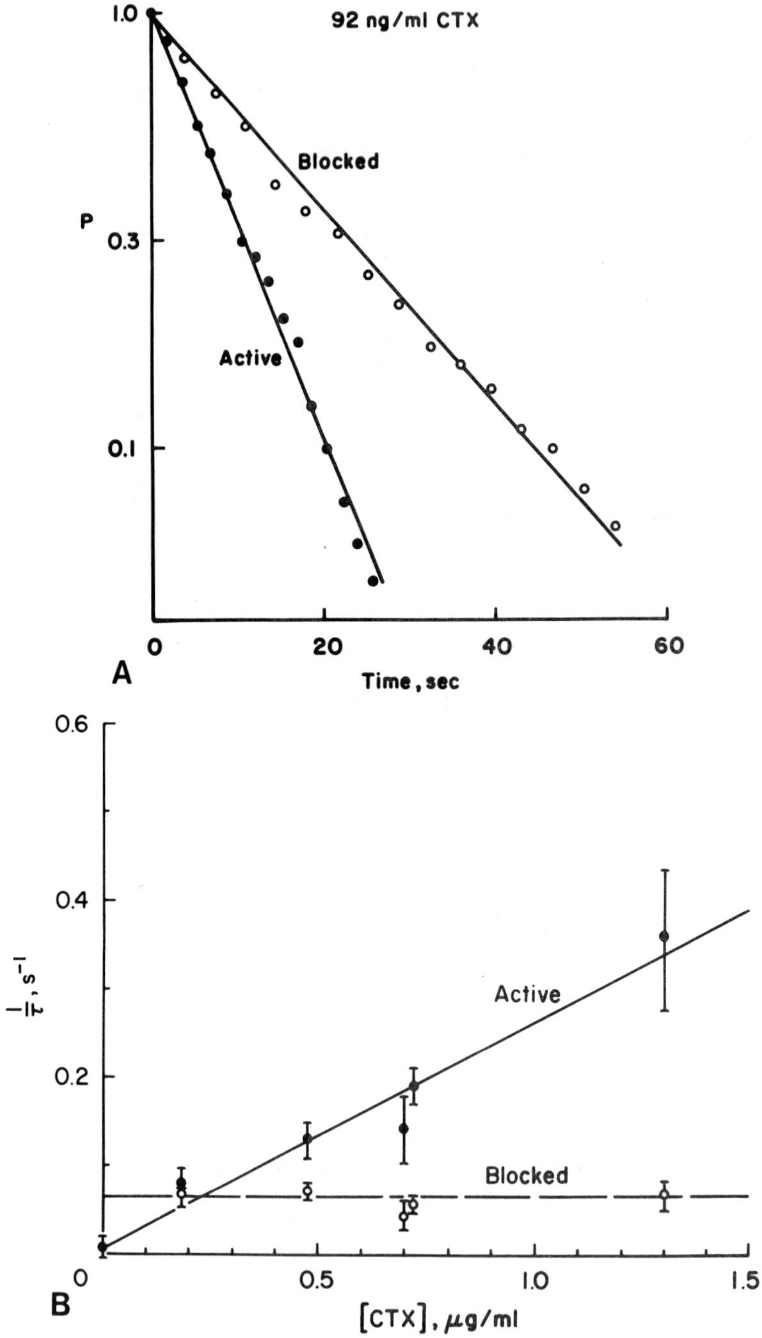

Figure 15. Kinetic characteristics of charybdotoxin blockage of the T-tubule CaK channel. A: Probability distributions of blocked and active dwell times at +50 mV. B: Active and blocked rates for CTX blockade. (From C. Miller, unpublished data.)

active state are effectively distributed as single exponentials (Fig. 15A). Also consistent with Scheme 4, the reciprocal of the mean active time, $1/\bar{\tau}_a = k_{on}$ [CTX], increases linearly with [CTX], and $1/\bar{\tau}_b$ is independent of [CTX] (Fig. 15B). Both $\bar{\tau}_a$ and $\bar{\tau}_b$ are essentially independent of voltage. From Fig. 15B, and assuming that CTX is only 30% pure, we (Miller *et al.*, 1985) obtained a $K_d \sim 10$ nM. Slow blockade of the type induced by CTX is also seen when Na^+ channels incorporated into planar bilayers are treated with TTX (see Chapter 14 by French *et al.*). However, the TTX blocking reaction is voltage dependent (French *et al.*, 1984; Moczydlowski *et al.*, 1984).

Although the specificity of CTX has not been characterized fully, several channels are insensitive to it. These include batrachotoxin-activated Na^+ channels from rat muscle, K^+ channels from sarcoplasmic reticulum, and K^+ channels from squid axon. However, CTX shows essentially the same blocking characteristics when directed against a maxi CaK channel from smooth muscle incorporated into bilayers (R. Latorre, unpublished results). Thus, CTX may prove to be a valuable tool in purifying and reconstituting maxi CaK channels.

7.2 Apamin Blocks an Unidentified Class of CaK Channels

Apamin is a bee venom toxin of 18 amino acids containing two disulfide bridges. In 1981, Burgess *et al.* reported that apamin was able to inhibit Ca^{2+}-induced K^+ fluxes in mammalian hepatocytes in the nanomolar range. However, the toxin is unable to inhibit the CaK channels in red blood cells. These results suggest that the CaK channel differs in these two tissues. Moreover, Vergara (1983) was unable to find any effect of apamin in the T-tubule maxi CaK channel at concentrations 100-fold larger than that used in hepatocytes.

A more detailed study of the effect of apamin on CaK channels has been done more recently by Romey and Lazdunski (1984). Using rat muscle cells in culture, they found that maxi CaK channels (Pallota and Magleby, 1981) coexist with a Ca^{2+}-activated slow outward current. Apamain has no effect on the maxi CaK channels but completely abolishes the slow outward current and the long-lasting afterhyperpolarization that follows an action potential. Curiously, Romey and Lazdunski (1984) were unable to record by means of the patch-clamp technique the activity of these apamin-sensitive channels. I find this a puzzling observation, since the density of apamin receptors is 7000 per cell, which, although low, is not lower than the density of red blood cell CaK channels, which have already been characterized. There are at least three explanations for this result: (1) the channel conductance is too low to be recorded; (2) since the experiments were made in excised patches, it is possible that this channel needs another modulator besides Ca^{2+} to open; and (3) the putative channel is not a channel but a carrier.

8. Phospholipid Surface Charge Effects

One of the main advantages of incorporating a biological channel into artificial membranes is that it opens the possibility of studying the effect of lipid composition on the different channel properties. In particular, in order to gain insight into the role of phospholipid surface charge on T-tubule maxi CaK channel conduction and gating, we incorporated channels into PE and PS membranes (Moczydlowski *et al.*, 1985). The fixed charge of PS produces a negative electrostatic surface potential in the aqueous phase adjacent to the membrane. This negative surface potential causes the local accumulation of cations and depletion of anions in solution, which in turn results in a diffuse double layer at equilibrium (reviewed by McLaughlin, 1977).

In the absence of any specific effects of the phospholipid membrane composition, we expect that the accumulation of K^+ and Ca^{2+} near the surface of the PS membrane will increase channel conductance and shift the curve of probability of opening versus voltage towards more negative voltages. However, biological channels are proteins of an appreciable size, which can protrude into the aqueous phases, leaving their ends well insulated from the lipid bilayer surface. Therefore, the fraction of the surface potential that the conduction system and the Ca^{2+} sites "sense" depends on how far from the bilayer surface those mechanisms are.

We found that at low K^+ concentrations, where the Debye length is large, the channel conductance is higher in PS membranes but approached the same conductance observed in PE at large K^+ concentrations. On the other hand, at the same Ca^{2+} concentration and voltage, the probability of opening is significantly greater in PS than PE. Using Gouy–Chapman–Stern theory, a theory that has been well tested in bilayers (McLaughlin *et al.*, 1981), we calculated the concentrations of K^+ and Ca^{2+} at various distances from the PS membrane surface. These calculations suggest that both the Ca^{2+} site(s) and the conduction machinery sense a similar fraction of the surface potential. This fraction is equivalent to the local electrostatic potential at 0.9 nm away from the surface of the PS bilayer. This may mean that gating and conduction machineries are located in the same domain of the channel protein. This hypothesis is consistent with the effects of external K^+ on sodium blockade and on the changes of probability of opening induced by Na^+ (Marty, 1983b). In the absence of structural information about the channel protein, these conclusions are, of course, highly speculative.

9. A Personal View of Maxi CaK Channels

I look on the maxi CaK channels as a large integral membrane protein with Ca^{2+} sites and conduction system somewhat removed from the lipid bilayer surface. The conduction process takes place in a short tunnel. Based on the QA ion blockade characteristics, it should have a maximum internal diameter smaller than TMA (~0.6 nm). This tunnel is predicted not to have a diameter of less than 0.5 nm; otherwise its convergence resistance would be too large to account for the large channel conductance. The tunnel possesses, at both ends, large mouths that make the voltage fall in a very short distance, and conduction takes place in a multiion occupancy mode. These characteristics allow the channel to attain its large conductance. Local concentration of K^+ can also be increased by the presence of fixed charges or dipoles at the channel mouths. Given the different natures of *cis* and *trans* QA ion blockade, the channel mouths are different from a structural point of view.

The outer part of the tunnel lets K^+, Rb^+, and Cs^+ enter but not Na^+. The inner part lets all these cations enter. As pointed out by Yellen (1984a), these conduction properties may imply that the selectivity filter is not located in a single identifiable place. However, Ba^{2+} appears to bind to a single site. For the delayed rectifier, it has been proposed that Ba^{2+}, because of its size ($r_{Ba^{2+}} = 0.135$ nm; $r_{K^+} = 0.133$ nm), is able to reach the selectivity filter and, because of its divalent nature, bind very tightly to it.

At least two basic models are able to explain the channel voltage dependence. In one, all the voltage dependence arises from a voltage-dependent Ca^{2+}-binding reaction, and in the other, only the close-open transitions are voltage dependent (e.g., Methfessel and Boheim, 1982). In this latter case, one can imagine that Ca^{2+} binding (voltage independent) modulates the number of gating charges available to be displaced by the applied electric field. Given the complexity of channel kinetics, it is difficult at present to say with any certainty which is the origin of the voltage-dependent process.

ACKNOWLEDGMENTS. I am grateful to Drs. O. Alvarez, X. Cecchi, D. Wolff, and C. Vergara for sharing with me the love for Ca^{2+}-activated K^+ channels. This work was supported in part by the Fondo Nacional de Investigacion Cientifica, Proyecto 1299, the Departmento de Desarrollo de la Investigacion, University of Chile, Proyecto B-1985-8413, and by a grant of The Tinker Foundation, Inc. of New York to the Centro de Estudios Cientificos de Santiago.

References

Andersen, O. S., 1982, Ion movement through gramicidin A channels: Studies on the diffusion-controlled association step, *Biophys. J.* **41**:147–165.

Andersen, O. S., and Procopio, J., 1980, Ion movement through gramicidin A channels: On the importance of aqueous diffusion resistance and ion–water interactions, *Acta Physiol. Scand. Suppl.* **481**:27–35.

Armstrong, C., 1975, Ionic pores, gates and gating currents, *Q. Rev. Biophys.* **1**:179–210.

Armstrong, C., and Hille, B., 1972, The inner quaternary ammonium ion receptor in potassium channel of the node of Ranvier, *J. Gen. Physiol.* **59**:388–400.

Armstrong, C., Swenson, R. P., and Taylor, S. R., 1982, Block of the squid axon K^+ channels by internally and externally applied Ba^{2+} ions, *J. Gen. Physiol.* **80**:663–682.

Barrett, J. N., Barrett, E. F., and Dribin, L. B., 1981, Calcium-dependent slow potassium conductance in rat skeletal myotubes, *Devl. Biol.* **82**:258–266.

Barrett, J. N., Magleby, K. L., and Pallota, B. S., 1982, Properties of single calcium-activated potassium channels in cultured rat muscle, *J. Physiol.* (*Lond.*) **331**:211–230.

Bezanilla, F., and Armstrong, C., 1972, Negative conductance caused by entry of sodium and cesium ions into the potassium channels of the squid axon, *J. Gen. Physiol.* **60**:588–608.

Blatz, A. I., and Magleby, K. L., 1984, Ion conductance and selectivity of single calcium-activated potassium channels in cultured rat muscle, *J. Gen. Physiol.* **84:**1–23.

Burgess, G. M., Claret, M., and Jenkinson, D. H., 1981, Effects of quinine and apamin on the calcium-dependent potassium permeability of mammalian hepatocytes and red cells, *J. Physiol.* (*Lond.*) **317:**67–90.

Colqhoun, D., Neher, E., Reuter, H., and Stevens, C. F., 1981, Inward current channels activated by intracellular Ca in cultured cardiac cells, *Nature* **294:**752–754.

Cook, D. L., Ikeuchi, M., and Fuyimoto, W. Y., 1984, Lowering of pH_1 inhibits Ca^{2+}-activated K^+ channels in pancreatic B-cells, *Nature* **311:**269–271.

Coronado, R., and Latorre, R., 1983, Phospholipid bilayers made from monolayers on patch-clamp pipettes, *Biophys. J.* **43:**231–236.

Coronado, R., Rosenberg, R., and Miller, C., 1980, Ionic selectivity, saturation and block in a K^+ selective channel from sarcoplasmic reticulum, *J. Gen. Physiol.* **76:**425–446.

De Peyer, J. E., Cachelin, A. B., Levitan, I. B., and Reuter, H., 1982, Ca^{2+}-activated K^+ conductance in internally perfused snail neurones is inhanced by protein phosphorylation, *Proc. Natl. Acad. Sci. U.S.A.* **79:**4207–4211.

Eisenman, G., and Horn, R., 1983, Ionic selectivity revisited: The role of kinetic and equilibrium processes in ion permeability through channels, *J. Membr. Biol.* **76:**197–225.

Ewald, D. A., Williams, A., and Levitan, I. B., 1985, Modulation of single Ca^{2+}-dependent K^+-channel activity by protein phosphorylation, *Nature* **315:**503–506.

Fink, R., and Luttgau, H. C., 1976, An evaluation of the membrane constants and the potassium conductance in metabolically exhausted fibres, *J. Physiol.* (*Lond.*) **263:**215–238.

Fitzhugh, R., 1983, Statistical properties of the asymmetric random signal, with applications to single-channel analysis, *Math. Biosci.* **64:**75–89.

French, R. J., and Shoukimas, J. J., 1981, Blockage of squid axon potassium conductance by internal tetra-n-alkylammonium ions of various sizes, *Biophys. J.* **34:**271–292.

French, R. J., Worley, J. F., and Kreuger, B. K., 1984, Block by saxitoxin of sodium channels incorporated into planar bilayers, *Biophys. J.* **45:**301–310.

Gallin, K., 1984, Calcium and voltage-activated potassium channels in human macrophages, *Biophys. J.* **46:**821–825.

Gárdos, G., 1958, The function of calcium in the potassium permeability of human erythrocytes, *Biochim. Biophys. Acta* **30:**653–654.

Goldman, D. E., 1943, Potential, impedance and rectification in membranes, *J. Gen. Physiol.* **27:**37–60.

Gorman, A. L. F., Woolum, J. C., and Cornwall, M. C., 1982, Selectivity of Ca^{2+}-activated and light-dependent K^+ channels for monovalent cations, *Biophys. J.* **38:**319–322.

Gration, K. A. F., Lambert, J. J., Ramsey, R., and Usherwood, P. N. R., 1981, Non-random openings and concentration-dependent lifetimes of glutamate-gated channels in muscle membranes, *Nature* **291:**423–425.

Grygorczyk, R., and Schwarz, W., 1983, Properties of the Ca^{2+}-activated K^+ conductance of human red cells as revealed by the patch-clamp technique, *Cell Calcium* **4:**499–510.

Hagglund, J., Enos, B., and Eisenman, G., 1979, Multi-site, multi-barrier, multi-occupancy models for the electrical behavior of single filing channels like those of gramicidin, *Brain Res. Bull.* **4:**154–158.

Hagglund, J., Eisenman, G., and Sandblom, J. P., 1984, Single-salt behavior of a symmetrical 4-site channel with barriers at its middle and ends, *Bull. Math. Biol.* **46:**41–80.

Hagiwara, S., and Takahashi, K., 1974, The anomalous rectification and cation selectivity of the membrane of a starfish egg cell, *J. Membr. Biol.* **18:**61–80.

Hamill, O. P., 1983, Potassium and chloride channels in red blood cells, in: *Single Channel Recording* (B. Sackmann and E. Neher, eds.), pp. 451–471, Plenum Press, New York.

Hidalgo, C., Donoso, P., Liberona, J. L., Parra, C., Riquelme, G., and Jaimovich, E., 1985, Properties of transverse tubule membranes isolated from frog muscle, *Biophys. J.* **47:**58a.

Hille, B., 1973, Potassium channels in myelinated nerve: Selectivity to small cations, *J. Gen. Physiol.* **61:**669–686.

Hille, B., 1975, Ionic selectivity of Na and K channels in nerve membranes, in: *Membranes: A Series of Advances*, Vol. 3 (G. Eisenman, ed.), pp. 255–323, Marcel Dekker, New York.

Hille, B., 1984, *Ionic Channels of Excitable Membranes*, Sinauer, Sunderland, Massachusetts.

Hille, B., and Schwarz, W., 1978, Potassium channels as multi-ion single-file pores, *J. Gen. Physiol.* **72**:409–442.

Hodgkin, A. L. and Huxley, A. F., 1952, Currents carried by sodium and potassium ions through the membrane of the giant axon of *Loligo, J. Physiol.* (*Lond.*) **116**:449–472.

Hodgkin, A. L., and Keynes, R. D., 1955, The potassium permeability of a giant nerve fiber, *J. Physiol.* (*Lond.*) **128**:61–88.

Hunter, M., Lopes, A. G., Boulpaep, E. L., and Giebisch, G. H., 1984, Single channel recording of calcium-activated potassium channels in the apical membrane of rabbit cortical collecting tubules, *Proc. Natl. Acad. Sci. U.S.A.* **81**:4237–4239.

Krueger, B. K., French, R. J., Blaustein, M. B., and Worley, J. F., 1982, Incorporation of Ca^{2+}-activated K^+ channels, from rat brain, into planar bilayers, *Biophys. J.* **37**:170a.

Latorre, R., and Miller, C., 1983, Conduction and selectivity in potassium channels, *J. Membr. Biol.* **71**:11–30.

Latorre, R., Vergara, C., and Hidalgo, C., 1982 Reconstitution in planar lipid bilayers of a Ca^{2+}-dependent Ka^+ channel from transverse tubule membranes isolated from rabbit skeletal muscle, *Proc. Natl. Acad. Sci. U.S.A.* **77**:7484–7486.

Latorre, R., Vergara, C., and Moczydlowski, E., 1983, Properties of a Ca^{2+}-activated K^+ channel in a reconstituted system, *Cell Calcium* **4**:343–357.

Latorre, R., Alvarez, O., Cecchi, X., and Vergara, C., 1985, Properties of reconstituted ion channels, *Annu. Rev. Biophys. Chem.* **14**:79–111.

Läuger, P., 1973, Ion transport through pores: A rate theory analysis, *Biochim. Biophys. Acta.* **311**:423–441.

Läger, P., 1976, Diffusion-limited ion flow through pores, *Biochim. Biophys. Acta* **455**:493–509.

Lux, H. D., Neher, E., and Marty, A., 1981, Single-channel activity associated with calcium dependent outward current in *Helix pomatia, Pfluegers Arch.* **389**:293–295.

Magleby, K. L., and McManus, O. B., 1985, Adjacent state analysis of the large conductance calcium-activated potassium channel in cultured rat muscle indicates that open states with short mean lifetimes are adjacent to shut states with long mean lifetimes and open states with long mean lifetimes are adjacent to shut states with short mean lifetimes, *Biophys. J.* **47**:1347a.

Magleby, K. L., and Pallota, B. S., 1983a, Calcium dependence of open and shut interval distributions from calcium-activated potassium channels in culture rat muscle, *J. Physiol. (Lond.)* **344**:585–604.

Magleby, K. L., and Pallota, B. S., 1983b, Burst kinetics of single calcium-activated potassium channels in culture rat muscle, *J. Physiol.* (*Lond.*) **344**:605–623.

Marty, A., 1981, Ca-dependent K channels with large unitary conductance in chromaffin cell membranes, *Nature* **291**:497–500.

Marty, A., 1983a, Ca^{2+}-dependent Ka^+ channels with large unitary conductance, *Trends Neurochem.* **6**:262–265.

Marty, A., 1983b, Blocking of large unitary calcium-dependent potassium currents, *Pfluegers Arch.* **396**:179–181.

Maruyama, Y., and Petersen, O. H., 1982, Cholecystokinin activation of single-channel currents is mediated by internal messenger in pancreatic acinar cells, *Nature* **300**:61–63.

Maruyama, Y., Gallagher, D. V., and Petersen, O. H., 1983, Voltage and calcium-activated potassium channels in baso-lateral acinar cell membranes of mammalian salivary glands, *Nature* **302**:827–829.

McLaughlin, S., 1977, Electrostatic potentials at membrane–solution interfaces, *Curr. Top. Membr. Transport* **9**:71–144.

McLaughlin, S., Mulrine, G. A. N., Gresalfi, T., Vaio, G., and McLaughlin, A., 1981, Adsorption of divalent cations to bilayer membranes containing phosphatidylserine, *J. Gen. Physiol.* **77**:445–473.

McManus, O. B., and Magleby, K. L., 1985, The large conductance calcium-activated potassium channel in cultured rat muscle has at least three open states of similar conductance and six shut states, *Biophys. J.* **47**:137a.

Meech, A. W., 1978, Calcium-dependent potassium activation in nervous tissues, *Annu. Rev. Biophys. Bioeng.* **7**:1–18.

Methfessel, C., and Boheim, G., 1982, The gating of single calcium-dependent potassium channels is described by an activation/blockade mechanism, *Biophys. Struct. Mech.* **9**:35–60.

Miller, C., 1978, Voltage-gated cation conductance channel from sarcoplasmic reticulum. Steady-state electrical properties, *J. Membr. Biol.* **40**:1–23.

Miller, C., 1982, *Bis*-quaternary ammonium blockers as structural probes of the sarcoplasmic reticulum K^+ channel, *J. Gen. Physiol.* **79**:869–891.

Miller, C., Moczydlowski, E., Latorre, R., and Phillips, M., 1985, Charybdotoxin, a protein inhibitor of single Ca^{2+}-activated K^+ channels from mammalian skeletal muscle, *Nature* **313**:316–318.

Moczydlowski, E., and Latorre, R., 1983, Gating kinetics of Ca^{2+}-activated K^+ channels from rat muscle incorporated into plan : lipid bilayers: Evidence for two voltage-dependent Ca^{2+} binding reactions, *J. Gen. Physiol.* **82**:511–542.

Moczydlowski, E., Garber, S,. and Miller, C., 1984, Batrachotoxin-activated Na^+ channels in planar lipid bilayers. Competition of tetrodotoxin block by Na^+, *J. Gen. Physiol.* **84**:665–686.

Moczydlowski, E., Alvarez, O., Vergara, C., and Latorre, R., 1985, Effect of phospholipid surface charge on the conductance and gating of a Ca^{2+}-activated K^+ channel in planar lipid bilayers, *J. Membr. Biol.* **83**:273–282.

Neher, E., and Steinback, J. H., 1978, Local anaesthetics transiently block currents through single acetylcholine-receptor channels, *J. Physiol.* (*Lond.*) **277**:153–176.

Pallota, B. S., and Magleby, L., 1981, Single channel recordings of Ca^{2+}-activated K^+ currents in rat muscle cell culture, *Nature* **293**:471–474.

Passow, H., 1983, Ca^{2+}-activated channels in erythrocytes and excitable cells, *Annu. Rev. Physiol.* **45**:359–374.

Petersen, O. H., and Maruyama, Y., 1984, Calcium-activated potassium channels and their role in secretion, *Nature.* **307**:693–696.

Reuter, H., and Stevens, C. F., 1980, Ion conductance and ion selectivity of potassium channels in snail neurons, *J. Membr. Biol.* **57**:103–118.

Romey, G., and Lazdunski, M., 1984, The coexistence in rat muscle cells of two distinct classes of Ca^{2+}-dependent K^+ channels with different pharmacological properties and different physiological functions, *Biochem. Biophys. Res. Commun.* **118**:669–674.

Rosemblatt, M., Hidalgo, C., Vergara, C., and Ikemoto, N., 1981, Immunological and biochemical properties of transverse tubule membranes isolated from rabbit skeletal muscle, *J. Biol. Chem.* **256**:8140–8148.

Tank, D. W., Miller, C., and Webb, W. W., 1982, Isolated-patch recordings from liposomes containing functionally reconstituted chloride channels from *Torpedo* electroplax, *Proc. Natl. Acad. Sci. U.S.A.* **79**:7749–7753.

Vergara, C., 1983, Characterization of a Ca^{2+}-Activated K^+ Channel from Skeletal Muscle Membranes in Artificial Bilayers, Ph.D. Dissertation, Harvard University, Cambridge, Massachusetts.

Vergara, C., and Latorre, R., 1983, Kinetics of Ca^{2+}-activated K^+ channels from rabbit muscle incorporated into planar bilayers: Evidence for a Ca^{2+} and Ba^{2+} blockade, *J. Gen. Physiol.* **82**:543–568.

Vergara, C., Moczydlowski, E., and Latorre, R., 1984, Conduction, blockade and gating in a Ca^{2+}-activated K^+ channel incorporated into planar bilayers, *Biophys. J.* **45**:73–76.

Walsh, J. V., and Singer, J. J., 1983, Ca^{++}-activated K^+ channels in vertebrate smooth muscle cells, *Cell Calcium* **4**:321–330.

White, M., and Miller, C., 1979, A voltage-gated anion channel from the electric organ of *Torpedo californía*, *J. Biol. Chem.* **254**:10161–10166.

Wibo, M., Delfosse, I., and Godfraind, J., 1982, Sidedness of plasma membrane vesicles insolated from intestinal smooth muscle, *Arch. Int. Pharmacodyn. Ther.* **258**:331–332.

Wolff, D., Cecchi, X., Naranjo, D., Alvarez, O., and Latorre, R., 1985, Cation selectivity and Cs^+ blockade in Ca^{2+}-activated channels from rabbit intestinal smooth muscle, *Biophys. J.* **47**:386a.

Wong, B. S., Lecar, H., and Adler, M., 1982, Single calcium-dependent potassium channels in clonal anterior pituitary cells, *Biophys. J.* **39**:313–317.
Woodhull, A. M., 1973, Ionic blockade of sodium channels in nerve, *J. Gen. Physiol.* **61**:687–708.
Yellen, G., 1982, Single Ca^{2+}-activated non-selective cation channels in neuroblastoma, *Nature* **296**:357–359.
Yellen, G., 1984a, Ionic permeation and blockade in Ca^{2+}-activated K^+ channels of bovine chromaffin cells, *J. Gen. Physiol.* **84**:157–186.
Yellen, G., 1984b, Relief of Na^+ block of Ca^{2+}-activated K^+ channels by external cations, *J. Gen. Physiol.* **84**:187–199.

Chapter 18

THE SARCOPLASMIC RETICULUM POTASSIUM CHANNEL

Lipid Effects

JOAN BELL

1. Introduction

The conduction, gating, and pharmacological properties of the sarcoplasmic reticulum (SR) K^+ channel have been well characterized (Coronado *et al.*, 1980; Coronado and Miller, 1980; Labarca *et al.*, 1980; Miller, 1982). It is a relatively "simple" channel in that it exists in only two states, open or closed, with quite slow (seconds) transitions between the two, is gated by voltage only, and does not inactivate. It thus provides an excellent vehicle for studying possible lipid effects on a channel protein. In turn, the artificial bilayer system is the ideal means for studying lipid effects because both the membrane composition and the ionic milieu can be determined and manipulated by the experimenter. The topic of this chapter is the role of the electrical charge of the bilayer in the observed conduction characteristics of the SR K^+ channel in artificial bilayers.

Choosing lipid compositions with which to make bilayers has generally been governed by practical considerations such as ease of bilayer formation, mechanical and electrical stability, and "fusability." However, it has become increasingly obvious that the lipid composition of the bilayer can affect the gating and/or conduction parameters of both bacterial ionophores and eucaryotic ion channels (Muller and Finkelstein, 1972; Apell *et al.*, 1979; Bell and Miller, 1984). These observations suggest that the choice of bilayer lipids should be based not only on technical concerns but also on knowledge of the possible effects of specific lipids on channels. In fact, as is the case with the SR K^+ channel, these effects can be used to elucidate structural and functional characteristics of the channel.

Figure 1 shows the composition of biologically occurring phospholipids that are often used in artifical bilayers, as well as one phospholipid analogue that was used specifically to study the effects of surface charge on the SR K^+ channel. Phosphatidylethanolamine (PE) and phosphatidylcholine (PC) are zwitterions,

JOAN BELL • Department of Physiology and Biophysics, Laboratory of Human Reproduction and Reproductive Biology, Harvard Medical School, Boston, Massachusetts 02115.

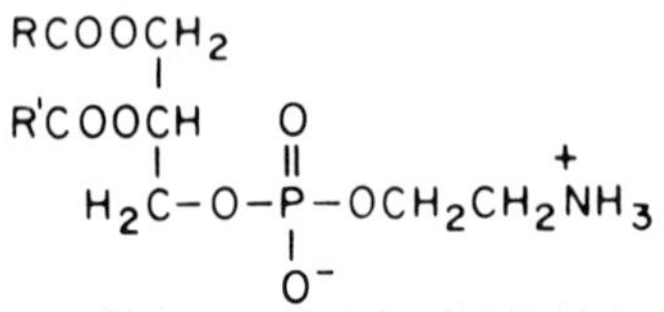

Phosphatidylethanolamine
(PE)

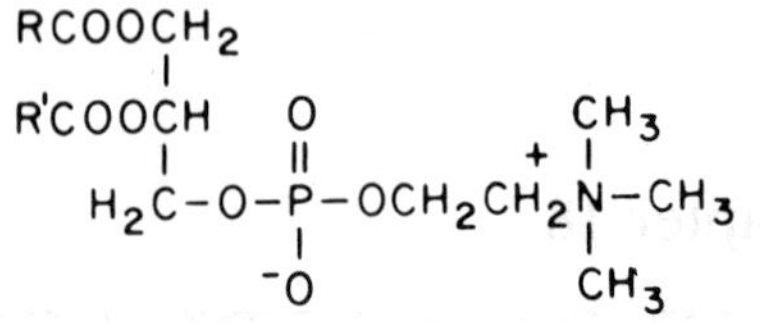

Phosphatidylcholine
(PC)

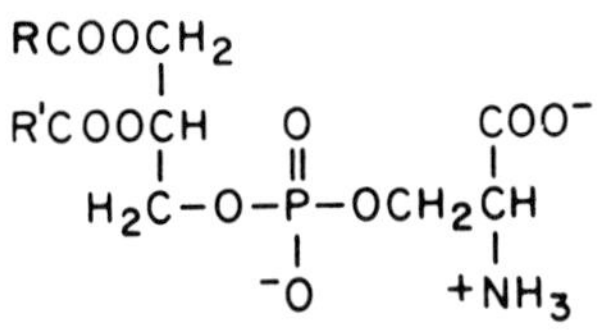

Phosphatidylserine
(PS)

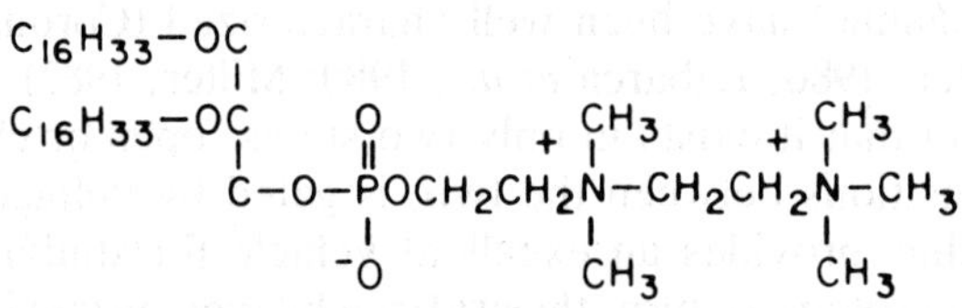

1,2-Dihexadecylphosphatidyldicholine
(DHPDC)

Figure 1. Lipid compositions. The composition of the most commonly used bilayer lipids is indicated, as is the composition of a synthetic positively charged lipid analogue, DHPDC (generously donated by Dr. O. Boheim). R and R′ represent fatty acids bound to a glycerol backbone via ester linkages. In the case of DHPDC, the aklyl side chains are linked via either bonds.

bearing no net charge, whereas phosphatidylserine (PS) has a negative charge, and the analogue, 1,2-dihexadecylphosphatidyldicholine (DHPDC) is positively charged. A bilayer composed either entirely or partially of charged phospholipids has a net "surface charge," which results in a surface potential at the membrane surface relative to the bulk solution bathing the membrane. A negatively charged surface has a surface potential that attracts cations and repels anions such that the cation concentration close to the membrane is higher and the anion concentration lower than in the bulk solution. The opposite situation occurs at positively charged surfaces. The result is that a membrane protein embedded in a charged bilayer may see ion concentrations different from those in the bulk solution, whereas a protein in a neutral membrane will see the same concentrations as those in the bulk solution. These concepts are outlined schematically in Fig. 2.

Therefore, under a given set of buffer conditions, a channel may have a higher

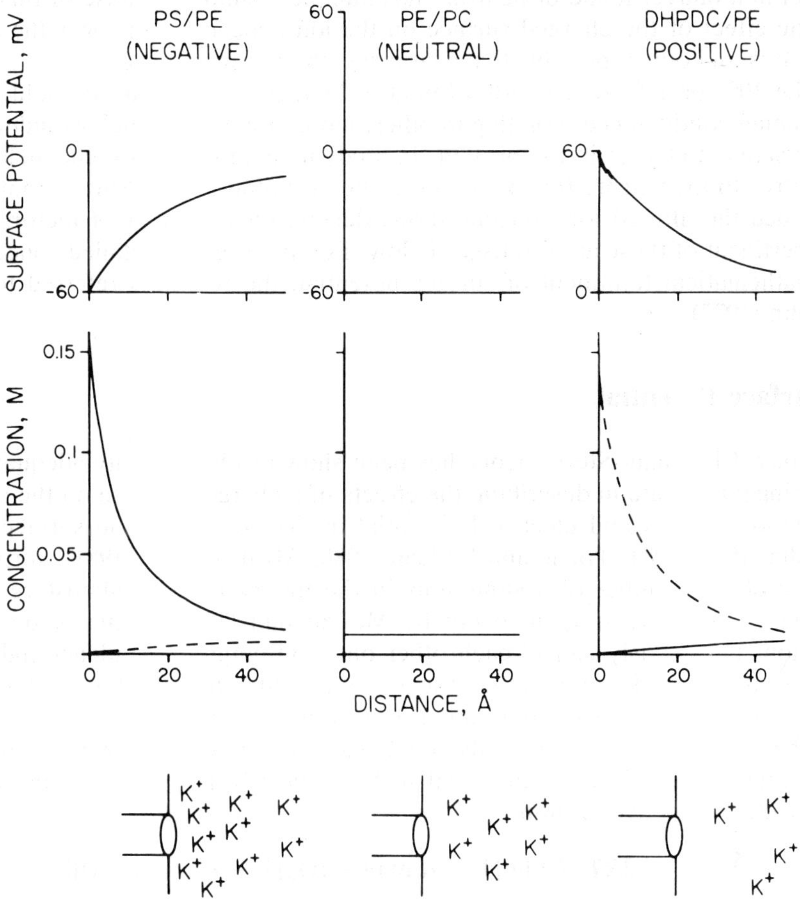

Figure 2. Effect of surface charge on the electrostatic potential and ion distributions near a phospholipid bilayer. In the PS/PE bilayer, there is a surface potential of −60 mV, which decays with distance into the solution to 0 mV. The PE/PC bilayer has no surface potential, whereas the DHPDC/PE bilayer has a positive surface potential. The cation and anion distributions with distance are shown in the second part of the figure, where the dashed line represents the cation concentration and the solid line the anion concentration. In the PE/PC membranes, the lines are superimposed because the ion concentrations are constant with distance from the surface. The curves were calculated according to Gouy–Chapman–Stern theory (see text), with a value of 1 e/700 Å^2 as the charge density in the PS/PE and DHPDC/PE bilayers. These values correspond to a 10% charge density assuming a molecular area per phospholipid of 1/70 Å^2 (Loosely-Millman, 1982). The electrolyte concentration was assumed to 0.01 M KCl with a binding constant of K^+ to PS of 0.15 M^{-1} (Eisenberg *et al.*, 1979). The bottom part of the figure is a highly schematic cartoon of a channel mouth flush with the bilayer and the K^+ levels near the mouth. It is meant to illustrate that a channel embedded in a PS/PE membrane would see a higher K^+ concentration than one in a PE/PC bilayer, whereas a channel in a DHPDC/PE membrane would see a lower concentration, even though in all cases the bulk K^+ level is the same.

or lower conductance in a bilayer made of charged phospholipids than the same channel in a bilayer made of neutral phospholipids, simply because of this electrostatic effect of the charged surface on the ion concentrations near the membrane. It is, of course, possible that a large membrane surface charge (i.e., greater than the 10% or 15% that is often found in biological membranes) could affect the channel conductance or gating in other, more direct ways such as inducing a conformational change in the protein. In studying lipid effects on channels, it is imperative to investigate this possibility. The magnitude of the surface potential and hence the altered ion concentrations depend on a number of factors. The most pertinent of these are discussed below. For an excellent detailed conceptual and mathematical treatment of surface potential, the reader is referred to McLaughlin (1977).

2. Surface Potential

Gouy–Chapman–Stern theory has been shown to be not only adequate but surprisingly accurate in describing the effects of a charged surface on the spatial distribution of ions and electrical potential in the vicinity of the surface (McLaughlin, 1970, 1983; Torrie and Valleau, 1980; Alvarez *et al.*, 1983), surprising because of the number of assumptions in the theory that are almost certainly incorrect. However, as pointed out by McLaughlin (1983), some of these assumptions fortuitously cancel each other out, and empirical evidence indicates that the others simply do not affect the accuracy with which the theory describes the electrostatic and ionic environment near a phospholipid bilayer.

The basis of the theory is the nonlinearized Poisson–Boltzman equation, which relates the change in potential with distance from the membrane to the bulk electrolyte concentration:

$$\Psi(x) = (2RT/zF)\ \ln\{[1 + \alpha \exp(-\kappa x)]/[1 - \alpha \exp(-\kappa x)]\} \tag{1}$$

where $\Psi(x)$ is the potential at distance x from the surface, R, T, F, and z have their usual meanings, and κ is

$$\kappa = \{2F^2C(\infty)/\epsilon_r\epsilon RT\}^{1/2} \tag{2}$$

Here, $C(\infty)$ is the bulk ionic strength, and ϵ_r and ϵ_o are the dielectric of water and the permittivity of free space, respectively. Also

$$\alpha = \{\exp[F\Psi(0)/2RT] - 1\}/\{\exp[F\Psi(0)/2RT] + 1\} \tag{3}$$

where $\Psi(0)$ is the surface potential.

The Grahame equation allows us to relate the surface potential to the fixed charge density:

$$\sigma = \{8C(\infty)\epsilon_r\epsilon_o RT\}^{1/2}\ \sinh[zF\Psi(0)/2RT] \tag{4}$$

where σ is the fixed charge density.

The ion concentrations are described by a Boltzmann distribution such that

$$C^+(x) = C(\infty)\exp - (zF\Psi(x)/RT) \qquad (5)$$
$$C^-(x) = C(\infty)\exp(zF\Psi(x)/RT)$$

where $C^+(x)$ and $C^-(x)$ are cation and anion concentrations at distance x.

Thus, by knowing the charge density, one can calculate the surface potential and the surface ion concentrations as well as the potential and ion concentrations at any distance from the membrane surface.

According to equation 4, the surface potential becomes smaller as the ionic strength is increased. This decrease in potential results from screening of the fixed charges by the ions in solution. The potential decreases because the distance between the fixed and mobile charges decreases, i.e., the "Debye length" ($1/\kappa$) decreases. Thus, at high ion concentrations, the surface potential is smaller, the ratio of the surface to bulk ion concentrations approaches unity, and the potential falls off more steeply with distance from the membrane than at low ion concentrations. However, in addition to screening the fixed charges, the ions in solution can actually bind to and neutralize the fixed charges, thus actually decreasing the charge density. Stern's contribution to the theory was to include a Langmuir adsorption isotherm, such that

$$\sigma_{\text{reduc}} = \sigma/[1 + K_aC_i(\infty)\exp(-F\Psi(0)/RT] \qquad (6)$$

where σ is the charge density without any binding, σ_{reduc} the reduced charge density, and K_a the association constant of the ion, i, for the fixed charge. The Stern equation thus includes both screening and binding effects on surface potential:

$$\sinh(F\Psi(0)/2RT) = \sigma/[1 + K_aC_i(\infty)\exp(-F\Psi(0)/RT][8\epsilon_r\epsilon_oRTC(\infty)]^{1/2} \qquad (7)$$

The consequences of the above relationships are multifarious. First, not only the ion concentrations but also the ionic species determine the magnitude of the surface potential and surface ion concentrations. Second, because of binding, the charge density in the membrane is no longer expected to be that of the solution used to make the bilayer. For example, a bilayer made from a solution of 65% PS/35% PE in 50 mM KCl will have an actual charge density of 45% using the experimentally determined K_a of K$^+$ to PS of 0.15 M^{-1} (Eisenberg *et al.*, 1979). This reduced charge density results in a calculated surface K$^+$ concentration of 3M rather than the 6M expected from a charge density of 65%.

Third, because of the valency term, z, in the above equations, multivalent ionsare much more potent that monovalent ions in screening surface potential. In addition, the binding constants for many multivalent ions to PS or phosphatidylglycerol (PG) are one to two orders of magnitude larger than are the monovalent binding constants. The binding constants of K$^+$, Na$^+$, and Li$^+$ to PS are 0.15, 0.6, and 0.8 M^{-1}, respectively, whereas those for Mg^{2+}, Ca^{2+}, and Ba^{2+}

are 8, 12, and 20 M^{-1}, respectively (Eisenberg *et al.*, 1979; McLaughlin *et al.*, 1981). Therefore, even trace amounts of multivalent ions are very important in determining the magnitude of the surface potential, particularly at low (<0.1 M) monovalent ion concentrations.

In fact, under most experimental conditions, the association constants of all the ions in solution (including buffers such as tris, MOPS, etc.) for the lipids in the bilayer are not known. Therefore, calculation of the surface potential from the charge density of the lipid mixture used to make the bilayer, even when one is using a single species of lipid (say, 100% PS), might not provide the most accurate value of the potential. When the true value of the surface potential is required, it is probably best to measure it directly under the conditions used and then to calculate the charge density. If at two or three ionic strengths the same charge density is calculated, then one can be fairly confident that all relevant variables are accounted for.

Fourth, negatively charged membranes will attract any species of cation in solution, *including protons*. In a bilayer containing as little as 15% PS, the surface pH will therefore be 6.2 when the bulk pH is 7.2. This observation makes it very important to understand the interactions of protons with a channel in neutral bilayers before looking at the channel in charged bilayers. For the SR K^+ channel, the pH of the bulk solution makes a large difference in the conductance in PS membranes. The data in the charged membranes could be well fit with Gouy–Chapman–Stern theory only after determining that protons competitively inhibit the K^+ conductance with a K_i of 0.5 μM and including that K_i in the calculations (see Section 4).

3. Measuring Surface Potential

In our study of phospholipid surface charge on conduction of the SR K^+ channel, we measured the surface potential by two independent means. The first

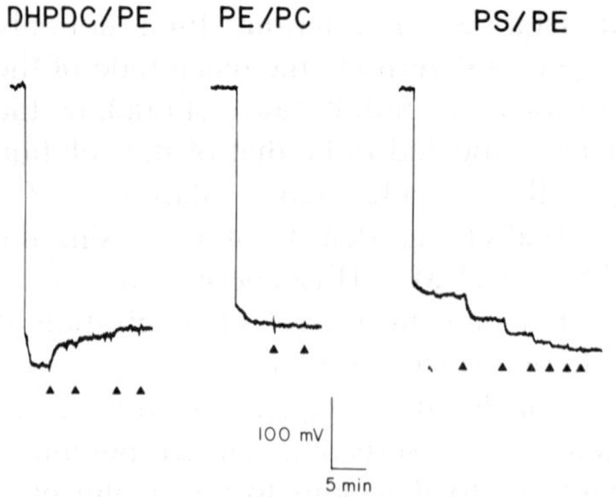

Figure 3. Measurements of potentials across phospholipid monolayers. In all traces, the large downward deflection is the change in potential (more positive) across the air–water interface when lipid is added to the surface of the solution. Thus, the potential before spreading of the monolayer is considered zero. The surface potential, Ψ(0), is the difference in potentials measured before the addition of any LiCl and after the final addition. Triangles mark successive additions of LiCl to the solution. All lipids were obtained from Avanti Biochemicals, Inc. (Birmingham, AL). Percentages refer to molar ratios in the lipid solution. DHPDC/PE: 25% DHPDC/75% PE, 50 mM K^+-gluconate, pH 7.2; final [LiCl] = 700 mM; measured Ψ(0) = −55 mV. PE/PC: 80% PE/20% PC, 50 mM K^+-gluconate, pH 7.2; final [LiCl] = 700 mM; measured Ψ(0) < −10 mV. PS/PE: 70% PS/30% PE, 100 mM K^+-gluconate, pH 7.2; final [LiCl] = 1.2 M; measured Ψ(0) = −80 mV.

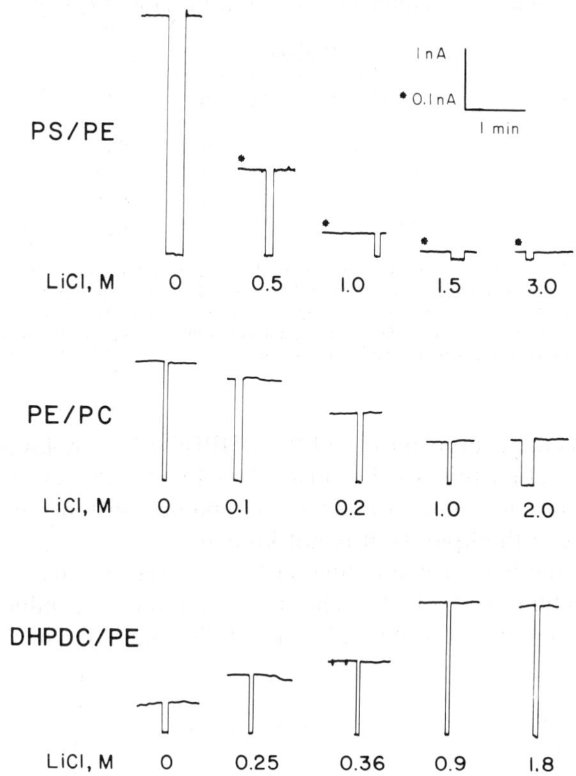

Figure 4. Nonactin currents as a function of LiCl concentration in negative, neutral, and positively charged artificial bilayers. Traces of nonactin-induced membrane currents are shown. Membranes were formed by the technique of Montal and Mueller (1972). In all cases the downward deflections are evoked by flipping a positive holding potential (+10 mV) to zero. Breaks between traces reflect the time needed to approach a new steady-state conductance after a LiCl addition. PS/PE: 70% PS/30% PE, 12 mM K$^+$-gluconate. PE/PC: 80% PE/20% PC, 22 mM K$^+$-gluconate. DHPDC/PE: 25% DHPDC/75% PE, 100 mM K$^+$-gluconate.

was to measure the potential across a monolayer of the lipid spread on the relevant buffer solution using a polonium–air electrode (Ohki and Sauve, 1978). The electrode over a lipid monolayer measures the sum of the surface and dipole potentials of the lipid. Raising the ionic strength changes the surface potential but does not affect the dipole potential; hence, the difference between the potentials measured at various low (2–200 mM) and high (1–2 M) electrolyte concentrations is taken to be the surface potential in the low-salt solution. Figure 3 shows the result of such an experiment using a PS/PE (70%/30%) lipid solution, a PE/PC (80%/20%) lipid solution, and a DHPDC/PE (25%/75%) lipid solution. In these experiments LiCl was used to neutralize the charge because it has a high affinity for PS (K_a = 0.8 M^{-1}). The measured potentials of the PS/PE and PE/PC monolayers indicated charge densities quite close to those expected (after correcting for

Table I. Determination of Surface Charge Density[a]

Lipid	Method: Monolayer	Method: Nonactin	Average	Percent[b]
80% PE/20% PC	<0.04	0.08 ± 0.02	0.06 ± 0.03	4
70% PS/30% PE	0.85 ± 0.01	0.96 ± 0.2	0.9 ± 0.2	64
15% PS/85% PE	—	0.21 ± 0.02	0.21 ± 0.02	15
25% DHPDC/75% PE	0.19 ± 0.07	0.18 ± 0.07	0.19 ± 0.05	13

[a] Measurements of surface charge are reported as charges/nm^2. Data represent mean ± S E of at least two determinations at two to five different K^+ gluconate concentrations. The charge density of the 15% PS/85% PE was measured with the nonactin method only.

[b] Apparent percentage of charged lipid in the monolayer or bilayer calculated from the measured charge density assuming a molecular area of 0.7 nm^2.

binding), whereas the charge density of the DHPDC/PE monolayers was only half that expected. Whether this was because of binding by Cl^- or gluconate$^-$ to the lipid (which was not accounted for since the binding constants are not known) or simply impurities in the lipid itself is not known.

The second method of estimating surface charge density was the nonactin method (McLaughlin *et al.*, 1970). The nonactin-induced conductance in a lipid bilayer depends on surface potential in the following way (in the limit of zero voltage):

$$\Psi_0 = -RT/F \ln(G^*/G_0)$$

where G^* is the membrane conductance induced by the nonactin–K^+ complex before the addition of inert electrolyte, in this case LiCl, and G_0 is the membrane conductance after the surface potential has been neutralized by adding high concentrations of LiCl. These experiments are shown in Fig. 4. In the PS/PE membranes, the K^+ conductance decreases as the LiCl causes a decrease in the negative surface potential and hence a decrease in surface K^+ concentration. In PE/PC membranes, although no affect of the LiCl addition is expected, there is a slight decrease in the conductance as LiCl is added. In the DHPDC/PE membranes, the LiCl causes an increase in the conductance as the resulting decrease in positive surface potential allows an increase in the surface K^+ concentration. Table I compiles the results from the two different methods, and it is apparent that the agreement is quite good.

4. Effect of Surface Charge on the Conductance of the SR K^+ Channel

Once we knew the surface potential and charge density, it was then reasonable to investigate and attempt to quantify the effect of surface charge, if any, on the channel. Figure 5 plots the single-channel conductance of the SR K^+ channel versus the K^+ concentration in neutral and negatively charged bilayers at pH 7.2

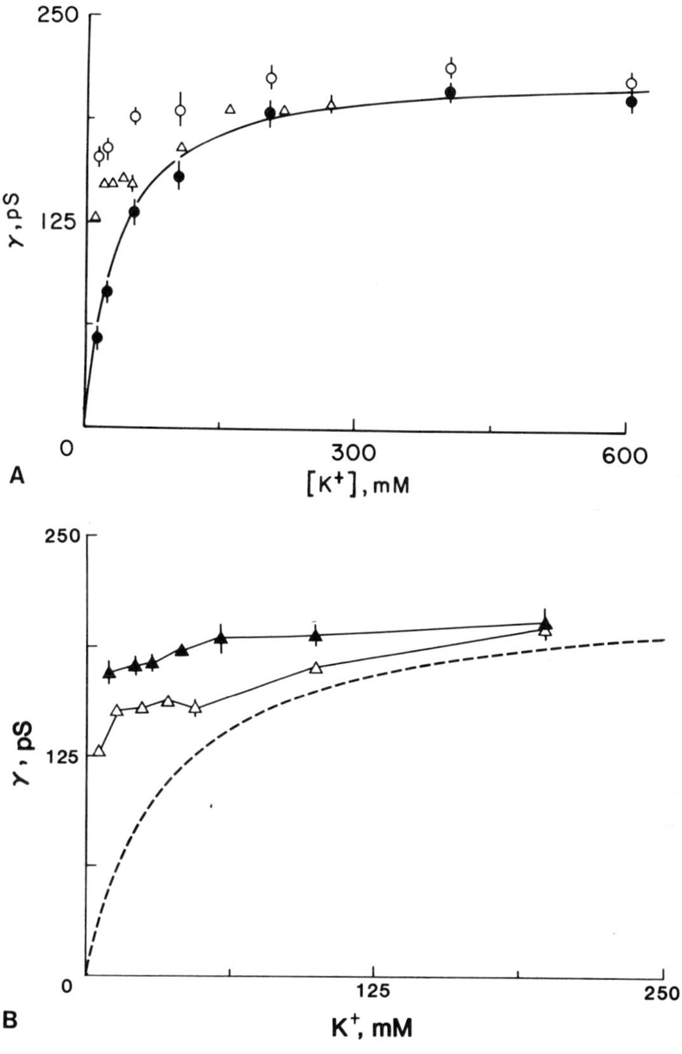

Figure 5. Single-channel conductance as a function of K$^+$ concentration and pH in neutral and negatively charged membranes. Sarcoplasmic reticulum vesicles were prepared according to Miller and Rosenberg (1979) and fused into bilayers of the desired lipid composition. A: Single-channel conductance in bilayers made from PE/PC (filled circles), 70% PS/30% PE (open circles), and 15% PS/85% PE (open triangles). The pH in all cases was 7.2. The solid line is a rectangular hyperbola drawn with a K_M of 40 mM and $V_{\max}$ of 220 pS. B: Single-channel conductance in bilayers made from 15% PS/85% PE at pH 7.2 (open triangles) and pH 8.5 (closed triangles). Dashed line indicates conductances expected in a neutral membrane. C: Single-channel conductance in bilayers made from 70% PS/30% PE at pH 7.2 (open circles) and pH 8.5 (closed circles). Dashed line has same meaning as in B.

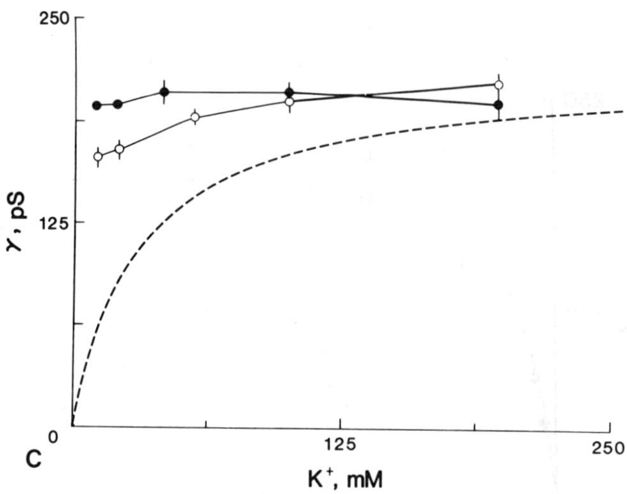

Figure 5. (*continued*)

and pH 8.5. As previously shown, in neutral membranes the conductance obeys Michaelis–Menten kinetics with a K_M for K^+ for 40 mM and V_{max} of 220 pS (Coronado *et al.*, 1980). However, in membranes containing both 15% and 65% negative charge, the single-channel conductance deviated from such simple kinetics at both pHs. Although the maximum conductances were the same as in neutral membranes, the conductances at lower K^+ concentrations (<0.15 M) were higher than in neutral membranes. In addition, the conductances in the negatively charged membranes were higher at pH 8.5 than at pH 7.2. The increase in conductance when the pH was raised was considerably larger than that expected (and observed) in a neutral bilayer considering the measured K_i for protons of 0.5 μM.

Because bilayers containing the positively charged lipid DHPDC were extremely difficult to make, data at only two K^+ concentrations were obtained from channels incorporated into these membranes. At 100 mM K^+, the conductances in PE/PC and DHPDC/PE membranes were not different, whereas at 50 mM K^+, the conductance was less in the positively charged membrane (data not shown). This behavior is precisely that expected from an electrostatic effect of the charged lipid on the K^+ concentration at the membrane surface. In PS-containing membranes at low K^+ concentrations, the surface K^+ remains high, and thus the conductance remains high, because of the surface potential. As the bulk K^+ concentration is raised, the conductance becomes independent of the surface (and bulk) K^+ concentration because the conductance has saturated.

The pH effect is also consistent with an electrostatic mechanism. In a PS membrane, the proton concentration as well as the K^+ concentration at the surface of the membrane will be higher than that in the bulk solution. The high surface proton concentration tends to decrease the conductance, whereas the high K^+ concentration tends to increase it. Raising the bulk pH causes a decrease in the-proton concentration near the channel mouth but does not affect the K^+ con-

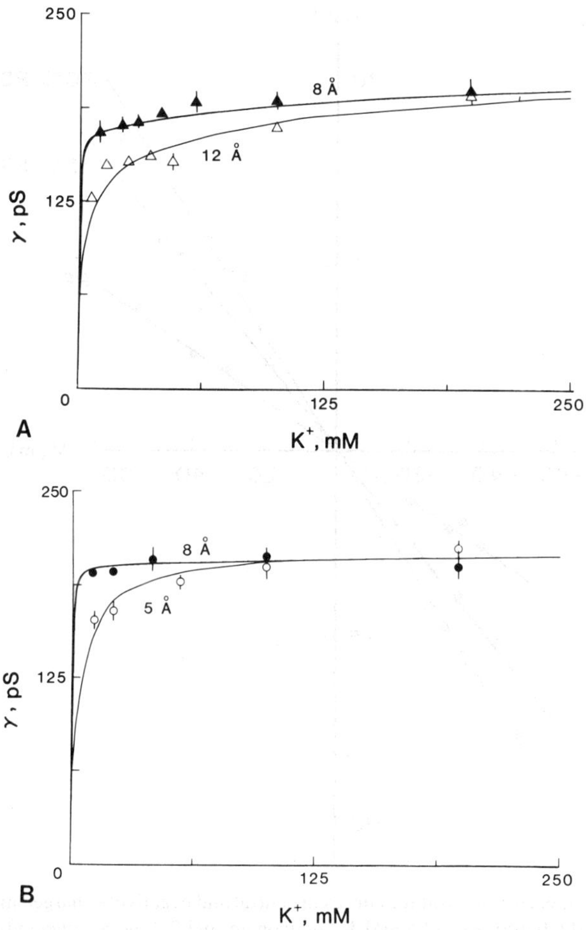

Figure 6. Single-channel conductance versus K^+ concentration and Gouy–Chapman–Stern theory. Data are from Fig. 5 and are fit with Gouy–Chapman–Stern theory assuming the channel mouth is isolated the indicated distance from the bilayer surface. A: Conductances in 15% PS/85% PE bilayers at pH 7.2 (open triangles) and 8.5 (closed triangles). B: Conductances in 70% PS/30% PE bilayers at pH 7.2 (open circles) and 8.5 (closed circles).

centration, resulting in the observed larger conductance at high pH. The effect is greater than that expected in a neutral membrane because at pH 7.2, the surface pH in the charged membranes is close to or lower than the pK_i. Raising the pH thus makes a considerable difference in the amount of inhibition expected.

Another explanation for the results is that the surface charge has somehow altered the structure of the open channel, resulting in the aberrant conductances. One control for an experiment of this type would be to add high concentrations of an inert electrolyte to neutralize the surface potential and see if the conductance approaches that in a neutral bilayer. Unfortunately, for the SR K^+ channel there does not seem to be such a cation. Of the 20 cations tested, all either permeate

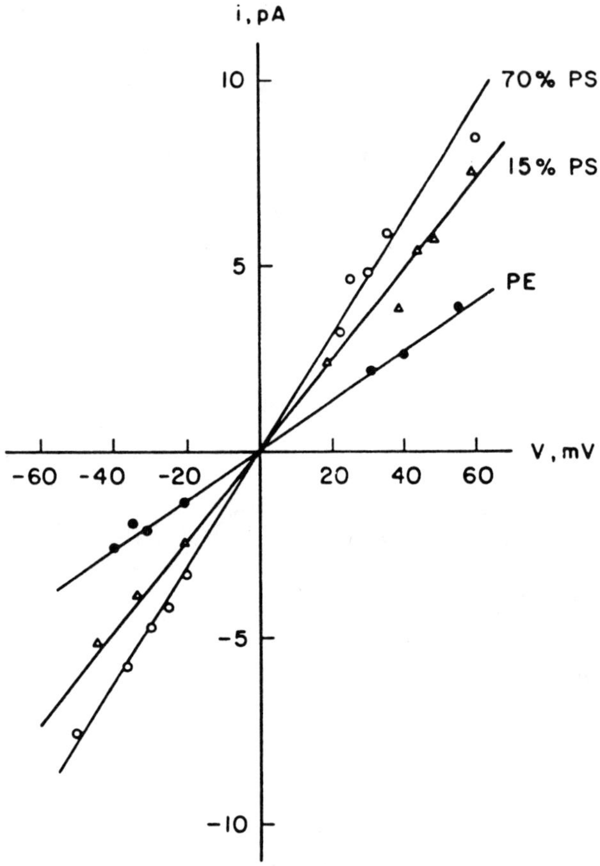

Figure 7. Open-channel current–voltage curves in neutral and negatively charged membranes. Records of channel current fluctuations in 12.5 mM K^+-gluconate, pH 7.2, were collected at different holding voltages.

or block the channel (Coronado and Miller, 1982). However, because the Na^+, K^+, and Li^+ maximum conductances are identical in PE/PC and PS/PE membranes, and because the Na^+/K^+ permeability ratio is the same in the two membranes, it appears that the channel itself is not altered by the presence of surface charge (data not shown) (Bell and Miller, 1984). The role of surface potential in increasing the single-channel conductance of the SR K^+ channel at low K^+ concentrations is therefore most easily explained by an electrostatic effect of the lipids on the ion concentrations, notably K^+ and H^+, near the mouth of the channel.

If the effect really is electrostatic, then the data should fit the conductances expected from Gouy–Chapman–Stern theory. However, examination of the data reveals that the effect of PS is less than that expected if the channel were sensing the K^+ concentration immediately adjacent to the bilayer surface. In the 65% PS membrane, the theory predicts that the conductances at both pH 7.2 and 8.5 remain saturated over the entire K^+ range tested. This prediction is also true for the conductances in the 15% PS membranes at pH 8.5, whereas at pH 7.2 the conductances at low K^+ are expected to drop only slightly. In both types of membranes at both pHs, the conductances at low K^+ decrease more than is expected from the theory. A reasonable explanation for this deviation is that the channel protein extends some distance into the aqueous solution on either side of the membrane. As seen in Fig. 2, a channel entryway that is flush with the membrane surface will sample the ionic environment immediately adjacent to the surface. However, if the protein either extends into the solution some distance x or is surrounded by a neutral annulus of radius x, then the channel will not experience the surface ion concentrations but the concentrations at that distance. The theory allows us to calculate the K^+ and H^+ concentrations at any distance from the bilayer, so we can determine this distance of isolation from the bilayer surface. Figure 6 shows that the data are fit quite well by assuming a distance of approximately 10 Å between the channel mouth and the surface of the lipid bilayer. This distance appears to be symmetrical on either side of the membrane as seen from the single-channel current–voltage curve in Fig. 7. If the channel were not symmetrical, the two sides of the channel would sense different K^+ concentrations, and the current–voltage curve would not be symmetrical around zero.

Although a characteristic distance of isolation may be obtained in this fashion, no geometrical conclusions about the mouth of the channel can be made. As mentioned above, many structural configurations could result in the same degree of isolation of the channel mouth from the surface ion concentrations of the bilayer. With these techniques, more precise elucidation of channel structure is simply not possible.

5. Surface Charge and Ion Channels

We have shown that the conduction mechanism of the SR K^+ channel can sense the surface potential of the surrounding membrane in a manner consistent with a purely electrostatic effect of the lipids on the local ion concentrations as described by Gouy–Chapman–Stern theory. The results indicate that both ends of the channel are close to but not flush with the membrane surface. It is clear that this essentially structural conclusion is based on channel activity under specific conditions. These results are entirely consistent with other structural inferences that have been made about the channel on the basis of blocker interactions, streaming potentials, and theoretical considerations of the channel conduction characteristics (Coronado and Miller, 1982; Miller, 1982; Latorre and Miller, 1983). Until more direct structural assays are available, these types of experiments seem to provide a reasonable means for exploring functional–structural relationships.

References

Alvarez, O. M., Brodwick, M., Latorre, R., McLaughlin, A., McLaughlin, S., and Szabo, G., 1983, Large divalent cations and the electrostatic potential adjacent to phospholipid bilayer membranes, *Biophys. J.* **44**:333–350.

Apell, H. J., Bamberg, E., and Läuger, P., 1979, Effects of surface charge on the conductance of the gramicidin channel, *Biochim. Biophys. Acta* **552**:369–378.

Bell, J., and Miller, C., 1984, Effects of phospholipid surface charge on ion conduction in the K^+ channel of sarcoplasmic reticulum, *Biophys. J.* **45**:279–287.

Coronado, R., and Miller, C., 1980, Decamethonium and hexamethonium block K^+ channels of sarcoplasmic reticulum, *Nature* **288**:495–497.

Coronado, R., and Miller, C., 1982, Conduction and block by organic cations in a K^+-selective channel from sarcoplasmic reticulum incorporated into planar phospholipid bilayers, *J. Gen. Physiol.* **79**:529–547.

Coronado, R., Rosenberg, R., and Miller, C., 1980, Ionic selectivity, saturation, and block in a K^+-selective channel from sarcoplasmic reticulum, *J. Gen. Physiol.* **76**:425–446.

Eisenberg, M., Gresalfi, T., Riccio, T., and McLaughlin, S., 1979, Adsorption of monovalent cations to bilayer membranes containing negative phospholipids, *Biochemistry* **18**:5213–5223.

Labarca, P., Coronado, R., and Miller, C., 1980, Thermodynamic and kinetic studies of the gating behavior of a K-selective channel from the sarcoplasmic reticulum membrane, *J. Gen. Physiol.* **76**:396–424.

Latorre, R., and Miller, C., 1983, Conduction and selectivity in potassium channels, *J. Membr. Biol.* **71**:11–30.

Loosley-Millman, M. E., Rand, R. P., and Parsegian, V. A., 1982, Effects of monovalent ion binding and screening and measured electrostatic forces between charged phospholipid bilayers, *Biophys. J.* **40**:221–232.

McLaughlin, S., 1977, Electrostatic potentials at membrane–solution interfaces, *Curr. Top. Membr. Transport* **9**:71–144.

McLaughlin, S., 1983, Experimental test of the assumptions inherent in the Gouy–Chapman–Stern theory of the diffuse double layer, in: *Physical Chemistry of Transmembrane Ion Motions* (G. Spach, ed.), pp. 69–76, Elsevier, Amsterdam.

McLaughlin, S., Szabo, G., Eisenman, G., and Ciani, S. M., 1970, Surface charge and the conductance of phospholipid membranes, *Proc. Natl. Acad. Sci. U.S.A.* **67**:1268–1275.

McLaughlin, S., Mulrine, N., Gresalfi, T., Vaio, G., and McLaughlin, A., 1978, Adsorption of divalent cations to bilayer membranes containing phosphatidylserine, **77**:445–473.

McLaughlin, S., Mulrine, N., Gresalfi, T., Vaio, G., and McLaughlin, A., 1981, Adsorption of divalent cations to bilayer membranes containing phosphatidylserine, *J. Gen. Phys.* **77**:445–473.

Miller, C., 1982, Coupling of water and ion fluxes in a K^+ channel from sarcoplasmic reticulum, *Biophys. J.* 227–230.

Miller, C., and Rosenberg, R., 1979, A voltage-gated cation conductance channel from sarcoplasmic reticulum. Effects of transition metal ions, *Biochemistry* **18**:1138–1145.

Montal, M., and Mueller, P., 1972, Formation of bimolecular membranes from lipid monolayers and a study of their electrical properties, *Proc. Natl. Acad. U.S.A.* **69**:3561–3566.

Muller, R., and Finkelstein, A., 1972, The effect of surface charge on the voltage-dependent conductance induced in thin lipid membranes by monazomycin, *J. Gen. Physiol.* **60**:285–306.

Ohki, S., and Sauve, U., 1978, Surface potential of phosphatidylserine monolayers. I. Divalent cation effect, *Biochim. Biophys. Acta* **511**:377–387.

Torrie, G. M., and Valleau, J. P., 1980, Electrical double layers. I. Monte Carlo study of a uniformly charged surface, *J. Chem. Phys.* **73**:5807–5816.

Chapter 19

CHARACTERIZATION OF DIHYDROPYRIDINE-SENSITIVE CALCIUM CHANNELS FROM PURIFIED SKELETAL MUSCLE TRANSVERSE TUBULES

ROBERTO CORONADO and HUBERT AFFOLTER

1. Introduction

The activity of calcium channels contributes to a large number of cellular functions, of which the best known are neurotransmitter release, pacemaking, motility, and muscle contraction (Reuter, 1983; Hille, 1984). Yet many more roles may be envisioned, as suggested by the fact that almost all cells, procaryotes and eucaryotes, have well-defined calcium currents (see Hille, 1984). More than any other channel, the calcium channel is presently the target of vigorous chemical and clinical research aming to understand how antiarrhythmic and antihypertensive drugs, such as dihydropyridines, modulate, block, or antagonize calcium-channel function (Janis *et al.*, 1985).

Inasmuch as most of these agents are lipophilic in nature, and their site of action may compromise the phospholipid phase (Glossmann *et al.*, 1982; Hescheler *et al.*, 1982), a study of calcium channels in planar lipid films seems particularly attractive. Indeed, in the past few years, this technique has been found highly suitable for the analysis of pharmacologically active drugs at the molecular level (Coronado, 1985; Miller, 1984). Recording of cellular calcium channels in planar lipid bilayers, however, presents several challenges. In the tissues of most interest, calcium channels are present in minute quantities, perhaps at surface densities far too low to be considered as reasonable sources of channels. Cultured rat cardiac cells, for example, have 0.1–0.5 calcium channels/μm^2, and chromaffin cells have 5–15 calcium channels/μm^2 (see Reuter, 1983). These numbers must be compared with densities of 500 acetylcholine receptor channels/μm^2 in *Torpedo*

ROBERTO CORONADO and HUBERT AFFOLTER • Department of Pharmacology, University of North Carolina at Chapel Hill, Chapel Hill, North Carolina 27514. Present addresses: R. C., Department of Physiology and Molecular Biophysics, Baylor College of Medicine, Houston, Texas 77030; H.A., Department of Physiology, Yale University School of Medicine, New Haven, Connecticut 06510.

electric organ or 250 sodium channels/μm^2 in electric eel, two sources routinely used for biochemical work (see Coronado, 1985; Miller, 1984).

Conceivably related to the low density is the generalized observation that calcium currents are notoriously labile and difficult to control, especially under cell-free conditions. A second challenge is purely technical. Calcium channels display kinetics that are fast and complex (Brown *et al.*, 1984a, 1984c; Lux and Brown, 1984). Activation kinetics, which occur in tens of milliseconds, are followed by an equally fast turning-off or inactivation process. Thus, out of necessity, calcium-channel currents must be collected using a pulse protocol in which voltage-elicited opening events are captured before inactivation renders the channel closed. Likewise, the unitary calcium current is remarkably small and never exceeds 1–2 pA in amplitude even when measured under saturation and optimized driving force conditions (Brown *et al.*, 1984a,b; Hess *et al.*, 1984; Kokubun and Reuter, 1984; Nowycky *et al.*, 1985). These two characteristics, currents that are small in amplitude and brief in duration, place the calcium channel somewhat beyond the dynamic range detectable in planar bilayers.

Recently, there have been two important developments that overcome the inherent difficulties of recording calcium channels *in vitro*. One is the development of biochemical preparations of muscle membranes highly enriched in calcium channel markers (Fosset *et al.*, 1983; Kawamoto *et al.*, 1984; Kirley and Schwartz, 1984; Rosemblatt *et al.*, 1981). Membranes topologically derived from the transverse tubular system of skeletal muscle (T-tubules) may contain 10- to 1000-fold higher antagonist-binding protein than any other membrane in which calcium-channel markers have been measured (Glossmann *et al.*, 1982). The second is the discovery of dihydropyridine drug molecules that stabilize the calcium channel in the open conformation (Schramm *et al.*, 1983; Towart and Schramm, 1984; Brown *et al.*, 1984b; Freedman and Miller, 1984; Kokubun and Reuter, 1984; Nowycky *et al.*, 1985). These "agonists" induce calcium channels to remain open for hundreds of milliseconds.

Making use of these advances, we recently described a strategy to record agonist-induced single calcium channels from rat skeletal muscle T-tubules in planar bilayers (Affolter and Coronado, 1985a,b; Coronado and Affolter, 1985a,b). Calcium channels were identified as such based on (1) a sensitivity to dihydropyridine agonists and antagonists; (2) a selectivity for divalent cations with the exclusion of Mg^{2+}; and (3) a steady-state voltage-dependent kinetics. This chapter reviews the basic properties by which T-tubule calcium channels are recognized and presents results from areas where planar bilayer research can make an important impact in understanding calcium-channel behavior.

2. What Calcium Channel Agonists and Antagonists Do

The structures of a few familiar calcium agonists and antagonists are shown in Fig. 1. The term calcium antagonist was introduced by Fleckenstein *et al.* (1984) to identify a group of molecules that acted by reversibly sealing calcium channels in the membrane of the myocardial cell. Inorganic cations such as Mn^{2+}, Co^{2+},

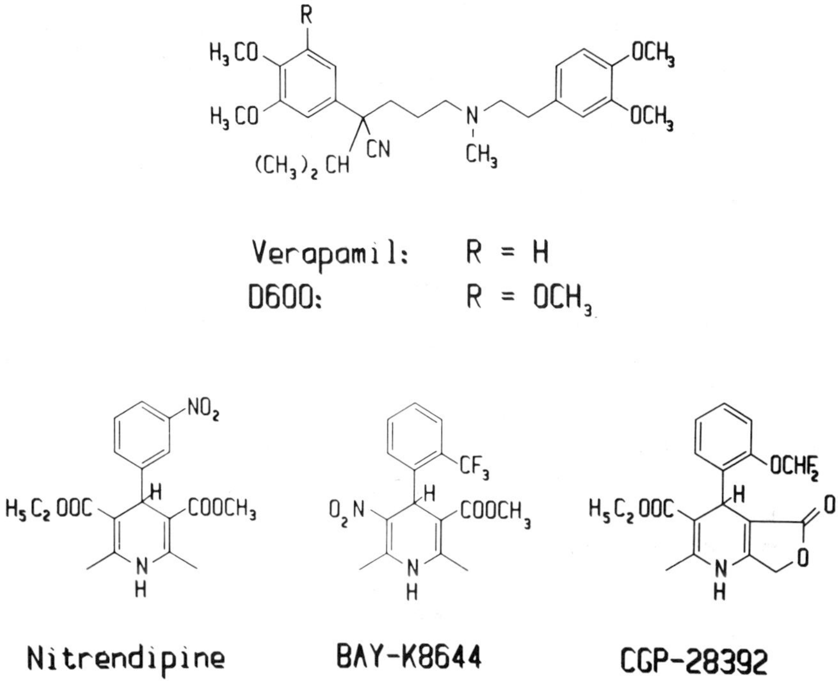

Figure 1. Structure of calcium channel antagonists and agonists: aralkylamine antagonists (verapamil, D600), dihydropyridine antagonists (nitrendipine), and dihydropyridine agonists (Bay-K8644, CGP-28392).

Cd^{2+}, or La^{3+} also fall in this category and may be regarded as unspecific because they are competitive ions that interfere with conducting divalent ions.

Two specific antagonist groups are best known: the aralkylamine series ("verapamil"-type), having a charged tertiary amine, and the dihydropyridine series ("nitrendipine"-type), without an essential basic group. Both series block divalent ion fluxes through calcium channels, but at different sites. Thus, bound radiolabels from each series can be displaced only by analogues from the same series, whereas antagonists of the other series are more than 1000-fold less active (Ehlert *et al.*, 1982; Garcia *et al.*, 1984). Extracellular verapamil or D600 (methoxyverapamil, gallopramil) at concentrations of 0.1 μM to 1 μM in heart and 10 μM to 40 μM in skeletal muscle suppresses calcium inward currents (Palade and Almers, 1985; Hescheler *et al.*, 1982; Lee and Tsien, 1983). The main effect is to slow down activation and the recovery from inactivation.

The site of blockade by D600, as shown by Hescheler *et al.* (1982), is located on the cytoplasmic end of the channel. Thus, it was suggested that the neutral form of externally added D600 (7% uncharged at pH 7.4) must partition into the lipid phase to reach its site of action. Conversely, D890, a permanently charged derivative of D600 that is impermeant through the lipid membrane, inhibits when added to the intracellular but not extracellular side (Hescheler *et al.*, 1982).

Blockade by nitrendipine-type antagonists reduces peak inward calcium current and speeds its decay. This has been interpreted as drug blockade of open channels (Lee and Tsien, 1983). The IC_{50}s are approximately the same in heart and skeletal muscle, varying from 0.1 μM to 10 μM (Palade and Almers, 1985; Lee and Tsien, 1983; Bean, 1984). In single-channel patch-clamp recordings, the antagonism of dihydropyridines has proven to be complex. Nitrendipine, for example, increases the number of null sweeps without activity, apparently stabilizing the channel in a closed conformation (Hess *et al.*, 1984). Yet, at low concentrations, it increases the mean open time, thus behaving as a partial agonist.

True agonists such as Bay-K8644 and CGP-28392 increase heart contractility and divalent cation influx over a wide range of concentrations (Schramm *et al.*, 1983; Towart and Schramm, 1984; Freedman and Miller, 1984; Hess *et al.*, 1984; Kokubun and Reuter, 1984; Brown *et al.*, 1984b; Nowycky *et al.*, 1985). In heart cells and cultured neurons, agonists increase whole-cell current by two- to tenfold. Agonist-induced activity can easily be recorded under steady-state 0-mV conditions, implying that the drug alters in a yet unknown way the natural inactivation kinetics. Dose–response curves of Bay-K8644 have been shown to be shifted by nitrendipine, a strong indication that agonists and antagonists are competitive in nature (Freedman and Miller, 1984). However, the site of action of dihydropyridines (i.e., intracellular versus extracellular) has not been resolved to our knowledge. The agonist-mediated increase in calcium current has been accounted for by the appearance of a population of calcium channels that have open lifetimes 10- to 100-fold longer than control channels. Agonists apparently have little effect on single-channel amplitudes and do not alter selectivity for divalent cations. A second effect of agonists is to induce bursts of open events, each burst lasting several seconds (Nowycky *et al.*, 1985).

3. Muscle T-Tubules: The Ideal Source of Calcium Channels

The scheme of isolation of skeletal muscle T-tubules is shown in Fig. 2. An ideal source of membranes for planar bilayer work would be one in which single-channel data obtained by reconstitution in planar bilayers could be integrated with radioligand-binding data, isotopic flux measurements, and biochemical manipulation of the channel protein, all in the same preparation. In the case of calcium channels, these goals leave few alternatives other than purified skeletal muscle T-tubules.

Soon after the discovery of calcium currents in skeletal muscle (Sanchez and Stefani, 1978), Almers *et al.* (1981) showed that these currents originated predominately in the T-tubule system. The T-system comprises a network of tubules, that relay action potentials generated at the muscle surface deep into the muscle fiber. Thanks to the effort of several groups, it is now clear that membranes collected from fragmented muscle separate into fractions of various densities and

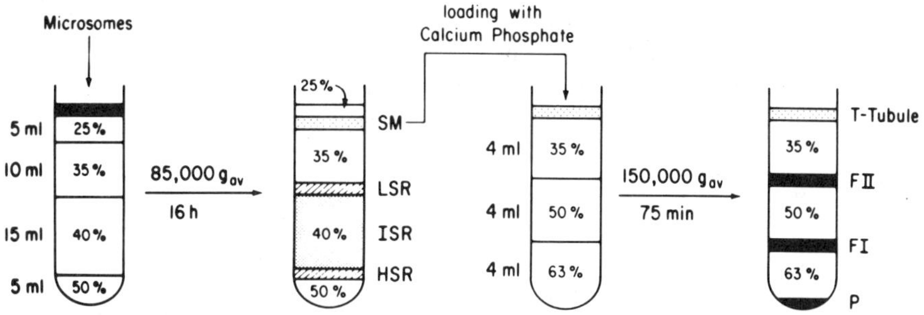

Figure 2. Biochemical purification of skeletal muscle T-tubules. Outline of procedure of Rosemblatt *et al.* (1981) used for the isolation of T-tubule membranes. Microsomes from rabbit skeletal white muscles (rat back and leg muscles in our work) are centrifuged in discontinuous sucrose gradients to isolate the light SM fraction from the various sarcoplasmic reticulum (LSR, ISR, HSR) fractions. The SM is further purified by incubation in Ca^{2+}-phosphate loading buffers. T-tubules are collected from the top of a second discontinuous gradient centrifugation. Active calcium channels have been found in either SM or T-tubule fractions but not in SR fractions.

morphologies. Their topological origin in the intact muscle has been traced back through the aid of enzyme markers, immunologic staining, lipid constituents, structural features, and many other advanced biochemical methods. Rosemblatt *et al.* (1981) (see Fig. 2A) developed a vesicular preparation of rabbit skeletal muscle T-tubules, whose origin has been verified by the following criteria. (1) Immunofluorescence studies using goat antibodies against purified T-tubules have shown that the antibodies label T-tubules but not the surface or the sarcoplasmic reticulum of muscle slices. Conversely, antibodies against surface membranes do not label T-tubules. (2) T-tubule membranes have a much higher cholesterol and phospholipid content than sarcoplasmic reticulum and the surface membrane. (3) T-tubule membranes have a much higher level of Mg^{2+}-ATPase than surface membranes.

Fosset *et al.* (1983) subjected these membranes to a binding study using radiolabeled nitrendipine as a calcium channel antagonist ligand. Quite surprisingly, the maximum binding capacity was found to be the highest ever reached in purified membranes: 50 pmol/mg protein. This was later confirmed by others (Kawamoto *et al.*, 1984). In heart, smooth muscle, and brain microsomes, nitrendipine is bound no more than 0.1–0.5 pmol/mg protein (Glossmann *et al.*, 1982), 50- to 100-fold lower than in T-tubules. In complete agreement with the electrophysiological measurements that locate calcium channels in the T-system (Almers *et al.*, 1981), the binding capacity of T-tubules is 30 times larger than that of surface membranes. This is a remarkable enrichment of receptors by any radioligand-binding standard. For example, in the same preparation, nitrendipine binding is 70-fold higher than TTX binding to sodium channels and 12,000-fold higher than apamine binding to $K^+(Ca^{2+})$ channels.

However, not all nitrendipine receptors may represent functional calcium channels, given the significant discrepancies between electrophysiological IC_{50}s and K_ds of radiolabels (Glossmann *et al.*, 1982; Lee and Tsien, 1983). Still, the nitrendipine-binding receptor retains much of the competitive and noncompetitive relationships among dihydropyridines, aralkylamines, and inorganic calcium-channel antagonists, which can be rationalized from the pharmacology of calcium channels *in vivo* (Murphy *et al.*, 1983; Curtis and Catterall, 1983, 1984; Kirley and Schwartz, 1984; Borsotto *et al.*, 1984; Campbell *et al.*, 1984). In a detergent–micellar form, the nitrendipine-binding protein behaves as a large multisubunit glycoprotein with a sedimentation value of 20 S and an estimated molecular weight above 200,000 (Curtis and Catterall, 1983, 1984; Borsotto *et al.*, 1984).

4. How Are T-Tubule Calcium Single Channels Recognized?

Figure 3 shows the activity of several T-tubule channels recorded in bilayers under various ionic conditions. A remarkable property of purified T-tubules is that channel activity is completely preserved during the fractionation of muscle and collection of membranes, a process that consumes 14–18 hr (Rosemblatt *et al.*, 1981). Latorre *et al.* (1982) showed, for example, the activity of $K^+(Ca^{2+})$ channels, and more recently, Mocdzydlowski (1985; see also other chapters in this volume) has characterized the batrachotoxin-opened sodium channel from the same preparation. In addition to calcium channels, our T-tubule membranes contain chloride channels, which can be recorded regularly.

Even though this variety of activity is quite rewarding, it creates a problem, since the calcium-channel current is only a fraction of a picoampere, approximately 20 times smaller than that of $K^+(Ca^{2+})$ channels. Signal isolation then becomes a problem, in the same way as it has been a problem in recordings in whole cells. It thus became clear that instead of trying to mimic the intracellular and extracellular solutions, the bilayer solutions should be dictated by the need to eliminate the undesirable single-channel currents. A standard set of ionic conditions and protocol to record T-tubule calcium channels is described in Fig. 4. In order to eliminate $K^+(Ca^{2+})$ channels, measurements are done in Na^+ instead of K^+. For the same reason, we routinely use Ba^{2+} and not Ca^{2+} as current carrier. Since under our recording conditions $E_{Ba^{2+}} \ll -200$ mV (nominally infinite), $E_{Na^+} = 0$ mV, $E_{Cl^-} = +40$ mV, Ba^{2+}-selective channels can easily be identified at zero holding potential (HP) as upward deflections (positive current). At this same voltage, chloride channels, recorded in one out of ten experiments, are identified as negative (downward) deflections that reverse at +35 mV.

Figure 4A shows barium current through single calcium channels in the absence of agonist. Events without agonist are brief and labile, since channels tend to close irreversibly within 30–45 sec after being recorded for the first time. This spontaneous activity is only present in Ba^{2+} media but is totally absent in Ca^{2+} or Sr^{2+}. However, when the fusion process is done in the presence of micromolar CGP-28392 (Fig. 4B) or Bay-K8644 (Fig. 4C), channel activity is stabilized with any of these three divalent cations, and recording time can easily be extended over 60–100 min at HP = 0 mV. Under the ionic conditions described in Fig. 4, agonist-induced channels reverse at voltages more negative than −50 mV. The

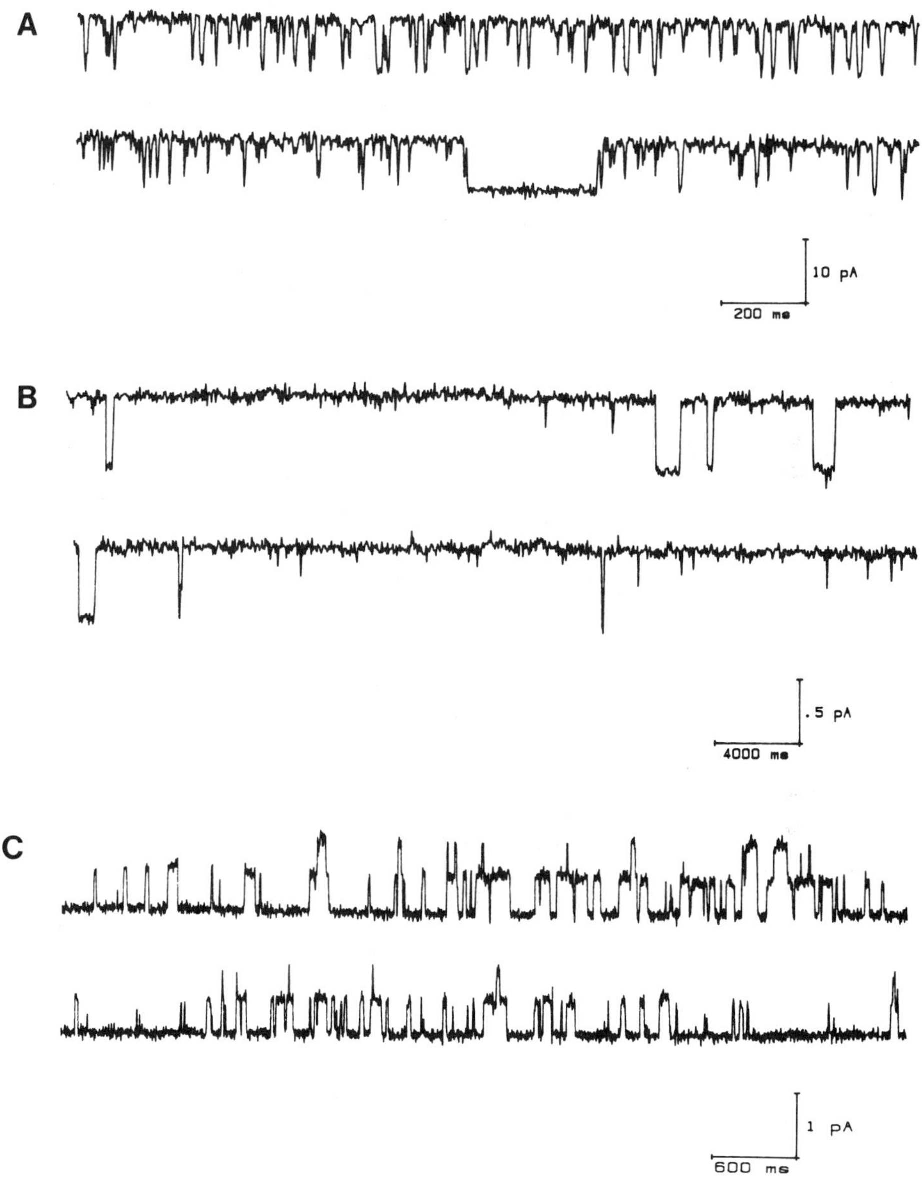

Figure 3. Functional channels of purified T-tubules in planar bilayers. A: $K(Ca^{2+})$ channels at HP = -50 mV in symmetrical 0.2 M KCl, 10 mM HEPES–tris, pH 7.0, 10 mM *cis* $CaCl_2$. B: Batrachotoxin-activated Na^+ channels at HP = $+50$ mV recorded in *cis–trans* 5 μM bactrachotoxin, 0.5 M NaCl, 10 mM HEPES–tris, pH 7.2. C: Chloride channels at HP = 0 mV recorded 0.2 M *cis* $BaCl_2$, 0.1 M *trans* $BaCl_2$, 10 mM HEPES–tris, pH 7.0. All openings are shown upwards.

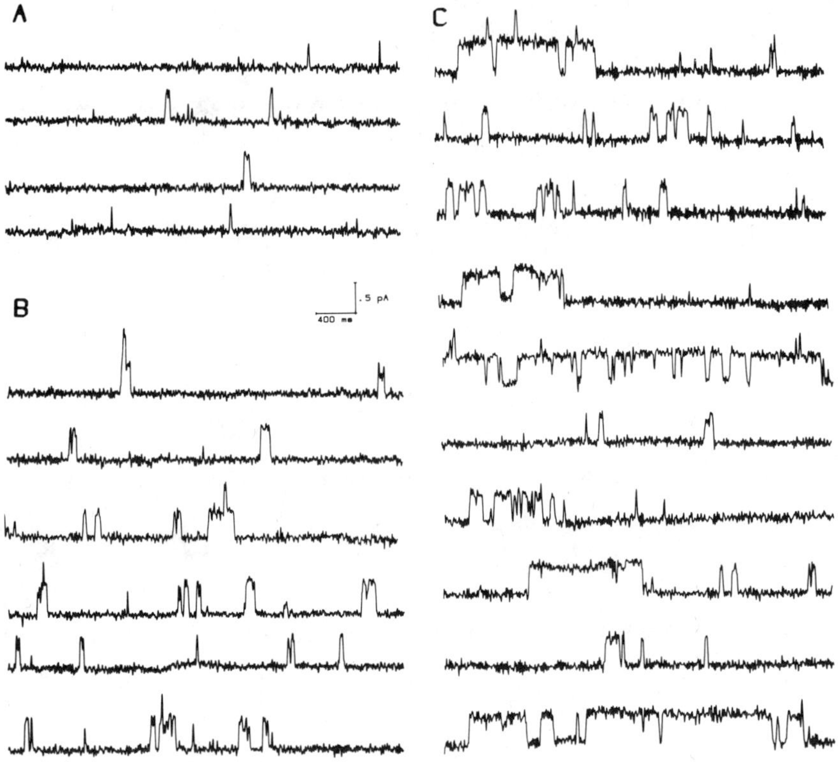

Figure 4. Barium current through single calcium channels activated by agonists. A: Records at HP = 0 mV in the absence of agonists; bilayer was formed in symmetrical *cis–trans* solution containing 0.05 M NaCl, 0.1 mM EGTA, 10 mM HEPES–tris, pH 7.0. Afterwards, *cis* $BaCl_2$ was raised to 0.1 M, and 100 μg of vesicular protein was added to the *cis* side. B,C: Records at HP = 0 mV in 3 μM CGP-28392 (B) or 3 μM Bay-K8644 (C) using the same solutions described in A. The bilayer was incubated for 5 min with agonist just prior to addition of *T*-tubule protein. Agonist was added from a concentrated solution (3 mM) in ethanol. Time mark (400 msec) and current mark (0.5 pA) is the same for all records. Holding potential (HP) is injected into the cup side (*cis* side), and the bath side (*trans* side) is maintained at virtual ground. (From Affolter and Coronado, 1985b.)

most remarkable feature of the agonist-induced activity is the appearance of long channel openings, which tend to cluster into bursts lasting 1–5 sec.

Whether the agonist acts on the channels in the native membrane or once channels are inserted into the bilayer is an issue that is difficult to resolve. Bay-K8644 or CGP-28392 are highly lipophilic molecules and will partition into both vesicular and planar bilayer lipid phases once the compound is added to the aqueous phase.

Of the many options now available (see Miller, 1984), we insert calcium channels by fusion of vesicles into Mueller–Rudin membranes (see Mueller and Rudin, 1969). At the expense of sacrificing time resolution, which is relatively low, Mueller–Rudin films offer an enormous stability in baseline current because of the $>10^{12}$-Ω seal resistance of these membranes. This simplifies computer analysis.

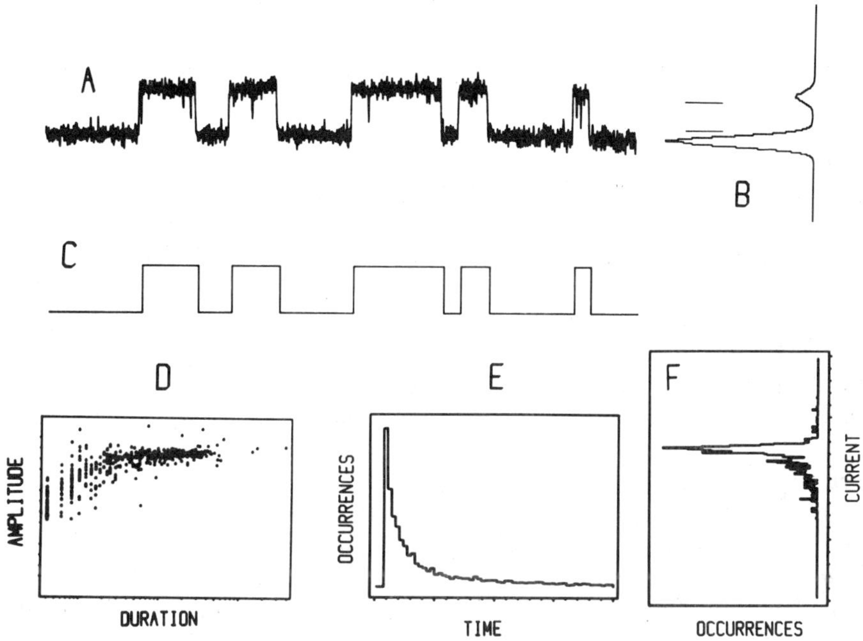

Figure 5. Computer handling of single-channel data. Raw data (A) is digitized, and a peak-current histogram (B) is taken to locate baseline and open-current levels. Two threshold detectors (see marks in histogram B), positioned at 1 SD from the mean baseline peak and 1 SD from the mean unitary current peak, define the open–closed transitions given in C. The sequence of transitions and the mean amplitude and duration of each transition are stored as a random-access file, which is later used to construct duration-versus-amplitude plots (D), lifetime histograms (E), and open-channel amplitude distributions (F).

Fusion is chosen primarily because T-tubule vesicles, like other muscle membrane fractions (Miller, 1978), are highly "fusiogenic" to planar films. Thus, we eliminate a typical source of irreproducibility in planar bilayers, which is inconsistency in channel insertion. Fusion conditions are met by adding vesicles to the hyperosmotic divalent *cis* solution and by casting planar bilayers from a 1:1 mixture of the neutral lipid phosphatidylethanolamine and the acidic lipid phosphatidylserine.

5. Computer Handling of Single-Channel Data

A session of single-channel data analysis is outlined in Fig. 5. A recurrent problem is the correct separation of baseline and unitary current levels, particularly in cases of records with a relatively low signal-to-noise ratio. Unfortunately, our records of calcium channels are a vivid example of such a problem. A second related difficulty arises from small but discernible time-dependent changes in baseline currents, which may introduce serious errors in the event-counting routines.

We minimized the former problem by counting open transitions with a dual-threshold detector program and by eliminating events whose amplitudes are clearly attenuated (in our case, all events <10 msec). The latter is solved by recording channels in small 100- to 200-pF Mueller–Rudin bilayers with bare resistances $>10^{12}\ \Omega$. In these membranes, drift is less than 0.1 pA over a 2-hr period.

Records taken on FM tape are filtered at 0.1–0.4 kHz (−3 dB point from eight-pole Bessel filter) and digitized on playback at 0.5–2.0 points/msec. Raw data are stored as files of 100,000 points (200 kbytes of IBM-PCXT/AT computer memory). For massive storage of raw data, removable hard disk cartridges are quite useful. Bernoulli disks (Vufax, Inc. GA) in use in our laboratory, for example, have a capacity of 10 Mbytes and can hold 49 raw data files. After construction of peak-current histograms to determine the mean and standard deviation (SD) of the baseline current and unitary current peaks, two threshold detectors are placed between baseline and open-peak current (see Fig. 5). One detector (discr 1) is placed at 1 SD from the mean base-line current, and the second detector (discr 2) is placed at 1 SD from the mean single channel unitary current. Open events are defined as transitions that cross discr 1 and discr 2 and remain above discr 2 for two or more consecutive points. Likewise, closed events must remain below discr 1 for two or more points to be considered as such. This provision takes care of most spurious noise spikes, which are unavoidable in records at high gain.

Within our programs, part machine language and part compiled BASIC (available on request to H.A.), it takes 20 sec to construct a peak-current histogram and 120 sec to complete a file of processed data containing the sequence of threshold-defined open and closed segments, their mean amplitudes, and individual durations. The advantages of constructing a processed data file are: (1) the file is now 20 kbytes instead of the 200 kbytes of the raw data file; (2) any histograms, including sequence-dependent histograms, can be obtained from this file in <30 sec; and (3) the relationship between amplitude and duration of events can be inspected so that attenuated events can be detected and eliminated.

6. Agonists Induce Long-Term Calcium-Channel Openings

Figure 6A shows lifetime distributions of open channels. In the absence of agonist, open events can be approximately fitted by a single exponential with a time constant, t_s = 20 msec. In CGP-28392 and Bay-K8644, distributions are biexponential and have been fitted with time constants: t_s = 16 msec, t_l = 125 msec and t_s = 28 msec, t_l = 186 msec, respectively. Thus, the agonists induce a distinctive population of stable open channels, o_l, that remain open approximately ten times longer than the population of brief channels, o_s, recorded in the absence of agonists.

The lack of long events in the absence of agonists is most clearly seen in Fig. 6B, where the tails of the distributions have been expanded to show bin content for $t > 100$ msec. A summary of open and closed lifetimes is shown in Table I.

Burst formation in the presence of agonist is a significant feature of the chan-

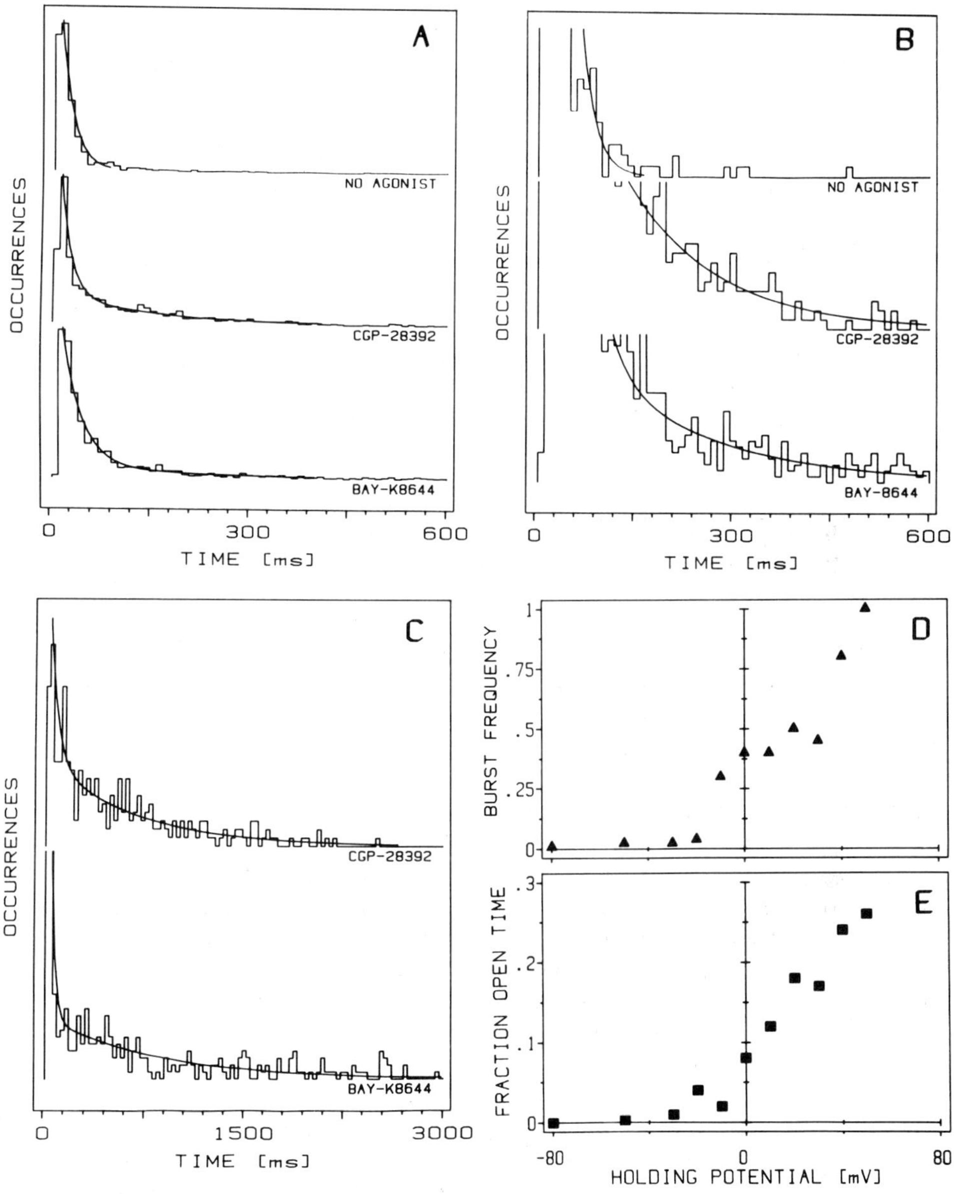

Figure 6. Open lifetime, burst distribution, and voltage dependence of agonist-induced events. A: Open-channel histograms at HP = 0 mV fitted for all events >10 msec. Fitted exponentials are given in Table I. B: Same histograms described, with *y*-axis (occurrences) amplified five times to show events at the tail of the distributions. C: Distribution of identified bursts in Bay-K8644 or CGP-28392 at HP = 0 mV. D: Frequency of single bursts measured as bursts/sec (labeled burst frequency), plotted as a function of HP. E: The fraction of total digitized points at each voltage (100,000 points at 2 msec/point) that fall under the open-peak current (labeled fraction open time), is plotted as a function of HP. All data in D and E are from a single record of barium current through single calcium channels with Bay-K8644. (From Affolter and Coronado, 1985b.)

Table I. Lifetime Distributions of T-Tubule Calcium Channels Incorporated into Planar Bilayers[a]

	t_s (msec)	t_l (msec)	f_l	n
Open events				
Control	20	—	—	557
Bay-K8644	28	186	0.15	1680
CGP-28392	16	125	0.17	948
Closed events				
Control	230	—	—	555
Bay-K8644	35	380	0.25	1683
CGP-28392	25	450	0.18	940
Burst events				
Bay-K8644	56	670	0.39	272
CGP-28392	27	820	0.10	353

[a] t_s, t_l correspond to short and long components of exponential distributions fitted by the sum of two exponential terms; f_l corresponds to the fraction of the population of total events fitted by t_l; n is the number of computer-defined events used in the fit. Open and closed events are defined by two-threshold criterion. Burst events are defined as a sequence of transitions closed–open–closed. . . in which the first and last closed are 100 msec or longer.

nel, as can be observed from the records. This has been confirmed quantitatively, since closed distributions are biexponential (Table I) and probability of channel opening is conditioned by the previous occurrence of a closing event. Figure 6C shows that burst distributions are also biexponential and have been fitted with $t_s = 56$ msec, $t_l = 670$ msec for CGP-28392 and $t_s = 27$ msec, $t\sim = 820$ msec for Bay-K8644. Both the number of bursts formed per unit time (Fig. 6D) and the fraction of time that channels spent open (Fig. 6E) are functions of voltage, increasing with *cis*-positive holdings potentials. In the range HP = -30 mV to HP = $+20$ mV, burst frequency increases e-fold per 15 mV; the fraction open time increases e-fold per 25 mV. In spite of the fact that the fraction of open time increases 20-fold from HP = -50 mV to HP = $+30$, the absolute value is low, seldom rising above 0.3 at positive holding potentials.

Thus, channel opening and bursting are voltage-dependent events of low probability, even in the presence of Bay-K8644. Even though agonist-induced long events represent, at most, 15% of the population of open-channel lifetimes (Table I), a simple calculation shows that this would be sufficient to generate a fourfold increase in the fraction of open time at HP = 0 mV. From data of Table I, the mean open time of the whole population (t_o) increases from 20 msec (no agonist) to 67 msec (Bay-K8644 data), and the mean closed time of the whole population (t_c) decreases from 230 msec (no agonist) to 120 msec (with agonist). Thus, the fraction of open time, $P = (1 + t_c/t_o)^{-1}$, increases from $P = 0.08$ (no agonist to $P = 0.35$ (with agonist) or 4.4-fold.

In heart cells at similar voltages, Bay-K8644 increases calcium current two- to fourfold (Hess *et al.*, 1984; Kokubun and Reuter, 1984). A significant difference relative to the whole-cell data is found in the time scale of agonist-induced events. The reported mean open time of Bay-K8644-induced events is 20 msec in heart cells (Hess *et al.*, 1984, Kokubun and Reuter, 1984; Brown *et al.*, 1984b) and 60 msec in neurons (Nowycky *et al.*, 1985) instead of 186 msec as measured here in T-tubules (Table I) at same drug concentration and comparable depolarization. In lieu of other explanations, we consider this difference to be a reflection of the known intrinsically slow kinetics of calcium channels in skeletal muscle (Sanchez and Stefani, 1978; Stanfield, 1977; Almers *et al.*, 1984). Interestingly, mean burst duration of Bay-K8644-induced channels in cultured neurons and T-tubules (Table I) is approximately the same.

7. T-Tubule Calcium Channels Are Selective for Divalents but Not for Mg^{2+}

Figure 7A compares the amplitude distribution of open events in the presence and absence of agonist. For Bay-K8644, short events (20 msec $< o_s <$ 50 msec) and long events (200 msec $< o_l <$ 1000 msec) are shown separately. The windows

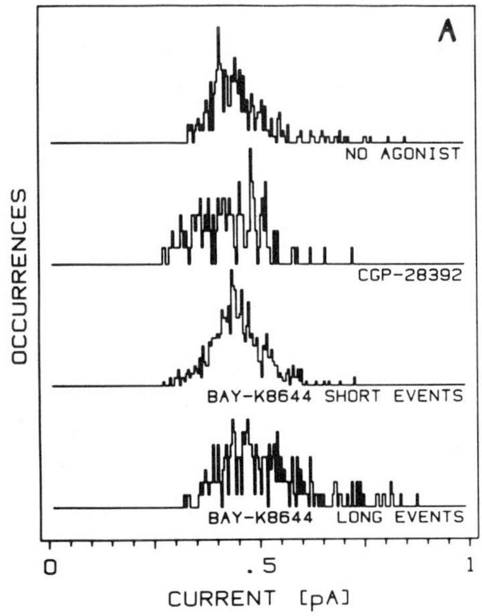

Figure 7. Amplitude distribution, divalent cation permeation, and reversal of Bay-K8644-induced single calcium channels. A: Amplitude distribution of open-channel currents at HP = 0 mV in *trans* 0.05 M NaCl–*cis* 0.05 M NaCl + 0.1 M $BaCl_2$ in the absence or presence of Bay-K8644 or CGP-28392. Events without agonist or CGP-28392 correspond to total openings >20 msec. Short events in Bay-K8644 are from the duration interval 20 msec $> t >$ 50 msec, and long events from the interval 200 msec $> t >$ 1000 msec. B: Bar histogram of mean and standard deviation of Bay-K8644-induced channel currents at HP = 0 mV measured in *trans* 0.05 M NaCl–*cis* 0.05 M NaCl + 0.1 XCl_2 (X = Ba^{2+}, Sr^{2+}, Ca^{2+}, Mg^{2+}). The Mg^{2+} currents cannot be measured accurately; the bar (0.05 pA) is an estimate for the upper limit. C: Reversal of single-channel currents under biionic *trans* 0.05 M NaCl–*cis* 0.05 M $BaCl_2$ (filled squares) or *trans* 0.05 M $BaCl_2$–*cis* 0.05 M NaCl (filled triangles). Reversal potential for this latter condition is +58 mV. (From Affolter and Coronado, 1985b).

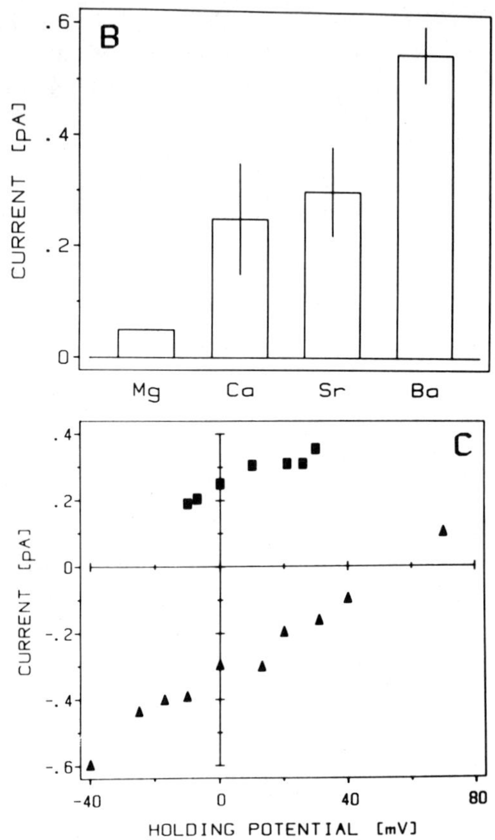

Figure 7. (*continued*)

chosen to sample short and long events seem justified, since we calculate with probability $P > 0.9$ that events <50 msec belong to the population fitted by t_s and events >200 msec belong to the population fitted by t_l. Measured under the same conditions, all open channels have the same mean open current of approximately 0.5 pA at HP = 0 mV. Also, in all cases, reversal occurs at a potential more negative than −50 mV (not shown). Thus, only the channel lifetime and not the pore selectivity is modified by the dihydropyridine agonist.

Shown in Fig. 7B are means and S.D.s of single-channel currents at HP = 0 mV after *cis* Ba^{2+} is replaced by one of the other group IIa divalent cations. Slope conductance at HP = −50 mV follows the sequence Ba^{2+} = 20 pS > Sr^{2+} = 12 pS > Ca^{2+} = 10 pS ≫ Mg^{2+} <2 pS. The high Ba^{2+} conductance and lack of any significant Mg^{2+} conductance indicates that the channel is endowed with permeation properties that are characteristic of calcium channels recorded in whole cells (Kostyuk *et al.*, 1983; Fukushima and Hagiwara, 1983; Hess and Tsien, 1984; Almers and McCleskey, 1984).

Discrimination against alkali monovalent cations is shown in Fig. 7C for the bionic pair 0.05 M Na^+/0.05 M Ba^{2+} measured from activity induced by Bay-K8644. By making $E_{Ba^{2+}}$ positive (*trans* Ba^{2+}/*cis* Na^+), it is possible to show a genuine reversal at +58 mV. This reversal corresponds to a permeability ratio $P_{Ba^{2+}}/P_{Na^+} = 25$ from the constant-field equation solved with $P_{Cl^-} = 0$. The lack of anion permeation is shown in Fig. 11.

8. Nitrendipine Blockade Lacks a Marked Sidedness

Described in Fig. 8 is the antagonistic action of nitrendipine on the T-tubule calcium channel. Blockade of Bay-K-8644-induced activity by nitrendipine was an expected finding given the competitive nature of these two drugs (Freedman and Miller, 1984). The overall effect of nitrendipine consists in a reduction of the probability of channel opening and, consequently, the appearance of long silent periods. Surprisingly, the records shown in Fig. 8 indicate that nitrendipine blockade lacks a marked sidedness, inasmuch as prolonged exposure of the channel to either *cis* or *trans* nitrendipine diminishes activity. This could indicate that the drug gains access to the channel after partitioning and diffusing through the bilayer lipid phase, similar to the block of calcium channels by aralkylamine-type antagonists (Hescheler *et al.*, 1982).

The two most significant features of nitrendipine block are shown in the window plots of Fig. 8 (bottom). In these experiments, control records and records after drug additions are digitized and split into 300-msec windows or frames. For each consecutive window in time, the plot shows the fraction of time spent open during that particular time interval, P_o. Blanks corresponds to frames without activity. Control records show events of high ($P_o > 0.9$) and low ($P_o < 0.4$) probability, which is a typical behavior of calcium channels (Hess *et al.*, 1984; Nowycky *et al.*, 1985). As a function of nitrendipine concentration, there is an increase in the number of empty frames. This indicates that nitrendipine diminishes the pool of active channels. At the same time, those channels that are available to open do so, but briefly, since P_o decreases progressively. At *cis–trans* 1 μM nitrendipine, P_o is usually <0.2.

Figure 9 shows how nitrendipine affects the various kinetic components controlling open-channel probability. Measurements were made in the absence of agonists, taking advantage of the fact that in high *cis* Na^+, low *cis* Ba^{2+}, spontaneous activity is enhanced (Affolter and Coronado, 1985a; Coronado and Affolter, 1985a). Consecutive additions of micromolar *cis* and *trans* nitrendipine (1) decrease the mean open time and simultaneously increase the closed time interval between open events; (2) decrease the duration of the burst unit; and (3) decrease the mean amplitude of the single-channel current. This effect probably results from limitations in frequency response secondary to the decrease in mean open time.

Figure 10 shows blockade of T-tubule calcium channels by the aralkylamine-type antagonist D600. Concentrations that induce block are higher than reported for cardiac tissues (Lee and Tsien, 1983) but totally consistent with the pharmacological profile gathered by Palade and Almers (1985) for the frog skeletal

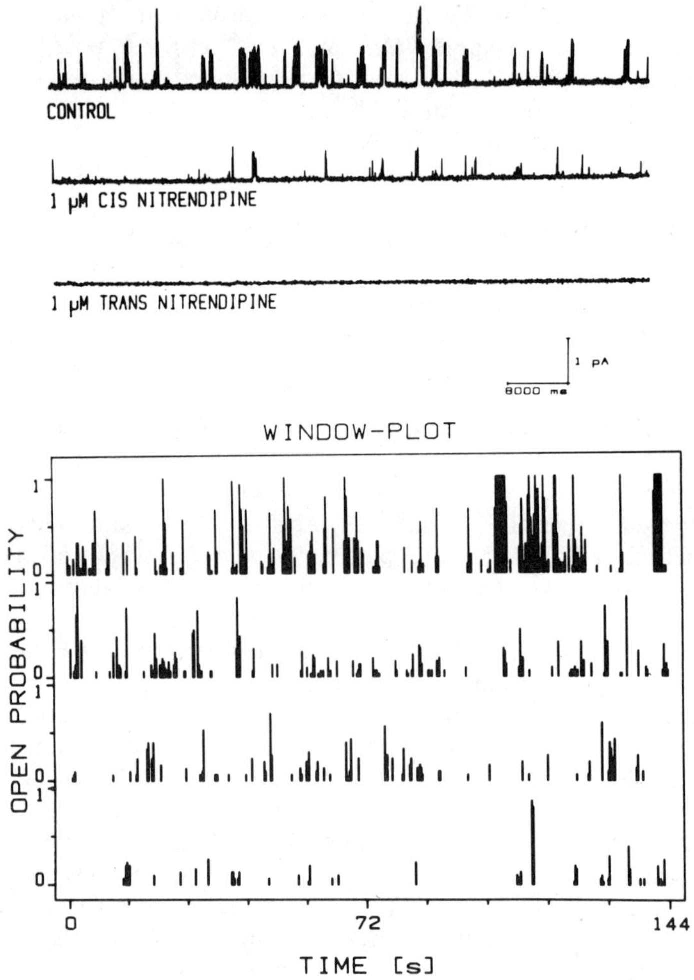

Figure 8. Nitrendipine blockade of mixed barium–sodium currents through single calcium channels. Top: Control records compressed in time at HP = 0 mV prior to addition of 1 μM *cis* nitrendipine and subsequent addition of 1 μM *trans* nitrendipine. Ionic solution was *cis* 0.25 M NaCl, 0.01 M $BaCl_2$, 10 mM HEPES–tris, pH 7.0; *trans* 0.05 M NaCl, 0.1 mM EGTA, 10 mM HEPES–tris pH 7.0. Bottom: The fraction of open time (P_0) was calculated in each consecutive 300-msec window segment. From top to bottom, window plots correspond to control, 0.25 μM, 0.5 μM, and 1 μM nitrendipine added to *cis* and *trans* sides. $t = 0$ in each trace corresponds to 60 sec after addition of nitrendipine. Recordings were made in *trans* 0.05 M NaCl, *cis* 0.25 M NaCl, 0.01 M $BaCl_2$ in absence of agonists. Empty frames correspond to segments without opening events.

muscle calcium channel. Qualitatively, D600 induces the same kind of kinetic changes as observed with nitrendipine. However, significant differences not yet studied in bilayers may be expected from the voltage dependence of these two antagonists. Figure 10A shows single-channel records compressed in time prior to and after *cis* addition of 40 μM D600. *Trans* block occurs at similar concen-

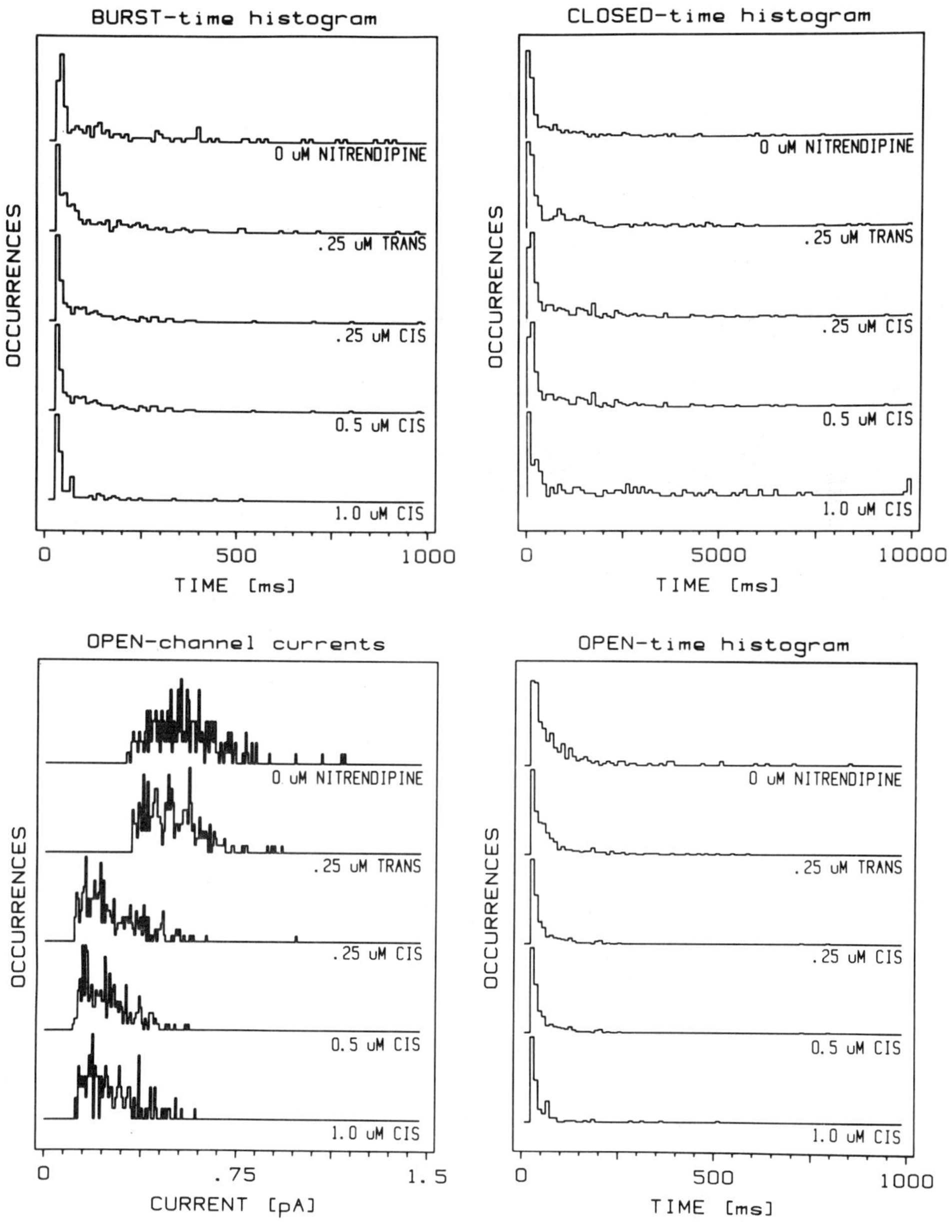

Figure 9. Kinetics of nitrendipine-blocked T-tubule calcium channels. Histograms after consecutive additions of *cis* and *trans* nitrendipine in *trans* 0.05 M NaCl, *cis* 0.25 M NaCl, 0.01 M $BaCl_2$. Open-channel currents correspond to mean values of events defined by the two-threshold criterion (see Fig. 5). Bursts are defined as sequences of open–closed transitions preceded and ended by closed events >100 msec. All histograms are from the same set of records.

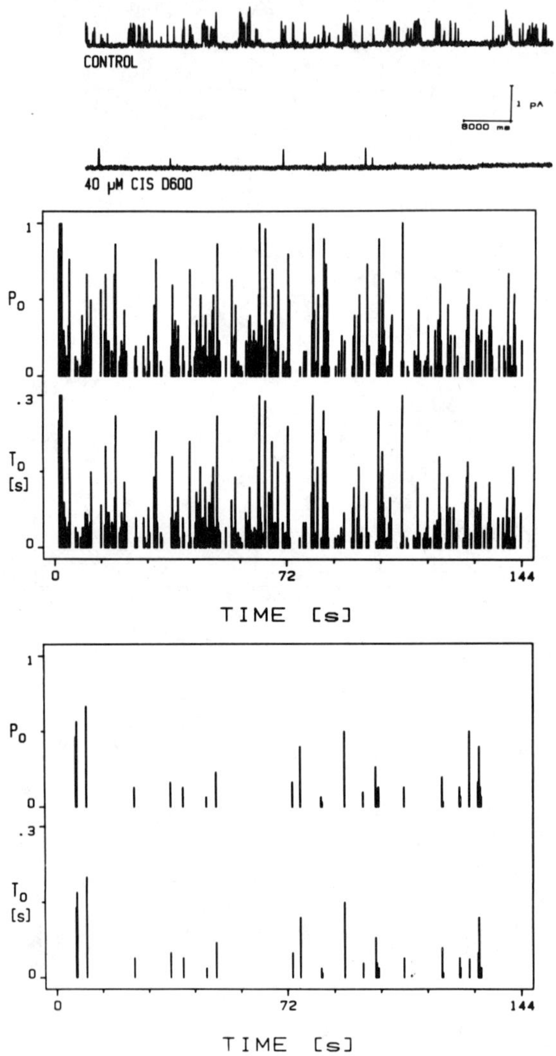

Figure 10. Blockade of T-tubule calcium channels by D600. Top: Records compressed in time of control and after *cis* addition of 40 μM D600. Measurements were made in 0.05 M NaCl *trans*, 0.25 M *cis*, 0.01 M $BaCl_2$. D600 was added from a 2 mM aqueous stock solution. Middle: For each 300-msec window segment of control record, the plot shows the fraction of open time (P_0) and mean open time (t_0). Bottom: Window plot of P_0 and t_0 for each 300-msec segment after *cis* addition of 40 μM D600; $t = 0$ corresponds to 45 sec after drug addition. Blanks correspond to windows without openings.

trations (not shown). As shown in the window plots of Fig. 10 (middle and bottom), D600 decreases both the frequency and duration of open events.

9. Sodium Permeation through T-Tubule Calcium Channels

Perhaps the most surprising finding that can only be explained qualitatively at present is shown in Fig. 11A. In low-*cis*-divalent (10 mM Ba^{2+}), high-*cis*-monovalent (0.25 M Na^+) solutions, calcium channels are capable of transporting Na^+ or K^+ at saturating rates higher than for divalent ions alone. That monovalent current is transported via calcium channels is demonstrated in pharmacological

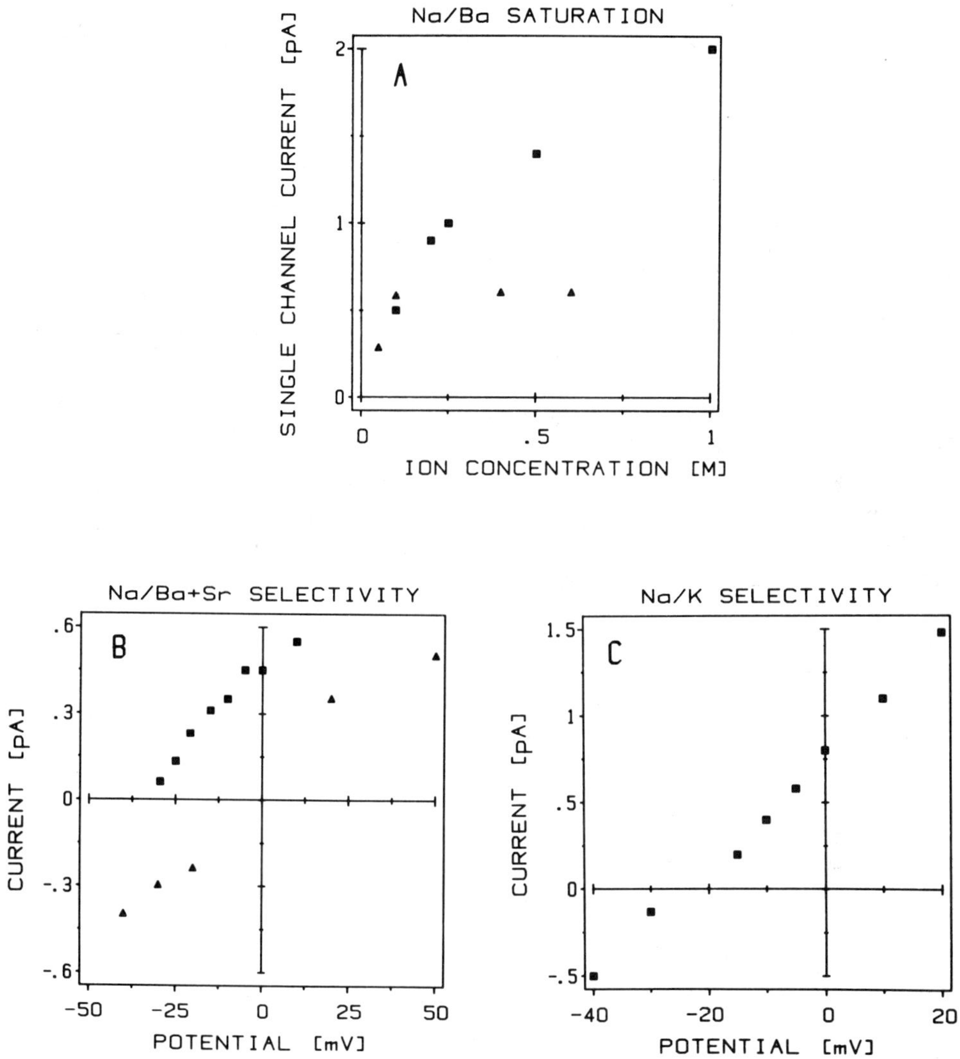

Figure 11. Saturation of calcium single-channel currents in sodium and barium. A: Amplitude of single-channel currents at HP = 0 mV measured in *trans* 0.05 M NaCl, *cis* 0.05 M NaCl + indicated *cis* molar concentration of $BaCl_2$ (triangles) or *trans* 0.05 M NaCl, *cis* 0.01 M $BaCl_2$ + indicated *cis* molar concentration of NaCl (squares). Concentrations are not corrected for ionic activities. The Ba^{2+} curve corresponds to activity elicited by *cis* 3 μM Bay-K8644. The Na^+ curve was obtained without agonist. B: Current–voltage curves in mixed Na^+, Ba^{2+}, and Sr^{2+} conditions. In *trans* 0.05 M NaCl, *cis* 0.25 M NaCl, $E_{rev} = -32$ mV ($E_{Na} = -35$ mV). Then additions of 0.01 M $BaCl_2$ and 0.01 M $SrCl_2$ were made to the *cis* and *trans* sides, respectively. In mixed monovalent–divalent solutions, the new reversal is $E_{rev} = -3$ mV ($E_{divalent} = 0$ mV). C: Current–voltage curves in biionic Na^+/K^+ solutions containing *trans* 0.125 M KCl, *cis*, 0.25 M NaCl. Reversal occurs at −22.5 mV, corresponding to a permeability ratio $P_{Na}/P_K = 1.4$.

experiments (Fig. 9) and by current–voltage curves in Fig. 11B. As shown in the latter, in high *cis* Na^+ and in the absence of divalent cations, channels aim at a reversal identical to E_{Na^+}. Addition of divalents to the *cis* and *trans* solutions decreases the single-channel currents and shifts reversal to $E_{divalent}$.

A secondary effect of divalents on monovalent-conducting calcium channels is a decrease in mean open time and frequency of events (not shown). Figure 11C shows that not only sodium but also potassium ions can permeate at equal rates. Thus, in the absence of conducting divalent ions, the channel becomes cationic nonspecific. The loss of selectivity by calcium channels when bathed in divalent-free solutions is a longstanding observation that has been clarified only recently (Almers and McCleskey, 1984; Hess and Tsien, 1984). In cardiac and skeletal muscle, the explanation needs only to invoke (1) a tighter binding for divalents over monovalents inside the calcium channel; (2) adherence of the calcium channel to the laws of ion occupancy and translocation described for mutliion single-file pores; and (3) ion repulsion inside the pore to boost the single-channel conductance in high-divalent solutions. Thus, a tighter binding for divalents would insure that in the presence of both kinds of ions the channel would be occupied more often by divalents than by monovalents. But this alone would not promote conduction, since the channel under these conditions remains actually blocked by divalents. The problem of translocation is solved by placing several binding sites in series so that electrostatic repulsions among ions when two or more sites are occupied at the same time weaken their respective ion–site binding energies. Whether the conduction hypothesis is valid in our case is not known, but, nevertheless, it is a very attractive model that can be thoroughly tested and expanded with the advantages offered by the planar bilayer method.

10. Conclusions

The behavior of T-tubule calcium channels in planar bilayers, including responsiveness to dihydropyridine agonists and antagonists, selectivity for divalents (but not Mg^{2+}), and voltage-dependent kinetics, indicates that channels remain functional, albeit unstable in absence of agonist, after membrane purification. Features such as the loss of selectivity in the absence of divalents strongly suggest that the channel is closely related to those recorded *in vivo*. This is not an assumption that can be taken for granted given the real difficulties encountered in maintaining whole-cell calcium currents fully active and the growing belief that calcium channels are intimately linked to cellular metabolism (Reuter, 1983). Most if not all of this cellular control of calcium channels would actually be lost when membranes are purified. How does the *T*-tubule channel survive? This question will probably remain without an answer until channel-mediated fluxes can be elicited in the native vesicular preparation, where many of these biochemical inquiries can be settled.

The demonstration that we have provided, however, of dihydropyridine agonist and antagonist action *in vitro* opens several possibilities for the use of these drugs in flux assays. Questions that the planar bilayer technique can approach quite successfully are those related to the modulation exerted by the bilayer phospholipid on the action of dihydropyridines. Are the dihydropyridine agonist and antagonist sites buried in the lipid membrane? If so, how does drug action depend on membrane composition? Is the modulated receptor hypothesis developed for drug–receptor interactions in sodium channels (Hille, 1977) applicable, for example, to blockade by nitrendipine or D600? These and other questions are ripe to receive answers in this system.

ACKNOWLEDGMENTS. Work in this laboratory is supported by NIH grant RO1 GM-32824. H.A. was partially supported by a fellowship from Swiss National Funds.

References

Affolter, H., and Coronado, R., 1985a, Planar bilayer recording of single calcium channels from purified muscle transverse tubules, *Biophys. J.* **47**:434a.

Affolter, H., and Coronado, R., 1985b, Agonists Bay-K8644 and CGP-28392 open calcium channels reconstituted from skeletal muscle transverse tubules, *Biophys. J.* **48**:341–347.

Almers, W., Fink, R., and Palade, P. T., 1981, Calcium depletion in frog muscle tubules: The decline of calcium current under maintained depolarization. *J. Physiol.* **312**:177–207.

Almers, W., and McCleskey, E. W., 1984, Non-selective conductance in calcium channels of frog muscle: Calcium selectivity in a single-file pore, *J. Physiol.* **353**:585–608.

Bean, B. P., 1984, Nitrendipine block of cardiac calcium channels: High affinity binding to the inactivated state, *Proc. Natl. Acad. Sci. U.S.A.* **81**:6388–6392.

Borsotto, M., Norman, R. I., Fosset, M., and Lazdunski, M., 1984, Solubilization of the nitrendipine receptor of skeletal muscle transverse tubule membranes: Interactions with specific inhibitors of the voltage-dependent calcium channel. *Eur. J. Biochem.* **142**:449–455.

Brown, A. M., Wilson, D. L., and Lux, H. D., 1984a, Activation of calcium channels, *Biophys. J.* **45**:125–127.

Brown, A. M., Kunze, D. L., and Yatani, A., 1984b, The agonist effect of dihydropyridines on Ca channels, *Nature* **311**:570–572.

Brown, A. M., Lux, H. D., and Wilson, D. L., 1984c, Activation and inactivation of single calcium channels in snail neurons, *J. Gen. Physiol.* **83**:751–769.

Campbell, K. P., Lipshutz, G. M., and Denney, G. H., 1984, Direct photoaffinity labeling of the high affinity nitrendipine-binding site in subcellular membrane fractions isolated from canine myocardium, *J. Biol. Chem* **259**:5384–5387.

Coronado, R., 1985, Recent advances in planar phospholipid bilayer techniques for monitoring ion channels, *Ann. Rev. Biophys. Biophys. Chem.* (in press).

Coronado, R., and Affolter, H., 1985a, Kinetics of dihydropyridine-sensitive single calcium channels from purified muscle transverse tubules, *Biophys. J.* **47**:434a.

Coronado, R., and Affolter, H., 1985b, Insulation of the conduction pathway of muscle transverse tubule calcium channels from the surface charge of bilayer phospholipid, *J. Gen. Physiol.* (in press).

Curtis, B. M., and Catterall, W. A., 1983, Solubilization of the calcium antagonist receptor from rat brain, *J. Biol. Chem.* **258**:7280–7283.

Curtis, B. M., and Catterall, W. A., 1984, Purification of the calcium antagonist receptor of the voltage-sensitive calcium channel from skeletal muscle transverse tubules, *Biochemistry* **23**:2113–2117.

Ehlert, F. J., Roeskes, W. R., Itoga, E., and Yamamura, H. I., 1982, The binding of [^{3}H]-nitrendipine to receptors for Ca^{2+} channel antagonists in the heart, cerebral cortex and ileum of rats, *Life Sci.* **30**:2191–2202.

Fleckenstein, A., 1984, Calcium antagonism: History and prospects for a multifaceted pharmacodynamic principle, in *Calcium Antagonists and Cardiovascular Disease*, (L. H. Opie, ed.), pp 9–28, Raven Press, New York.

Fosset, M., Jaimovich, E., Delpont, E., and Lazdunski, M., 1983, Nitrendipine receptors in skeletal muscle, *J. Biol. Chem.* **258:**6086–6092.

Freedman, S. B., and Miller, R. J., 1984, Calcium channel activation: A different type of drug action, *Proc. Natl. Acad. Sci. U.S.A.* **81:**5580–5583.

Fukushima, Y., and Hagiwara, S., 1983, Voltage-gated Ca^{2+} channel in mouse myeloma cells. *Proc. Natl. Acad. Sci. U.S.A.* **80:**2240–2243.

Garcia, M. L., Trumble, M. J., Reuben, J. P., and Kaczorowski, G. J., 1984, Characterization of verapamil binding sites in cardiac membrane vesicles, *J. Biol. Chem.* **259:**15013–15016.

Glossmann, H., Ferrsy, D. R., Leubbecke, F., Mewes, R., and Hoffmann, F., 1982, Calcium channels: Direct identification with radioligand binding studies, *Trends in Pharmacol. Sci.* **3:**431–437.

Hescheler, J., Pelzer, D., Trube, G., and Trautwein, W., 1982, Does the organic calcium channel blocker D600 act from the inside or outside on the cardiac cell membrane? *Pflugers Arch.* **393:**287–291.

Hess, P., and Tsien, R. W., 1984, Mechanism of ion permeation through calcium channels, *Nature* **309:**453–456.

Hess, P., Lansman, J. B., and Tsien, R. W., 1984, Different modes of calcium channel gating favoured by Ca agonists and antagonists, *Nature* **11:**538–544.

Hille, B., 1977, The pH-dependent rate of action of local anesthetics on the node of Ranvier, *J. Gen. Physiol.* **69:**475–496.

Hille, B., 1984, Evolution and Diversity, in *Ionic Channels of Excitable Membranes* (B. Hille, ed.), pp. 371–383. Sinauer Associates, Sutherland, Massachusetts.

Janis, R. A., Bellemann, P., Sarmiento, J. G., and Triggle, 1985, The dihydropyridine receptor in *IX Bayer Symposium. Cardiovascular Effects of Dihydropyridine-Type Calcium Antagonists and Agonists* (A. Fleckenstein and C. Breemen, eds.), Springer-Verlag, New York.

Kawamoto, R., Brant, N., and Caswell, A., 1984, Localization and solubilization of nitrendipine binding sites in skeletal muscle, *Biophys. J.* **45:**206a.

Kirley, T. L., and Schwartz, A., 1984, Nimodipine and nitrendipine binding to rabbit skeletal t-tubular membranes: Effects of D-cis-Diltiazem (DTZ) and attempts to solubilize and affinity label the putative calcium channel, *Biophys. J.* **45:**396a.

Kokubun, S., and Reuter, H., 1984, Dihydropyridine derivatives prolong the open state of Ca channels in cultured cardiac cells, *Proc. Natl. Acad. Sci. U.S.A.* **81:**4824–4827.

Kostyuk, P. G., Mironov, S. L., and Shuba, Y. M., 1983, Two ion-selecting filters in the calcium channel of the somatic membrane of mollusc neurons, *J. Memb. Biol.* **76:**83–93.

Latorre, R., Vergara, C., and Hidalgo, C., 1982, Reconstitution in planar bilayers of a Ca^{2+}-dependent K^+ channel from transverse tubule membranes isolated from rabbit skeletal muscle, *Proc. Natl. Acad. Sci. U.S.A.* **79:**805–809.

Lee, K. S., and Tsien, R. W., 1983, Mechanism of calcium channel blockade by verapamil, D600, diltiazem and nitrendipine in single dialysed heart cells, *Nature* **302:**790–794.

Lux, H. D., and Brown, A. M., 1984, Patch and whole cell calcium currents recorded simultaneously in snail neurons, *J. Gen. Physiol.* **83:**727–750.

Miller, C., 1978, Voltage-gated cation conductance channel from fragmented sarcoplasmic reticulum: Steady-state electrical properties, *J. Memb. Biol.* **40:**1–23.

Miller, C., 1984, Integral membrane channels: Studies in model membranes, *Physiol. Rev.* **63:**1209–1242.

Moczydlowski, E., 1985, Na-channel block by u-conotoxin GIIIA: A peptide toxin specific for skeletal muscle, *Biophys. J.* **47:**190a.

Mueller, P., and Rudin, D. O., 1969, Translocators in bimolecular lipid membranes: Their role in dissipative and conservative bioenergy transductions, *Curr. Top. Bioenerg.* **3:**157–249.

Murphy, K. M. M., Gould, R. J., Largent, B. L., and Snyder, S., 1983, A unitary mechanism of calcium antagonist drug action, *Proc. Natl. Acad. Sci. U.S.A.* **80:**860–864.

Nowycky, M. C., Fox, A. P., and Tsien, R. W., 1985, Novel mode of neuronal calcium channel gating and its promotion by the dihydropyridine calcium agonist Bay K 8644, *Proc. Natl. Acad. Sci. U.S.A.* **82:**2178–2182.

Palade, P. T., and Almers, W., 1985, Slow calcium and potassium currents in frog skeletal muscle: Their relationship and pharmacological properties, *Pflugers Arch.* (in press).

Reuter, H., 1983, Calcium channel modulation by neurotransmitters, enzymes and drugs, *Nature* **301:**569–574.

Rosemblatt, M., Hidalgo, C., Vergara, C., and Ikemoto, I., 1981, Immunological and biochemical properties of transverse tubule membranes isolated from rabbit skeletal muscle, *J. Biol. Chem.* **256:**8140–8148.

Sanchez, J. A., and Stefani, E., 1978, Inward calcium current in twitch muscle fibres of the frog, *J. Physiol.* **283:**197–209.

Schramm, M., Thomas, G., Towart, R., and Franckowiak, G., 1983, Novel dihydropyridines with positive ionotropic action through activation of Ca^{+2} channel, *Nature* **303:**535–537.

Stanfield, P. R., 1977, A calcium dependent inward current in frog skeletal muscle fibres, *Pflugers Arch.* **368:**267–270.

Towart, R., and Schramm, M., 1984, Recent advances in the pharmacology of the calcium channel, *Trends Pharmacol. Sci.* **5:**111–113.

Chapter 20

CALCIUM CHANNELS

MARK T. NELSON

1. Introduction

Intracellular calcium ions play a critical role in many tissues. In many cases, a rise in intracellular calcium is the trigger for one or more cellular functions such as release of neurotransmitters from nerve terminals, control of fertilization, and contraction of most types of muscle cells including skeletal, smooth, and cardiac (Byerly and Hagiwara, 1981). Calcium ions enter cells by moving down their electrochemical gradient through channels in specific transmembrane proteins. In most cases, these channels are voltage gated and open when the membrane is depolarized. In some nerve and muscle cells, an inward calcium current can actually account for regenerative spikes. The calcium channel is a major site for regulation by drugs and hormones. For example, in heart and invertebrate neurons, neurotransmitters, apparently acting via a cAMP–protein phosphorylation cascade, have been shown to regulate calcium channels in the target cells (Reuter, 1983). A number of substances (the so-called "organic calcium antagonists") have been shown to modify calcium channel activity in excitable cells (Triggle and Janis, 1984). These blockers (e.g., verapamil, nifedipine, and diltiazem) have been used clinically to treat certain cardiac arrhythmias, angina, and hypertension (see Triggle and Janis, 1984). The dihydropyridines appear to be the most exciting of these organic calcium antagonists because of their high potency and their potential as therapeutic agents.

Calcium currents have been difficult to study for the following reasons (see Hagiwara and Byerly, 1981 for details): (1) problems in separating calcium currents from other currents; (2) calcium that enters cells through calcium channels activates other conductances; (3) lack of good voltage control in preparations with strong Ca^{2+} currents; (4) inability to control the solute composition including the level of free calcium on the inside of the cell; and (5) inaccessibility of membrane structures that have important calcium currents (e.g., mammalian nerve terminals).

Recently, it has been possible to measure currents through single calcium channels (Tsien, 1983), thereby overcoming problems 1–3. Although patch-clamp

MARK T. NELSON • Department of Pharmacology, University of Miami School of Medicine, Miami, Florida 33101.

techniques have the distinct advantages of high resolution and preservation of the topology of channels in the cell membrane, the study of single Ca^{2+} channels has been limited by three important factors. (1) In the cell-attached configuration, neither the solution in the patch-clamp pipette nor that on the inside of the cell membrane can be readily manipulated, thereby hampering ion-substitution and channel-blocking experiments. In the outside-out configuration or the inside-out configuration, the experimenter has ready access to one side of the membrane but not to both sides of the channel in the patch. Although it has been possible to measure single-calcium-channel currents in outside-out patches (Carbone and Lux, 1984), thus far it has been difficult to make maintained recordings of single calcium channels in excised patches in the inside-out configuration (cf. Cavalie *et al.*, 1983). (2) Because membrane surface charges can have a pronounced effect on the measured conductance and kinetic properties of calcium channels (Wilson *et al.*, 1983), it is desirable to have control of membrane lipid composition. (3) Single ion channels in small or inaccessible membrane structures (i.e., nerve terminals, transverse tubule membranes) cannot be studied.

An alternative approach for studying the properties of single calcium channels is to incorporate these channels into planar lipid bilayers. Because the bilayer system permits direct control over the solutions on both sides of the membrane and over the lipid composition, it is well suited for investigation of the ion-permeation properties and the pharmacological regulation of calcium channels. During the last 2 years, calcium channels from brain synaptosomes (Nelson, 1983; Nelson *et al.*, 1984), *Paramecium* cilia (Ehrlich *et al.*, 1984), skeletal muscle transverse tubule membranes (Coronado and Affolter, 1985), and skeletal muscle sarcoplasmic reticulum membranes (Orozco *et al.*, 1985; Smith *et al.*, 1985) have been incorporated into planar lipid bilayers. Furthermore, the bilayer technique may provide the means for linking the functional properties of the purified calcium channel to structural elements of the channel molecule.

The goal of this chapter is to supply a guide for investigators who wish to incorporate calcium channels from native membranes into bilayers. The following section gives a general strategy for incorporating and identifying single calcium channels in planar lipid bilayers. Section 3 will show how this approach was used to identify a voltage-dependent calcium channel from rat brain synaptosomes.

2. Hunting for Calcium Channels

This section has five parts. The first three parts (2.1–2.3) describe a strategy for incorporating calcium channels from the biological material of interest into a planar lipid bilayer. Section 2.4 deals with the strategies for detecting small currents through calcium channels after they have been incorporated into a planar lipid bilayer that has large intrinsic membrane current noise. The last part (Section 2.5) describes criteria for the classification of measured current fluctuations as resulting from calcium channel activity.

2.1. Choice of Preparation

The first step is to choose a cell type or membrane structure that has a reasonably high density of calcium channels. Thus far, calcium channels have been incorporated into planar lipid bilayers from membrane preparations with high densities of calcium channels: nerve terminals (synaptosomes), skeletal muscle transverse tubule membranes, and *Paramecium* cilia.

The median eminence of the rat brain was chosen as a source of calcium channels becauses it is rich in nerve terminals and virtually devoid of neuronal cell bodies (Moore and Johnston, 1982). Nerve terminals are thought to have a high density of calcium channels. The membranes are prepared from 10–15 rat brain median eminences by a centrifugation procedure similar to that used for preparing rat whole-brain synaptosomes (Krueger *et al.*, 1979; Nelson *et al.*, 1984) with the omission of the sucrose gradient step. The synaptosome preparation has been shown to contain pinched-off, resealed nerve terminals (Blaustein *et al.*, 1977). These "resealed" nerve terminals are physiologically functional: they maintain resting potentials (Blaustein and Goldring, 1975), release neurotransmitters on depolarization (cf. Drapeau and Nachshen, 1984), and have functional sodium channels (Krueger *et al.*, 1983) and calcium channels (Nachshen and Blaustein, 1980). These "synaptosome" preparations were stored in low-concentration aliquots (about 0.1 mg/ml 0.4 M sucrose) at −77°C, these aliquots were further diluted to about 0.05 mg/ml before addition to a divalent-cation-containing solution that bathes one side of the bilayer. This procedure was used in order to minimize divalent-cation-induced aggregation of membranes.

2.2. Bilayers

A simple rule of thumb is to use the smallest possible lipid bilayer into which channels will incorporate with a reasonable frequency. Thus far, calcium channels have been incorporated into decane-containing planar lipid bilayers made of phosphatidylethanolamine (PE) and PE/phosphatidylserine (PS) mixtures derived from bovine brain (Avanti Polar Lipids, Birmingham, AL). To increase the probability of incorporating a functional calcium channel into the bilayer, I recommend that in preliminary experiments, medium-sized bilayers (hole diameter 150–250 μm) be used. For a bilayer painted on a 250-μm hole, the intrinsic membrane current noise is substantial, about 1 pA at 500 Hz. With single-calcium-channel currents being no greater than 1.5 pA, it is necessary to filter at 50–100 Hz in order to measure distinct current fluctuations. This heavy filtering clearly will remove many important fast fluctuations.

When calcium channels have been incorporated on a regular basis into medium-sized bilayers, then one can attempt to incorporate them into smaller bilayers. A simple procedure that I have used successfully to guarantee incorporation into small bilayers is first to incorporate channels into a 250-μm bilayer and then to transfer the membrane vesicle solution to a chamber that has a smaller hole (e.g., 100 μm). The calcium channel from brain synaptosomes has been studied in painted bilayers of 100 to 250-μm diameter.

2.3. Incorporation of Biological Material

Constituents of biological membranes such as ion channels can be incorporated into planar lipid bilayers by fusion of the biological membrane fragments with the planar bilayer (Cohen *et al.*, 1982). The incorporation of ion channels may occur by a process that is similar to exocytosis of secretory vesicles, with the result that the vesicle membrane becomes incorporated into the planar bilayer along with membrane channels in the vesicle. Although evidence for this mechanism has been provided by studies of liposome fusion with bilayers, the exact mechanism of incorporation of calcium channels is not known. A consequence of this fusion mechanism is that, provided all vesicles that incorporate are of the same sidedness, all channels will be incorporated with same orientation. In support of such a mechanism, the calcium channels from synaptosomes incorporate into bilayers with one orientation. This method has the advantages that the ion channels do not pass through an air–water interface and that the channels incorporate with a uniform orientation. Addition of the membrane vesicle suspension (final concentration 0.1–1.0 μg protein/ml solution) to one side of a preformed bilayer resulted in current fluctuations. To facilitate incorporation of membrane vesicles from other preparations, the surfaces of the membrane vesicles may need to be enzymatically "cleaned" by collagenase or other enzymes (analogous to the preparation of cell surfaces for patch clamping).

Recently, calcium channels from skeletal muscle (Affolter and Coronado, 1985) have been incorporated into bilayers in the tips of micropipettes by forming planar bilayers by the apposition of a monolayer of purified phospholipids and a monolayer made from the native membrane vesicles with all their intrinsic proteins (cf. Schlinder and Quast, 1980). This technique has the advantage of high current and time resolution (because of the small membrane surface area). It has the following disadvantages: the channel must pass through an air–water interface, thereby introducing the possibility of denaturation, and access to one side of the channel is lost.

2.4. Detecting Currents through a Single Calcium Channel: Small Currents among a Large Membrane Current Background

The rate of divalent cation movement through a single calcium channel can be as fast as about 4.5 million ions per second, corresponding to a single-channel current of 1.5 pA. The current noise of a bilayer made on 250-μm hole is about 0.5–1.0 pA at 500 Hz. Furthermore, the lifetime of the conducting state of the calcium channel can be brief and dependent on the transmembrane voltage (cf. Tsien, 1983). Therefore, to detect a small current among a large background, the single-channel current and mean conducting time must be maximized, and the background noise minimized. This section gives a general procedure for detecting calcium-channel current fluctuations in a planar lipid bilayer.

2.4.1. Step 1: Increase the Single-Channel Current

For preliminary experiments, barium ions should be used as the charge carrier for the following reasons (cf. Tsien, 1983): (1) barium ions usually move through single calcium channels faster than other divalent cations (cf. Nelson *et al.*, 1984); (2) barium ions block current flow through many types of cation channels that may complicate or obscure single-calcium-channel current measurements; (3) barium ions, unlike calcium ions, do not activate other conductances; (4) barium ions do not promote calcium-dependent inactivation. To maximize the single-channel current, high divalent cation concentrations should be used (single-channel currents tend to saturate above 100 mM divalent cation concentrations).

2.4.2. Step 2: Increase the Mean Conducting Time

2.4.2a. Via Voltage. The probabilty of a calcium channel being open is voltage dependent. A voltage protocol should be employed to find the voltage range in which the mean open (or burst) duration is easily resolvable (cf. Nelson *et al.*, 1984).

2.4.2b. Via Calcium Agonists. If there is reason to believe that the calcium channel that is under investigation is sensitive to the calcium agonist Bay K 8644, then this substance should be also included in the incubation medium. Bay K 8644 has been shown to increase the frequency of long-lived channel events (cf. Hess *et al.*, 1984; Brown *et al.*, 1984).

2.4.2c. Via Temperature. By lowering the temperature of the bathing solution, it may be possible to slow down the kinetics.

2.5. Criteria for a Calcium Channel

After one has identified a barium-conducting channel incorporated into a planar lipid bilayer, what criteria must be fulfilled in order to label this channel a "calcium channel." Unfortunately, a tetrodotoxin-equivalent toxin for the calcium channel has not as yet been found. Moreover, as described below, the properties of calcium channels vary substantially from cell type to cell type. In fact, recent evidence suggests that multiple types of calcium channels with substantially different properties can exist in the same cell membrane (Nowycky *et al.*, 1985). These observations suggest that the criteria for identifying a calcium channel must be made general enough to encompass these variations in channel properties.

The following general properties seem to be shared by all voltage-dependent calcium channels:

1. Permeability: Calcium, barium, and strontium ions permeate, with single-

channel conductances generally being the greatest with barium ions as the charge carrier

2. Single-channel conductances are small. However, they appear to vary from about 5 to 25 pS (for about 100 mM barium ions)
3. Selects for divalent cations over monovalent ions and magnesium
4. Voltage-dependent activation: The steepness and position of activation curves also appear to vary
5. Current flow through the open channel can be blocked by inorganic ions such as cadmium and lanthanum. The order and affinities of the blocking ions for the calcium channel also appear to vary from tissue to tissue (cf. Hagiwara and Byerly, 1981).

These five "common" properties can vary, but the following properties exhibit substantial differences among calcium channel types (Tsien, 1983):

1. Inhibition of calcium currents by organic agents such as the dihydropyridines, D600, and diltiazem range from none to potent. These agents appear to exert their actions by affecting the probability of the channel being open and not by blocking ion flow through the open channel (Hess *et al.*, 1984)
2. Inactivation. Inactivation of calcium currents varies from none to substantial. For calcium currents that inactivate, inactivation can be either voltage dependent or calcium dependent (Tsien, 1983)
3. Kinetics: multiple gating modes. Calcium channels can exhibit large differences in their closing and opening rates. In fact, a single type of calcium channel can exhibit substantial spontaneous shifts in gating (Hess *et al.*, 1984).

3. Single Calcium Channels from Brain Synaptosomes

By an approach similar to the one described in Section 2, single calcium channels from brain synaptosomes were identified and characterized. This section describes results indicating that these channels fulfill the criteria for being calcium channels.

3.1. Ion Permeation and Selectivity

After the exposure of a planar lipid bilayer to membrane vesicles from rat brain synaptosomes, current fluctuations such as the ones shown in Fig. 1 were observed. These unitary current fluctuations appear to be the result of the opening and closing of single calcium channels. Calcium, barium, and strontium ions can permeate this channel (Fig. 1) (Nelson *et al.*, 1984), thereby satisfying criterion 1 for the identification of calcium channels. In addition, manganese ions can also permeate this channel (Nelson, 1984; Drapeau and Nachshen, 1984) (Fig. 2). The single-channel conductances were small (criterion 2), with barium (8 pS) > calcium = strontium (5–6 pS) > manganese (4 pS). In the presence of these permeant divalent cations, cadmium, lanthanum, potassium, and sodium ions did not have measurable permeation rates (a divalent cation flux rate of about 300,000 ions per second or 0.1 pA is about the limit of resolution).

The calcium channel appears to select for certain divalent cations over other ions not by simple rejection of undesired ions but by preferential binding of one ion type over another (cf. Hess and Tsien, 1984). Therefore, a simple comparison of the single-channel conductances of the channel with a variety of permeant ions as charge carriers does not directly indicate what the relative selectivities are. Relative selectivities can, however, be determined from the effects of mixtures of ions on the reversal potentials (E_{rev}) of the single-channel currents. The reversal potentials of currents through the single calcium channels from synaptosomes were dependent on the permeant divalent cations and were not affected by monovalent ions (Nelson *et al.*, 1984) (criterion 3 for a calcium channel).

An important feature of calcium channels is that they select for permeant divalent cations over magnesium ions. In the presence of identical concentrations of permeant divalent cations, the single-channel current–voltage relationship was linear and extrapolated to 0 mV. The replacement of the *cis* side by a magnesium–calcium mixture caused E_{rev} to shift from O mV to +48 mV. The equilibrium potential for magnesium (E_{Mg}) was nominally minus infinity, and the calcium equilibrium potential was calculated to be +58 mV. If the channel was magnesium selective, then E_{rev} would shift from O mV in the negative direction, and if the channel was calcium selective, then E_{rev} would be between O mV and +58 mV. E_{rev} was determined to be +48 mV, suggesting that the channel is highly selective for calcium ions over magnesium ions. An exact selectivity ratio was not determined because the reversal potentials were estimated by linear extrapolation, and E_{Mg} was undefined (minus affinity). Furthermore, if the channel indeed contains two divalent cation binding sites, then the selectivity ratio would depend on the ratios of the permeant ions on both sides of the channel (see Hille, 1984).

3.2. Channel Gating

3.2.1. Voltage Dependence

One property common to all calcium channels is that depolarizing the cell membrane increases the probability of the channel being open. For the synaptosomal calcium channel, membrane depolarization decreased the mean closed time and increased the mean conducting (open) time (Nelson *et al.*, 1984) (criterion 4). The open times of the channel were punctuated with fast, incompletely resolved closures, and, therefore, at higher resolution, these open times may be analogous to "burst" times.

Figure 2A illustrates the effect of voltage on a single calcium channel. At −120 mV, current fluctuations were not observed (bottom trace). When the voltage was stepped to +120 mV, current fluctuations were observed (top trace). The

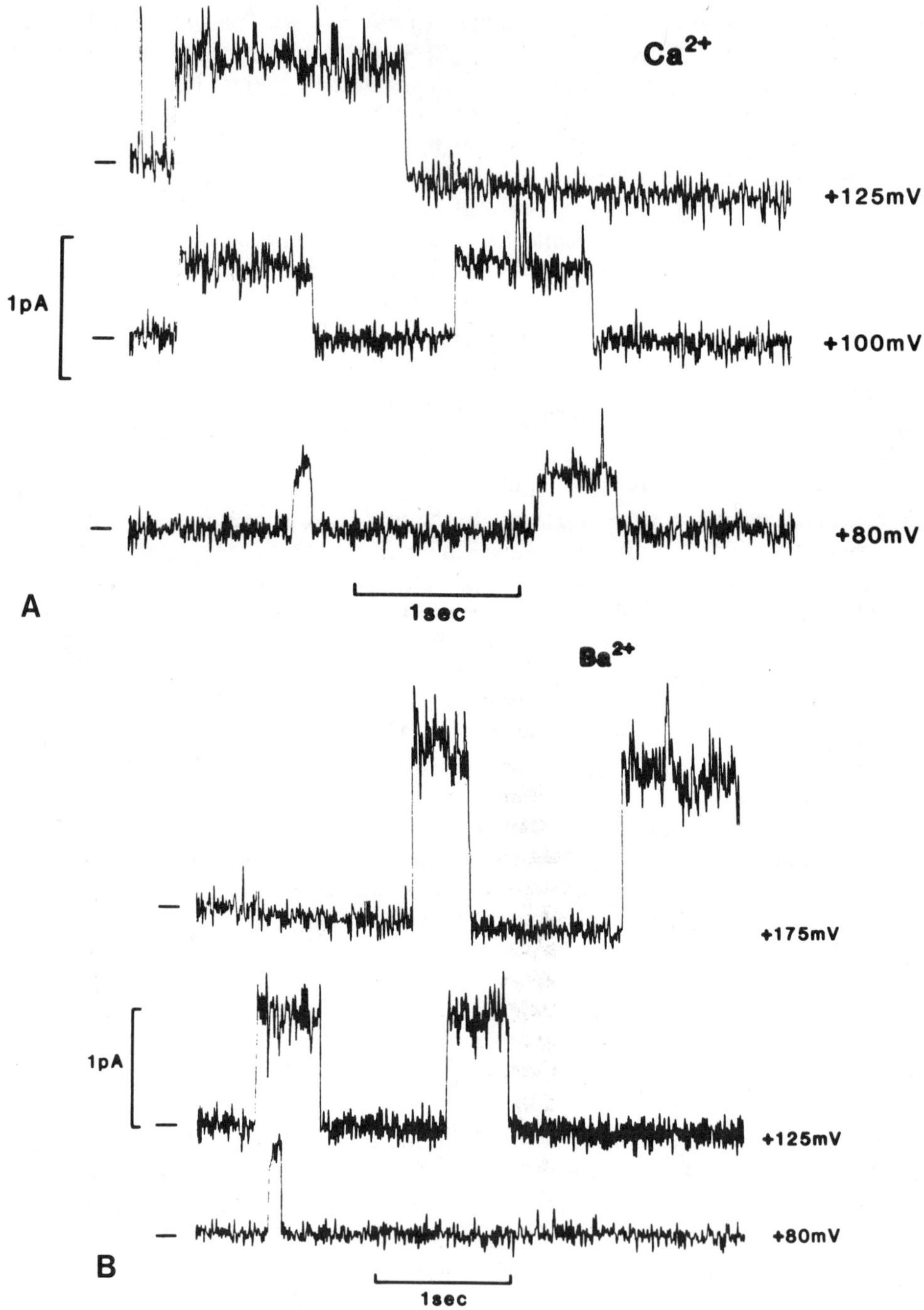

Figure 1. Currents through single calcium channels from brain synaptosomes. A: Calcium ion movement through single channels. Single-channel conductance was about 5 pS and independent of voltage. B: Barium ion movement through single channels. The single-channel conductance was about 9 pS. Note that even at voltages of +175 mV, the channel was not open all of the time. C: Strontium ion movement through single channels. The single-channel conductance was about 6 pS. In all experiments there were identical concentrations of divalent cations (250 mM) on both sides of the membrane; i.e., the single-channel current reversal potential was at 0 mV.

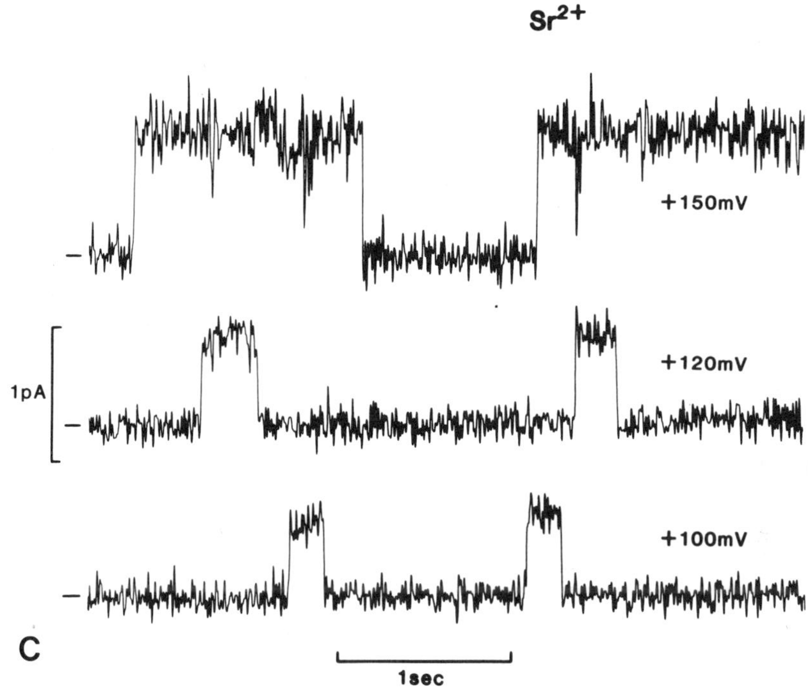

Figure 1. (*continued*)

difference between zero current and the current level at −120 mV (−0.8 pA) corresponds to the current flow across the bare bilayer in response to a 120-mV change in voltage. Figure 2B shows the effect of voltage on the probability of a channel being open (*P*). The relationship can be fit with a curve derived from the Boltzmann equation assuming an "equivalent gating charge" of 1.5 (range being 1.5–2.0). The equilibrium constant for the open–close reaction changed e-fold with 16 mV. Although the voltage–*P* relationship could be fit with a Boltzmann function, it was not possible to drive *P* to 1 even with voltages as high as +245 mV (see Fig. 1B; cf. Cavalie *et al.*, 1983). The significance of this observation is not clear. However, it may suggest that the kinetic scheme of this calcium channel is more complicated than suspected.

The synaptosomal calcium channels under the experimental conditions used did not appear to inactivate. This property was not investigated in detail. In general, calcium channels that are involved in excitation–secretion coupling do not exhibit substantial inactivation (Nachshen and Blaustein, 1980; Tsien, 1983).

3.2.2. Ion Dependence

The single-channel conductances are dependent on the nature of the permeating ion (Nelson *et al.*, 1984; Nelson, 1984). An interesting finding was that

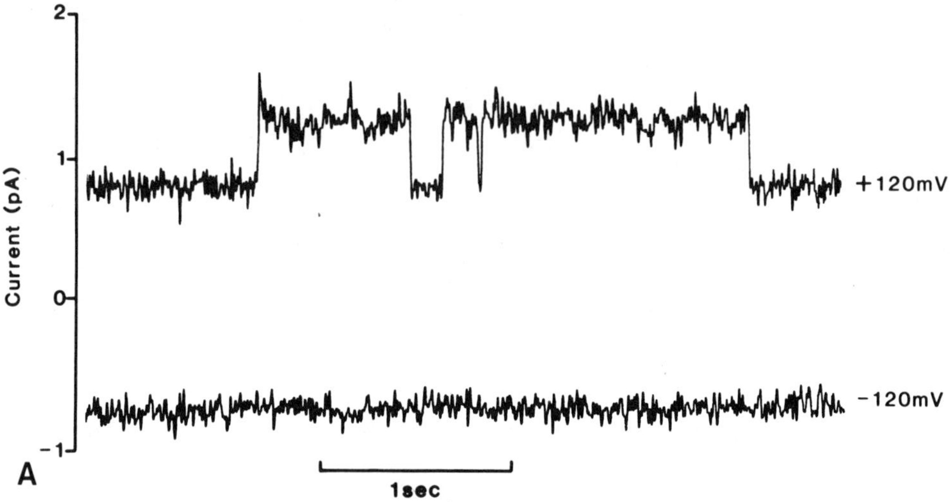

Figure 2. Voltage dependence of single calcium channels. A: Original record of current fluctuations at −120 mV and +120 mV. Manganese was the charge carrier. B: The relationship between the probability of the channel being open (*P*) and voltage. The membrane had two functional channels in it, and *P* was determined from the opening and closing rate constants. At +190 mV, *P* was determined from the mean current (*I*), where $P = (I)/(i)(N)$ (*N* is the number of functional channels, and *i* is the unitary current). The solid line was derived from the Boltzmann function, $P = 1/(1 + \exp(qF(V - V^*)/RT)$, where *q*, the "equivalent gating charge," was equal to 1.5 and was determined from the linearization of the equation in C, V^* was the voltage at which the opening and closing rates were equal. C: The logarithmic transform of the dissociation constant (*K*) of the close–open reaction of the channel. *K* = (mean open time)/(mean closed time) or $(1 - P)/(P)$. The opening and closing rates were equal at about +162 mV (i.e., $K = 1$) and *K* changes e-fold for about 16 mV change in membrane potential.

the mean open (burst) times of the synaptosomal calcium channel also depended on the nature of the permeant divalent cation. The order of single-channel conductances (Ba > Ca = Sr > Mn) was opposite that of the mean open times. This apparent relationship between ion permeation and mean open times suggests that a common site(s) regulates these processes. However, the nature of the charge carrier did not affect the mean closed times. Furthermore, the effect depended on the ion on the charge-carrying side of the channel and not on the nature of the ion on the opposite side. This evidence suggests that the site of action of the permeant ion on the mean open time is in the channel. A similar relationship between single potassium-channel conductances and mean open times has recently been reported (Spruce *et al.*, 1984).

One possible explanation for such ion dependence is that the calcium channel rapidly fluctuates between two nonzero conducting states or conformations (cf. Läuger, 1983). Each of these conformations would have a different lifetime. Therefore, the measured unitary current and mean conducting time of a single channel would be the average of time spent in each conducting conformation. By interacting with its binding site in the channel, a permeant ion could possibly affect

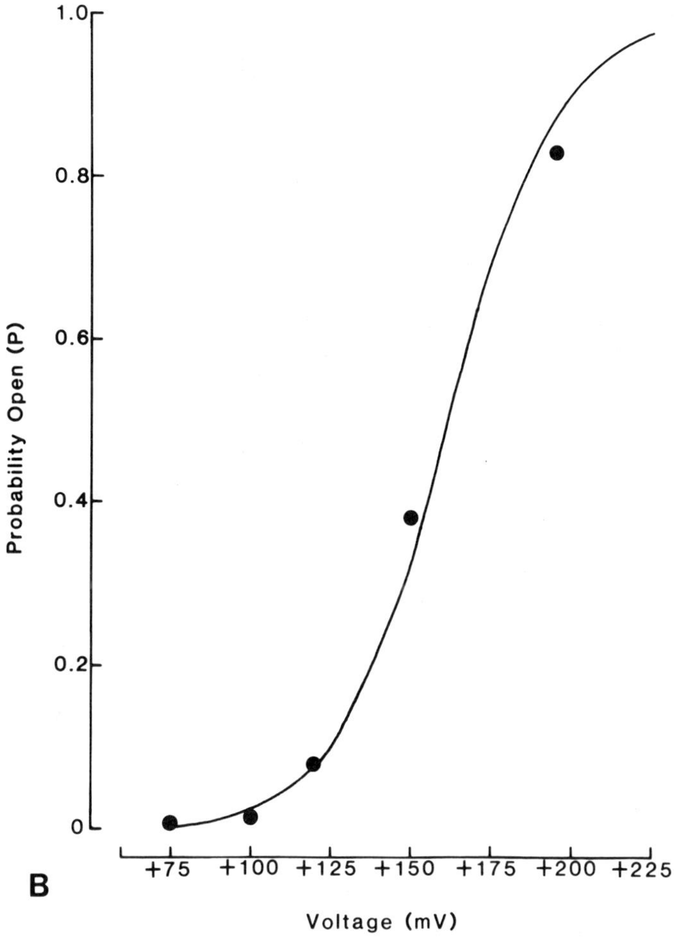

B

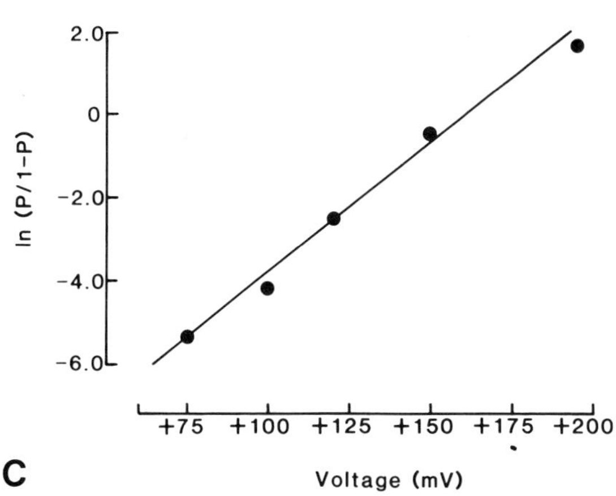

C

Figure 2. (*continued*)

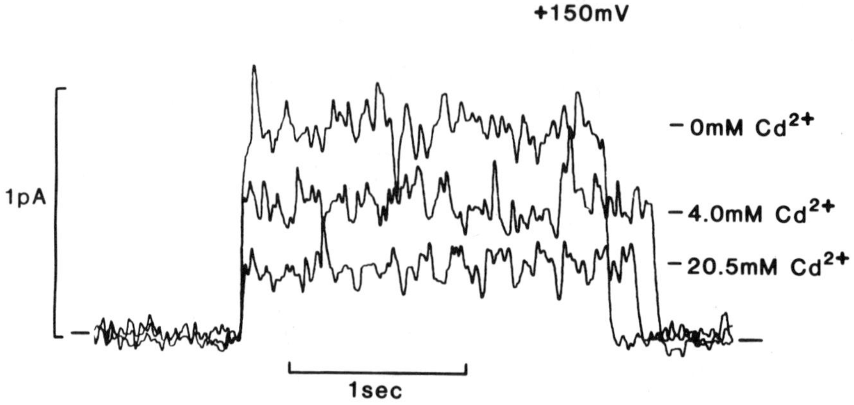

Figure 3. Reduction of single-channel currents by cadmium. Comparison of single-channel currents with 0 Cd, 4.0 mM Cd, and 20.5 mM Cd in the *cis* side. Current records taken in the presence of each concentration of Cd were superimposed. Calcium (250 mM) was the charge carrier. Filtered at 30 Hz.

the distribution between these two conformations and thereby affect both the apparent single-channel current and the mean conducting time. The effects of ion concentration, ionic strength, and temperature on these properties may provide evidence for or against such a model.

3.3. Block

Another property common to all calcium channels studied to date is that they can be blocked by multivalent cations such as cadmium, cobalt, or lanthanum (criterion 5). The inorganic blocking cations appear to compete with the permeant ions for a common binding site (Hagiwara and Byerly, 1981). In the presence of permeating ions, lanthanum and cadmium, when added to the *cis* side of the bilayer, reduced the single-channel current in a dose-dependent manner (Fig. 3; see also Figure 4 from Nelson *et al.*, 1984, for an example of lanthanum block). The block by these ions is explained by a mechanism whereby one cadmium ion binds to a single site in the open state of the channel and is unable to move through the channel (Nelson, 1985). While bound to this site, cadmium prevents ion flow through the channel. The mean block time of the channel by a cadmium ion has been estimated to be about 25 μsec (i.e., far below the resolution of the recording system presently employed). Although these results indicate that a single site accessible from the *cis* side regulates ion flow through the channel, they do not prove that this site is the same as the one(s) that the permeant ions bind on their passage through the channel. To provide evidence that cadmium ions interact with the permeation site, the effects of the nature of permeant ion on the block by cadmium were investigated.

The nature of permeant ion did not affect the maximal block possible (100%), but it did affect the concentration of cadmium ions needed for 50% reduction of

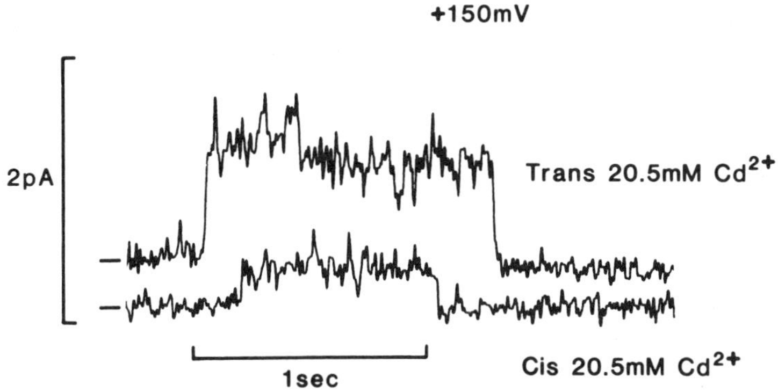

Figure 4. Sidedness of cadmium block. Cadmium ions (20.5 mM), when added to the *trans* side, reduced the single-channel current by 16% and, when added to the *cis* side, reduced the single current by 72%.

the single-channel current ($K_{0.5}$) with the following order: $K_{0.5}$(Mn) > $K_{0.5}$(Ca = Sr) > $K_{0.5}$(Ba). This order was inverse to that of the single-channel currents and consistent with cadmium ions competing with the permeating ions for a common site in the channel. Furthermore, cadmium ions blocked current flow when added to either side of the channel. However, the block was more potent from the *cis* side (Fig. 4). This difference in efficacy of block could be caused by a voltage dependence of the blocking reaction.

To resolve the discrete block and unblock events, more potent blockers (i.e., ones with longer mean blocked times) such as lanthanum ions must be used, and the membrane current noise must be reduced by using small bilayers. If discrete block and unblock events cannot be discerned with some blocking ions, then it may be possible to analyze the current fluctuations that have been truncated by the recording system (cf. "flickery" block, Yellen, 1984; for an example see Figure 4 of Nelson *et al.*, 1984). The order of blocking affinity of inorganic ions for the synaptosomal calcium channel was lanthanum > cadmium > manganese ≫ magnesium, sodium, potassium. Although all types of calcium channels can be blocked by these polyvalent ions, the relative order of efficacy of block has been reported to vary (cf. Hagiwara and Byerly, 1981). If this proposition is found to be true at the single-channel level, then this would suggest that there are fundamental differences between the permeation sites in calcium channels in different tissues. Recent evidence suggests that the on-rate constant of cadmium may be fast (10^9 M^{-1} sec^{-1}) and essentially diffusion limited (Lansman *et al.*, 1985). The off-rate constants for cadmium ions are much slower, about 10^3–10^4 sec^{-1}.

Isotopic fluxes through synaptosomal calcium channels are sensitive to certain organic calcium-channel blockers (Nachshen and Blaustein, 1979). In accord with these observations, single calcium channel activity from synaptosomes in a bilayer was affected by 5 μm D600. In agreement with the macroscopic current data from skeletal muscle (Almers and McCleskey, 1984), micromolar concentrations of nitrendipine affected single calcium channels from skeletal muscle incorporated into planar lipid bilayers (Coronado and Affolter, 1985), indicating that the nitrendipine site of the calcium channel can also be incorporated into planar lipid bilayers.

4. Conclusions and Prospects

To date, calcium channels from brain synaptosomes, *Paramecium* cilia, skeletal muscle transverse tubule membranes, and skeletal muscle sarcoplasmic reticulum have been incorporated into planar lipid bilayers. These studies constitute the first single-calcium-channel recordings from these tissues. The future should see an increase in the number of investigators studying calcium channels from a variety of tissues incorporated into planar lipid bilayers. As mentioned above, a number of these calcium channels may not be ordinarily accessible to direct electrical measurement. The bilayer system should prove particularly useful for the investigation of the ion permeation and blocking properties of calcium channels. Easy access to both sides of the calcium channel will facilitate studies of the effects of pharmacological and modulating agents on the channel's properties.

A single-channel experiment can unambiguously identify a channel as being a calcium channel. However, to determine that a calcium channel in the bilayer is indeed derived from the membrane under investigation requires a detailed comparison of the channel's properties in the bilayer and in the intact system. When possible, the properties of the single calcium channels in the bilayer and in the intact membrane patch should be compared under identical experimental conditions. For further verification of the channel's identity, single-channel properties should also be compared with the macroscopic properties.

Nitrendipine, a "calcium antagonist," has been used as a high-affinity ligand in order to purify a protein that may be the voltage-dependent calcium channel (Curtis and Catterall, 1984). However, because there can be large differences between the binding affinity of nitrendipine for these proteins and nitrendipine's potency as a functional inhibitor of calcium movements, the possibility still remains that this protein is not the calcium channel (cf. Bean, 1984). One viable way of demonstrating that this protein is indeed the calcium channel is to reconstitute the nitrendipine-binding protein into planar lipid bilayers. Furthermore, if the reconstitution proved successful, then the bilayer system could be used to determine the effects of structural modifications of the purified calcium channel on its single-channel behavior.

Acknowledgment. Supported by The American Heart Association.

References

Affolter, H., and Coronado, R., 1985, Planar bilayer recording of single calcium channels from purified muscle transverse tubules, *Biophys. J.* **47**:434a.

Almers, W., and McCleskey, E. W., 1984, Nonselective conductance in calcium channels of frog muscle: Calcium selectivity in a single-file pore, *J. Physiol.* (*Lond.*) **353**:585–608.
Bean, B., 1984, Nitrendipine block of cardiac calcium channels: High-affinity binding to the inactivated state, *Proc. Natl. Acad. Sci., U.S.A.* **81**:6388–6392.
Blaustein, M. P., and Goldring, J. M., 1975, Membrane potentials in pinched-off presynaptic nerve terminals monitored with a fluorescent probe: Evidence that synaptosomes have potassium diffusional potentials, *J. Physiol.* (*Lond.*) **247**:589–615.
Blaustein, M. P., Kendrick, N. C., Fried, R. C., and Ratzlaff, R. W., 1977, Calcium metabolism at the mammalian presynaptic nerve terminal: Lessons from the synaptosome, in *Society for Neuroscience Symposia*, Volume 2 (M. Cowan and J. A. Ferrendelli, eds.) pp. 172–194, Society for Neuroscience, Bethesda, Maryland.
Brown, A., Kunze, D., and Yatani, A., 1984, The agonist effect of dihydropyridines on Ca channels, *Nature* **311**:570–572.
Carbone, E., and Lux, H., 1984, A low voltage-activated, fully inactivating Ca channel in vertebrate sensory neurones, *Nature* **310**:501–502.
Cavalie, A., Ochi, R., Pelzer, D., and Trautwein, W., 1983, Elementary currents through Ca channels in guinea pig myocytes, *Pflugers Archiv* **398**:284–297.
Cohen, F. S., Akabas, M. H., and Finkelstein, A., 1982, Osmotic swelling of phospholipid vesicles causes them to fuse with a planar lipid bilayer membrane, *Science* **217**:458–460.
Coronado, R., and Affolter, H., 1985, Kinetics of dihydropyridine-sensitive single calcium channels from purified muscle transverse tubules, *Biophys. J.* **47**:434a.
Curtis, B., and Catterall, W., 1984, Purification of the calcium antagonist receptor of the voltage-sensitive calcium channel from skeletal muscle transverse tubules, *Biochemistry* **23**:2113–2118.
Drapeau, P., and Nachshen, D., 1984, Manganese fluxes and manganese-dependent neurotransmitter release in presynaptic nerve endings isolated from rat brain, *J. Physiol.* **348**:493–510.
Ehrlich, B., Finkelstein, A., Forte, M., and Kung, C., 1984, Voltage-dependent calcium channels from *Paramecium* cilia incorporated into planar lipid bilayers, *Science* **225**:427–428.
Hagiwara, S., and Byerly, L., 1981, Calcium channel, *Ann. Rev. Neurosci.* **4**:69–125.
Hess, P., Lansman, J., and Tsien, R. W., 1984, Different modes of Ca channel gating behaviour favoured by dihydropyridine Ca agonists and antagonists, *Nature* **311**:538–544.
Hess, P., and Tsien, R. W., 1984, Mechanism of ion permeation through calcium channels, *Nature* **309**:453–456.
Hille, B., 1984, Ionic channels in excitable membranes, Sinauer, Sunderland, Massachusetts.
Krueger, B. K., Ratzlaff, R. W., Strichartz, G. R., and Blaustein, M. P., 1979. Saxitoxin, binding to synaptosomes, membranes, and solubilized binding sites from rat brain, *J. Membr. Biol.* **50**:287–310.
Krueger, B. K., Worley III, J. F., and French, R. J., 1983, Single sodium channels from rat brain incorporated into planar lipid bilayer membranes, *Nature* (*Lond.*) **303**:172–175.
Lansman, J., Hess, P., and Tsien, R. W., 1985, Direct measurement of entry and exit rates for calcium ions in single calcium channels, *Biophys. J.* **47**:67a.
Läuger, P., 1983, Conformational transitions of ionic channels, in *Single-Channel Recording* (B. Sakmann and E. Neher, eds.), pp. 177–188, Plenum, New York.
Moore, K. E., and Johnston, C. A., 1982, The median eminence: Aminergic control mechanisms, in *Neuroendocrine perspectives*, Volume 1 (E. E. Mueller and R. M. Macleod, eds.), pp. 25–68, Elsevier Biomedical Press, New York.
Nachshen, D. A., and Blaustein, M. P., 1979, The effects of some organic calcium antagonists on calcium influx in presynaptic nerve terminals, *Mol. Pharmacol.* **16**:579–586.
Nachshen, D. A., and Blaustein, M. P., 1980, Some properties of potassium-stimulated calcium influx in presynaptic nerve endings, *J. Gen. Physiol.* **76**:709–728.
Nelson, M. T., 1983, Single calcium-channel current measurements from brain synaptosomes in planar lipid bilayers, *Soc. Neurosci. Abstr.* **9**.
Nelson, M. T., French, R. J., and Krueger, B. K., 1984, Voltage-dependent calcium channels from brain incorporated into planar lipid bilayers, *Nature* **308**:77–80.
Nelson, M. T., 1984, Permeant ions affect both the conductance and closing rate of single calcium channels from rat brain in planar bilayers, *J. Physiol.* **357**:58P.

Nelson, M. T., 1985, Divalent cation interactions with single calcium channels from rat brain: Evidence for two sites, *Biophys. J.* **47**:67a.

Nowycky, M., Fox, A. P., and Tsien, R. W., 1985, Three types of calcium channels in chick dorsal root ganglion cells, *Biophys. J.* **47**:67a.

Orozco, C., Suarez-Isla, B. A., Froehlich, J., and Heller, P., 1985, Calcium channels in sarcoplasmic reticulum membranes, *Biophys. J.* **47**:57a.

Reuter, H., 1983, Calcium channel modulation by neurotransmitters, enzymes and drugs, *Nature* **301**:569–574.

Schlindler, H., and Quast, U., 1980, Functional acetylcholine receptor from *Torpedo marmorata* in planar membranes, *Proc. Natl. Acad. Sci. U.S.A.* **77**:3052–3056.

Smith, J., Coronado, R., and Meissner, G., 1985, A nucleotide-stimulated calcium conducting channel from sarcoplasmic reticulum incorporated into planar lipid bilayers, *Biophys. J.* **47**:451a.

Spruce, A. E., Standen, N. B., Stanfield, P. R., and Wilson, S. W., 1984, Rubidium ions prolong potassium channel opening in frog skeletal sarcolemma, *J. Physiol.* (*Lond.*) **357**:45P.

Triggle, D., and Janis, R., 1984, Calcium channel antagonists: New perspectives from the radioligand binding assay, in: *Modern Methods in Pharmacology*, pp. 1–28, Alan Liss, New York.

Tsien, R. W., 1983, Calcium channels in excitable cell membranes, *Ann. Rev. Physiol.* **45**:341–358.

Wilson, D. L., Morimoto, K., Tsuda, Y., and Brown, A., 1983, Interaction between calcium ions and surface charge as it relates to calcium currents, *J. Membr. Biol.* **72**:309–324.

Yellen, G., 1984, Ionic permeation and blockade in Ca-activated K channels of bovine chromaffin cells, *J. Gen. Physiol.* **84**:157–187.

Chapter 21

PHOSPHORYLATION OF A RECONSTITUTED POTASSIUM CHANNEL

IRWIN B. LEVITAN

1. Introduction

Ion channels in biological membranes are dynamic entities. They of course act as pores through which ions can move down their electrochemical gradients, but more than that, many of them are regulatable pores. Channels can exist in more than one functional state, and rapid transitions between these states can be regulated by the voltage across the membrane or by the binding of a small molecule to a specific receptor that is intimately associated with the channel. Furthermore, it is now becoming evident that these rapid transitions are themselves subject to longer-term modulation, which may last for seconds, minutes, or even hours.

What mechanisms might give rise to long-lasting modulation of ion-channel activity in excitable cells? Cyclic AMP (cAMP) and calcium ions are well established as second messengers in the regulation of carbohydrate metabolism, where they modulate the activity of certain enzymes by causing them to be phosphorylated (for a recent review see Cohen, 1982). It has long been suspected that they might play a similar role in the regulation of ion channels, and evidence from several laboratories has confirmed that this is the case. Most studies of long-term channel modulation have utilized voltage-clamp techniques, and inferences about channel activity have been drawn from macroscopic current measurements. More recently, membrane patch recording and channel reconstitution have been used to obtain more direct measures of ion-channel activity and its regulation. In this chapter I discuss results from my laboratory that indicate that reconstitution is a powerful tool to investigate the molecular details of the metabolic modulation of ion channels.

1.1. Background

Beginning about the late 1960s, it became fashionable to put cyclic nucleotides on or into nerve and muscle cells and ask whether this did anything to their

IRWIN B. LEVITAN • Graduate Department of Biochemistry, Brandeis University, Waltham, Massachusetts 02254.

electrical activity. By the end of the 1970s, it had become clear that cAMP can indeed alter membrane properties (for a recent review see Siegelbaum and Tsien, 1983), and the search began for the molecular mechanisms of the actions of cAMP. On the one hand, voltage-clamp approaches were used to identify the ionic conductances that are altered by cAMP, and one fact that emerged is that the cAMP-modulated conductance is cell, laboratory, and investigator dependent; that is, there is no single ionic conductance that is the universal target for cAMP action, but rather the conductance that is affected varies from cell to cell, even within the same organism. This rather surprising and little understood finding has stood the test of time, and indeed, the list of ionic currents that are modulated by cAMP continues to grow.

The second major approach has been to investigate the biochemical basis of cAMP's action and, more specifically, to ask whether its effects are mediated by the action of a cAMP-dependent protein kinase. Several laboratories have introduced the active catalytic subunit of the kinase directly into nerve or muscle cells and have found that it can modulate membrane properties in a way predictable from the known actions of cAMP on these same cells (Kaczmarek *et al.*, 1980; Castellucci *et al.*, 1980; DePeyer *et al.*, 1982; Osterrieder *et al.*, 1982; Alkon *et al.*, 1983). A complementary approach to this pharmacological use of exogenous catalytic subunit is to inject intracellularly a naturally occurring highly specific inhibitor of cAMP-dependent protein kinase. Experiments of this type have established a role for endogenous kinase in the physiological actions of serotonin on several different molluscan neurons (Adams and Levitan, 1982; Castellucci *et al.*, 1982).

1.2. Single-Channel Analysis

More recently, the techniques of single-channel analysis have been applied to the problem of long-term modulation. One important advance has been the introduction of a general test for involvement of an intracellular second messenger in a physiological response. Since the gigohm seal between a patch electrode and a cell membrane is a lateral diffusion barrier, neurotransmitter applied to the extracellular medium outside a patch electrode does not have access to receptors located in the patch of membrane under the electrode. Thus, if neurotransmitter outside the electrode affects channels in the membrane patch, this must result from occupation of receptors outside the patch, and the only way these receptors can communicate with the channels in the patch is by means of some diffusible intracellular messenger. This elegant technique has been used to demonstrate a role for cAMP in the serotonin-induced closing of individual K^+ channels in *Aplysia* sensory neurons (Siegelbaum *et al.*, 1982), and for Ca^{2+} in the effects of acetylcholine and cholecystokinin on nonspecific cation channels in pancreatic acinar cells (Maruyama and Petersen, 1982).

Patch recording has also been used to examine further the role of protein phosphorylation in channel modulation. By detaching patches from cells in the "inside-out" configuration (Hamill *et al.*, 1981), one can gain access to the cytoplasmic membrane surface and investigate the effects of applying the catalytic subunit of cAMP-dependent protein kinase together with ATP to the membrane. It has been found that phosphorylation of detached patches in this way leads to closure of single serotonin-sensitive K^+ channels from *Aplysia* sensory neurons (Shuster *et al.*, 1985) and to activation of Ca^{2+}-dependent K^+ channels in *Helix* neurons (Ewald *et al.*, 1985). These results indicate that the phosphorylation target is something that comes away with the patch of membrane when it is detached from the cell. As I detail below, this approach recently has been extended to reconstituted Ca^{2+}-dependent K^+ channels from *Helix*.

2. Reconstitution of Channels from Molluscan Nervous System: Methodology

Our initial efforts to reconstitute ion channels from *Helix* and *Aplysia* nervous system were almost too successful. Crude membrane vesicle preparations from the ganglia of these creatures fuse extremely readily with large planar bilayers, and one of the major problems has been to obtain single-channel membranes and to pick out the channel of interest from the several different kinds that have been observed. As is evident from the results described below, this problem has largely been circumvented by the use of much smaller bilayers formed on the tip of patch electrodes.

2.1. Preparation of Membrane Vesicles

We routinely prepare crude membrane vesicles for channel reconstitution by homogenizing ganglia, identified groups of neurons, or single identified neurons in a low-ionic-strength buffer (e.g., 20 mM MOPS or HEPES) at neutral pH. The homogenate is centrifuged at 10,000 *g* for 15 min to remove connective tissue, nuclei, and large membrane fragments, and the supernatant is recentrifuged at 100,000 *g* for 60–90 min. The pellet from this second centrifugation is taken as the source of membrane vesicles for reconstitution.

2.2. Large Planar Bilayers

For reconstitution by fusion with large planar bilayers, the membrane pellet is resuspended in 20 mM MOPS or HEPES containing 1 M sucrose. The preparation is allowed to sit at 0°C for 12–36 hr before being added to the *cis* side of a standard bilayer chamber (see Alvarez, Chapter 5, this volume). Usually about 0.5–5 μg of membrane protein is added to the 3-ml chamber. The membranes are formed by painting a mixture of phosphatidylethanolamine (PE) and phosphatidylserine (PS) (3:1) in decane over the hole in the partition between the *cis* and *trans* sides. Fusion of channel-containing vesicles, measured by the change in

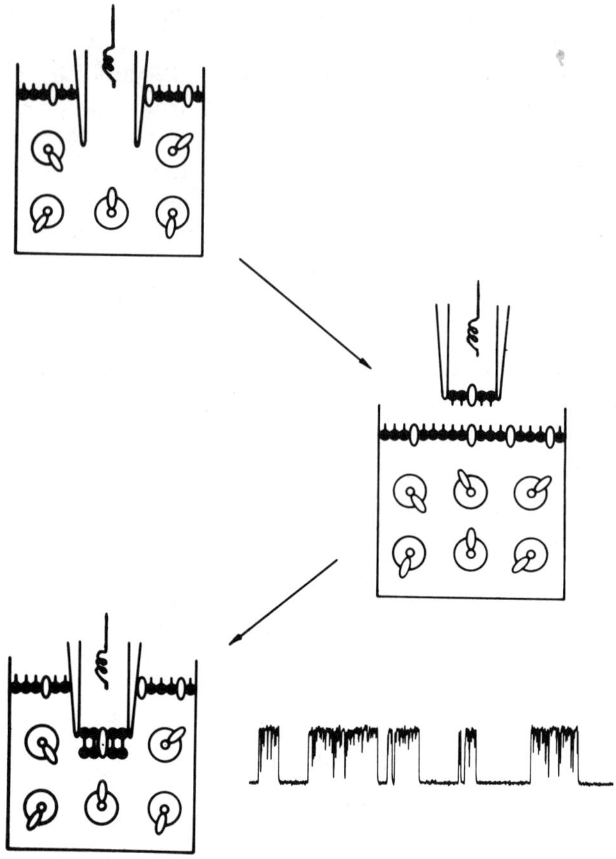

Figure 1. Formation of a bilayer on the tip of a patch electrode. A frozen and thawed vesicle/phospholipid mixture is placed in a small chamber, and a lipid monolayer containing ion channels (and other membrane proteins) forms spontaneously at the air–liquid interface. A fire-polished patch electrode is inserted through the interface while positive pressure is applied (top). The pressure is released, and the electrode is withdrawn through the interface to pick up a monolayer on the tip (middle). When it is reinserted, it picks up a second monolayer to form a bilayer (bottom left), and if there is an ion channel in the bilayer, single-channel currents can be recorded (bottom right). (For similar pictures see Wilmsen *et al.*, 1983; Suarez-Isla *et al.*, 1983; Coronado and Latorre, 1983.)

conductance across the bilayer, usually occurs within seconds or minutes; on rare occasions a transmembrane osmotic gradient of KCl or urea was used to promote fusion. Both the *cis* and *trans* sides of the bilayer chamber contain KCl and no other added ion (except Ca^{2+} where specifically noted). We have not yet carried out experiments under ionic conditions that would reveal the presence of channels selective for ions other than Cl^- or K^+.

2.3. Bilayers on Patch Electrodes

We have also reconstituted channels from molluscan neurons in much smaller bilayers formed on the tips of patch electrodes. This "tip-dip" method, which is described in Fig. 1, was introduced by Wilmsen *et al.* (1983) and has been used very effectively by several other workers (Suarez-Isla *et al.*, 1983; Coronado and Latorre, 1983). For our experiments, the membrane pellet after the 100,000-*g* centrifugation is resuspended in 20 mM MOPS or HEPES containing an appropriate concentration of KCl. This is then mixed with a sonicated suspension of PE:PS (3:1) in the same buffer to give a lipid/protein ratio of about 10/1. Then Ca^{2+} is added to a final concentration of 0.1 mM, and the mixture is frozen and thawed three times prior to its addition to a small chamber in which the bilayer is formed, as described in Fig. 1. After formation of the bilayer, the electrode can be transferred to adjacent chambers that contain different bathing media. With minor modifications of the technique and the use of very small chambers, we have been successful in preliminary experiments in reconstituting channels from individual identified nerve cells.

3. Phosphorylation of Reconstituted Ca^{2+}-Dependent K^+ Channels

The impetus to undertake these experiments came from our earlier finding that intracellular perfusion of the catalytic subunit of cAMP-dependent protein kinase in certain *Helix* neurons can selectively enhance the Ca^{2+}-dependent K^+ current in these cells (DePeyer *et al.*, 1982). Subsequent experiments with detached membrane patches (described above) suggested that the phosphorylation target is not a cytoplasmic protein but is something that remains asociated with the membrane when the patch is pulled off the cell. To extend this approach and ask whether the phosphorylation target is the channel itself, we attempted to identify and phosphorylate Ca^{2+}-dependent K^+ channels in a reconstituted system.

3.1. Large Planar Bilayers

As indicated above, *Helix* and *Aplysia* membrane vesicles fuse readily with planar bilayers, and with KCl on both sides of the bilayer we have observed at least four channels that can be distinguished one from another on the basis of their single-channel conductances and gating kinetics. Examples of several of these channels are shown in Fig. 2. When there is a KCl gradient across the membrane, virtually all of the single-channel currents reverse at the K^+ equilibrium potential, indicating that they select strongly for K^+ over Cl^-; on only one occasion have we observed what appears to be a Cl^- channel from these vesicles. At least one of the K^+ channels is Ca^{2+} dependent, since its gating is dependent on the presence of Ca^{2+} on the *cis* side of the membrane (Fig. 3). About 20% of the time this channel appears to insert "backwards" and requires Ca^{2+} on the *trans* side. When the catalytic subunit of cAMP-dependent protein kinase, together with Mg-ATP, is added to the *cis* side of the bilayer, there is a marked change in the gating kinetics of the Ca^{2+}-dependent K^+ channel (Fig. 3). The channel remains open much more of the time after phosphorylation, and the gating

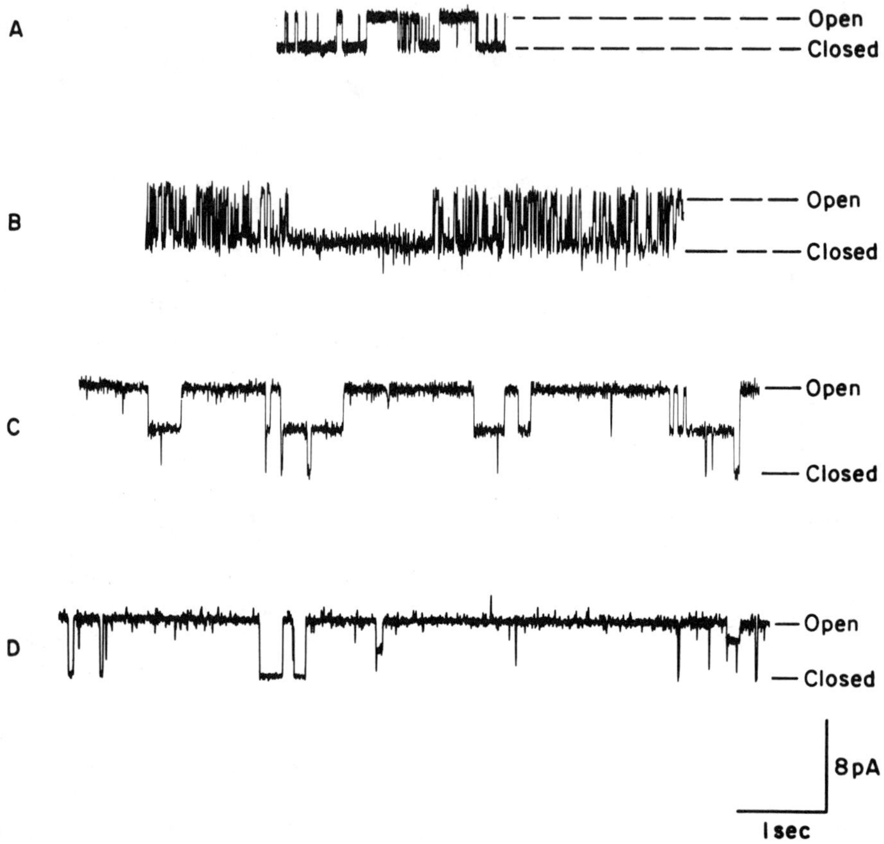

Figure 2. Examples of reconstituted channels from *Helix* and *Aplysia* ganglia. Each of these traces was recorded at +40 mV from a large planar bilayer with 200 mM KCl in both the *cis* and *trans* chambers. Note the differences in gating kinetics and amplitudes of the single-channel currents and apparent channel substates in C and D. When there was a KCl gradient across the bilayer, each of these single-channel currents reversed at the K^+ equilibrium potential. Membrane vesicles from *Helix* (A–C) and *Aplysia* (D) ganglia. Bandwidth 300 Hz.

appears to be less dependent on Ca^{2+} (Fig. 3). On the one occasion that we attempted to phosphorylate one of the channels that had inserted backwards, the kinase was effective on the *trans* but not on the *cis* side.

3.2. Tip-Dip Bilayers

With the large planar bilayers, it was rare to obtain a single-channel membrane. Although steps were taken to limit vesicle fusion, single fusions often dumped several channels into the bilayer, and it was difficult to identify the channel of interest among them. To circumvent this difficulty, we carried out the experiment with much smaller bilayers formed on the tip of a patch electrode.

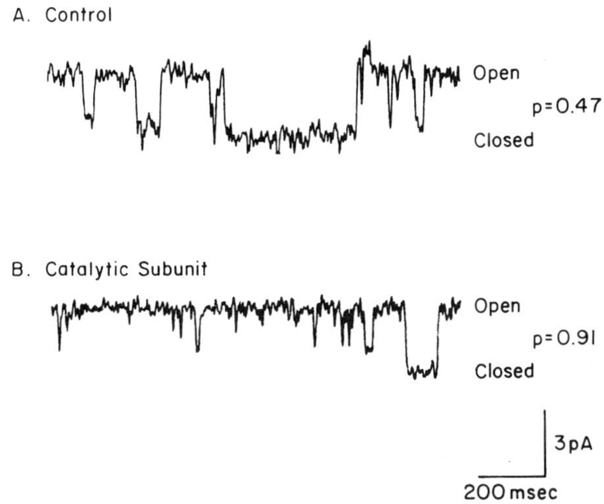

Figure 3. Phosphorylation of a Ca^{2+} dependent K^+ channel in a large planar bilayer. This channel was shown to be Ca^{2+} dependent by addition of EGTA to the *cis* chamber (not shown). When 0.1 μM of the catalytic subunit of cAMP-dependent protein kinase was added to the *cis* chamber, the normal gating of the channel (10 μM free Ca^{2+}, +40 mV) changed dramatically. A: Control. B: Kinase. The *P* value to the right of each trace is the probability of the channel being open, calculated from >20 sec of record. Bandwidth 300 Hz.

This has the added advantage of providing lower-noise recordings because of the small size of the membrane. On the other hand, the channel must be able to survive an air–liquid interface (Fig. 1), something that many proteins are reluctant to do. However, we have been fortunate in that we often obtain membranes that contain only a single Ca^{2+}-dependent K^+ channel under our experimental conditions. Again, channel insertion into the bilayer appears to be asymmetric in that the Ca^{2+}-sensitive site is usually accessible from the bathing medium outside the electrode (Fig. 4). When the kinase catalytic subunit and ATP are added to this bathing medium, the proportion of time that the channel spends in the open state increases dramatically (Fig. 4). This effect is not observed when heat-treated or DTNB-inactivated kinase is used (Ewald *et al.*, 1985). The magnitude of the change in channel gating is illustrated in Table I, which gives the probability of the channel being open (at +40 mV) under various experimental conditions. Similar changes in probability were observed over a wide range of membrane voltages (Ewald *et al.*, 1985).

4. Summary and Conclusions

An understanding of neuromodulation will require studies at many levels of complexity, from the cellular physiology of neuronal networks to the molecular mechanisms involved in regulation of ion-channel activity. Channel reconstitution approaches allow investigation at the most basic of these levels, the biochemical

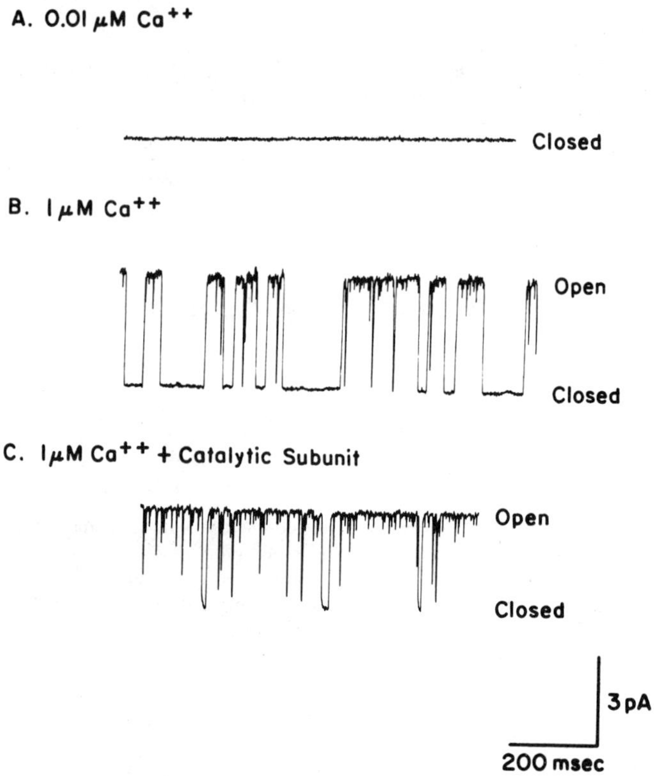

Figure 4. Phosphorylation of a Ca^{2+} dependent K^+ channel in a bilayer on the tip of a patch electrode. This channel was virtually never open at 0.01 μM free Ca^{2+} (A) and was open 15–20% of the time at 1 μM free Ca^{2+} (B). After phosphorylation with the kinase, the channel closed only rarely (C). All at +40 mV. Note the excellent signal-to-noise characteristics of the recording (compare with Figs. 2 and 3). Bandwidth 1 kHz. (Modified from Ewald *et al.*, 1985.)

Table I. Probability (P) of the Ca^{2+}-Dependent K^+ Channel Being Open at +40 mV Under Various Experimental Conditions[a]

Treatment	P
10^{-8} M Ca^{2+}	0.03
10^{-6} M Ca^{2+}	0.19
10^{-6} M Ca^{2+} + inactive kinase	0.17
10^{-6} M Ca^{2+} + active kinase	0.86

[a] The P values were calculated from at least 30 sec of record containing hundreds of opening and closing transitions.

and biophysical mechanisms of channel modulation. These approaches are still in their infancy: single-channel gating model analysis has yet to be applied to these phenomena, and the biochemistry of the phosphorylation reaction is only now beginning to be studied in detail. Nevertheless, at least one significant conclusion can be drawn from the limited experiments to date. Since proteins are effectively at infinite dilution in the artificial bilayers, it can be concluded that the phosphorylation target must be some protein that travels with the channel in the bilayer; that is, the target is either the channel protein itself or something intimately associated with the channel. Thus, the reconstitution experiments already have provided a major new insight into mechanisms of neuromodulation and surely will continue to do so.

ACKNOWLEDGMENTS. Most of this work was done in collaboration with Drs. Douglas Ewald and Alan Williams. Supported by NIH grant NS17910.

References

Adams, W. B., and Levitan, I. B., 1982, Intracellular injection of protein kinase inhibitor blocks the serotonin-induced increase in K^+ conductance in *Aplysia* neuron R15, *Proc. Natl. Acad. Sci. U.S.A.* **79:**3877–3880.

Alkon, D. L., Acosta-Urquidi, J., Olds, J., Kuzma, G., and Neary, J. T., 1983, Protein kinase injection reduces voltage-dependent potassium currents, *Science* **219:**303–306.

Castellucci, V. F., Kandel, E. R., Schwartz, J. H., Wilson, F. D., Nairn, A. C., and Greengard, P., 1980, Intracellular injection of the catalytic subunit of cyclic AMP-dependent protein kinase simulates facilitation of transmitter release underlying behavioral sensitization in *Aplysia, Proc. Natl. Acad. Sci. U.S.A.* **77:**7492–7496.

Castellucci, V. F., Nairn, A., Greengard, P., Schwartz, J. H., and Kandel, E. R., 1982, Inhibitor of adenosine 3′:5′-monophosphate-dependent protein kinase blocks presynaptic facilitation in *Aplysia, J. Neurosci.* **2:**1673–1681.

Cohen, P., 1982, The role of protein phosphorylation in neural and hormonal control of cellular activity, *Nature* **296:**613–620.

Coronado, R., and Latorre, R., 1983, Phospholipid bilayers made from monolayers on patch-clamp pipettes, *Biophys. J.,* **43:**231–236.

DePeyer, J. E., Cachelin, A. B., Levitan, I. B., and Reuter, H., 1982, Ca^{2+}-Activated K^+ conductance in internally perfused snail neurons is enhanced by protein phosphorylation, *Proc. Natl. Acad. Sci. U.S.A.* **79:**4207–4211.

Ewald, D., Williams, A., and Levitan, I. B., 1985, Modulation of single Ca^{2+}-dependent K^+ channel activity by protein phosphorylation, *Nature* **315:**503–506.

Hamill, O. P., Marty, A., Neher, E., Sakmann, B., and Sigworth, F. J., 1981, Improved patch-clamp techniques for high-resolution current recording from cells and cell-free membrane patches, *Pfluegers Arch.* **391:**85–100.

Kaczmarek, L. K., Jennings, K. R., Struwasser, F., Nairn, A. C., Walter, U., Wilson, F. D., and Greengard, P., 1980, Microinjection of catalytic subunit of cyclic AMP-dependent protein kinase enhances calcium action potentials of bag cell neurons in cell culture, *Proc. Natl. Acad. Sci. U.S.A.* **77:**7487–7491.

Maruyama, Y., and Petersen, O. H., 1982, Cholecystokinin activation of single-channel currents is mediated by internal messenger in pancreatic acinar cells, *Nature* **300:**61–63.

Osterrieder, W., Brum, G., Hescheler, J., Trautwein, W., Flockerzi, V., and Hofmann, F., 1982, Injection of subunits of cyclic AMP-dependent protein kinase into cardiac myocytes modulates Ca^{2+} current, *Nature* **298:**576–578.

Shuster, M., Camardo, J., Siegelbaum, S., and Kandel, E. R., 1985, Cyclic AMP-dependent protein kinase closes the serotonin-sensitive K^+ channels of *Aplysia* sensory neurons in cell-free membrane patches, *Nature* **313**:392–395.

Siegelbaum, S. A., and Tsien, R. W., 1983, Modulation of gated ion channels as a mode of transmitter action, *Trends Neurosci.* **6**:307–313.

Siegelbaum, S. A., Camardo, J. S., and Kandel, E. R., 1982, Serotonin and cyclic AMP close single K^+ channels in *Aplysia* sensory neurones, *Nature* **299**:413–417.

Suarez-Isla, B. A., Wan, K., Lindstrom, J., and Montal, M., 1983, Single-channel recordings from purified acetylcholine receptors reconstituted in bilayers formed at the tip of patch pipets, *Biochemistry* **22**:2319–2323.

Wilmsen, U., Methfessel, C., Hanke, W., and Boheim, G., 1983, Channel current fluctuation studies with solvent-free lipid bilayers using Neher–Sakmann pipettes, in: *Physical Chemistry of Transmembrane Ion Motions*, pp. 479–485, Elsevier, Amsterdam.

Chapter 22

VOLTAGE GATING IN VDAC

Toward a Molecular Mechanism

MARCO COLOMBINI

1. Background on Voltage-Dependent Anion-Selective Channels

The reconstitution of transport systems into phospholipid bilayer membranes was originally undertaken in order to gain new insights into the structure and mechanism of action of these molecular machines. This goal is being achieved as evidenced by the contents of this book. This chapter deals with some of the results obtained from studies performed on a reconstituted channel-forming protein isolated from mitochondria. It focuses on the results that yield information on the molecular mechanism that causes this channel to be voltage-gated.

1.1. Biological Function

In describing the structure or function of mitochondria, the outer membrane is often discussed almost as an afterthought, a brief interlude to the main event, the inner membrane. The outer membrane almost seems superfluous, its function nebulous. Some have even proposed that it is not related to the inner membrane at all but to the endoplasmic reticulum (Ernster and Kuylenstierna, 1970). Perhaps it is merely a descendant of the membrane used to engulf the original symbiote. The outer membrane often seems to be in the way. Small molecules must cross it to travel from the cytoplasm to the inner membrane. Mitochondrial proteins synthesized by cytoplasmic ribosomes must cross the outer membrane in order to be incorporated into the functional inner membrane. It is clear that the outer membrane is located at the crossroads of a greal deal of traffic, but does that make this structure trivial?

In the past few years it has become clear (Colombini, 1979; Zalman *et al.*, 1980; Freitag *et al.*, 1982; Roos *et al.*, 1982; Mannella and Colombini, 1984) that much of the molecular traffic crossing the outer membrane does so via the voltage-dependent anion-selective channel-forming protein called VDAC [also referred

MARCO COLOMBINI • Laboratories of Cell Biology, Department of Zoology, University of Maryland, College Park, Maryland 20742.

to as mitochondrial porin (Freitag *et al.*, 1982; Linden *et al.*, 1982a) in some circles because it bears some resemblance (Colombini, 1980a; Zalman *et al.*, 1980) to the family of bacterial channel formers collectively referred to as porins*] Although it had been known for many years that the outer membrane's permeability to molecules depended on their size (Werkheiser and Bartley, 1957; Pfaff *et al.*, 1968; Wojtczak and Zaluska, 1969), the large size of many of the permeating species led many investigators to view the outer membrane as a leaky structure of little importance to the mitochondrion's function.

The channels isolated from the outer membrane are not only large enough to account for the permeability of the membrane to nonprotein molecules (Colombini, 1980b; Zalman *et al.*, 1980; Mannella, 1982) but have also been shown to be controllable (Schein *et al.*, 1976; Colombini, 1979). This raises the possibility that, *in vivo*, these channels may be regulated. The ability to regulate the permeability of the outer membrane would mean that the function of the inner membrane could depend on the outer membrane's permeability.

1.2. Properties

The properties of VDAC have been highly conserved throughout mitochondrial evolution. Large channels are visible in electron micrographs of negatively stained mitochondrial outer membranes from mung beans to rats (Parsons *et al.*, 1965; Mannella *et al.*, 1984). Voltage-gated anion-selective channels have been isolated from rat, paramecium, yeast, and even corn mitochondria (Colombini, 1979; Smack and Colombini, 1985). Their single-channel properties are remarkably similar. This includes single-channel conductance, ion selectivity, and the characteristics of their voltage dependence. Although mitochondrial RNAs and mitochondrial proteins such as cytochrome oxidase have evolved (Tzagoloff, 1982), VDAC's characteristics seem to have remained largely unchanged, indicating that strong selective pressures exist that retain these properties.

VDAC channels are highly conductive, 14 nS in 4 M KCl (Fig. 1). There is essentially no sign of saturation of the channel's ability to conduct KCl even at this high salt concentration (inset to Fig. 1). Although both cations and anions permeate the channel, a distinct preference for anions is evident (Colombini, 1980a). Each channel exhibits one maximal conducting state, known as the open state, and a variety of lower conductance states, referred to as closed states. Figure 2B shows the extent of the variability of the conductance of the more

* The use of the term mitochondrial porin is unfortunate in that it implies that porin and VDAC are very similar molecules. The only strong similarities are the polypeptide molecular weight and their location in an outer membrane. As for the former, connexin also has a very similar polypeptide molecular weight. The differences between porin and VDAC are many. They include the single-channel size, selectivity, voltage dependence, quaternary structure, and the fact that one is found in procaryotes and the other in eucaryotes. In addition, other channel formers whose properties are much more similar to VDAC's have been discovered on the surface membrane of some cells (e.g., Blatz and Magleby, 1983; Nelson *et al.*, 1984). To call all these channels porin would render the term almost as general as the terms channel and pore.

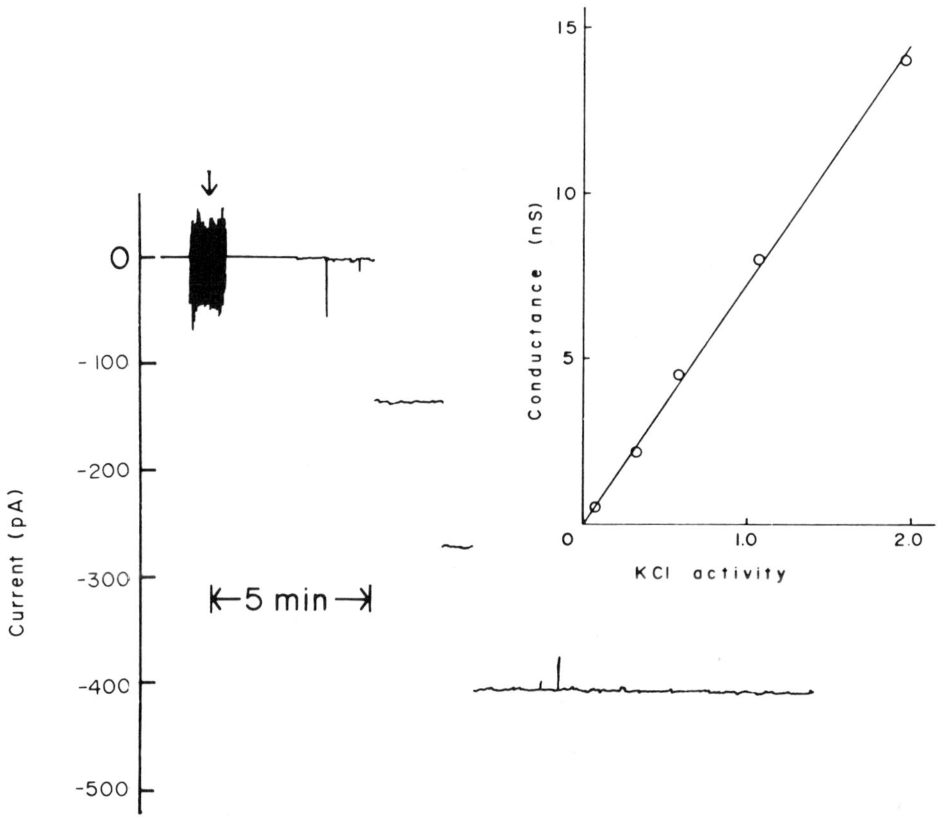

Figure 1. Current through VDAC channels in 4 M KCl. At the point indicated by the arrow, 5 μl of a rat liver VDAC-containing solution (dissolved in 0.7% Triton X-100) was added to the 4 M KCl solution bathing a planar bilayer made with soybean phospholipids as previously described (Schein *et al.*, 1976). The transmembrane voltage was clamped at − 10 mV, and the current was monitored. The three current discontinuities represent three channels inserting into the membrane. The inset is a plot of single-channel conductance as a function of KCl activity. Each point represents the most probable channel conductance from a histogram of single-channel conductances measured at that particular KCl activity.

probable closed states occupied by rat VDAC. A single VDAC can occupy a large variety of closed states although it does show a preference for a particular conductance level. Figure 2A shows the states occupied by a typical rat liver VDAC, and these do vary from channel to channel.

Channel closure occurs when either positive or negative fields (referring to the direction of the electric field vector) are applied to the channel-containing membrane. Figure 3 shows the probability of finding an individual channel open as a function of applied electrical potential. Each channel switches its conductance in at least two regions in the electrical potential range. One occurs between 10 and 30 mV, and the other between − 10 and − 30 mV. That these two conductance-

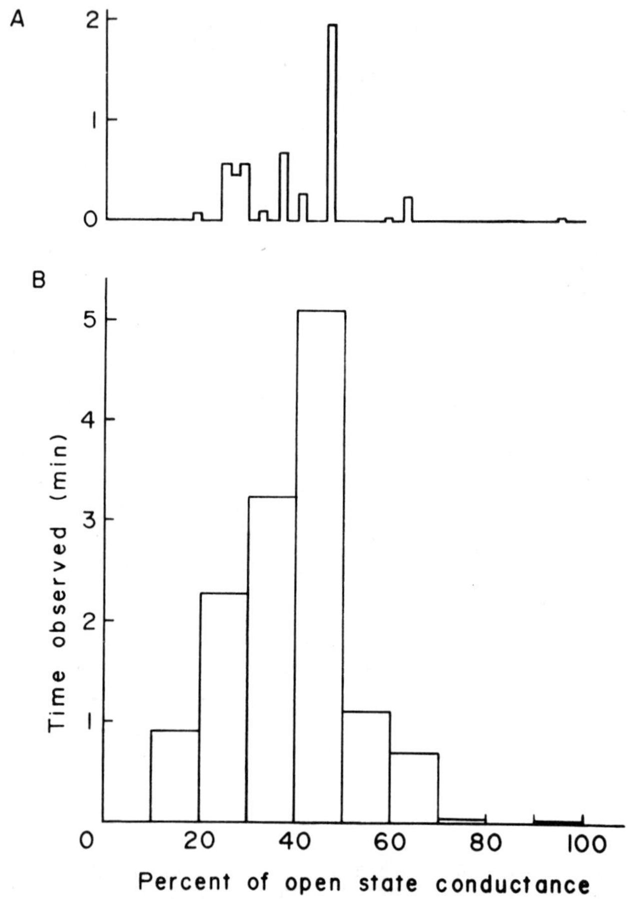

Figure 2. Histograms of VDAC conductances in the "closed state." The current flowing through a single VDAC channel was monitored while the voltage was varied between plus and minus 60 mV (the aqueous solution was 1 M KCl, 5 mM $CaCl_2$). The channel was open at low fields and closed at high fields. The total amount of time the channel was observed in a given closed state (defined as a given conductance level) was recorded. Panel A represents the results obtained for one channel, and panel B the pooled results from measurements made on a number of single channels.

switching regions represent two distinct voltage-gating mechanisms is indicated by the results shown in Fig. 4. The channels were first closed at +50 mV, and then the field was reversed at such a rate (response time <50 μsec) that very few channels would have been able to open [channels open at zero potential in the millisecond time scale (Colombini, 1979)]. The conductance still returned to a value equivalent to the opening of all the channels followed by relatively slow channel closure. Thus, the conformations achieved at positive electric fields differ from those achieved at negative fields. Each channel can therefore occupy one open state and any of two sets of closed states, one set at positive fields and the other at negative fields.

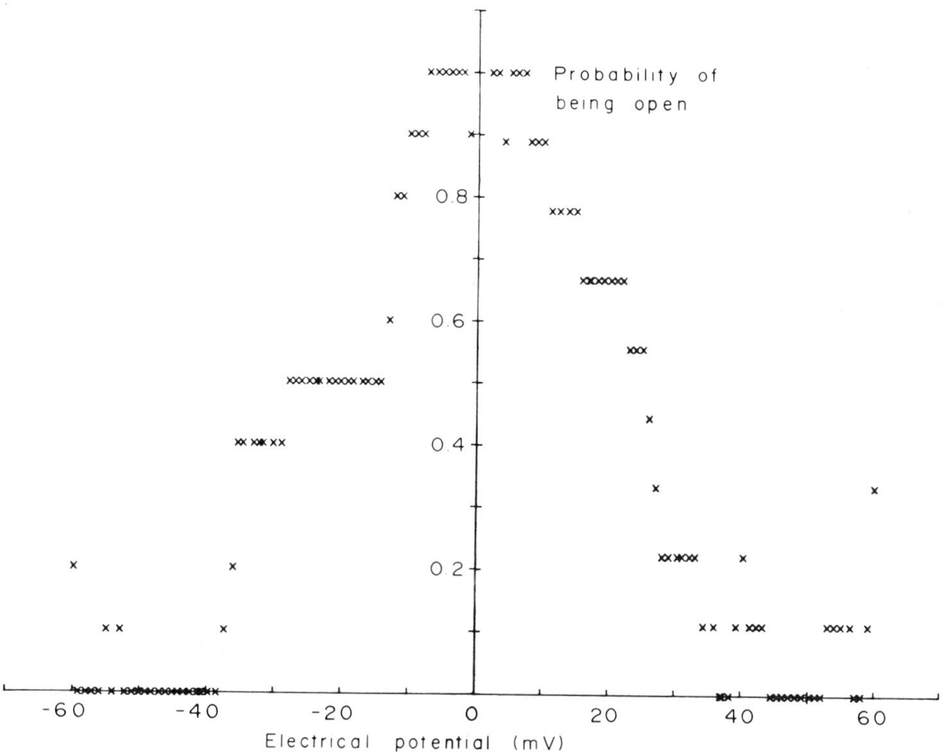

Figure 3. Probability of finding a channel in the open state as a function of applied electrical potential. Asolectin planar phospholipid membranes made by the Montal–Mueller method in the presence of 1 M KCl, 5 mM $CaCl_2$, 50 mM MOPS (Na^+ salt, pH 7.5) were used. A single rat liver VDAC channel was inserted, and its voltage-gating properties were analyzed using a triangular voltage wave at 1.2 mV/sec (voltage-clamp conditions as in Schein *et al.*, 1976).

1.3. Structure

Each channel is composed of polypeptide chains whose molecular weight is 29,000 to 35,000 depending on the source of the material and the SDS polyacrylamide gel system used (Zalman *et al.*, 1980; Linden *et al.*, 1982a; Roos *et al.*, 1982; Freitag *et al.*, 1982; Colombini, 1983; Nakashima *et al.*, 1985). Linden and Gellerfors (1983) showed that the detergent-solubilized entity was a dimer, and it was known (Colombini, 1980a; Roos *et al.*, 1982) that the amount of VDAC conductance in planar phospholipid membranes was directly proportional to the amount of detergent-solubilized VDAC added to the aqueous phase. Hence, one detergent-solubilized entity forms one channel, and each channel is composed of two polypeptide chains.* The channels form simple cylindrical pores that span

* Despite comments to the contrary in a recent review article, this finding is not inconsistent with the apparent molecular weight of the detergent–VDAC complex of 110,000 (Colombini, 1980a). It is well known that membrane proteins bind a great deal of detergent when solubilized in detergent. Indeed, this was found to be true for VDAC (Linden and Gellerfors, 1983).

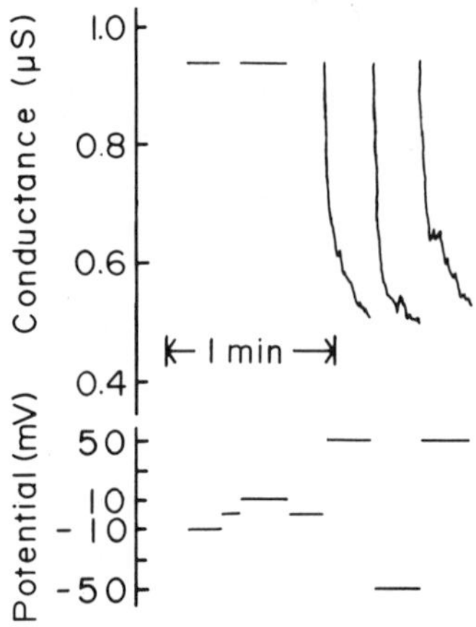

Figure 4. Evidence that the closed state achieved at positive electric fields differs from that achieved at negative fields. The membrane was made as indicated in Fig. 3 except that the medium was unbuffered. The lower trace shows the potential at which the membrane was clamped, and the upper trace shows the membrane conductance. (The conductance could not be measured directly when the potential was zero.) The time required to change the voltage was less than 50 μsec. *N. crassa* VDAC was used in this experiment.

the membrane (Mannella *et al.*, 1984, and Fig. 10). In some outer membranes (Mannella *et al.*, 1983), the channels aggregate to form predominently triplets and multiples of triplets. The basic triplet unit is sickle shaped. Thus, the arrays have a low degree of symmetry, possibly indicating that the channels are present in different conformational forms. Perhaps the channel in the center of the triplet is in a different conformation from the channels at the ends.

The VDAC channels form large aqueous pores. Their permeability to nonelectrolytes is consistent with a pore radius (for the open state) of 20 Å (Colombini, 1980b). The ordered arrays formed by these channels in plant outer mitochondrial membranes allowed Mannella and Bonner (1975) to obtain X-ray diffraction patterns. Some interpretations (Mannella and Ratkowski, 1979) of these patterns agree with the 20 Å estimate for the channel's radius, but others indicate a smaller pore radius (Mannella, 1981). Computer-filtered images of negatively stained outer membranes containing arrays of channels indicate a smaller radius, 12 to 15 Å (Mannella, 1982). Recent measurements of the flux of [^{14}C]glucose through VDAC are consistent with a slightly larger radius (23 Å, Kulkarni, 1984).

2. Voltage Gating: Expectations and Predictions

2.1. Ways of Achieving Channel Closure

Channel closure is usually monitored functionally rather than structurally. It is detected as a reduction in the current flow through a channel. This reduction could be partial or complete. In the case of VDAC it is partial. Since what is being monitored is a flow of ions, this flow can be reduced either by changing the electrostatic environment within the channel or by obstructing the channel. These two modes will be referred to as an electrostatic block and a steric block, respectively. Although they are not mutually exclusive, these two modes can be distinguished experimentally.

One can look for a steric block by measuring the channel's permeability to nonelectrolytes in the open and closed state. A sizable change in the nonelectrolyte permeability would argue strongly for a steric block, whereas the lack of such a change would indicate an electrostatic block. The presence of an electrostatic block might be indicated by a change in selectivity between anions and cations associated with channel closure. Although such a selectivity change indicates that the electrostatic environment within the channel has changed, it does not necessarily follow that this change is responsible for channel closure (i.e., the conductance decrease).

The conformational changes associated with channel closure could be localized to one region of the channel or extend over a large portion of the pore. In the case of a steric block, the difference between uniform constriction and localized constriction could be detected by measuring changes in the pore volume associated with channel closure and changes in the effective pore radius with closure.

2.2. Ways of Detecting and Responding to the Electric Field

It is generally agreed that the simplest way for a protein to sense the electric field is either to use charged residues on the protein or to rely on large dipoles formed by domains in the protein or the protein as a whole. Voltage dependence would result if channel closure were coupled to (1) the movement of charges on the protein relative to the electric field; or (2) a change in the dipole moment in the direction of the field and/or a change in the intensity of the field at the dipole. The charges or dipoles can be referred to as the sensors that allow the channel to detect and respond to the electric field. A large energy change at the sensor by one or both of these general mechanisms (about 400 J per 1-mV change in transmembrane potential difference) would produce a steep voltage dependence.

Using these general mechanisms, one can devise a myriad of specific mechanisms to account for voltage gating in a particular channel. Although this has been done by many investigators, only a few of these detailed mechanisms are supported by compelling experimental observations. The best-understood voltage-dependent channels are some of the ectocytic channel formers (those that come from the environment and insert into a cell's membranes). Two of these, alamethicin and monazomycin, are rod-shaped molecules that tend to adsorb to the surface of membranes. In the presence of an electric field, some are driven into the membrane (the dipoles align with the field), where they aggregate to form barrel-stave channels (Eisenberg *et al.*, 1973; Heyer *et al.*, 1976). Recently, a different mechanism has been proposed to explain the voltage gating of alamethicin (Hall *et al.*, 1984).

For the endocytic channel formers (channels inserted by a cell into its own membranes), the specific molecular mechanisms are less well defined and have, in general, much less experimental support. The large size and complexity of the proteins that form these channels coupled with the small amount of detailed structural information have made progress toward a molecular mechanism rather slow.

However, an interesting pattern has emerged. A number of endocytic channels are open at low electric fields and closed at high fields. Some are closed at both high positive and high negative fields (sometimes one switching region is referred to as inactivation). Perhaps a common fundamental mechanism is operating in many of these systems. It is interesting to note that a number of voltage-gated ectocytic channels function in exactly the opposite manner, i.e., they are most conductive at high fields. Perhaps they are designed to selectively insert into living cells by inserting into cells that maintain a transmembrane voltage difference.

3. Observations on VDAC

3.1. The Nature of the Barrier

There were early indications that an electrostatic barrier might be operating in VDAC. The channels did not close completely, and one might expect that an electrostatic barrier might not be able to block ion flow totally (especially in vew of the large size of VDAC's pore). More importantly, on closure, the ion selectivity of the channels changed from weakly anion selective to nonselective or slightly cation selective. Thus, the electrostatic environment within the channel did change when the channel closed.

In order to distinguish between a steric and an electrostatic barrier, one must obtain an estimate of the effective radius of a channel in the open and closed state. Although some use conductance measurements to estimate pore radii, these are clearly inappropriate to distinguish between the two types of barriers. Only the measurement of nonelectrolyte fluxes will yield the needed information.

In the case of VDAC, nonelectrolyte fluxes were used to estimate the pore radius in two ways. The first and most straightforward way was to determine the largest nonelectrolyte that could penetrate the pore and the smallest that could not penetrate. The second was to measure the flux of a small nonelectrolyte and determine what size cylindrical pore would account for the observed flux per channel.

The first approach yielded an estimated radius of 20 Å for the open state (Colombini, 1980b). The technique, which was developed for VDAC, involves measuring fluxes through channels reconstituted into liposomes. It is easy, does not require the use of radioisotopes, and can and has been applied to other channels (for example, diphtheria toxin, Kagan *et al.*, 1981). However, it does not allow one easily to apply electrical potentials across the channels, and thus it proved very difficult to use the same method to determine the radius of VDAC's closed state.

The use of the second approach has yielded the answer despite difficulties encountered in converting fluxes into an estimate of the radius. VDAC channels were inserted into planar bilayer membranes, and [^{14}C]glucose fluxes were measured through channels held in either the open or closed state by controlling the electric field (Kulkarni, 1984). A marked drop in glucose flux occured on channel closure (Fig. 5). At -10 mV, the channels were all open, whereas at potentials

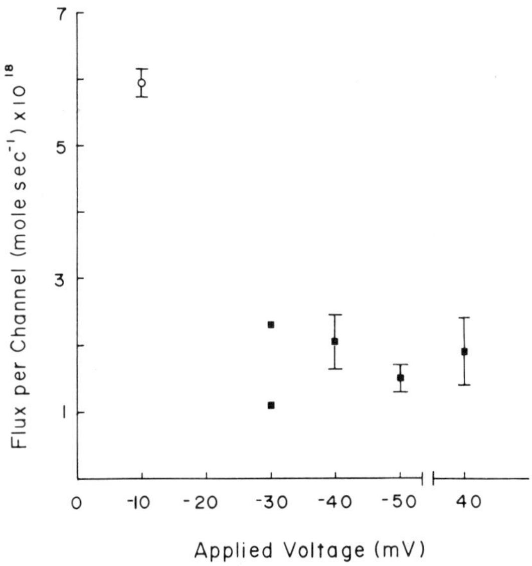

Figure 5. Glucose flux through VDAC as a function of the applied electrical potential. These results were taken from Kulkarni (1984). The membranes were made as indicated in Fig. 3 except that the aqueous medium was 1.0 M NaCl, 5 mM $CaCl_2$ on the *cis* side (the potential refers to this side) and 0.1 M NaCl, 5 mM $CaCl_2$ on the *trans* side. [^{14}C]Glucose was added to the *trans* side, and its appearance on the *cis* side was measured. The open symbol represents glucose flux through channels in the open state; the filled symbols are for the closed state. The data at −30 mV were corrected for flux through channels that had not closed. The error bars indicate the standard error of the mean.

of higher magnitude, the channels were closed. Note that the direction of the current had little effect on the glucose flux. The electric-field-induced conductance drop was accompanied by a drop in glucose flux that was independent of the sign and magnitude of the field (as long as the field was sufficient to close the channels). The decrease in glucose flux was comparable in magnitude to the conductance decrease. Thus, a steric block is almost certainly a major factor in VDAC closure.

Estimates of pore radius from these fluxes required some understanding of the way the experiments were performed. (An osmotic gradient was present resulting in bulk flow through the channel.) Analysis of the results yielded a change in pore radius from 23 Å for the open channel to 18 Å for the closed channel. These values seem to be on the high side compared to results obtained by other techniques.

3.2. The Nature of the Sensor

As discussed above, the sensor in the channel could either be charges or dipoles or perhaps both. If charges are involved, then it should be possible to neutralize them and thus eliminate the voltage dependence. This was done both by titrating the charges and by modifying them chemically. Both types of experiments were performed on VDAC channels after they had been inserted into planar phospholipid membranes (Bowen *et al.*, 1985; Doring and Colombini, 1985a). Thus, the voltage dependence of the channels could be measured before and after the specific treatment.

Chemical modification of VDAC with reagents that modify either amino groups or arginine groups was very revealing. Arginine reagents had no effect

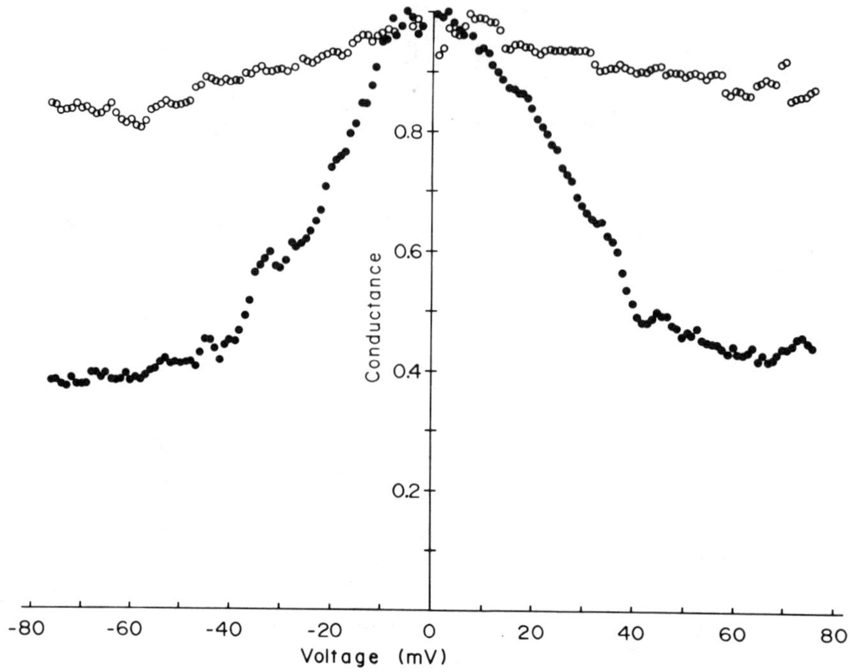

Figure 6. The voltage dependence of VDAC before (filled symbols) and after (open symbols) succinic anhydride modification. (These results are from Doring and Colombini, 1985a, and are reprinted with the permission of the *Journal of Membrane Biology*.) For each set of points, the conductance was normalized to the highest observed conductance. These results were obtained with rat liver VDAC.

(Bowen *et al.*, 1985), but succinic anhydride (a reagent that reacts with amino groups and generates a free carboxyl group, Habeeb *et al.*, 1958) had pronounced effects (Doring and Colombini, 1985a). VDAC's voltage dependence was reduced by treatment with anhydride (Fig. 6). Along with a drop in the steepness of the voltage dependence (measured by the parameter n), the voltage at which half the channels would be closed (V_o) increased (Fig. 7). This is expected if the conformational energies (i.e., the voltage-independent components of the energy levels) of the open and closed states remained unaltered as the channels were modified. A reduction in the steepness of the voltage dependence means that the channel does not change its energy level as much for a given change in the electric field. Thus, a higher field should be needed to compensate for conformational energy differences between different states of the channel. This was indeed the case. The drop in n and the rise in V_o almost exactly compensated. Hence, the energy difference between the states in the absence of an electric field (nFV_o) changed little with anhydride modification (Fig. 7). These results strongly support the contention that succinic anhydride modified the gating charges.

When the medium's pH was raised, VDAC's voltage dependence diminished in a dose-dependent manner (Bowen *et al.*, 1985). As with the anhydride treat-

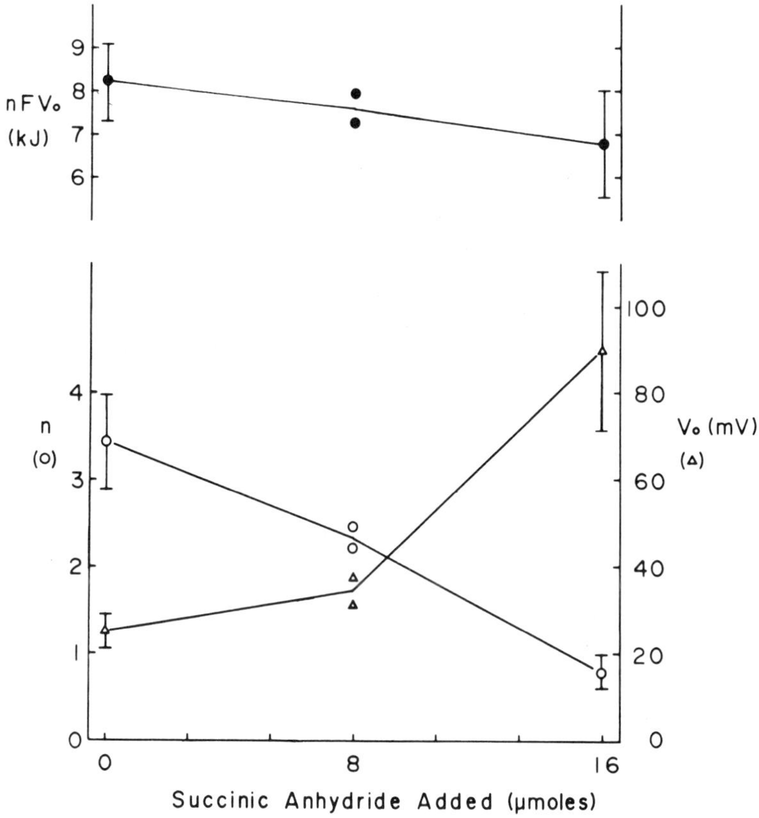

Figure 7. Quantitation of the reduction of VDAC's voltage dependence by succinic anhydride. (These results are from Doring and Colombini, 1985a, and are reprinted with the permission of the *Journal of Membrane Biology*.) Results such as those shown in Fig. 5 were analyzed to obtain the steepness of the voltage dependence (measured by the parameter n, open circles), the voltage at which half the channels were closed (V_0, triangles), and the energy difference between the open and closed states in the absence of a field (nFV_0, where F is the Faraday constant).

ment, the steepness of the voltage dependence, n, decreased (Fig. 8), whereas the voltage at which half the channels were closed, V_o, increased. Once again there was only a small decrease in the energy difference between the states (in the absence of a field) as the pH was raised from neutrality to pH 11.

The short-term exposure of VDAC to alkaline pH turned out to be a surprisingly mild treatment. The conductance of the channels in the open and closed states did not change very much for short exposures to the high pH. The kinetics were slowed down somewhat. However, on the whole, the major effect was a reduction in voltage dependence. The mean apparent pK of the titratable groups was 10.6.

The pK of the titratable groups and the modification of these groups by succinic anhydride strongly indicate that the gating charges are amino groups, prob-

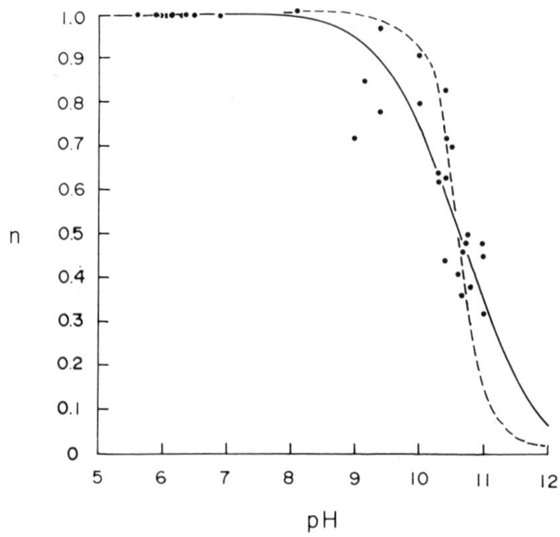

Figure 8. Reduction in VDAC voltage dependence at elevated pH. (These results are from Bowen *et al.*, 1985 and are reprinted with the permission of the Journal of Membrane Biology.) Rat liver VDAC was used. The steepness of the change in conductance with voltage was measured (and expressed by the parameter *n*) prior to and after the pH was elevated. The value of *n* was normalized to its value measured prior to the elevation of the medium pH. The dotted line represents a best fit (pK = 10.6) assuming five gating charges with identical pKs. The solid line is a best fit (pK 9.6 to 11.6) assuming five gating charges with pKs spread evenly over the pK range.

ably lysine groups. If uncharged dipoles are involved in the voltage dependence, their contribution seems to be minor.

3.3. The Nature of the Selectivity Filter

The large size of VDAC's pore (20 Å) and its high conductance (7.5 nS per molal for KCl) make it unlikely that most of the ions crossing the channel associate with the channel's walls for any significant period of time. Hence, the channel's weak anion selectivity is probably the result of the electric field created by positive charges on the walls of the pore rather than ion-binding sites. Support for this conjecture has come from chemical modification experiments.

Modification of the channel with succinic anhydride results in a dramatic selectivity shift from anion selectivity to cation selectivity (Doring and Colombini, 1985a). The ratio of the ion permeabilities changed almost tenfold. For LiCl, it changed from 3 to 1, Cl^- over Li^+, to 3 to 1, Li^+ over Cl^-. Since the succinic anhydride reaction essentially converts an amino group to a carboxyl group, if the anhydride reacted with amino groups in the channel's interior, then the net charge of the pore could have been radically altered.

3.4. Dynamics of the Gating Process

Since succinic anhydride appears to react with VDAC's gating charges, it was used as a probe to determine the location of these charges and whether the location changed with VDAC's conformational state (Doring and Colombini, 1985b). Channels were inserted into the membrane from the aqueous phase on

Table I. Voltage Dependence of VDAC Modified in the Closed State[a]

	Positive field[b]		Negative field[b]	
	Cis[c] (%)	*Trans* (%)	*Cis* (%)	*Trans* (%)
Voltage dependence				
Unchanged		50		
Small change	75			9
Lost		50	70	
Channels remained closed			30	91

[a] Data from Doring and Colombini (1985b).
[b] Positive field indicates that the *cis* side was at a positive electrical potential relative to the *trans* side. Negative field signifies the *cis* side was negative.
[c] The *cis* side of the membrane was that to which VDAC was added.

one side of the membrane (the *cis* side). The accessibility of the charges to succinic anhydride was tested by adding the reagent to either the *cis* or *trans* side. With channels in their open state, no differential reactivity was detected with *cis* or *trans* reagent addition. However, a marked asymmetry was found when anhydride was added to closed VDAC channels.

The channels in the planar phospholipid membranes were held in the closed state by means of an electric field. Both positive and negative fields were used to examine the reactivity of both types of closed states. Table I summarizes the results. Succinic anhydride was most effective in eliminating the voltage-gating property if it was added to the negative side of the membrane. Hence, on channel closure, the gating charges become more accessible to one side of the membrane than to the other side. The side to which the charges become more accessible differs depending on whether the closed state was achieved with a positive or negative field. It is gratifying that the positive gating charges became more accessible to the negative side of the membrane.

4. Interpretations

4.1. How Many Gating Mechanisms Are There in VDAC?

A population of VDAC channels will close when both positive and negative fields are applied. This is also the case for a single channel (Fig. 3). Thus, there are two voltage-dependent switching regions, which indicate two gating mechanisms. The alternative that the channel might be driven into the same closed state at both positive and negative fields was discounted by the experiment described in Section 1.2. However, is it possible that one gating mechanism might account for closure at both positive and negative fields much as the same restoring force in a spring accounts for resistance to displacement for both positive and negative forces?

The succinic anhydride modification experiments revealed not only an asymmetry in the location of the gating charges for VDAC in different conformations but also a remarkable symmetry in the voltage dependence of the modified VDAC despite asymmetric addition of the modifier (Doring and Colombini, 1985b). All modification protocols modified the ability of channels to close in response to both positive and negative fields to about the same extent. It certainly is possible that two sets of gating charges are present in the channel and both are modified to the same extent in most of the experiments despite the asymmetric additon. Indeed, in a few cases, the modified channels did behave asymmetrically, i.e., closed preferentially with applied fields in one direction. Once in a great while, unmodified channels have been observed to rectify strongly: they conducted and closed with electric fields in one direction but conducted extremely poorly in the opposite direction. However, in view of the consistent modification of both gating mechanisms by succinic anhydride, it is more likely that only one set of gating charges exists to account for closure at both positive and negative fields. Thus, one sensor (the gating charges) could operate two gates, one functioning at positive fields and the other at negative fields. The observed asymmetries might then result from altered or defective gates rather than two different sensors.

4.2. Are the Selectivity Filter and the Voltage Sensor One and the Same?

Succinic anhydride modification reversed VDAC's selectivity, indicating that the charge on the amino groups modified by the anhydride has a strong influence on VDAC's selectivity. Thus, it is likely that these groups are located at the mouth of the channel or line the walls of the aqueous pore. Is it possible that the gating charges are located in the same place, and, thus, perhaps one set of charges is performing both functions.

Both the titration experiments and the anhydride modification experiments are consistent with amino groups being the gating charges. The experiments in which anhydride was added asymmetrically indicate that in the closed state, these charges are preferentially accessible from the negative-potential side of the membrane. Hence, on channel closure, either the charges are moved across the membrane or the gate restricts access to the charges from the positive side of the membrane. If the charges crossed the membrane through a low-dielectric-constant region, this unfavorable process would probably be compensated for by being coupled to a very favorable conformational change. If the charges were neutralized by raising the pH, the closing rates should increase dramatically. This was not observed. Thus, it is likely that if the gating charges must cross the membrane when the channels close, they do so through a high-dielectric-constant medium such as the pore of the channel. On the other hand, if the accessibility change is caused by the way the gate restricts access to the gating charges, these charges must be located in the mouth of the pore or in the pore itself. If so, then they could serve as a selectivity filter as well as gating charges.

4.3. Proposed Mechanism

Although the following mechanism is tentative and needs to be tested further, it is supported by a great deal of experimental evidence. The mechanism must

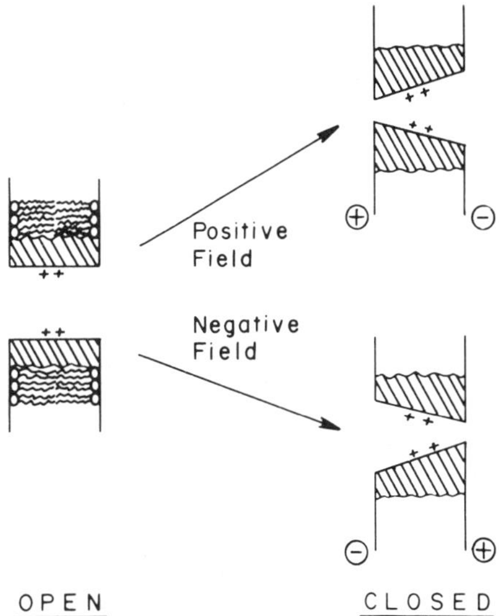

Figure 9. Proposed voltage-gating mechanism for the channel former VDAC. The hatched areas represent the channel-forming protein in a membrane. The positive charges within the pore are ionized lysine groups. Although only four are depicted in these cross-sectional views, there are probably more than four charges on the walls of the pore. The application of an electric field causes the channel to constrict on the positive side of the membrane.

account for one open state and a different closed state at each of positive and negative electric fields, the existence of multiple closed states at each field, the steric block, the positive gating charges, the positive charges conferring selectivity, the differential accessibility of the gating charges in the closed state, one set of gating charges being responsible for two closed states, and the voltage-dependent energy difference between the conformational states.

Figure 9 shows the proposed channel structure and how it might change in response to applied electric fields. The gating charges are located on the walls of the pore, where they also control the channel's selectivity. In the closed state, the gating charges are more accessible from the negative side. Note that it is easy to see how multiple closed states could occur. On closing, the channel might constrict to different extents depending on the different ways in which the protein residues pack. The voltage dependence would arise from two components: (1) the positively charged channel wall is aligning with the electric field, thus increasing the effective dipole moment of the channel in the direction of the field; (2) the constriction at the mouth of the pore would change the effective electrical potential at the gating charges, thus resulting in the movement of the charges through an electrical potential difference (Colombini, 1984). These effects could account for closure at both positive and negative fields using the same gating charges.

The proposed mechanism is consistent with VDAC's kinetic properties (Schein *et al.*, 1976; Colombini, 1979), rapid opening kinetics (milliseconds) and slow closing kinetics (seconds). Pore constriction associated with channel closure should require some rearrangement of protein side chains in order to pack them in closer proximity. Channel opening, on the other hand, should involve the spreading apart of protein groups and could therefore be a faster process. If VDAC is held in a closed conformation for long periods of time, the channel seems to

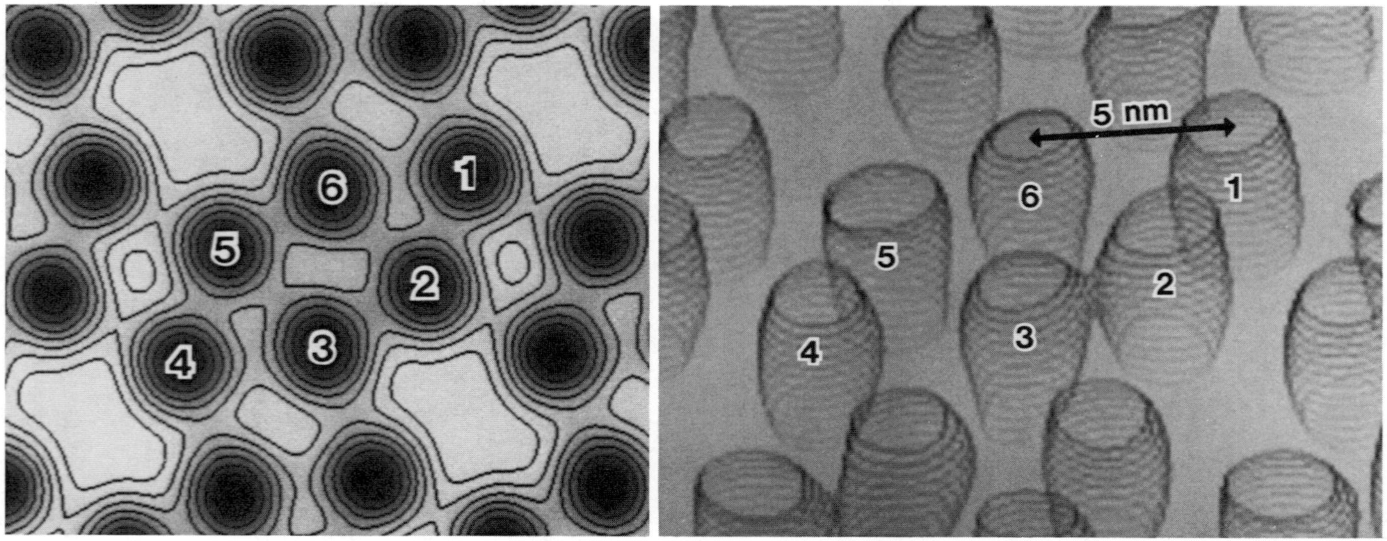

Figure 10. Crystalline arrays of VDAC channels in outer membranes of *N. crassa* mitochondria (for more information see Mannella *et al.*, 1984). The membranes had been negatively stained with phosphotungstate, and so the contour lines represent the density of this material (probably visualizing VDAC's aqueous pore). The left panel is a two-dimensional view through the membrane with two sets of VDAC triplets labeled 1-2-3 and 4-5-6. This group of six is repeated, forming a parallelogram planar lattice. The right panel is an oblique view of a three-dimensional reconstruction of the structure showing how the stain is distributed through the membrane (5.4 nm thick). (Figure courtesy of Dr. Carmen A. Mannella.)

adapt to this conformation so that the rate of opening slows down. This could be explained in terms of rearrangement of the protein groups in the more cramped closed state, which may occur with time, resulting in lower free energy states. For the protein to leave the new lower energy states would require more energy and therefore more time.

This mechanism is also consistent with VDAC's known structure as reported by Mannella (1982) and Mannella *et al.* (1984). Three-dimensional reconstruction of VDAC's pore from electron micrographs taken of negatively stained VDAC arrays (Fig. 10) reveal that each channel forms a simple cylindrical pore (not the complex structure of porin). The channels are organized into triplets in the *N. crassa* outer mitochondrial membranes (Mannella *et al.*, 1983), and these triplets form the ordered arrays. No attempt has been made to modify the model to account for the triplet structure of the arrays, since there is no evidence that the triplet structure has any bearing on VDAC's voltage dependence.

Volume changes of VDAC's pore were estimated by Zimmerberg and Parsegian (1984). They indicate that, on closure, VDAC's pore reduces its volume by 20,000 to 35,000 $Å^3$ (J. Zimmerberg, personal communication). This number is rather large and might indicate that the pore is reduced in diameter uniformally rather than constricting at one end (Zimmerberg and Parsegian, 1984). However, if the numbers listed above (Section 3.1) for the change in VDAC's effective radius on closure are used to estimate a volume change assuming uniform constriction (and assuming a pore 54 Å long, Mannella *et al.*, 1984), a value of 34,800 $Å^3$ is obtained. On the other hand, if only one end constricted, forming a conically shaped structure (as indicated in Fig. 9), a volume change of 18,000 $Å^3$ would result. The uncertainties in the estimates are too great to make firm conclusions.

5. Outlook for the Future

The molecular mechanism underlying the voltage dependence of channels is an important unsolved problem. The work on VDAC has led to a plausible mechanism underlying the phenomenology. The proposed conformational changes are quite consistent with the known behavior of proteins in general. However, more work needs to be done to confirm or rule out this mechanism.

More information is needed on the structure of the protein in general and on the structure of the open and closed states in particular. Electron microscopy and other biophysical techniques have yielded and will yield some of this information. Although the open state is readily available, the structure of the closed state is more difficult to study because an electric field is normally needed to achieve this state. Indirect means of achieving the closed state are being pursued (e.g., chemical modification, allosteric modifiers, and osmotic pressure).

Since VDAC's primary sequence is available, it is possible to try to fold the polypeptide chain using the rules developed for other membrane proteins. VDAC's small size and large pore radius should greatly reduce the number of possible conformations. On the other hand, unlike membrane proteins, VDAC's secondary structure seems to be mainly β (P. S. Mangan and M. Colombini unpublished results), making the folding process more difficult. Nevertheless, the possible ways in which the polypeptide chain might fold should indicate whether the proposed voltage-gating mechanism is viable. Since the VDAC gene is cloned, the generation of mutants should also prove to be an effective way of testing the proposed mechanism.

These approaches, in conjunction with electrophysiological studies on the reconstituted channel, should lead to a detailed understanding of VDAC's voltage-gating mechanism. As the voltage-gating mechanisms of other endocytic channels are studied, a pattern may emerge, or perhaps classes of voltage-gating mechanisms might come to light. It is quite possible that VDAC's gating mechanism might be used only by large channel formers and that a very different mechanism is operating in the case of narrow channels.

Other important aspects of VDAC's function have been only hinted at in this chapter. One of these is VDAC's ability to bind hexokinase (Fiek *et al.*, 1982; Linden *et al.*, 1982b; Nakashima *et al.*, 1985) and perhaps other enzymes (Ostlund *et al.*, 1983). Thus, VDAC is probably right in the midst of some very interesting cell biology.

References

Blatz, A. L., and Magleby, K. L., 1983, Single voltage-dependent chloride-selective channels of large conductance in cultured rat muscle, *Biophys. J.* **43:**237–241.

Bowen, K. A., Tam, K., and Colombini, M., 1985, Evidence for titratable gating charges controlling the voltage-dependence of the outer mitochondrial membrane channel, VDAC, *J. Memb. Biol.* **86:**51–59.

Colombini, M., 1979, A candidate for the permeability pathway of the outer mitochondrial membrane, *Nature* **279:**643–645.

Colombini, M., 1980a, Structure and mode of action of a voltage-dependent anion-selective channel (VDAC) located in the outer mitochondrial membrane, *Ann. N.Y. Acad. Sci.* **341:**552–563.

Colombini, M., 1980b, The pore size and properties of channels from mitochondria isolated from *Neurospora crassa, J. Memb. Biol.* **53:**79–84.

Colombini, M., 1983, Purification of VDAC (voltage-dependent anion-selective channel) from rat liver mitochondria, *J. Memb. Biol.* **74:**115–121.

Colombini, M., 1984, A novel mechanism for voltage control of channel conductance, *J. Theor. Biol.* **110:**559–567.

Doring, C., and Colombini, M., 1984, On the nature of the molecular mechanism underlying the voltage dependence of the channel-forming protein, VDAC, *Biophys. J.* **45:**44–46.

Doring, C., and Colombini, M., 1985a, Voltage dependence and ion selectivity of the mitochondrial channel, VDAC, are modified by succinic anhydride, *J. Memb. Biol.* **83:**81–86.

Doring, C., and Colombini, M., 1985b, The mitochondrial voltage-dependent channel, VDAC, is modified asymmetrically by succinic anhydride, *J. Memb. Biol.* **83:**87–94.

Eisenberg, M., Hall, J. E., and Mead, C. A., 1973, The nature of the voltage-dependent conductance induced by alamethicin in black lipid membranes, *J. Memb. Biol.* **14:**143–176.

Ernster, L., and Kuylenstierna, B., 1970, Outer membrane of mitochondira, in: *Membranes of Mitochondria and Chloroplasts* (E. Racker, ed.), p. 205, Van Nostrand Reinhold, New York.

Fiek, C., Benz, R., Roos, N., and Brdiczka, D., 1982, Evidence for identity between the hexokinase binding protein and the mitochondrial porin in the outer membrane of rat liver mitochondria, *Biochim. Biophys. Acta* **688:**429–440.

Freitag, H., Neupert, W., and Benz, R., 1982, Purification and characterization of a pore protein of the outer mitochondrial membrane from *Neurospora crassa, Eur. J. Biochem.* **123:**629–636.

Habeeb, A. F. S. A., Cassidy, H. G., and Singer, S. J., 1958, Molecular structural effects produced in proteins by reaction with succinic anhydride, *Biochim. Biophys. Acta* **29**:587–593.

Hall, J. E., Vodyanoy, I., Balasubramanian, T. M., and Marshall, G. R., 1984, Alamethicin, a rich model for channel behavior, *Biophys. J.* **45**:233–247.

Heyer, E. J., Muller, R. U., and Finkelstein, A., 1976, Inactivation of monazomycin-induced voltage-dependent conductance in thin lipid membranes, *J. Gen. Physiol.* **67**:731–748.

Kagan, B. L., Finkelstein, A., and Colombini, M., 1981, Diphtheria toxin fragment forms large pores in phospholipid bilayer membranes, *Proc. Natl. Acad. Sci. U.S.A.* **78**:4950–4954.

Kulkarni, S. V., 1984, Sizing the closed state of the mitochondrial channel, VDAC, Thesis, Department of Zoology, University of Maryland, College Park, Maryland.

Linden, M., and Gellerfors, P., 1983, Hydrodynamic properties of porin isolated from other membranes of rat liver mitochondria, *Biochim. Biophys. Acta* **736**:125–129.

Linden, M., Gellerfors, P., and Nelson, B. D., 1982a, Purification of a protein having pore forming activity from the rat liver mitochondrial outer membrane, *Biochem. J.* **208**:77–82.

Linden, M., Gellerfors, P., and Nelson, B. D., 1982b, Pore protein and the hexokinase-binding protein from the outer membrane of rat liver mitochondria are identical, *FEBS Lett.* **141**:189–192.

Mannella, C. A., 1981, Structure of the outer mitochondrial membrane. Detection of in-plane subunit structure, *Biochim. Biophys. Acta* **645**:33–40.

Mannella, C. A., 1982, Structure of the outer mitochondrial membrane: Ordered arrays of porelike subunits in outer-membrane fractions from *Neurospora crassa* mitochondria, *J. Cell Biol.* **94**:680–687.

Mannella, C. A., and Bonner, W. D., Jr., 1975, X-ray diffraction from oriented outer mitochondrial membranes, *Biochim. Biophys. Acta* **413**:226–233.

Mannella, C. A., and Colombini, M., 1984, Evidence that the crystalline arrays in the outer membrane of *Neurospora* mitochondria are composed of the channel protein, VDAC, *Biochim. Biophys. Acta* **774**:206–214.

Mannella, C. A., and Ratkowski, A. J., 1979, Structure of the pore complex in the outer membrane of mitochondria, *J. Cell Biol.* **83**:270a.

Mannella, C. A., Colombini, M., and Frank, J., 1983, Structural and functional evidence for multiple channel complexes in the outer membrane of *Neurospora crassa* mitochondria, *Proc. Natl. Acad. Sci. U.S.A.* **80**:2243–2247.

Mannella, C. A., Radermacher, M., and Frank, J., 1984, Three-dimensional structure of mitochondrial outer-membrane channels from fungus and liver, *Proc. Annu. EMSA Meet.* **42**:644–645.

Nakashima, R. A., Mangan, P. S., and Pedersen, P. L., 1985, Evidence for a relationship between hexokinase-binding and the mitochondrial VDAC protein in AS-30D hepatoma cells: DCCD labels the pore protein and inhibits hexokinase binding, *Biophys. J.* **47**:238a.

Nelson, D. J., Tang, J. M., and Palmer, L. G., 1984, Single-channel recordings of apical membrane chloride conductance in A6 epithelial cells, *J. Memb. Biol.* **80**:81–89.

Ostlund, A. K., Gohring, U., Krause, J., and Brdiczka, D., 1983, The binding of glycerol kinase to the outer-membrane of rat liver mitochondria. Its importance in metabolic regulation, *Biochem. Med.* **30**:231–245.

Parsons, D. F., Bonner, W. D., Jr., and Verboon, J. G., 1965, Electron microscopy of isolated plant mitochondria and plastids using both thin-section and negative staining techniques, *Can. J. Bot.* **43**:647–655.

Pfaff, E., Klingenberg, E., Ritt, E., and Vogell, W., 1968, Korrelation des unspezifisch permeablen mitochondrialen Raumes mit dem "Intermembrane-Raum," *Eur. J. Biochem.* **5**:222–232.

Roos, N., Benz, R., and Brdiczka, D., 1982, Identification and characterization of the pore-forming protein in the outer membrane of rat liver mitochondria, *Biochim. Biophys. Acta* **686**:204–214.

Schein, S. J., Colombini, M., and Finkelstein, A., 1976, Reconstitution in planar lipid bilayers of a voltage-dependent anion-selective channel obtained from *Paramecium* mitochondria, *J. Memb. Biol.* **30**:99–120.

Smack, D. P., and Colombini, M., 1985, Voltage-dependent channels found in the membrane fraction of corn mitochondria, *Plant Physiol.* (in press).

Tzagoloff, A., 1982, *Mitochondria*, pp. 115, 307–317, Plenum Press, New York.

Werkheiser, W. C., and Bartley, W., 1957, The study of steady-state concentrations of internal solutes of mitochondria by rapid centrifugal transfer to a fixation medium, *Biochem. J.* **66**:79–91.
Wojtczak, L., and Zaluska, H., 1969, On the permeability of the outer mitochondrial membrane to cytochrome *c*. I. Studies on whole mitochondria, *Biochim. Biophys. Acta* **193**:64–72.
Zalman, L. S., Nikaido, H., and Kagawa, Y., 1980, Mitochondrial outer membrane contains a protein producing nonspecific diffusion channels, *J. Biol. Chem.* **255**:1771–1774.
Zimmerberg, J., and Parsegian, V. A., 1984, Polymer-inaccessible space as a measure of the volume change during channel opening and closing: VDAC channels, *Biophys. J.* **45**:59a.

Chapter 23

ANALYSIS AND CHEMICAL MODIFICATION OF BACTERIAL PORINS

ROLAND BENZ

1. Introduction

The cell envelope of gram-negative bacteria such as *Escherichia coli, Salmonella typhimurium,* and *Pseudomonas aeruginosa* consists of three different layers, the outer membrane, the peptidoglycan (murein) layer, and the inner membrane (Nikaido, 1979a; Beveridge, 1981). The inner membrane acts as a real diffusion barrier and contains, in addition to the respiration chain and the H^+-ATPase, a large number of different transport systems for substrates. The peptidoglycan or murein layer consists of repeating N-acetylglucosaminyl-N-acetylmuramyl dimers, which form a macromolecule surrounding the cell (Beveridge, 1981). This layer protects the cells from lysis. The outer membrane is a molecular filter with defined exclusion limits for hydrophilic substrates (Nikaido and Nakae, 1979). In enteric bacteria, the outer membrane acts also as a barrier for hydrophobic compounds such as detergents (Nikaido, 1979b). The active substrates of the molecular sieving properties for hydrophilic substances are a major class of proteins called matrix proteins (Rosenbusch, 1974) or porins (Nakae, 1976).

The porins have a molecular weight between 30,000 and 50,000 and normally form trimers in the outer membrane that are, in many cases, stable in sodium dodecylsulfate (Benz, 1985). Porins have been isolated from the outer membrane of a variety of gram-negative bacteria (Nakae, 1975, 1976; Hancock *et al.*, 1979, 1982; Nikaido and Rosenberg, 1981; Luckey and Nikaido, 1980a,b). Experiments on reconstituted vesicles in the presence of porins from enteric bacteria have shown that the vesicle membranes become permeable for sugars and amino acids. The exclusion limit for the permeation of substrates in the case of *E. coli* and *S. typhimurium* is around 600 to 800 daltons, whereas hydrophilic substances with molecular weights up to 6000 can penetrate vesicles reconstituted in the presence of a porin from *P. aeruginosa* (Nakae, 1975, 1976; Hancock *et al.*, 1979).

Experiments on reconstituted vesicles provide valuable information on the

ROLAND BENZ • Department of Biology, University of Konstanz, D-7750 Konstanz, West Germany.

size of the porin pores. In this respect, the liposome swelling assay is of considerable interest for the detection of a solute specificity of the pores (Nikaido and Rosenberg, 1981). More detailed information about the pore interior and especially about the selectivity of the pores can be obtained from experiments on lipid bilayer membranes (Benz, 1985). Furthermore, porin trimers are very stable, and it is possible to perform chemical modifications. The reconstitution of native and chemically modified porins into lipid bilayer membranes and the study of their properties are described in this chapter. Most features of the porin pores are consistent with the assumption that they form large water-filled channels. However, there are also exceptions, such as the protein P of *P. aeruginosa* (Benz *et al.*, 1983) and the maltoporin of *E. coli* (Luckey and Nikaido, 1980a,b), which both form solute-specific channels.

2. Isolation and Reconstitution of Bacterial Porins

2.1. Isolation and Purification

The porins of most gram-negative bacteria are usually tightly associated with the peptidoglycan layer (Rosenbusch, 1974). This allows a rapid isolation procedure (Nikaido, 1983). The cells are harvested and washed. Then they are resuspended in a small volume and passed three times through a French pressure cell. The cell envelope containing the two membranes and the peptidoglycan are obtained by centrifugation. Most components of the outer and the inner membranes are removed by washing the pellet with a buffer containing 2% sodium dodecylsulfate (SDS) at elevated temperature. The SDS-insoluble part should only contain a few proteins associated with or covalently bound to the peptidoglycan layer.

There are two different methods to remove the porins from the peptidoglycan layer. The first uses the enzymatic degradation of the murein by lysozyme and trypsin (Nakae, 1975). This method has the disadvantage that the porin is obtained in aggregates of large molecular weight (Nakae, 1975; Benz *et al.*, 1978b). The second more recent method uses a salt extraction procedure with a buffer containing 1% SDS, 0.4 M NaCl, and 5 mM ethylendiamine tetraacetic acid (EDTA) (Nakamura and Mizushima, 1976; Tokunaga *et al.*, 1979; Nikaido, 1983). The porins are obtained by gel filtration.

Gram-negative bacteria in many cases have more than one porin in the outer membrane (Benz, 1985). Here it is the best to use mutant strains that have one porin in the outer membrane. Such mutants are available in a large number for enteric bacteria such as *E. coli* or *S. typhimurium*. The pore-forming activity of the porin solutions remains constant for several months at 4°C. In lyophilized form, the protein remains active for at least 1 year stored at −20°C (Benz *et al.*, 1978b).

2.2. Reconstitution of Porins into Planar Bilayers

Three different methods have successfully been used to reconstitute porins from gram-negative bacteria into lipid bilayer membranes. The first method uses

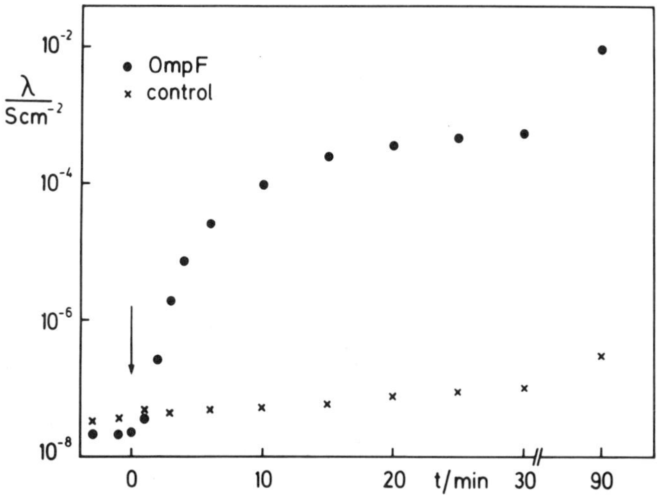

Figure 1. Specific membrane conductance as a function of time after the addition of 20 ng/ml OmpF trimers of *E. coli* to a black membrane from diphtanoylphosphatidylcholine/*n*-decane (arrow). The crosses represent a control experiment in which only 5 μg/ml Triton X-100 was added to another membrane. 1 M KCl, 25°C.

the addition of detergent-solubilized protein to the aqueous phase bathing a membrane (Benz *et al.*, 1978b; Young *et al.*, 1983). This method has been used in many studies of porin pores. It is described in more detail below. Second, the lipid bilayer membrane is formed from reconstituted vesicles (Schindler and Rosenbusch, 1978); third, the porin is inserted into the planar bilayer via fusion of the reconstituted vesicles with the membrane (Cohen *et al.*, 1982). The latter two methods are not described here because they are given in full detail in other chapters of this book (see Chapters 6 and 8).

The addition of porins to the aqueous phase bathing a membrane is the simplest and most successful reconstitution method. Figure 1 shows a typical reconstitution experiment of this type. Porin OmpF from *E. coli* was added in a final concentration of 20 ng/ml to a black membrane from diphytanoylphosphatidylcholine/*n*-decane. After an initial lag, presumably because of diffusion of the protein through unstirred layers, the conductance increased by about four orders of magnitude within about 30 min. Only a slight additional increase (as compared with the initial one) occurred after that time (Fig. 1). It has to be noted, however, that in the presence of all porins, the membrane conductance kept increasing until membrane breakage. The addition of detergent alone had only a small effect on the membrane conductance (see control in Fig. 1). The use of different detergents, e.g., Triton X-100, cholate, or octylglycoside, had only a small influence on porin reconstitution, although it seemed that the exchange of an ionic detergent for a neutral one increased the yield of the reconstitution, presumably because of the reduction of the surface charge of the protein–detergent complex. Porin trimers

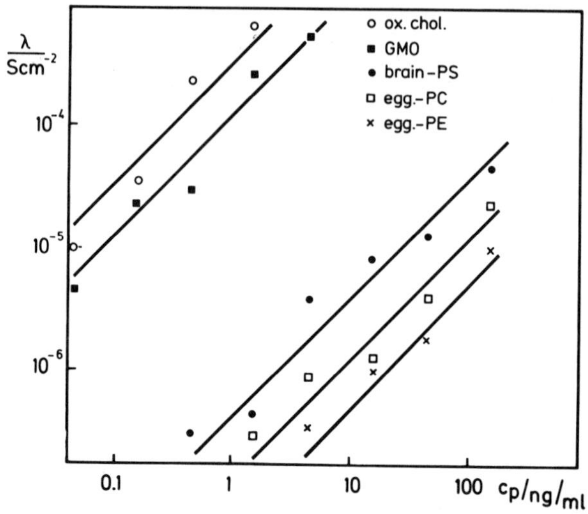

Figure 2. Specific membrane conductance λ as a function of the concentration of a mixture of porins of *Salmonella typhimurium* (OmpC, OmpF, and OmpD) in the aqueous phase; 1 M KCl; T = 25°C. Less than 5 μg/ml SDS was present in the aqueous solutions. The membranes were formed from different lipids dissolved in *n*-decane. The results were obtained 20 min after blackening of the membranes.

dissociated into monomers were found to be inactive in reconstitution experiments with lipid bilayer membranes and vesicles (Nakae *et al.*, 1979).

Since in most bilayer experiments in the presence of porins, a steady conductance level could not be reached, the dependence of the conductance on various parameters was difficult to determine. A meaningful comparison was possible, however, using the conductance level at a fixed time after the addition of the protein. Since most of the conductance increase was over after 20 to 30 min, this time was used for a comparison (Benz *et al.*, 1978b, 1980; Benz and Hancock, 1981). Figure 2 shows the influence of membrane composition on the incorporation of a mixture of different porins from *S. typhimurium* (OmpF, OmpC, and OmpD) into the membrane. Lipids with small polar head groups like monoolein and oxidized cholesterol allow a higher yield of reconstitution than the more common lipids. A similar "lipid specificity" was also found in reconstitution experiments with porins from other gram-negative bacteria (Benz *et al.*, 1978b; Benz and Hancock, 1981).

The incorporation of many conductance units into the lipid bilayer membranes, as described above, allows the investigation of the current–voltage behavior of the porin pores. Our own experiments were in most but not all cases consistent with the assumption that the porins form voltage-independent channels; i.e., the observed current was a linear function of the applied voltage up to 100 mV for porins of *E. coli* (Benz *et al.*, 1978b), *S. typhimurium* (Benz *et al.*, 1980), and *P. aeruginosa* (Benz and Hancock, 1981). Only a porin from the outer membrane of the cyanobacterium *Anabaena variabilis* switched at 100 mV transmembrane potential to substates (Benz and Böhme, 1985).

Schindler and Rosenbusch (1978) reported voltage-gated porin pores from *E. coli* in their experiments using reconstituted vesicles for bilayer formation (see Chapter 8 for the reconstitution method). A voltage dependence was also observed for a porin (protein I) of *Neisseria gonorrhoeae* outer membrane (Young *et al.*, 1983). The physiological role of the voltage dependence of the bacterial porins is, however, difficult to understand in these cases. The outer membrane of *E. coli* contains about 10^5 pores (Nikaido, 1979a), which corresponds to a channel density of 10^{12} pores/cm^2. The time constant of the membrane is very small under normal conditions (10 nsec at 0.1 M ionic strength; Benz *et al.*, 1980). This means that any membrane potential will decay in a short time, and Donnan potentials are far too small to reach 100 mV across the outer membrane (Stock *et al.*, 1977). In fact, no indication for the voltage control of the outer membrane permeability has been observed *in vivo* (Nikaido and Vaara, 1985), and this is consistent with the assumption that the outer membrane simply forms a molecular filter.

3. Analysis of Bacterial Porins

Most known porins from outer bacterial membranes form so-called "general diffusion pores." These pores are not highly structured, and they are only weakly anion or cation selective and do not contain a selectivity filter for solutes of any kind (Benz, 1984, 1985). The general diffusion pores can simply be treated as if they form a cylinder with diameter d and length l in the outer membrane (Benz *et al.*, 1978b). The outer membranes of certain gram-negative bacteria contain, in addition to the general diffusion pores, inducible channels that show a high specificity for ions or for uncharged molecules. Examples of these specific channels are the anion channel of the *P. aeruginosa* outer membrane and the maltoporin (LamB) of the *E. coli* outer membrane. The latter channel facilitates the diffusion of maltose and maltodextrins (Benz *et al.*, 1986b; Luckey and Nikaido, 1980a,b).

3.1. General Diffusion Porins

3.1.1. Single-Channel Analysis

The addition of small amounts (1–10 ng/ml) of porins to the aqueous phase on one or both sides of a black membrane of small surface area resulted in a stepwise increase of the membrane conductance. These conductance steps are specific to the presence of the proteins and were not observed when only the detergent dodecylsulfate or Triton X-100 was added to the aqueous phase. Figure 3 shows an experiment using porin OmpC from *E. coli*. All steps were directed upward. Terminating (downward) steps were only rarely observed during experiments with many different porins and membranes of long lifetimes (Benz *et al.*,

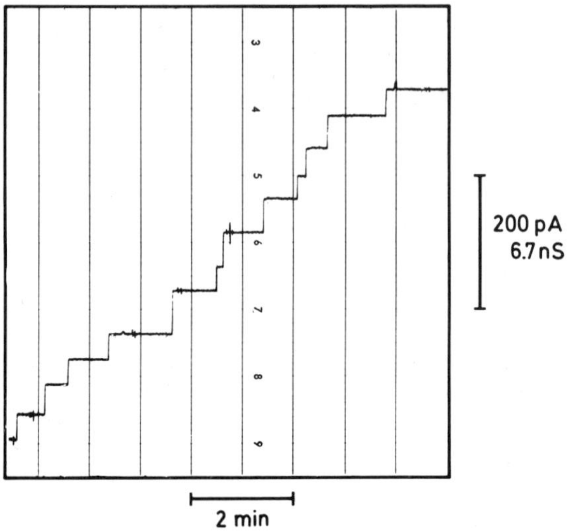

Figure 3. Stepwise increase of the membrane current in the presence of 2 ng/ml OmpC of *E. coli* K-12 added to the aqueous phase containing 1 M KCl; T = 25°C. The membrane was formed from diphytanoylphosphatidylcholine/*n*-decane; V_m = 10 mV.

1978b). Similar results were also found for the porins from *S. typhimurium* and *P. aeruginosa* (Benz *et al.*, 1980; Benz and Hancock, 1981). Only porin F pores observed in the presence of total outer membrane from *P. aeruginosa* (Hancock and Nikaido, 1978) had a much shorter lifetime, on the order of 50–100 msec, as compared with purified porin F, although the absolute level of the pore conductance was identical in both cases.

The conductance fluctuations measured in the presence of general diffusion porins were not uniform in size but were distributed over a range of about a factor of two to three from the smallest to the largest conductance step (Benz *et al.*, 1979). Furthermore, multiple conductance steps were observed when porin oligomers were added to the aqueous phase (Benz *et al.*, 1980). It is interesting to note that the specific porins gave a narrow distribution of conductance fluctuations (Hancock *et al.*, 1983).

Single-channel measurements in the presence of general diffusion porins were performed with a variety of different salts and salt concentrations (Benz *et al.*, 1979, 1980; Benz and Hancock, 1981; Benz and Böhme, 1985). For all general diffusion pores studied so far, the average pore conductance, $\overline{\Lambda}$, was a linear function of the specific conductance, σ, of the aqueous phase; i.e., the ratio $\overline{\Lambda}/\sigma$ varied only little in contrast to variations of $\overline{\Lambda}$ by two orders of magnitude (see Table I). A linear relationship is also shown in Fig. 4, where the average pore conductance, $\overline{\Lambda}$, of the F-porin from *P. aeruginsa* is given as a function of σ. The same was found for the porins from *S. typhimurium* (Benz *et al.*, 1980) and *E. coli* (Benz *et al.*, 1979, dotted line in Fig. 4), although $\overline{\Lambda}/\sigma$ is larger for *P. aeruginosa* porin F. This may be explained by a larger diameter of the *P. aeruginosa* pore (Benz and Hancock, 1981), which is also consistent with the liposome permeability assay using carbohydrates of different sizes (Hancock and Nikaido, 1978) and the liposome swelling method (Yoshimura *et al.*, 1983).

Table I. Single-Channel Conductance $\bar{\Lambda}$ of the OmpF Pore of *S. typhimurium* for Different Salt Solutions

Salt	*c* (M)	$\bar{\Lambda}$ (nS)	$\bar{\Lambda}/\sigma$ (10^{-8} cm)[a]
LiCl	1.0	0.9	1.3
NaCl	1.0	1.8	2.1
KCl	1.0	2.4	2.1
$MgCl_2$	0.5	0.62	1.0
K_2SO_4	0.5	1.4	1.8
$MgSO_4$	0.5	0.32	1.0
$Tris^+$ $HEPES^-$	0.5	0.088	1.2
$N(C_2H_5)_4^+$ $HEPES^-$	0.5	0.045	0.94

[a] σ is the specific conductance of the salt solutions of the concentration *c*. Taken from Benz *et al.* (1980).

The linear relationship between specific aqueous conductance σ and the average pore conductance $\bar{\Lambda}$ observed for the general diffusion pores suggests that the ions move inside the pore as they do in an aqueous environment; i.e., the pores are hydrophilic inside. This observation allows us to give a rough estimate of the size of the general diffusion pores. Assuming that the porin pores are filled with a solution of the same specific conductance as the external solution, and assuming a cylindrical pore with a length of 6 nm (Garavito *et al.*, 1983; Dorset *et al.*, 1983), the average pore diameter d (= $2r$) may be calculated according to:

$$\bar{\Lambda} = \sigma \pi r^2 / l \qquad (1)$$

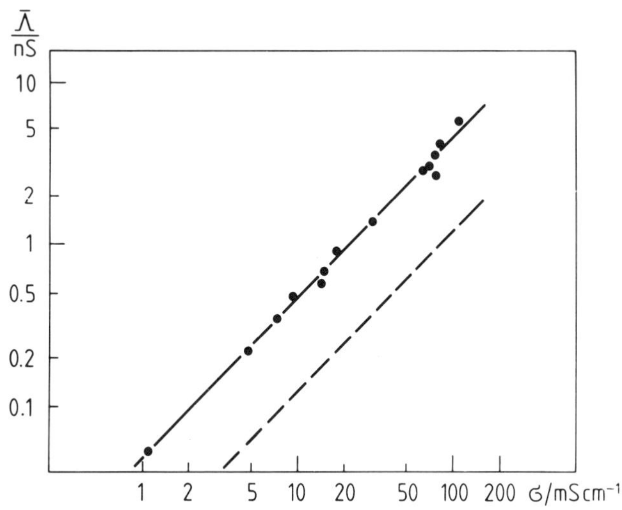

Figure 4. Average single-channel conductance $\bar{\Lambda}$ of protein F from *P. aeruginosa* given as a function of the specific conductance σ of the corresponding aqueous salt solution. The broken line reflects the relationship between $\bar{\Lambda}$ and σ for *E. coli* and *S. typhimurium* porins.

Table II. Single-Channel Conductance in 1 M KCl and Diameter of Different Gram-Negative Bacterial General Diffusion Porins[a]

Porin	$\bar{\Lambda}$ (nS)	*d* (nm)
E. coli (Benz *et al.*, 1985)		
OmpF	1.9	1.1
OmpC	1.5	1.0
PhoE	1.8	1.1
K	1.5	1.0
NmpC	1.3	1.0
S. typhimurium (Benz *et al.*, 1980)		
OmpF	2.2	1.2
OmpC	2.4	1.3
OmpD	2.5	1.3
P. aeruginosa (Benz and Hancock, 1981)		
F	5.6	2.0

[a] The diameter of the pores is calculated from equation 1 assuming a length of the pores of 6 nm and a specific conductance of the pore interior of 110 mS cm^{-1}.

Table II shows the diameters of a number of general diffusion pores from different gram-negative bacteria calculated from the conductance of pores in 1 M KCl (σ = 110 mS cm^{-1}). The pore diameters given in Table II show excellent agreement with those calculated from the exclusion molecular weight (Nakae, 1975, 1976) and the liposome swelling assay (Nikaido and Rosenberg, 1983).

3.1.2. Specificity of General Diffusion Pores

Further information on the structure of general diffusion pores may be obtained by the study of the ionic selectivity of the pores using zero-current potential measurements. Figure 5 shows the results of such experiments obtained for different *E. coli* porins in membranes from diphytanoylphosphatidylcholine/*n*-decane. Most porins were found to be cation selective (Benz *et al.*, 1985). In those cases, a positive membrane potential was observed on the more dilute side. For two porins (PhoE and NmpC), the more dilute side became negative. From the measured V_m and the concentration gradient c''/c' across the membrane, the ratio P_c/P_a of the permeabilities (P_c for cations and P_a for anions) was calculated using the Goldman–Hodgkin–Katz equation:

$$V_m = (RT/F) \ln \frac{P_c c'' + P_a c'}{P_c c' + P_a c''} \qquad (2)$$

where R, T, and F have their usual meanings.

Table III shows the zero-current membrane potentials for the different *E. coli* porins. OmpF and OmpC are present in the *E. coli* strain K-12. Their expres-

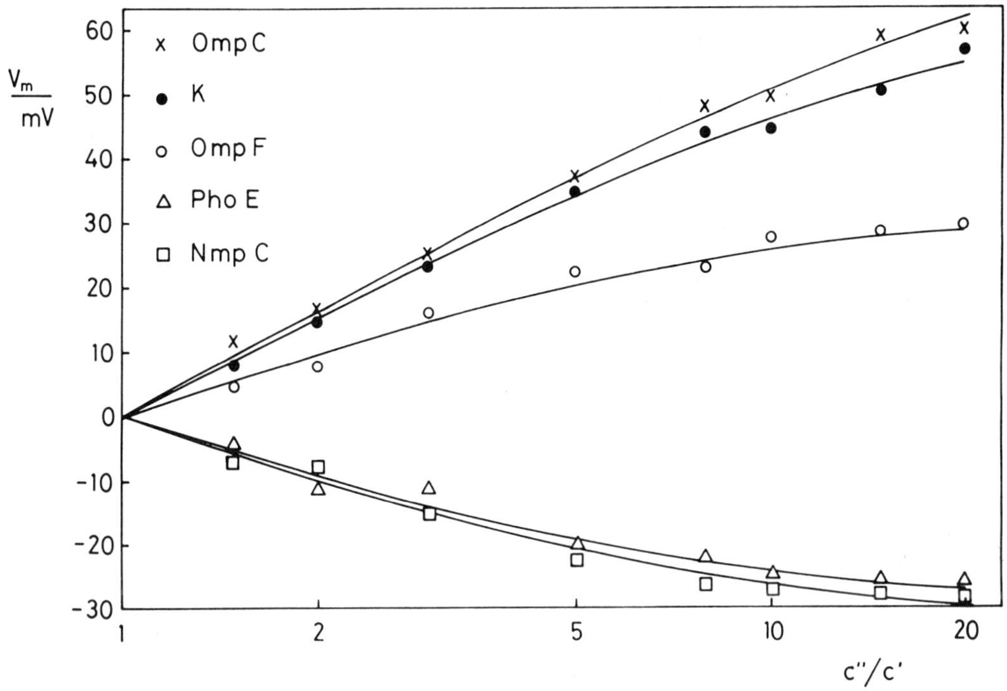

Figure 5. Zero-current membrane potentials V_m of membranes of diphytanoylphosphatidylcholine/*n*-decane in the presence of five different porins from *E. coli* as a function of the KCl gradient c''/c' across the membranes. The potential corresponds to the more dilute side of the membrane. The lines were drawn according to equation 2 with the following values for P_c/P_a: 26 (OmpC), 12 (K), 3.6 (OmpF), 0.3 (PhoE), and 0.26 (NmpC).

Table III. Zero-Current Membrane Potentials V_m for Different *E. coli* Porins in the Presence of a Tenfold KCl Gradient[a]

Porin	V_m (mV)	P_c/P_a
OmpF	26	3.6
OmpC	50	26
PhoE	−24	0.30
NmpC	−26	0.27
K	46	15

[a] V_m is the electrical potential of the dilute side (10^{-2} M) minus the potential on the concentrated side (10^{-1} M). The ratios of the permeabilities P_c(cation) and P_a (anion) was calculated from equation 3. Taken from Benz *et al.* (1985).

sion is regulated by the osmolarity of the growth media (van Alphen and Lugtenberg, 1978). The PhoE porin is induced in *E. coli* under conditions of phosphate deprivation (Tommassen and Lugtenberg, 1980). Although this porin has a certain preference for negatively charged solutes, it still forms a general diffusion pore (Nikaido and Rosenberg, 1983; Benz *et al.*, 1984a). NmpC appears in revertants of porin-deficient mutants, whereas porin K was found in *E. coli* strains that form capsules (Benz *et al.*, 1985). Charged groups on the surface of the protein or inside the pores are presumably responsible for the observed selectivity. This is supported by the pH dependence of the selectivity (Benz *et al.*, 1979) and by the observation that additional lysines appear in the primary sequence of PhoE as compared with the largely homologous OmpC and OmpF sequences (Mizuno *et al.*, 1983).

3.2. Properties of Specific Porins

3.2.1. Porin P from *P. aeruginosa*

The outer membrane of *P. aeruginosa* has very special sieving properties, which make this organism quite resistant to most antibiotics (Angus *et al.*, 1981; Hancock *et al.*, 1982). Part of these sieving properties arise from the limited permeability of the porin F channel. The outer membrane may also contain under normal growth conditions, another porin protein P with a molecular weight of 48,000. This porin is induced when *P. aeruginosa* is grown on media with low phosphate concentrations together with an alkaline phosphatase (Hancock *et al.*, 1982). Protein P forms highly anion-selective channels in lipid bilayer membranes (Benz *et al.*, 1983). Zero-current membrane potentials had a Nernstian slope ($P_c/P_a \ll 1$), and they indicated that the protein P channel is at least 100 times more permeable for chloride than potassium.

The current steps of the porin P channel are found to be fairly homogeneous as compared with the steps measured in the presence of the general diffusion porins. Figure 6 shows conductance steps observed with protein P on a diphytanoylphosphatidylcholine/*n*-decane membrane. All conductance steps were directed upward, indicating a similar long lifetime to the general diffusion porins. The lifetime of the pores was found to be independent of voltage, and linear current–voltage relationships were observed up to 200 mV for the protein P channel.

The cation present in the aqueous salt solutions had no influence on the single-channel conductance. In 100 mM chloride solution $\bar{\Lambda}$ was always around 160 pS irrespective of the size and charge of the cations. A much larger variation of the single-channel conductance $\bar{\Lambda}$ was observed when the anions were varied. Table IV shows $\bar{\Lambda}$ for a variety of different 100 mM salt solutions that contain the same cation (potassium) but different anions. The protein P pore has the largest single channel conductance in 0.1 M chloride solution (Benz *et al.*, 1986a). Among the halides, the selectivity follows the Eisenman sequence AVI (Eisenman, 1965). A variety of other anions also permeate through the channel, whereas in the presence

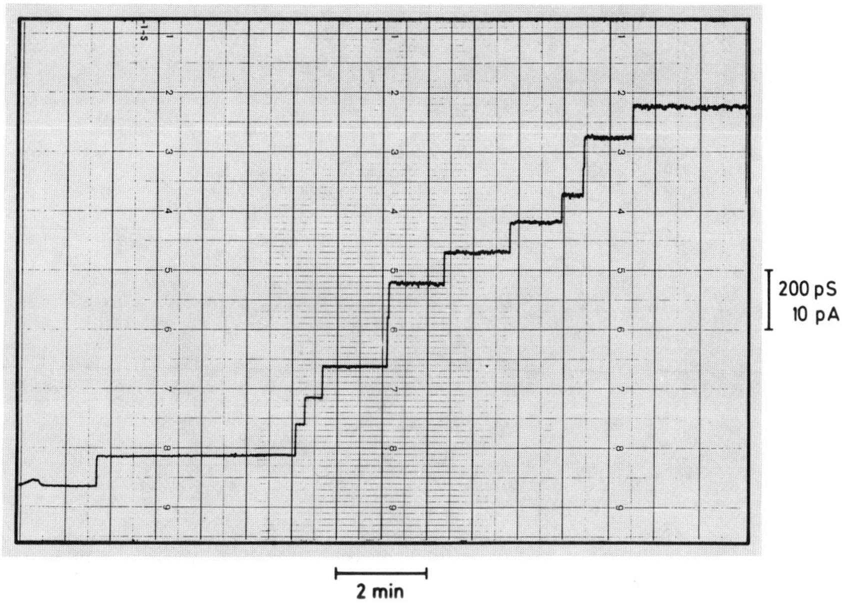

Figure 6. Single-channel record of a diphytanoylphosphatidylcholine/*n*-decane membrane in the presence of 20 ng/ml protein P of *P. aeruginosa* and 30 mM KCl in the aqueous phase. A voltage of 50 mV was applied through calomel electrodes with salt bridges; T = 25°C.

of salts with large organic anions, such as $HEPES^-$ (N-2-hydroxyethylipiperazine-N′-2-ethanesulfonic acid), no conductance fluctuations could be observed at all.

The data given in Table IV suggest that the porin P channel of *P. aeruginosa* contains a selectivity filter with a diameter of 0.6 nm and at least one positive

Table IV. Single-Channel Conductance $\bar{\Lambda}$ of the Protein P Channel in Different 0.1 M Salt Solutions[a]

Anion	$\bar{\Lambda}$ (pS)	*R* (nm)
F^-	136	1.36
Cl^-	159	1.81
Br^-	86	1.95
I^-	13	2.16
NO_2^-	69	
NO_3^-	24	
$HCOO^-$	82	
$HEPES^-$	<2	

[a] The corresponding cation was in all cases potassium. *R* is the ion radius according to Pauling (1960). Taken from Benz *et al.* (1986a).

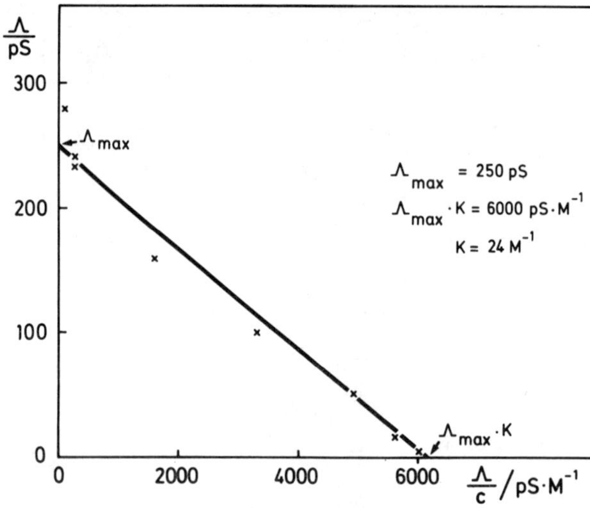

Figure 7. Eadie–Hofstee plot of the average single-channel conductance $\overline{\Lambda}$ as a function of the chloride concentration in the aqueous phase; T = 25°C.

charge (Benz *et al.*, 1986a). The binding of anions to this selectivity filter presumably accounts for the concentration dependence of the single-channel conductance (Benz, 1984). Figure 7 shows an Eadie–Hofstee plot of the single-channel conductance as a function of Λ/c (c is chloride concentration). The data of Fig. 7 could be fitted easily assuming single occupancy of the pore (i.e., one binding site for anions inside the pore) using the following equation (Läuger, 1973):

$$\Lambda(c) = \Lambda_{\max} K \cdot c/(1 + K \cdot c) \tag{3}$$

with the binding constant $K = 24\ \mathrm{M}^{-1}$ and a maximum conductance of 250 pS.

3.2.2. Maltoporin from *E. coli*

Besides the normal porins OmpF and OmpC, the outer membrane of *E. coli* K-12 contains the LamB protein or maltoporin when the cells grow on maltose or maltodextrins (Szmelcman and Hofnung, 1975). This protein is different from the general diffusion porins and does not show any immunologic cross reaction with them (Overbeeke *et al.*, 1980). The results of liposome swelling experiments in the presence of maltoporin showed that the rate of penetration of maltose was much larger than that of sucrose or lactose despite similar molecular weights (Luckey and Nikaido, 1980a,b). This facilitated diffusion has been explained by the assumption of a specific binding site for maltose and maltodextrins inside the channel.

Lipid bilayer experiments are not very well suited to perform permeability

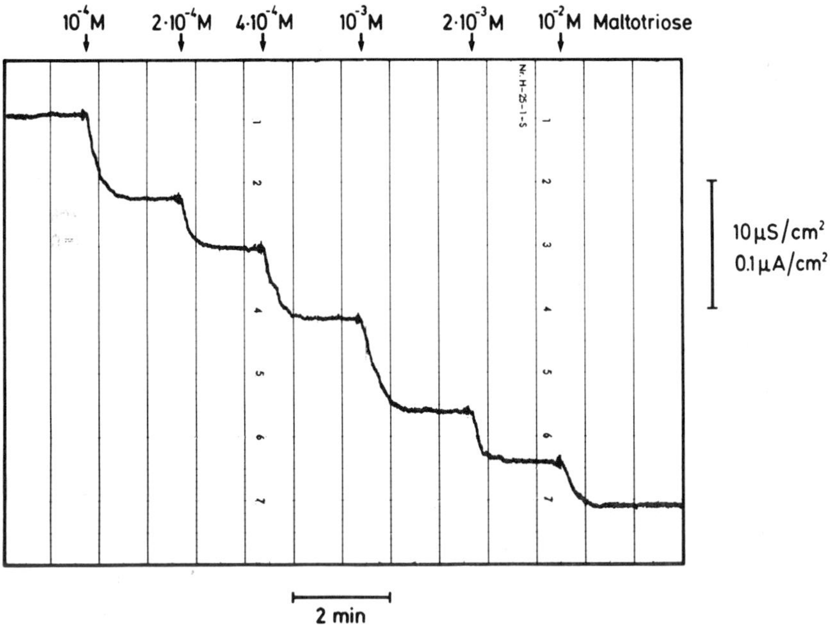

Figure 8. Influence of increasing concentrations of maltotriose on the macroscopic conductance of a diphytanoylphosphatidylcholine/*n*-decane membrane containing maltoporin from *E. coli*. The aqueous phase contained 1 M KCl and 40 ng/ml maltoporin; V_m = 10 mV; T = 25°C.

measurements with uncharged solutes. This should also be the case for the study of maltoporin. However, lipid bilayer experiments are able to give insight into the binding of maltose and maltodextrins to this pore. Maltoporin has a single-channel conductance of 160 pS in 1 M KCl. The addition of maltotriose reduces the single-channel conductance, and large concentrations completely block the pore (Benz *et al.*, 1986b). In fact, it is possible to titrate the maltoporin conductance with maltose and maltodextrins and to calculate the binding constant of these substances to the channel interior. Figure 8 shows an experiment of this type. LamB was added to a black lipid bilayer membrane, and the increase of the conductance was measured as a function of time. Thirty minutes after the addition, the membrane conductance was almost stationary. Different concentrations of maltotriose were added to the aqueous phase with stirring (arrows). The membrane conductance decreased as a function of the maltotriose concentration (Fig. 8). From the conductance decrease a binding constant of about $2 \cdot 10^3$ M^{-1} for maltotriose to the pore interior can be calculated (Benz *et al.*, 1986b). This value shows excellent agreement with the binding constant of maltotriose to whole cells (Ferenci *et al.*, 1980). These results indicate that it is possible to study the specificity of channels for neutral solutes in lipid bilayer membranes.

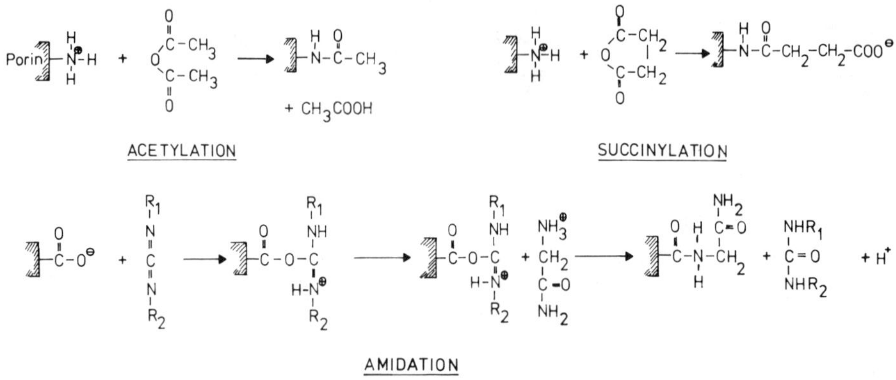

Figure 9. Chemical modification of the porins.

4. Chemical Modification of Bacterial Porins

Figure 9 shows three different possibilities for the chemical modification of charged groups attached to a protein. Acetylation and succinylation were performed using acetyl anhydride at pH 7 and succinic anhydride at pH 8 (Tokunaga *et al.*, 1981). Acetylation blocks a positively charged group, whereas succinylation converts an amino group into a carboxyl group, i.e., a positive charge into a negative charge. The amidation of carboxyl groups is performed in a two-step reaction with 1-ethyl-3-(3-dimethyl-aminopropyl) carbodiimide in the presence of glycineamide, which neutralizes the negative charge (Benz *et al.*, 1984c). Porins are very stable, and the trimeric structure is retained after chemical modification. Only intense amidation leads to dissociation of OmpF trimers into monomers (Tokunaga *et al.*, 1981).

4.1. Effect on the General Diffusion Porins

The pores formed by the general diffusion porins in lipid bilayer membranes can be described as hollow cylinders in which the ions have the same mobility sequence as in the aqueous phase. Provided that the chemical modification does not change the structure of the pores, it should not influence their single-channel conductance in KCl because K^+ and Cl^- have about the same aqueous mobility [limiting molar conductivities 73.5 mS/cmM and 76.4 mS/cmM, respectively (Castellan, 1983)]. In fact, chemical modification had only a small influence on the single-channel conductance of OmpF and PhoE in 1 M KCl (Benz *et al.*, 1984c; Darveau *et al.*, 1984). A somewhat larger effect is expected when a mobile cation is combined with a less mobile anion and vice versa. In these cases, a much larger influence of the modification of charged groups was observed. The PhoE pore in 0.5 M $MgCl_2$ had a single-channel conductance of 930 pS, whereas $\bar{\Lambda}$ for acetylated

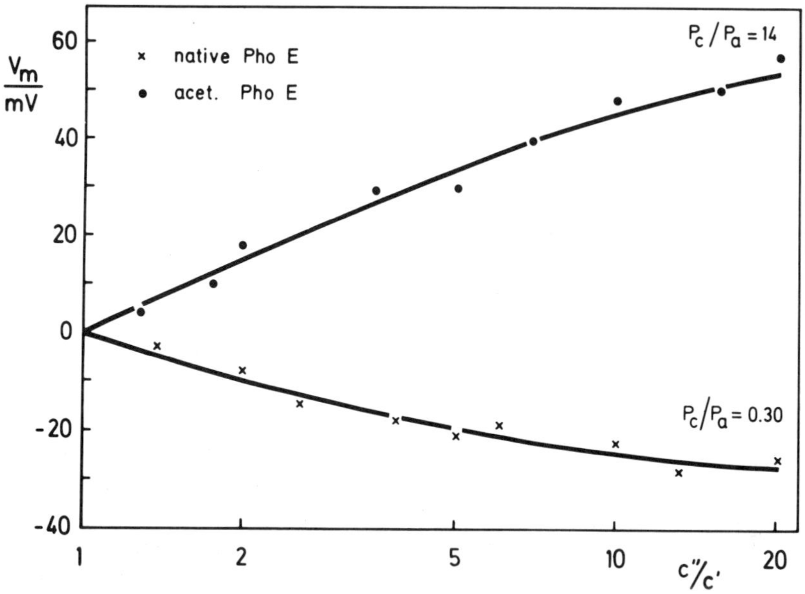

Figure 10. Zero-current membrane potentials V_m of membranes from diphytanoylphosphatidylcholine/ *n*-decane in the presence of native and acetylated PhoE as a function of the KCl gradient c''/c' across the membranes. The potential corresponds to the more dilute side of the membrane. The lines were drawn according to equation 2 with the indicated values for P_c/P_a; T = 25°C.

PhoE was only 550 pS (Darveau *et al.*, 1984). This result indicated that the chemical modification has an influence on the ionic selectivity of the PhoE porin.

Figure 10 shows zero-current membrane potentials of native and acetylated PhoE. Whereas the more dilute side becomes more negative for increasing salt gradients in the case of the native porin, the acetylated PhoE forms cation-selective pores; i.e., the more dilute side is positive. The results analyzed according to the Goldman–Hodgkin–Katz equation suggested that the native PhoE channel exhibited a 3.3-fold preference for Cl^- over K^+. With acetylated porin PhoE in

Table V. Zero-Current Membrane Potentials V_m of Acetylated PhoE in the Presence of a Tenfold Salt Gradient[a]

Salt	V_m (mV)	P_c/P_a Acetylated	P_c/P_a Native	Increase of cation selectivity
LiCl	24	3.3	0.11	30
KCl	45	14	0.30	46
KF	50	24	0.65	30

[a] The relative permeabilities of cations and anions were calculated from equation 2. Taken from Darveau *et al.* (1984).

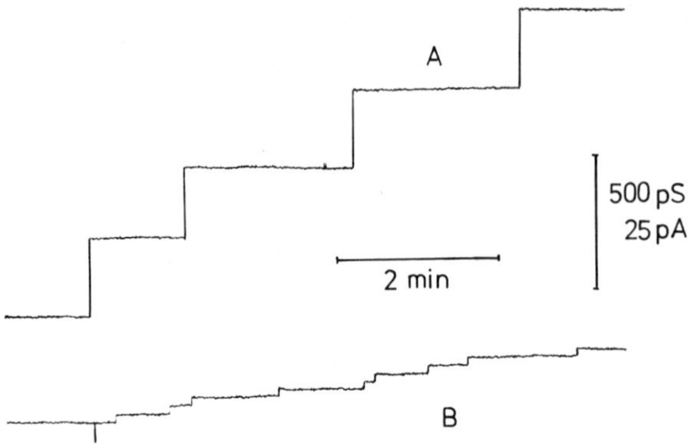

Figure 11. Single-channel records of current steps in the presence of native protein P (A) and acetylated protein P (B). The aqueous phase contained 1 M KCl, pH 6, and 10 ng/ml protein. V_m = 50 mV; T = 25°C.

the membrane, the membrane showed a 14-fold preference of K^+ over Cl^-. This 46-fold change of selectivity is consistent with the selectivity change observed for other salts (see Table V). The actual numerical differences in the relative permeabilities of cations and anions of the acetylated PhoE channel reflected the relative mobilities of the anions and cations (Darveau *et al.*, 1984). The change of the selectivity after the chemical modification is consistent with the observation that PhoE is an acidic protein (Overbeeke *et al.*, 1983). The pore presumably contains an excess of positive charges before the acetylation. The acetylation blocks the positive charges, which results in a change of the selectivity.

4.2. Effect on Porin P of *P. aeruginosa*

Porin P of *P. aeruginosa* forms highly anion-selective channels with a small selectivity filter (see Section 3.2). The nature of the filter was probed by acetylation and succinylation of the amino groups of protein P. This resulted in small changes in the mobility of protein P on dodecylsulfate polyacrylamide gels (Hancock *et al.*, 1983), although all the protein was present as trimeric aggregates after chemical modification. Succinylation totally blocked ion movement through the channel. Acetylation caused a reduction of the single-channel conductance. Figure 11 shows the effect of acetylation on the single-channel conductance in 1 M KCl solution. In addition, the anion selectivity was drastically reduced, and the saturation of the single-channel conductance with increasing anion concentration was no longer observed. It is suggested that lysines located at the selectivity filter of the pore are responsible for the channel selectivity (Benz *et al.*, 1984b,c; Hancock *et al.*, 1983).

5. Conclusion

Porins from the outer membrane of gram-negative bacteria form ion-permeable channels in lipid bilayer membranes. The simplest method for reconstitution consists of the addition of the porin either from a detergent solution or an osmotic shock solution directly to the aqueous phase (Benz *et al.*, 1978a,b). The insertion of the protein aggregates into the membranes in this reconstitution method is presumably controlled by hydrophobic forces because the rate of insertion shows no dependence on the ionic strength of the aqueous phase (Benz *et al.*, 1978b, 1980). The reconstitution rate was found to be strongly dependent on the lipid used for membrane formation. The properties of the single unit, however, were found to be independent of the lipid environment (Benz *et al.*, 1980). The insertion of the porin trimers into the lipid bilayer membranes may be controlled by a kinetic process that is facilitated by lipids with small polar head groups.

Most porins form obviously large water-filled pores in lipid bilayer membranes and in the outer membrane of gram-negative bacteria (Benz, 1985). These general diffusion pores are permeable to the large organic ions $Tris^+$ and $HEPES^-$ without detectable interaction with the pore interior. The pore conductance shows no saturation for increasing salt concentration. Furthermore, the current–voltage characteristic was ohmic, which is also expected for a wide, unselective channel. The diameters of the general diffusion pores calculated from the single-channel conductance show excellent agreement with the results of the reconstituted vesicles, in which hydrophilic solutes of molecular weights up to 600 (*E. coli*, Nakae, 1976), 700 (*S. typhimurium*, Nakae, 1975), and 6000 (*P. aeruginosa*, Hancock *et al.*, 1979) were found to be permeable through the porin pores. A similar good agreement was found for a comparison between the lipid bilayer experiments and the liposome swelling assay (Yoshimura *et al.*, 1983; Nikaido and Rosenberg, 1983).

The properties of the porin P channel of *P. aeruginosa* outer membrane and of the maltoporin of *E. coli* are quite different from those of the other porins. Whereas most porins have only a limited selectivity for solutes, the porin P channel appears highly selective for anions, and the maltoporin is specific for maltose and maltodextrins. Both pores presumably have a binding site and a selectivity filter with a diameter around 0.6 nm inside the pore, compared with a diameter of at least 1.1 nm for the general diffusion pores. Taking these findings together, it is obvious that the further study of the properties of maltoporin and porin P will be of special interest.

The outer membrane of gram-negative bacteria acts as a molecular filter for hydrophilic solutes (Di Rienzo *et al.*, 1978; Nikaido, 1979a). The results obtained from reconstitution experiments support this view. In particular, we observed no gating of the porin pores up to 100 mV, and we found that almost all pores had lifetimes of the order of minutes. Only the porin F from *P. aeruginosa* showed a much shorter lifetime (50–100msec) when total outer membrane, instead of purified porin, was present in the lipid bilayer experiments (Benz and Hancock, 1981). The short lifetime of protein F pores is presumably related to the high antibiotic resistance of this organism (Angus *et al.*, 1981). In any case, the vesicle permeability assay (Hancock and Nikaido, 1978; Hancock *et al.*, 1979; Nakae, 1975, 1976) and the liposome swelling assay (Nikaido and Rosenberg, 1981) are consistent with the presence of open pores in the absence of applied voltage, a fact that supports the results of our own lipid bilayer experiments.

The question of whether one porin trimer contains three pores (Nikaido, 1979b; Schindler and Rosenbusch, 1978; Lugtenberg and van Alphen, 1983) or only one pore (Benz *et al.*, 1978b; Di Rienzo *et al.*, 1978) was controversal in the past. Our own lipid bilayer experiments were always consistent with one pore in a trimer. More recently, OmpF trimers of *E. coli* B have been crystallized using a polyethylene glycol or salt-generated two-phase system (Garavito *et al.*, 1983). The maximum dimensions as analyzed by X-ray diffraction are 8 nm·5.5 nm. The pore-forming complex contains a large amount of β-pleated sheets oriented perpendicular to the membrane plane (Garavito *et al.*, 1983), which is consistent with earlier studies using circular dichroism measurements (Tokunaga *et al.*, 1979; Nakae *et al.*, 1979). A more detailed picture of the pore structure was obtained by electron microscopic analysis of OmpF protein embedded into lipid vesicles. The results of these studies suggest that the pore in the OmpF trimer is basically one channel with three openings facing the outside of the cell (Dorset *et al.*, 1983). The three channels merge approximately in the middle of the membrane in a central region with a diameter around 1 nm (Dorset *et al.*, 1984).

The primary sequences of three porins of *E. coli* strains B and K-12 (OmpF, OmpC, and PhoE) have been determined (Chen *et al.*, 1979; Inokuchi *et al.*, 1982; Mizuno *et al.*, 1983; Overbeeke *et al.*, 1983). The amino acid composition of the porins is not particularly hydrophobic, and only a limited number of subsequences with more than three consecutive hydrophobic amino acids are found in these proteins. The comparison of the primary sequences of the three porins shows that their structural genes presumably evolved from a common ancestral gene (Mizuno *et al.*, 1983). The cation selectivity of the OmpF and the OmpC pores presumably arises from the excess of negatively charged amino acids (Inokuchi *et al.*, 1982; Mizuno *et al.*, 1983), whereas the anion selectivity of the PhoE pore may be caused by the appearance of additional lysines in the primary sequence (Overbeeke *et al.*, 1983). The primary sequence of the maltoporin is not related to that of the general diffusion porins OmpF, OmpC, and PhoE (Clément and Hofnung, 1981).

The study of porin channels from the outer membrane of gram-negative bacteria offers an elegant way to investigate the structure–function relationship of pores. Porin trimers are very stable and allow chemical modification without damage to their tertiary and quaternary structures (Tokunaga *et al.*, 1981; Benz *et al.*, 1984c), which finally leads to a selectivity change of the porins (Darveau *et al.*, 1984). The close relationship of the anion-selective porin PhoE and the cation-selective porins OmpF and OmpC allows the construction of hybrid genes and of hybrid pores (Tommassen *et al.*, 1984). The study of the hybrid pores in lipid bilayer membranes would make it possible to localize the amino acids responsible for the selectivity of the pores (Tommassen *et al.*, 1985). Also of further interest is the study of the mechanism of anion translocation through the protein P channel from *P. aeruginosa*.

ACKNOWLEDGMENTS. This work was supported by grants of the Deutsche Forschungsgemeinschaft (Be 865/3-1 and Be 865/3-2).

References

Angus, B. L., Carey, A. M., Caron, D. A., Kropinski, A. M. B., and Hancock, R. E. W., 1981 Outer membrane permeability in *Pseudomonas aeruginosa*: Comparison of a wild-type with an antibiotic-supersusceptible mutant, *Antimicrob. Agents Chemother.* **21**:299–309.

Benz, R., 1984, Structure and selectivity of porin channels, in: *Current Topics in Membranes and Transport*, Vol. 21 (E. D. Stein, ed.), pp. 199–219, Academic Press, San Diego.

Benz, R., 1985, Porin from bacterial and mitochondrial outer membranes, *CRC Crit. Rev. Biochem.* **19**:145–190.

Benz, R., and Böhme, H., 1985, Pore formation by an outer membrane protein of the cyanobacterium *Anabaena variabilis, Biochim. Biophys. Acta* **812**:286–292.

Benz, R., and Hancock, R. E. W., 1981, Properties of the large ion permeable pores formed from protein F of *Pseudomonas aeruginosa* in lipid bilayer membranes, *Biochim. Biophys. Acta* **646**:298–308.

Benz, R., Boehler-Kohler, B. A., Dieterle, R., and Boos, W., 1978a, Porin activity in the osmotic shock fluid of *Escherichia coli, J. Bacteriol.* **135**:1080–1090.

Benz, R., Janko, K., Boos, W., and Läuger, P., 1978b, Formation of large ion-permeable membrane channels by the matrix protein (porin) of *Escherichia coli, Biochim. Biophys. Acta* **511**:305–319.

Benz, R., Janko, K., and Läuger, P., 1979, Ionic selectivty of pores formed by the matrix protein (porin) of *Escherichia coli, Biochim. Biophys. Acta* **551**:238–247.

Benz, R., Ishii, J., and Nakae, T., 1980, Determination of ion permeability through the channels made of porins from the outer membrane of *Salmonella typhimurium* in lipid bilayer membranes, *J. Memb. Biol.* **56**:19–29.

Benz, R., Gimple, M., Poole, K., and Hancock, R. E. W., 1983, An anion selective channel from the *Pseudomonas aeruginosa* outer membrane, *Biochim. Biophys. Acta* **730**:387–390.

Benz, R., Darveau, R. P., and Hancock, R. E. W., 1984a, Outer membrane protein PhoE from *Escherichia coli* forms anion-selective pores in lipid bilayer membranes, *Eur. J. Biochem.* **140**:319–324.

Benz, R., Poole, K., and Hancock, R. E. W., 1984b, Characterization and chemical modification of small anion specific channels formed in lipid bilayer membranes by outer membrane protein of *Pseudomonas aeruginosa, Biophys. J.* **45**:81–82.

Benz, R., Tokunaga, H., and Nakae, T., 1984c, Properties of chemically modified porin from *Escherichia coli* in lipid bilayer membranes, *Biochim. Biophys. Acta* **769**:348–356.

Benz, R., Schmid, A., and Hancock, R. E. W., 1985, Ion selectivity of gramnegative bacterial porins, *J. Bacteriol.* **162**:722–727.

Benz, R., Poole, K., and Hancock, R. E. W., 1986a, Mechanism of ion translocation through the anion selective channel of *Pseudomonas aeruginosa* outer membrane, *J. Gen. Physiol.* (in press).

Benz, R., Schmid, A., Nakae, T., and Vos-Scheperkeuter, G., 1986b, Pore formation by LamB preparations of *E. coli* in lipid bilayer membranes, *J. Bacteriol.* (in press).

Beveridge, T. J., 1981, Ultrastructure, chemistry and function of the bacterial wall, *Int. Rev. Cytol.* **72**:229–317.

Castellan, G. W., 1983, *Physical Chemistry. The Electrical Current in Ionic Solutions*, pp. 769–780, Addison-Wesley, Reading.

Chen, R., Kramer, C., Schmidmayer, W., and Henning, U., 1979, Primary structure of major outer membrane protein I of *Escherichia coli* B, *Proc. Natl. Acad. Sci. U.S.A.* **76**:5014–5017.

Clément, J. M., and Hofnung, M., 1981, Gene sequence of the λ receptor, an outer membrane protein of *Escherichia coli* K12, *Cell* **27**:507–514.

Cohen, F. S., Akabas, M. H., and Finkelstein, A., 1982, Osmotic swelling of phospholipid vesicles causes them to fuse with a planar phospholipid bilayer membrane, *Science* **217**:458–460.

Darveau, R. P., Hancock, R. E. W., and Benz, R., 1984, Chemical modification of the anion selectivity of the PhoE porin from the *Escherichia coli* outer membrane, *Biochim. Biophys. Acta* **774**:67–74.

Di Rienzo, J. M., Nakamura, K., and Inouye, M., 1978, The outer membrane of gram-negative bacteria: Biosynthesis, assembly and function, *Annu. Rev. Biochem.* **47**:481–532.

Dorset, D. L., Engel, A., Häner, M., Massalski, A., and Rosenbusch, J. P., 1983, Two-dimensional crystal packing of matrix porin. A channel forming protein in *Escherichia coli* outer membrane, *J. Mol. Biol.* **165**:701–710.

Dorset, D. L., Engel, A., Massalski, A., and Rosenbusch, J. P., 1984, Three dimensional structure of a membrane pore. Electron microscopical analysis of *Escherichia coli* outer membrane matrix porin, *Biophys. J.* **45**:128–129.

Eisenman, G., 1965, Some elementary factors involved in specific ion permeation. *Excerpta Medica Int. Cong. Ser.* **87**:489–506.

Ferenci, T., Schwentorat, M., Ullrich, S., and Vilmart, J., 1980, Lambda receptor in the outer membrane of *Escherichia coli* as a binding protein for maltodextrins and starch polysaccharides, *J. Bacteriol.* **142**:521–526.

Garavito, R. M., Jenkins, J. A., Jansonius, J. N., Karlsson, R., and Rosenbusch, J. P., 1983, X-ray diffraction analysis of matrix porin, an integral membrane protein A from *Escherichia coli* outer membranes, *J. Mol. Biol.* **164**:313–327.

Hancock, R. E. W., and Nikaido, H., 1978, Outer membranes of gram-negative bacteria XIX. Isolation from *Pseudomonas aeruginosa* PA 01 and use in reconstitution and definition of permeability barrier, *J. Bacteriol.* **136**:381–390.

Hancock, R. E. W., Decad, G. M., and Nikaido, H., 1979, Identification of the protein producing transmembrane diffusion pores in the outer membrane of *Pseudomonas aeruginosa* PA O1, *Biochim. Biophys. Acta* **554**:323–331.

Hancock, R. E. W., Poole, R. K., and Benz, R., 1982, Outer membrane protein P of *Pseudomonas aeruginosa* regulation by phosphate deficiency and formation of small anion-specific channels in lipid bilayer membranes, *J. Bacteriol.* **150**:730–738.

Hancock, R. E. W., Poole, R. K., Gimple, M., and Benz, R., 1983, Modification of the conductance, selectivity, and concentration-dependent saturation of *Pseudomonas aeruginosa* protein P channels by chemical acetylation, *Biochim. Biophys. Acta* **735**:137–144.

Inokuchi, K., Muthoh, N., Matsuyama, S., and Mizushima, S., 1982, Primary structure of the OmpF gene that codes for a major outer membrane protein of *Escherichia coli* K12, *Nucleic Acids Res.* **10**:6957–6968.

Läuger, P., 1973, Ion transport through pores: A rate-theory analysis *Biochim. Biophys. Acta* **311**:423–441.

Luckey, M., and Nikaido, H., 1980a, Diffusion of solutes through channels produced by phage lambda receptor protein of *Escherichia coli*: Inhibition by higher oligosaccharides of maltose series, *Biochem. Biophys. Res. Commun.* **93**:166–171.

Luckey, M., and Nikaido, H., 1980b, Specificity of diffusion channels produced by λ phage receptor protein of *Escherichia coli, Proc. Natl. Acad. Sci. U.S.A.* **77**:167–171.

Lugtenberg, B., and van Alphen, L., 1983, Molecular architecture and functioning of the outer membrane of *Escherichia coli* and other gram-negative bacteria, *Biochim. Biophys. Acta* **737**:51–115.

Mizuno, T., Chou, M.-Y., and Inouye, M., 1983, A comparative study on the genes for three porins of the *Escherichia coli* outer membrane: DNA sequence of the osmoregulated ompC gene, *J. Biol. Chem.* **258**:6932–6940.

Nakae, T., 1975, Outer membrane of *Salmonella typhimurium*: Reconstitution of sucrose-permeable membrane vesicles, *Biochem. Biophys. Res. Commun.* **64**:1224–1230.

Nakae, T., 1976, Identification of the outer membrane protein of *Escherichia coli* that produces transmembrane channels in reconstituted vesicle membranes, *Biochem. Biophys. Res. Commun.* **71**:877–889.

Nakae, T., Ishii, J., and Tokunaga, M., 1979, Subunit structure of functional porin oligomers that form permeability channels in the outer membrane of *Escherichia coli, J. Biol. Chem.* **254**:1457–1461.

Nakamura, K., and Mizushima, S., 1976, Effects of heating in dodecyl sulfate solution on the conformation and electrophoretic mobility of isolated major outer membrane proteins from *Escherichia coli* K12, *J. Biochem.* **80**:1411–1422.

Nikaido, H., 1979a, Permeability of the outer membrane of bacteria, *Angew. Chem.* **18**:337–350.

Nikaido, H., 1979b, Nonspecific transport through the outer membrane, in: *Bacterial Outer Membranes* (M. Inouye, ed.), pp. 367–407, Wiley-Interscience, New York.

Nikaido, H., 1983, Proteins forming large channels from bacterial and mitochondrial outer membranes: Porins and phage lambda receptor protein, *Methods. Enzymol.* **97**:85–100.

Nikaido, H., and Nakae, T., 1979, The outer membrane of gram-negative bacteria, *Adv. Microb. Physiol.* **20**:163–250.

Nikaido, H., and Rosenberg, E. Y., 1981, Effect of solute size on diffusion rates through the transmembrane pores of the outer membrane of *Escherichia coli, J. Gen. Physiol.* **77**:121–135.

Nikaido, H., and Rosenberg, E. Y., 1983, Porin channels in *Escherichia coli*: Studies with liposomes reconstituted from purified proteins, *J. Bacteriol.* **153**:241–252.

Nikaido, H., and Vaara, M., 1985, Molecular basis of bacterial outer membrane permeability, *Microbiol. Rev.* **49**:1–32.

Nikaido, H., Rosenberg, E. Y., and Foulds, J., 1983, Porin channels in *Escherichia coli*: Studies with β-lactams in intact cells, *J. Bacteriol.* **153**:232–240.

Overbeeke, N., Bergmans, H., van Mansfeld, F., and Lugtenberg, B., 1983, Complete nucleotide sequence of phoE, the structural gene for the phosphate limitation inducible outer membrane pore protein of *Escherichia coli* K12, *J. Mol. Biol.* **163**:513–532.

Overbeeke, N., van Scharrenburg, G., and Zugtenberg, B., 1980, Antigenic relationships between pore proteins of *Escherichia coli* K12, *Eur. J. Biochem.* **110**:247–254.

Pauling, L., 1960, *The Nature of the Chemical Bond*, Cornell University Press, Ithaca.

Rosenbusch, J. P., 1974, Characterization of the major envelope protein from *Escherichia coli, J. Biol. Chem.* **249**:8019–8029.

Schindler, H., and Rosenbusch, J. P., 1978, Matrix protein of *Escherichia coli* outer membranes forms voltage-controlled channels in lipid bilayers, *Proc. Natl. Acad. Sci. U.S.A.* **75**:3751–3755.

Stock, J. B., Rauch, B., and Roseman, S., 1977, Periplasmic space in *Salmonella typhimurium* and *Escherichia coli, J. Biol. Chem.* **252**:7850–7861.

Szmelcman, S., and Hofnung, M., 1975, Maltose transport in *Escherichia coli* K12: Involvement of the bacteriophage lambda receptor, *J. Bacteriol.* **124**:112–118.

Tokunaga, M., Tokunaga, H., Okajima, Y., and Nakae, T., 1979, Characterization of porins from the outer membrane of *Salmonella typhimurium*. 2. Physical properties of the functional oligomeric aggregates, *Eur. J. Biochem.* **95**:441–448.

Tokunaga, H., Tokunaga, M., and Nakae, T., 1981, Permeability properties of chemically modified porin trimers from *Escherichia coli* B, *J. Biol. Chem.* **256**:8024–8029.

Tommassen, J., and Lugtenberg, B., 1980, Outer membrane protein e of *Escherichia coli* K-12 is coregulated with alkaline phosphatase, *J. Bacteriol.* **143**:151–157.

Tommassen, J., Pugsley, A. P., Korteland, J., Verbakel, J., and Lugtenberg, B., 1984, Gene encoding a hybrid ompF–PhoE pore protein in the outer membrane of *Escherichia coli* K12, *Mol. Gen. Genet.* **197**:503–508.

Tommassen, J., van der Ley, P., van Zeijl, M., and Agterberg, M., 1985, Localization of functional domains in *E. coli* K-12 outer membrane porins with the aid of hybrid genes obtained by *in vivo* recombination, *EMBO J.* **4**:1583–1587.

van Alphen, L., and Lugtenberg, B., 1978, Influence of osmolarity of the growth medium on the outer membrane protein pattern of *Escherichia coli, J. Bacteriol.* **131**:623–630.

Yoshimura, F., Zalman, L. S., and Nikaido, H., 1983, Purification and properties of *Pseudomonas aeruginosa* porin, *J. Biol. Chem.* **258**:2308–2314.

Young, J. D.-E., Blake, M., Mauro, A., and Cohn, Z. A., 1983, Properties of the major outer membrane protein from *Neisseria gonorrhoeae* incorporated into model lipid membranes, *Proc. Natl. Acad. Sci. U.S.A.* **80**:3831–3835.

INDEX